Texts in Mathematics

Volume 2

Adventures in Formalism

Volume 1
An Introduction to Discrete Dynamical Systems and their General Solutions
F. Oliveira-Pinto

Volume 2
Adventures in Formalism
Craig Smoryński

Texts in Mathematics Series Editor
Dov Gabbay   dov.gabbay@kcl.ac.uk

Adventures in Formalism

Craig Smoryński

ISBN 978-1-84890-060-8

College Publications
Scientific Director: Dov Gabbay
Managing Director: Jane Spurr
Department of Computer Science
King's College London, Strand, London WC2R 2LS, UK

http://www.collegepublications.co.uk

Cover designed by Laraine Welch
Printed by Lightning Source, Milton Keynes, UK

Contents

Preface

Formalism

This is a book about formalism in mathematics. What is that? Well, if you mention the word "formalism" to a mathematician, he will probably respond with David Hilbert, with mathematics as a game played with meaningless symbols, and maybe even with the litany of the three philosophies of mathematics—logicism, intuitionism, and formalism. A mathematical logician might think along different lines and describe *a* formalism, i.e. a formal system. The mathematician with a broader historical background will think of Leonhard Euler, "the greatest formalist of the hundred-year period following the invention of the calculus"[1]. So, which response would be correct?

Out of curiosity I pulled Samuel Johnson's *A Dictionary of the English Language* off the shelf. The word "formalism" is not in it. But "formal", "formalist", "formality", "formalize" and "formally" are. "Formalist" has a curious definition:

> One who practises external ceremony; one who prefers appearance to reality; one who seems what he is not.

The sample quotation Johnson supplies from South's *Sermons* illustrates this nicely:

> A grave, stanch, skilfully managed face, set upon a grasping aspiring mind, having got many a sly *formalist* the reputation of a primitive and severe piety.

A formalist stresses form over content; indeed, to him content and reality are irrelevant. The formalist mathematician manipulates symbols *formally*, not worrying himself about the meanings— if any— of the symbols. This doesn't really describe the arch-formalists Euler and Hilbert, though one can recognise this behaviour (minus the deceptive component) in the work of some early 20th

---

[1] Jerome H. Manheim, *The Genesis of Point Set Topology*, Pergamon Press, Oxford, and The Macmillan Company, New York, 1964, p. 9.

century mathematical logicians: Bertrand Russell and his *Axiom of Reducibility*, Alonzo Church and the λ-calculus, and Willard van Orman Quine and his *New Foundations*. But this is hardly the kind of formalism I wish to discuss.

Johnson, by the way, gives four meanings of "formally", two of which are

1. According to established rules, methods, ceremonies or rites.
2. Ceremoniously; stifly; precisely.

These reflect in part some of the 7 meanings he supplies for the word "formal":

1. Ceremonious; solemn; precise; exact to affectation.
2. Done according to established rules and methods; not irregular; not sudden; not extemporaneous.
4. External; having the appearance but not the essence.

I don't know if there is any good definition of formalism in mathematics written down anywhere. Full of expectation, I checked W. Bynum, E.J. Brown, and Roy Porter, *Dictionary of the History of Science*[2], a work more interesting than its title implies as its entries are not only full sentences, but paragraphs and, occasionally, essays. Under "formalism", however, I found only the brief entry:

formalism. *See* philosophies of mathematics.

Following the "link" brings one to 4 short paragraphs, half of one concluding:

Meanwhile, D. Hilbert (1862 - 1943) advocated 'formalism', where mathematical systems are studied for properties such as consistency and completeness.

Art historians define, e.g., "mannerism" with greater care and accuracy than this.

I decided to check some history books. First was Morris Kline's magnum opus[3]. The index points to an 8 page discussion of Hilbert's foundational views. This offers a fuller explanation, but only of one narrow aspect of formalism. One of the standard classroom textbooks[4] also begins with David Hilbert, but in its non-chronological order, soon arrives at George Peacock:

Peacock is best known for his textbook *Treatise of Algebra*, an early precursor of the formalistic approach algebra would take on later in the century.[5]

The book does offer a little information about Peacock's approach, and much earlier it uses the word "formal" in discussing Euler, but except in connexion with Hilbert's formalism it offers no definition of "formalism" itself.

---

[2] Princeton University Press, Princeton, 1981.
[3] *Mathematical Thought from Ancient to Modern Times*, Oxford University Press, New York, 1972.
[4] David M. Burton, *The History of Mathematics; An Introduction*, 6th edition, McGraw-Hill, New York, 2005.
[5] *Ibid.*, p. 627. The "of" in the title of title of Peacock's book is a typo and should read "on".

In light of the results of my exhaustive, in depth research on the matter, I have drawn the conclusion that no one has bothered to properly define *formalism* in mathematics in a way that goes beyond Hilbert's assumed philosophy. I therefore claim the prerogative and offer the following:

0.1 Definition. Formalism *is the practice of mathematics that is* formal *in the sense that it is*

1. *precise; exact to affectation; or*
2. *done according to established rules; or*
3. *having the appearance if not the essence.*

The attentive reader will notice some slight modifications of Johnson's definition here. The most significant (other than renumbering) is the replacement of "but" by "if" in the last clause; for, in doing mathematics formally, one is not denying there is an essence underlying one's activity, but merely ignoring such.

I identify three types of formalism, not by the three clauses of the Definition, but rather by the various motivations behind the formality. In *Type I Formalism* one forges ahead doing formal manipulations of objects or symbols in total unconcern for the validity of one's manipulations, either because one is blissfully unaware of the lack of any underlying essence or because one's manipulations are purely heuristic— reasoning by analogy, so to speak. Probably the most famous examples of this are the early use of $\sqrt{-1}$ when one had no idea what it was, and, a modern favourite, the physicist's use of the Dirac δ-function. I discuss this type of formalism in Chapter II.

Type II Formalism occurs when one simply replaces an intuitive notion by a precise, formally defined concept. The quintessential example is the notion of limit. Type II formalism is the topic of Chapter III.

Finally, *Type III Formalism*, or the modern axiomatic approach, replaces an intuitive notion, not by a precisely defined concept, but by a precise delineation of properties that one is allowed to use in dealing with the notion. This is discussed in Chapter IV.

I precede these discussions with Chapter I in which I discuss a single example— the notion of an infinite sum— which has been subjected to all three types of formalism at a most accessible level.

Finally, in Chapter V, I round out my presentation with a little history of the negative reactions to formalism. The mildest and one of the most natural reactions is to the sort of inevitable excess that caused me to retain Johnson's phrase "exact to affectation". More serious objections are that formalism leaves no room for creativity and turns mathematics into a meaningless game, suggesting I restore the original wording to Johnson's clauses 2 and 4 of his definition of "formal".

When I was approximately half finished writing this book I acquired a copy of Klaus Volkert, *Die Krise der Anschauung*[6] [*The Crisis of Intuition*]. In a footnote on page XIX we read

---
[6] Vandenhoeck & Ruprecht, Göttingen, 1986.

The term "formalism" shows up repeatedly in the history of mathematics with varying meanings. Characteristic of all "formalisms" is the emphasis on syntax as opposed to semantics, of form over content. One can distinguish three meanings.

1. Formalism in the above described sense: so long as the meaning of a sign (e.g. $\sqrt{-1}$) is still unexplained (in the example: so long as the question, "Is there a root of -1?", is unanswered), the application of the accompanying marks (in the example these are material images of the form "$\sqrt{-1}$") is established exclusively through syntactic rules (for example, if the product $(\sqrt{-1})(\sqrt{-1})$ occurs, one can always replace it by -1).

2. Formalism in the sense of a "computation procedure" (for example, the formalism of linear algebra).

3. Formalism in the narrow terminological sense of a formal system, as goes back to Hilbert.

The latter two of his meanings are of no interest here, and the first is a bit broad. Thus, for the purposes of the present book I will stick with my definition and trinity of types.

The title of Volkert's book is borrowed from a paper of the same title by Hans Hahn[7], which also provides the impetus for the book. The central theme thereof is how Type II formal definitions of concepts of continuity, curve, dimension give rise to the construction of "monsters"— continuous but nowhere differentiable functions, space-filling curves, etc.— the existence of which violates the intuition and would not be conceivable on a merely intuitive notion of the concepts involved. Both Hahn and Volkert discuss the philosophical aspects of the formalism/intuition divide, Volkert naturally going into greater detail, covering the history, and bringing the subject more up-to-date. The book by Manheim cited in our first footnote discusses the importance of such teratological constructions from a more mathematical point of view but, being some 20 years older than Volkert's book, is not as up to date.

Alternate Introduction (On the Contents)

Chapter I is a microcosmic reflexion of the rest of the book, introducing and illustrating with a single example the three types of formalism. It does not possess the impact of a multitude of examples, but offers a quick entry to the subject. And embedded in the discussion are some matters of importance later.

The heyday of Type I formalism was during the 17th and 18th centuries, but Type I formalism extends well beyond these centuries. Chapter II discusses nu-

---

[7] In: *Krise und Neuaufbau in den exakten Wissenschaften. Fünf Wiener Vorträge*, F. Deuticke, Leipzig and Vienna, 1933. The paper appears in English translation, "The crisis in intuition", in: Hans Hahn, *Empiricism, Logic and Mathematics; Philosophical Papers*, D. Reidel Publishing Co., Dordrecht, 1980.

merous examples, beginning with Euclid, continuing through the Renaissance, pausing a while in the 17th and 18th centuries, and finally emerging in the 20th. One might be surprised to find Euclid here, his name having been synonymous with rigour for so long. But his definitions were hardly Type II formal definitions, his terms vaguely defined, and his axiomatic practice not as strictly adhered to as at present. One finds definite Type I formalism in *The Elements*. That said, Type I formalism really took off during the Renaissance and grew rapidly after the discovery of the Calculus as mathematicians, dissatisfied with the Greeks' perceived habit of hiding their methods worked out methods of their own in an explosion of ingenuity. It was truly an adventurous undertaking, resulting in most wonderful results via the most fantastical and unreliable reasonings. The mathematics of the period raise eyebrows today, but never cease to entertain and fascinate. The practice of unbridled Type I formalism died out among mathematicians during the 19th century with the rise of Type II formalism, and the closing example of Chapter II, the Dirac δ-function, comes from a 20th century physicist and the successful application of its use troubled mathematicians for years.

Much of 19th century mathematics can be described as a backlash or retrenchment followed by consolidation. The field had grown rapidly and in importance. But there were exceptions to theorems, students to teach, and matters to be sorted out. In the 19th century the foundations of mathematics slowly emerged and Type II formalism began to take over. The infinitistic methods of the Calculus were the most problematic and drew the most attention, but philosophical issues also arose: What were negative numbers, these things that were "less than nothing"? And what about imaginary numbers? Such problems were resolved largely through Type II definitions—replacements for vague notions of limit and continuity, convergence, area, and even of the various classes of numbers—negative, imaginary and complex, and ultimately of the real numbers themselves. The Type II definitions of workaday notions of limit, continuity, etc., are familiar from the Calculus and, but for some of the history, I can simply refer the reader to any Calculus textbook, or, in the cases where the term being defined has several inequivalent candidates for a Type II definition (continuity *vs.* uniform continuity, convergence *vs.* absolute convergence *vs.* uniform convergence) to any textbook on the theory of the Calculus. I do discuss some of these concepts in Chapter III as they arise, but the main emphasis in this Chapter is the Type II definition/construction of the numbers themselves.

I don't know if it has gone out of fashion or if it never was universally taught, but when I was a student there was a course on the foundations of the real numbers, starting with an axiomatic treatment of the positive integers and proceeding through various constructions to lay the foundations of the sets of all integers, the rationals, the real numbers, and occasionally the complex numbers. There have been a number of textbooks devoted to this subject and perhaps more are still being written. I note, however, that I have taught at universities where such a course was not in the catalogue. There is good reason for this: the various constructions are incorporated individually in other courses.

I have chosen to devote the major portion of Chapter III, about a third of the book, to the constructions of these various number systems. I justify my treatment on the fullness of the account, both in mathematical and historical detail, as well as the presentation of the use of Type II concepts to provide foundations for some Type I derivations, and the inclusion of some material that is not that widely known.

One byproduct of the replacement of vague Type I notions by formal Type II definitions is a precision that not only allows one to rigorously prove theorems under specifiable conditions, but even on occasion to construct counterexamples when the conditions are not met. The results can be highly non-intuitive and have led to something of a backlash. The final section preceding the exercises to Chapter III discusses the first truly spectacular such construction, namely that of a continuous function which is nowhere differentiable.

Chapter IV discusses the axiomatic method, what it is, where axioms come from, etc.

Finally, Chapter V discusses the reaction to formalism. This includes the backlash to things like nowhere differentiable functions, as well as other aspects of dissatisfaction with Type II formalism in particular. The discussion is political, pædagogical, psychological, and scientific.

Background of the Reader

I have tried to keep the mathematical prerequisites to a minimum, but the issues raised simply do not arise in elementary mathematics. The book presupposes a knowledge of the Calculus, but, presumably, not of the foundations of the Calculus. American textbooks in upper division mathematics courses used to state as their prerequisite "a course in Calculus or the equivalent mathematical maturity". The standard course in Calculus ought to provide the necessary mathematical background for the present book, but, with many watered-down textbooks on the market and many instructors loath to subject the student to not easily digestible material, the Calculus course may no longer be assumed to supply the undefinable quality referred to as "mathematical maturity". I am inclined to believe the reader who has made an honest attempt to understand the concepts and proofs in a rigorous course in the Calculus will have no trouble with most of the sequel[8]; but he who has skipped over the conceptual material in favour of merely learning the algorithms and applications of the Calculus, will likely find the book a rather eccentric collection of odds and ends— a broad, but not very deep, haphazard assortment of topics constituting a sort of wordy introduction to some of the concepts of higher mathematics.

All in all, my attempt to write a book accessible to a wider audience has failed. But it should be suited, if not for college sophomores fresh out of the Calculus, at least for juniors who have had one additional course providing the

---

[8] Chapters I to IV grow successively in the level of sophistication required, and Chapter IV deals in more generalities and has fewer concrete details than the previous chapters.

hypothetical mathematical maturity. And I would hope there is enough novelty in the choice of topics to interest the more advanced reader as well.

Acknowledgements

There are three people I owe a large debt of gratitude to— my friends Jimmie Johnson, Dirk van Dalen, and Eckart Menzler-Trott. The three of them pulled me out of a long period of inactivity, the causes of which I need not bore the reader with, and contributed in one way or another to the writing of this book. Jimmie Johnson arranged for me to teach part time at a local community college and, while that was even less intellectually stimulating than it sounds, he did one time assign me to teach the History of Mathematics. The result of that course (and a course I taught years ago, thanks to my friend Bob Gold at *The Ohio State University*) was my book, *History of Mathematics; A Supplement*[9]. Chapter I of the present work is in part a continuation of that book.

Dirk van Dalen also awakened me from my slumbers when he suggested me as the translator for Eckart Menzler-Trott's biography of Gerhard Gentzen[10]. Van Dalen also provided the initial inspiration for the present work years ago when I was looking into the evolution of Hilbert's Programme[11] by pointing out to me that the criticism of formalism as turning mathematics into a meaningless game did not originate with Hermann Weyl but went back much earlier. While I didn't pursue the matter further at the time, I did file it away in the back of my mind for later consideration.

Eckart Menzler-Trott, through his visits to Westmont and our joint pilgrimage to Helsinki to meet with Jan von Plato, has also been a powerful stimulant, as well as a vital resource. More than one passage in this book has been rewritten because of remarks he has made or material he has sent me. And my debt to him in Chapter V should be obvious.

In like manner my newer friend Robert Murray Jones has forced me to make some changes in my exposition through his criticism and procurement of additional source material.

Speaking of resources, I wish also to acknowledge the additional material assistance of Fred Thulin and Ulrich Kohlenbach in the acquisition of materials.

I also wish to thank an old friend, Robert Tragesser, whose enthusiastic response provided much needed encouragement. While I don't think I am one of Augustus de Morgan's paradoxers, my views are often enough at odds with those of others, that such reassurance I am on the right path serves as an anti-inhibitive.

---

[9] Springer-Verlag, New York, 2007.

[10] *Gentzens Problem: Mathematische Logik im nationalsozialistischen Deutschland*, Birkhäuser Verlag, Basel, 2001; English translation by Craig Smoryński and Edward Griffor, *Logic's Lost Genius: The Life of Gerhard Gentzen*, American Mathematical Society, Providence (Rhode Island), 2007.

[11] This is most accessible in its modified form as an appendix to the English translation of the Gentzen biography cited in the previous footnote.

On a less personal note, the first work cited in this book— Jerome Manheim's *The Genesis of Point Set Topology*— was, along with some of the material in the Gentzen biography, the spark that set me off writing the present work.It is a pleasure to acknowledge these influences.

And, of course, my debt to the scholarship of Detlef Laugwitz, particularly in Chapter III, is so obvious as to barely require explicit mention.

It is a pleasure to acknowledge these influences.

I

Infinite Series

1 Grandi's Series

An infinite sequence of real numbers,

$$a_0, a_1, \ldots, \tag{1}$$

may be thought of as a succession of numbers, or, in modern mathematical terms as a function,

$$a : \mathbb{N} \to \mathbb{R}$$

of the set $\mathbb{N} = \{0, 1, \ldots\}$ of natural numbers[1] into the set $\mathbb{R}$ of real numbers.

Given a sequence (1) the finite sums

$$\sum_{i=0}^{n} a_i = a_0 + a_1 + \ldots + a_n$$

have obvious meanings and automatically inherit some properties from the arithmetic of the reals, e.g.

$$r \sum_{i=0}^{n} a_i = r(a_0 + a_1 + \ldots + a_n)$$

$$= ra_0 + ra_1 + \ldots + ra_n = \sum_{i=0}^{n} ra_i \tag{2}$$

and

---

[1] Many mathematicians begin the natural numbers at 1, a practice conforming better to the meaning of the word "natural" as only three cultures are known to have invented 0— the Hindus, the late Sumerians, and the Mayans. Logicians however regard the set of natural numbers as the first infinite ordinal number, where, following Johann von Neumann, an ordinal number is the set of its predecessors. Such a sequence begins with the empty set, which is identified with 0. Being a logician, I thus include 0 among the natural numbers.

$$\sum_{i=0}^{n} a_i + \sum_{i=0}^{n} b_i = (a_0 + a_1 + \ldots + a_n) + (b_0 + b_1 + \ldots + b_n)$$
$$= (a_0 + b_0) + (a_1 + b_1) + \ldots + (a_n + b_n)$$
$$= \sum_{i=0}^{n} (a_i + b_i). \tag{3}$$

But what happens if one adds all the elements of the sequence, i.e. what is the sum of the series, as the sequence is called when its elements are added:

$$\sum_{i=0}^{\infty} a_i = a_0 + a_1 + \ldots ? \tag{4}$$

Better stated: what do we mean by the sum (4)? For, asking *what* the sum is could be construed as asking for its value, which can sometimes be found formally even if one doesn't know what it means.

A case in point is the so-called Grandi's series given by choosing $a_i = (-1)^i$:

$$\sum_{i=0}^{\infty} (-1)^i = 1 - 1 + 1 - 1 + \ldots \tag{5}$$

This series was first introduced by Jacob Bernoulli in 1696 who summed it by considering the more general series,

$$\frac{l}{m+n} = \frac{l}{m}\left(1 + \frac{n}{m}\right)^{-1} = \frac{l}{m} - \frac{ln}{m^2} + \frac{ln^2}{m^3} - \ldots \tag{6}$$

and obtaining, for $m = n$:

$$\frac{l}{2m} = \frac{l}{m} - \frac{l}{m} + \frac{l}{m} - \ldots$$

and, for $l = m = n = 1$,

$$\frac{1}{2} = 1 - 1 + 1 - \ldots \tag{7}$$

He declared this to be a "not inelegant paradox", which it truly is. This is not yet a mathematical contradiction, but it does contradict our expectations: a sum of integers ought not to be a fraction. It becomes an outright contradiction if we compare the result of (7) with the results of summing (5) by re-associating the terms:

$$\sum_{i=0}^{\infty} (-1)^i = (1 - 1) + (1 - 1) + \ldots$$
$$= 0 + 0 + \ldots = 0 \tag{8}$$

or,

$$\sum_{i=0}^{\infty} (-1)^i = 1 + (-1 + 1) + (-1 + 1) + \ldots$$

$$= 1 + 0 + 0 + \ldots = 1 \tag{9}$$

The Italian mathematician Guido Grandi published a book in 1703 in which he used the expansion

$$\frac{1}{1+x} = 1 - x + x^2 - x^3 + \ldots \tag{10}$$

to obtain (7). Here let me quote Carl Boyer:

> As an example of the addition of differentials to give a finite magnitude, Grandi referred to the paradoxical result $1-1+1-1+\ldots = 0+0+\ldots = \frac{1}{2}$. This, he suggested to Leibniz, could be compared with the mysteries of the Christian religion and with the creation of the world by which an absolutely infinite force created something out of absolutely nothing.[2]

Leibniz agreed with the result, but not with the derivation. Instead he looked at the sequence $s_0, s_1, \ldots$ of partial sums,

$$s_n = \sum_{i=0}^{n} (-1)^i = \begin{cases} 1, & n \text{ is even} \\ 0, & n \text{ is odd.} \end{cases}$$

To him it was a question of probability: 0 and 1 are equally likely, so the expected value[3] is $\frac{1}{2} \cdot 0 + \frac{1}{2} \cdot 1 = \frac{1}{2}$. He reported on this in a letter to Christian Wolf published in the *Acta Eruditorum* in 1713.[4] Wolf concurred and also concluded

$$1 - 2 + 4 - 8 + \ldots = \frac{1}{3} \tag{11}$$

and

$$1 - 3 + 9 - 27 + \ldots = \frac{1}{4},$$

---

[2] Carl Boyer, *The History of the Calculus and Its Conceptual Development*, Dover, New York, 1959, p. 241.

[3] The expected value of an "experiment" is just the weighted average of the outcomes, with the weights supplied by the probabilities of the outcomes. One thinks of the probability of an outcome as the *relative frequency* or percentage of times the outcome occurs under repetitions of the experiment. Thus, if there are two outcomes with probabilities p and q, respectively and with respective values P and Q, then after n trials the outcomes will occur approximately np and nq times, respectively. One's overall "winnings" will be $npP + nqQ$, so that the average winning or *expected value* of a single experiment will be

$$\frac{1}{n}(npP + nqQ) = pP + qQ.$$

[4] I do not know Latin and am relying on secondary sources. Most general history texts don't have much to say on this subject. The exception is Morris Kline, *Mathematical Thought from Ancient to Modern Times*, Oxford University Press, NY, 1972. A more specialised reference is Richard Reiff, *Geschichte der unendlichen Reihen*, Verlag der H. Laupp'schen Buchhandlung, Tübingen, 1889.

both by extending Leibniz's argument.[5] Leibniz countered that the terms of a series must be decreasing in size or, as with (5), they must be the limit of a series with decreasing terms as exemplified by letting x approach 1 from below in

$$\frac{1}{1+x} = 1 - x + x^2 - x^3 + \ldots$$

Euler got involved around 1730. To quote Boyer again,

> The lack of care with which Euler handled the infinite is evidenced also in his use of divergent series. As Leibniz had suggested that $1 - 1 + 1 - 1 + \ldots = \frac{1}{2}$, so Euler held that from $\frac{1}{(1+1)^2} = \frac{1}{4}$ one could conclude that $1 - 3 + 5 - 7 + \ldots = 0$. Numerous similar examples of divergent series are to be found in his work.[6]

Euler's approach to the Grandi series began with the expansion

$$\frac{1}{1-x} = 1 + x + x^2 + x^3 + \ldots,$$

which he also used to justify (11) and

$$1 + 2 + 4 + 8 + \ldots = -1, \tag{12}$$

which last is even more startling than Grandi's result.

David Burton's popular textbook on the history of mathematics passes the following resoundingly negative judgment:

> By paying insufficient heed to questions of convergence, Euler was sometimes led into absurdities. There is the occasion when he used the binomial theorem to expand $\frac{1}{(1+x)^2} = (1+x)^{-2}$ as
>
> $$\frac{1}{(1+x)^2} = 1 - 2x + 3x^2 - 4x^3 + \cdots,$$
>
> and then set $x = -1$ to obtain
>
> $$\infty = 1 + 2 + 3 + 4 + \cdots$$
>
> Since he regarded infinity as a number (the reciprocal of 0), this sum caused no wonderment. From the series $\frac{1}{1-x} = 1 + x + x^2 + x^3 + \cdots$, with x replaced by 2, Euler found the strange equation
>
> $$-1 = 1 + 2 + 4 + 8 + \cdots,$$
>
> where positive integers are added to produce a negative value. Observing that the terms of the series exceed the corresponding ones of the earlier series, he argued that -1 is larger than infinity. In a further flight

---

[5] Kline, *op. cit.*, p. 446. Both formulæ obviously follow from (10). I do not see how to obtain these sums probabilistically.

[6] *Op. cit.*, p. 246.

of fancy, Euler contended that infinity separates positive and negative numbers, much as 0 does.[7]

This delightful passage brings to mind the prose of Eric Temple Bell and it is refreshing to find it in a textbook. Bell himself doesn't have too much to say about all of this. In his general history, he says

> The Bernoullis, James (1654 - 1705) and John[8] (1667 - 1748) (Swiss), recognised the divergence of the harmonic series[9] in 1689; and James (1696) noted "the not-inelegant paradox" that $1/(1-x)$ when expanded as $1+x+x^2+\cdots$ gives $1-1+1-\cdots = \frac{1}{2}$ for $x = -1$, a revelation which Leibniz and others attempted to justify by the theory of probability, but which the devout Euler humbly accepted. Less eminent mathematicians applied similar considerations to theology. Thus, by rearranging the terms of a divergent series, it was readily proved that $0 = 1$, and hence that God had created the universe out of nothing. Even Leibniz, in a philosophical mood, drew similar conclusions from the binary scale of notation.[10]

In his biographical account of Euler, he says

> One peculiarity of Euler's analysis must be mentioned in passing, as it was largely responsible for one of the main currents of mathematics in the nineteenth century. This was his recognition that unless an infinite series is *convergent* it is unsafe to use. For example, by long division we find
> $$\frac{1}{x-1} = \frac{1}{x} + \frac{1}{x^2} + \frac{1}{x^3} + \frac{1}{x^4} + \cdots,$$
> the series continuing indefinitely. In this put $x = \frac{1}{2}$. Then
> $$-2 = 2 + 2^2 + 2^3 + 2^4 + \ldots$$
> $$= 2 + 4 + 8 + 16 + \ldots$$
>
> The study of *convergence*...shows us how to avoid absurdities like this... The curious thing is that although Euler recognized the necessity for caution in dealing with *infinite* processes, he failed to observe it in much of his own work. His faith in analysis was so great that he would sometimes seek a preposterous "explanation" to make a patent absurdity respectable.[11]

---

[7] David Burton, *History of Mathematics; An Introduction*, 6th edition, McGraw-Hill, New York, 2005, pp. 533 - 534.

[8] I.e., Jacob and Johann, respectively.

[9] The series $\sum_{i=1}^{\infty} \frac{1}{i} = 1 + \frac{1}{2} + \frac{1}{3} + \ldots$ is called the harmonic series. It doesn't converge— cf. section 3, below.

[10] Eric Temple Bell, *Development of Mathematics*, McGraw-Hill, New York, 1940, p. 473.

[11] Eric Temple Bell, *Men of Mathematics*, Simon and Schuster, New York, 1937, pp. 151 - 152.

Euler must have been one of Bell's heroes, for the tone is disappointingly mild coming from one of history's more outspoken critics. Where it not for the line beginning "The curious thing...", I would consider the two quotes mutually contradictory. As it is, it presents Euler's behaviour as schizoid, something not unknown in the history of science, particularly among those individuals on the borderline between magic and science.

Morris Kline clarifies our Eulerian puzzle admirably:

> In writing to Goldbach on August 7, 1745, Euler refers to Bernoulli's argument that divergent series such as
>
> $$+1 - 2 + 6 - 24 + 120 - 720 + \ldots$$
>
> have no sum but says that these series have a definite *value*. He notes that we should not use the term "sum" because this refers to actual addition. He then states the general principle which explains what he means by a definite value. He points out that the divergent series comes from finite algebraic expressions and then says that the value of the series is the *value of the algebraic expression from which the series comes*.[12]

Euler corresponded with Nikolaus Bernoulli on the subject in 1743 and the latter suggested this value would not be unique if the expansions of two different functions happened to coincide. Two years later, in his already mentioned letter to Goldbach, Euler wrote, "Bernoulli gives no examples and I do not believe it possible that the same series could come from two truly different algebraic expressions. Hence it follows unquestionably that any series, divergent or convergent, has a definite sum or value".[13]

Almost 40 years later, the Frenchman Jean-Charles Callet came up with the example,

$$\frac{1 + x + \ldots + x^{m-1}}{1 + x + \ldots + x^{n-1}} = \frac{1 - x^m}{1 - x^n}$$
$$= 1 - x^m + x^n - x^{n+m} + x^{2n} - \ldots, \quad (13)$$

whence on setting $x = 1$,

$$\frac{m}{n} = 1 - 1 + 1 - 1 + 1 - \ldots \quad (14)$$

Lagrange countered that (14) does not follow from (13) because (13) does not list all the terms in the full power series. For $m = 3$ and $n = 5$, (13) would read

$$1 + 0x + 0x^2 - x^3 + 0x^4 + x^5 + 0x^6 + 0x^7 - x^8 + 0x^9 + x^{10} + \ldots,$$

whence (14) would thus be

$$1 + 0 + 0 - 1 + 0 + 1 + 0 + 0 - 1 + 0 + 1 + 0 + \ldots,$$

---

[12] *Op. cit.*, pp. 462 - 463.
[13] *Ibid.* p. 463.

and, following Leibniz, the most probable value of *this* series would be $\frac{3}{5}$.

Another half century would pass before Bernard Bolzano would take up the baton and criticise a proof of (7) published in 1830 by the almost anonymous "M.R.S.". Bolzano's critique is most interesting and we shall have a look at it later. For now we have a lot to consider. In the immediately following section we will have a look at the most absurd element of the preceding material and account for it. We follow this in section 3 with a discussion of the standard solution to the question of the meaning of an infinite sum (4) and where Bernoulli's, Grandi's, *et al*'s arguments break down. The topic will then turn to Bolzano's discussion of infinite sums in sections 4 and 5. Sections 6 and 7 consider more general notions of summation where Bernoulli, Grandi, *et al* would not be perceived to be wrong. Finally, we make a brief attempt to see what this all means in section 8.

2 Preposterously Absurd Flights of Fancy

The title of this section comes from the epithets hurled on Euler by Burton and Bell. They may have been intended for equations (7), (11), (12) and their kind; but, of all that was described in the preceding section, they seem to apply most fittingly to Grandi's linking of the Biblical theory of creation with the paradoxical equation,

$$0 + 0 + \ldots = 1 - 1 + 1 - 1 + \ldots = \frac{1}{2}. \tag{15}$$

The separation between science and theology has not always been as total as it almost is today. In the 17th century, Newton's teacher Isaac Barrow was an ordained minister, as were all permanent faculty at Cambridge who did not get a special exemption. Newton got the exemption, but was nevertheless a religious man and wrote a great deal on theology. Grandi worked in a monastery before moving up academically to the University of Pisa. Euler was a religious man[14], as was Leibniz. And Bolzano, in the 19th century, was a priest. I don't know the temporal or geographic extent of the requirement that professors took religious orders, but I can report that in the mid-19th century, the decision not to enforce it in the newly founded Cornell University was a matter of quite some controversy:

---

[14] Recall the apocryphal anecdote of his having been called upon to silence Denis Diderot, whose atheistic utterances at court in St. Petersburg were annoying everyone. Euler stood up and said, "Monsieur!

$$\frac{a + b^n}{n} = x$$

therefore God exists; respond!" (Augustus de Morgan, *A Budget of Paradoxes*, 2 volumes, 2nd edition, Open Court Publishing Company, Chicago, 1915, volume II, pp. 3 - 4.)

Opposition began at once. In the State Legislature it confronted us at every turn, and it was soon in full blaze throughout the State— from the good Protestant bishop who proclaimed that all professors should be in holy orders, since to the Church alone was given the command, "Go, teach all nations," to the zealous priest who published a charge that Goldwin Smith— a profoundly Christian scholar— had come to Cornell in order to inculcate the "infidelity of the *Westminster Review*"; and from the eminent divine who went from city to city denouncing the "atheistic and pantheistic tendencies" of the proposed education, to the perfervid minister who informed a denominational synod that Agassiz, the last great opponent of Darwin, and a devout theist, was "preaching Darwinism and atheism" in the new institution.[15]

In these modern times when it is illegal in the United States to even think about God in a neighbourhood that might be frequented by children who attend public school, all of this may seem a bit strange. Well, as Al Jolson said, you ain't heard nothing yet! In the early centuries of Christianity there arose the theory of the two Books God gave man to guide him in his faith. One was the Book of Scripture, i.e. the *Bible*. The other was the Book of Nature, for in Nature God planted clues to lead us to or remind us of religious truth. Bestiaries, or books of beasts, which lasted well into the middle ages provide an excellent illustration of the interpretation of the Book of Nature. One of the more popular of these was *Physiologus*, wherein we read:

> The third attribute of the lion is this: when the lioness brings forth her young, she brings it forth dead. But the Lioness watches over her cubs until the third day, when the father comes and roars and breathes in its face and wakens it. So did God and Father of the Universe waken the first born of all creation, our Lord Jesus Christ, his Son, from the dead. Well now spoke Jacob when he said of Judah the Lion's whelp: "Who shall rouse him up?"[16]

There is at least one religious or moral inference to be drawn from the behaviour of each animal. *Physiologus* ends with the polypus[17]:

> We must not pass over the deceitfulness, cunning, and trickery of the fish which is called Polypus. For on whatever stone or rock in the sea he fixes himself he makes his colour resemble the thing he hangs upon; and the fishes, while they stupidly swim round him, fancy he is dead

---

[15] Andrew White, *A History of the Warfare of Science with Theology in Christendom*, 2 volumes, 15th printing, D. Appleton and Company, New York, 1905, vol. I, pp. vii - viii. (The first printing was 1896.) White's book drew some serious criticism. James Walsh, a professor of the History of Medicine countered with *The Popes and Science; The History of the Papal Relations to Science During the Middle Ages and Down to Our Own Time*, Fordham University Press, New York, 1908.

[16] William Rose (ed.), *The Epic of the Beast. Consisting of English Translations of* THE HISTORY OF REYNARD THE FOX *and* PHYSIOLOGUS, George Routledge & Sons, Ltd., London, and E.P. Dutton & Company, New York, nd, p. 188.

[17] I.e., the octopus.

and gone, because his colour does not distinguish him from the rock. Then they approach the stone so close as to touch him: thus he finds his food brought to him through his own artifice.

So are they framed who hide their cunning under a bright appearance and change their minds to please their rulers and adapt their deeds to their liking so that to outward appearance they seem to obey cheerfully—. The righteous man in whom is no deceit is like Jacob, a plain man (*Gen.* 25, 27) in whom is no deceit. And of such was it said God maketh the solitary to dwell in a house. (*Ps.* 68, 7).[18]

A number of centuries and some degree of rationality separate *Physiologus* from Grandi and Leibniz. The gap between them and the mediæval scholar Alexander Neckham (1157 - 1217) is somewhat narrower. Neckham gave an interesting moral interpretation of the Horn angle (that is the angle between a circle and one of its tangent lines). Neckham started with the observation that if two circles share a tangent, the smaller circle appears to

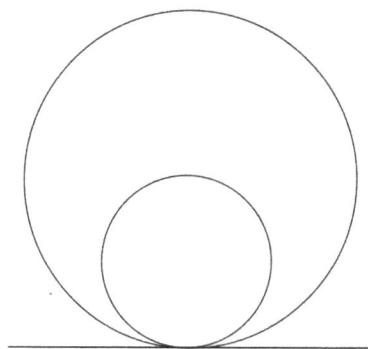

Figure 1

have the larger Horn angle. (Cf. *Figure 1.*) Now these circles represent the image to the eye of two equal sized balls placed at different distances from the observer. The nearer ball will look like the larger circle and the more distant like the smaller circle.

> However, the smaller the circle appears, the greater appears the angle of tangency: therefore the more remote the angle of tangency, the larger it appears to be. Similarly, the further the acquaintance of a powerful man is from being achieved, the more worthy of praise is he considered to be. [However,] having become the friend of the powerful man, so much less desirable will his friendship appear to you.
>
> Nevertheless, to some people it seems that this argument is invalid. The smaller the circumference the larger the angle of tangency appears to be, since as the angle under which the circle appears decreases because of its elongation from the eye, so the angle under which the angle of tangency appears decreased as it is removed [from the eye]. But how? Under what angle is the angle of tangency viewed? Likewise a straight rod appears bent in water, which is customarily attributed to reflection[19] of the rays from the surface of the water. [Now,] waters represent tribulations and the straight rod good works. Thus the works of the just, who are vexed by the tribulations, are often regarded as

---

[18] *Ibid.*, pp. 249 - 250.

[19] He should have written "refraction".

bent, although they are [actually] straight. Furthermore, the man who is in a dark place sees a man standing in the light, but not vice versa; in the same way, unimportant people, whose fortune is dark, perceive the deeds of important people, but not vice versa.[20]

If we keep the concept of the Book of Nature before us, Neckham's remarks are less nonsensical than they at first seem. God arranged things so that nature would teach us religious and moral lessons. Some natural occurrences, like the still birth of lion cubs and their three day wait for life, have obvious interpretations. The meanings of others have to be ferreted out and we see Neckham explaining the significance of one optical illusion.

The links between mathematics and theology do not stop with Neckham. Augustus de Morgan cites several works on the subject in his *Budget*[21]. In chronological order they are:

John Craig, *Theologiæ Christianne Principia Mathematica*, London, 1699.[22]

Richard Jack, *Mathematical principles of theology, or the existence of God geometrically demonstrated*, London, 1747.[23]

Anonymous, *Mathematical Illustrations of Doctrine*, London, 1855.[24]

A contemporary of Leibniz and Grandi, Craig was a Scotsman— "Scotchman" according to de Morgan— and a competent mathematician, who published various works on the Calculus and tried to introduce Leibniz's differential calculus into England. In the book in question he assumed that suspicions against historical evidence increased in proportion to the square of time and calculated that Christianity would come to an end by the year 3150. On the basis of the scriptural query, "When the Son of Man cometh, shall he find faith on the earth?", he concluded 3150 to be the earliest possible date of the Second Coming.

Here I absolutely must digress from de Morgan to quote W.W. Rouse Ball on John Craig. After discussing Craig's positive contributions, he starts a new paragraph:

It is however much easier to obtain a lasting reputation by eccentricity than by merit; and hundreds who never heard of Craig's work on

---

[20] Edward Grant, *A Source Book in Medieval Science*, Harvard University Press, Cambridge (Mass.), 1974, p. 381.

[21] Augustus de Morgan, *op. cit.*

[22] *Ibid.*, vol. I, pp. 129 - 130. A second, posthumous edition appeared in 1755 and a partial English translation was published anonymously in 1964. Finally, in 1991 there appeared a full translation complete with historical background and biographical information on Craig: Richard Nash, *John Craige's Mathematical Principles of Christian Theology* (Southern Illinois University Press, Carbondale and Edwardsville). The book consists of 6 chapters, the first 2 devoted to the analysis of evidence as cited by de Morgan, the last 4 to a variant of Pascal's wager, which wager will be described shortly.

[23] *Ibid.*, vol. I, p. 149.

[24] *Ibid.*, vol. II, pp. 70 - 74.

fluxions know of him as the author of *Theologia Christianae principia mathematica* published in 1699. He here starts with the hypothesis[25] that evidence transmitted through successive generations diminishes in credibility as the square of the time. The general idea was due to the Mahommedan apologists, who enunciated it as an axiom, and then argued that as the evidence for the Christian miracles daily grows weaker a time must come when they will have no evidential value, whence the necessity of another prophet. Curiously enough Craig's formulæ shew that the oral evidence would by itself have become worthless in the eighth century, which is not so very far removed from the date of Mahommed's death (632). He asserts that the gospel evidence will cease to have any value in the year 3150. He then quotes a text to shew that at the second coming faith will not be quite extinct among men: and hence the world must come to an end before 3150.[26] This was reprinted abroad, and seriously answered by many divines; but most of his opponents were better theologians than mathematicians, and would have been wiser if they had contented themselves with denying his axioms.[27]

De Morgan was not favourably impressed by Jack's book and describes it as neither theology nor geometry, but "a logical vagary". On this basis, I think we can discount Jack as a "kook", a theological equivalent of a circle squarer or angle trisector.

The anonymous author of the third book was not unknown to de Morgan, who reports, "The author has published mathematical works with his name," but declines to give it, respecting the author's anonymity. The work is represented as a short one of only 16 pages and ultimately derives from Pascal's Wager, published posthumously in 1670. The argument— that it is safer to believe in God than not to— goes back, in fact, to Arnobius (fl. c. 300 A.D.) whose Latin remarks are cited from Pierre Bayle's citation and appear in English translation by David Eugene Smith, de Morgan's posthumous American editor:

> But Christ himself does not prove what he promises. It is true. For, as I have said, there cannot be any absolute proof of future events. Therefore since it is a condition of future events that they cannot be grasped or comprehended by any efforts of anticipation, is it not more reasonable, out of the two alternatives that are uncertain and that are hanging in doubtful expectation, to give credence to the one that gives some hope rather to the one that offers none at all? For in the former case there is no danger if, as is said to threaten, it becomes empty and void; while in the latter case the danger is greatest, that is, the loss of salvation, if when the time comes it is found that it was not a falsehood.

---

[25] Ball here footnotes a reference to de Morgan's account in the *Budget*.

[26] Ball is not alone in reading Craig as placing the Second Coming before 3150 instead of after it. Nash, *op. cit.*, p. 89, attributes such to a mistranslation.

[27] W.W. Rouse Ball, *A History of the Study of Mathematics at Cambridge*, Cambridge University Press, Cambridge, 1889, p. 78.

Pascal's Wager is a more famous, somewhat bloated version of this framed in the language of gambling, for Pascal was one of the founders of probability theory.

One usually formulates Pascal's Wager in terms of expected value. God exists with a probability which, all things being equal, we take to be $\frac{1}{2}$.[28] Believing in him will cost you some finite amount (F) (time spent praying, tithes to the Church, etc.), not believing in him won't cost you anything in this life. If God exists and you believe in him your reward will be infinite ($+\infty$); but if you don't believe in him your punishment will be infinite ($-\infty$). If God doesn't exist, there is no reward or punishment. So, if you choose to believe, your expected value is

$$E = \frac{1}{2}(+\infty) + \frac{1}{2}(-F) = +\infty,$$

and if you choose not to believe, your expected value is

$$\frac{1}{2}(-\infty) + \frac{1}{2}(0) = -\infty.$$

Thus, you should choose to believe.

De Morgan points out one of the flaws[29] in this argument: choosing to believe for the sake of infinite gain is not likely to be smiled upon:

> I wonder whether Pascal's curious imagination ever presented to him in sleep his convert, in a future state, shaken out of a red-hot dice-box upon a red-hot hazard-table[30], as perhaps he might have been, if Dante had been the later of the two.

So we see a long tradition from early Christendom to the middle of the 19th century linking natural science with theology. As Grandi and Leibniz are between the limits of the connexions cited and coeval with the theologically inclined Newton and Craig, their seeing such between their work and the Creation is not really remarkable. They were from a more rational era than Neckham or the author of *Physiologus* and it is unlikely they were searching for theological interpretations of mathematics. But they were learned scholars of broad interests and the analogy between (15) and the Biblical account would have immediately suggested itself. Bell's version— let me repeat a portion of one of the passages quoted in the previous section:

---

[28] Any nonzero probability will work. The argument should thus be convincing to any agnostic. It is not valid, however, to any devout atheist who takes this probability to be 0.

[29] Another flaw is the formal nature of the Wager. The expected value is intuitively explained as the average value one would expect from performing a repeatable experiment, where probability is more-or-less a relative frequency of occurrence. (Cf. footnote 3 above.) The probability of God's existence is a subjective estimate of one's faith, not a relative frequency— He either exists or doesn't, not with some frequency as more and more people die. The calculation is a purely formal manipulation with nothing behind it.

[30] I.e. a table on which to roll the dice. The word "hazard" was often used in describing games of chance, i.e. gambling.

Thus, by rearranging the terms of a divergent series, it was readily proved that $0 = 1$, and hence that God had created the universe out of nothing. Even Leibniz, in a philosophical mood, drew similar conclusions from the binary scale of notation.

This is much stronger than Boyer's remark cited even earlier that Grandi "suggested to Leibniz" that equation (15) "could be compared with the mysteries of the Christian religion and with the creation of the world by which an absolutely infinite force created something out of absolutely nothing". That they would see a comparison is perfectly natural. That one could add a bunch of 0's and come up with something ought to at least remind anyone of God's creation of the universe out of nothing— whether one believes the Biblical theory or not: all that is necessary to see the parallel is knowledge of the story of the Creation, not a belief in it. Bell's version, that Grandi proved that God created the universe and Leibniz "drew similar conclusions", does indeed point to a preposterously absurd flight of fancy. But I doubt it is true and think Bell's account the flight of fancy. For one thing, the best that (15) can logically do is remove the obvious objection that one cannot create something out of nothing; it proves that it is not impossible for God to have done so, not that he did it. And I think it is safe to assume both men intelligent enough to have realised this.

I don't have Grandi's book before me, but I do have Leibniz's works on the binary number system[31] and do not find him attempting to prove that God created the world from nothing. Rather, he states in a letter of 29 March 1698 that binary arithmetic "is a beautiful sign of the perpetual creation of things from nothingness and their dependence on God". Eric Temple Bell is notorious for not getting all of his facts straight and I think we can safely discount his overstated remarks.[32] That said, it must be admitted that in another work Leibniz did, in fact, attempt to prove the existence of God, but this was a separate matter, one of philosophical argumentation, not a conclusion drawn from a few mathematical results.

Incidentally, in the 1768 publication of Leibniz's letter of 19 March 1698 there is a footnote stating that the immediate cause of the correspondence with Leibniz was the author's having read a work, *Mathesis, ejusque Indoles, Theologia applicanda* [roughly: *Mathematics and Its Methods Applied to Theology*].

Given all this background, I don't think we should find the link between (15) and the Biblical account of the Creation all that bizarre. It should also suggest to us that the other "absurdities" referred to by Burton and Bell in their rush to judgement may also not be so absurd.

---

[31] *Herrn von Leibniz' Rechnung mit Null und Eins*, 3rd edition, Siemens Aktiengesellschaft, West Berlin, 1979.

[32] Note too that none of the arguments cited in section 1 involved the *rearrangment* of terms in a series. Grandi did some re-association, but that was it.

3 Geometric Progressions and Limits

The extracts cited in section 1 above did not say whether formulæ (6) and (10) were derived from Newton's Binomial Theorem, ordinary long division, or by summing the geometric progression. All three methods were known at the time, but summing the progression is the simplest and oldest of these three possibilities.

A *geometric progression* is formally defined to be an infinite sequence of numbers generated by starting with a given number a and multiplying successively by another fixed number r, thus:

$$a, ar, ar^2, \ldots$$

One of the problems of the Rhind Papyrus shows that the ancient Egyptians knew how to sum the first $n+1$ terms of a geometric progression[33], and Euclid gave a geometric derivation of the result in the *Elements* (Proposition IX-35). Today, this is easily done algebraically. One starts with the sum of the first $n+1$ elements,

$$s_n = \sum_{i=0}^{n} ar^i = a + ar + \ldots + ar^n, \tag{16}$$

and multiplies by r:

$$s_n r = (a + ar + \ldots + ar^n)r \tag{17}$$
$$= ar + ar^2 + \ldots + ar^{n+1}, \text{ by distributivity.} \tag{18}$$

It follows by commutativity, associativity, and distributivity (of -1), that

$$s_n r - s_n = ar^{n+1} - a,$$

whence

$$s_n(r - 1) = a(r^{n+1} - 1)$$
$$s_n = a\frac{r^{n+1} - 1}{r - 1}, \text{ if } r \neq 1. \tag{19}$$

This trick can be applied to the entire geometric series, as the progression is called when it is summed. Let

$$s = \sum_{i=0}^{\infty} ar^i = a + ar + \ldots \tag{20}$$

Then

$$sr = (a + ar + \ldots)r \tag{21}$$

---

[33] Richard J. Gillings, *Mathematics in the Time of the Pharaohs*, Dover, New York, 1982, pp. 166 - 170. The original edition was published by MIT Press, 1972.

$$= ar + ar^2 + \ldots, \tag{22}$$

whence

$$sr - s = (ar + ar^2 + \ldots) - (a + ar + \ldots) \tag{23}$$
$$= (ar + ar^2 + \ldots) + (-a - ar - \ldots) \tag{24}$$
$$= (-a - ar - \ldots) + (ar + ar^2 + \ldots) \tag{25}$$
$$= (-a + ar) + (-ar + ar^2) + \ldots \tag{26}$$
$$= -a + (ar - ar) + (ar^2 - ar^2) + \ldots \tag{27}$$
$$= -a + 0 + 0 + \ldots$$
$$= -a,$$

and, finally,

$$s(r - 1) = -a$$
$$s = -\frac{a}{r-1} = \frac{a}{1-r}, \text{ if } r \neq 1. \tag{28}$$

In school we first learn (28) as the means of converting a repeating decimal into an ordinary fraction, as for example,

$$.\overline{3} = .333\ldots = \frac{3}{10} + \frac{3}{100} + \ldots$$

Here, $a = \frac{3}{10}, r = \frac{1}{10}$, and

$$s = \frac{3/10}{1 - 1/10} = \frac{3/10}{9/10} = \frac{3}{9} = \frac{1}{3}.$$

A more difficult one:

$$.\overline{142857} = \frac{142857/1000000}{1 - 1/1000000} = \frac{142857}{999999} = \frac{1}{7},$$

as one may check by performing the long division. Finally, of course, our teacher dazzles us with

$$.\overline{9} = .999\ldots = \frac{9/10}{1 - 1/10} = \frac{9/10}{9/10} = 1.$$

That there is something wrong with the derivation of (28) is demonstrated probably first in a Calculus course when the teacher might decide to plug some values like $a = 1$ and $r = -1, -2, -3$ or 2 into the formula to obtain

$$1 - 1 + 1 - 1 + \ldots = \frac{1}{2}$$
$$1 - 2 + 4 - 8 + \ldots = \frac{1}{3}$$
$$1 - 3 + 9 - 27 + \ldots = \frac{1}{4}$$

$$1 + 2 + 4 + 8 + \ldots = -1.$$

The difference between the successful and unsuccessful applications of (28) is readily clarified by consulting the partial sums s_n and formula (19). Fix $a = 1$, and observe

$$s_n = \frac{r^{n+1} - 1}{r - 1} = \frac{1}{r-1} r^{n+1} - \frac{1}{r-1}.$$

For $|r| < 1$, we write this as

$$s_n = \frac{1}{1-r} - \frac{1}{1-r} r^{n+1},$$

which is very close to $1/(1-r)$, the error being a constant $1/(1-r)$ times r^{n+1} which is very small for large n. Thus the sums, $s_0, s_1, \ldots$ get closer and closer to $1/(1 - r)$. For $|r| > 1$, however, r^{n+1} is very large in absolute value and gets larger and larger with n. The sequence $s_0, s_1, \ldots$ does not settle down, but gets larger and larger in absolute value. And, for $r = \pm 1$, we ignore (19) and observe that, for $r = 1$,

$$s_n = n + 1,$$

while, for $r = -1$,

$$s_n = \begin{cases} 1, & n \text{ is even} \\ 0, & n \text{ is odd.} \end{cases}$$

In the one case s_n grows without bound, while in the other it alternates between 1 and 0.

We can summarise this state of affairs thus:

3.1 Theorem. *In the real numbers, the infinite series $\sum_{i=0}^{\infty} r^i$ converges to $\frac{1}{1-r}$ for $|r| < 1$ and diverges for $|r| \geq 1$.*

We haven't explained the terms "converge" and "diverge", but given the previous discussion, their meanings and the truth of the theorem should be clear in the present case. Indeed, in the 14th century, Nicole Oresme (c. 1325 - 1382) stated a form of Theorem 3.1 for positive r very clearly:

> Secondly, it must be noted that if an addition were made to infinity by proportional parts in a ratio of equality or of greater inequality [i.e., a geometric progression with $r \geq 1$], the whole would become infinite; if, however, this addition should be made [by proportional parts] in a ratio of lesser inequality [i.e., $0 < r < 1$] the whole would never become infinite, even if the addition continued into infinity.[34]

This quotation is from a work, *Quæstiones super geometriam Euclides* [*Questions on the geometry of Euclid*], written about 1350. In the translated selection from which the above quote is taken, Oresme shows he not only knows the sum of the geometric progression to exist when $0 < r < 1$, but he also knows how

---

[34] Edward Grant, *A Source Book in Medieval Science*, Harvard University Press, Cambridge (Mass), 1974, p. 134.

to find this sum. Perhaps even more impressive, he showed the divergence of the *harmonic series*

$$\sum_{i=1}^{\infty} \frac{1}{i} = \frac{1}{1} + \frac{1}{2} + \frac{1}{3} + \dots,$$

by grouping terms as follows:

$1 > \frac{1}{2}$

$\frac{1}{2} \geq \frac{1}{2}$

$\frac{1}{3} + \frac{1}{4} > \frac{1}{4} + \frac{1}{4} = \frac{1}{2}$

$\frac{1}{5} + \frac{1}{6} + \frac{1}{7} + \frac{1}{8} > \frac{1}{8} + \frac{1}{8} + \frac{1}{8} + \frac{1}{8} = \frac{4}{8} = \frac{1}{2}$

$\vdots$

whence

$$\sum_{i=1}^{\infty} \frac{1}{i} > \frac{1}{2} + \frac{1}{2} + \dots = \infty.$$

Oresme also makes clear that not all of his contemporaries understood series as well as he did. Indeed, Edward Grant, in introducing the selection from Oresme in his source book, states that infinite series were much talked about in Oresme's day because of the paradoxes they give rise to.[35] As evidenced by our discussion in section 1, these paradoxes were not dispelled by the recognition that some series summed to definite numbers, while others diverged to infinity. Indeed, because of the introduction of negative numbers, new paradoxes like Bernoulli's paradox of the Grandi series emerged. It wasn't until the 19th century that infinite sums were sufficiently understood that they ceased to be paradoxical. The first step in this direction was the definition of the sum of an infinite series as the limit of the sequence of partial sums:

3.2 Definition. *Let* $a_0, a_1, \dots$ *be an infinite sequence. For each natural number* $n \geq 0$, *let*

$$s_n = \sum_{i=0}^{n} a_i.$$

---

[35] Even in modern times these paradoxes continue to amuse. On my bookshelf is a pleasant little volume written for popular consumption: Eugene P. Northrop, *Riddles in Mathematics; A Book of Paradoxes*, D. van Nostrand Company, New York, 1944. It devotes half a chapter ("7. Outward Bound (*Paradoxes of the Infinite*)") to obtaining strange sums for infinite series. An even slimmer volume, Patrick Hughes and George Brent, *Vicious Circles and Infinity; A Panoply of Paradoxes*, Doubleday & Company, Inc., New York, 1975, is devoted more to philosophy than to mathematics, and, in particular, has nothing to say about infinite series. But it does testify once again to the ongoing fascination with paradoxes, and it also informs us that Northrop's book was reprinted in London in 1967. Books like Douglas Hofstadter's *Gödel, Escher, Bach: an Eternal Golden Braid* (Harvester Press, Hassocks (Sussex), 1979) or Rudolph von Bitter-Rucker's *Infinity and the Mind* (Birkhäuser, Boston, 1982), which are still in print, demonstrate the continued interest in paradoxes of the infinite, though the nature of the paradoxes stressed has changed from the numerical to the linguistic and logical.

We define

$$\sum_{i=0}^{\infty} a_i = \lim_{n \to \infty} s_n$$

if this limit exists.

This, of course, presupposes one to have defined the notion of the limit of a sequence:

3.3 Definition. *Let $a_0, a_1, \ldots$ be an infinite sequence. A real number L is the* limit *of the sequence, written*

$$\lim_{n \to \infty} a_n = L,$$

if for any $\epsilon > 0$, however small, there is a number n_0 such that whenever $n > n_0$ we have

$$|a_n - L| < \epsilon.$$

When such a number L exists, the sequence is said to *converge* to L and to be *convergent*; if no such L exists, the sequence *diverges* and is called *divergent*.

Definition 3.3 may appear strange if one has not seen it before. But it can be explained in fairly simple terms. The sequence

$$1, \frac{1}{2}, \frac{1}{3}, \frac{1}{4}, \ldots \tag{29}$$

keeps getting smaller and smaller, the term $a_n = \frac{1}{n+1}$ getting closer and closer to 0. But it is also getting closer and closer to -1. One reason 0 is the limit and -1 is not is that the sequence not only approaches 0, but gets as close as we want to 0. If we specify how close we want to get by choosing some tolerance $\epsilon > 0$, we can find n_0 such that

$$|a_{n_0} - 0| < \epsilon,$$

which we cannot do for -1:

$$|a_n - (-1)| > 1, \text{ for all } n.$$

Now, this much also holds for the sequence,

$$1, \frac{1}{2}, 1, \frac{1}{3}, 1, \frac{1}{4}, 1, \ldots$$

This gets as close as we want to 0, but at every other step it bounces away. Sequence (29), however, gets close and stays close: given $\epsilon > 0$, for all $n > 1/\epsilon$,

$$|a_n - 0| < \epsilon.$$

Thus, Definition 3.3 states, in formal terms, that whatever degree of closeness we specify, by going far enough out in the sequence we can get that close to the limit L and stay that close if we proceed even farther out in the sequence.

Using these definitions, our remarks on the preceding Theorem 3.1 are readily turned into a formal proof thereof. Perhaps most interesting, however, are the proofs of divergence for $r = \pm 1$.

$\underline{r = 1}$. If a limit existed for $\sum_{i=0}^{\infty} 1$, then by definition, the sequence of partial sums $s_n = 1 + \ldots + 1 = n + 1$ would converge to some number L. Choose $\epsilon = \frac{1}{2}$ and n_0 such that for any $n > n_0$

$$|s_n - L| < \frac{1}{2}.$$

But observe

$$
\begin{aligned}
1 = |n + 2 - (n + 1)| &= |s_{n+1} - s_n| \\
&= |s_{n+1} - L + L - s_n| \\
&\leq |s_{n+1} - L| + |L - s_n| \\
&\leq |s_{n+1} - L| + |s_n - L| \\
&< \frac{1}{2} + \frac{1}{2} = 1, \text{ for } n > n_0.
\end{aligned}
$$

The contradiction, that $1 < 1$, means our assumption of convergence was false.

$\underline{r = -1}$. Let $\sum_{i=0}^{\infty}(-1)^i$ converge to L. The partial sums

$$
s_n = \begin{cases} 1, & n \text{ is even} \\ 0, & n \text{ is odd} \end{cases}
$$

again satisfy

$$1 = |s_{n+1} - s_n|$$

and s_n, s_{n+1} cannot both be close to a given L.

On the positive side, one proves several lemmas.

3.4 Lemma. *Let $a_0, a_1, \ldots$ be an infinite sequence. If L_1, L_2 are real numbers such that*

$$\lim_{n \to \infty} a_n = L_1 \quad and \quad \lim_{n \to \infty} a_n = L_2, \tag{30}$$

then $L_1 = L_2$, i.e. limits, if they exist, are unique.

Proof. Suppose (30) holds and $L_1 \neq L_2$. Choose $\epsilon = |L_1 - L_2|$ and find n_1, n_2 such that

$$\text{for } n > n_1, \quad |a_n - L_1| < \frac{\epsilon}{2}$$

$$\text{for } n > n_2, \quad |a_n - L_2| < \frac{\epsilon}{2},$$

and note that, for any $n > n_0 = \max\{n_1, n_2\}$, one has

$$
\begin{aligned}
|L_1 - L_2| = |L_1 - a_n + a_n - L_2| \\
\leq |L_1 - a_n| + |a_n - L_2| \\
< \frac{\epsilon}{2} + \frac{\epsilon}{2} = \epsilon = |L_1 - L_2|,
\end{aligned}
$$

a contradiction. Thus $L_1 = L_2$. □

3.5 Lemma. *Let $\sum_{i=0}^{\infty} a_i$ and $\sum_{i=0}^{\infty} b_i$ converge. Then $\sum_{i=0}^{\infty}(a_i + b_i)$ converges and*

$$\sum_{i=0}^{\infty}(a_i + b_i) = \sum_{i=0}^{\infty} a_i + \sum_{i=0}^{\infty} b_i.$$

Proof. Let $\sum_{i=0}^{\infty} a_i = L_1$ and $\sum_{i=0}^{\infty} b_i = L_2$. Choose $\epsilon > 0$ and n_1, n_2 such that

$$\text{for } n > n_1, \quad \left| \sum_{i=0}^{n} a_i - L_1 \right| < \frac{\epsilon}{2}$$

$$\text{for } n > n_2, \quad \left| \sum_{i=0}^{n} b_i - L_2 \right| < \frac{\epsilon}{2},$$

and observe, for $n > n_0 = \max\{n_1, n_2\}$,

$$\left| \sum_{i=0}^{n}(a_i + b_i) - (L_1 + L_2) \right| = \left| \sum_{i=0}^{n} a_i - L_1 + \sum_{i=0}^{n} b_i - L_2 \right|$$

$$\leq \left| \sum_{i=0}^{n} a_i - L_1 \right| + \left| \sum_{i=0}^{n} b_i - L_2 \right|$$

$$< \frac{\epsilon}{2} + \frac{\epsilon}{2} = \epsilon. \qquad \square$$

3.6 Lemma. *Let $\sum_{i=0}^{\infty} a_i$ converge and let r be any real number. Then $\sum_{i=0}^{\infty} r a_i$ converges and*

$$\sum_{i=0}^{\infty} r a_i = r \sum_{i=0}^{\infty} a_i.$$

Proof. Assume first that $r \neq 0$. Let $\sum_{i=0}^{\infty} a_i = L$. Finally, let $\epsilon > 0$ be given and suppose n_1 is such that

$$\text{for } n > n_1, \quad \left| \sum_{i=0}^{n} a_i - L \right| < \frac{\epsilon}{|r|}.$$

Then, for $n > n_0 = n_1$,

$$\left| \sum_{i=0}^{n} r a_i - rL \right| = \left| r \sum_{i=0}^{n} a_i - rL \right| = |r| \cdot \left| \sum_{i=0}^{n} a_i - L \right| < |r| \cdot \frac{\epsilon}{|r|} = \epsilon.$$

For $r = 0$, the multiplied series is $\sum_{i=0}^{\infty} 0$ with all partial sums equal to 0, while $0 \cdot \sum_{i=0}^{\infty} a_i$ is $0 \cdot L = 0$, and

$$\left| \sum_{i=0}^{n} 0 - 0 \right| < \epsilon$$

for any $\epsilon > 0$ and any $n \geq 0$. $\qquad \square$

3.7 Remark. For $r \neq 0$ the converse holds: if $\sum_{i=0}^{\infty} ra_i$ converges, then so does $\sum_{i=0}^{\infty} a_i$ and

$$r \sum_{i=0}^{\infty} a_i = \sum_{i=0}^{\infty} ra_i.$$

One sees this simply by multiplying $\sum_{i=0}^{\infty} ra_i$ by $1/r$.

Lemma 3.4 allows us to speak of *the* sum when it exists. Lemmas 3.5 and 3.6 are the *linearity* conditions that occur all over mathematics. With them we can almost go back to our derivation of (28) summing the geometric progression and justify the steps. Assuming the sum (20) exists as a limit, Lemma 3.6 justifies the passage from (21) to (22). The passage from (22) to (23) is just the substitution of equals, and Lemma 3.6 is again applied to conclude (24). The commutativity of finite sums yields (25) and Lemma 3.5 yields (26). The rest of the derivation is a re-association, which still needs some justification.

The sequences of (26) and (27),

$$(-a + ar) + (-ar + ar^2) + \dots \tag{26'}$$

$$-a + (ar - ar) + (ar^2 - ar^2) + \dots \tag{27'}$$

are both re-associations of the unassociated series

$$-a + ar - ar + ar^2 - ar^2 + \dots \tag{31}$$

If we look at the sequence of partial sums of (31),

$$s_0 = -a$$
$$s_1 = -a + ar$$
$$s_2 = -a + ar - ar$$
$$\text{etc.,}$$

we see that the partial sums of (26′) are

$$s_1, s_3, s_5, \dots$$

while those of (27′) are

$$s_0, s_2, s_4, \dots$$

In short, they are subsequences of the sequence of partial sums of (31).

3.8 Lemma. *Suppose $a_0, a_1, \dots$ is an infinite sequence of real numbers and $\lim_{n \to \infty} a_n = L$. Let $j_0 < j_1 < \dots$ be an infinite sequence of non-negative integers and define a new sequence $b_0, b_1, \dots$ by setting $b_n = a_{j_n}$. Then the sequence $b_0, b_1, \dots$ converges to L, i.e.*

$$\lim_{n \to \infty} b_n = \lim_{n \to \infty} a_n.$$

Proof. Let $\epsilon > 0$. Choose n_0 such that, for $n > n_0$,

$$|a_n - L| < \epsilon.$$

Now $b_n = a_{j_n}$ and $j_n \geq n$, so

$$|b_n - L| = |a_{j_n} - L| < \epsilon. \qquad \square$$

Now the transition from line (26) to (27) in our earlier derivation is justified by noting that (26′) and (27′) both equal the limit of series (31)— *provided* this limit exists. At this stage, the only way of seeing this is to show directly that the limit is $-a$. A more elegant approach to deriving (28) might be to go from (25) to (27) in a slightly different manner using the following Lemma instead of Lemma 3.8:

3.9 Lemma. *Let $a_0, a_1, \ldots$ be an infinite sequence and define a sequence $b_0, b_1, \ldots$ by*

$$b_0 = 0$$
$$b_{n+1} = a_n.$$

Then $\sum_{i=0}^{\infty} b_i$ exists iff $\sum_{i=0}^{\infty} a_i$ exists and

$$\sum_{i=0}^{\infty} b_i = \sum_{i=0}^{\infty} a_i.$$

The sequence b is obtained from the sequence a simply by shoving a 0 in front of the latter sequence, whence the Lemma says

$$0 + a_0 + a_1 + \ldots = a_0 + a_1 + \ldots$$

Proof of the Lemma. Assume first that $\sum_{i=0}^{\infty} a_i$ exists. Let $\sum_{i=0}^{\infty} a_i = L$. Let $\epsilon > 0$ be given, and choose n_1 such that for $n > n_1$,

$$\left| \sum_{i=0}^{n} a_i - L \right| < \epsilon.$$

Note that

$$\left| \sum_{i=0}^{n} b_i - L \right| = \left| 0 + \sum_{i=0}^{n-1} a_i - L \right|$$

$$= \left| \sum_{i=0}^{n-1} a_i - L \right|$$

$$< \epsilon$$

provided $n - 1 > n_1$. So, if we choose $n_0 = n_1 + 1$, we have

$$\left| \sum_{i=0}^{n} b_i - L \right| < \epsilon$$

for $n > n_0$.

The converse argument, that if $\sum_{i=0}^{\infty} b_i$ exists, then so does $\sum_{i=0}^{\infty} a_i$ and the two limits are equal is treated similarly. □

We can now justify all the steps in summing the geometric progression algebraically in case the series converges: For

$$s = \sum_{i=0}^{\infty} ar^i$$

we have

$$
\begin{aligned}
sr &= (a + ar + \ldots)r \\
&= ar + ar^2 + \ldots, \text{ by Lemma 3.6} \\
sr - s &= (ar + ar^2 + \ldots) - (a + ar + \ldots) \\
&= (ar + ar^2 + \ldots) + (-a - ar - \ldots), \text{ by Lemma 3.6} \\
&= (-a - ar - \ldots) + (ar + ar^2 + \ldots) \\
&= (-a - ar - \ldots) + (0 + ar + ar^2 + \ldots), \text{ by Lemma 3.9} \\
&= (-a + 0) + (-ar + ar) + \ldots, \text{ by Lemma 3.5} \\
&= -a + 0 + 0 + \ldots && (32) \\
&= -a && (33) \\
s(r - 1) &= -a \\
s &= \frac{-a}{r - 1} = \frac{a}{1 - r}, \text{ if } r \neq 1.
\end{aligned}
$$

All of the unexplained transitions, other than the step from (32) to (33), are justified by ordinary algebra applied to finite quantities. The passage from (32) to (33) seems too obvious to require justification, but if one wishes to prove it, note that all the partial sums of (32) are $-a$, whence $-a$ is the limit of the series. We can isolate this as a lemma:

3.10 Lemma. *Let $a_0, a_1, \ldots$ be a sequence and suppose $a_n = 0$ for all $n > n_0$. Then*

$$\sum_{i=0}^{\infty} a_i = \sum_{i=0}^{n_0} a_i.$$

Proof. Note that for $n > n_0$, the n-th partial sum is

$$s_n = \sum_{i=0}^{n} a_i = \sum_{i=0}^{n_0} a_i + a_{n_0+1} + \ldots + a_n = \sum_{i=0}^{n_0} a_i + 0 + \ldots + 0 = \sum_{i=0}^{n_0} a_i,$$

whence for $n > n_0$,

$$\left| s_n - \sum_{i=0}^{n_0} a_i \right| = 0 < \epsilon. \qquad\qquad \square$$

So we see that, when the geometric series converges, all the steps used in solving for the sum of the progression are justified and we get the correct answer. Of course, we get this information and more— namely, a determination of when the series converges— from Theorem 3.1. Nonetheless, as we will see later, there is still some interest in our justification of the algebraic determination of the sum.

4 Bernard Bolzano on Infinite Sums

Occasionally in the history of science there are individuals who make tremendous discoveries which go unnoticed. The individual is forgotten until the discovery is made anew, the result enters the mainstream, and eventually someone uncovers the individual's priority and he is duly celebrated. The most famous example is probably Gregor Mendel and his discovery of the laws of heredity. In mathematics, Grassmann and his theory of vectors comes to mind. And then there is Bernard Bolzano.

Bolzano was a Bohemian priest, working in the mathematical backwaters of what is now the Czech Republic. His publications on the foundations of analysis went largely unnoticed and were overshadowed by the subsequent rediscovery and publication of his results by Augustin Louis Cauchy working in Paris, a major mathematical centre. Today, Bolzano's name is known to all mathematicians through i) his share of formal credit for the eponymous Bolzano-Weierstrass Theorem, by which every bounded sequence contains a convergent subsequence, ii) where authorship is mentioned, the Intermediate Value Theorem, and iii) his construction of an everywhere nondifferentiable function. Books on the general history of mathematics mention him briefly on the way to discussing Cauchy, or they mention him in passing as a precursor to Georg Cantor and the theory of sets. These are fair representations. In the case of set theory, Bolzano saw more clearly than any of his predecessors what was going on, but unlike Richard Dedekind and Georg Cantor, he did not develop the subject very far mathematically. As to the foundations of analysis, he had our modern notions of limit, continuity, Cauchy-convergence of sequences, the Bolzano-Weierstrass Theorem, and a proof of the Intermediate Value Theorem. These are undeniably great achievements, but they had little impact because of the obscurity of his publications and, in a different sense, the obscurity of his writing. Hans Hahn, a competent mathematician and expert in real analysis, and native German speaker to boot, had this to say about Bolzano's definition of infinite sums in annotating one of Bolzano's works:

The definition of the concept of a sum given here is so abstract, and of such little clarity, that it is difficult to determine its exact meaning. It could mean the following...[36]

Several sections after defining the sum, Bolzano proves in a footnote spread across the bottoms of three pages that the geometric series $a + ae + ae^2 + \ldots$ converges to $\frac{a}{1-e}$ for $0 < e < 1$. Hahn criticises the proof thus:

What is said here about sums of infinitely many magnitudes, in particular the calculation worked through in footnote *), does not agree with the modern view of this situation. Probably on the basis of the summation concept introduced in §5, Bolzano believed himself justified to consider without further ado a sum of infinitely many summands. We have already seen that this concept of sum is too indefinite, to be able to begin to do anything with it. One will thus have objected to B's proof that there is no precise concept tied to the sums occurring in it, and that the calculations carried out (for example factoring out an individual term) are not grounded on anything.[37]

No less an authority on the infinite than Georg Cantor declared much[38] of what Bolzano wrote about infinite sums to be "haltlos und irrig" [groundless and incorrect][39]. Julian Lowell Coolidge summed up the situation with what appears to be the greatest of Bolzano's paradoxes of the infinite:

I cannot help expressing surprise that the man who first introduced the correct definition of a convergent series should not have grasped the idea of defining a series as a limit...[40]

That Bolzano could introduce the correct definition of a convergent series and yet not deem a series to *be* a limit is not really that paradoxical. As "one of the fathers" of the arithmetisation of analysis[41], Bolzano was at the beginning of the transition of analysis from geometry to arithmetic, of the replacement of vague intuitive notions by precise formally defined concepts. He stood at the threshold and only partially crossed over it. Bob van Rootselaar says in his

---

[36] Bernard Bolzano, *Paradoxien des Unendlichen*, 2nd edition, Verlag von Felix Meiner, Leipzig, 1920, p. 134. (The interior lists 1920 as the date of publication, while the cover of my copy states 1921. It is usually cited as having been published in 1920, so I choose that date here.)

[37] *Ibid.*, p. 137.

[38] Pp. 29 - 33 of *Paradoxien des Unendlichen*.

[39] Georg Cantor, "Über unendliche, lineare Punktmannichfaltigkeiten", part 5, *Mathematische Annalen* 21, pp. 545 -591; here: p. 561.

[40] Julian Lowell Coolidge, *The Mathematics of Great Amateurs*, Dover Publications, Inc., New York, 1963, p. 202. The book was originally published by Oxford University Press in 1949 and a new edition annotated by Jeremy Gray was published by that Press in 1990.

[41] Felix Klein, *Vorlesungen über die Entwicklung der Mathematik im 19. Jahrhundert*, Springer-Verlag, Berlin, 1926, volume I, p. 56. The two volumes were re-issued in a single volume edition by Springer in 1979.

entry on Bolzano in *The Dictionary of Scientific Biography* that Bolzano was at the borderline between intension and extension. Recall Euler's, to modern eyes mysterious or even mystifying, distinction between the sum and the value of an infinite series and his remark on the value of the algebraic expression from which the series comes. The infinite series was not just a sequence of numbers to be added up, but was an expansion, following a definite rule, of an algebraic expression. The sum might be determined as a limit of a sequence of partial sums, and so depended extensionally on the numbers, but the value was determined by the rule for generating the given sequence. Bolzano similarly had a dual view of infinite series. The meaning was determined by adding the numbers all at once; the value was the limit of a sequence of finite additions.

Bolzano's definition of a sum reads more like the definition of set theoretic union:

> As is known there are also classes [Inbegriffe], whose parts themselves are yet collected together, i.e. are once again classes. Among these are also such... for which nothing in their essential nature changes if we interpret the parts of the parts as parts of the whole. I call them sums, with a word borrowed from mathematics. Because that is just the concept of a sum that $A + (B + C) = A + B + C$ must be the case.[42]

As I've already reported, Hahn finds this explanation lacking in clarity. He does, however, point to several sections of Bolzano's *Wissenschaftslehre*, a four volume work that has been abridged and translated into English a couple of times. The translation by Rolf George only summarises §84, "Concepts of Sets and Sums":

> B. introduces the notion of a set as a class where the manner of connection between elements is not specified. Sets fall into two kinds: those where the parts of parts are themselves parts of the class, and those where this is not the case. He calls the former "sums". "It follows from the concept of a sum that it is not changed if the order of its parts is changed, and that it is not changed if one of the parts is replaced by the parts of that part."
>
> In a note he contends that "the meaning which the mathematicians connect with the word 'sum' when they claim to get a sum whenever they conjoin a couple of expressions with a '+' is the same as mine. It is perhaps different in the sense that they apply the concept only to *magnitude*— as their science requires. However, they take the word in an entirely different sense when they undertake to find the sum of a set of given number-expressions, for instance $1 + \frac{1}{2} + \frac{1}{4} + \frac{1}{8} + ad$ *infinitum*. In this case they mean by 'sum' the simplest possible expression that is equivalent to the given set of number expressions."[43]

---

[42] *Op. cit.*, p. 4. I'd be willing to bet that I've mistranslated this last sentence were it not for the fact that Hahn points out in his note on p. 134 that the equation is the associative law and holds for multiplication as well.

[43] Rolf George (ed. and translator), *Bernard Bolzano, Theory of Science*, University of California Press, Berkeley and Los Angeles, 1972, p. 127.

Hahn had read the whole of Bolzano's discussion of sums and still found the notion unclear. So if this summary doesn't exactly clarify the situation, we should not blame ourselves. This poses an embarrassing problem for me as I want to report on what Bolzano had to say about infinite series, but I do not understand it all because I do not know what he means by a sum. I can only say, "Look, he could mean something like *this*, and if *this* is what he means, here is what he does with it and it makes sense."

If we consider a line segment AB, we can subdivide it into two segments AC and CB and regard AB as the sum of AC and CB. If AC is further subdivided into AD and DC, we can replace AC by AD and DC and regard AB as the sum of AD, DC and CB. Moreover, associativity is automatic:

$$AB = (AD + DC) + CB = AD + DC + CB.$$

If we have a whole bunch of intervals laid out one after another,

$$A_0A_1, A_1A_2, \ldots,$$

their sum will be another, possibly infinitely long, segment. If the length of segment A_iA_{i+1} is a_i, the length of the sum of the segments will be the sum of the lengths of the segments, $\sum_{i=0}^{\infty} a_i$. Viewed this way, the sum is given to us in one step. It is the length of a line segment, not necessarily the limit of the sequence of partial sums.

Similarly, one could consider infinitely many forces acting on an object. The sum total of all those forces is determined by its effect on the object without recourse to taking a limit.

If we bear this in mind the following passage from §32 of *Paradoxien des Unendlichen*, nonsensical on first reading, makes some small sense:

> In 1830 someone signing himself M.R.S. attempted to prove in Gergonne's *Annales de Mathématique* (vol. 20, no. 12) that the well-known infinite series
> $$a - a + a - a + a - a + \ldots \; ad \; inf.$$
> has the value $\frac{a}{2}$; to do this he set this value $= x$, believed he could conclude that
> $$x = a - a + a - a + \ldots \; ad \; inf. = a - (a - a + a - a + \ldots \; ad \; inf.)$$
> and the series enclosed in parentheses is identified to that being calculated, and could again be set $= x$, whence
> $$x = a - x$$
> and that
> $$x = \frac{a}{2}.$$

The error here is not hidden deep. The series in parentheses obviously no longer has the same list of members as the first that was set $= x$;

but rather it is missing the first a. Its value must thus, if it can be pretended at all, be given as $x - a$; which however would only yield the identity

$$x = a + x - a.^{44}$$

At first sight this critique makes no sense at all. But if we think of the sum as collecting all the pieces, like assembling all the line segments or combining all the forces, we must admit that the series $-a + a - a + \ldots$ is indeed missing the first term of x. Hahn's annotation merely points out that today we explain away the calculation by pointing to the divergence of the series and the non-existence of x.[45]

Once we realise that x doesn't exist, Bolzano's criticism just seems to be beside the point. If, however, we don't realise this, it does seem to be a way out of the paradox. To complicate matters, Bolzano goes on[46] to prove that x doesn't exist, or, in his words, the sum $a - a + a - \ldots$ is an "objectless number-expression", i.e. an algebraic expression that doesn't refer to anything. He does this by recalling that it is in the nature of a sum that one can re-order its parts and group them however we please, and then noting that

$$(a - a) + (a - a) + \ldots = 0$$
$$a + (-a + a) + \ldots = a$$
$$-a + (a - a) + \ldots = -a,$$

the third sum obtained by permuting adjacent pairs and then grouping. It follows that $a - a + a - \ldots$ is not a sum in his sense. Hahn's annotation[47] explains the notion of absolute convergence and remarks that any series, such as

$$1 - \frac{1}{2} + \frac{1}{3} - \frac{1}{4} + \ldots \tag{34}$$

which is convergent but not *absolutely convergent*[48] can be made to converge to any given limit or to diverge by a suitable rearrangement of its terms.

The divergence of the sum of the absolute values of the terms of (34),

$$1 + \frac{1}{2} + \frac{1}{3} + \frac{1}{4} + \ldots, \tag{35}$$

as pointed out in §3, above, was first shown by Oresme in the 14th century. It had also been rediscovered by one of the Bernoullis. The convergence of (34) is easy to prove by appeal to the Bolzano-Weierstrass Theorem. Actually calculating the sum ($\ln 2$, i.e. the logarithm of 2 to the base $e = 2.71828\ldots$) requires more work. I think it safe to assume Bolzano would have been aware of the

---

[44] *Op. cit.*, p. 49.
[45] *Ibid.*, p. 145.
[46] *Ibid.*, pp. 50 - 51.
[47] *Ibid.*, p. 146.
[48] The notion of absolute convergence will be formally defined and discussed in the next section.

convergence of the series (34) and the divergence of (35). Whether it occurred to him that (34) could, like the Grandi series, be rearranged to give other limits and thus is not properly a sum in his sense I cannot say. Bolzano died in 1847, his *Paradoxien des Unendlichen* was first published in 1851, and the general result referred to by Hahn was proven by Bernhard Riemann in 1854 (and published some time later). However, in 1837 P.G.L. Dirichlet had published examples of convergent series whose sums could be altered by rearrangement.

Mathematically, Bolzano is on very shaky ground as soon as his series contains terms of opposite signs. In 1839, when Dirichlet was still rearranging the terms of non-absolutely convergent series to obtain different limits, he also proved that, for an absolutely convergent series, the rearrangements always yielded the same limits. Thus, from a modern standpoint we might be tempted to point to absolutely convergent series as Bolzano's sums. But this doesn't really capture his intent. In a set theoretic union, one only adds elements together; one performs no deletions. A mathematical sum $a_0 + a_1 + \ldots$ as the magnitude of the union of non-overlapping sets with individual non-negative magnitudes $a_0, a_1, \ldots$ makes good intuitive sense. Since the union is given in no particular order, the magnitude of the union ought not to matter on how the terms in the sum are associated or ordered. There is a genuine conception behind Bolzano's use of sums. He did not clarify it very well, but it would appear that it applies to sums of non-negative terms and the numerical value of his sum coincides with the limit of the sequence of partial sums. In short, in the arithmetic domain, Bolzano's notion of an infinite sum is just a more restricted notion of sum than the modern one.

If we bear all of this in mind, we can read Bolzano's proof that the sum of the geometric progression[49],

$$s = 1 + e + e^2 + \ldots, \quad 0 < e < 1,$$

is $\frac{1}{1-e}$ and not be mystified by it. He starts by noting that for any positive integer n,

$$\begin{aligned}
s &= 1 + e + e^2 + \ldots + e^{n-1} + e^n + e^{n+1} + \ldots \\
&= \frac{1-e^n}{1-e} + e^n + e^{n+1} + \ldots \\
&= \frac{1-e^n}{1-e} + P_1,
\end{aligned} \tag{36}$$

where P_1 denotes the sum

$$e^n + e^{n+1} + \ldots$$

P_1 depends on e and n. He now notes that

$$P_1 = e^n + e^{n+1} + \ldots = e^n[1 + e + \ldots]$$

and warns that the sum within the brackets,

---

[49] *Ibid.*, pp. 24 - 26.

$$[1 + e + \ldots],$$

looks like the original sum s, but it isn't the same because, although both sums are infinite, it has n fewer terms than does s. All we can say about the expression within the brackets is that it is $s - P_2$, where we may assume P_2 is a positive quantity depending on n.

As Hahn notes in his annotated remark cited earlier, the calculations here are "not grounded on anything". Bolzano has offered no justification for assuming the distributive law,

$$e^n + e^{n+1} + \ldots = e^n[1 + e + \ldots].$$

However, if we are thinking of sums of lengths, we can offer an intuitive justification on the basis of change of scale: if, for example, $a_0, a_1, \ldots$ are measures in feet, we can measure the whole in inches by summing the conversions

$$12a_0 + 12a_1 + \ldots$$

or converting the sum

$$12(a_0 + a_1 + \ldots),$$

and the results should be the same. So we can let this objection slide.

If we think of $1, e, e^2, \ldots$ as being the lengths of intervals $I_0, I_1, \ldots$, then indeed the expression within the brackets depends on the intervals $I_n, I_{n+1}, \ldots$ and not on all of $I_0, I_1, \ldots$; but the quantities are not the lengths of $I_n, I_{n+1}, \ldots$, but $1/e^n$ times these lengths. These values turn out to be exactly the lengths of $I_0, I_1, \ldots$ and the expression is equal to s and not just $s - P_2$ for P_2 positive. If we don't realise $P_2 = 0$, there is no reason here to assume P_2 positive and not negative. If it were negative, the rest of the argument would fail.

Continuing his proof, Bolzano concludes

$$s = \frac{1 - e^n}{1 - e} + e^n(s - P_2) = \frac{1 - e^n}{1 - e} + e^n s - e^n P_2, \tag{37}$$

whence

$$s(1 - e^n) = \frac{1 - e^n}{1 - e} - e^n P_2, \tag{38}$$

and finally

$$s = \frac{1}{1 - e} - \frac{e^n}{1 - e^n} P_2. \tag{39}$$

Combining this last with (36),

$$\frac{1 - e^n}{1 - e} + P_1 = \frac{1}{1 - e} - \frac{e^n}{1 - e^n} P_2$$

$$-\frac{e^n}{1 - e} + P_1 = -\frac{e^n}{1 - e^n} P_2$$

$$P_1 + \frac{e^n}{1 - e^n} P_2 = \frac{e^n}{1 - e}.$$

Now the right-hand side can be made arbitrarily small by choosing n large. Hence P_1 and $\frac{e^n}{1-e^n}P_2$, being positive and adding up to $\frac{e^n}{1-e}$, which is also positive, can also be made small by choosing n large, whence each of (36) and (39) yields $s = \frac{1}{1-e}$.

Hahn[50] completed his remarks on this proof by citing the formal definition of convergence and giving a correct proof that the limit of the partial sums is $\frac{1}{1-e}$. Hahn's proof is much simpler and seems to be more to the point:

$$a + ae + \ldots + ae^{n-1} = a\frac{1 - e^n}{1 - e},$$

but, for $|e| < 1$,

$$\lim_{n\to\infty} e^n = 0,$$

whence

$$\lim_{n\to\infty} (a + ae + \ldots + ae^{n-1}) = \frac{a}{1 - e}. \tag{40}$$

A look at (36),

$$s = \frac{1 - e^n}{1 - e} + P_1,$$

brings to light the difference in what was attempted. Bolzano wanted to show not only that

$$\lim_{n\to\infty} (1 + e + \ldots + e^{n-1}) = \frac{1}{1 - e},$$

but that

$$s = \lim_{n\to\infty} (1 + e + \ldots + e^{n-1}),$$

i.e. that the remainder gradually disappears:

$$\lim_{n\to\infty} P_1 = 0.$$

Hahn is allowed to ignore P_1.

Bolzano's proof fails at his task. Embedded in it are all the steps needed to show (40), but he cannot show that this limit equals s without assuming s to exist (as he does at the outset) and is finite (the passage from (37) to (38) is only justified numerically if s is finite). If one ignores the intensional aspect— i.e., worrying about which classes the sums come from— and extensionally assumes that the sum depends only on the terms in it, so that the sum inside the brackets equals s and the red herring P_2 drops out of the picture, the derivation after (36) proceeds as follows:

$$s = \frac{1 - e^n}{1 - e} + e^n[1 + e + \ldots]$$

$$= \frac{1 - e^n}{1 - e} + e^n s \tag{41}$$

$$s(1 - e^n) = \frac{1 - e^n}{1 - e} \tag{42}$$

---

[50] *Ibid.*, pp. 137 -139.

$$s = \frac{1}{1-e},$$

where, again, in the passage from (41) to (42) one *assumes* s exists and is finite. No use of the assumption $|e| < 1$ need be made[51] and this argument is no more valid for such e than it was for $e = -1$.

I don't see anything in the *Paradoxien des Unendlichen* that allows one to conclude the existence of any infinite sum that contains more than finitely many nonzero terms. Bolzano does say that in the finite case his notion agrees with ordinary addition. He states that the terms of a sum can be re-associated and rearranged. He implicitly uses some arithmetic laws like distributivity. And he assumes that, if one truncates a series, the sum of the original series equals the sum of the truncated portion added to the finite sum of the deleted terms. Now all of these properties hold for the more primitive notion of sum given by adding the nonzero terms if only finitely many terms are nonzero and declaring the series divergent otherwise. Bolzano's explanation of the notion of sum is too vague to rule out this primitive notion of summation. Hence Bolzano cannot prove that any but a finite series has a sum. I will explain this in greater detail later; for now I want to turn to a sharper explanation of Bolzano's sums.

5 Bolzano Summation and Absolute Convergence

Given an infinite sequence $a_0, a_1, \ldots$ one can more readily write down the expressions

$$a_0 + a_1 + \ldots \qquad \text{or} \qquad \sum_{i=0}^{\infty} a_i$$

and start manipulating them than one can explain what they mean. The standard definition of the infinite sum as the limit of the sequence of partial sums is only one possible explanation, as I hope to have indicated in discussing Bolzano's conception of summation. I wish to do this more definitively in the present section by presenting a very precise definition of a notion of summation that fits Bolzano's intuitions well. The key, as already hinted a few pages back, is to begin by restricting our attention to sequences of non-negative real numbers.

Bolzano's contributions to the foundations of analysis include the recognition of the Least Upper Bound Principle.

5.1 Definition. *Let X be a non-empty set of real numbers. A number B is an* upper bound *for X if*

$$\text{for all } x \in X, \ x \le B.$$

B is a least upper bound *for X if*
i. B is an upper bound for X; and
ii. if B' is any upper bound for X, then $B \le B'$.

---

[51] One can justify the last step by dividing both sides of (42) by $1 - e^n$, or by using the assumption $|e| < 1$ and taking the limit.

Note that, if B_1 and B_2 are least upper bounds for X, then by 5.1.ii, $B_1 \leq B_2$ and $B_2 \leq B_1$, i.e. $B_1 = B_2$. Thus, least upper bounds, when they exist, are unique.

5.2 Theorem (Least Upper Bound Principle). *Let X be a nonempty set of real numbers. If X has an upper bound, then X has a least upper bound.*

Theorem 5.2 can either be taken as an axiom or be proven on the basis of another axiom or, if one has in mind one of the constructions of the set of real numbers from the rationals, on the basis of that construction. Let us accept it as an axiom here.

5.3 Remark. Dual to the Least Upper Bound Principle is a Greatest Lower Bound Principle. It does not need to be assumed as an additional axiom as it reduces, via the order reversing property of negation, to the Least Upper Bound Principle: If a set X has a lower bound, say a, then $-a$ is an upper bound for $-X = \{-x | x \in X\}$. Choose B to be the least upper bound for this set and notice that $-B$ is the greatest lower bound for $--X = \{--x | x \in X\} = \{x | x \in X\} = X$. Results about least upper bounds are fairly automatically accompanied by dual results about greatest lower bounds. These are often left unstated, or stated without proof as they usually reduce to the corresponding results about least upper bounds by inserting minus signs in the proper places.

We can use the least upper bound principle to define the sums of some infinite series as follows.

5.4 Definition. *Let $a : a_0, a_1, \ldots$ be a sequence of non-negative real numbers. We define $S_B(a)$ to be the least upper bound of the set of finite sums,*

$$\left\{ \sum_{i \in Y} a_i \mid \emptyset \neq Y \subseteq \mathbb{N}, \ Y \text{ finite} \right\},$$

of elements of the sequence, if this least upper bound exists. Further, when $S_B(a)$ exists, we say a is Bolzano summable *and call $S_B(a)$ the* Bolzano sum *of a.*

5.5 Lemma. *Let $a : a_0, a_1, \ldots$ be a sequence of non-negative real numbers. $S_B(a)$ exists iff $\lim_{n \to \infty} \sum_{i=0}^{n} a_i$ exists and, if these exist, then*

$$S_B(a) = \lim_{n \to \infty} \sum_{i=0}^{n} a_i.$$

Proof. Let

$$s_n = \sum_{i=0}^{n} a_i$$

be the $n+1$-st partial sum. Because the a_i's are all non-negative, the sequence of partial sums is monotone:

$$s_0 \leq s_1 \leq s_2 \leq \ldots$$

Suppose first that $L = \lim_{n \to \infty} s_n$ exists. By monotonicity, each $s_n \leq L$. (*Exercise.* Prove this using Definitions 3.2 and 3.3.) For any finite set $Y = \{i_0, \ldots, i_m\} \subseteq \mathbb{N}$, if n is the maximum of the elements of Y, we have

$$\sum_{i \in Y} a_i = a_{i_0} + \ldots + a_{i_m} \leq a_0 + \ldots + a_n \leq L.$$

Thus the set of finite sums of elements of the sequence a is bounded and has a least upper bound $S_B(a)$.

Let $B = S_B(a)$. Then $B \leq L$. If $B \neq L$, then $B = L - \epsilon$ for some $\epsilon > 0$. For large enough n, however, we have

$$|s_n - L| < \epsilon,$$

i.e.

$$L - \epsilon < s_n < L + \epsilon.$$

This would make $s_n > B$, contrary to the definition of B. Thus $B = L$.

If we assume $B = S_B(a)$ exists, we prove that $L = \lim_{n \to \infty} s_n$ exists by showing directly that B is the limit. Let $\epsilon > 0$. $B - \epsilon$ is less than B and thus cannot be an upper bound of the set of finite sums as B is the least such bound. Hence, for some finite $Y \subseteq \mathbb{N}$,

$$B - \epsilon < \sum_{i \in Y} a_i.$$

As before, for some n_0,

$$\sum_{i \in Y} a_i < s_{n_0},$$

whence

$$B - \epsilon < s_{n_0}.$$

Because the sequence is monotone, for any $n > n_0$, we have $s_n \geq s_{n_0}$, whence

$$B - \epsilon < s_{n_0}$$
$$< s_n,$$

and, because B is an upper bound, $s_n \leq B$. Thus

$$B - \epsilon < s_n \leq B < B + \epsilon,$$

i.e.

$$|s_n - B| < \epsilon$$

for all $n > n_0$. $\square$

For sequences of non-negative real numbers, S_B seems to offer a good explanation, or interpretation, of Bolzano's vague notion of summation. That it agrees with the now-familiar definition of a sum as the limit is further testimony to its naturalness. Other testimony in the form of additional properties of sums that S_B can be shown to possess is ready at hand. The prime example of such a property is linearity:

5.6 Lemma (Linearity Properties). *Let a, b be Bolzano summable sequences of non-negative real numbers and let r be a non-negative real number. Then:*
i. $a + b$ is Bolzano summable and $S_B(a + b) = S_B(a) + S_B(b)$.
ii. ra is Bolzano summable and $S_B(ra) = rS_B(a)$.

Proof. i. Assume a, b are Bolzano summable.

$$S_B(a) + S_B(b) = \sum_{i=0}^{\infty} a_i + \sum_{i=0}^{\infty} b_i, \text{ by Lemma 5.5}$$

$$= \sum_{i=0}^{\infty} (a_i + b_i), \text{ by Lemma 3.5}$$

$$= S_B(a + b), \text{ by Lemma 5.5,}$$

where we implicitly invoke the existential conclusions of the named lemmas in concluding successively the existence of $\sum a_i$ and $\sum b_i$, $\sum (a_i + b_i)$, and finally $S_B(a + b)$.

ii. Similarly, if a is Bolzano summable and $r \geq 0$,

$$rS_B(a) = r \sum_{i=0}^{\infty} a_i$$

$$= \sum_{i=0}^{\infty} (ra_i)$$

$$= S_B(ra),$$

by appeal to Lemmas 5.5, 3.6, and 5.5.   □

It is worth noting that Lemma 5.6 can be proven directly from the definition of S_B without the detour via convergence supplied by Lemma 5.5.

Alternate Proof of Lemma 5.6. i. Suppose a, b are Bolzano summable and consider a finite sum of elements of the sequence $a + b$:

$$\sum_{i \in I} (a + b)_i = \sum_{i \in I} (a_i + b_i)$$

$$= \sum_{i \in I} a_i + \sum_{i \in I} b_i, \ I \subseteq \mathbb{N}, I \text{ finite}$$

$$\leq S_B(a) + S_B(b).$$

It follows that $S_B(a) + S_B(b)$ is an upper bound on the sums $\sum_{i \in I} (a + b)_i$ and thus is not smaller than the least upper bound of such sums:

$$S_B(a + b) \leq S_B(a) + S_B(b).$$

Suppose, by way of contradiction, the inequality were strict. Then, for some $\epsilon > 0$,

$$S_B(a) + S_B(b) = S_B(a + b) + \epsilon.$$

Now $S_B(a) - \epsilon/2 < S_B(a)$ and thus is not an upper bound on the set of sums $\sum_{i \in I} a_i$. Choose some finite set J such that

$$S_B(a) - \epsilon/2 < \sum_{i \in J} a_i.$$

Similarly, choose a finite set K such that

$$S_B(b) - \epsilon/2 < \sum_{i \in K} b_i,$$

and set $I = J \cup K$. Then

$$S_B(a) - \epsilon/2 \ < \ \sum_{i \in J} a_i \ \leq \ \sum_{i \in I} a_i$$

$$S_B(b) - \epsilon/2 \ < \ \sum_{i \in K} b_i \ \leq \ \sum_{i \in I} b_i$$

Hence

$$S_B(a + b) < S_B(a) + S_B(b) - \epsilon < \sum_{i \in I} (a_i + b_i)$$

contrary to the definition of $S_B(a + b)$.

    ii. Similar.   □

5.7 Remark. Because we are dealing with sequences of non-negative numbers, a converse to the existential portion of Lemma 5.6.i holds: if $a + b$ is Bolzano summable, then so are a and b. For, any finite sum $\sum_{i \in I} a_i$ or $\sum_{i \in I} b_i$ is bounded above by $\sum_{i \in I}(a_i + b_i)$, whence by $S_B(a + b)$ and the sets of such finite sums are bounded and have least upper bounds.

    Notice that when $r = 0$, Lemma 5.6.ii yields

$$S_B(0, 0, \ldots) = 0.$$

This seems obvious enough as not to merit being singled out for attention or even to require a proof, but as shown by Grandi's supposed proof of the existence of God, wherein he derived

$$0 + 0 + \ldots = \frac{1}{2}, \tag{43}$$

(page 3, above), when venturing into unknown territory even the obvious must be verified. Of course, (43) is readily dismissed, but Lagrange's demonstration that the insertion of 0's can affect the value of a divergent series (page 7, above) carries some weight. Thus, I pause to consider the effect of 0's on the Bolzano sum of a sequence.

5.8 Lemma (Eliminability of 0). *Let $a : a_0, a_1, \ldots$ be a sequence of non-negative real numbers, and let*

$$J : j_0 < j_1 < \dots$$

be a (possibly finite) sequence of natural numbers such that the subsequence

$$b : a_{j_0}, a_{j_1}, \dots$$

includes every nonzero element of a.

i.   If $J : j_0 < j_1 < \dots < j_{n-1}$ is finite, a is Bolzano summable and

$$S_B(a) = a_{j_0} + \dots + a_{j_{n-1}}.$$

ii. (In particular) if J is empty, $S_B(a) = 0$, i.e.

$$S_B(0, 0, 0, \dots) = 0.$$

iii. If J is infinite, then a is Bolzano summable iff b is Bolzano summable. Moreover, when these equivalent conditions hold,

$$S_B(a) = S_B(b).$$

   Proof. The proofs of the three parts are pretty much the same, so I will just present that of assertion iii.

   Let $a : a_0, a_1, \dots$ be given, $J = \{j_0 < j_1 < \dots\}$ an infinite set of non-negative integers, and define b by $b_i = a_{j_i}$. Let $I \subseteq \mathbb{N}$ be finite and consider a sum

$$\sum_{i \in I} a_i. \tag{44}$$

Let

$$I^+ = \{i \in I \mid i \in J\}, \quad I^- = \{i \in I \mid i \notin J\},$$

and note

$$\sum_{i \in I} a_i = \sum_{i \in I^+} a_i + \sum_{i \in I^-} a_i$$

$$= \sum_{i \in I^+} a_i + \sum_{i \in I^-} 0$$

$$= \sum_{i \in I^+} a_i. \tag{45}$$

Now, each $i \in I^+$ being in J,

$$\sum_{i \in I^+} a_i = \sum_{k \in K} a_{j_k}, \text{ for some finite } K \subseteq \mathbb{N}$$

$$= \sum_{k \in K} b_k. \tag{46}$$

Combining this last with (45), we see that every finite sum (44) equals a finite sum of the form (46). Because b is a subsequence of a, every sum of the form (46) is already of the form (44), it follows that the sets of values of finite sums of elements of the sequences a and b coincide. Hence they either are both bounded or unbounded and, in the former case, share the same upper bounds—in particular, the same least upper bound: $S_B(a) = S_B(b)$.   □

5.9 Corollary (Truncation Property). *Let $a : a_0, a_1, \ldots$ be an infinite sequence of non-negative real numbers and let k be a positive integer. Define a sequence b by*

$$b_i = a_{k+i}.$$

Then: a is Bolzano summable iff b is Bolzano summable. Moreover, when these conditions hold, $S_B(a) = a_0 + \ldots + a_{k-1} + S_B(b)$.

Proof. Let $a : a_0, a_1, \ldots$ and $k \geq 0$ be given, and define

$$b : a_k, a_{k+1}, \ldots$$

Define the auxiliary sequences

$$b' : 0, 0, \ldots, 0, a_k, a_{k+1}, \ldots$$
$$c' : a_0, \ldots, a_{k-1}, 0, 0, \ldots$$

By Lemma 5.8, c' is Bolzano summable and

$$S_B(c') = a_0 + \ldots + a_{k-1}. \tag{47}$$

Also by the Lemma, b is Bolzano summable iff b' is and when they are,

$$S_B(b') = S_B(b). \tag{48}$$

But Lemma 5.6.i and Remark 5.7 tell us that a is Bolzano summable iff b' and c' are Bolzano summable, i.e. iff b' is Bolzano summable, i.e. iff b is Bolzano summable, and moreover that when they are so summable,

$$S_B(a) = S_B(b') + S_B(c').$$

Combining this with (47) and (48) yields the Corollary.   □

The properties of Bolzano summability presented are natural extensions to infinite sums of properties of finite sums, so natural that one might easily use some of them— as Bolzano did— without justification or even mention. Two properties Bolzano was explicit on were the associative property and the nondependence of the sum on the order. This latter property is easy to state formally.

5.10 Lemma (Rearrangement Property). *If $\pi : \mathbb{N} \to \mathbb{N}$ is a permutation[52] and a sequence b is defined from a sequence a by*

$$b_i = a_{\pi(i)},$$

then a is Bolzano summable iff b is Bolzano summable and, when these sequences are summable, $S_B(a) = S_B(b)$.

---

[52] A *permutation* of a set X is a function $\tau : X \to X$ that maps X one-to-one (distinct elements map to distinct elements) onto (every element is mapped onto) itself. The formally defined concept completely captures the notion of a rearrangement.

For, the finite sums of b_i's are finite sums of a_j's and vice versa, whence the sets of such sums coincide and share a common least upper bound.

The associative property as described by Bolzano is that one can replace a collection of parts of a sum by their sum. We might interpret this as follows: If a is a Bolzano summable sequence and b is a subsequence, the sum of a equals the sum of b plus the sum of the remaining terms of the sequence. Because 0's don't matter, we can further interpret this as follows:

5.11 Lemma (Associative Property). *Let $a : a_0, a_1, \ldots$ be a sequence of non-negative real numbers and let $J \subseteq \mathbb{N}$. Define sequences b_i, c_i by*

$$b_i = \begin{cases} a_i, & i \in J \\ 0, & i \notin J \end{cases}, \quad c_i = \begin{cases} 0, & i \in J \\ a_i, & i \notin J \end{cases}.$$

Then: $S_B(a)$ exists iff $S_B(b)$ and $S_B(c)$ both exist. Moreover, when these equivalent conditions hold,

$$S_B(a) = S_B(b) + S_B(c).$$

Proof. For a, b, c as stated, we have $a = b + c$. Lemma 5.6.i applies.   □

For sequences with no negative terms, our interpretation of Bolzano summability offers a good interpretation of the infinite sum. It does coincide with the notion of the sum as the limit of the finite sums, and it allows of simple proofs of the basic properties one would expect sums to have. The only thing missing is applicability to sequences with negative terms. We can see this easily enough by considering the sequence,

$$a : -1, -\frac{1}{2}, -\frac{1}{4}, -\frac{1}{8}, \ldots,$$

for which we should expect the value -2, the negative of the sum of the geometric progression,

$$1, \frac{1}{2}, \frac{1}{4}, \frac{1}{8}, \ldots$$

We get instead

$$S_B(a) = 0.$$

For, all finite sums are negative, whence 0 is an upper bound. For any $\epsilon > 0$, we can see that $-\epsilon$ is not an upper bound by choosing n so large that

$$\frac{1}{2^n} < \epsilon,$$

i.e.

$$-\epsilon < -\frac{1}{2^n}.$$

As the right-hand side of this last inequality is a term of the sequence, hence a finite sum of such terms, no number less than 0 is an upper bound for these sums and 0 is indeed the least upper bound.

Similarly, if b is the sequence

$$-2, 1, \frac{1}{2}, \frac{1}{4}, \dots,$$

we should expect

$$S_B(b) = -2 + \left(1 + \frac{1}{2} + \frac{1}{4} + \dots\right) = -2 + 2 = 0,$$

but we would in fact get $S_B(b) = 2$.

With a little care, however, the extension of S_B to sequences with mixed terms is not difficult. If a has no positive terms, then $-a$ has no negative terms and we know what $S_B(-a)$ is. Assuming linearity,

$$-S_B(-a) = S_B(--a) = S_B(a)$$

and we have our definition of $S_B(a)$ for sequences a of non-positive numbers:

5.12 Definition. *Let* $a : a_0, a_1, \dots$ *be a sequence of non-positive real numbers. Define* $b = -a$ *by* $b_i = -a_i$ *for all* $i \in \mathbb{N}$. *We define*

$$S_B(a) = -S_B(b),$$

if this number exists.

Thus,

$$S_B\left(-1, -\frac{1}{2}, -\frac{1}{4}, -\frac{1}{8}, \dots\right) = -S_B\left(1, \frac{1}{2}, \frac{1}{4}, \frac{1}{8}, \dots\right)$$
$$= -2,$$

as expected.

We could equally well have defined $S_B(a)$ to be the greatest lower bound of the finite sums, in more direct analogy to Definition 5.4. The two definitions are equivalent because the negation of numbers reverses order.

The extension to arbitrary sequences is determined by assuming the Associative Property to carry over:

5.13 Definition. *Let* $a : a_0, a_1, \dots$ *be a sequence of real numbers and define* a^+, a^- *by*

$$a_i^+ = \begin{cases} a_i, & \text{if } a_i \geq 0 \\ 0, & \text{if } a_i < 0 \end{cases}, \quad a_i^- = \begin{cases} 0, & \text{if } a_i \geq 0 \\ a_i, & \text{if } a_i < 0 \end{cases}.$$

We define

$$S_B(a) = S_B(a^+) + S_B(a^-),$$

provided both numbers on the right exist.

But for the retention and introduction of 0's, we can think of a^+ as the sequence of positive terms of a— the *positive part* of a if you will— and a^- as the *negative part*. If the Association Property is to generalise to arbitrary sequences, it will be the case that we can split the sequence into positive and negative parts.

When we extend Bolzano summation in this way, the basic properties— Linearity, Eliminability of 0, Truncatability, Rearrangability, and Associativity— still hold. The proofs of these properties for Bolzano sums of sequences of mixed terms are not repeats of the earlier proofs, but algebraic reductions. First one concludes from a property's validity for sequences of non-negative real numbers that it holds for sequences of non-positive numbers via Definition 5.12, and then one combines these via Definition 5.13 to conclude the property's validity for sequences in general.

The first of these steps is fairly trivial. For example, if a, b are sequences of non-positive real numbers, then $-(a+b), -a, -b$ are sequences of non-negative real numbers and $-(a + b) = -a + -b$, whence

$$
\begin{aligned}
S_B(a + b) &= -S_B(-(a + b)), \text{ by Definition 5.12} \\
&= -S_B(-a + -b) \\
&= -S_B(-a) + -S_B(-b), \text{ by Lemma 5.6.i} \\
&= S_B(a) + S_B(b), \text{ by Definition 5.12.}
\end{aligned}
$$

Or, to verify the Rearrangement Property, let $\pi : \mathbb{N} \to \mathbb{N}$ be a permutation, a a sequence of non-positive real numbers, and b the permuted sequence: $b_i = a_{\pi(i)}$. Then:

$$
\begin{aligned}
S_B(b) &= -S_B(-b), \text{ by Definition 5.12} \\
&= -S_B(-a), \text{ by Lemma 5.10} \\
&= S_B(a), \text{ by Definition 5.12.}
\end{aligned}
$$

The second step, the extension to sequences with mixed terms, can range from being equally trivial to involving some bookkeeping with the indices or the associated subsequences. The Rearrangement Property, for example, is proven by taking a sequence a, splitting it in two,

$$
a = a^+ + a^-,
$$

defining the permuted sequence b and splitting it in two,

$$
b = b^+ + b^-,
$$

and noting that b^+ is just a permutation of a^+ and b^- a permutation of a^-. Thus,

$$
\begin{aligned}
S_B(a) &= S_B(a^+) + S_B(a^-) \\
&= S_B(b^+) + S_B(b^-) \\
&= S_B(b).
\end{aligned}
$$

I shall leave to the reader the details of the proofs that all these properties of Bolzano summation for sequences of non-negative numbers extend to the summation of mixed sequences. For the purposes of the present book, the statement of the properties and the flavour of the proofs suffice.

What I wish to finish this section with is a discussion of the relation of Bolzano summability to the usual definition of the sum of an infinite series as a limit. We know that, for a sequence a of non-negative real numbers, a is Bolzano summable iff $\sum a_i$ converges and, should these equivalent conditions hold, $S_B(a) = \sum a_i$. Multiplying by -1, one easily sees that this also holds for sequences of non-positive real numbers. It does not hold in general, however, as is illustrated by the convergent *alternating harmonic series* of (34):

$$S_B\left(1, -\frac{1}{2}, \frac{1}{3}, -\frac{1}{4}, \ldots\right) = S_B\left(1, \frac{1}{3}, \frac{1}{5}, \ldots\right) + S_B\left(-\frac{1}{2}, -\frac{1}{4}, -\frac{1}{6}, \ldots\right)$$

$$= S_B\left(1, \frac{1}{3}, \frac{1}{5}, \ldots\right) - S_B\left(\frac{1}{2}, \frac{1}{4}, \frac{1}{6}, \ldots\right),$$

by definition— provided both summands on the right exist. But Lemma 5.5 applies and

$$S_B\left(1, \frac{1}{3}, \frac{1}{5}, \ldots\right) = \sum_{i=0}^{\infty} \frac{1}{2i + 1}$$

$$S_B\left(\frac{1}{2}, \frac{1}{4}, \frac{1}{6}, \ldots\right) = \sum_{i=0}^{\infty} \frac{1}{2i + 2},$$

provided the series on the right converge. But neither does: If the second series converged, then so would

$$2\sum_{i=0}^{\infty} \frac{1}{2i + 2} = \sum_{i=0}^{\infty} \frac{1}{i + 1},$$

which is the harmonic series shown earlier to diverge. Moreover, a term by term comparison shows the first series to dominate the second, whence it too diverges to infinity.

The Bolzano summable sequences form a proper subset of the set of sequences that give rise to convergent series, and, if a is Bolzano summable then

$$S_B(a) = \sum_{i=0}^{\infty} a_i.$$

Moreover, there is a characterisation of Bolzano summable sequences in terms of convergence.

5.14 Definition. *A series $\sum_{i=0}^{\infty} a_i$ of real numbers is* absolutely convergent *if $\sum_{i=0}^{\infty} |a_i|$ converges. A sequence is* conditionally convergent *if it is convergent, but not absolutely convergent.*

As we have not introduced an adverb further modifying "convergent" to describe those series that are absolutely convergent but not convergent, one should guess that there are no such things. This is indeed the case: absolute convergence implies convergence. We will prove this shortly. First, however, let us prove the following.

5.15 Theorem. *Let a be a sequence of real numbers. a is Bolzano summable iff $\sum a_i$ converges absolutely.*

Proof. Write $|a|$ for the sequence of absolute values of members of a. Let a^+, a^- be the positive and negative parts of a, and observe

$$a \text{ is Bolzano summable } \quad \text{iff} \quad S_B(a^+), S_B(a^-) \text{ exist, by definition}$$
$$\text{iff} \quad S_B(a^+), S_B(-a^-) \text{ exist, by definition}$$
$$\text{iff} \quad S_B(|a|) \text{ exists, by Lemma 5.11}$$
$$\text{iff} \quad \sum_{i=0}^{\infty} |a_i| \text{ converges, by Lemma 5.5.} \qquad \square$$

5.16 Theorem. *Every absolutely convergent series is convergent.*

Proof. Repeating part of the proof of Theorem 5.15,

$$\sum_{i=0}^{\infty} a_i \text{ is absolutely convergent} \Rightarrow S_B(a) \text{ exists}$$
$$\Rightarrow S_B(a^+), S_B(a^-) \text{ exist}$$
$$\Rightarrow S_B(a^+), S_B(-a^-) \text{ exist}$$
$$\Rightarrow \sum_{i=0}^{\infty} a_i^+, \sum_{i=0}^{\infty} -a_i^- \text{ converge}$$
$$\Rightarrow \sum_{i=0}^{\infty} a_i^+, \sum_{i=0}^{\infty} a_i^- \text{ converge}$$
$$\Rightarrow \sum_{i=0}^{\infty} a_i^+ + \sum_{i=0}^{\infty} a_i^- \text{ converges}$$
$$\Rightarrow \sum_{i=0}^{\infty} a_i \text{ converges},$$

since $a = a^+ + a^-$. $\qquad \square$

5.17 Corollary. *Let $a : a_0, a_1, \ldots$ be Bolzano summable. Then*

$$S_B(a) = \sum_{i=0}^{\infty} a_i.$$

Proof. Split a into its positive and negative parts and observe:

$$S_B(a) = S_B(a^+) + S_B(a^-)$$

$$= \sum_{i=0}^{\infty} a_i^+ + \sum_{i=0}^{\infty} a_i^-$$

$$= \sum_{i=0}^{\infty} a_i. \qquad \square$$

Digression

The proof given that absolute convergence implies convergence, though not absent from the textbooks, is not the one usually presented. The standard proof, though not necessary for the purposes of the present chapter, is not entirely irrelevant for the book and I take the liberty of digressing to outline this proof briefly.

One starts with a series $a : a_0, a_1, \ldots$ that is assumed absolutely convergent:

$$\sum_{i=0}^{\infty} |a_i| = M < \infty.$$

One then considers the partial sums

$$s_n = \sum_{i=0}^{n} a_i$$

and notes that

$$|s_n - s_{n+p}| = |a_{n+1} + \ldots + a_{n+p}|$$

$$\leq |a_{n+1}| + \ldots + |a_{n+p}|$$

$$\leq \sum_{i=n+1}^{\infty} |a_i|$$

$$\leq \left| M - \sum_{i=0}^{n} |a_i| \right|$$

$$< \epsilon,$$

when n is sufficiently large, i.e. the sequence of partial sums is a *Cauchy sequence*:

5.18 Definition. *A sequence* $b_0, b_1, \ldots$ *of real numbers is said to be* Cauchy convergent *and is called a* Cauchy sequence *if, for every* $\epsilon > 0$, *there is a number* n_0 *such that for all* $m, n > n_0$ *one has* $|b_m - b_n| < \epsilon$.

One very quickly sees that every convergent series is Cauchy convergent: Let $\lim_{n \to \infty} b_n = L$, choose $\epsilon > 0$, and let n_0 be so large that for all $k > n_0$,

$$|b_k - L| < \frac{\epsilon}{2}.$$

For any $m, n > n_0$ we have

$$\begin{aligned}
|b_m - b_n| &= |b_m - L + L - b_n| \\
&\leq |b_m - L| + |L - b_n| \\
&< \frac{\epsilon}{2} + \frac{\epsilon}{2} = \epsilon.
\end{aligned}$$

The converse is somewhat deeper:

5.19 Theorem. *Every Cauchy sequence of real numbers is convergent.*

Accepting this Theorem, the proof of Theorem 5.16 was completed the moment we showed the partial sums to be a Cauchy sequence.

The notion of a Cauchy sequence was discovered by several mathematicians: José Anastácio da Cunha, Bernard Bolzano, and Augustin Louis Cauchy. A Portuguese, da Cunha published his attempts to make Calculus more rigorous in *Principios mathemáticas*, a sort of serial encyclopædia, in the 1780's and finally in book form in 1790. This was translated into French in 1811, but neither edition was widely circulated, it was excessively concise, and it is reported that his treatment of Cauchy convergence was mistranslated in the more accessible French version.

In 1817 Bolzano stated the result clearly in his paper, "Rein analytischer Beweis des Lehrsatzes, daß zwischen je zwey Werthen, die ein entgegengesetztes Resultat gewähren, wenigstens eine reelle Wurzel der Gleichung liege"[Purely analytic proof of the theorem, that between any two values receiving opposite results, at least one root of the equation lies], which paper of some 60 pages was devoted to proving the Intermediate Value Theorem. Bolzano gave a proof of Theorem 5.19 that doesn't quite satisfy. We will have a little more to say about this in Chapter III, below (pp. 260 - 261).

Neither da Cunha's nor Bolzano's works brought the concept of a Cauchy sequence to the attention of the mathematical public. This was done by Augustin Louis Cauchy in his textbook *Cours d'analyse* of 1821. Cauchy states Theorem 5.19 without proof, thus more-or-less accepting it as an axiom.

The modern approach to the real numbers is either to characterise them axiomatically or to construct them from the rational numbers via some infinitistic construction. We will discuss the latter in Chapter III, below. As to the axiomatic approach, the key axiom is a *completeness axiom* asserting there to be no gap in the real number line. There are various ways of formulating completeness. One is the Least Upper Bound Principle; another is Theorem 5.19. Each can be derived from the other. As I am more inclined to consider the Least Upper Bound Principle to be intuitively evident than I am to consider the same of the convergence of all Cauchy sequences, I would opt for assuming the former as an axiom and deriving the latter from it.

I have said before that this is not intended as a textbook in mathematics and there is no immediate reason to include a proof of Theorem 5.19 here. I shall nonetheless do so with an eye to one criticism of formalism to be discussed later in Chapter V. The reader may choose to skip the proof for now.

Proof of Theorem 5.19. Let $b_0, b_1, \ldots$ be Cauchy convergent. Choose n_0 large enough so that, for all $m, n > n_0$,

$$|b_m - b_n| < 1.$$

Then $M = \max\{|b_0|, \ldots, |b_{n_0}|, |b_{n_0+1}| + 1\}$ bounds the sequence: Let b_m be given. If $m \le n_0$, then $|b_m| \le M$ because $|b_m|$ is listed in the set $\{|b_0|, \ldots, |b_{n_0}|, |b_{n_0+1}| + 1\}$ of which M is the maximum. If $m > n_0$,

$$\begin{aligned}|b_m| &= |b_m - b_{n_0+1} + b_{n_0+1}| \\ &\le |b_m - b_{n_0+1}| + |b_{n_0+1}| \\ &\le 1 + |b_{n_0+1}| \le M.\end{aligned}$$

It follows that $b_0, b_1, \ldots$ is bounded. Define

$$X = \{x \in \mathbb{R} \mid \text{for all but finitely many } n, \, x \le b_n\}.$$

X is nonempty and has an upper bound. Let B be the least upper bound of X and let $\epsilon > 0$.

Because $B - \epsilon$ is not an upper bound on X, there is some $x \in X$ for which $B - \epsilon < x$. Now, if $x \le b_n$, then $B - \epsilon < b_n$ and we see that, for all but finitely many n, $B - \epsilon < b_n$, i.e. there is an n_1 such that for all $n > n_1$, $B - \epsilon < b_n$.

The number $B + \epsilon/2$ is greater than any element of X because it is greater than B, which is greater than or equal to every element thereof. It follows that there are infinitely many elements b_n of the sequence for which $b_n < B + \epsilon/2$. But the sequence is Cauchy convergent, so there is a number n_2 such that for all $m, n > n_2$, $|b_m - b_n| < \epsilon/2$. If we choose $m > n_2$ so that $b_m < B + \epsilon/2$, we see that for any $n > n_2$,

$$b_n < b_m + |b_m - b_n| < B + \epsilon/2 + \epsilon/2 = B + \epsilon.$$

Thus, for $n > n_0 = \max\{n_1, n_2\}$, we have

$$B - \epsilon < b_n < B + \epsilon,$$

i.e. $-\epsilon < b_n - B < \epsilon$, i.e. $|b_n - B| < \epsilon$. Thus, $B = \lim_{n \to \infty} b_n$. □

This is probably a good place to mention another completeness property of the real numbers, one closely associated with the name Bolzano.

5.20 Theorem (Bolzano-Weierstrass Theorem). *Every bounded sequence of real numbers has a convergent subsequence.*

Proof The proof follows the same lines as the proof of Theorem 5.19, but has an added complication. We start with a bounded sequence $b_0, b_1, \ldots$, the set

$$X = \{x \in \mathbb{R} \mid \text{for all but finitely many } n, \, x \le b_n\},$$

and its least upper bound B as before.

For any $\epsilon > 0$, $B - \epsilon$ is not an upper bound of X and it follows that there is a number n_ϵ such that for all $n > n_\epsilon$, $B - \epsilon < b_n$. This will hold for any subsequence $b_{i_0}, b_{i_1}, \ldots$ of the original sequence: $k > n_\epsilon \Rightarrow B - \epsilon < b_{i_k}$.

The extra complication arises in trying to bound the subsequence below $B + \epsilon$. Because $B + \epsilon \notin X$, there are infinitely many terms b_n of the sequence for which $b_n < B + \epsilon$. If we now generate a subsequence, we will indeed have

$$B - \epsilon < b_{i_k} < B + \epsilon \tag{49}$$

for all sufficiently large k, but as the sequence $b_{i_0}, b_{i_1}, \ldots$ depends on ϵ, we will not guarantee the same holds for, say, $\epsilon/2$: it could happen that

$$B - \epsilon < B + \frac{\epsilon}{2} < b_{i_k}$$

for all k. We must handle all ϵ's at once.

We inductively define a sequence $i_0 < i_1 < i_2 < \ldots$ by stages in such a way as to guarantee, for all $k > 0$,

$$B - \frac{1}{k} < b_{i_k} < B + \frac{1}{k}. \tag{50}$$

If we can do this, we are done. For, given any $\epsilon > 0$, for all $k > 1/\epsilon$, we have

$$B - \epsilon < B - \frac{1}{k} < b_{i_k} < B + \frac{1}{k} < B + \epsilon,$$

i.e. $|B - b_{i_k}| < \epsilon$ for all sufficiently large i_k, i.e. $\lim_{k \to \infty} b_{i_k} = B$.

At stage 0 we simply take $i_0 = 0$.

At stage $k + 1$, we note again that $B + 1/(k + 1) \notin X$, whence there are infinitely many terms b_n of the sequence satisfying

$$B - \frac{1}{k + 1} < b_n < B + \frac{1}{k + 1}.$$

Choose i_{k+1} to be such an $n > i_k$. (For the sake of definiteness, one might choose i_{k+1} to be the least such n.) Thus,

$$B - \frac{1}{k + 1} < b_{i_{k+1}} < B + \frac{1}{k + 1},$$

and condition (50) has been met. $\qquad\square$

5.21 Remark. Such infinitary constructions take some getting used to. It might be easier to imagine that one has generated for each $\epsilon = 1/(k + 1)$ a subsequence $b_{i_0(\epsilon)}, b_{i_1(\epsilon)}. \ldots$ satisfying (49):

$$B - \epsilon < b_{i_k(\epsilon)} < B + \epsilon$$

for all sufficiently large k, and that one has the sequences arranged in a large array:

$$b_{i_0(1)}, \quad b_{i_1(1)}, \quad b_{i_2(1)}, \quad \cdots$$
$$b_{i_0(\frac{1}{2})}, \quad b_{i_1(\frac{1}{2})}, \quad b_{i_2(\frac{1}{2})}, \quad \cdots$$
$$b_{i_0(\frac{1}{3})}, \quad b_{i_1(\frac{1}{3})}, \quad b_{i_2(\frac{1}{3})}, \quad \cdots$$
$$\vdots$$

One could then diagonalise on this array by defining, say, $i_0 = i_0(1), i_1 = i_1(1/2), i_2 = i_2(1/3), \ldots$ As one wants $i_0 < i_1 < i_2 < \ldots$, one has to modify this slightly by choosing $i_0 = i_0(1)$ and after this i_{k+1} to be the first element of the sequence

$$i_0\left(\frac{1}{k+1}\right), i_1\left(\frac{1}{k+1}\right), \ldots$$

greater than i_k. At the blackboard, where I can wave my hands and point at the array, I would present the proof in this manner; in print, I find the proof as formally presented with its single sequence more congenial. Both proofs use the same idea; only the representations differ.

It is worth noting that the Bolzano-Weierstrass Theorem quickly implies the convergence of Cauchy convergent sequences.

Proof of Theorem 5.19 from Theorem 5.20. Let $b_0, b_1, \ldots$ be Cauchy convergent. As in the direct proof of Theorem 5.19, the sequence $b_0, b_1, \ldots$ is bounded. By the Bolzano-Weierstrass Theorem it has a convergent subsequence, $b_{i_0}, b_{i_1}, \ldots$ Let L be the limit of this subsequence, let $\epsilon > 0$, choose n_1 large enough so that

$$i_k > n_1 \Rightarrow |b_{i_k} - L| < \frac{\epsilon}{2},$$

and n_2 so that

$$m, n > n_2 \Rightarrow |b_m - b_n| < \frac{\epsilon}{2}.$$

Let $n_0 = \max\{n_1, n_2\}$ and choose $n > n_0, m = i_k > n_0$, and observe

$$|b_n - L| = |b_n - b_m + b_m - L|$$
$$\leq |b_n - b_m| + |b_m - L|$$
$$\leq |b_n - b_{i_k}| + |b_{i_k} - L|$$
$$< \frac{\epsilon}{2} + \frac{\epsilon}{2} = \epsilon. \qquad \square$$

6 Generalised Sums; Examples

Viewed abstractly, the assignment of a value to serve as the sum of the elements of a sequence is the application of a function S mapping some subset of the set $\mathbb{R}^{\mathbb{N}}$ of infinite sequences of real numbers to the set $\mathbb{R}$ of real numbers. Obviously, this is too abstract a view; not every function $S : X \subseteq \mathbb{R}^{\mathbb{N}} \to \mathbb{R}$ merits being considered a notion of summation. What must such a function satisfy to gain such consideration? We have established some properties of the classical notion of summation as the limit of the sequence of its finite sums, as well as some properties of Bolzano summation. Are these sufficient? Are they necessary? I don't propose to answer these questions here, but merely to ask them and to provide for one's contemplation of the solution a small stock of sum functions S and a list of some properties these functions may or may not possess.

Let us begin with a brief list of sum functions. There is no better place to start with than the finite sums:

6.1 Example. A sequence $a : \mathbb{N} \to \mathbb{R}$ has *finite support* if there is an $n_0 \in \mathbb{N}$ such that $a_n = 0$ for all $n > n_0$. For such a sequence a, we define

$$S_F(a) = a_0 + \ldots + a_{n_0}.$$

In words, $S_F(a)$ is defined just in case a has only finitely many nonzero elements and equals the sum of these elements. For any other sequence a, $S_F(a)$ is undefined. This is a perfectly reasonable explanation of an infinite sum: $\sum_{i=0}^{\infty} a_i$ exists iff we can actually add all the elements of the sequence, which we cannot actually do if infinitely many of them are nonzero.

6.2 Example. The usual definition of a series as the limit of the sequence of partial sums,

$$S_\Sigma(a) = \sum_{i=0}^{\infty} a_i = \lim_{n \to \infty} \sum_{i=0}^{n} a_i,$$

has as domain the set of convergent series.

This is a somewhat broader definition of summation, and has many of the properties of S_F, but not all.

6.3 Example. Bolzano summation S_B as defined in §5 above has the set of absolutely convergent series as domain.

S_B is defined for fewer sequences than is S_Σ, but it has some properties of S_F that S_Σ does not.

There are two additional examples going back to Leibniz that we have touched on already. The first of these is:

6.4 Example. Let $a : a_0, a_1, \ldots$ be a sequence of real numbers, and let $s_0, s_1, \ldots$ be the associated sequence of partial sums,

$$s_n = \sum_{i=0}^{n} a_i.$$

Define the sequence $t_0, t_1, \ldots$ of *averages* of the finite sums by

$$t_n = \frac{1}{n+1}(s_0 + \ldots + s_n).$$

We define

$$S_A(a) = \lim_{n \to \infty} t_n,$$

provided this limit exists.

The domain of S_A properly includes the set of convergent sequences (cf. Lemma 6.7, below), as exemplified by the Grandi sequence,

$$g : 1, -1, 1, -1, \ldots$$

The sequence of partial sums of g,

$$s : 1, 0, 1, 0, \ldots$$

does not converge, but the sequence t of averages,

$$t : 1, \frac{1}{2}, \frac{2}{3}, \frac{2}{4} = \frac{1}{2}, \frac{3}{5}, \frac{3}{6} = \frac{1}{2}, \ldots$$

does converge: In general,

$$t_{2n} = \frac{n+1}{2n+1}, \quad t_{2n+1} = \frac{n+1}{2n+2} = \frac{1}{2},$$

and it is clear that

$$\lim_{n \to \infty} t_n = \frac{1}{2}.$$

Otto Hölder generalised the definition of S_A as follows: If the sequence of averages doesn't converge, try the sequence of averages of averages, then the sequence of averages of these, etc. For example, he offered the sequence

$$h : 1, -2, 3, -4, 5, -6, \ldots \tag{51}$$

The partial sums of this are

$$s : 1, -1, 2, -2, 3, -3 \ldots$$

and their averages are

$$t : 1, 0, \frac{2}{3}, 0, \frac{3}{5}, 0 \ldots$$

In general

$$t_{2n} = \frac{n+1}{2n+1}, \quad t_{2n+1} = 0$$

and

$$\lim_{n \to \infty} t_{2n} = \frac{1}{2}, \quad \lim_{n \to \infty} t_{2n+1} = 0,$$

whence $t_0, t_1, \ldots$ does not converge. The n-th term of the average of the t_is,

$$u_n = \frac{1}{n+1} (t_0 + \ldots + t_n),$$

is not so easy to determine, but, as we will shortly see, the sequence of these converges to the average of the limits of the subsequences $t_0, t_2, \ldots$ and $t_1, t_3, \ldots$, which is $\frac{1}{4}$.

Formal manipulation offers (heuristic) support for assigning the value $\frac{1}{4}$ to h as follows: Performing long division or developing the Taylor series (without paying close attention to issues of convergence), we see that

$$\frac{1}{(1+x)^2} = 1 - 2x + 3x^2 - 4x^3 + \ldots \tag{52}$$

whence substituting 1 for x yields

$$\frac{1}{4} = \frac{1}{2^2} = 1 - 2 + 3 - 4 + \ldots$$

Alternatively, subtract the Grandi sequence g from h:

$$
\begin{array}{rrrrrl}
1 & -2 & 3 & -4 & 5 & \ldots \\
1 & -1 & 1 & -1 & 1 & \ldots \\
\hline
0 & -1 & 2 & -3 & 4 & \ldots
\end{array}
$$

Thus, if we abuse notation and let g, h denote their respective sums, $h - g = -h$, whence $2h = g = \frac{1}{2}$ and $h = \frac{1}{4}$.

Hölder's generalisation iterated Leibniz's averaging technique. To state this formally let us introduce a minor bit of notation.

6.5 Definition. *Let $s : s_0, s_1, \ldots$ be a sequence. The sequence $A(s)$ of average values of s is the sequence $t_0, t_1, \ldots$ given by*

$$t_n = \frac{1}{n+1} \big(s_0 + \ldots + s_n \big).$$

6.6 Example. Let k be a positive integer. A sequence a is *(H,k)-summable* if the k-fold application of A to the sequence $s_0, s_1, \ldots$ of partial sums of a, $A^k(s)$, is a convergent sequence. If a is (H, k)-summable, we write $S_{H,k}(a)$ for this limit. We say a is *Hölder summable* if a is (H, k)-summable for some k, and we define

$$S_H(a) = S_{H,k}(a)$$

for such k.

Note that $(H,1)$-summability is just the average summability of Example 6.4 and $S_{H,1}(a) = S_A(a)$ whenever these are defined.

The definition of S_H presupposes some consistency to the values given by the sums: Should a be (H, k)-summable and (H, m)-summable, then $S_{H,k}(a) = S_{H,m}(a)$. This is readily established.

6.7 Lemma. *Let $s : s_0, s_1, \ldots$ be a convergent sequence with limit L and let $t_0, t_1, \ldots$ be the sequence $A(s)$ of averages of s. Then: $\lim_{n \to \infty} t_n = L$.*

Proof. Choose $\epsilon > 0$.
For any n, note that

$$|t_n - L| = \left| \frac{1}{n+1} \sum_{i=0}^{n} s_i - L \right|$$

$$= \frac{1}{n+1} \left| \sum_{i=0}^{n} (s_i - L) \right|$$

$$\leq \frac{1}{n+1} \sum_{i=0}^{n} |s_i - L|. \tag{53}$$

Because L is the limit of the s_i's, we can choose n_0 large enough so that $|s_i - L| < \epsilon/2$ for all $i > n_0$. This suggests assuming $n > n_0$ and breaking the sum in (53) into two parts:

$$|t_n - L| \leq \frac{1}{n+1} \sum_{i=0}^{n_0} |s_i - L| + \frac{1}{n+1} \sum_{i=n_0+1}^{n} |s_i - L|$$

$$< \frac{1}{n+1} \sum_{i=0}^{n_0} |s_i - L| + \frac{1}{n+1} \sum_{i=n_0+1}^{n} \frac{\epsilon}{2}$$

$$< \frac{1}{n+1} \sum_{i=0}^{n_0} |s_i - L| + \frac{n - n_0}{n+1} \cdot \frac{\epsilon}{2}$$

$$< \frac{1}{n+1} \sum_{i=0}^{n_0} |s_i - L| + \frac{\epsilon}{2}. \tag{54}$$

As n_0 is a fixed finite number, there are only finitely many values

$$|s_0 - L|, |s_1 - L|, \ldots, |s_{n_0} - L|$$

occurring in (54) and they have a maximum, say M. From (54) it follows that

$$|t_n - L| < \frac{1}{n+1} \sum_{i=0}^{n_0} M + \frac{\epsilon}{2}$$

$$< \frac{n_0 + 1}{n+1} M + \frac{\epsilon}{2}. \tag{55}$$

Adding the assumption

$$\frac{n_0 + 1}{n+1} M < \frac{\epsilon}{2},$$

i.e.

$$n + 1 > \frac{2(n_0 + 1)M}{\epsilon},$$

will allow us to conclude

$$|t_n - L| < \frac{\epsilon}{2} + \frac{\epsilon}{2} = \epsilon. \qquad \square$$

This Lemma has several immediate consequences for us. First, if $\sum_{i=0}^{\infty} a_i$ converges, then $S_A(a)$ exists and $S_A(a) = \sum_{i=0}^{\infty} a_i$. Second, if $S_{H,k}(a) =$

$\lim_{n\to\infty}(A^k(s))_n$ exists, then so does $S_{H,k+1}(a) = \lim_{n\to\infty}(A^{k+1}(s))_n$, and $S_{H,k+1}(a) = S_{H,k}(a)$. In particular, $S_H(a)$ is unambiguously defined.

We can also apply it and some elementary results on limits to the sequence h of (51) to verify rigorously that $S_{H,2}(h) = \frac{1}{4}$. Recall that the sequence $t_0, t_1, \ldots$ of averages of the partial sums satisfied

$$t_{2n} = \frac{n+1}{2n+1}, \quad t_{2n+1} = 0.$$

If we now consider

$$u_n = \frac{1}{n+1}(t_0 + \ldots + t_n)$$
$$= \frac{1}{n+1}(t_0 + t_2 \ldots + t_{2[n/2]}),$$

where $[n/2]$ denotes the greatest integer $\leq n/2$. But by the Lemma,

$$\lim_{n\to\infty} \frac{t_0 + t_2 \ldots + t_{2[n/2]}}{[n/2]+1} = \lim_{n\to\infty} t_{2n} = \frac{1}{2}.$$

Hence

$$u_n \approx \frac{1}{2} \cdot \frac{t_0 + t_2 \ldots + t_{2[n/2]}}{[n/2]+1} \to \frac{1}{2} \cdot \frac{1}{2} = \frac{1}{4}.$$

The compatibility of these sums is reassuring. The compatibility with Leibniz's second method of summing divergent series is awesome.

6.8 Example. Let $a : a_0, a_1, \ldots$ be a sequence of real numbers and suppose for all x with $|x| < 1$ the series,

$$\sum_{i=0}^{\infty} a_i x^i,$$

converges. We define

$$S_L(a) = \lim_{x\to 1^-} \sum_{i=0}^{\infty} a_i x^i,$$

provided this limit exists.[53]

For the Grandi sequence g one has for $|x| < 1$,

$$1 - x + x^2 - x^3 + \ldots = \frac{1}{1+x},$$

whence

$$S_L(g) = \lim_{x\to 1^-} (1 - x + x^2 - x^3 + \ldots)$$

---

[53] The notation $\lim_{x\to 1^-} f(x)$ means the limit of $f(x)$ as x approaches 1 from below. The formal definition reads: $\lim_{x\to 1^-} f(x) = L$ iff for any $\epsilon > 0$ there is a $\delta > 0$ such that, for all x satisfying $1 - \delta < x < 1$, one has $|f(x) - L| < \epsilon$.

$$= \lim_{x \to 1^-} \frac{1}{1+x} = \frac{1}{2}.$$

Similarly for the Hölder sequence h, for $|x| < 1$ one has

$$1 - 2x + 3x^2 - 4x^3 + \ldots = \frac{1}{(1+x)^2},$$

whence

$$S_L(h) = \lim_{x \to 1^-} (1 - 2x + 3x^2 - 4x^3 + \ldots)$$
$$= \lim_{x \to 1^-} \frac{1}{(1+x)^2} = \frac{1}{2^2} = \frac{1}{4}.$$

Leibniz introduced S_L in his letter to Christian Wolf in connexion with his criticism of Wolf's summations of some strange series (cf. page 3, above). In this letter he also stated, without proof, the following:

6.9 Theorem. *Let $a : a_0, a_1, \ldots$ be a sequence of real numbers. If $S_A(a)$ exists, then so does $S_L(a)$ and $S_L(a) = S_A(a)$.*

The first partial proof was supplied by Niels Henrik Abel[54], who showed that $S_L(a)$ existed and equalled $S_\Sigma(a)$ when the latter existed. The full result was first proven in 1880 by Georg Frobenius[55]. Two years later, Hölder[56] introduced his generalisation of Leibniz's averaging technique and proved the corresponding generalisation of Frobenius's result:

6.10 Theorem. *Let $a : a_0, a_1, \ldots$ be a sequence of real numbers. If $S_H(a)$ exists, then so does $S_L(a)$ and $S_L(a) = S_H(a)$.*

Both proofs are complicated extensions of the proof of Lemma 6.7. Because he has to prove a far more general result, Hölder organises the details more carefully and his proof of Theorem 6.9 is more readable and I refer the curious reader to Hölder's paper for the proof, which I shall not give here.[57] What I wish to do instead is to enumerate and discuss some of the properties these functions have which make them defensible notions of summation.

The properties of a sum function we have thus far considered are the following:

---

[54] N.H. Abel, "Untersuchungen über die Reihe $1 + \frac{m}{1}x + \frac{m \cdot (m-1)}{1 \cdot 2}x^2 + \frac{m \cdot (m-1)(m-2)}{1 \cdot 2 \cdot 3}x^3 \ldots$", *Journal für die reine und angewandte Mathematik* 1 (1826), pp. 311 - 339.

[55] G. Frobenius, "Ueber die *Leibnitz*sche Reihe", *Journal für die reine und angewandte Mathematik* 89 (1880), pp. 262 - 264.

[56] O. Hölder, "Grenzwerthe von Reihen an der Convergenzgrenze", *Mathematische Annalen* 20 (1882), pp. 535 -549.

[57] A more all-encompassing reference, readily available in English in a Dover reprint, is Konrad Knopp, *Theory and Application of Infinite Series*, 2nd English edition, Blackie & Sons Ltd., London, 1951, pp. 488f.

Agreement With Finite Sums S agrees with finite sums if, for any $a \in \mathbb{R}^{\mathbb{N}}$ of finite support, $S(a)$ is the sum of the nonzero elements of a, i.e. $S(a) = S_F(a)$.

Linearity S is linear if it satisfies the two conditions for any $a, b \in \mathbb{R}^{\mathbb{N}}$ and any $r \in \mathbb{R}$:

i. *L1.* If $S(a)$ and $S(b)$ are defined, then so is $S(a + b)$ and

$$S(a + b) = S(a) + S(b);$$

ii. *L2.* If $S(a)$ is defined, then so is $S(ra)$ and

$$S(ra) = rS(a).$$

Truncation/Prefixing For any $a \in \mathbb{R}^{\mathbb{N}}$ and any $n \in \mathbb{N}$, $S(a)$ is defined iff $S(a_{n+1}, a_{n+2}, \ldots)$ is defined and

$$S(a_0, a_1, \ldots) = a_0 + \ldots + a_n + S(a_{n+1}, a_{n+2}, \ldots).$$

Irrelevance of 0 Let $a : a_0, a_1, \ldots$ be a sequence of real numbers and let

$$J : j_0 < j_1 < \ldots$$

be a (possibly finite) sequence of natural numbers such that the subsequence

$$b : a_{j_0}, a_{j_1}, \ldots$$

includes every nonzero element of a.

i. If $J = \{j_0, \ldots, j_{n-1}\}$ is finite,

$$S(a) = a_{j_0} + \ldots + a_{j_{n-1}};$$

ii. If J is infinite, $S(a)$ exists iff $S(b)$ exists, and when they do, $S(a) = S(b)$.

Rearrangement Let $a \in \mathbb{R}^{\mathbb{N}}$, $\pi : \mathbb{N} \to \mathbb{N}$ any permutation, and let b be defined by

$$b_i = a_{\pi(i)}.$$

$S(a)$ exists iff $S(b)$ exists and, when they do, $S(a) = S(b)$.

Associativity Let $a : a_0, a_1, \ldots$ be a sequence of non-negative real numbers and let $J \subseteq \mathbb{N}$. Define sequences b, c by

$$b_i = \begin{cases} a_i, & i \in J \\ 0, & i \notin J \end{cases}, \quad c_i = \begin{cases} 0, & i \in J \\ a_i, & i \notin J \end{cases}.$$

Then: $S(a)$ exists iff $S(b)$ and $S(c)$ both exist and, when they do,

$$S(a) = S(b) + S(c).$$

Should these not all appear completely familiar, it is because I have stated them here in their final, general forms.

6.11 Theorem. *The functions $S_F, S_B, S_\Sigma, S_A = S_{H,1}, S_{H,2}, \ldots, S_H, S_L$ all satisfy Agreement with Finite Sums, Linearity, and the Truncation/Prefixing properties. Of these functions, only S_F, S_B and S_Σ satisfy Irrelevance of 0, and only S_F and S_B satisfy Rearrangement and Associativity.*

We have already proven Agreement with Finite Sums for S_Σ (Lemma 3.10) and for S_B (Lemma 5.8.i). It is a special case of the Irrelevance of 0, but should be mentioned separately because it holds more generally than does Irrelevance of 0. It is valid for all of the sum functions we have defined. For, the sequences of finite support are included in the domains of each of these functions, and we have established that S_B extends S_F, S_Σ extends S_B, etc.; in symbols:

$$S_F \subseteq S_B \subseteq S_\Sigma \subseteq S_A = S_{H,1} \subseteq S_{H,2} \subseteq \ldots \subseteq S_H \subseteq S_L.$$

We would not expect every sum function to lie somewhere in this chain, or even to be compatible with all the elements of it. It would seem, however, that agreeing with S_F on all sequences of finite support is an absolutely minimal condition to impose on a function S before allowing it to be called a sum.

Insofar as it deals with two of the most fundamental properties of addition, Linearity is another necessary condition any S must satisfy before it can be called an infinite sum. Both subproperties $L1$ (the same mixture of commutativity and associativity we use when we add a column of numbers digit by digit instead of entry by entry) and $L2$ (the distributive law) are so obvious that everyone used them without thinking before any formal definition of summation of infinite series was offered.

We proved linearity for S_Σ (Lemmas 3.5 and 3.6) and S_B (Lemma 5.6) already. That it holds for S_F is obvious, as it is finite summation which suggested the properties. From the formal point of view, however, one must give a proof. (Remember the words "rite" and "ceremonious" in Johnson's definition of "formally" on page vi, above. And, of course, I kept the phrase "exact to affectation" in Definition 0.1 of the Preface.) Besides, the linearity of S_F is not a result about finite sums, but about sums of infinite sequences all but finitely many elements of which are 0. That this summation behaves like finite summation as it was intended to requires formal verification. (This is one of the criticisms of formalism. By replacing intuitive concepts by formal "equivalents", what was once obvious and is still trivial now requires a proof like a genuine theorem.)

The proof for S_F is very simple. If a, b have finite support, then so do $a + b$ and ra for any $r \in \mathbb{R}$. Thus $S_F(a + b)$ and $S_F(ra)$ are defined. Suppose n_0 is large enough so that $a_n = b_n = 0$ for $n > n_0$. By definition,

$$S_F(a + b) = \sum_{i=0}^{n_0}(a_i + b_i) = \sum_{i=0}^{n_0} a_i + \sum_{i=0}^{n_0} b_i = S_F(a) + S_F(b)$$

$$S_F(ra) = \sum_{i=0}^{n_0} ra_i = r \sum_{i=0}^{n_0} a_i = rS_F(a).$$

The proof of linearity for S_A is only a little more complex. Suppose $S_A(a)$ and $S_A(b)$ exist.

$$S_A(a) + S_A(b) = \lim_{n\to\infty} \frac{1}{n+1} \sum_{i=0}^{n} \left(\sum_{j=0}^{i} a_j \right) + \lim_{n\to\infty} \frac{1}{n+1} \sum_{i=0}^{n} \left(\sum_{j=0}^{i} b_j \right)$$

$$= \lim_{n\to\infty} \left[\frac{1}{n+1} \sum_{i=0}^{n} \left(\sum_{j=0}^{i} a_j \right) + \frac{1}{n+1} \sum_{i=0}^{n} \left(\sum_{j=0}^{i} b_j \right) \right]$$

$$= \lim_{n\to\infty} \frac{1}{n+1} \left[\sum_{i=0}^{n} \left(\sum_{j=0}^{i} a_j \right) + \sum_{i=0}^{n} \left(\sum_{j=0}^{i} b_j \right) \right]$$

$$= \lim_{n\to\infty} \frac{1}{n+1} \sum_{i=0}^{n} \left(\sum_{j=0}^{i} a_j + \sum_{j=0}^{i} b_j \right)$$

$$= \lim_{n\to\infty} \frac{1}{n+1} \sum_{i=0}^{n} \left(\sum_{j=0}^{i} \left(a_j + b_j \right) \right)$$

$$= S_A(a+b).$$

The proof that $S_A(ra)$ is defined and equal to $rS_A(a)$ is similar, as are the proofs for $S_{H,k}$ for $k > 1$ and I omit these.

The proof of the linearity of S_H is a reduction to the linearity of all the $S_{H,k}$'s. Suppose $S_H(a)$ and $S_H(b)$ exist. Then there are k_1, k_2 such that $S_{H,k_1}(a), S_{H,k_2}(b)$ exist and

$$S_H(a) = S_{H,k_1}(a), \quad S_H(b) = S_{H,k_2}(b).$$

Choose $k = \max\{k_1, k_2\}$ and observe

$$S_H(a) + S_H(b) = S_{H,k_1}(a) + S_{H,k_2}(b)$$
$$= S_{H,k}(a) + S_{H,k}(b) = S_{H,k}(a+b) = S_H(a+b),$$

and similarly,

$$rS_H(a) = rS_{H,k_1}(a) = S_{H,k_1}(ra) = S_H(ra).$$

Finally, for S_L, one has

$$S_L(a) + S_L(b) = \lim_{x\to 1^-} \sum_{i=0}^{\infty} a_i x^i + \lim_{x\to 1^-} \sum_{i=0}^{\infty} b_i x^i$$

$$= \lim_{x\to 1^-} \left(\sum_{i=0}^{\infty} a_i x^i + \sum_{i=0}^{\infty} b_i x^i \right)$$

$$= \lim_{x\to 1^-} \sum_{i=0}^{\infty} (a_i x^i + b_i x^i)$$

$$= \lim_{x \to 1^-} \sum_{i=0}^{\infty} (a_i + b_i)x^i = S_L(a + b),$$

and similarly $rS_L(a) = S_L(ra)$.

The validity of the Truncation/Prefixing property for S_F is obvious and we proved it already for S_B (Lemma 5.9). For S_Σ we proved only a special case of it in Lemma 3.9. The ambitious reader can go back to the proof and modify it to obtain the full result. Alternatively, he can wait until the next section where it is proven that the special case yields the full result.

To prove the Truncation/Prefixing property for S_A, note first that it suffices to prove the property for prefixing or truncating a single entry, as multiple items can be handled by simple repetition. Now, if a is $a_0, a_1, \ldots$, we have

$$S_A(a) = \lim_{n \to \infty} \frac{1}{n+1} \sum_{i=0}^{n} \sum_{j=0}^{i} a_j$$

$$= \lim_{n \to \infty} \frac{1}{n+1} \sum_{i=0}^{n} \left(a_0 + \sum_{j=1}^{i} a_j \right)$$

$$= \lim_{n \to \infty} \frac{1}{n+1} \left((n+1)a_0 + \sum_{i=1}^{n} \sum_{j=1}^{i} a_j \right)$$

$$= a_0 + \lim_{n \to \infty} \frac{1}{n+1} \sum_{i=1}^{n} \sum_{j=1}^{i} a_j$$

$$= a_0 + \lim_{n \to \infty} \left(\frac{n}{n+1} \cdot \frac{1}{n} \sum_{i=1}^{n} \sum_{j=1}^{i} a_j \right)$$

$$= a_0 + \lim_{n \to \infty} \left(\frac{1}{n} \sum_{i=1}^{n} \sum_{j=1}^{i} a_j \right)$$

$$= a_0 + S_A(a_1, a_2, \ldots),$$

where, except for the first and last equations which are the definitions of S_A, each equation asserts equivalence of the existence of the limits in question as well as equality of the existing numbers. The proofs for $S_{H,k}$ are along the same lines, but are of course notationally messier and I omit them. The result for S_H reduces to the results for the individual $S_{H,k}$'s.

Finally, to establish the result for S_L, assume $\sum_{i=0}^{\infty} a_i x^i$ converges for all $|x| < 1$ and note:

$$\lim_{x \to 1^-} \sum_{i=0}^{\infty} a_i x^i = \lim_{x \to 1^-} \left(a_0 + \sum_{i=1}^{\infty} a_i x^i \right)$$

$$= a_0 + \lim_{x \to 1^-} \sum_{i=1}^{\infty} a_i x^i$$

$$= a_0 + \lim_{x \to 1^-} \left(x \cdot \sum_{i=1}^{\infty} a_i x^{i-1} \right)$$

$$= a_0 + \lim_{x \to 1^-} x \cdot \lim_{x \to 1^-} \sum_{i=1}^{\infty} a_i x^{i-1}$$

$$= a_0 + 1 \cdot S_L(a_1, a_2, \ldots).$$

The Irrelevance of 0 property holds for S_F and we proved it for S_B (Lemma 5.8). Moreover, the standard ϵ-n_0 argument will show S_Σ to have this property as well. Of greater interest is its failure for $S_A, S_{H,2}, \ldots, S_H, S_L$. This we have already seen with Lagrange's discussion of Callet and the Grandi series (page 7, above): there is a sequence c obtained by inserting 0's into the Grandi sequence g for which

$$S_A(c) = 3/5 \neq 1/2 = S_A(g).$$

This counterexample holds for all extensions, $S_{H,2}, \ldots, S_H, S_L$ of S_A.

Rearrangement and Associativity obviously hold for S_F and we proved these for S_B (Lemmas 5.10 and 5.11, respectively).

The failure of the Rearrangement property for S_Σ was demonstrated by Dirichlet and Riemann in as universal a sense as possible: if a is conditionally convergent, one can find permutations $\pi, \rho : \mathbb{N} \to \mathbb{N}$ such that for b, c defined by

$$b_i = a_{\pi(i)}, \quad c_i = a_{\rho(i)},$$

one has $\sum b_i$ convergent but $S_\Sigma(b) \neq S_\Sigma(a)$ and $\sum c_i$ not even convergent. The pairs a, b remain counterexamples for $S_{H,2}, \ldots, S_H, S_L$.

The failure of Associativity for S_Σ is also as universal as possible. If a is only conditionally convergent, the separation of a into its positive and negative parts provides a quick counterexample: $S_\Sigma(a)$ is defined, but neither $S_\Sigma(a^+)$ nor $S_\Sigma(a^-)$ exists.

This failure carries all the way from $S_A = S_{H,1}$ to S_L as the following Lemma demonstrates.

6.12 Lemma. *Let $b : b_0, b_1, \ldots$ be a sequence of non-negative real numbers and suppose $\sum_{i=0}^{\infty} b_i$ diverges. Then $S_L(b)$ does not exist.*

Proof. Suppose to the contrary that $\lim_{x \to 1^-} \sum_{i=0}^{\infty} b_i x^i = L$ exists. Choose n so large that

$$\sum_{i=0}^{n} b_i > L + 1.$$

Now $\sum_{i=0}^{n} b_i x^i$ is a polynomial, whence it is continuous and there is a $\delta > 0$ such that, for all x,

$$1 - \delta < x < 1 \Rightarrow 0 < \sum_{i=0}^{n} b_i - \sum_{i=0}^{n} b_i x^i < \frac{1}{2}.$$

Thus,

$$\sum_{i=0}^{n} b_i x^i > \sum_{i=0}^{n} b_i - \frac{1}{2} > L + 1 - \frac{1}{2} = L + \frac{1}{2},$$

is bounded away from its limit, a contradiction. □

7 Generalised Sums; An Axiomatic Approach

Linearity is a central concept of modern mathematics and we should expand on it. One is first introduced to the concept of linearity in one's study of the Calculus with the differentiation rules:

$$\frac{d}{dx}(f(x) + g(x)) = \frac{d}{dx}f(x) + \frac{d}{dx}g(x)$$

$$\frac{d}{dx}rf(x) = r\frac{d}{dx}f(x)$$

and the linearity of the integral:

$$\int (f(x) + g(x))dx = \int f(x)dx + \int g(x)dx$$

$$\int rf(x)dx = r\int f(x)dx.$$

In many American colleges, the next course a mathematics major is likely to take after Calculus is Linear Algebra, the two fundamental concepts of which are the notions of a *vector space* and a *linear transformation*. A vector space is basically a collection of objects that can be added and subtracted, and multiplied by real numbers. The set of real numbers forms a vector space. In Multivariable Calculus and Vector Analysis, one learns that

$$\mathbb{R}^n = \{(x_1, \ldots, x_n) \mid x_1, \ldots, x_n \in \mathbb{R}\}$$

is a vector space when one defines addition by

$$(x_1, \ldots, x_n) + (y_1, \ldots, y_n) = (x_1 + y_1, \ldots, x_n + y_n)$$

and *scalar multiplication* by

$$r(x_1, \ldots, x_n) = (rx_1, \ldots, rx_n).$$

Either in the Linear Algebra course or in a later Analysis course one learns that the set of, say, continuous functions defined on a given interval I form a vector space when one defines

$$(f + g)(x) = f(x) + g(x), \qquad (rf)(x) = r \cdot f(x).$$

And, in Differential Equations, one learns that the set of solutions to a homogeneous linear differential equation forms a vector space under these operations.

Aside from $\mathbb{R}$ itself, the vector spaces of interest to us here are $\mathbb{R}^{\mathbb{N}}$ and its subspaces. The closure clauses of the linearity conditions (if $S(a)$ and $S(b)$ are defined, then $S(a+b)$ and $S(ra)$ are defined) guarantee that the domains of our sum functions $S_F, S_B, S_\Sigma, \ldots$ are all vector spaces.

The formal definition of a vector space stipulates the properties that *vector addition* and *scalar multiplication* must satisfy.

7.1 Definition. *A set V together with binary functions*

$$+ : V \times V \to V,$$

$$\cdot : \mathbb{R} \times V \to V,$$

and a designated element $\mathbf{0} \in V$, constitutes a vector space *if the following conditions are satisfied:*
i.   *for all $u, v \in V, u + v = v + u$*
ii.   *for all $u, v, w \in V, u + (v + w) = (u + v) + w$*
iii. *for all $v \in V, v + \mathbf{0} = \mathbf{0} + v = v$*
iv.   *for all $v \in V, v + (-1)v = \mathbf{0}$*
v.   *for all $u, v \in V, r \in \mathbb{R}, r(u + v) = ru + rv$*
vi.   *for all $v \in V, q, r \in \mathbb{R}, (q + r)v = qv + rv$*
vii. *for all $v \in V, q, r \in \mathbb{R}, q(rv) = (qr)v.$*

7.2 Remark. Perhaps the above definition becomes more elegant if one completely separates the conditions on vector addition and scalar multiplication by replacing iv by
iv′. for all $v \in V$, there is an element $v^* \in V$ such that $v + v^* = \mathbf{0}$.
When one does this, however, one must add some axiom normalising the scalar multiplication, such as
viii.   for all $v \in V, 1v = v$; or
viii′. for all $v \in V$, there are $r \in \mathbb{R}, w \in V$ such that $v = rw$; or
viii″. for all $v \in V, v^* = (-1)v$.
Without some such condition, one cannot rule out the possibility that $rv = \mathbf{0}$ for all r, v.

One can make this definition more abstract by replacing $\mathbb{R}$ by $\mathbb{F}$, where $\mathbb{F}$ is any *field*, or (to use the archaic term) *domain of rationality*, such as the sets $\mathbb{Q}$ and $\mathbb{C}$ of rational and complex numbers, respectively. Or, one can make it more concrete by considering only *subspaces* of a given vector space.

7.3 Lemma. *Let $V_0 \subseteq V$ be nonempty, where V is a vector space. If*
i.   *V_0 is closed under addition: $u, v \in V_0 \Rightarrow u + v \in V_0$; and*
ii. *V_0 is closed under scalar multiplication: $v \in V_0, r \in \mathbb{R} \Rightarrow rv \in V_0$,*
then: V_0 is a vector space under the operations inherited from V.

Proof. All of the conditions of Definition 7.1 are identities and are valid in any subset of V for which the terms are defined. By assumption, V_0 is closed under addition and scalar multiplication. Moreover, $\mathbf{0} \in V_0$ because $\mathbf{0} = 0v$ for any $v \in V_0$, as the reader may easily verify. □

By this Lemma, the closure of the domains of S_F, S_B, etc. under addition and multiplication by real numbers shows these domains to be vector spaces.

7.4 Example. Define

$$V_F = dom(S_F), \quad V_B = dom(S_B), \quad V_\Sigma = dom(S_\Sigma), \quad \text{etc.}$$

For $X = F, B, \Sigma, \ldots, L$, the set V_X is a vector space.

The second fundamental concept of Linear Algebra is that of a linear transformation:

7.5 Definition. *Let V, W be vector spaces. A function $T : V \to W$ is a* linear transformation *if*
i. for all $u, v \in V, T(u + v) = T(u) + T(v)$; and
ii. for all $v \in V, r \in \mathbb{R}, T(rv) = rT(v)$.

7.6 Example. For $X = F, B, \Sigma, \ldots, L$, the function $S_X : V_X \to \mathbb{R}$ is a linear transformation.

This is just a restatement of the linearity of these functions. Some additional elementary examples are:

7.7 Examples. i. The projection $P_i : \mathbb{R}^{\mathbb{N}} \to \mathbb{R}$ defined by $P_i(a) = a_i$ is a linear transformation.
ii.   The cumulative sum $C : \mathbb{R}^{\mathbb{N}} \to \mathbb{R}$ defined by

$$C(a) = s, \quad \text{where } s_n = \sum_{i=0}^{n} a_i,$$

is a linear transformation.
iii. The function $A : \mathbb{R}^{\mathbb{N}} \to \mathbb{R}$ of Definition 6.5,

$$A(s) = t, \quad \text{where } t_n = \frac{1}{n+1}(s_0 + \ldots + s_n),$$

is a linear transformation.
iv. Let $V \subseteq \mathbb{R}^{\mathbb{N}}$ be the set of convergent sequences and define, for $a \in V$,

$$L(a) = \lim_{n \to \infty} a_n.$$

L is a linear transformation.
v.   Let $I = (-1, 1)$ and V the set of continuous functions $f : I \to \mathbb{R}$ for which $\lim_{x \to 1^-} f(x)$ exists and define $L_2 : V \to \mathbb{R}$ by

$$L_2(f) = \lim_{x \to 1^-} f(x).$$

L_2 is a linear transformation.

To establish these, note that part i is really nothing more than the definition of $a + b$ and ra: the i-th terms of these sequences are $a_i + b_i$ and ra_i, respectively.

The second example makes a non-trivial claim, but the proof is not difficult. Suppose $a, b \in \mathbb{R}^{\mathbb{N}}$ and consider the n-th term of $C(a) + C(b)$:

$$(C(a) + C(b))_n = (C(a))_n + (C(b))_n$$

$$= \sum_{i=0}^{n} a_i + \sum_{i=0}^{n} b_i = \sum_{i=0}^{n} (a_i + b_i)$$

$$= \sum_{i=0}^{n} (a + b)_i = (C(a + b))_n.$$

Multiplication by a real number is handled similarly:

$$(C(ra))_n = \sum_{i=0}^{n} (ra)_i = \sum_{i=0}^{n} ra_i = r \sum_{i=0}^{n} a_i = r(C(a))_n.$$

This mode of proof is slightly inefficient. For this reason, one often combines the two linearity conditions— *preservation of addition* and *preservation of scalar multiplication*— into a single condition of *preservation of linear combination*.

7.8 Lemma. *Let V, W be vector spaces. A function $T : V \to W$ is a linear transformation iff for all $u, v \in V$ and all $q, r \in \mathbb{R}$, one has*

$$T(qu + rv) = qT(u) + rT(v). \tag{56}$$

Proof. Assume first that T is a linear transformation and u, v, q, r are given. Observe,

$$T(qu + rv) = T(qu) + T(rv), \quad \text{by Definition 7.5.i}$$
$$= qT(u) + rT(v), \quad \text{by Definition 7.5.ii.}$$

Conversely, if T satisfies (56), then

$$T(u + v) = T(1u + 1v) = 1T(u) + 1T(v) = T(u) + T(v)$$

and

$$T(rv) = T(rv + \mathbf{0}) = T(rv + 0v) = rT(v) + 0T(v) = rT(v) + \mathbf{0} = rT(v). \qquad \square$$

We illustrate the use of this Lemma in verifying that the function A is a linear transformation. Let $a, b \in \mathbb{R}^{\mathbb{N}}, q, r \in \mathbb{R}$ and observe

$$(A(qa + rb))_n = \frac{1}{n + 1} \sum_{i=0}^{n} (qa + rb)_i$$

$$= \frac{1}{n + 1} \sum_{i=0}^{n} (qa_i + rb_i)$$

$$= \frac{1}{n + 1} \left[q \sum_{i=0}^{n} a_i + r \sum_{i=0}^{n} b_i \right]$$

$$= q\frac{1}{n+1}\sum_{i=0}^{n}a_i + r\frac{1}{n+1}\sum_{i=0}^{n}b_i$$
$$= q(A(a))_n + r(A(b))_n,$$

whence $A(qa + rb) = qA(a) + rA(b)$.

The proofs that the limits iv and v are linear transformations are essentially the same as the proof of linearity of S_Σ. Using Lemma 7.8, however, we use one slightly more complicated ϵ-estimate instead of two slightly easier ones. To prove iv, let $a, b \in \mathbb{R}^\mathbb{N}$ have limits A, B, respectively, and assume $q, r \in \mathbb{R}$. Let $\epsilon > 0$ and consider

$$|qa_n + rb_n - (qA + rB)| = |qa_n - qA + rb_n - rB|$$
$$\leq |qa_n - qA| + |rb_n - rB|$$
$$\leq |q| \cdot |a_n - A| + |r| \cdot |b_n - B|. \qquad (57)$$

Suppose neither $|q|$ nor $|r|$ is 0. Find n_0 so that, for all $n > n_0$,

$$|a_n - A| < \frac{\epsilon}{2|q|}, \qquad |b_n - B| < \frac{\epsilon}{2|r|}.$$

Then from (57) we have

$$|qa_n + rb_n - (qA + rB)| < |q| \cdot \frac{\epsilon}{2|q|} + |r| \cdot \frac{\epsilon}{2|r|}$$
$$< \frac{\epsilon}{2} + \frac{\epsilon}{2} = \epsilon,$$

and we see that

$$\lim_{n\to\infty}(qa_n + rb_n) = qA + rB = q\lim_{n\to\infty}a_n + r\lim_{n\to\infty}b_n.$$

If one of $|q|$ and $|r|$ is 0, the term containing it on the right-hand side of (57) is 0 and thus less than $\epsilon/2$ for all n.

I leave the verification that the limit of v is linear to the reader.

The transformations of Examples 7.7 were not chosen at random. Notice that

$$S_{H,1}(a) = S_A(a) = L(A(C(a))) \quad \text{and} \quad S_{H,2}(a) = L(A(A(C(a)))),$$

more generally,

$$S_{H,k}(a) = L(A^k(C(a))),$$

and

$$S_L(a) = L_2\big(S_\Sigma(a(x))\big), \quad \text{for a power series } a(x).$$

The proof we gave of the linearity of S_A essentially amounts to applying the following:

7.9 Lemma. *i. Let $T_0 : U \to V$, $T_1 : V \to W$ be linear transformations. The composition $T_1 \circ T_0 : U \to W$ is a linear transformation.*
ii. Let $T_0 : U \to V$, $T_1 : V_0 \to W$ be linear transformations, where $V_0 \subseteq V$ is itself a vector space. Then $U_0 = \{u \in U | T_0(u) \in V_0\}$ is a vector space and $T_1 \circ T_0 : U_0 \to W$ is a linear transformation.

Proof. i. If $u, v \in U$ and $q, r \in \mathbb{R}$, then $qu + rv \in U$ and $(T_1 \circ T_0)(qu + rv)$ is defined. Moreover,

$$
\begin{aligned}
(T_1 \circ T_0)(qu + rv) &= T_1(T_0(qu + rv)) \\
&= T_1(qT_0(u) + rT_0(v)), \text{ by linearity of } T_0 \\
&= qT_1(T_0(u)) + rT_1(T_0(v)), \text{ by linearity of } T_1 \\
&= q(T_1 \circ T_0)(u) + r(T_1 \circ T_0)(v).
\end{aligned}
$$

ii. Let $T_0 : U \to V$ and let $V_0 \subseteq V$ also be a vector space, and define $U_0 = \{u \in U | T_0(u) \in V_0\}$. Now $T_0(\mathbf{0}) = \mathbf{0}$ (for, $T_0(\mathbf{0}) = T_0(0 \cdot \mathbf{0}) = 0 T_0(\mathbf{0}) = \mathbf{0}$), so U_0 is not empty. If $u, v \in U_0$, then $T_0(u), T_0(v) \in V_0$, whence

$$T_0(u + v) = T_0(u) + T_0(v) \in V_0, \text{ and thus } u + v \in U_0,$$

$$T_0(qu) = qT_0(u) \in V_0, \text{ and thus } qu \in U_0.$$

It follows that U_0 is a vector space.

The rest of part ii now follows from part i because $T_0 : U_0 \to V_0$, and $T_1 : V_0 \to W$ are linear transformations.   □

7.10 Definition. *Let $V \subseteq \mathbb{R}^{\mathbb{N}}$. A function $S : V \to \mathbb{R}$ is a* weak sum function, *or a* summation function in the weak sense, *if*
i. V is a vector space and S is a linear transformation; and
ii. $V_F \subseteq V$ and, for all $a \in V_F$, $S(a) = S_F(a)$.

In other words, S is a summation function in the weak sense if it has the linearity property and agrees with finite sums. The formulation of the present definition reminds us explicitly that the domain of S, by being a vector space, has some structure too, in this case the closure properties of Lemma 7.3.

I use the adjective "weak" in describing this notion of a summation function because it makes the weakest demands on the function that still impose some sum-like properties. Greater familiarity with Linear Algebra provides a multitude of examples that might suggest the present definition to be too broad. We turn now to some rudimentary considerations in the construction of such examples.

7.11 Definition. *Let $U \subseteq V$ be vector spaces and let $v_0 \in V$ be such that $v_0 \notin U$. We define*

$$U[v_0] = \{u + rv_0 | u \in U \text{ and } r \in \mathbb{R}\}.$$

$U[v_0]$ is again a vector space and any linear transformation on U can be extended to one on $U[v_0]$. If we start with $U \supseteq V_F$ and a transformation $T : U \to \mathbb{R}$ that extends S_F, i.e. if T is a weak sum function on U, then any extension of T to $U[v_0]$ is also a weak sum function. This is how we can construct simple and plentiful examples of weak sum functions. Before proving this, however, let us consider some examples of U, V, v_0.

7.12 Examples. We let $V = \mathbb{R}^{\mathbb{N}}$ in each of the following:

i. $U = V_F$, $v_0 : 1, \frac{1}{2}, \frac{1}{4}, \frac{1}{8}, \dots$

ii. $U = V_{\Sigma}$, $v_0 = g : 1, -1, 1, -1, \dots$

iii. $U = V_{\Sigma}$, $v_0 = e : 1, 2, 4, 8, \dots$

iv. $U = V_{\Sigma}[g]$, $v_0 = h : 1, -2, 3, -4, \dots$

v. $U = V_{\Sigma}[e]$, $v_0 = w : 1, -2, 4, -8, \dots$

The crucial thing about each of these examples is the verification that $v_0 \notin U$. For i-iii this is easy enough. V_F is the set of sequences in $\mathbb{R}^{\mathbb{N}}$ with only finitely many nonzero terms and the geometric progression in i had infinitely many such. V_{Σ} is the set of sequences defining convergent series, and the series constructed from g and e do not converge. For iv and v we have to work a little harder.

First, that the spaces U in 7.12.iv and 7.12.v are indeed vector spaces will be established shortly. To see that $h \notin V_{\Sigma}[g]$, note that every element of $V_{\Sigma}[g]$ has the form

$$a + rg : a_0 + r, a_1 - r, a_2 + r, a_3 - r, \dots$$

for some sequence $a : a_0, a_1, \dots$ for which $\sum_{i=0}^{\infty} a_i$ converges and some real number r. Because $\sum_{i=0}^{\infty} a_i$ converges, $\lim_{n \to \infty} a_n = 0$ and $a + rg$ is bounded: there are m, M such that

$$m < a_n \pm r < M$$

for all n. This is not the case for h.

There is no difference in size between the sequence $e : 1, 2, 4, 8, \dots$ which Euler summed to -1 (page 4, above) and $w : 1, -2, 4, -8, \dots$ which Wolf summed to $\frac{1}{3}$ (page 3, above). But the one alternates in sign and the other doesn't. This behaviour is unaffected on multiplication by a nonzero r and is unaffected from some point on if one adds the elements of a convergent series: Every element of $V_{\Sigma}[e]$ is of the form

$$b : a_0 + r, a_1 + 2r, a_2 + 3r, \dots$$

If $r \neq 0$, there is some n_0 such that $|a_n| < |r|$ for all $n > n_0$. But then b_n has the same sign as r for all $n > n_0$, i.e. the sign remains fixed from b_{n_0} on. This is not the case for the alternating sequence $w : 1, -2, 4, -8, \dots$

As I said, we want to use these spaces to construct some simple weak sum functions. We do that via the following Theorem.

7.13 Theorem. *Let $U \subseteq V$ be vector spaces and let $v_0 \in V$ be such that $v_0 \notin U$. Then:*

i. $U[v_0] \subseteq V$ is a vector space under the operations of V.

ii. Every element of $U[v_0]$ is uniquely representable in the form $u + rv_0$: if $u_1 + r_1v_0 = u_2 + r_2v_0$, then $u_1 = u_2$ and $r_1 = r_2$.

iii. Let $T : U \to W$ be a linear transformation and let $w_0 \in W$ be arbitrary. There is a linear transformation $T' : U[v_0] \to W$ satisfying

a. for all $u \in U$, $T'(u) = T(u)$,

b. $T'(v_0) = w_0$.

Proof. The proofs are very simple. Let U, V, v_0 be given as stated.

i. $U[v_0]$ contains $v_0 = \mathbf{0} + 1v_0$ and thus is not empty. Let $u_1 + r_1v_0, u_2 + r_2v_0 \in U[v_0]$ and observe

$$(u_1 + r_1v_0) + (u_2 + r_2v_0) = (u_1 + u_2) + (r_1 + r_2)v_0 \in U[v_0]$$

$$r(u_1 + r_1v_0) = ru_1 + rr_1v_0 \in U[v_0].$$

Thus, by Lemma 7.3, $U[v_0]$ is a vector space.

ii. Suppose u_1, u_2, r_1, r_2 are given such that

$$u_1 + r_1v_0 = u_2 + r_2v_0.$$

Then

$$u_1 - u_2 = (r_2 - r_1)v_0 \tag{58}$$

and, if $r_1 \neq r_2$, we would have

$$v_0 = \frac{1}{r_2 - r_1}(u_1 - u_2) \in U,$$

which is assumed not to be the case. Thus $r_1 = r_2$ and the right-hand side of (58) is $\mathbf{0}$, i.e. $u_1 - u_2 = \mathbf{0}$, i.e. $u_1 = u_2$.

iii. Given $T : U \to W$, $w_0 \in W$, define T' on $U[v_0]$ as follows. Each $v \in U[v_0]$ has a unique representation

$$v = u + rv_0.$$

Set

$$T'(v) = T(u) + rw_0.$$

Given $v_1 = u_1 + r_1v_0, v_2 = u_2 + r_2v_0 \in U[v_0]$, and $q_1, q_2 \in \mathbb{R}$, observe

$$
\begin{aligned}
T'(q_1v_1 + q_2v_2) &= T'(q_1(u_1 + r_1v_0) + q_2(u_2 + r_2v_0)) \\
&= T'(q_1u_1 + q_2u_2 + (q_1r_1 + q_2r_2)v_0) \\
&= T(q_1u_1 + q_2u_2) + (q_1r_1 + q_2r_2)w_0, \text{ by definiton} \\
&= q_1T(u_1) + q_2T(u_2) + q_1r_1w_0 + q_2r_2w_0, \text{ by linearity of } T \\
&= q_1(T(u_1) + r_1w_0) + q_2(T(u_2) + r_2w_0) \\
&= q_1T'(v_1) + q_2T'(v_2), \text{ by definition.}
\end{aligned}
$$

But this last is just the criterion for linearity established in Lemma 7.8.   □

As I said, we can apply this Theorem to the spaces of Examples 7.12.

7.14 Examples. i. Extend S_F to $V_F[1, \frac{1}{2}, \frac{1}{4}, \frac{1}{8}, \ldots]$ by setting

$$S\left(1, \frac{1}{2}, \frac{1}{4}, \frac{1}{8}, \ldots\right) = 5.$$

That is, there is a weak sum function giving the wrong value to the geometric progression $1, \frac{1}{2}, \frac{1}{4}, \frac{1}{8}, \ldots$ This cannot happen if we start with S_Σ defined on V_Σ, as $(1, \frac{1}{2}, \frac{1}{4}, \frac{1}{8}, \ldots) \in V_\Sigma$. But a divergent geometric progression can be given any value we choose. In particular,

ii. Starting with $S = S_\Sigma$ on $U = V_\Sigma$, by the Theorem we can extend this to $S' : V_\Sigma[g] \to \mathbb{R}$ choosing any value we please for g. We can give g the Bernoulli-Grandi value of $\frac{1}{2}$, or one of the plausible values $0, 1$ determined by pairing adjacent elements of the sequence, or we can choose a value randomly. The conditions of weak summability are too weak to make the decision for us.

iii. The same goes for the progression $e : 1, 2, 4, 8, \ldots$ We are free to follow Euler's lead in extending S_Σ to $V_\Sigma[e]$ by defining $S(e) = -1$ or not as we please.

iv - v. The point here is that we can iterate the process. We can start by letting $S = S_\Sigma$ on V_Σ, extend it to $V_\Sigma[g]$ by choosing $S'(g) = r_1$, and then extending this to $(V_\Sigma[g])[h]$ by choosing $S''(h) = r_2$. We can choose $r_1 = \frac{1}{2}$ and $r_2 = \frac{1}{4}$ in accordance with S_H or not: we can give either, both, or neither their "correct" values.

The iteration of 7.12.iv - v can be carried well beyond the finite— into the *transfinite*, to use the set theoretic lingo— and one can prove the following:

7.15 Theorem. *Let $U \subseteq V$ be vector spaces, and let $T : U \to W$ be a linear transformation. T can be extended to a linear transformation $T' : V \to W$.*

7.16 Corollary. *There is a weak sum function $S \supseteq S_\Sigma$ defined for all sequences in $\mathbb{R}^{\mathbb{N}}$.*

The proofs of these results are, of course, beyond the scope of the present book as they require a knowledge of Set Theory usually reserved for the graduate level.

Back in §3 we demonstrated the correctness of the calculation of the sum of a geometric progression

$$a, ar, ar^2, \ldots$$

for $|r| < 1$, i.e. on assumption that the sum existed. And in §6 (following (52)) we determined the sum of the Hölder sequence h by assuming it and the Grandi sequence g to have sums. In these derivations we used one additional property of ordinary summation, namely a special case of the Truncation/Prefixing property asserting that we could prefix a 0 to a sequence without affecting the summability or the sum of the sequence.

7.17 Definition. *The* shift *and* truncation *operators on $\mathbb{R}^{\mathbb{N}}$ are the functions $\sigma, \tau : \mathbb{R}^{\mathbb{N}} \to \mathbb{R}^{\mathbb{N}}$ defined by*

$$\sigma(a) = (0, a_0, a_1, \ldots)$$
$$\tau(a) = (a_1, a_2, \ldots)$$

for $a : a_0, a_1, \ldots$ More formally,

$$\sigma(a) = b, \ where \ b_n = \begin{cases} 0, & n = 0 \\ a_{n-1}, & n > 0 \end{cases}$$

$$\tau(a) = c, \ where \ c_n = a_{n+1}.$$

Note that σ, τ are linear transformations. (*Exercise!*)

7.18 Lemma. *Let $S : V \to \mathbb{R}$ be a weak sum function. S has the Truncation/Prefixing property iff*
i. V is closed under σ, τ; and
ii. for all $a \in V$, $S(a) = S(\sigma(a))$.

Proof. Closure under σ, τ and the equation $S(a) = S(\sigma(a))$ are just specialisations of the Truncation/Prefixing property. Thus, assume i, ii and let $a \in V$, say

$$a : a_0, a_1, \ldots$$

To verify closure under truncation, note that

$$(a_{n+1}, a_{n+2}, \ldots) = \tau^{n+1}(a) \in V.$$

For closure under prefixing, note that

$$(b_0, \ldots, b_n, a_0, a_1, \ldots) = (b_0, \ldots, b_n, 0, 0, \ldots) + \sigma^n(a) \in V,$$

because $(b_0, \ldots, b_n, 0, 0, \ldots) \in V_F \subseteq V$, $\sigma^n(a) \in V$, and V is a vector space.
Finally,

$$S(a_0, a_1, \ldots, a_n, a_{n+1}, \ldots) = S(a_0, \ldots, a_n, 0, 0, \ldots) + S(0, \ldots, 0, a_{n+1}, \ldots)$$
$$= a_0 + \ldots + a_n + S(0, \ldots, 0, a_{n+1}, \ldots)$$
$$= a_0 + \ldots + a_n + S(a_{n+1}, \ldots),$$

the last step obtained by applying ii n times to $(a_{n+1}, \ldots)$. $\square$

7.19 Definition. *Let $V \subseteq \mathbb{R}^{\mathbb{N}}$. A linear transformation $S : V \to \mathbb{R}$ is a* general sum function, *or* generalised sum, *if*
i. $V \supseteq V_F$ is a vector space closed under σ, τ; and
ii. a. for all $a \in V_F$, $S(a) = S_F(a)$
    b. for all $a \in V$, $S(a) = S(\sigma(a))$.

In other words, a generalised sum is a weak sum function that also possesses the Truncation/Prefixing property. All the sums $S_F, S_B, S_\Sigma, \ldots, S_L$ introduced in §6 are generalised sums.

The ability to prefix a 0 without changing the sum imposes some restrictions on the possible values that S can assign to certain sequences.

7.20 Theorem. *Let $S : V \to \mathbb{R}$ be a generalised sum. Consider the geometric progression,*

$$G : 1, r, r^2, \ldots$$

i. If $r \neq 1$ and $S(G)$ is defined, then $S(G) = \frac{1}{1-r}$.
ii. If $r = 1$, then $S(G)$ is not defined.

Proof. i. Observe

$$\begin{aligned}
G &= (1, 0, 0, \ldots) + (0, r, r^2, \ldots) \\
&= (1, 0, 0, \ldots) + r(0, 1, r, \ldots) \\
&= (1, 0, 0, \ldots) + r\sigma(G),
\end{aligned}$$

whence

$$S(G) = S(1, 0, 0, \ldots) + rS(\sigma(G)) = 1 + rS(G)$$

and

$$(1 - r)S(G) = 1$$

from which follows $S(G) = \frac{1}{1-r}$ so long as $r \neq 1$.
    ii. Suppose $S(1, 1, 1, \ldots)$ existed. Then

$$S(0, 1, 1, \ldots) = S(\sigma(1, 1, \ldots)) = S(1, 1, 1, \ldots).$$

But $(1, 1, 1, \ldots) - (0, 1, 1, \ldots) = (1, 0, 0, \ldots)$ and

$$\begin{aligned}
0 = S(1, 1, 1, \ldots) - S(1, 1, 1, \ldots) &= S(1, 1, 1, \ldots) - S(0, 1, 1, \ldots) \\
&= S(1, 0, 0, \ldots) = 1,
\end{aligned}$$

a contradiction.   □

Note that Theorem 7.20 does not guarantee the existence of $S(G)$, only that the usual algebraic value is the only possible one for a generalised sum.

7.21 Lemma. *Let $S : V \to \mathbb{R}$ be a generalised sum, let $r \neq 1$, and let G be the geometric progression $1, r, r^2, \ldots$ There is an extension $S' : V[G] \to \mathbb{R}$ which is also a generalised sum.*

Proof. If $G \in V$, there is nothing to prove. If $G \notin V$, extend S to $V[G]$ by defining

$$S'(G) = \frac{1}{1 - r}$$

in accordance with Theorem 7.13. S' is a linear transformation.
    Now

$$\begin{aligned}
\sigma(G) &= (0, 1, r, r^2, \ldots) \\
&= (1/r, 1, r, r^2, \ldots) - (1/r, 0, 0, 0, \ldots) \\
&= \frac{1}{r}G - (1/r, 0, 0, 0, \ldots) \in V[G],
\end{aligned}$$

whence $S'(\sigma(G))$ is defined. Similarly, $\tau(G) \in V[G]$.

Moreover,

$$
\begin{aligned}
S'(\sigma(G)) &= \frac{1}{r}S'(G) - \frac{1}{r} \\
&= \frac{1}{r} \cdot \frac{1}{1-r} - \frac{1}{r} = \frac{1-1+r}{r(1-r)} \\
&= \frac{1}{1-r} = S'(G). \qquad (59)
\end{aligned}
$$

Let $v = u + qG \in V[G]$ for some $u \in V, q \in \mathbb{R}$. then

$$
\begin{aligned}
S'(\sigma(v)) &= S'(\sigma(u + qG)) \\
&= S'(\sigma(u) + q\sigma(G)), \text{ by linearity of } \sigma \\
&= S'(\sigma(u)) + qS'(\sigma(G)) \\
&= S'(u) + qS'(G), \text{ by (59)} \\
&= S'(u + qG) = S'(v).
\end{aligned}
$$

Thus, S' is a generalised sum. □

In Set Theory application of Lemma 7.20 is iterated through the transfinite to prove the following:

7.22 Theorem. *Let $S : V \to \mathbb{R}$ be a generalised sum. There is a generalised sum extending S that is defined for all non-constant geometric progressions.*

The specialisation of Lemma 7.20 to $r = \pm 1$ also tells us something about arithmetic and polynomial progressions.

7.23 Theorem. *Let $P(X)$ be a non-zero polynomial and define sequences*

$$
p : P(0), P(1), P(2), P(3), \ldots
$$

$$
p^* : P(0), -P(1), P(2), -P(3), \ldots
$$

i. There is a number q such that, for any generalised sum S for which $S(p^*)$ is defined, $S(p^*) = q$.
ii. For no generalised sum S is $S(p)$ defined.

Proof. The proof is by induction on the degree $d \geq 0$ of P.

Basis. $P(X) = q$ is a constant for some $q \neq 0$. Then $p^* = qg$ and $p = q|g|$ and the result follows from Theorem 7.20 for $r = -1$ and $r = 1$, respectively.

Induction Step. Assume P of degree $d + 1$, for $d \geq 0$.

The key to reducing the result to the case of degree d is the *difference operator* Δ defined by

$$
\Delta f(x) = f(x+1) - f(x).
$$

The difference operator is a discrete analogue to the differentiation operator and similarly reduces the degree of a polynomial by 1. Observe that

$$p - \sigma(p) = (P(0) - 0, P(1) - P(0), P(2) - P(1), \ldots$$
$$= (P(0), \Delta P(0), \Delta P(1), \ldots)$$
$$= (P(0), Q(0), Q(1), Q(2), \ldots)$$

for some polynomial Q of degree d, whence the induction hypothesis tells us that

$$\tau(p - \sigma(p)) = (Q(0), Q(1), Q(2), \ldots)$$

has no sum under S, whence neither can p.

Similarly,

$$p^* + \sigma(p^*) = (P(0), -\Delta P(0), \Delta P(1), -\Delta P(2), \ldots)$$
$$= (P(0), R(0), -R(1), R(2), \ldots)$$

for some polynomial R of degree d. The induction hypothesis applies and there is some value r which $S(R(0), -R(1), R(2), -R(3), \ldots)$ must assume if it is defined and, as before, if $S(p^*)$ is defined, we have

$$2S(p^*) = P(0) + r$$

and

$$S(p^*) = \frac{P(0) + r}{2}. \qquad \square$$

As an exercise, I suggest the reader verify the value $\frac{1}{4}$ for the Hölder sequence h (51), or determine the value of

$$S(1, -4, 9, -16, \ldots),$$

assuming it to exist.

As is the case with geometric progressions, the positive half of Theorem 7.23 only tells us what the sum must be if it exists and not that it exists. It does not take much thought to conjecture that $S_{H,m}(p^*)$ exists for $m > d$, where d is the degree of P, whence S_H would sum all alternating polynomial progressions. I leave the exploration of this matter to the advanced reader.

It is a little strange that the geometric progression,

$$1 + 2 + 4 + 8 + \ldots,$$

which grows very rapidly, could have a sum and yet the more slowly growing arithmetic progression,

$$1 + 2 + 3 + 4 + \ldots,$$

could not. It is also a bit strange that the sum should be negative. We could accept, with Euler, that ∞ separates the positive and negative numbers and that our geometric progression grows so rapidly that it managed to jump over ∞ to land among the negative numbers. Or, one could refuse to accept this intuition and attempt to rule out the possibility by considering another property of finite summation, namely order preservation, which we may state in the following form:

Order Preservation If $a : a_0, a_1, \ldots$ are all non-negative, then $S(a) \geq 0$.

7.24 Definition. *A generalised sum $S : V \to \mathbb{R}$ satisfying the Order Preservation property is called an* order preserving generalised sum.

It is not hard to see that $S_F, S_B, S_\Sigma, \ldots, S_L$ are all order preserving generalised sums. (It suffices to prove this to be the case for S_L, which I leave as an easy exercise for the reader.) Taking these as typifying the sorts of summation functions we are interested in, we might be tempted to go back and redefine the notion of generalised sum to include order preservation among its defining clauses. My preference is not to do this because, unlike the other properties considered, order preservation has nothing to do with the mechanics of summation, but is more a property of the real numbers than of summation— as is seen by contemplating addition of complex numbers or modular arithmetic. Thus I present order preservation as a property of a special kind of generalised sum and not as a defining property thereof.

Returning to the point at hand, order preservation rectifies the situation of a geometric progression overstepping infinity:

7.25 Theorem. *Let S be an order preserving generalised sum and let $G : 1, r, r^2, \ldots$ be a geometric progression. If $r > 1$, then $S(G)$ is undefined.*

For, if $S(G)$ were defined, then by Theorem 7.20, $S(G)$ would be $\frac{1}{1-r}$, which is negative for $r > 1$, and order preservation would be violated.

We can do better.

7.26 Theorem. *Let $a : a_0, a_1, \ldots$ be a sequence of non-negative real numbers and S an order preserving generalised sum. If $S(a)$ is defined, then so is $S_\Sigma(a)$ and $S_\Sigma(a) \le S(a)$.*

Proof. Suppose a is given and $S(a) = M$. If $S_\Sigma(a)$ is undefined, or is defined but greater than $S(a)$, then there is a number n_0 such that

$$\sum_{i=0}^{n_0} a_i > M. \tag{60}$$

But

$$S(a) = \sum_{i=0}^{n_0} a_i + S(a_{n_0+1}, a_{n_0+2}, \ldots), \text{ by Truncation}$$

$$\ge \sum_{i=0}^{n_0} a_i + 0, \text{ by Order Preservation}$$

$$> M, \text{ by (60)},$$

a contradiction.   □

In particular, the Theorem shows that if $S_\Sigma(a)$ is undefined, then so is $S(a)$. This has an interesting corollary.

7.27 Corollary. *If S is an order preserving generalised sum that satisfies the Associative property and $a \in \mathbb{R}^{\mathbb{N}}$ is such that $S(a)$ exists, then $S_B(a)$ exists. In particular, no order preserving generalised sum properly extending S_B can satisfy Associativity.*

Proof. Suppose S has the two properties in question, $S(a)$ is defined, and a^+, a^- are the positive and negative parts of a, respectively. By Associativity, $S(a^+)$ and $S(a^-)$ are defined and, by Theorem 7.26, $S_\Sigma(a^+)$ and $S_\Sigma(a^-)$ exist. But then, by the proof of Theorem 5.15, $\sum_{i=0}^{\infty} a_i$ is absolutely convergent and $S_B(a)$ exists. □

Let me finish this section with one more result, a partial reversal of the inequality of Theorem 7.26.

7.28 Theorem. *Let S be an order preserving generalised sum, the domain of which includes all convergent geometric series. Suppose $a : a_0, a_1, \ldots$ is a sequence of non-negative real numbers, and suppose there is a geometric progression,*

$$q, qr, qr^2, \ldots, \quad q > 0, \quad 0 < r < 1,$$

and a number n_0 such that, for all $n > n_0$, $a_n \le qr^n$. If $S(a)$ is defined, then $S(a) = S_\Sigma(a)$.

Stated more succinctly, if $\sum_{i=0}^{\infty} a_i$ converges at least as rapidly as some convergent geometric progression, S is defined for all convergent geometric progressions, and $S(a)$ is defined, then $S(a) = \sum_{i=0}^{\infty} a_i$.

Proof. Suppose $\sum_{i=0}^{\infty} a_i = L$. Let $\epsilon > 0$ be given.

Choose $n > n_0$ large enough so that

$$\sum_{i=n}^{\infty} qr^i < \epsilon. \qquad (61)$$

Now

$$S(a) = \sum_{i=0}^{n} a_i + S(a_{n+1}, a_{n+2}, \ldots)$$

$$\le \sum_{i=0}^{n} a_i + \sum_{i=n+1}^{\infty} qr^i$$

$$< \sum_{i=0}^{n} a_i + \epsilon, \text{ by (61)}$$

$$< L + \epsilon. \qquad (62)$$

But $L \le S(a)$ by Theorem 7.26, which combined with (62) yields

$$L \le S(a) < L + \epsilon.$$

As $\epsilon > 0$ was arbitrary, $L \le S(a) \le L$, i.e. $S(a) = L$. □

8 Formal Summary

From a strictly mathematical point of view, our selection of topics in this chapter must seem a bit eccentric. In a mathematical treatment of sequences and series, one would follow the definitions of limit with numerous examples, proofs of convergence by comparison tests, the definition of the product of two series and the rôle played by absolute convergence in showing that the sum of a product of two series converges to the product of the two sums. One would also introduce Taylor series and their composition, and perhaps even discuss S_A and the rôle it plays in the theory of Fourier series. Our present goal, however, is not strictly mathematical. We did not introduce infinite sums to study them for their own sake, but for use as an illustration of formalism in action.

Summing the geometric series algebraically is a prime example of Type I Formalism, the mindless manipulation of expressions following established rules valid in another context. Presumably one's intuition for an infinite sum is adding an infinite column of figures and noticing that the running sums are settling down. This pretty much means writing

$$\sum_{i=0}^{\infty} a_i = a_0 + a_1 + \ldots = \ldots (((a_0 + a_1) + a_2) + \ldots,$$

i.e. left associating. At any finite stage one has completed a sum of the form,

$$\sum_{i=0}^{n} a_i,$$

and one is looking to the limiting behaviour of these partial sums. In the algebraic determination of

$$1 + r + r^2 + \ldots, \tag{63}$$

however, one is associating the terms differently:

$$1 + (r + (r^2 + \ldots, \tag{64}$$

which, calculationally, infinitely often requires an infinite sum to have been completed before the finitary operation can be applied. One then assumes the infinite distributive law,

$$r + (r^2 + (r^3 + \ldots = r(1 + (r + (r^2 + \ldots,$$

in order to rewrite (63) as

$$1 + r + r^2 + \ldots = 1 + r(1 + r + r^2 + \ldots)$$

and solve the equation

$$x = 1 + rx.$$

When $|r| < 1$, the procedure works: it gives the "correct" value for the infinite sum. When $r = -1$ or $|r| > 1$, it also gives a value, but a paradoxical one for

$r = -1$, and a value one is more likely to deem "wrong" than paradoxical for $|r| > 1$, especially if r is positive.

It might be instructive to consider a couple of other infinite expressions. First, consider Jacob Bernoulli's mixture of summing and taking square roots:

$$\alpha = \sqrt{1 + \sqrt{1 + \sqrt{1 + \ldots}}} \tag{65}$$

Computationally, this is analogous to (64): There is no first step. We cannot take the square root until we have added 1 to a square root which we cannot take until... Algebraically, however, we see that

$$\alpha = \sqrt{1 + \alpha},$$

whence α is a solution to the equation

$$x = \sqrt{1 + x}, \tag{66}$$

i.e.

$$x^2 = 1 + x,$$

and the Quadratic Formula quickly yields

$$x = \frac{1 \pm \sqrt{5}}{2},$$

and, since α is (implicitly) assumed positive,

$$\alpha = \frac{1 + \sqrt{5}}{2}.$$

One may try to give computational significance to (65) by calculating the "partial" results got by ignoring all but the first, first two, first three, etc. roots:

$$\sqrt{1}, \sqrt{1 + \sqrt{1}}, \sqrt{1 + \sqrt{1 + \sqrt{1}}}, \ldots$$

and hoping these approach a limit. The best way to see that they do is to notice that the function

$$f(x) = \sqrt{1 + x}$$

is a *contraction map* for $x > 0$: if $x, y > 0$,

$$|f(x) - f(y)| = |f'(\theta)| \cdot |x - y|, \text{ some } \theta \text{ between } x, y$$
$$= \frac{|x - y|}{2\sqrt{1 + \theta}}$$
$$< \frac{|x - y|}{2\sqrt{1}} = \frac{1}{2}|x - y|,$$

since $\theta > 0$. It can now be shown that, if x_0 is any positive real number and one defines $x_1, x_2 \ldots$ by

$$x_{n+1} = f(x_n),$$

then the sequence is a Cauchy sequence, whence $\beta = \lim_{n\to\infty} x_n$ exists and $\beta = f(\beta) = \sqrt{1+\beta}$, i.e. β is a root of (66), i.e. $\beta = \alpha$.

Another example of Bernoulli's is given by the expression

$$\beta = \sqrt{a\sqrt{a\sqrt{\cdots}}},$$

where a is a positive real number. Again, there is no first step in the computation. One can, however, notice that

$$\beta = \sqrt{a\beta}$$

and determine algebraically that $\beta = a$ or $\beta = 0$. There is no particular reason to assume one value or the other. However,

$$\beta = \sqrt{a}\sqrt{\sqrt{a\sqrt{\cdots}}} = \sqrt{a} \cdot \sqrt{\sqrt{a}} \cdot \sqrt{\sqrt{\sqrt{a}}} \cdots$$
$$= a^{\frac{1}{2}} \cdot a^{\frac{1}{4}} \cdot a^{\frac{1}{8}} \cdots$$
$$= a^{\frac{1}{2}+\frac{1}{4}+\frac{1}{8}\cdots} = a^1 = a,$$

all of this assuming one can continue the distribution of the radical across the infinite product and that the exponent laws hold when infinitely many terms are involved. And, numerically, one can consider the sequence of truncated terms,

$$\sqrt{a}, \sqrt{a\sqrt{a}}, \sqrt{a\sqrt{a\sqrt{a}}}, \ldots,$$

and verify its convergence to a for positive real numbers a.

A less successful example, discovered by logicians in the 1950's, deals with infinitary expressions in logic. Let "$A \to B$" read "A implies B" and consider, for any assertion B, the infinite sentence

$$A : \quad \ldots \to B) \to B) \to B. \tag{67}$$

It is not really possible to compute the truth value of A directly from that of B, but we can determine logically that A must be true. For, notice that A is of the form $A \to B$. If A were false, then, since a false assertion implies anything, $A \to B$ is true. But this is just A. Hence, if A is false, A is true, and we conclude A cannot be false. Thus, it is true as claimed. And here is where the trouble begins. Because A is true and $A \to B$ ($= A$) is true, it follows that B is also true. But B was an arbitrary statement, whence we conclude that every statement is true!

These successes and failures are typical of Type I Formalism. One applies computation rules or lines of reasoning valid in one domain to objects of another or to expressions that might or might not denote objects, often without thinking. So long as the results are correct or reasonable, one continues to do so. When one starts getting paradoxical results, one has a problem to which one can respond in various ways.

The most extreme response is simple rejection. Infinitely iterated implications like (67) have no applications, so we can simply declare them meaningless and drop the subject. When the applications are meaningful enough, however, one doesn't want to do this. In the early days of the Calculus, from just before Newton until the days of Bolzano and Cauchy, the predominant approach was to proceed formally on the theory, or hope, that the intuitions of experienced mathematicians would lead them safely through the minefields of paradox. One can declare the formal argument to be mere heuristic used to obtain results, which can later be supplied with genuine proofs that make no reference to the formal heuristic. This was the approach of Archimedes who, however, did not insist on Eudoxian proofs because his mechanical method sometimes led to paradoxes, but for the simpler reason that it provided no proofs of its own.

Finally, one can try to account for the success of the formalistic reasoning where it succeeds and remove the paradoxicality where it fails. This is where Types II and III Formalism come in. Bernoulli's "not inelegant paradox" is quickly resolved by the standard definition of an infinite sum as the limit of the sequence of partial sums: the formula for the sum of an infinite geometric progression can only be established when the series converges— and it doesn't converge for $r = -1$. Like the rejection of infinitely iterated implication as meaningless, one could simply reject divergent series.

We do not reject divergent series for several reasons. First, if we see the likes of Leibniz and Euler finding values for

$$1 - 1 + 1 - 1 + \dots \quad \text{and} \quad 1 + 2 + 4 + 8 + \dots,$$

then we must consider the possibility that there is something to these sums. Second, it has turned out that Leibniz's technique of averaging the sums yields usable results.[58] And third, the definition of the sum of a series we learn in the Calculus is not a formal characterisation of an intuitive notion, but *a* formal sharpening of a rather vague intuitive notion, and not the only one. This is perhaps best exemplified by pointing to S_B as a sharpening of the intuitive notion of the infinite sum as the result of an unordered accumulation. Indeed, if there is such a thing as *the* correct definition of an infinite sum we might ask if it is given by the limit of a convergent series and absolute convergence is an extra condition introduced for technical reasons, or if an infnite sum is the limit or Bolzano sum of an absolutely convergent series and the summation of convergent series is merely an extension of this. Or, is the "correct" definition yet something else?

The fact that the Type II formal concept is a replacement for and not necessarily an equivalent to the intuitive concept has some immediate and not necessarily pleasant consequences. The most immediate of these is the necessity

---

[58] Most famous is probably Leopold Fejér's theorem in Fourier Analysis: If a is the Fourier series of a function f at a point of continuity x_0 of f, then $S_A(a)$ exists and $S_A(a) = f(x_0)$.

of proving "obvious" results by possibly unusual or unmotivated proofs. A classic example of this is the proof of the uniqueness of the limit (Lemma 3.4). Intuitively, the result is obvious: if a sequence tends to a certain number as a limit, it cannot tend to a different number as well. But Lemma 3.4 is not dealing with the intuitive notion, rather it deals with a formal replacement and we must show that this formal replacement has all the properties we might expect of the notion of limit.

This verification of the adequacy of one's formal definitions (as with Lemmas 3.4 - 3.6 and 3.8 - 3.10), though necessary, often has the appearance of belabouring the obvious. Pædagogically it is bad because it draws attention away from intuition, and the skills it emphasises, though valuable, are at a low level... But this is a rant best saved for Chapter V, where we consider some of the negative reactions to formalism in mathematics.

What is left for us here is to say a few words about Type III Formalism, the modern axiomatic method. In the traditional axiomatic method, one begins with a definite structure and a list of axioms that are intuitively obvious properties of the structure. In the modern approach, one starts with the axioms and the structure is to be determined by them. This is overstating things a bit. What we've done here is start with a structure— the collection of infinite sequences of real numbers and some vague notion of summing them when possible. Now this structure is not very definite. So we can give some axioms for infinite sums and hope these lead to some insight into a correct definition of summation. In this we have been singularly unsuccessful. But it is not always so.

A remarkably successful application of Type III Formalism was made in 1382 by Oresme in defining fractional exponents. Hermann Hankel declared this to be the first application of a general methodological principle he called the *Principle of the Permanence of Formal Laws*[59]. This principle, which may be taken as either an axiom or an heuristic principle, says that if we have some structure defined on one set of objects and wish to extend it, we should do so in such a way that the universal laws valid in the original domain remain valid in the extension. In essence, Oresme extended the use of positive integral exponents to positive rational ones (for a positive base) by assuming the law,

$$(a^m)^n = a^{(mn)},$$

which holds for positive real a and positive integral m, n, remains valid when m, n are positive rational numbers. If that is the case, then

$$(a^{\frac{1}{n}})^n = a^{(\frac{1}{n} \cdot n)} = a^1 = a,$$

and $a^{1/n}$ is the n-th root of a. And the equation

---

[59] H. Hankel, *Geschichte der Mathematik in Alterthum und Mittelalter*, B.G. Teubner Verlag, Leipzig, 1874, pp. 350 - 351. Hankel referred to his *Theorie der complexe Zahlensysteme*, Leipzig, 1867, p. 10, for an earlier enunciation of the principle, which in fact was already stated under the name, the Principle of the Permanence of Equivalent Forms, by the English mathematician George Peacock around 1830. We shall have more to say about Peacock in Chapter IV.

$$a^{\frac{m}{n}} = a^{\left(\frac{1}{n} \cdot m\right)} = \left(a^{\frac{1}{n}}\right)^m$$

or

$$a^{\frac{m}{n}} = a^{\left(m \cdot \frac{1}{n}\right)} = \left(a^m\right)^{\frac{1}{n}}$$

determines $a^{\frac{m}{n}}$ for any positive real number a and any positive rational number m/n. Eventually, one would use the further law,

$$a^{m+n} = a^m \cdot a^n,$$

to conclude that $a^0 = 1$ because

$$a^m = a^{m+0} = a^m \cdot a^0,$$

and that $a^{-m} = 1/a^m$ because

$$1 = a^0 = a^{m-m} = a^m \cdot a^{-m}.$$

Our Type III approach to infinite sums has not been so successful. But it does further illustrate that the question of what such a thing is is not so clear-cut as our familiarity with the idea of an infinite series as the limit of the partial sums which led Coolidge to utter those words on page 25 that we might so readily accept, that it is surprising Bolzano "should not have grasped the idea of defining a series as a limit". There are other possibilities; to be sure, they may be obscured by the limit definition, but they are there.

With this, we take leave of infinite sums. We will come back to them later, but for now we change gears and move on in Chapter II to consider some classical examples of formalism, some even wilder than Grandi's series.

9 Exercises

We begin with a few simple exercises.

9.1 Exercise. (Nikolaus Bernoulli) Obviously

$$\frac{1}{1-2} = \frac{1}{1-1-1} = -1.$$

Perform the long division formally to show

$$\frac{1}{1-1-1} = 1 + 1 + 2 + 3 + 5 + 8 + 11 + \ldots,$$

where each summand after the first two is the sum of the two preceding terms[60]. Similarly, observe

$$\frac{1}{1-2} = 1 + 2 + 4 + 8 + \ldots$$

Conclude that Order Preservation does not hold for these sums.

---

[60] The sequence $1, 1, 3, 5, \ldots$ of numbers is none other than the famous Fibonacci sequence of Leonardo of Pisa, aka Fibonacci.

9.2 Exercise. (Continued Fractions) Consider

$$\alpha = 1 + \cfrac{1}{1 + \cfrac{1}{1 + \dots}}$$

i. Find a value for α algebraically by observing $\alpha = 1 + 1/\alpha$.
ii. Find a value for α by considering the sequence of ever larger truncations:

$$1, 1 + \frac{1}{1}, 1 + \cfrac{1}{1 + \frac{1}{1}}, \dots$$

[It might be more convenient to write

$$x_0 = 1, \quad x_{n+1} = 1 + \frac{1}{x_n}$$

and consider the sequence $x_0, x_1, x_2, \dots$.]

9.3 Exercise. (Infinite Products) Write

$$\prod_{i=0}^{\infty} a_i = a_0 \cdot a_1 \cdot a_2 \cdots$$

and consider the product

$$\alpha = \prod_{i=0}^{\infty} 2^{(-1)^n} = 2 \cdot \frac{1}{2} \cdot 2 \cdot \frac{1}{2} \cdot 2 \cdots$$

i. Assuming the analogues of Linearity, Truncation/Prefixing, and Agreement with Finite Sums,

$$\prod(a_i b_i) = \prod a_i \cdot \prod b_i, \quad \prod a_i = 1 \cdot a_0 \cdot a_1 \cdots, \quad \prod 1 = 1,$$

show: $\alpha(\frac{1}{2}\alpha) = 1$, whence $\alpha = \sqrt{2}$.
ii. Consider the partial products,

$$p_n = \prod_{i=0}^{n} a_i$$

for $a_{2i} = 2$, $a_{2i+1} = \frac{1}{2}$. What is the expected value of these products? What value should α receive?
iii. Assuming formally that

$$\log \prod a_i = \sum \log a_i,$$

find a value for $\log \alpha$ and thus a value for α.

9.4 Exercise. (Another Infinite Product) Let

$$\alpha = a(a(a(\dots$$

Show algebraically that $\alpha = 1$ or $\alpha = 0$.

9.5 Exercise. (Iterated Subtraction) Let

$$\alpha = 1 - (1 - (1 - (\dots$$

i.   Determine α algebraically.
ii. Examine the sequence of truncated results,

$$1, 1 - 1, 1 - (1 - 1), \dots$$

[I.e., consider the sequence

$$x_0 = 1, \quad x_{n+1} = 1 - x_n.]$$

What value do you derive for α?
iii. What can be said about

$$\beta = \dots - 1) - 1) - 1?$$

9.6 Exercise. (Iterated Division) Consider

$$\alpha = 2 \div (2 \div (2 \div (\cdots$$

$$\beta = \cdots \div 2) \div 2) \div 2.$$

i.   Find α algebraically.
ii. Consider the sequence $x_0 = 2, x_{n+1} = 2/x_n$. Does this suggest a value for α?
iii. Find β algebraically.
iv. Find a value for β using the sequence $x_0 = 2, x_{n+1} = x_n/2$.

9.7 Exercise. (Iterated Roots and Divisions) Let $a > 0$ and consider

$$\alpha = \sqrt{a \div \sqrt{a \div \sqrt{a \div \sqrt{\cdots}}}}$$

i. Find a value for α algebraically by noticing $\alpha = \sqrt{a/\alpha}$.
ii. Consider the sequence $x_0 = \sqrt{a}$, $x_{n+1} = \sqrt{a/x_n}$. Set $a = 2$ and use your calculator to determine the limit of the sequence.
iii. Show that, for the sequence x_n of part ii, we have

$$x_n = a^{e_n},$$

where

$$e_0 = \frac{1}{2}, \quad e_{n+1} = \frac{1}{2} - \frac{1}{2}e_n.$$

Show:

$$e_n = \frac{1}{3} + \frac{1}{6}\left(-\frac{1}{2}\right)^n$$

and conclude

$$\lim_{n \to \infty} x_n = \lim_{n \to \infty} a^{e_n} = a^{\lim e_n} = a^{1/3}.$$

9.8 Exercise. (Iterated Roots and Divisions) For positive integers n,

$$\sqrt[n]{a} = a^{1/n}.$$

Extend the notation to arbitrary rational numbers so that, in particular,

$$\sqrt[1/2]{a} = a^2.$$

Consider, for $a > 0$,

$$\alpha = \sqrt[1/2]{a \div \sqrt[1/2]{a \div \sqrt[1/2]{\cdots}}}.$$

Repeat Exercise 9.7 to the extent to which it is possible.

Our next exercise requires a little more mathematical sophistication.

9.9 Exercise. (Monotone Convergence Theorem) A special case of the Bolzano-Weierstrass Theorem is the Monotone Convergence Theorem asserting that every bounded monotone sequence converges, i.e., if $b_0 \le b_1 \le \ldots \le A$, then $\lim_{n \to \infty} b_n$ exists.
i.   Prove the Monotone Convergence Theorem by reduction to Lemma 5.5.
ii. Prove the Monotone Convergence Theorem by mimicking the proof of Lemma 5.5 or the proofs of Theorems 5.19 and 5.20.
iii. (For advanced undergraduates and above). Derive the Bolzano-Weierstrass Theorem from the Monotone Convergence Theorem as follows: Let $b_0, b_1, \ldots$ be a bounded sequence and define

$$c_n = \text{ least upper bound of } \{b_n, b_{n+1}, \ldots\}.$$

Show: c is convergent and some subsequence of b converges to the same limit at c.

Our next exercise requires a great deal more mathematical sophistication.

9.10 Exercise. (Euler Transformation) Recall from page 71 the definition of the difference operator Δ:

$$(\Delta a)_n = a_{n+1} - a_n.$$

Note that $\Delta a = \tau(a) - a$, write $\Delta = \tau - 1$ and apply the Binomial Theorem to obtain

$$\Delta^n = (\tau - 1)^n = \sum_{i=0}^{n} \binom{n}{i} (-1)^{n-i} \tau^i.$$

Conclude

$$(\Delta^n a)_0 = \sum_{i=0}^{n} \binom{n}{i} (-1)^{n-i} a_i.$$

i.   Derive Euler's formula,

$$\sum_{n=0}^{\infty}(-1)^n a_n = \sum_{n=0}^{\infty}(-1)^n \frac{(\Delta^n a)_0}{2^{n+1}}, \tag{68}$$

formally without regard to issues of convergence by first showing the coefficient of a_k in (68) to be

$$\sum_{n=k}^{\infty}\frac{(-1)^n}{2^{n+1}}\binom{n}{k}(-1)^{n-k} = (-1)^k\sum_{n=k}^{\infty}\frac{\binom{n}{k}}{2^{n+1}},$$

then observing that for $k = 0$ this is

$$\sum_{n=0}^{\infty}\frac{1}{2^{n+1}} = 1,$$

and then using the identity,

$$\binom{n+1}{k+1} = \binom{n}{k+1} + \binom{n}{k},$$

to show the coefficient of a_{k+1} to be minus that of a_k.

ii.   Apply Euler's formula to the Grandi sequence $g = \sum(-1)^n a_n$, where $a_n = 1$. Do the same for $\sum(-2)^n$, i.e. $a_n = 2^n$.

iii.   Noting that for $a_n = \frac{1}{n+1}$

$$\Delta^k a_n = (-1)^k \frac{k!}{(n+1)\cdots(n+k+1)},$$

apply Euler's transformation(68) to the alternating harmonic series,

$$1 - \frac{1}{2} + \frac{1}{3} - \frac{1}{4} + \cdots$$

Sum the first 20 terms of the original series and the transformed series and compare the results with $\ln 2$. Which is converging more rapidly? [Note: With the modern calculators, this is not as arduous a task as it sounds. On the *TI-83 Plus*, for example, one enters

$$Y_1 = (-1)^\wedge X/(X+1)$$

in the Equation Editor for the original series and then the corresponding expression for the transformed series as Y_2. One then quits the editor and stores

$$seq(Y_1(X), X, 0, 20), \quad seq(Y_2(X), X, 0, 20),$$

in lists L_1, L_2, respectively. A final trick: store

$$cumSum(L_1), \quad cumSum(L_2)$$

in lists L_3, L_4, respectively, and compare these two lists in the statistical List Editor. Finally, calculating

$$L_3(20), \quad L_4(20), \quad \ln 2$$

or

$$sum(L_1), \quad sum(L_2), \quad \ln 2$$

will make the final comparison. With some facility with the calculator, this takes only a minute or two. Euler, of course, would have had to do all of this by hand— unless he was already blind, in which case he would have performed all the calculations in his head.]

iv.   (For graduate students and above) Show that, if $\sum (-1)^n a_n$ converges, then so does the transformed series and (68) holds.

v.   (Ditto) For any sequence $b : b_0, b_1, b_2, \ldots$, write $b_n = (-1)^n a_n$ and define S_E by

$$S_E(b) = \sum_{n=0}^{\infty} (-1)^n \frac{(\Delta^n a)_0}{2^{n+1}},$$

provided it converges. Discuss S_E with respect to the summation properties of §6. How does S_E relate to the other summation functions discussed? How does it handle geometric progressions?

Finally, for those more interested in history, an historical exercise:

9.11 Exercise. (Cesàro Summability) No discussion of generalised sums is complete without mention of Ernesto Cesàro. Look him up and write a short report on his work on generalised sums.

II

Classic Examples of Formalism

1 Euclid and His Magnitudes

The origins of mathematics are lost alongside the origins of writing. But we can imagine the so-called counting numbers $1, 2, 3, 4 \ldots$ as having had their origin in actual counting. The positive rational numbers would have arisen from more general mensuration problems involving subdividable quantities. For example, some problems preserved from ancient Egypt concern rationing, where the number of loaves of bread to be distributed is not evenly divisible by the number of those among whom the bread is to be distributed, and one must deal with parts of a loaf. Positive real numbers, like $\sqrt{2}$, would slip in unnoticed as one became more mathematically sophisticated. The Greeks went beyond mathematics and delved into the metaphysics of the subject.

One can, of course, not speak of Greek mathematics in the singular. There was the early Pythagorean approach by which all was number and geometry was based on number. Then came the classical period, best exemplified by Euclid, where all was geometry and the only numbers mentioned were whole numbers. And, finally, there are the great applied mathematicians, Hero, Archimedes, and Ptolemy, who, regardless of metaphysical-ontological principles, dealt with real numbers much as we do now. That said, we can paint a simplified picture of the Greek view of number as follows: *Number* meant counting number: $1, 2, 3, 4, \ldots$[1] Rational numbers were not numbers, i.e. actual objects, but *ratios*— relations between numbers. Mathematics was studied by the leisure class as philosophy, so the needs of commercial arithmetic could safely be ignored, thus allowing one to thus demote the rationals. The Pythagoreans elevated oriental number mysticism to an official ontology: all was number, and all relations were thus ratios. Pythagoras had had phenomenal success with this in music and soon he and his followers were applying this to geometry.

The centrality of the notion of similarity and the fact that mensuration formulæ can be rewritten as proportions with no constant of proportionality stated combine to make this a plausible approach. However, the principle that

---

[1] Actually, 1 was the *unit* that generated the *numbers* (i.e. multiplicities) $2, 3, 4 \ldots$ The distinction serves no mathematical purpose and we can ignore it.

all objects are numbers and all relations ratios can be used as more than motivation. It has consequences. According to it, the ratios appealed to in dealing
with, say, similar triangles all have operational significance. If AB and CD are
line segments and their ratio is m/n, then dividing AB into m equal parts and
CD into n such results in line segments of a common length, a unit that can
be used to measure AB and CD by, whereby AB is m units in length and
CD is n units. This *commensurability* of any two line segments can be used to
prove geometric theorems.[2] Unfortunately, not all pairs of line segments turn
out to be commensurable; the ratio of the diagonal to the side of a square, for
example, is $\sqrt{2}$, which is not rational.[3]

The existence of incommensurable line segments presents one with two immediate problems. First, if AB and CD have no common measure, they have
no ratio: we don't know what we mean by the comparison $AB : CD$. Second, assuming proportions to have some meaning we are not aware of, how do
we deal with them? We know the laws of fractions, i.e. proportions between
numbers; what are the laws of proportions between possibly incommensurable
objects?

The situation is ripe for Type I Formalism: Acknowledge one doesn't know
what a proportion is, but assume proportions satisfy the usual properties that
rational proportions satisfy:

$$\frac{a}{b} = \frac{c}{d} \quad \text{iff} \quad \frac{a}{c} = \frac{b}{d}$$
$$\frac{a}{b} = \frac{a+a}{b+b}$$

etc. Presumably this was done. If we assume a bit more, namely some basic
properties of similar triangles, and cross-multiplication, i.e.,

$$\frac{a}{b} = \frac{c}{d} \quad \text{iff} \quad ad = bc,$$

where the interpretation of ad and bc would be the area of the rectangles of
sides a, d and b, c, respectively, (assuming a, b, c, d to represent lengths), then
we get a remarkably simple version of Euclid's proof of the Pythagorean Theorem: Let ABC be a right triangle with right angle at C, draw the squares on
the sides of ABC, and drop a perpendicular from C to the far side of the square
on the hypotenuse, as in *Figure 1*. Comparing the angles we see that triangles

---

[2] I cite a couple of examples in my *History of Mathematics; A Supplement* (Springer-
Verlag, New York, 2007), borrowed from Arthur Gittleman, *History of Mathematics*
(Merrill Publishing Company, 1975), who in turn took them from Thomas Heath's
annotated edition of Euclid's *Elements* (available in paperback from Dover).

[3] Again, one can consult any of the three works cited in the previous footnote for a
more extensive discussion of this point.

ABC, ACK and CBK are similar, whence

$$\frac{AC}{AK} = \frac{AB}{AC}, \quad \frac{BC}{BK} = \frac{AB}{BC},$$

i.e.

$$AC \cdot AC = AB \cdot AK,$$
$$BC \cdot BC = AB \cdot BK,$$

that is, the area of the square $ACHI = AC \cdot AC$ is the area of the rectangle $ADJK$ of sides $AD = AB$ and AK, and the area of the square $BFGC$ is the same as that of the rectangle $BKJE$ of sides $BE = AB$ and BK. But these add up to the square $ADEB$ of side AB. Hence

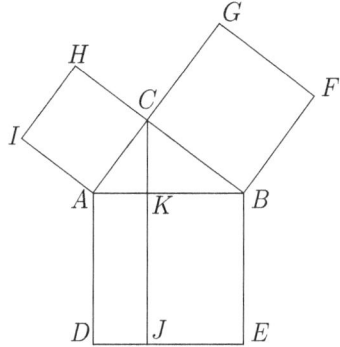

Figure 1

$$AB^2 = AC^2 + BC^2.$$

This may have been too many conscious assumptions for the philosophically minded Greeks, and Plato lent his weight to the demand for a satisfactory foundation of geometry. This was given by Plato's younger contemporary Eudoxus, who provided a Type II formalistic definition of the proportion of two magnitudes. The theory of proportions developed by Eudoxus is presented in full, with applications, by Euclid in the *Elements*. The Eudoxian theory bears looking into. Right now, however, I want to consider Euclid and his treatment of magnitudes.

Euclid's treatment of magnitude is probably the conceptually most puzzling aspect of the *Elements*. Magnitude is something like length or area, but without the number. There is no number because there is no unit. Once you choose the unit, a number is determined. Thus a ruler becomes 12 inches long when the inch is chosen as unit and 1 foot long when the foot is chosen. But, however you measure it, the ruler has a certain length and even without the numbers magnitudes can be compared. Thus one can assert that a yardstick is longer than a ruler and even that it is 3 times as long, regardless of the choice of measure.

Euclid discusses magnitudes and the theory of proportions in Book V. Unlike points, lines, planes, and solids, he offers no definition of magnitude, only what it means for one magnitude to be a part or a multiple of another, when two magnitudes stand in the same ratio, when one ratio is greater than another, etc. He then proceeds to derive 25 propositions about magnitudes, 20 of them dealing with proportions. These latter include, for example, Proposition 8,

$$a > b \quad \Rightarrow \quad \frac{a}{c} > \frac{b}{c} \quad \& \quad \frac{c}{a} < \frac{c}{b},$$

which will pop up shortly, and Proposition 16,

$$\frac{a}{b} = \frac{c}{d} \quad \Rightarrow \quad \frac{a}{c} = \frac{b}{d},$$

which Euclid oddly does not appeal to at the most obvious place. In Heath's edition, the demonstrations are accompanied by illustrations using line segments, suggesting Euclid is proving these propositions for linear magnitudes. However, a cursory reading of a few randomly chosen proofs reveals the proofs to be fully general and the lines are used merely to focus the mind.

Today we would choose a unit and thus consider magnitudes to be positive real numbers. Euclid did not do so. He does not even assume each pair of magnitudes to have a ratio, defining two magnitudes a, b to have a ratio in case some multiple ma is greater than b and some multiple nb greater than a, where m, n are positive integers. This is certainly a bit mysterious, but the mystery can be cleared up by observing that, if for some positive integers m, n

$$b < ma \quad \text{and} \quad a < nb,$$

then

$$\frac{1}{m} < \frac{a}{b} < n,$$

and the ratio lies between two rational numbers. If one assumes the Archimedean Axiom, that such m, n exist for any given pair a, b of magnitudes, then the definition of the equality of two ratios $a/b, c/d$ reduces to the assertion that the ratios split the positive rational numbers identically: for any $q \in \mathbb{Q}, q > 0$,

$$q < \frac{a}{b} \quad \text{iff} \quad q < \frac{c}{d}$$

$$q = \frac{a}{b} \quad \text{iff} \quad q = \frac{c}{d}$$

$$q > \frac{a}{b} \quad \text{iff} \quad q > \frac{c}{d}.$$

Moreover, the Euclidean definition of inequality of ratios asserts then that

$$\frac{a}{b} < \frac{c}{d} \quad \text{iff} \quad \text{for some } q \in \mathbb{Q} \text{ with } 0 < q, \ \frac{a}{b} \leq q < \frac{c}{d}.$$

Euclid did not explicitly[4] assume the Archimedean Axiom, i.e. that any two magnitudes possess a ratio. This means he does not rule out the possibility of

---

[4] He assumes the Archimedean axiom implicitly in the proof of Proposition V-8. If a is a magnitude and ε is infinitesimal relative to a, then

infinite or infinitesimal magnitudes. Some authors explain this by pointing to Proposition III-16 characterising the tangent to the circle at a point on the circle as the line perpendicular to the diameter at that point. Euclid proved that any other line passing through that point had to intersect the circle at another point. Thus no line can be squeezed between a circle and its tangent, and thus the angle made between the circle and its tangent— the so-called Horn angle[5] (see page 9, above)— is less than any acute rectilineal angle (see

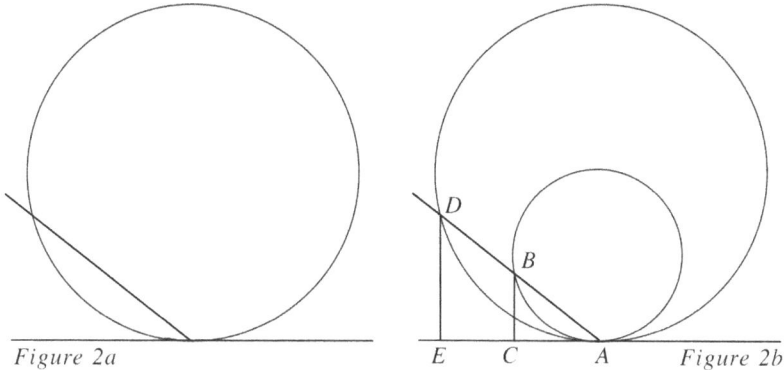

Figure 2a E C A Figure 2b

Figure 2a). If h is the magnitude of the Horn angle and a that of an ordinary angle, then $\frac{1}{n}a$ is also an ordinary angle and, as it cannot be squeezed in there, it is not less than h, i.e. $nh \not> a$. Thus a and h have no ratio. Nowadays, of course, we define h numerically by defining the measure of the angle between a line and a curve to be the angle between the given line and the line tangent to the curve at the point of intersection, which is 0 when the given line is the tangent line.

The Horn angle puzzled mathematicians for centuries and is a fine formalistic enigma. The fact is that we know what the angle between two lines is and how to measure it consistently in several ways. We can draw a circle centered at the vertex and measure the length of arc cut by the angle, or the area of the sector determined by the two lines. The measures vary with the radius of the circle, but the ratios of arc length to radius and sector area to square of

$$a + \varepsilon > a \quad \text{and} \quad 2a > a + \varepsilon,$$

whence a and $a + \varepsilon$ have a ratio. By V-8,

$$\frac{a+\varepsilon}{a} > \frac{a}{a} \quad \text{and} \quad \frac{a}{a} > \frac{a}{a+\varepsilon}.$$

The first of these is true, but the latter is not: $a/a = 1$ and $a/(a+\varepsilon)$ is infinitesimally close to 1, whence there is no rational q satisfying

$$\frac{a}{a+\varepsilon} \leq q < \frac{a}{a}.$$

---

[5] It and other angles formed by curves tangent at a point are also called *contingency angles* or *contact angles*.

the radius do not. We can do neither of these with the Horn angle. Euclid's conclusion from the inability to fit a rectilineal angle inside the region of a Horn angle that the horn angle is smaller in magnitude than the magnitudes of any rectilineal angle is purely formal, for what the magnitude of the Horn angle is or should be is left undetermined.

Consider again Alexander Neckham's remark that the Horn angle of the larger circle appears smaller than that of the smaller circle. This is more than appearance. The region determined by the Horn angle of the smaller circle cannot be fitted into the region determined by the Horn angle of the larger circle. By Euclid's reasoning, the magnitude of the Horn angle of the larger circle is less than that of the Horn angle of the smaller circle. But look now at *Figure 2b* combining Euclid's and Neckham's figures. Here we have dropped two perpendiculars BC and DE from the points of intersection of the line ABD with the two circles to the common tangent of the circles. This creates two "triangles" given on the one hand by the arc $\widehat{AB}$ and the segments BC and CA, and, on the other hand, by the arc $\widehat{AD}$ and the segments DE and EA. These "triangles" $\widehat{ABC}$ and $\widehat{ADE}$ are evidently "similar" and, reasoning formally, the angles $\widehat{BAC}$ and $\widehat{DAE}$, i.e the two Horn angles, must be equal in magnitude.

Combining Neckham's observation and Euclid's reasoning with our latest conclusion, we see unerringly that every Horn angle is greater than itself in magnitude! Type I Formalism often leads to contradictory conclusions which are only cleared up by taking a step back, realising one doesn't know what one is talking about, and either shrugging one's shoulders and walking away or finding the undefined source of confusion and repairing it with one or more Type II formal definitions. In the present case, someone eventually realised that if one defines the angle between two curves to be the angle between the tangents formed at the point of intersection, then the contradictions vanish: the Horn angle has magnitude 0 and there is no more to be said.

Actually, there is a great deal more to be said. The modern solution to the problem of the Horn angle is based on a Type II formal definition of angle and the magnitude thereof. It is not the only possible one. Detlef Laugwitz has shown[6] how to construct a notion of magnitude for angles in such a way that Horn angles have nonzero (and thus necessarily non-real) magnitudes. While the details may be a little grubby, the basic idea is fairly simple. Suppose we are given two curves forming a curvilinear angle ABC as in *Figure 3a* below.

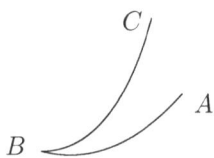

Figure 3a

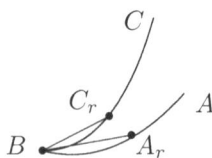

Figure 3b

6 "Die Messung von Kontingenzwinkeln", *Journal für die reine und angewandte Mathematik* 245 (1970), pp. 133 - 142.

If we intersect these curves with a circle of radius $r > 0$ centered at B in points A_r, C_r, respectively, we obtain a rectilineal angle $A_r B C_r$ approximating ABC, as in *Figure 3b*, above. Our usual modern choice is, essentially to take the limit of (the numerical values of) these angles as the numerical value for the angle ABC:

$$\text{magnitude of } ABC = \lim_{r \to 0} A_r B C_r.$$

Laugwitz captures the notion of ultimate angle differently by declaring two angles to be *equivalent* if ultimately they are the same:

$$ABC \cong A'B'C' \quad \text{iff} \quad \begin{array}{l} \text{for some } r_0 > 0, \text{ one has} \\ \text{for all } r \text{ with } 0 \le r < r_0, \\ A_r B C_r = A'_r B' C'_r. \end{array}$$

$$ABC \prec A'B'C' \quad \text{iff} \quad \begin{array}{l} \text{for some } r_0 > 0, \text{ one has} \\ \text{for all } r \text{ with } 0 \le r < r_0, \\ A_r B C_r < A'_r B' C'_r. \end{array}$$

This will not totally order the angular magnitudes, as the angles $A_r B C_r$ and $A'_r B' C'_r$ may oscillate. But it is a good enough definition to allow one to verify anew Euclid's assertion that Horn angles are smaller than every rectilineal angle, as well as to model Neckham's optical illusion that the Horn angle of a smaller circle is larger than that of a larger one.

First, note that if ABC is a rectilineal angle of ordinary measure θ, then every angle $A_r B C_r$ has measure θ and the same must hold for all sufficiently small r for any angle $A'B'C'$ equivalent to ABC. We thus identify the magnitude $[ABC]$ with the real number θ.

As for the Horn angle, imagine the circle as being tangent to the origin and centered at the point $(0, k)$. The equation of the circle is

$$X^2 + (Y - k)^2 = k^2. \tag{1}$$

The point A_r of intersection with the circle

$$X^2 + Y^2 = r^2, \tag{2}$$

of radius r centered at the origin is readily determined by subtracting equation (2) from equation (1):

$$(Y - k)^2 - Y^2 = k^2 - r^2$$
$$-2Yk + k^2 = k^2 - r^2$$
$$Y = \frac{-r^2}{-2k} = \frac{r^2}{2k}.$$

The sine of the angle α_r determined by $A_r B C_r$ will thus be $(r^2/2k)/r = r/2k$, whence

$$\alpha_r = \arcsin(r/2k).$$

For any $\theta > 0$, we have $\alpha_r < \theta$ so long as $r < 2k\theta$, whence

$$\alpha = [ABC] < \theta,$$

and Euclid's assertion is verified. As for Neckham's, note that if $k > m$, then

$$\frac{r}{2k} < \frac{r}{2m}$$

and, as x is very close to $\sin x$ for x close to 0, one has

$$\arcsin \frac{r}{2k} < \arcsin \frac{r}{2m}$$

for sufficiently small r. Thus, the Horn angle of the circle with radius k is less than that of the circle of radius m.

The Horn angle is not the only formal element of Euclid's treatment of magnitudes. There is a major formalistic assumption in his use of the fourth proportional in proving the theorem on the areas of circles.

The problem of the fourth proportional is simple to state. Given three magnitudes, a, b, c, find a fourth one d completing the proportion:

$$\frac{a}{b} = \frac{c}{d}.$$

The sought for magnitude d is called the fourth proportional. Euclid takes pains to show how this can be done for linear magnitudes with his Proposition VI-12. Given line segments of magnitudes a, b, c, respectively, he shows how to construct a segment the magnitude of which is the fourth proportional. This is an easy enough construction. Let, e.g., b be greater than a and construct a triangle of sides a, c as in *Figure 4a* and extend AB to some point D for which

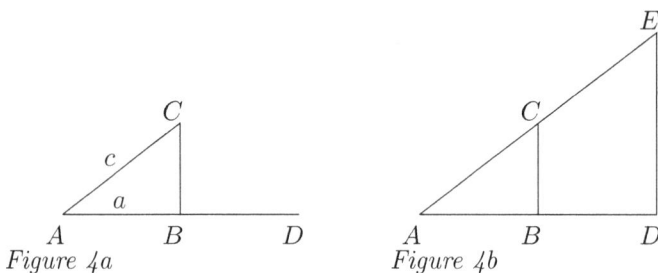

Figure 4a   *Figure 4b*

$AD = b$. If one now places a line parallel to BC at D and extends AC until it meets this parallel at some point E, as in *Figure 4b*, then the triangles ABC, ADE will be similar and

$$\frac{AB}{AD} = \frac{AC}{AE}, \quad \text{i.e.} \quad \frac{a}{b} = \frac{c}{AE}$$

and AE is the fourth proportional. Several books later, where he wants to determine the area of a circle, he implicitly assumes given two squares S_1, S_2, and a circle C, the existence of a figure F yielding the fourth proportional:

$$\frac{S_1}{S_2} = \frac{C}{F}.$$

Should he be allowed this assumption?

Clearly Euclid has assumed the fourth proportional, which exists for magnitudes of linear segments, to exist for areas as well. Is this merely a formal argument, or does he feel somehow justified in carrying over the conclusion? *If* lengths and areas are measured by the same magnitude, and *if* he assumes for any magnitude the existence of a figure of that magnitude, then his conclusion is justified: Given any magnitudes a, b, c, there are line segments of these magnitudes. The magnitude of their fourth proportional d thus exists. If one takes a, b, c to be the magnitudes of S_1, S_2, C, respectively, then the existence assumption yields a figure F of the magnitude d standing in the fourth proportion to a, b, c. However, Euclid nowhere explicitly makes these assumptions and, in the present case, the only other justification I can see is the actual presentation of F, which in this case assumes the theorem that the ratio of the areas of two circles is the same ratio as that of the squares on their respective diameters— his proof of which result presupposes the existence of the figure F.[7]

Now, the existence of a figure of a given magnitude is a continuity assumption of the sort overlooked by the Greeks and, like the fact that a straight line entering a triangle must come out again, may be taken as an unconscious assumption. If so, why did he prove explicitly that every rectangle can be squared?

Euclid's formulation of the theorem on the area of a circle is this: If C_1, C_2 are circles of diameters d_1, d_2, respectively, then

$$\frac{C_1}{C_2} = \frac{d_1^2}{d_2^2},$$

where d_i^2 denotes the area of the square with side d_i. Oddly enough, he does not apply Proposition V-16 mentioned earlier to conclude

$$\frac{C_1}{d_1^2} = \frac{C_2}{d_2^2}$$

and verify the existence of the constant π $(= 4C_1/d_1^2)$. Is this because of the inhomogeneity of the ratio of a circle to a square as opposed to the homogeneous comparisons of circle to circle or square to square? Maybe circles and squares have different types of magnitudes?

The continuity result does not follow from Euclid's axioms, even assuming the Archimedean Axiom. Indeed, familiarity with abstract algebra allows one to construct a model of Euclidean geometry in which linear segments and circular arcs do not have the same magnitudes, and the areas of squares and circles are measured by different magnitudes. Moreover, the model, while it may not

---

[7] Of course, this doesn't mean that all proofs presuppose this existence. Bartel van der Wærden, *Science Awakening*, Oxford University Press, New York, 1961, asserts, "At the beginning of this proof, the existence of a fourth proportional is tacitly assumed; the same thing takes place in all the analogous proofs in Book XII. This assumption is not necessary; a somewhat subtler formulation of the proof could have avoided it." (p. 185).

invalidate the existence of the fourth proportional, does bring this existence into question.

1.1 Definition. *A real or complex number α is called an* algebraic number *if it is the zero of some non-constant polynomial with rational coefficients. In other words, α is an algebraic number if there is a polynomial,*

$$P(X) = q_n X^n + q_{n-1} X^{n-1} + \ldots + q_1 X + q_0, \quad q_n \neq 0, \quad q_0, \ldots q_n \in \mathbb{Q},$$

such that $P(\alpha) = 0$. If an algebraic number α is a real number, we say that α is an algebraic real number *or, more simply, an* algebraic real.

Every rational number is an algebraic real: If $q \in \mathbb{Q}$, then q is a zero of $P(X) = X - q$. Not every algebraic real is rational, as familiarly exemplified by $\sqrt{2}$ which is an irrational zero of $P(X) = X^2 - 2$. And, not every real number is an algebraic number. This fact goes deeper and it is very difficult to prove a specific number is not algebraic. Charles Hermite proved that e is not algebraic in a paper published in 1873. In 1882 Ferdinand Lindemann proved that π is not algebraic. This proof was simplified and generalised by Carl Weierstrass in 1885 and the basic results for e and π proven more simply by David Hilbert in 1893.[8]

The algebraic reals form a number system in their own right. They form what is called a *field*: they are closed under addition, subtraction, multiplication, and division by nonzero algebraic reals. Moreover, if $P(X)$ is a polynomial with real algebraic coefficients, then any real zero of P is itself an algebraic real, i.e. it is the zero of some polynomial $Q(X)$ (of possibly higher degree) with rational coefficients.[9] In particular, if α is a positive algebraic real and it has square root $\sqrt{\alpha}$, this is a zero of the polynomial

$$P(X) = X^2 - \alpha,$$

whence $\sqrt{\alpha}$ is also an algebraic real, i.e. the field of algebraic reals is closed under taking square roots of positive numbers.

Our interest in the algebraic reals here is that they have enough closure properties that we can use them to construct a model of geometry by simply doing coordinate geometry with all coordinates restricted to being algebraic real numbers. To this end, we define a point (α, β) of the real plane $\mathbb{R}^2$ to be an *algebraic point* just in case both coordinates α, β are algebraic reals. The *algebraic plane* is the set of algebraic points in $\mathbb{R}^2$. An *algebraic line* is the set of all algebraic points lying on a line in $\mathbb{R}^2$ that contains at least two

---

[8] These seminal papers are reprinted in L. Berggren, J. Borwein, and P. Borwein, *Pi; A Source Book*, 3rd ed., Springer-Verlag, New York, 2004. The papers are untranslated and unannotated, and, moreover, the inking leaves something to be desired. Thus, I refer the interested but possibly timid reader to Charles Hadlock, *Field Theory and Its Classical Problems*, Mathematical Association of America, 1978, for a gentler exposition of the proof.

[9] These are not trivial facts. I refer the reader to Hadlock's book cited in the preceding footnote for the proofs.

algebraic points. An *algebraic line segment* is the set of all algebraic points of an ordinary line segment determined by algebraic endpoints. Algebraic triangles, rectangles, etc. are figures composed of algebraic line segments. An *algebraic circle* is determined by choosing an algebraic point to serve as centre and an algebraic number as radius; it is then the set of algebraic points at the given distance from the given centre, i.e. the algebraic points of the ordinary circle with the given centre and radius. And, finally, *algebraic circular arcs* are determined by choosing two algebraic points on an algebraic circle to serve as endpoints and consists of all the algebraic points lying on the ordinary arc determined by the given endpoints.

It is a routine, but not particularly rewarding, matter to show that an algebraic line can also be defined as the set of algebraic points satisfying a linear equation,

$$AX + BY + C = 0, \quad A, B \text{ not both } 0,$$

with A, B, C algebraic real numbers; and algebraic circles are the sets of algebraic solutions to equations,

$$AX^2 + BXY + AY^2 + DX + EY + F = 0, \quad A \neq 0,$$

with A, B, D, E, F algebraic reals.

Our description of the model is complete as soon as we define what we mean by magnitude in the model. We define the magnitude of an algebraic line, circular arc, rectangle, circle, etc. to be the length or area we usually assign to the object. Thus, for example, the magnitude of the line segment connecting the algebraic points (α_1, β_1) and (α_2, β_2) is just the usual distance between the points,

$$\sqrt{(\alpha_2 - \alpha_1)^2 + (\beta_2 - \beta_1)^2},$$

which is an algebraic real. A rectangle with algebraic sides will have as area the product of the lengths of the sides— again an algebraic real. Perimeters of algebraic polygons, being finite sums of algebraic reals, are themselves algebraic real numbers as well. It can further be shown that the areas of all algebraic polygons are algebraic reals.

With circles, the situation is different. The circumference of the circle of radius r is $2\pi r$ and, if r is algebraic, this circumference cannot be algebraic as otherwise

$$\pi = \frac{2\pi r}{2r},$$

being the ratio of two algebraic numbers, would also be algebraic. Thus the circumference of every algebraic circle is a non-algebraic number of the form $2\pi r$ for an algebraic real number r. Similarly, every algebraic circle has a non-algebraic area of the form πr^2 for an algebraic real number r. The magnitudes of line segments do not exhaust all the magnitudes around and one cannot conclude the existence of a figure whose magnitude is in the fourth proportion to three given magnitudes from Euclid's Proposition on magnitudes of line segments.

Indeed there is a question for which I do not know the answer:

1.2 Query. *Is there a figure composed piecewise of algebraic circular arcs and algebraic line segments, the length or area of which completes the proportion*

$$\frac{1}{2\pi} = \frac{2\pi}{x}?$$

I note that 1 is the magnitude of the line segment of length 1 and 2π the circumference of a circle of radius 1. The quantity $x = 4\pi^2$ is not the length of any algebraic line segment, the area of any algebraic polygon, nor the circumference or area of any algebraic circle. It is conceivably the length of some algebraic circular arc or the sum of such, or even the area of some strange figure concocted out of algebraic circular segments and sectors with, perhaps, some algebraic polygons gluing the pieces together. My knowledge of *transcendental numbers*, as non-algebraic numbers are called, is too weak to allow me to answer my query. Thus, I cannot point to this model as providing a counterexample to the existence of a figure standing in the fourth proportion to some magnitudes, but I can point to it as demonstrating that Euclid hasn't proved such existence by proving that a line segment can be found standing in the fourth proportion to these line segments.

Euclid's proof, depending on the formal application of a result established for line segments and lengths to two dimensional figures and areas, is made correct by introducing the real numbers and continuity axioms. This would not be done until over two thousand years had passed. We shall similarly postpone our discussion of this, but only until the next chapter.

2 Galileo and the Infinite

The notion of the infinite is vague enough that it can lead to paradox without the use of extra accessories like additions, subtractions, etc. The word itself means not finite, and "finite" derives from the latin verb *finio* meaning to limit, to bound, to end, etc. "Fin" pops up in the same sense in such words as "finish" and "finally". The ancient Greek word for the infinite, *apeiron* ($\alpha\pi\varepsilon\iota\rho o\nu$) similarly means unbounded[10] or indefinite[11]. European philosophy was introduced to the concept of the apeiron by Anaximander, who regarded it as some primordial stuff from which everything derives and to which everything returns. At the same time it was some sort of deity. Perhaps one could compare it with the later pan-theistic views of God as some all encompassing abstract substance[12]. God is infinite; nothing is greater than God; etc. In mathematics, the meaning of infinity is a bit narrower and "not finite", "not bounded", "unending" describe well its use in pre-Galilean times.

---

[10] Th.G. Sinnige, *Matter and Infinity in the Presocratic Schools and Plato*, van Gorcum & Comp. N.V., Assen, 1968, p. 16.

[11] Edward A. Maziarz and Thomas Greenwood, *Greek Mathematical Philosophy*, Frederick Unger Publishing Co., New York, 1968, p. 6.

[12] Spinoza's *Ethics* comes to mind.

On apparently linguistic grounds Aristotle declared the actual infinite not to exist; for, the "infinite is that which, regarded as a quantity, can always be increased by something outside it"[13]. I think this means that, by definition, the infinite is unending, i.e. incomplete. Infinity for Aristotle was an unactualised potentiality.

By Aristotle's day there were better reasons to question the actual existence of the infinite in mathematics. In Aristotle's *Mechanica*, a work of dubious authorship, but considered contemporaneous or almost contemporaneous with Aristotle, there is the paradox now known as *Aristotle's Wheel*. If one takes a wheel and rolls it on the ground once, the length of the path traced by the wheel will equal the circumference of the wheel. But, if one attaches a small wheel to a large one at the hub, the paths of both wheels will have the same

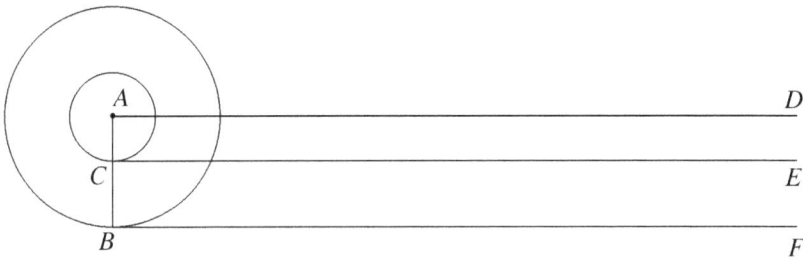

Figure 5

length, as in *Figure 5*. As the large wheel rolls from B to F, each point on its circumference touching the line BF only once, the small wheel will roll from C to E with each point on its circumference touching the line CE only once. But CE has the same length as BF; yet CE must equal the circumference of the small circle and BF the larger circumference of the larger circle.

Nowadays we gloss over the paradox with a few well-chosen words about cardinality and measure not being the same. We might even add a word or two about there being a good deal of sliding going on with the smaller wheel, though if hard pressed one might have a bit of difficulty pinpointing where the sliding occurs: each point of the smaller wheel touches the line CE in exactly one point and not throughout an interval. Perhaps this is why one likes to emphasise that the "same" problem already arises in comparing the two wheels without rolling them. The radii of the larger wheel set up a one-to-one correspondence between the points on the circumferences of the two circles.[14] The comparison is not entirely apt; there is a bit more to this paradox than just the confusion between cardinality and measure and it has a long history.[15]

Aristotle and the wheel named after him date to the 4th century BC. In the 5th century AD, Proclus reported on another, simpler paradox of infinity:

---

[13] Quoted in Sinnige, *op. cit.*, pp. 150 - 151.

[14] Bernard Bolzano, *Paradoxien des Unendlichen*, Verlag von Felix Meiner, Leipzig, 1921, p. 83.

[15] Cf. Israel E. Drabkin, "Aristotle's Wheel: Notes on the history of a paradox", *Osiris* 9 (1950), pp. 162 - 198.

But if from one diameter two semicircles are produced, and if an indefinite number of diameters can be drawn through the center, it will follow that the number of semicircles is twice infinity.[16]

That is, by observing that every diameter of a circle defines two semicircles, one concludes there to be twice as many semicircles as diameters. So, if the number of diameters is infinite, the number of semicircles is twice infinity. The contradiction here is that "twice infinity" sounds larger than infinity and yet, what could be larger than infinity?

This paradox, involving only the cardinal aspect of infinity and not measure, is perhaps a little purer; certainly it is a little simpler. And it is easier to dispel. Today we do so by formally defining what the cardinality of a set is, as well as what the sum of two infinite cardinals is, and then showing that the double of an infinite cardinal number is no larger than the cardinal itself. For later reference I sketch this here.

The first step is to define two sets A and B to have the *same cardinality* if there is a *one-to-one correspondence* between A and B, i.e. a function $f : A \to B$ that is *one-to-one* and *onto*:

i. for all $a, b \in A$, if $a \neq b$, then $f(a) \neq f(b)$;

ii. for all $b \in B$, there is some $a \in A$ for which $f(a) = b$.

The abstraction of a *cardinal number* shared by two sets of the same cardinality is a set theoretic subtlety that need not concern us here. Suffice it to say that in modern axiomatic set theory several Type II definitions have been given.

The sum $\mathfrak{a} + \mathfrak{b}$ of two cardinal numbers $\mathfrak{a}$ and $\mathfrak{b}$ is defined to be the cardinal number of the union of two disjoint sets A and B of cardinalities $\mathfrak{a}$ and $\mathfrak{b}$, respectively. It can be shown fairly easily that the sum is well-defined, i.e. it does not depend on which pair A, B of sets of cardinalities $\mathfrak{a}, \mathfrak{b}$ are chosen so long as they have these cardinalities and are disjoint.

The final step, showing that an infinite cardinal $\mathfrak{a}$ and its double $\mathfrak{a} + \mathfrak{a}$ are equal, is not trivial, but there are easy examples. Let $\mathbb{N}$ be the set of natural numbers, $\mathbb{E}$ the set of even natural numbers, and $\mathbb{O}$ the set of odd natural numbers. These sets all have the same cardinality, for $\mathbb{N}$ is mapped one-to-one onto $\mathbb{E}$ by the map $n \mapsto 2n$, and it is mapped one-to-one onto $\mathbb{O}$ via $n \mapsto 2n+1$. If $\mathfrak{a}$ is their common cardinal number, we have

$$\mathfrak{a} + \mathfrak{a} = \text{cardinality of } \mathbb{E} \cup \mathbb{O}$$
$$= \text{cardinality of } \mathbb{N}$$
$$= \mathfrak{a}.$$

As I say, this is the modern solution to the paradox. The Greek solution, taught by Aristotle, was to deny actual existence to the infinite. Proclus continues his remark thus:

This difficulty is alleged by some persons against the indefinite divisibility of magnitudes. We reply that a magnitude is indefinitely divisible,

---

[16] Glenn Morrow (ed.), *Proclus; A Commentary on the First Book of Euclid's Elements*, Princeton University Press, Princeton, 1970, p. 125.

but not into an infinite number of parts. The latter statement makes an infinite number actual, the former merely potential; the latter assigns existence to the infinite, the other only genesis. With one diameter, then, two semicircles come into being, and the diameters will never be infinite in number, even though they can be taken indefinitely. So the number of semicircles will never be twice infinity; those that are produced at any time will be twice a finite number; for the diameters taken at any time are always finite in number.[17]

There are a couple of things to be explained here. One is the distinction between actual and potential infinity, and the other is the indefinite divisibility of magnitudes.

The Greek distinction between potential and actual infinity is nowadays most commonly explained by reference to constructive mathematics. We imagine ourselves constructing objects in time. A line consisting of an infinity of points could not be constructed in a finite period, but a container in which to place points as they are constructed, or a rule specifying which points will belong to the line when they are constructed can be constructed in a finite amount of time. We can, for example, construct a line by giving its equation, say, $2x+3y = 5$. When we construct the line we do not construct an infinity of points comprising the line; only the line, or rule, itself has been constructed. We can then begin to construct points on the line, e.g. $(1, 1), (4, -1), (7, -3), (10, -5)$, etc. At any time we have constructed only finitely many points on the line. We do not have infinitely many points on the line; that is, we do not have an *actual* infinity. However, the number of possible points on the line is *potentially* infinite insofar as there is no bound on the number of points we can construct by some future date.

The view that the infinite is only potential and not actual would be defended in general mathematics as late as the 19th century by Carl Friedrich Gauss, and it would be maintained by a few constructivists in the 20th century. But the actual infinite was making inroads into mathematics already in the Middle Ages, and became something to be reckoned with during the Renaissance. The most fruitful manifestation of the infinite in this latter period was in the form of indivisibles.

The notion of an indivisible goes back to the Greeks and predates Aristotle. The natural name to assign to them is Democritus, whose early atomic theory held that matter was ultimately composed of small indivisible parts. In geometry one would likewise consider the line as composed of indivisibles, though whether these should be points, infinitesimally small line segments of non-zero length, or very small line segments of definite positive length is not clear. The texts do not survive; all we have are mentions by commentators who may or may not have understood what they were commenting on, and thus may or may not have represented the views of the atomists faithfully.

---

[17] *Ibid.*

A case in point is Antiphon whose quadrature of the circle is referred to but not described by Aristotle in his *Physics* and described for the first time by Themistius in his *Commentary on Aristotle's Physics*:

> ... Antiphon, who inscribed an equilateral triangle in the circle, and on each of the sides set up another triangle, an isosceles triangle with its vertex on the circumference of the circle, and continued this process, thinking that at some time he would make the side of the last triangle, although a straight line, coincide with the circumference.[18]

No one knows exactly what to make of this. Did Antiphon believe that for some large n the $3 \cdot 2^n$-gon so constructed would coincide with the circle? Was the use of the word "last" a misconception of Themistius and is Antiphon a forerunner of Eudoxus and the method of exhaustion? Or, did Antiphon have in mind the final result after the doubling the number of sides of the inscribed polygon had been carried out an infinite number of times— a polygon of infinitely many infinitesimally small sides that coincided with the circle?

Through carefully distinguishing potential from actual infinity, the potentially infinite divisibility of a line from the actual infinite division of such, Aristotle effectively banished indivisibles from mathematical practice for nearly two millennia. The subsequent development of the method of exhaustion in Greek mathematics makes viewing Antiphon as its anticipator the favoured interpretation of what he did, but all three interpretations are plausible.

By Galileo's day, indivisibles were making a comeback. These were heroic times in European mathematics. By the middle of the 16th century, the cubic equation had been solved. This was the first piece of mathematics that in no way stemmed from the Greeks. And by the 1630s, the period I wish to discuss, Archimedes had been translated and mastered and European mathematicians, impatient with the sterile logic of Greek geometric proofs, were forging ahead using their own heuristic methods. Kepler had used indivisibles and now Galileo and Bonaventura Cavalieri, a student of one of Galileo's students[19], were doing the same.

Galileo used indivisibles for various purposes, some of which might not occur to us with our experience of them (if any) coming from the use of infinitesimals in Calculus. Some of his physical explanations in his *Discourses and Mathematical Demonstrations Concerning Two New Sciences*[20] (1638) depend on his view of matter as being composed of infinitely many infinitely small indivisibles

---

[18] Ivor Thomas, *Selections Illustrating the History of Greek Mathematics, vol. I: From Thales to Euclid*, Harvard University Press, Cambridge (Mass.), 1939, and William Heineman Ltd., London, 1939, pp. 311 - 313.

[19] Cavalieri was taught by Benedetto Castelli, a former student of Galileo. Cavalieri was eventually introduced to Galileo and they entered into a lengthy correspondence, Cavalieri considering himself a disciple of Galileo. In some history books Cavalieri is referred to as a student of Galileo, which is a fair description under an extended meaning of the word "student". Authors who call him a pupil of Galileo's, however, are way off base.

[20] The traditional English translation, *Dialogues Concerning the Two New Sciences*, by Henry Crew and Alfonso de Salvio, first published by Macmillan, New York,

interspersed with infinitely many empty spaces. He illustrates his belief in the existence of such empty spaces via his solution of the paradox of Aristotle's Wheel.

To explain away the apparent paradox, Galileo suggested replacing the circular wheels by hexagonal ones as in *Figure 6*. As one starts to rotate

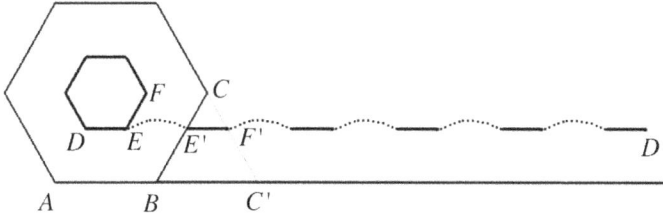

Figure 6

the large wheel ABC, it does not revolve around the centre O, but pivots about B, lifting the small wheel. By the time the large wheel comes to rest with side BC covering interval BC', all of the points of the wheel other than B itself will have swept out arcs of circles centered at B. The side EF of the small wheel will have been dragged through the air before landing on segment $E'F'$ of the line DD' corresponding to CE of *Figure 5*.

The same is true of concentric wheels of any number n of sides. The line traced out by the larger wheel will have length equalling the perimeter of the larger wheel, and the line traced out by the smaller wheel will consist of n line segments totalling in length the perimeter of the small wheel along with their $n-1$ separating gaps. Galileo concludes

So in the case of the circles, polygons having an infinitude of sides, the line traversed by the continuously distributed infinitude of sides is in the greater circle equal to the line laid down by the infinitude of sides in the smaller circle but with the exception that these latter alternate with empty spaces; and since the sides are not finite in number but infinite, so also are the intervening empty spaces not finite but infinite. The line traversed by the larger circle consists then of an infinite number of points which completely fill it; while that which is traced by the smaller

1924, was reissued by Northwestern University, Evanston and Chicago, 1939. It was incorporated into the Great Books of the Western World, published by the Encyclopædia Britannica in 1952 and was also published by Dover. Some relevant mathematical passages are reproduced in Dirk Struik, *A Source Book in Mathematics, 1200 - 1800*, Harvard University Press, Cambridge (Mass.), 1969. A new translation by Galileo scholar Stillman Drake appeared in 1974 (*Two New Sciences*, University of Wisconsin Press, Madison, 1974). The relevant passages of Drake's translation are reproduced in Ronald Calinger (ed.), *Classics of Mathematics*, Moore Publishing Company, Oak Park (Ill.), 1982. As the Great Books edition is likely to be the most accessible edition, I give page references to it. That said, I would still recommend consulting the printing in Drake or Calinger for the footnotes. I also recommend the short discussion of Galileo given in Margaret Baron, *The Origins of the Infinitesimal Calculus*, Pergamon Press, Oxford, 1969, pp. 118 - 120.

circle consists of an infinite number of points which leave empty spaces and only partly fill the line. And here I wish you to observe that after dividing and resolving a line into a finite number of parts, that is, into a number which can be counted, it is not possible to arrange them again into a greater length than that which they occupied when they formed a *continuum* and were connected without the interposition of as many empty spaces. But if we consider the line resolved into an infinite number of infinitely small and indivisible parts, we shall be able to conceive the line extended indefinitely by the interposition, not of a finite, but of an infinite number of infinitely small indivisible empty spaces.[21]

Here there is no ambiguity, no question as to what is meant as with Antiphon's inscribed polygons. Galileo said the circle was a polygon with infinitely many sides and he meant it. The *Discourses*, like his earlier *Dialogue Concerning the Two Chief World Systems*[22], is largely written in the form of a dialogue among three participants— Salviati, Sagredo, and Simplicio. Salviati and Sagredo represent Galileo's views and are named after two of Galileo's friends. Simplicio represents the Aristotelian viewpoint and is named after Simplicius, a noted commentator on Aristotle. In the *Discourses*, Simplicio attempts to raise the potential/actual distinction regarding infinity and Salviati effectively shoots it down. Certainly, he says, one is only dealing with potentiality in trying to construct infinitely many indivisibles by constructing polygons of ever more sides, but one does construct such in a single step by taking a compass and drawing a circle.[23]

It is almost impossible to form a fair judgement of Galileo's solution. As a mathematical logician I am sufficiently well-versed in Nonstandard Analysis to accept the circle as a polygon with infinitely many infinitesimal sides. However, Nonstandard Analysis is a bit more devious than Galileo. The circle is not itself an infinitely many sided polygon, but is the *standard part* of such a polygon. In fact, it is the standard part of infinitely many such polygons. So such polygons can be used to model Aristotle's Wheel and explain the discrepancy between the length CE and the circumference of the small circle in *Figure 5*, but his assumption that the mere act of drawing a circle produces a polygon of infinitely many indivisible sides is not justified. The circle drawn is the standard part of infinitely many such polygons and Galileo has not explained how to determine which one he has singled out.

Of course, such criticism as I have given is unfair. It says his Type I result is incorrect according to my favoured Type II conception, and resembles the criticisms of the summations of certain series reported on in Chapter I. The correct criticism is that Galileo's use of indivisibles was not grounded on anything. He doesn't tell us what indivisibles are or what their properties are. Look at *Figure 6* again, but imagine the hexagon standing in for one of Galileo's infinitely

---

[21] Galileo, p. 141.

[22] This is his comparison of the Ptolemaic and Copernican systems, the work that brought him before the Inquisition.

[23] *Ibid.*, p. 151.

many sided polygons. Both DE and AB are indivisibles, but they don't have the same lengths. In my image, I gave the small hexagon a diameter 40% the diameter of the larger one, but suppose it were 50%. That would make DE half the size of AB. Is AB still indivisible or can it be divided in half? How does it change the picture if the larger polygon has twice as many sides as the smaller?[24] Maybe one cannot compare sizes of indivisibles because of their elasticity or fuzziness or...? Or, perhaps, Galileo does not intend "indivisible" to mean "incapable of being divided" but merely infinitely small?

Galileo's disciple Cavalieri also applied indivisibles, but his image of them was completely different. For Galileo, an indivisible was the same dimension as the object being considered: a line was composed of infinitely many indivisible line segments, an area of infinitely many indivisible areas, and a solid of infinitely many indivisible volume elements. For Cavalieri, an indivisible was an element of lower dimension: a line was composed of points, an area of line segments, and a solid of 2-dimensional cross-sections. The length of a line was the sum of its points; the area of a plane figure the sum of its linear elements; and the volume of a solid the sum of its cross-sections.

Cavalieri never explained what these sums where or how to calculate them, but he did explain how to use this conception to derive areas and volumes of certain figures from the known areas and volumes of other figures by applying a theorem now known as *Cavalieri's Principle*:

> **Theorem**. If between the same parallels any two plane figures are constructed, and if in them, any straight lines being drawn equidistant from the parallels, the included portions of any one of these lines are equal, the plane figures are also equal to one another; and if between the same parallel planes any solid figures are constructed, and if in them, any planes being drawn equidistant from the parallel planes, the included plane figures out of any one of the planes so drawn are equal, the solid figures are likewise equal to one another.[25]

A simple application of Cavalieri's Principle is the proof that triangles of equal base and height have the same area. Consider two such triangles ABC and $A'B'C'$ as in *Figure 7*, below. If we assume A, B, A', B' to lie in a

---

[24] In set theory, as we've already mentioned, doubling the infinite makes no change in cardinality, but in Nonstandard Analysis it does.

[25] This is quoted from Struik's *Source Book* (*op.cit.*), p. 210, where in turn it is quoted from G.W. Evans, "Cavalieri's theorem in his own words", *American Mathematical Monthly* 24 (1917), pp. 447 - 451. Evans's article is a loose translation of the relevant passage of Cavalieri's *Geometria indivisibilibus continuorum* of 1635. Another translation by Mary Walker is to be found in David Eugene Smith, *A Source Book in Mathematics*, 1929, reprinted by Dover, New York, 1959. Both English versions include not only the statement of the Theorem, but also its attempted proof, essentially a superposition argument.

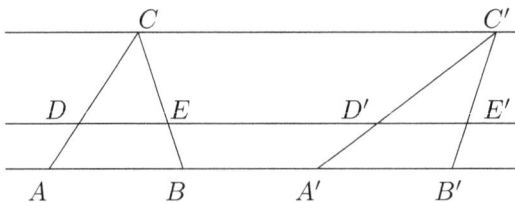

Figure 7

straight line, then the assumption of equal height (say, H) makes the line CC' parallel to $ABA'B'$. If we draw another parallel line between these two, it will intersect the triangles in points D, E, D', E' as pictured. The triangles DEC and $D'E'C'$ have equal height (say h). Moreover, the triangles DEC and ABC are similar, as are $D'E'C'$ and $A'B'C'$. Thus

$$\frac{DE}{AB} = \frac{h}{H} = \frac{D'E'}{A'B'}.$$

Assuming $AB = A'B'$ we have

$$DE = \frac{h}{H}AB = \frac{h}{H}A'B' = D'E'.$$

Thus the cross-sections of the two figures given by lines parallel to the base are equal and by Cavalieri's Principle the triangles have equal area.

As mentioned, Cavalieri, in anticipation of integration, thought of the area as the sum of line segments. In this case the areas of the two figures are given by the same sums and hence must be equal. There is a danger in this conception, as he himself was aware. For, he wrote to another of Galileo's disciples, Evangelista Torricelli about the following paradox[26]: Consider the non-isosceles triangle ABC of *Figure 8*, below. Draw a line parallel to BC

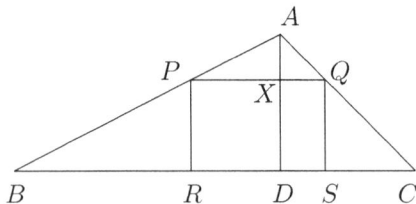

Figure 8

intersecting the sides AB, AC in points P, Q, respectively. Drop perpendiculars from these points to obtain equal line segments PR and QS. If one sums over these segments one will conclude the triangles ABD and ACD to be equal in area, which, of course, they are not.

Cavalieri's way out of the paradox was to say that if, e.g., AB were twice the length of AC, and AC had 100 points, then AB would have 200 points. Presumably, he concludes the area ABD to be twice that of ACD, which we

---

[26] Struik, *op.cit.*, pp. 218 - 219.

would see to be the case by summing not the equal segments PR and QS, but the segments PX and QX, which stand in the ratio 2:1.

Cavalieri's solution yields the correct result, but only by applying another formalist paradox— the misidentification of measure and cardinality. From the assumption that AB is twice as long as AC, he concludes that (although the parallel lines PQ set up a one-to-one correspondence between the points of AB and AC) the line AB has twice as many points as AC— it is "twice infinity" in Proclus's lingo.

I suppose Galileo would have explained the matter by pointing to an infinite number of indivisible gaps in the line AB. Cavalieri at some point generalised his Principle to allow one to conclude that if the cross-sections (in *Figure 8* the segments PX and QX) always stood in a constant ratio (say, 2:1) then the areas (in this case the triangles ABD and ACD) stood in the same ratio. One wonders: if we accept Galileo's gaps, could we accept the result? For, the parallels determined by the points Q of AC are skipping the gaps and must not cover half the area of triangle ABD.

Indivisibles are intuitive, useful heuristic devices and one can go a long way with them. But they do raise more questions than they answer, and without the proper care can lead one astray. It is no wonder they fell into disrepute and were discarded after the Calculus became algorithmic and the ϵ-δ techniques were introduced. The former eliminated the need for the heuristic use of indivisibles, and the latter provided genuine rigour which indivisibles could not give.

To get back on course, I wish to discuss what is called *Galileo's Paradox* by Bolzano or his editors[27], for here he makes implicit use of Cavalieri's Principle.

First, I should explain about the three participants of the dialogues of the *Discourses*. Sagredo is the host. He is an intelligent layman, open to reason and capable of understanding. Salviati is sort of the honoured guest. He is the brilliant philosopher who wins over Sagredo with his demonstrations. Both Sagredo and Salviati speak for Galileo. Simplicio is the Aristotelian philosopher who lacks mathematics. He plays the foil, but not so stupidly as he had done in the earlier *Dialogue Concerning the Two Chief World Systems*, the work that got Galileo into trouble.

Galileo's Paradox is presented by Salviati:

> And this I shall do by showing you two equal surfaces, together with two equal solids located upon these same surfaces as bases, all four of which diminish continuously and uniformly in such a way that their remainders always preserve equality among themselves, and finally both the surfaces and solids terminate their previous constant equality by degenerating, the one solid and the one surface into a very long line, the other solid and the other surface into a single point; that is, the latter to one point, the former to an infinite number of points.[28]

---

[27] I have the second edition of Bolzano's *Paradoxien des Unendlichen* and do not find the phrase "Paradoxie von Galilei" in the text or in the table of contents, but only as the header topping page 89.

[28] Galileo, p. 142.

Sagredo responds, "This proposition appears to me wonderful indeed; but let us hear the explanation and demonstration".[29] Salviati responds with a geometric construction. He starts with a semicircle and isosceles triangle inscribed in a rectangle as in *Figure 9*, below. He then drops the perpendicular

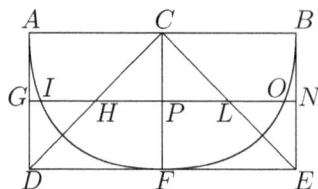
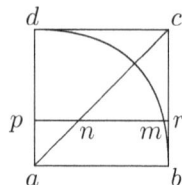

Figure 9   Figure 10

CF and revolves the whole around it. There result three solids of revolution: A cylinder generated by revolving the rectangle $ADEB$, a hemisphere generated by revolving the semicircle AFB, and a cone generated by revolving the triangle CDE. The cone is one of the solids he promised. The other is the "round bowl" obtained by hollowing out the cylinder through the removal of the hemisphere. Nowadays we would describe it as the solid of revolution obtained by revolving the region $ADFEBFA$ or even just ADF around the axis CF.

In *Figure 9*, an extra line, GN is drawn parallel to DE. This represents a plane parallel to the base of the cone and bowl. Salviati now remarks that cross-sections obtained by intersecting this plane with the two solids, that is with the cone and the bowl, are equal in area. Hence the cone and the bowl have equal volumes. In fact, the same is true for the pairs of solids resulting by removing everything from the bowl and cone below the given plane, i.e. for the smaller cone obtained by revolving the triangle CHL around the axis CP and the "round razor" obtained by revolving one or both of the figures AGI and BNO around CP.[30]

As one raises the plane represented by GN towards that represented by AB, the equality of the two solids is "continuously and uniformly preserved".

> And as the cutting plane comes near the top, the two solids (always equal) as well as their bases (areas which are also equal) finally vanish, one pair of them degenerating into the circumference of a circle, the other into a single point, namely, the upper edge of the bowl and the apex of the cone. Now, since as these solids diminish equality is maintained between them up to the very last, we are justified in saying that, at the very extreme and final end of this diminution, they are still equal and that one is not infinitely greater than the other. It appears therefore that we may equate the circumference of a large circle to a single point. And this which is true of the solids is true also of the surfaces which form their bases; for these also preserve equality between

---

[29] After the demonstration, Sagredo finds that "This demonstration is ingenious and the inferences drawn from it are remarkable". Galileo evidently appreciated good work!

[30] All this is done without mentioning Cavalieri, probably for reasons to be explained later.

themselves throughout their diminution and in the end vanish, the one
into the circumference of a circle, the other into a single point. Shall
we not then call them equal seeing that they are the last traces and
remnants of equal magnitudes? Note also that, even if these vessels
were large enough to contain immense celestial hemispheres, both their
upper edges and the apexes of the cones therein contained would always remain equal and would vanish, the former into circles having the
dimensions of the largest celestial orbits, the latter into single points.
Hence in conformity with the preceding we may say that all circumferences of circles, however different, are equal to each other, and are each
equal to a single point.[31]

Following this Sagredo asks for the geometric proofs of the equality of the
cross-sectional areas and the truncated volumes. Salviati partially obliges by
showing the equality of the cross-sectional areas and refers to a book by Luca
Valerio, "the Archimedes of our age", for the proof of the equality of the solids,
adding that, "for our purpose, it suffices to have seen that the above-mentioned
surfaces are always equal and that, as they keep on diminishing uniformly, they
degenerate, the one into a single point, the other into the circumference of a
circle larger than any assignable". Galileo had, of course, already derived this
result for the solids from that for the surfaces by implicit appeal to Cavalieri's
Principle. That he does not make this explicit may be due to a preference
for the classical methods of proof he uses later in the *Discourses*, to the fact
that, as stated, the solids are not necessary for making his point, or to the
scrupulousness of his attribution of the figures and result to Luca Valerio, who
had put the equality of the solids to a different use.

Indeed, in Bolzano's account of Galileo's Paradox, given in §46 of *Paradoxien des Unendlichen*[32], no mention is made of the solids at all. Bolzano draws
a slightly different picture and labels things differently (*Figure 10*, above), but
the situation is the same. Bolzano's picture is just a vertical flip of the right
half of Galileo's picture (*Figure 9*) with different lettering. The surfaces (in a
plane perpendicular to the diagram) one is interested in are the circle centered
at p of radius pn and the ring centered at p obtained by deleting the circle
centered at p of radius pm from the circle centered at p of radius pr. The area
of the circle is $\pi(pn)^2$ and that of the ring is $\pi(pr)^2 - \pi(pm)^2$. Bolzano then
states blithely that

$$\pi(pn)^2 = \pi(pr)^2 - \pi(pm)^2, \tag{3}$$

and Hans Hahn, in his annotation[33], explains that $pr = ab = am$, whence

$$(pr)^2 = (am)^2 = (pa)^2 + (pm)^2, \tag{4}$$

by the Pythagorean Theorem. But the angle pan is the same as pac, i.e. half a
right angle, whence $pa = pn$ and (4) becomes

---

[31] *Ibid.*, p. 143.
[32] Pp. 88 - 91.
[33] *Ibid.*, pp. 149 - 150.

$$(pr)^2 = (pn)^2 + (pm)^2,$$

and (3) follows. Galileo similarly invokes the Pythagorean Theorem. Once one has the figures, a more modern approach would be to ignore the surfaces and simply calculate the volumes of the solids of revolution using the tools of the Calculus.

Incidentally, Hahn finishes his annotation with the dismissive remark,

> This however is neither false nor paradoxical; rather, it is trivial, because they each have surface area 0. The appearance of a paradox comes about only through the erroneous conception, that the content of a point set is a measure for the quantity of points contained in it.[34]

Bolzano was certainly aware of the distinction between measure and quantity, having already in §20 of the book set up a one-to-one correspondence between the intervals $[0, 5]$ and $[0, 12]$ although the one is definitely larger in measure than the other. He also expressed the opinion that Galileo probably only introduced his paradox to stimulate thought.[35]

It is tempting to illustrate the identification of measure and cardinality by pointing to Cavalieri's dissolution of his paradoxical calculation of the areas of the triangles of *Figure 8* above. This was, however, not a real proof but a reasoning by analogy. Cavalieri in fact went so far as to compare the vertical lines to threads, 200 in the left triangle and 100 in the right. Some mediæval philosophers, in attempting to fix the nature of the continuum, did however appeal explicitly to the identification of magnitude and cardinality. Around the turn of the 11th to 12th centuries, the Arab scholar al-Ghazzālī, known as Algazel in the West, refuted Democritean atoms in geometry as follows[36]: If the plane consists of extended atoms, we could construct a square measuring 4 atoms by 4 atoms. Each of the diagonals and sides of the square would consist of 4 atoms. Hence the diagonal equals the side, which is impossible because one proves in geometry the diagonal of the square to be larger than the side of the square. Al-Ghazzālī would not be the last to use this proof.

There were, by the way, 5 distinct views of the composition of the continuum in the 14th century. The traditional Aristotelian view was that the line was a potentially infinitely divisible whole, not constructed of points. This was the predominant belief in pre-Galilean times. Democritus believed in corporeal indivisibles— extended atoms. Pythagoras and Plato had favoured a line segment consisting of finitely many points, a point differing from Democritean atoms in that it had no dimension. A fourth view was that a line segment consisted of infinitely many adjacent points. And finally, the currently accepted view, is that the line consisted of infinitely many densely packed points. Galileo's polygonal circles seem to suggest a 6th possibility: infinitely many adjacent indivisibles other than points.

---

[34] *Ibid.*, p. 150.

[35] *Ibid.*, p. 88.

[36] Edward Grant, *A Source Book in Medieval Science*, Harvard University Press, Cambridge (Mass), 1974, p. 316.

Following the presentation of his paradox, Galileo remarks through the voice of Salviati:

> One of the main objections urged against this building up of continuous quantities out of indivisible quantities is that the addition of one indivisible to another cannot produce a divisible, for if this were so it would render the indivisible divisible. Thus, if two indivisibles, say two points, can be united to form a quantity, say a divisible line, then an even more divisible line might be formed by the union of three, five, seven, or any other odd number of points. Since however these lines can be cut into two equal parts, it becomes possible to cut the indivisible which lies exactly in the middle of the line. In answer to this and other objections of the same type we reply that a divisible magnitude cannot be constructed out of two or ten or a hundred or a thousand indivisibles, but requires an infinite number of them.[37]

Simplicio responds with

> Here a difficulty presents itself which seems to me insoluble. Since it is clear that we may have one line greater than another, each containing an infinite number of points, we are forced to admit that, within one and the same class, we may have something greater than infinity, because the infinity of points in the long line is greater than the infinity of points in the short line. This assigning to an infinite quantity a value greater than infinity is quite beyond my comprehension.[38]

Salviati comes back with one of the most lucid remarks on the infinite to be found in Galileo:

> This is one of the difficulties which arise when we attempt, with our finite minds, to discuss the infinite, assigning to it those properties which we give to the finite and limited; but this I think is wrong, for we cannot speak of infinite quantities as being the one greater or less than or equal to another. To prove this I have in mind an argument...[39]

I like Salviati's remark. It offers as clear a diagnosis of a case of Type I Formalism as one could hope to find in the literature: difficulties arise when we assign properties of one class (in this case the finite) to another (the infinite). Simplicio's difficulty seems to be threefold. First, there is the measure vs. cardinality problem: a longer line must have more points than a shorter. Second, there is the proper inclusion problem: if a set X properly extends Y, then X has more points than Y and so the number of such points must be

---

[37] Galileo, p. 144. This seems a bit confused. The best explanation seems to be that "divisible" means potentially infinitely divisible and "indivisible" means not capable of being divided at all. This makes the two notions opposite extremes, not negations of each other. For example, the sum of two indivisibles is capable of being divided once, but not twice, and is thus neither a divisible nor an indivisible.

[38] *Ibid.*

[39] *Ibid.*

greater. Third, there is the belief that infinity is an absolute; nothing can be greater than infinity. Salviati has supposedly dealt with the first of these in discussing Aristotle's Wheel. He now proposes to attack the latter two problems by showing that the notion of "greater than", familiar of finite collections, does not apply to infinite collections. Attempting to apply it is a Type I formalistic error:

> If I should ask further how many squares there are one might reply truly that there are as many as the corresponding number of roots, since every square has its own root and every root its own square, while no square has more than one root and no root more than one square.[40]

Simplicio agrees and Salviati continues:

> But if I inquire how many roots there are, it cannot be denied that there are as many as there are numbers because every number is a root of some square. This being granted we must say that there are as many squares as there are numbers because they are just as numerous as their roots, and all the numbers are roots. Yet at the outset we said that there are many more numbers than squares, since the larger portion of them are not squares. Not only so, but the proportionate number of squares diminishes as we pass to larger numbers. Thus up to 100 we have 10 squares, that is, the squares constitute 1/10 part of all the numbers; up to 10000, we find only 1/100 part to be squares; and up to a million only 1/1000 part; on the other hand in an infinite number, if one could conceive of such a thing, he would be forced to admit that there are as many squares as there are numbers taken together.[41]

Now Sagredo asks, "What then must one conclude under those circumstances?" Salviati replies

> So far as I can see we can only infer that the totality of all numbers is infinite, that the number of squares is infinite, and that the number of their roots is infinite; neither is the number of squares less than the totality of all numbers, nor the latter greater than the former; and finally the attributes "equal", "greater", and "less", are not applicable to infinite, but only to finite, quantities. When therefore Simplicio introduces several lines of different lengths and asks me how it is possible that the longer ones do not contain more points than the shorter, I answer him that one line does not contain more or less or just as many points as another, but that each line contains an infinite number. Or if I had replied to him that the points in one line were equal in number to the squares, in another, greater than the totality of numbers; and in the little one, as many as the number of cubes, might I not, indeed, have

---

[40] *Ibid.* It is very tempting to read into this Galileo's recognition of the existence of a one-to-one correspondence as the basis for deciding equicardinality. If we read on however we find he never made this identification.

[41] *Ibid.*

satisfied him by thus placing more points in one line than in another and yet maintaining an infinite number in each?[42]

Even today this is a reasonable response. We cannot compare infinite quantities in the same way we can compare finite ones. Period. End of discussion. Our modern solution, using one-to-one correspondences to define the equality of cardinality of sets is a more sophisticated and more useful solution to the problem, but it is not more *correct*.

Galileo should have stopped here while he was ahead. What comes next is a bit confused. It does explain why Galileo referred to the one-to-one correspondence between the numbers (i.e., positive integers) and their squares instead of that between the numbers and, say, the even numbers. Sagredo now chimes in with a paradoxical observation:

> Pray stop a moment and let me add to what has already been said an idea which just occurs to me. If the preceding be true, it seems to me impossible to say either that one infinite number is greater than another or even that it is greater than a finite number, because if the finite number were greater than, say, a million it would follow that on passing from the million to higher and higher numbers we would be approaching the infinite; but this is not so; on the contrary, the larger the number to which we pass, the more we recede from infinity, because the greater the numbers the fewer are the squares contained in them; but the squares in infinity cannot be less than the totality of all the numbers, as we have just agreed; hence the approach to greater and greater numbers means a departure from infinity.[43]

Salviati agrees, saying

> And thus from your ingenious argument we are led to conclude that the attributes "larger", "smaller", and "equal" have no place either in comparing infinite quantities with each other or in comparing infinite with finite quantities.[44]

It is not the case that "the greater the numbers the fewer are the squares contained in them". The number of squares is not receding from infinity; it is simply proceeding there at an ever decreasing rate. Galileo's *Discourses* marked not only the beginning of a modern view of infinity, but was also the first deep study of motion and rates of change and Galileo's gaffe is that much more puzzling for this.

Galileo seems to be serious, however, for a little later Salviati says

> In the preceding discussion we concluded that, in an infinite number, it is necessary that the squares and cubes be as numerous as the totality of the natural numbers, because both of these are as numerous as their roots which constitute the totality of the natural numbers. Next we

---

[42] *Ibid.*, pp. 144 - 145.
[43] *Ibid.*, p. 145.
[44] *Ibid.*, p. 145.

saw that the larger the numbers taken the more sparsely distributed the cubes; therefore it is clear that the larger the numbers to which we pass the farther we recede from the infinite number; hence it follows that, since this process carries us farther and farther from the end sought, if on turning back we shall find that any number can be said to be infinite, it must be unity. Here indeed are satisfied all those conditions which are requisite for an infinite number; I mean that unity contains in itself as many squares as there are cubes and natural numbers.[45]

There follows some irrelevant remarks about {1} sharing properties of the sets of squares, cubes, and other powers, and the definitive conclusion:

Therefore we conclude that unity is the only infinite number. These are some of the marvels which our imagination cannot grasp and which should warn us against the serious error of those who attempt to discuss the infinite by assigning to it the same properties which we employ for the finite, the natures of the two having nothing in common.[46]

Galileo's conclusion that 1 is an infinite number fits in nicely with his earlier equating of a single point and a circle. That it is the only infinite number has to be rejected because if 1 point equates to a circle, 2 points must equate to 2 circles and thus to "twice infinity"...

Galileo was reportedly put off by Kepler's early Pythagoreanism.[47] Thus I won't suggest he was willing to accept the conclusion that 1 is an infinite number because it is the number of unity and is the generator of all the numbers, as Anaximander's apeiron was the source of all. It may have appealed to his religious beliefs, e.g. trinitarianism: There is but one God, namely 1) God the Father, 2) God the Son, and 3) God the Holy Ghost.[48]

I am not being facetious here[49]. The language "some of the marvels which our imagination cannot grasp" is reminiscent of pious homilies about how "the Lord works in mysterious ways". We can no more understand the infinite than we can understand God. The relation between God and the infinite is not the mere analogy between the Grandi sum and the biblical theory of creation that would be cited around the end of the century. The next great voice of the infinite, Bolzano, would be quite explicit on this point[50]. And Cantor felt an

---

[45] *Ibid.*, p. 146.

[46] *Ibid.*, pp. 146 - 147.

[47] Michael Sharratt, *Galileo, Decisive Innovator*, Blackwells, Oxford, 1994, p. 70.

[48] I am tempted to believe he was laying the groundwork for a defence of Bernard Shaw, who a few hundred years later would find even more God (in light of the trinity, singular and plural are the same here) in his entertaining little book, *The Adventures of the Black Girl in Her Search for God*. If so, however, it didn't work: one cleric was compelled to counter with *The Adventures of Gabriel in His Search for Mr. Shaw; A Modest Companion for Mr. Shaw's Black Girl*. But its author, W.R. Matthews, did not call for a burning at the stake as suffered by Giordano Bruno or the house arrest imposed on Galileo; he merely poked a little good-natured fun at GBS.

[49] Well, maybe I am in the preceding footnote.

[50] Bolzano, *op. cit.*, p. 8.

affinity to theology because of his "religious conviction, that scholastic philosophy and theology in particular provided starting points to a deeper discussion of
the problem of infinity"[51]. Indeed, Felix Klein said that "the scholastic specula-
tions... often turn out to be the most correct formulation of what we designate
today as set theory".[52]

But while religious convictions may stimulate interest in and discussion of
the infinite, it can also be a hindrance. Ascribing the infinite to God alongside
His ineffability, can prevent one from questioning such questionable results as
Galileo had derived. I am inclined to disagree with Bolzano's interpretation of
Galileo's intent. I don't think the latter merely meant to stimulate thought but
actually believed he had proven 1 to be infinite.

Two things, however, are certain. First, Galileo recognised that one could
not assume properties of the finite to carry over to the infinite. And, second, he
recognised, in some primitive sense, that sets in one-to-one correspondence with
each other had the same number of elements— whatever that should mean. He
did not take the extra step and reverse the implication. Bolzano did, but he did
not develop the concept mathematically, being content to show the limitations
of equicardinality: a set could have the same cardinality as a proper subset,
and, if the sets were line segments, they needn't have the same length.

The real breakthrough came with Georg Cantor, who founded a theory of
cardinal numbers, and proved for the first time that there are distinct infinite
cardinal numbers. There is an unwritten law that any introductory discussion
of infinite sets must include a proof of this result. As the reader has likely seen
the proof by diagonalising on a list of decimal expansions, I prefer to give a
variant of Cantor's earlier proof.

2.1 Theorem. *The sets $\mathbb{N}$ and $\mathbb{R}$ do not have the same cardinality.*

Cantor proved this via the following technical lemma:

2.2 Lemma. *Let $[a, b]$ be a closed interval ($a < b$), and suppose $c : c_0, c_1, \ldots$ is
a sequence of distinct real numbers. There is a real number $d \in [a, b]$ such that
d is not in the range of the sequence c.*

The Lemma yields Theorem 2.1. For, if $\mathbb{N}$ and $\mathbb{R}$ had the same cardinality,
there would be a function $c : \mathbb{N} \to \mathbb{R}$ which was one-to-one and onto. But, by
the Lemma, there is an element $d \in [0, 1]$ which is not in the range of c.

The idea behind the proof of Lemma 2.2 is very simple. Given the sequence
$c : c_0, c_1, \ldots$ and the interval $[a, b]$, define a sequence $I_0 \supseteq I_1 \supseteq \ldots$ of closed
intervals as follows: Choose $I_0 = [a, b]$. Given $I_0 \supseteq \ldots \supseteq I_n$, choose I_{n+1} to
be any closed subinterval of I_n which does not contain c_n. Because $c_n \notin I_{n+1}$,
c_n does not lie in the intersection $I = \bigcap_n I_n$ of the closed intervals. Thus,
any $d \in I$ satisfies the requirements of the Lemma. That such a d exists is
guaranteed by a completeness axiom of the type discussed in Chapter I (page
45, above):

---

[51] Walter Purkert and Hans Joachim Ilgauds, *Georg Cantor 1845 - 1918*, Birkhäuser
Verlag, Basel, 1987, p. 117.

[52] *Ibid.*

2.3 Theorem (Nested Interval Property). *Let $I_0 \supseteq I_1 \supseteq \ldots$ be a nested sequence of closed bounded intervals in $\mathbb{R}$. The intersection of this sequence is not empty, i.e. there is a number d such that, for all $n, d \in I_n$.*

Proof. Write $I_n = [a_n, b_n]$. Note that for all m, n we have

$$I_n \subseteq I_m \quad \text{or} \quad I_m \subseteq I_n,$$

according as $n \geq m$ or $n \leq m$ by the nesting of the intervals. Thus,

$$a_m < a_n < b_n < b_m \quad \text{or} \quad a_n < a_m < b_m < b_n.$$

Either way we see that $a_m < b_n$, whence b_n is an upper bound on the set $L = \{a_m | m \in \mathbb{N}\}$.

Let d be the least upper bound of L. For any n we have

$$a_n \leq d \leq b_n,$$

the first inequality holding because d is an upper bound for L and the latter because b_n is an upper bound. Thus, for all $n, d \in I_n$.   □

This completes the proof of Cantor's Theorem 2.1.

This has been something of a streamlined version of Cantor's proof. The practice in his day was to be more explicit in constructions, and the step where I said one should "choose I_{n+1} to be any closed subinterval of I_n which..." would be replaced by an explicit choice.[53] Beginners in higher mathematics may be more comfortable with a proof that is a little more explicit in its construction, even though the construction might complicate the proof somewhat. Cantor's explicit construction complicates the proof in the designation of I_{n+1}, as well as in the proof that c_n is not in the intersection.

He starts with an interval $[a, b]$ and a sequence $c : c_0, c_1, \ldots$ of distinct real numbers, and then sets out to define a sequence $I_0 \supseteq I_1 \supseteq \ldots$ of intervals. This is done inductively in stages.

Stage 0. Set $I_0 = [a, b]$ and define $a_0 = a, b_0 = b$.

---

[53] Adepts at set theory would also point to the apparent reliance on the Axiom of Choice, which allows one to combine all these individual choices of I_{n+1} into a single sequence $I_0, I_1, \ldots$ Cantor's proof was published in 1874 ("Über eine Eigenschaft des Inbegriffs aller reellen algebraischen Zahlen", *Journal für die reine und angewandte Mathematik* 77 (1874), pp. 258 - 262), decades before the Axiom was explicitly formulated or became an issue. In any event, its use here is unnecessary as one can choose I_{n+1} to be the first appropriate interval with rational endpoints under some enumeration of the set of real intervals with rational endpoints. Incidentally, the title of Cantor's paper translates to "On a property of the class of all real algebraic numbers". In this paper, Cantor first proves the set of real algebraic numbers to have the same cardinality as the set of natural numbers and then that the set of all real numbers does not have this cardinality. This gave a new proof of the existence of transcendental numbers. In fact, Cantor proved not merely that the cardinalities are different, but that the set of real numbers has a greater cardinality than the set of natural numbers and thus there are more transcendental than algebraic real numbers.

Stage $n + 1$. Given $I_n = [a_n, b_n]$. There are two cases.

Main case. There are at least two elements of the sequence $c_0, c_1, \ldots$ in the interior (a_n, b_n) of I_n. Choose the *first two*, c_m, c_k in the enumeration $c_0, c_1, \ldots$, and set

$$a_{n+1} = \min\{c_m, c_k\}, \quad b_{n+1} = \max\{c_m, c_k\}, \quad I_{n+1} = [a_{n+1}, b_{n+1}].$$

Auxiliary case. If there are not at least two elements of the sequence c in the interior of I_n, simply choose

$$a_{n+1} = a_n, \quad b_{n+1} = b_n, \quad I_{n+1} = I_n.$$

Note that, if at any stage $n+1$ in the construction the auxiliary case occurs, it will occur at every later stage and I_m will equal I_n for all $m > n$. The intersection of all the intervals will be I_n and its interior will contain at most one element c_k of the sequence $c_0, c_1, \ldots$ Any element d of the interior other than c_k will satisfy the demands of the Lemma.

If the main case occurs at every stage of the construction, one has an infinite nested sequence $I_0 \supseteq I_1 \supseteq \ldots$ of closed bounded intervals, which have a nonempty intersection by the Nested Interval Property. To see that any element d of this intersection satisfies Lemma 2.2, it suffices to show that no element of the sequence $c_0, c_1, \ldots$ is in the intersection, i.e. no c_n is in every I_m.

2.4 Lemma. *For all n, we have $c_0, \ldots, c_n \notin (a_{n+1}, b_{n+1})$.*

Proof. By induction on n.

Basis. If $c_0 \notin I_0$, then clearly $c_0 \notin (a_1, b_1) \subseteq I_0$. If, on the other hand, $c_0 \in I_0$, it is either an endpoint of the interval or an interior point. In the former case it is not in (a_1, b_1), because a_1, b_1 are interior points of I_0; in the latter case, c_0 is not in (a_1, b_1) because, being the first element of the list $c_0, c_1, \ldots$ in the interior of I_0, it is chosen to be one of a_1, b_1.

Induction step. From the assumption that $c_0, \ldots, c_n \notin (a_{n+1}, b_{n+1})$, it follows that $c_0, \ldots, c_n \notin (a_{n+2}, b_{n+2})$. The proof that $c_{n+1} \notin (a_{n+2}, b_{n+2})$ repeats the proof of the basis step. I leave the details to the reader to sort out.   □

Today one would opt for an expositionally more efficient construction. One needs to deal with closed intervals because the Nested Interval Property doesn't generally hold for open intervals. The introduction of the open intervals in Cantor's proof is an unnecessary complication. One has to think a bit (i.e. prove Lemma 2.4) to see that each c_n actually does fail to be in the final intersection. In his exposition[54], Herbert Meschkowski remained faithful to the abstract principles of Cantor's proof, but simplified the presentation through a simpler construction: At stage $n + 1$, one is given an interval $I_n = [a_n, b_n]$ and defines

$$I_n^1 = \left[a_n, a_n + \frac{1}{3}(b_n - a_n)\right]$$

---

[54] Herbert Meschkowski, *Hundert Jahre Mengenlehre*, Deutscher Taschenbuch Verlag, Munich, 1973, pp. 18 - 21.

$$I_n^2 = \left[a_n + \frac{1}{3}(b_n - a_n), a_n + \frac{2}{3}(b_n - a_n)\right]$$

$$I_n^3 = \left[a_n + \frac{2}{3}(b_n - a_n), b_n\right].$$

The number c_n cannot lie in all three of these intervals. Thus, if we define I_{n+1} to be the first of these not containing c_n, we automatically have $c_n \notin I_{n+1}$ and there is no need to prove an analogue to Lemma 2.4.

Meschkowski's version of the construction was probably suggested to him by another construction of Cantor's. The *Cantor Set*, or *Cantor Middle Third Set*, is obtained from the unit interval by successively deleting the middle thirds of intervals. Stated more explicitly, one starts with

$$C_0 = [0, 1]$$

and obtains C_1 by deleting the open interval $(\frac{1}{3}, \frac{2}{3})$:

$$C_1 = \left[0, \frac{1}{3}\right] \cup \left[\frac{2}{3}, 1\right].$$

One then deletes the middle thirds of the two intervals comprising C_1:

$$C_2 = \left[0, \frac{1}{9}\right] \cup \left[\frac{2}{9}, \frac{1}{3}\right] \cup \left[\frac{2}{3}, \frac{7}{9}\right] \cup \left[\frac{8}{9}, 1\right]$$

Continuing in this way, one generates a nested sequence $C_0 \supseteq C_1 \supseteq \ldots$ of closed bounded sets. The Cantor set is their intersection: $C = \bigcap_n C_n$. It can be shown that C has the same cardinality as the interval $[0, 1]$. However, the size of C can also measured. It is 1 (the length of $[0, 1]$) minus the sum of the lengths of all the deleted intervals. Now these deletions are at stage
1: 1 interval of length 1/3
2: 2 intervals of length 1/9
3: 4 intervals of length 1/27
and generally 2^{n-1} intervals of length $(1/3)^n$ at stage n. Thus altogether one has thrown away a set of measure

$$\frac{1}{3} + \frac{2}{3^2} + \frac{4}{3^3} + \ldots = \frac{1}{3}\left(1 + \frac{2}{3} + \left(\frac{2}{3}\right)^2 + \ldots\right)$$

$$= \frac{1}{3} \cdot \frac{1}{1 - \frac{2}{3}} = \frac{1}{3} \cdot \frac{3}{1} = 1,$$

and C has measure 0. This is about as strong a counterexample to the identification of cardinality and measure as one could hope for.[55]

---

[55] I have not defined what I mean here by the measure of a set. The reader may wish to consider this calculation as an exercise in Type I Formalism. I remark that there are good Type II definitions of the measure of a set of real numbers that apply to sets other than bounded intervals and finite unions of such, and that the above calculation is valid for these definitions.

I have now got far removed from the subject of Galileo. My only excuse for this is that I find these simple, but startling, results irresistibly appealing. For the reader who agrees, I have added some exercises at the end of this Chapter. For the rest, it is time to move on.

3 Taylor Series

3.1 Taylor Series Before Lagrange

A *power series* is a series of the form

$$s(x) = \sum_{k=0}^{\infty} a_k x^k, \tag{5}$$

or, more generally, of the form

$$s_a(x) = \sum_{k=0}^{\infty} a_k (x - a)^k. \tag{6}$$

Power series arose early in the history of the Calculus. Ordinary long division turned many rational functions into power series. We are already familiar with

$$\frac{1}{1+x} = 1 - x + x^2 - x^3 + \ldots \tag{7}$$

Even the more recalcitrant $1/x$ is amenable to such transformation:

$$\frac{1}{x} = \frac{1}{1 + (x-1)} = 1 - (x-1) + (x-1)^2 - (x-1)^3 + \ldots$$

Newton's precursors knew some other examples, perhaps most famously,

$$\tan^{-1} x = \int \frac{dx}{1 + x^2} = \int (1 - x^2 + x^4 - \ldots) dx$$
$$= C + x - \frac{x^3}{3} + \frac{x^5}{5} - \ldots$$
$$= x - \frac{x^3}{3} + \frac{x^5}{5} - \ldots, \tag{8}$$

where the constant of integration is determined to equal 0 by the identity $\tan^{-1} 0 = 0$.

The first general theorem on the representation of functions by expanding them into power series was due to Isaac Newton. This is his extension of the Binomial Theorem to the case of rational exponents.

The ordinary Binomial Theorem, already known some centuries earlier in China, owes its European discovery to Thomas Harriot and, more popularly, Blaise Pascal. It reads:

3.1 Theorem (Binomial Theorem). *For any $a, b \in \mathbb{R}$ and any positive integer n,*

$$(a + b)^n = \sum_{k=0}^{n} \binom{n}{k} a^{n-k} b^k, \tag{9}$$

where

$$\binom{n}{0} = 1, \quad \binom{n}{k} = \frac{n(n-1)\cdots(n-k+1)}{k(k-1)\cdots 1} \text{ for } k > 0.$$

Specialising $a = 1, b = x$ gives us a form more closely resembling (5):

$$(1 + x)^n = \sum_{k=0}^{n} \binom{n}{k} x^k. \tag{10}$$

In fact, since $\binom{n}{k} = 0$ for $k > n$ ($n - n$ is a factor of the numerator for such k), one can write this as a power series

$$(1 + x)^n = \sum_{k=0}^{\infty} \binom{n}{k} x^k. \tag{11}$$

Newton's great insight was that for $1 > x \geq -1$ (11) still holds when n is replaced by an arbitrary rational number:

3.2 Theorem (Newton's Binomial Theorem). *Let $1 > x \geq -1$ be a real number, and let q be any rational number. Then*

$$(1 + x)^q = \sum_{k=0}^{\infty} \binom{q}{k} x^k,$$

where

$$\binom{q}{0} = 1, \quad \binom{q}{k} = \frac{q(q-1)\cdots(q-k+1)}{k(k-1)\cdots 1} \text{ for } k > 0.$$

In a letter intended for Leibniz and written on 24 October 1676 to Henry Oldenburg, secretary of the Royal Society, Newton explained how he arrived at this result by interpolating terms in the sequences of coefficients of the series (11) for $n = 0, 1, 2, \ldots$[56] Newton cannot be said to have provided a proof, but he did make good use of the result.

One of Newton's applications was the derivation of a power series representation for the inverse sine:

---

[56] The essential parts of Newton's letter can be found in English translation in the source books of Dirk Struik, Ronald Calinger, and David Eugene Smith cited earlier in this chapter. It is also quoted in German translation in Richard Reiff, *Geschichte der unendlichen Reihen*, Verlag der H. Laupp'schen Buchhandlung, Tübingen, 1889, pp. 20 - 23. A few select quotes and a detailed analysis can be found in C.H. Edwards, Jr., *The Historical Development of the Calculus*, Springer-Verlag, New York, 1979, pp. 178 - 187.

$$\sin^{-1} x = \int \frac{dx}{\sqrt{1 - x^2}} = \int (1 - x^2)^{-1/2} dx$$

$$= \int \sum_{k=0}^{\infty} \binom{-1/2}{k} (-x^2)^k dx$$

$$= C + \sum_{k=0}^{\infty} (-1)^k \binom{-1/2}{k} \frac{x^{2k+1}}{2k+1}$$

$$= \sum_{k=0}^{\infty} (-1)^k \binom{-1/2}{k} \frac{x^{2k+1}}{2k+1}, \quad \text{since } \sin^{-1} 0 = 0$$

$$= x - \frac{-1/2}{1} \cdot \frac{x^3}{3} + \frac{(-1/2)(-3/2)}{2 \cdot 1} \cdot \frac{x^5}{5}$$

$$- \frac{(-1/2)(-3/2)(-5/2)}{3 \cdot 2 \cdot 1} \cdot \frac{x^7}{7} + \cdots$$

$$= \sum_{k=0}^{\infty} \frac{1 \cdot 3 \cdots (2k-1)}{2^k \cdot k!(2k+1)} x^{2k+1}, \tag{12}$$

where $0! = 1$, $k! = k(k-1)\cdots 1$ for $k > 0$ and $1 \cdot 3 \cdots (2k-1)$ is assumed to be 1 for $k = 0$.

From this one can find the power series representation for $\sin x$ by assuming the function written in the form (5), substituting it into the representation for $\sin^{-1}$,

$$x = \sin^{-1}(\sin x) = \sum_{k=0}^{\infty} \left(\frac{1 \cdot 3 \cdots (2k-1)}{2^k \cdot k!(2k+1)} \right) \left(\sum_{i=0}^{\infty} a_i x^i \right)^{2k+1},$$

expanding this expression and solving for $a_0, a_1, \ldots$, successively. Such a computation would not have daunted the likes of Newton and his contemporaries, and they would quickly have discerned a pattern to the coefficients $a_0, a_1, \ldots$ If I do not carry this out here, it is because Brook Taylor proved a theorem that for many functions made this massive computation unnecessary.

In modern terms, Taylor's Theorem says that under certain conditions a function f can be represented near a point a in its domain by a power series of the form

$$f(x) = f(a) + f'(a)(x - a) + \frac{f^{(2)}(a)}{2}(x - a)^2 + \cdots$$

$$= \sum_{k=0}^{\infty} \frac{f^{(k)}(a)}{k!}(x - a)^k, , \tag{13}$$

where $f^{(k)}$ denotes the k-th derivative of f (with $f^{(0)} = f$). Letting $h = x - a$, (13) can be written in a form more closely resembling (5):

$$f(a + h) = \sum_{k=0}^{\infty} \frac{f^{(k)}(a)}{k!} h^k. \tag{14}$$

In the special case where $a = 0$, we have $h = x$ and both (13) and (14) assume the special form

$$f(x) = \sum_{k=0}^{\infty} \frac{f^{(k)}(0)}{k!} x^k, \tag{15}$$

which more closely matches (5). Series of the form (13) and (14) are called the *Taylor series*, or the *Taylor series expansion* of f at a; the series (15) is called the *Maclaurin series (expansion)* of f.

Despite appearances, the Maclaurin series is no less general than the Taylor series. For, one can obtain the Taylor expansion of f at a by finding the Maclaurin expansion of $g(x) = f(x+a)$:

$$g(x) = \sum_{k=0}^{\infty} \frac{g^{(k)}(0)}{k!} x^k = \sum_{k=0}^{\infty} \frac{f^{(k)}(a)}{k!} x^k,$$

whence

$$f(x) = g(x-a) = \sum_{k=0}^{\infty} \frac{f^{(k)}(a)}{k!} (x-a)^k.$$

For this reason one can safely restrict one's attention to Maclaurin series in establishing results about Taylor series. If I do not do so, it is only to remain mildly faithful to the historical sources.

Before discussing what Taylor proved, let us pause to see how easily his formula yields the Maclaurin expansion of the sine function. For $f(x) = \sin x$ and any non-negative integer k, we have

$$f^{(4k)}(x) = \sin x,\ f^{(4k+1)}(x) = \cos x,\ f^{(4k+2)}(x) = -\sin x,\ f^{(4k+3)}(x) = -\cos x,$$

whence

$$f^{(4k)}(0) = 0,\quad f^{(4k+1)}(0) = 1,\quad f^{(4k+2)}(0) = 0,\quad f^{(4k+3)}(0) = -1,$$

and

$$\sin x = x - \frac{x^3}{3!} + \frac{x^5}{5!} - \frac{x^7}{7!} + \dots \tag{16}$$

This is much easier than first having to find the power series for $\sin^{-1} x$ and then (the step I did not do above) inverting the given series.

Similarly, the series for $f(x) = e^x$ is readily determined by the fact that $f^{(k)}(x) = e^x$ for all k:

$$e^x = 1 + x + \frac{x^2}{2!} + \frac{x^3}{3!} + \frac{x^4}{4!} + \dots \tag{17}$$

On the other hand, functions like $f(x) = \sin^{-1} x$ or $g(x) = \tan^{-1} x$, which are antiderivatives of rational functions, are more readily handled by expanding their derivatives into power series by long division and then integrating the series term-by-term. The reader might wish to verify this by calculating the first few derivatives of the inverse sine function and guessing the general term of the Maclaurin expansion from these.

Stating exactly what Taylor proved is a little difficult. The formulæ (13) and (14) are expressed as the equality of functions. He did not state his result in terms of functions, but rather in terms of a relation between variables y and x.[57] Curves were often presented either geometrically as loci or kinematically as the paths traced out by points with simultaneous horizontal and vertical movement. Taylor expressed his version of (13) and (14) kinematically in terms of Newtonian *fluxions*, i.e. in terms of the k-th derivatives of the two variable motions with respect to time. In doing so, he imposed the extra condition that x be a uniformly increasing function of time. That it be increasing guarantees that the path will not double back and thus y will be a function of x. That the increase be uniform pretty much means that the fluxions of y are the derivatives with respect to x.

With a slight change of notation, we could almost say that Taylor proved his eponymously named theorem in the form of equation (14). I say "almost" because such a statement does not explain what functions he can be said to have proven the equation for or what sense he gave to the equality. The modern interpretation of the equation would be that both sides exist for the same values of x and are equal when defined. This is known to be false and was known to be false in Taylor's day. (Consider (7).) This could be compared to equations involving rational functions, e.g.

$$\frac{x-1}{x^2-1} = \frac{1}{x+1}.$$

The two expressions are not defined for the same values of x; that on the right is defined for $x = 1$, while that on the left is not. What is true is that where both expressions are defined they have the same value. This, however, fails for (14), but this fact would not be known for over a century until 1822, when Augustin Louis Cauchy published the example,

$$f(x) = \begin{cases} e^{-1/x^2}, & x \neq 0, \\ 0, & x = 0 \end{cases},$$

which has derivatives of all orders at 0: $f^{(k)}(0) = 0$ for all k. Its Maclaurin series is thus $\sum_{k=0}^{\infty} 0x^k = 0$ and converges to $f(x)$ only at the point $x = 0$.

Today one either states carefully conditions under which the Taylor series converges to a given function and proves that result, or one defines the Taylor series expansion via (14) and writes something like

$$f(x) \sim \sum_{k=0}^{\infty} \frac{f^{(k)}(a)}{k!}(x-a)^k$$

or

---

[57] He denoted these by x and z, respectively. An English translation of the relevant passage of Taylor's book, *Methodus Incrementorum directa et inversa* (published 1716, but dated 1717), appears in Struik's source book (*op. cit.*) and is reprinted in Calinger's similar compilation (*op. cit.*) Both volumes contain historical remarks putting the result into historical perspective. Reiff's book on infinite series (*op. cit.*, pp. 80 - 88) gives an, in some ways, fuller historical treatment. I also recommend Edwards's history of the Calculus (*op. cit.*) for a detailed account of Taylor series.

$$f(x + h) \sim \sum_{k=0}^{\infty} \frac{f^{(k)}(x)}{k!} h^k$$

to indicate the series on the right to be the expansion of the function. In Taylor's day, the use of equality could mean something like this as we saw back in Chapter I when Euler spoke of the value of a divergent series. Or, perhaps it could simply mean that the right-hand side was the end result of an infinite expansion carried out formally.

Taylor's Theorem was not an immediate success. His book was not clearly written, and Taylor himself did not make much use of the result. According to Reiff, "Taylor himself did not fully recognise the significance of the formula, for its applications in the book are sparse and a treatment in the *Philosophical Transactions* on the numerical solution of equations is the only later writing of Taylor's known to us based on his formula."[58] The situation changed a quarter century later when the Scottish mathematician Colin Maclaurin published his *Treatise on Fluxions* (1742). Maclaurin's book lay greater emphasis on Taylor's series, in the form (15), and gave a simplified attempted proof that survives today. The proof implicitly rested on two assumptions: i) every function f can be represented in the form (5), and ii) a power series can be differentiated term-by-term. The first assumption is, of course, false, but the second assumption is true albeit not trivial. Nonetheless, Maclaurin's proof does show that, if a function has a power series representation of the form,

$$f(x) = a_0 + a_1 x + a_2 x^2 + \ldots, \tag{18}$$

then this series is given by (15):

$$a_k = \frac{f^{(k)}(0)}{k!}. \tag{19}$$

One shows this by successively differentiating (18):

$$f'(x) = a_1 + 2a_2 x + 3a_3 x^2 + \ldots$$
$$f^{(2)}(x) = 2a_2 + 3 \cdot 2a_3 x + 4 \cdot 3a_4 x^2 + \ldots$$

and, in general,

$$f^{(k)}(x) = k!a_k + (k+1)\cdots 2a_{k+1} x + (k+2)\cdots 3a_{k+2} x^2 + \ldots,$$

whence $f^{(k)}(0) = k!a_k$, i.e. (19) holds.

Maclaurin follows his derivation with the curious remark that "this proposition may be likewise deduced from the binomial theorem".[59] The proof that

---

[58] Reiff, *op. cit.*, p. 82.

[59] The relevant passages appear in Struik, *op. cit.*, and are reprinted in Calinger, *op. cit.* Reiff, *op. cit*, discusses Maclaurin's proof and his rôle in the history of Taylor's Theorem, but does not mention the Binomial Theorem in this connexion. The same is true of Edwards, *op. cit.*

suggests itself is a bit more modern than what Maclaurin probably had in mind, but abstractly it pretty much captures the spirit of Taylor's earlier proof.

Recall the difference operator Δ from page 71 and generalise it to an arbitrary increment by

$$\Delta_h(f)(x) = f(x+h) - f(x).$$

And, while we are at it, let us introduce the operators

$$E_h(f)(x) = f(x+h), \qquad I(f)(x) = f(x).$$

Define the sum of two operators F, G to be their "pointwise" sum,

$$(F+G)(f) = F(f) + G(f),$$

and their product to be their composition,

$$(FG)(f) = F(G(f)).$$

Notice that

$$E_{1h}(f)(x) = f(x+h) = E_h(f)(x)$$
$$E_{2h}(f)(x) = f(x+2h) = f((x+h)+h) = E_h(f)(x+h) = E_h(E_h(f))(x)$$

and generally,

$$E_{nh}(f)(x) = (E_h)^n(f)(x).$$

Note too that

$$E_h = I + \Delta_h.$$

Thus,

$$E_{nh} = (E_h)^n = (I + \Delta_h)^n$$
$$= \sum_{k=0}^{n} \binom{n}{k} I^{n-k} \Delta_h{}^k = \sum_{k=0}^{n} \binom{n}{k} \Delta_h{}^k, \tag{20}$$

by the Binomial Theorem, where $\Delta_h{}^0 = I^0 = I$.

Now suppose we are given f and x. Choose n to be any positive integer and subdivide the interval $[0, x]$ into n equal subintervals of length $h = x/n$. Thus $x = nh$ and

$$f(x) = E_x(f)(0) = E_{nh}(f)(0).$$

But (20) yields

$$E_x = E_{nh} = \sum_{k=0}^{n} \frac{n(n-1)\cdots(n-k+1)}{k!} \Delta_h{}^k$$
$$= \sum_{k=0}^{n} \frac{1 \cdot (1 - \frac{1}{n}) \cdots (1 - \frac{k-1}{n})}{k!} n^k \Delta_h{}^k$$
$$= \sum_{k=0}^{n} \frac{1 \cdot (1 - \frac{1}{n}) \cdots (1 - \frac{k-1}{n})}{k!} x^k \frac{\Delta_h{}^k}{h^k}, \tag{21}$$

since $n = x/h$. Letting n be infinite,

$$E_x = E_0^\infty = \sum_{k=0}^\infty \frac{1(1-\frac{1}{\infty})\cdots(1-\frac{k-1}{\infty})}{k!} x^k D^k \tag{22}$$

$$= \sum_{k=0}^\infty \frac{x^k}{k!} D^k, \tag{23}$$

where D is the differentiation operator. Thus

$$f(x) = E_x(f)(0) = \sum_{k=0}^\infty \frac{x^k}{k!} D^k f(0) = \sum_{k=0}^\infty \frac{f^{(k)}(0)}{k!} x^k.$$

The formal application of the Binomial Theorem is actually unproblematic. It can easily be justified rigorously in this case (Cf. Exercise 6.7 at the end of this chapter.) or dispensed with entirely as was done by Taylor. The real problem is the passage from (21) to (22), where infinitely many limits are being taken simultaneously. The handling of the individual limits is not hard. For fixed k, one shows using elementary techniques that

$$\lim_{n\to\infty} \frac{1(1-\frac{1}{n})\cdots(1-\frac{k-1}{n})}{k!} =$$
$$= \frac{(\lim_{n\to\infty} 1)(\lim_{n\to\infty}(1-\frac{1}{n}))\cdots(\lim_{n\to\infty}(1-\frac{k-1}{n}))}{k!}$$
$$= \frac{1\cdot 1\cdots 1}{k!} = \frac{1}{k!}.$$

The evaluation of

$$\lim_{h\to 0} \frac{\Delta_h{}^k f(0)}{h^k} = f^{(k)}(0)$$

requires an iterated application of L'Hôpital's Rule: If g, h are functions for which

$$\lim_{x\to a} g(x) = \lim_{x\to a} h(x) = 0,$$

then

$$\lim_{x\to a} \frac{g(x)}{h(x)} = \lim_{x\to a} \frac{g'(x)}{h'(x)},$$

provided the latter limit exists.

For the application in mind, one notes that $\Delta_h = E_h - I$, whence

$$\Delta_h{}^k = (E_h - I)^k = \sum_{i=0}^k \binom{k}{i} E^{k-i}(-I)^i,$$

and

$$\Delta_h{}^k f(0) = \sum_{i=0}^k (-1)^i \binom{k}{i} f((k-i)h).$$

The derivative with respect to h is

$$\sum_{i=0}^{k}(-1)^i(k-i)\binom{k}{i}f'((k-i)h) = \sum_{i=0}^{k-1}(-1)^i(k-i)\binom{k}{i}f'((k-i)h).$$

Thus, L'Hôpital's Rule yields

$$\lim_{h\to 0}\frac{\Delta_h^k f(0)}{h^k} = \lim_{h\to 0}\sum_{i=0}^{k-1}\frac{(-1)^i(k-i)\binom{k}{i}f'((k-i)h)}{k\cdot h^{k-1}}. \qquad (24)$$

But

$$\frac{k-i}{k}\binom{k}{i} = \frac{(k-i)k(k-1)\cdots(k-i+1)}{k\cdot i!}$$

$$= \frac{(k-1)\cdots(k-i+1)(k-i)}{i!} = \binom{k-1}{i},$$

whence (24) yields

$$\lim_{h\to 0}\frac{\Delta_h^k f(0)}{h^k} = \lim_{h\to 0}\frac{\sum_{i=0}^{k-1}(-1)^i\binom{k-1}{i}f'((k-i)h)}{h^{k-1}}$$

$$= \lim_{h\to 0}\frac{\Delta_h^{k-1}f'(0)}{h^{k-1}}.$$

Iteration shows that, for $i\le k$,

$$\lim_{h\to 0}\frac{\Delta_h^k f(0)}{h^k} = \lim_{h\to 0}\frac{\Delta_h^{k-i}f^{(k)}(0)}{h^{k-i}},$$

which equals $f^{(k)}(0)$ for $i=k$.

All of this is very nice but it still doesn't justify the step from (21) to (22). Indeed, Struik quotes Felix Klein's description of this passage as "a transition to the limit of extraordinary audacity"[60]. No modern textbook treatment of Taylor's Theorem attempts such an approach. The typical textbook proof today proceeds by estimating the difference

$$R_n(x,h) = f(x+h) - \sum_{k=0}^{n}\frac{f^{(k)}(x)}{k!}h^k.$$

The estimate of this difference was first given by Joseph-Louis Lagrange.

3.2 Lagrange

Taylor's and Maclaurin's proofs can safely be discussed *in vacuo*. They make sense with or without reference to their historical setting. Taylor's derivation is,

---

[60] *Op. cit.*, p. 332.

if nothing else, an heuristic determination of the coefficients[61], and Maclaurin's derivation is a correct proof that the coefficients of a convergent power series are the derivatives divided by the appropriate factorials. Sense can also be made of Lagrange's work without placing it in context, but one will not fully appreciate it.[62]

The thing that must be considered is the notion of function. There are a couple of good sources on the development of the notion of function. First, there is a pair of papers which I group together because of the similarity of title and common journal of publication:

A.P. Youschkevitch, "The concept of function up to the middle of the 19th century", *Archive for History of Exact Sciences* 16 (1976/77), pp. 37 - 85.

A.F. Monna, "The concept of function in the 19th and 20th centuries, in particular with regard to the discussions between Baire, Borel, and Lebesgue", *Archive for History of Exact Sciences* 9 (1972), pp. 57 - 84.

In the 1930s, Nikolai Nikolaievich Luzin published an article dealing with the subject in *The Great Soviet Encyclopedia*. Three quarters of a century later, it was published in 2 parts in English translation and several follow up articles appeared:

N. Luzin, "Function: Part I", *American Mathematical Monthly* 105 (1998), pp. 59 - 67; "Function: Part II", *American Mathematical Monthly* 105 (1998), pp. 263 - 270.

S.S. Demidov, "Two letters by N.N. Luzin to M. Ya. Vygodskiĭ", *American Mathematical Monthly* 107 (2000), pp. 64 - 82.

Detlef Laugwitz, "Comments on the paper 'Two letters by N.N. Luzin to M. Ya. Vygodskiĭ'" *American Mathematical Monthly* 107 (2000), pp. 267 - 276.

Jeremy Gray, "Weierstrass, Luzin and Intuition", *American Mathematical Monthly* 108 (2001), pp. 865 - 870.

The first pair of these papers is more than adequate for our present purposes; the second group, however, is quite relevant for the purposes of Chapter V, below, on reactions to formalism. I thus recommend all these references to the reader of more than average curiosity.

In the interest of a self-contained exposition, I cannot tell the reader to interrupt his or her reading and go read these papers. Nor can I reproduce in the space allotted all the relevant detail; I must perforce be brief and, in being so, recite claims that may seem a little odd without the supporting facts. Nonetheless, here in brief (or, in caricature) is the history of the notion of function in the 18th century: At the beginning of the century, the word "function", having been introduced into mathematics in a paper of Leibniz in 1692, was in use, but without a definite meaning. The correlation of a dependent variable y with an independent variable x was noted, but the emphasis of finding formulæ and

---

[61] Despite the currently fashionable historian's condescending mantra of "sufficient unto the rigour of the day", it can in no way be considered a proof.

[62] In *The Origins of Cauchy's Rigorous Calculus*, MIT Press, Cambridge (Mass), 1981, p. 117, Judith V. Grabiner summed it up nicely by stating, "Lagrange's work is almost, but not quite, acceptable by modern standards".

describing the law governing the correlation led to the notion of a function as being given by an *analytic expression*: $f(x) = \ldots x \ldots$ An analytic expression was an expression built up from variables and constants by addition, subtraction, multiplication, division, and eventually the taking of roots, infinite sums, etc. A function was *continuous* if it was given throughout by a single analytic expression; it was *discontinuous* or *mixed* if it was piecewise defined by different expressions on different intervals. Cauchy would later criticise this definition by pointing to the discontinuous function,

$$|x| = \begin{cases} x, & x \geq 0 \\ -x, & x < 0, \end{cases}$$

which became continuous when written $|x| = \sqrt{x^2}$. The study of counterexamples only began in the 19th century. The functions of the 18th century were fairly well-behaved. Except for finitely many singularities, any function encountered had derivatives of all orders and could be expanded into a power series around any but these few points. The power series might not always converge, but this wasn't really relevant. Even in applications, where numerical results were needed, convergence was not required; what was required was an estimate of the error in using the first few terms of a series.

One more thing: functions defined by analytic expressions could be expanded formally into power series and, by Taylor or Maclaurin, the series was completely determined by the behaviour of the function in a small neighbourhood of the point around which the function was being expanded. Two such *analytic functions* agreeing in such a neighbourhood were equal.

The period from the middle of the 16th century to the end of the 18th century was an heroic age of discovery in mathematics. Intuition and invention far outstripped logic and rigour. Essentially all of mathematics was Type I formalism, the unjustified, uncodified manipulation of symbols to get results which could be checked numerically but not proven. Lagrange reasoned that mathematics and its applications only needed analytic functions and could be founded upon them by using infinite series. One needn't appeal to the vague notion of limit and the oft-criticised use of infinitesimals. In 1797 his book *La Théorie des Fonctions analytiques* [*Theory of analytic functions*] was published.

In this book Lagrange attempted to prove all functions expandable into power series about all but a few exceptional points. Needless to say he failed. What he did do, however, was develop a theory of functions that are so expandable. In doing so, he made some strong assumptions. In particular, if

$$f(x+h) = \sum_{k=0}^{\infty} a_k h^k,$$

he assumed all the a_k's to be analytic functions of x: $a_0 = a_0(x) = f(x)$. He called the function $a_1(x)$ the *first derived function* of f, writing $a_1(x) = f'(x)$ (introducing the notation by the way). Being analytic itself, $a_1(x)$ would have its own derived function a_1', which he called the *second derived function* of f, $f^{(2)} = f'' = a_1'$. This, of course, would be continued down the line.

This definition served as the basis for a sort of inductive algebraic determination of the coefficients of the power series expansion of f, i.e. an algebraic "proof" of Taylor's Theorem. Starting from the assumption,

$$f(x+h) = \sum_{k=0}^{\infty} a_k(x)h^k, \tag{25}$$

Lagrange added another increment o:

$$f(x+(h+o)) = \sum_{k=0}^{\infty} a_k(x)(h+o)^k$$

$$= \sum_{k=0}^{\infty} a_k(x)\left(\binom{k}{0}h^k + \binom{k}{1}h^{k-1}o + \ldots + \binom{k}{k}o^k \right)$$

$$= \sum_{k=0}^{\infty} \binom{k}{0}a_k(x)h^k + \left(\sum_{k=1}^{\infty} \binom{k}{1}a_k(x)h^{k-1} \right) o +$$

$$+ \left(\sum_{k=2}^{\infty} \binom{k}{2}a_k(x)h^{k-2} \right) o^2 + \ldots$$

$$= f(x+h) + \left(a_1(x) + 2a_2(x)h + 3a_2(x)h^2 + \ldots \right) o +$$

$$+ \left(a_2(x) + 3a_3(x)h + 6a_4(x)h^2 + \ldots \right) o^2 + \ldots \tag{26}$$

But $x+(h+o) = (x+h)+o$, whence

$$f(x+(h+o)) = f((x+h)+o) = \sum_{k=0}^{\infty} a_k(x+h)o^k, \tag{27}$$

whence, comparing coefficients in (26) and (27), we see

$$a_1(x+h) = a_1(x) + 2a_2(x)h + 3a_2(x)h^2 + \ldots,$$

whence

$$f''(x) = a_1'(x) = 2a_2(x), \text{ i.e. } a_2(x) = \frac{f'(x)}{2}.$$

Again,

$$a_2(x+h) = a_2(x) + 3a_3(x)h + 6a_4(x)h^2 + \ldots,$$

whence

$$a_2'(x) = 3a_3(x), \text{ i.e. } a_3(x) = \frac{a_2'(x)}{3} = \frac{1}{3}\frac{a_1''(x)}{2} = \frac{f^{(3)}(x)}{3 \cdot 2}$$

etc.[63]

---

[63] English translations of Lagrange's exposition of the proof appear in Struik, *op. cit.*, pp. 389 - 395, and in John Fauvel and Jeremy Gray, *The History of Mathematics; A Reader*, Macmillan Education Ltd, London, 1987, pp. 561 - 562. It is also sketched in Reiff, *op. cit.*, p. 180, Edwards, *op. cit.*, pp. 246 - 247, and Jerome H. Manheim, *The Genesis of Point Set Topology*, Pergamon Press, Oxford, and The Macmillan Company, New York, 1964.

This can hardly be said to be an improvement on Maclaurin's proof. The most important contribution Lagrange made to Taylor's Theorem in *La Théorie des Fonctions analytiques* was his estimate of the error,

$$R_n(x, h) = f(x + h) - \sum_{k=0}^{n} \frac{f^{(k)}(x)}{k!} h^k, \tag{28}$$

in approximating $f(x+h)$ by using the first $n+1$ terms of the Taylor expansion of f at x. He returned to the subject in 1799 in presenting his *Leçons sur le Calcul des Fonctions* (published 1801) giving a different proof. The first of these proofs strikes me as a little more involved, so I will present the latter.[64]

In modern terms, Lagrange's proof yields the following:

3.3 Theorem (Taylor's Theorem with Remainder). *Let $f, f', f^{(2)}, \ldots,$ $f^{(n+1)}$ all be defined and continuous on a closed bounded interval I. Let $p, q \in I$ be such that $f^{(n+1)}(p)$ is the minimum and $f^{(n+1)}(q)$ the maximum value of $f^{(n+1)}$ on I. Then, for any $x, x + h \in I$, $f(x + h)$ lies between*

$$\sum_{k=0}^{n} \frac{f^{(k)}(x)}{k!} h^k + \frac{h^{n+1}}{(n+1)!} f^{(n+1)}(p)$$

and

$$\sum_{k=0}^{n} \frac{f^{(k)}(x)}{k!} h^k + \frac{h^{n+1}}{(n+1)!} f^{(n+1)}(q).$$

In other words, $R_n(x, h)$ lies between

$$\frac{h^{n+1}}{(n+1)!} f^{(n+1)}(p) \quad and \quad \frac{h^{n+1}}{(n+1)!} f^{(n+1)}(q).$$

3.4 Corollary. *Under the conditions of the Theorem, one has*

$$f(x + h) = \sum_{k=0}^{n} \frac{f^{(k)}(x)}{k!} h^k + \frac{h^{n+1}}{(n+1)!} f^{(n+1)}(x_0)$$

for some x_0 between x and $x + h$, i.e.

$$R_n(x, h) = \frac{h^{n+1}}{(n+1)!} f^{(n+1)}(x_0)$$

for some x_0 between x and $x + h$.

---

[64] Both of his books are rather old, but they were reprinted (*Fonctions analytiques* in the second, 1813 edition) in his collected works, which are accessible online. If one doesn't like to read French, the 1797 proof is discussed in German in Reiff, *op. cit.*, pp. 154 - 155, and in English in greater detail in Edwards, *op. cit.*, pp. 297 - 299; the proof from the *Leçons* appears in English translation in Jean-Luc Chabert, *A History of Algorithms; From the Pebble to the Microchip*, Springer-Verlag, Berlin, 1999, pp. 409 - 411.

For, by continuity of $f^{(n+1)}$, the Intermediate Value Theorem applies.[65]

Today we use Theorem 3.3 or Corollary 3.4 to prove the convergence of the Taylor series of the function to the function provided the derivatives do not grow too rapidly:

3.5 Corollary. *Let f be infinitely differentiable in a closed bounded interval I. If there is a uniform bound M on $|f^{(n)}(z)|$ for all $n \in \mathbb{N}$ and $z \in I$, then for each x, h for which $x, x + h \in I$, the Taylor series of f at x converges and*

$$f(x + h) = \sum_{k=0}^{\infty} \frac{f^{(k)}(x)}{k!} h^k.$$

Proof. Observe

$$\left| f(x + h) - \sum_{k=0}^{n} \frac{f^{(k)}(x)}{k!} h^k \right| = |R_n(x, h)| = \frac{|h|^{n+1}}{(n+1)!} \left| f^{(n+1)}(x_0) \right| \tag{29}$$

for some x_0 between x and $x + h$. But by assumption,

$$\frac{|h|^{n+1}}{(n+1)!} \left| f^{(n+1)}(x_0) \right| \leq \frac{|h|^{n+1}}{(n+1)!} M < \frac{B^{n+1} M}{(n+1)!}, \tag{30}$$

where B is, say, the length of I. Choose $n_0 > B$ and observe that for $n > n_0$,

$$\frac{B^{n+1} M}{(n+1)!} = \frac{B}{n+1} \cdot \frac{B}{n} \cdots \frac{B}{n_0 + 1} \cdot \frac{B^{n_0} M}{n_0!}$$

$$< \frac{B}{n_0 + 1} \cdot \frac{B}{n_0 + 1} \cdots \frac{B}{n_0 + 1} \cdot constant$$

$$< \left(\frac{B}{n_0 + 1} \right)^{n - n_0 + 1} \cdot constant, \tag{31}$$

which can be made as small as we like by choosing $n > n_0 + 1$ sufficiently large. Combining (29) - (31) shows

$$\lim_{n \to \infty} \sum_{k=0}^{n} \frac{f^{(k)}(x)}{k!} h^k = f(x + h). \qquad \square$$

From Corollary 3.5 we can conclude that the Taylor series of many functions converge to the functions. For example, for $f(x) = \sin x$, the derivatives are always $\pm \sin x$ or $\pm \cos x$, which are uniformly bounded in absolute value by $M = 1$ on any interval $[0, x]$ (or, if x is negative, $[x, 0]$). Substituting 0 for x

---

[65] The Intermediate Value Theorem was first proved for continuous functions by Bolzano in 1817. Before that it was fairly widely believed and, according to Monna's paper cited above (pp. 62 - 63), was often taken in a rather strong form as the definition of continuity— even after Bolzano and Cauchy had defined the notion "correctly", i.e. after they had given the modern definition. (Well, almost— cf. the discussion of Cauchy in Chapter III, section 6, below (pp. 361 - 373).)

and x for $x + h$ in the Corollary, we see that the Maclaurin series (16) for $\sin x$ converges to $\sin x$ for all $x \in \mathbb{R}$. Similarly, while the successive derivatives of $f(x) = e^x$ are not bounded on $\mathbb{R}$, they are uniformly bounded on each closed bounded interval $I \subseteq \mathbb{R}$. Hence the Maclaurin series (17) for e^x converges everywhere to e^x.

As for the proof of Theorem 3.3, Lagrange did not offer a full proof of the result but only for the cases $n = 0, 1, 2$ and positive h. This is certainly sufficient to convince anyone of the general truth of the theorem, but formally it is an incomplete ritual and falls short.

Let us first examine what he did. Let f, p, q, n be given as in the statement of the Theorem. Let $x, x + h$ be in the given interval and assume h is positive. By the Fundamental Theorem of Calculus,

$$f^{(n)}(x + h) - f^{(n)}(x) = \int_x^{x+h} f^{(n+1)}(t)dt.$$

But

$$hf^{(n+1)}(p) = \int_x^{x+h} f^{(n+1)}(p)dt$$

$$\leq \int_x^{x+h} f^{(n+1)}(t)dt$$

$$\leq \int_x^{x+h} f^{(n+1)}(q)dt = hf^{(n+1)}(q),$$

whence

$$f^{(n)}(x + h) - f^{(n)}(x) - hf^{(n+1)}(p) \geq 0$$

$$hf^{(n+1)}(q) + f^{(n)}(x) - f^{(n)}(x + h) \geq 0.$$

The primitives of these with respect to h are

$$f^{(n-1)}(x + h) - hf^{(n)}(x) - \frac{h^2}{2}f^{(n+1)}(p) + C_1$$

$$\frac{h^2}{2}f^{(n+1)}(q) + hf^{(n)}(x) - f^{(n-1)}(x + h) + C_2.$$

Choosing C_1, C_2 making these equal to 0 for $h = 0$, we get

$$C_1 = -f^{(n-1)}(x), \qquad C_2 = f^{(n-1)}(x),$$

i.e. we get the primitives,

$$f^{(n-1)}(x + h) - f^{(n-1)}(x) - hf^{(n)}(x) - \frac{h^2}{2}f^{(n+1)}(p)$$

$$\frac{h^2}{2}f^{(n+1)}(q) + hf^{(n)}(x) + f^{(n-1)}(x) - f^{(n-1)}(x + h),$$

with both of these quantities being non-negative for $h \geq 0$ as they are 0 at $h = 0$ and their derivatives are non-negative.

Continuing in this manner, one would show for $1 \leq m \leq n+1$,

$$\frac{f^{(n+1)}(p)}{m!}h^m \leq f^{(n+1-m)}(x+h) - \sum_{k=0}^{m-1}\frac{f^{(n+k+1-m)}(x)}{k!}h^k \leq \frac{f^{(n+1)}(q)}{m!}h^m \quad (32)$$

by induction on m, concluding the Theorem on setting $m = n+1$ in the case where h is positive. The changes for h negative, and indeed the verbal statement of the Theorem using the word "between" instead of inequalities, are brought about by the fact that when h is negative the inequalities in (32) reverse in going from m to $m+1$ (note that h^m changes sign). Note too that each line involving p is repeated, with the terms negated and inequalities reversed for q. All in all, Lagrange's proof is an inefficient organisation of its ideas. One can do better by moving the appeal to the Intermediate Value Theorem up from the Corollary following the proof to the first step of the proof: From

$$f^{(n)}(x+h) - f^{(n)}(x) = \int_x^{x+h} f^{(n+1)}(t)dt$$

we conclude via the Mean Value Theorem for Integrals (or the Intermediate Value Theorem applied to $hf^{(n+1)}(t)$),

$$f^{(n)}(x+h) - f^{(n)}(x) = hf^{(n+1)}(x_0) \quad (33)$$

for some x_0 between x and $x+h$.

3.6 Lemma. *For $1 \leq m \leq n+1$,*

$$f^{(n+1-m)}(x+h) - \sum_{k=0}^{m-1}\frac{f^{(m+k+1-m)}(x)}{k!}h^k = \frac{f^{(n+1)}(x_0)}{m!}h^m. \quad (34)$$

Note that no assumption on the positivity of h is made and that in place of two inequalities, a single equation now occurs.

Proof. By induction on m.

Basis. $m = 1$. This is (33).

Induction step. Assume (34) holds for m, replace h by t and integrate between 0 and h:

$$\frac{f^{(n+1)}(x_0)}{(m+1)!}h^{m+1} =$$

$$= f^{(n+1-m-1)}(x+h) - f^{(n+1-m-1)}(x) - \sum_{k=0}^{m-1}\frac{f^{(n+k+1-m)}(x)}{k!}\cdot\frac{h^{k+1}}{k+1}$$

$$= f^{(n+1-m-1)}(x+h) - f^{(n+1-m-1)}(x) - \sum_{k=1}^{m}\frac{f^{(n+k-m)}(x)}{k!}h^k$$

$$= f^{(n+1-m-1)}(x+h) - \sum_{k=0}^{m} \frac{f^{(n+k-m)}(x)}{k!} h^k. \qquad \square$$

And, of course, setting $m = n+1$ yields

$$f^{(0)}(x+h) - \sum_{k=0}^{n} \frac{f^{(n+k+1-n-1)(x)}}{k!} h^k = \frac{f^{(n+1)}(x_0)}{(n+1)!} h^{n+1},$$

i.e.

$$f(x+h) - \sum_{k=0}^{n} \frac{f^{(k)}(x)}{k!} h^k = \frac{f^{(n+1)}(x_0)}{(n+1)!} h^{n+1},$$

which is just Corollary 3.4 and which is equivalent to the Theorem.

4 Euler

When I was a student the list of all time great mathematicians had three names on it: Archimedes, Isaac Newton, and Carl Friedrich Gauss. Leonhard Euler's name was not on the list. In recent decades there has been a change of opinion and some would place Euler at the top of the list. He was certainly the most prolific mathematician who ever lived, his breadth the broadest, and his influence on succeeding generations among the greatest. However, he lived in an age of formalism and his formal proofs do not quite pass today's muster. Amazingly, some of the oddest of them are nonetheless almost correct— and, here I refer not just to the results but the proofs themselves. It is time for us to approach the master and examine some of his formal proofs.

4.1 The Binomial Theorem

As I mentioned in the preceding section, Newton did not prove his form of the Binomial Theorem. He derived the formula by an interpolation scheme, a good strategy for discovery but not a proof method. On and off throughout the 18th century there were attempts to prove the Theorem. Euler was no exception and tried several times to prove it. I shall describe his proof of 1774[66].

For any rational number p, let $[p]$ denote the power series

$$[p] = \sum_{k=0}^{\infty} \binom{p}{k} x^k. \qquad (35)$$

4.1 Lemma. *For all $p, q \in \mathbb{Q}$, $[p] \cdot [q] = [p+q]$.*

---

[66] Not knowing Latin, I've not consulted Euler's paper itself. For the present proof I rely on the sketch given by Klaus Volkert, *Geschichte der Analysis*, Bibliographisches Institut, Mannheim, 1988, p. 147.

Proof. Observe,

$$[p] \cdot [q] = \left(\sum_{k=0}^{\infty} \binom{p}{k} x^k \right) \left(\sum_{k=0}^{\infty} \binom{q}{k} x^k \right)$$

$$= \sum_{k=0}^{\infty} \left(\sum_{i=0}^{k} \binom{p}{i} \binom{q}{k-i} \right) x^k, \tag{36}$$

where we simply collect all the x^k terms. Now if we examine the coefficients of x^k, we see they are

$$\binom{p}{0}\binom{q}{k} + \binom{p}{1}\binom{q}{k-1} + \ldots + \binom{p}{k}\binom{q}{0}.$$

For $k = 0$, this is

$$\binom{p}{0}\binom{q}{0} = 1 = \binom{p+q}{0},$$

for $k = 1$,

$$\binom{p}{0}\binom{q}{1} + \binom{p}{1}\binom{q}{0} = 1 \cdot q + p \cdot 1 = p + q = \binom{p+q}{1},$$

and for $k = 2$,

$$\binom{p}{0}\binom{q}{2} + \binom{p}{1}\binom{q}{1} + \binom{p}{0}\binom{q}{2} = 1 \cdot \frac{q(q-1)}{2 \cdot 1} + p \cdot q + \frac{p(p-1)}{2 \cdot 1} \cdot 1$$

$$= \frac{q^2 - q + 2pq + p^2 - p}{2}$$

$$= \frac{(q+p)^2 - q - p}{2} = \frac{(q+p)(q+p-1)}{2}$$

$$= \binom{q+p}{2} = \binom{p+q}{2}.$$

Thus we see the correct pattern and it only remains to verify by induction on k that

$$\sum_{i=0}^{k} \binom{p}{i} \binom{q}{k-i} = \binom{p+q}{k}. \tag{37}$$

To prove (37), we first introduce a bit of notation. For any real number a and any positive integer j,

$$a^{(j)} = a(a-1) \cdots (a-j+1),$$

i.e. $a^{(j)}$ is the product of j factors beginning with a and each factor being 1 less than the previous one. We also define $a^{(0)} = 1$. Then

$$\binom{p}{i} = \frac{p(p-1) \cdots (p-i+1)}{i!} = \frac{p^{(i)}}{i!}, \quad \binom{q}{k-i} = \frac{q^{(k-i)}}{(k-i)!}$$

and

$$\binom{p}{i}\binom{q}{k-i} = \frac{p^{(i)}q^{(k-i)}}{i!(k-i)!} = \frac{k!}{i!(k-i)!} \cdot \frac{p^{(i)}q^{(k-i)}}{k!}$$
$$= \binom{k}{i}\frac{p^{(i)}q^{(k-i)}}{k!}.$$

Now, for the induction, note that

$$\binom{p+q}{k+1} = \frac{(p+q)^{(k+1)}}{(k+1)!} = \frac{(p+q)\cdots(p+q-k+1)(p+q-k)}{(k+1)k!}$$

$$= \frac{p+q-k}{k+1}\binom{p+q}{k}$$

$$= \frac{p+q-k}{k+1}\sum_{i=0}^{k}\binom{p}{i}\binom{q}{k-i}, \quad \text{by induction hypothesis}$$

$$= \frac{1}{k+1}\sum_{i=0}^{k}(p+q-k)\binom{k}{i}\frac{p^{(i)}q^{(k-i)}}{k!}$$

$$= \frac{1}{(k+1)!}\sum_{i=0}^{k}(p-i+q-(k-i))\binom{k}{i}p^{(i)}q^{(k-i)}$$

$$= \frac{1}{(k+1)!}\sum_{i=0}^{k}\binom{k}{i}\left(p^{(i+1)}q^{(k-i)}+p^{(i)}q^{(k-i+1)}\right)$$

$$= \frac{1}{(k+1)!}\sum_{i=0}^{k}\binom{k}{i}p^{(i+1)}q^{(k-i)} + \frac{1}{(k+1)!}\sum_{i=0}^{k}\binom{k}{i}p^{(i)}q^{(k+1-i)}$$

$$= \frac{1}{(k+1)!}\sum_{i=0}^{k-1}\binom{k}{i}p^{(i+1)}q^{(k-i)} + \frac{1}{(k+1)!}\binom{k}{k}p^{(k+1)}q^{(0)}$$

$$\quad + \frac{1}{(k+1)!}\binom{k}{0}p^{(0)}q^{(k+1)} + \frac{1}{(k+1)!}\sum_{i=1}^{k}\binom{k}{i}p^{(i)}q^{(k+1-i)}$$

$$= \frac{1}{(k+1)!}\sum_{i=1}^{k}\binom{k}{i-1}p^{(i)}q^{(k-i+1)} + \frac{1}{(k+1)!}\binom{k}{k}p^{(k+1)}q^{(0)}$$

$$\quad + \frac{1}{(k+1)!}\binom{k}{0}p^{(0)}q^{(k+1)} + \frac{1}{(k+1)!}\sum_{i=1}^{k}\binom{k}{i}p^{(i)}q^{(k+1-i)}$$

$$= \binom{k}{0}\frac{p^{(0)}q^{(k+1)}}{(k+1)!} + \sum_{i=1}^{k}\left(\binom{k}{i-1}+\binom{k}{i}\right)\frac{p^{(i)}q^{(k+1-i)}}{(k+1)!}$$

$$\quad + \binom{k}{k}\frac{p^{(k+1)}q^{(0)}}{(k+1)!}$$

$$= \binom{k+1}{0}\frac{p^{(0)}q^{(k+1)}}{(k+1)!} + \sum_{i=1}^{k}\binom{k+1}{i}\frac{p^{(i)}q^{(k+1-i)}}{(k+1)!}$$

$$+ \binom{k+1}{k+1} \frac{p^{(k+1)}q^{(0)}}{(k+1)!},$$

using the familiar identities,

$$\binom{k}{0} = 1 = \binom{k+1}{0}, \quad \binom{k}{i-1} + \binom{k}{i} = \binom{k+1}{i}, \quad \binom{k}{k} = 1 = \binom{k+1}{k+1}.$$

But the right-hand sum of our last equation is

$$\sum_{i=0}^{k+1} \binom{k+1}{i} \frac{p^{(i)}q^{(k+1-i)}}{(k+1)!} = \sum_{i=0}^{k+1} \binom{p}{i}\binom{q}{k+1-i},$$

as was to be shown.[67]

Now assume p is a positive rational number, $p = m/n$ where m, n are positive integers. Observe

$$[p]^n = [p] \cdots [p] = [np] = [m] = \sum_{k=0}^{\infty} \binom{m}{k} x^k = \sum_{k=0}^{m} \binom{m}{k} x^k = (1+x)^m,$$

whence

$$[p] = ((1+x)^m)^{\frac{1}{n}} = (1+x)^p.$$

If p is negative, note that the Lemma yields

$$[p] \cdot [|p|] = [p + |p|] = [0] = 1,$$

whence

$$[p] = \frac{1}{[|p|]} = \frac{1}{(1+x)^{|p|}} = (1+x)^p.$$

Thus, Newton's Binomial Theorem is proven for all rational exponents p.

This proof only fails in rigour in its lack of attention to matters of convergence. This was finally addressed half a century later by Niels Henrik Abel[68],

---

[67] Formula (37) is known as Vandermonde's Theorem. W. L. Ferrar, *A Text-Book of Convergence*, Oxford University Press, Oxford, 1938, pp. 94 - 95, presents a clever noncomputational proof which I reproduce as Exercise 6.9 at the end of this chapter.

[68] N.H. Abel, "Untersuchungen über die Reihe: $1 + \frac{m}{1}x + \frac{m \cdot (m-1)}{1 \cdot 2} \cdot x^2 + \frac{m \cdot (m-1) \cdot (m-2)}{1 \cdot 2 \cdot 3} \cdot x^3 + \dots$ u.s.w.", *Journal für die reine und angewandte Mathematik* 1 (1826), pp. 311 - 339. A few excerpts from the paper citing Abel's convergence and divergence criteria appear in English translation in Garrett Birkhoff, *A Source Book in Classical Analysis*, Harvard University Press, Cambridge (Mass), 1973, and are reprinted in Calinger's source book (*op. cit.*). Oddly enough, the title of Abel's paper is not given in either of these references.

In his textbook of 1821, *Cours d'analyse de l'école royale polytechnique*, Cauchy proved the Binomial Theorem for all real exponents. The proof is not generally considered correct because i) it depended on Cauchy's false theorem that the limit of a convergent sequence of continuous functions is continuous, and ii) at one point who showed that each series $[p]$ converges for $|x| < 1$, that the product of $[p]$ and $[q]$ is the product series given by (36), and that this equals $[p + q]$.

Euler preceded Lagrange chronologically and thus did not have Taylor's Theorem with the Lagrange form of the remainder to apply. This is just as well as it doesn't work here.[69] Lagrange's form for the remainder applies for $0 \leq x < 1$, but turns out not to apply when $-1 < x < 0$. For this, another estimate of the remainder due to Cauchy applies.[70] Another proof proceeds first by applying a convergence criterion to show $[p]$ converges for $-1 < x < 1$ and then differentiating the ratio

$$\frac{[p]}{(1 + x)^p}$$

with respect to x. Now

$$\frac{d[p]}{dx} = \frac{d}{dx} \sum_{k=0}^{\infty} \binom{p}{k} x^k = \sum_{k=0}^{\infty} \binom{p}{k} k x^{k-1}$$

$$= \sum_{k=1}^{\infty} \frac{p(p-1)\cdots(p-k+1)}{(k-1)\cdots 1} x^{k-1}$$

$$= p \sum_{k=1}^{\infty} \binom{p-1}{k-1} x^{k-1} = p \sum_{k=0}^{\infty} \binom{p-1}{k} x^k$$

$$= p[p-1].$$

Thus

$$\frac{d}{dx} \frac{[p]}{(1+x)^p} = \frac{p[p-1](1+x)^p - [p]p(1+x)^{p-1}}{(1+x)^{2p}}$$

$$= \frac{p(1+x)^{p-1}}{(1+x)^{2p}} \left((1+x)[p-1] - [p] \right). \qquad (38)$$

But

$$(1+x)[p-1] = \sum_{k=0}^{\infty} \binom{p-1}{k} x^k + \sum_{k=0}^{\infty} \binom{p-1}{k} x^{k+1}$$

in the argument he made a suspicious interchange of limits. Detlef Laugwitz, "Infinitely small quantities in Cauchy's textbooks", *Historia Mathematica* 14 (1987), pp. 258 - 274, especially pp. 266 - 267, argues that Cauchy's proof is correct if carried out not in the reals but in a system extending the reals by the addition of infinitely large and infinitely small quantities. See the concluding subsection of Chapter III, section 6, pp. 361 - 373, below, for a bit more information on Cauchy's proof.

[69] Cf. e.g. G.H. Hardy, *A Course in Pure Mathematics*, 10th edition, Cambridge University Press, Cambridge, 1952, p. 292. Hardy returns periodically to the binomial series in his book. Thus, he also presents the Euler-Abel proof (p. 387, p. 432).

[70] *Ibid.*, p. 328.

$$= \binom{p-1}{0} x^0 + \sum_{k=1}^{\infty} \binom{p-1}{k} x^k + \sum_{k=0}^{\infty} \binom{p-1}{k} x^{k+1}$$

$$= \binom{p-1}{0} x^0 + \sum_{k=1}^{\infty} \binom{p-1}{k} x^k + \sum_{k=1}^{\infty} \binom{p-1}{k-1} x^k$$

$$= \binom{p-1}{0} x^0 + \sum_{k=1}^{\infty} \binom{p}{k} x^k = [p].$$

Thus $(1+x)[p-1] - [p] = 0$ and by (38),

$$\frac{d}{dx} \frac{[p]}{(1+x)^p} = 0, \text{ for all } x \in (-1,1),$$

i.e. $[p] = K(1+x)^p$ for some constant K. Setting $x = 0$ shows $K = 1$, i.e. $[p] = (1+x)^p$ for $-1 < x < 1$.

This last proof, though lacking the motivation of Euler's proof, is favoured by many authors[71] insofar as it only requires knowledge of where $[p]$ converges, the termwise differentiability of power series, and the fact that a function whose derivative on an interval is always 0 must be constant.

4.2 Product Formula for $\sin x$

Let us return to Euler. Any discussion of formalism must include his treatment of the sine function from his *Introductio in Analysin Infinitorum*[72] of 1748. The book is divided, of course, into chapters, and each chapter further divided into numbered "articles". The numbers run consecutively throughout the book. All told there are 381 articles. The results on the sine function we wish to discuss are spread through several chapters in the book in the section from article 72 to article 167, that is, the latter parts of Chapter IV to the early parts of Chapter X.

The results I wish to present are Euler's product formula for sine and its application to the problem of summing the reciprocals of the squares. He first presented an heuristic argument for this result in a paper in 1734[73]. Here he pointed out that the zeros of the function,

---

[71] I find it in Ferrar, *op. cit.*, pp. 99 - 101; R.E. Johnson and F.L. Kiokemeister, *Calculus with Analytic Geometry*, 3rd. edition, Allyn and Bacon, Inc., Boston, 1964, pp. 446 - 447; Albert G. Fadell, *Vector Calculus with Differential Equations*, D. van Nostrand Company, New York, 1968, pp. 267 - 268; and James Stewart, *Single Variable Calculus; Concepts and Contexts*, 3rd edition, Thomson Brooks/Cole, Belmont (Cal.), 2005, p. 620 (Exercise 15). After finding all of these, I stopped checking for more.

[72] An English translation by John D. Blanton exists: Euler, *Introduction to the Analysis of the Infinite, Book I*, Springer-Verlag, New York, 1988.

[73] Leonhard Euler, "De summis serierum reciprocarum", *Commentarii academiæ scientarium Petropolitanæ* 7 (1734). Basically the same derivation was given by Johann Bernoulli some time after his brother Jacob died (1705). It was not published, however, until his collected works appeared in 1742. Cf. Reiff, *op. cit.*, pp. 60 - 62 for details.

$$f(x) = \frac{\sin x}{x} = 1 - \frac{x^2}{3!} + \frac{x^4}{5!} - \cdots$$

are exactly the numbers $\pm k\pi$ for $k \in \mathbb{N}, k \neq 0$, whence the "polynomial" f factors

$$f(x) = \prod_{k=0}^{\infty}(x - k\pi)(x + k\pi) = \prod_{k=0}^{\infty}(x^2 - k^2\pi^2). \tag{39}$$

Now, if one has a polynomial $P(z)$ with constant coefficient 1 and which factors into linear factors,

$$P(z) = a_0 + a_1 z + \ldots + a_n z^n = \prod_{i=1}^{n}(z - r_i),$$

then

$$a_0 = 1 = \prod_{j=1}^{n}(-r_j)$$

$$a_1 = \sum_{i=1}^{n}\prod_{j\neq i}(-r_j) = \sum_{i=1}^{n}\frac{-1}{r_i}\prod_{j=1}^{n}(-r_j) = \sum_{i=1}^{n}\frac{-a_0}{r_i} = \sum_{i=1}^{n}\frac{-1}{r_i}.$$

Letting $z = x^2$ in (39), with n now infinite, we have

$$\frac{-1}{3!} = a_1 = \sum_{k=1}^{\infty}\frac{-1}{k^2\pi^2},$$

whence

$$\frac{\pi^2}{6} = \sum_{k=0}^{\infty}\frac{1}{k^2}.$$

In the book, Euler starts from scratch and makes a couple of seemingly slight, but essential variations on this theme.

[Chapter IV, §72]. Euler states Newton's Binomial Theorem for rational exponents without proof and gives some sample expansions.

[Chapter VI]. Euler introduces exponentiation and logarithms. With respect to the discussion in Chapter I of the present work on the religious orientation of mathematicians of his day, I cannot resist citing one of his applications even though it is hardly relevant here. Example III (in §110) begins: "Since after the flood all men descended from a population of six, if we suppose that the population after two hundred years was 1,000,000, we would like to find the annual rate of growth." Take away the religious reference and you have a problem suitable for today's precalculus courses.

[Chapter VII, §114]. Euler takes ω to be an infinitely small number, and a a real number greater than 1. Because $a^0 = 1$, it follows that $a^\omega = 1 + \psi$ for some infinitesimal ψ. Let $\psi = k\omega$, so that

$$a^\omega = 1 + k\omega. \tag{40}$$

Euler illustrates this by choosing $a = 10, k\omega = 10^{-6}$, calculating

$$w = \log(1 + 10^{-6}) = .00000043429,$$

and concluding

$$k = \frac{10^{-6}}{\omega} = 2.30258.$$

He states

> We see that k is a finite number which depends on the value of the base a. If a different base had been chosen, then the logarithm of the same number $1 + k\omega$ will differ from the logarithm already given. It follows that a different value of k will result.

[This is hardly a proof. Note, however, that by (40),

$$k = \frac{a^\omega - 1}{\omega},$$

and k must thus equal the derivative of a^x at $x = 0$. From the Calculus, we know that $k = \ln a$, the natural logarithm of a. Euler's *Introductio* was a precalculus text, so he could make no use of this fact and just noted this finiteness in an offhand manner.]

[§115]. By the Binomial Theorem,

$$a^{j\omega} = (1 + k\omega)^j = 1 + \frac{j}{1}k\omega + \frac{j(j-1)}{1 \cdot 2}k^2\omega^2 + \frac{j(j-1)(j-2)}{1 \cdot 2 \cdot 3}k^3\omega^3 + \dots$$

If z is a finite number and we choose $j = z/\omega$, we have

$$a^z = \left(1 + \frac{kz}{j}\right)^j = 1 + \frac{1}{1}kz + \frac{1(j-1)}{1 \cdot 2j}k^2z^2 + \frac{1(j-1)(j-2)}{1 \cdot 2j \cdot 3j}k^3z^3$$
$$+ \frac{1(j-1)(j-2)(j-3)}{1 \cdot 2j \cdot 3j \cdot 4j}k^4z^4 + \dots$$

Note that, because z is finite and ω is infinitesimal, j must be infinite.

[§116]. I quote Euler:

> Since j is infinitely large, $\frac{j-1}{j} = 1$, and the larger the number we substitute for j, the closer the value of the fraction $\frac{j-1}{j}$ comes to 1. Therefore, if j is a number larger than any assignable number, then $\frac{j-1}{j}$ is equal to 1. For the same reason $\frac{j-2}{j} = 1$, $\frac{j-3}{j} = 1$, and so forth. It follows that $\frac{j-1}{2j} = \frac{1}{2}$, $\frac{j-2}{3j} = \frac{1}{3}$, $\frac{j-3}{4j} = \frac{1}{4}$, and so forth. When we substitute these values, we obtain

$$a^z = 1 + \frac{kz}{1} + \frac{k^2z^2}{1 \cdot 2} + \frac{k^3z^3}{1 \cdot 2 \cdot 3} + \frac{k^4z^4}{1 \cdot 2 \cdot 3 \cdot 4} + \dots$$

This should bring to mind Klein's description of Taylor's proof as "a transition to the limit of extraordinary audacity" as it is the same transition. The next few sections are devoted to deriving infinite series representations for logarithms.

[§122]. Euler defines e to be the base for which $k = 1$. Thus

$$e = 1 + \frac{1}{1} + \frac{1}{1 \cdot 2} + \frac{1}{1 \cdot 2 \cdot 3} + \frac{1}{1 \cdot 2 \cdot 3 \cdot 4} + \dots$$

and

$$e^z = 1 + \frac{z}{1} + \frac{z^2}{1 \cdot 2} + \frac{z^3}{1 \cdot 2 \cdot 3} + \frac{z^4}{1 \cdot 2 \cdot 3 \cdot 4} + \dots$$

[Chapter VIII]. Euler introduces the trigonometric functions, recalling without proof numerous identities. Most relevant are the addition formulæ [§128]:

$$\sin(y \pm z) = \sin y \cos z \pm \cos y \sin z$$
$$\cos(y \pm z) = \cos y \cos z \mp \sin y \sin z,$$

the parametrisation of the unit circle [§132]:

$$(\sin z)^2 + (\cos z)^2 = 1,$$

and its factorisation,

$$(\cos z + i \sin z)(\cos z - i \sin z) = 1.$$

Using these and the addition formulæ he derives

$$(\cos z \pm i \sin z)^n = \cos nz \pm i \sin nz$$

for $n = 2, 3$ and concludes it to hold in general. From this he concludes

$$\cos nz = \frac{(\cos z + i \sin z)^n + (\cos z - i \sin z)^n}{2} \tag{41}$$

$$\sin nz = \frac{(\cos z + i \sin z)^n - (\cos z - i \sin z)^n}{2i}, \tag{42}$$

which, when expanded, yield

$$\cos nz = (\cos z)^n - \frac{n(n-1)}{1 \cdot 2}(\cos z)^{n-2}(\sin z)^2$$
$$+ \frac{n(n-1)(n-2)(n-3)(n-4)}{1 \cdot 2 \cdot 3 \cdot 4}(\cos z)^{n-4}(\sin z)^4$$
$$- \frac{n(n-1)(n-2)(n-3)(n-4)(n-5)}{1 \cdot 2 \cdot 3 \cdot 4 \cdot 5 \cdot 6}(\cos z)^{n-6}(\sin z)^6 + \dots$$

and

$$\sin nz = \frac{n}{1}(\cos z)^{n-1}\sin z$$
$$- \frac{n(n-1)(n-2)}{1 \cdot 2 \cdot 3}(\cos z)^{n-3}(\sin z)^3$$

$$+ \frac{n(n-1)(n-2)(n-3)(n-4)}{1 \cdot 2 \cdot 3 \cdot 4 \cdot 5}(\cos z)^{n-5}(\sin z)^5 - \dots$$

[§134]. Noting that $\sin z = z$ and $\cos z = 1$ for infinitely small z, he chooses, for any finite number v, an infinite number n and $z = v/n$. Then, as in §116, he concludes

$$\cos v = 1 - \frac{v^2}{1 \cdot 2} + \frac{v^4}{1 \cdot 2 \cdot 3 \cdot 4} - \frac{v^6}{1 \cdot 2 \cdot 3 \cdot 4 \cdot 5 \cdot 6} + \dots$$

and

$$\sin v = x - \frac{v^3}{1 \cdot 2 \cdot 3} + \frac{v^5}{1 \cdot 2 \cdot 3 \cdot 4 \cdot 5} - \frac{v^7}{1 \cdot 2 \cdot 3 \cdot 4 \cdot 5 \cdot 6 \cdot 7} + \dots$$

Thus, without resorting to differentiation, he has derived the Taylor expansions for the sine and cosine functions. Nowadays, we would immediately apply these and the expansion of e^z from §122 to conclude

$$e^{iv} = \cos v + i \sin v.$$

Euler, however, proceeds differently.

[§138]. Going back to §133, Euler now chooses z infinitely small and j infinite so that $v = jz$, $\sin z = \frac{v}{j}$ and $\cos z = 1$. Then from (41) and (42) he concludes

$$\cos v = \frac{\left(1 + \frac{iv}{j}\right)^j + \left(1 - \frac{iv}{j}\right)^j}{2}$$

$$\sin v = \frac{\left(1 + \frac{iv}{j}\right)^j - \left(1 - \frac{iv}{j}\right)^j}{2i},$$

and thus

$$\cos v = \frac{e^{iv} + e^{-iv}}{2}, \quad \sin v = \frac{e^{iv} - e^{-iv}}{2i}.$$

From these he obtains

$$e^{iv} = \cos v + i \sin v.$$

Euler devotes the next few sections to manipulations of these series and and logs to derive

$$\frac{\pi}{4} = 1 - \frac{1}{3} + \frac{1}{5} - \frac{1}{7} + \dots$$

and such like. (Compare (8) above.)

[Chapter IX]. The chapter "On Trinomial Factors" starts out looking like it is headed in a different direction. Although there is not yet (1748) an adequate proof of the Fundamental Theorem of Algebra, Euler asserts that every polynomial in z with real coefficients factors completely into linear factors over the complex numbers and into linear and quadratic factors over the reals. He now sets out to find the quadratic trinomial factors, doing this in general terms in §§143 - 149. He gets down to specifics in §§150 - 151 where he finds the factors of $a^n + z^n$ and $a^n - z^n$, respectively. Today we go about this by writing the roots of these polynomials in the forms

$$z = a\rho_1 \quad \text{and} \quad z = a\rho_2,$$

respectively, where ρ_1, ρ_2 are n-th roots of $-1, 1$, respectively. Now each ρ_i can be taken as a power of a primitive n-th root, whence, for some k with $1 \leq k \leq n$,

$$\rho_1 = (e^{\frac{\pi i}{n}})^k, \quad \rho_2 = (e^{\frac{2\pi i}{n}})^k.$$

Thus

$$a^n + z^n = \prod_{k=1}^{n} \left(z - ae^{\frac{k\pi i}{n}} \right)$$

$$a^n - z^n = -(z^n - a^n) = -\prod_{k=1}^{n} \left(z - ae^{\frac{2k\pi i}{n}} \right).$$

One now has to consider the various cases in which n is even or odd. For convenience I consider only $a^n - z^n$ for $n = 2m + 1$ odd. We have

$$a^n - z^n = -(z - ae^0) \prod_{k=1}^{m} \left((z - ae^{\frac{2k\pi i}{n}})(z - ae^{\frac{2(n-k)\pi i}{n}}) \right)$$

$$= -(z - a) \prod_{k=1}^{m} \left((z - ae^{\frac{2k\pi i}{n}})(z - ae^{\frac{-2k\pi i}{n}}) \right)$$

$$= -(z - a) \prod_{k=1}^{m} \left(z^2 - az(e^{\frac{2k\pi i}{n}} + e^{\frac{-2k\pi i}{n}}) + a^2 \right)$$

$$= -(z - a) \prod_{k=1}^{m} \left(z^2 - 2az \cos \frac{2k\pi}{n} + a^2 \right)$$

$$= (a - z) \prod_{k=1}^{m} \left(z^2 - 2az \cos \frac{2k\pi}{n} + a^2 \right) \qquad (43)$$

[§155]. Euler applies this factorisation to

$$e^x - 1 = \left(1 + \frac{x}{j} \right)^j - 1$$

with j infinite. I skip this in favour of
    [§156]. Euler factors

$$e^x - e^{-x} = \left(1 + \frac{x}{j} \right)^j - \left(1 - \frac{x}{j} \right)^j.$$

Here $a = 1 + \frac{x}{j}, z = 1 - \frac{x}{j}, n = j = 2m + 1$, say, with j and m infinite. Equation (43) becomes

$$e^x - e^{-x} = \left(1 + \frac{x}{j} - \left(1 - \frac{x}{j} \right) \right)$$

$$\cdot \prod_{k=1}^{m} \left(\left(1+\frac{x}{j}\right)^2 - 2\left(1+\frac{x}{j}\right)\left(1-\frac{x}{j}\right)\cos\frac{2k\pi}{n} + \left(1-\frac{x}{j}\right)^2 \right)$$

$$= \frac{2x}{j} \prod_{k=1}^{m} \left(2 + \frac{2x^2}{j^2} - 2\left(1-\frac{x^2}{j^2}\right)\cos\frac{2k\pi}{n} \right)$$

$$= \frac{2x}{j} \prod_{k=1}^{m} \left(2 + \frac{2x^2}{j^2} - 2\left(1-\frac{x^2}{j^2}\right)\left(1-\frac{2k^2\pi^2}{j^2}\right) \right),$$

using the series expansion,

$$\cos\frac{2k\pi}{j} = 1 - \frac{(2k\pi)^2}{2j^2} + \dots \tag{44}$$

and dropping all terms after the second because, j being infinite, they are infinitely small in comparison to the second. But this yields

$$e^x - e^{-x} = \frac{2x}{j} \prod_{k=1}^{m} \left(2 + \frac{2x^2}{j^2} - 2\left(1-\frac{x^2}{j^2}-\frac{2k^2\pi^2}{j^2}+\frac{2k^2\pi^2x^2}{j^4}\right) \right)$$

$$= \frac{2x}{j} \prod_{k=1}^{m} \left(\frac{4x^2}{j^2} + \frac{4k^2\pi^2}{j^2} - \frac{4k^2\pi^2x^2}{j^4} \right).$$

Factoring $4k^2\pi^2/j^2$ out of the quadratic factor, Euler concludes

$$1 + \frac{x^2}{k^2\pi^2} - \frac{x^2}{j^2}$$

to be a factor of $e^x - e^{-x}$. Now, when one multiplies the m factors, the terms $-x^2/j^2$ only contribute an infinitesimal amount to the resulting sum and can be omitted. Thus

$$\frac{e^x - e^{-x}}{2} = x\left(1 + \frac{x^2}{3!} + \frac{x^4}{5!} + \frac{x^6}{7!} + \dots\right)$$

$$= x\prod_{k=1}^{m}\left(1 + \frac{x^2}{k^2\pi^2}\right)$$

$$= x\left(1 + \frac{x^2}{\pi^2}\right)\left(1 + \frac{x^2}{4\pi^2}\right)\left(1 + \frac{x^2}{9\pi^2}\right)\dots \tag{45}$$

[§157]. Euler repeats the exercise for $e^x + e^{-x}$.

[§158]. Euler now sets $x = iz$ in (45) to obtain

$$\sin z = \frac{e^{iz} - e^{-iz}}{2i} = z\left(1 - \frac{z^2}{\pi^2}\right)\left(1 - \frac{z^2}{4\pi^2}\right)\left(1 - \frac{z^2}{9\pi^2}\right)\dots,$$

his famous product formula for the sine.

[Chapter X, §165]. In a chapter titled "On the Use of the Discovered Factors to Sum Infinite Series", Euler says

If $1 + Az + Bz^2 + cz^3 + Dz^4 + \ldots = (1 + \alpha z)(1 + \beta z)(1 + \gamma z)(1 + \delta z) \cdots$,
then these factors, whether they be finite or infinite in number, must
produce the expression $1 + Az + Bz^2 + Cz^3 + Dz^4 + \ldots$, when they are
actually multiplied. It follows that the coefficient A is equal to the sum
$\alpha + \beta + \gamma + \delta + \epsilon + \cdots$. The coefficient B is equal to the sum of the products
taken two at a time. Hence $B = \alpha\beta + \alpha\gamma + \alpha\delta + \beta\gamma + \beta\delta + \gamma\delta + \cdots$.
Also the coefficient C is equal to the sum of the products taken three
at a time, namely $C = \alpha\beta\gamma + \alpha\beta\delta + \beta\gamma\delta + \alpha\gamma\delta + \cdots$. We also have D as
the sum of products taken four at a time, and E is the sum of products
taken five at a time, etc. All this is clear from ordinary algebra.

[§169]. He now applies this observation to the product formula (45) for the
hyperbolic sine:

$$\frac{e^x - e^{-x}}{2x} = 1 + \frac{x^2}{6} + \frac{x^4}{120} + \ldots = \prod_{k=1}^{\infty} \left(1 + \frac{x^2}{(k\pi)^2}\right),$$

by first setting $x^2 = \pi^2 z$:

$$1 + \frac{\pi^2}{6}z + \frac{\pi^4}{120}z^2 + \ldots = (1 + z)\left(1 + \frac{z}{4}\right)\left(1 + \frac{z}{9}\right)\cdots.$$

The result is his famous formula

$$\sum_{k=1}^{\infty} \frac{1}{k^2} = 1 + \frac{1}{4} + \frac{1}{9} + \ldots = \frac{\pi^2}{6}.$$

Following Bolzano-Cauchy-Weierstrass, until the advent of Nonstandard
Analysis, Euler's derivation of the product formula for sine and of a value
for the sum of the reciprocals of the squares was deemed purely heuristic. The
sum would be established via Fourier series and the product representation
considered established only by Weierstrass's product representation theorem
for analytic functions in Complex Analysis. When Nonstandard Analysis began
to be developed it was inevitable that Euler's proofs would be looked at anew
and given new life. W.A.J. Luxemburg tackled[74] the product formula for sine
and gave a correct nonstandard proof along the lines set forth by Euler. One
of the points that Luxemburg changed was Euler's replacement of $\cos(2k\pi/j)$
by $1 - 2k^2\pi^2/j^2$. In its place, Luxemburg uses the double angle formula

$$\cos\frac{2k\pi}{j} = \cos^2\frac{k\pi}{j} - \sin^2\frac{k\pi}{j} = 1 - 2\sin^2\frac{k\pi}{j} = 2\cos^2\frac{k\pi}{j} - 1$$

to write

$$2\left(1 + \frac{x^2}{j^2}\right) - 2\left(1 - \frac{x^2}{j^2}\right)\cos\frac{2k\pi}{j} = 2\left(1 - \cos\frac{2k\pi}{j}\right) + 2\left(1 + \cos\frac{2k\pi}{j}\right)\frac{x^2}{j^2}$$

[74] W.A.J. Luxemburg, "What is nonstandard analysis?", *American Mathematical Monthly* 80, no. 6 Part II (Papers in the foundations of mathematics), (1973), pp. 38 - 67.

$$= 4 \sin^2 \frac{k\pi}{j} + 4 \left(\cos^2 \frac{k\pi}{j} \right) \frac{x^2}{j^2}$$

$$= 4 \sin^2 \frac{k\pi}{j} \left(1 + \frac{x^2}{j^2 \tan^2(k\pi/j)} \right).$$

The factors

$$\sin^2 \frac{k\pi}{j} \quad \text{and} \quad 1 + \frac{x^2}{j^2 \tan^2(k\pi/j)}$$

of this are infinitesimally close to

$$1 \quad \text{and} \quad 1 + \frac{x^2}{k^2 \pi^2},$$

respectively. Eventually, Mark McKinzie and Curtis Tuckey provided rigorous justification for all the steps used by Euler, as he himself set them down.[75]

4.3 Euler-Maclaurin Summation Formula

I would like to finish our discussion of Euler with one of my favourite Eulerian results. This is the Euler-Maclaurin Summation Formula, the standard introductory proof of which in the textbooks is a marvel of Type I Formalism. The proof does not look at all correct and yet it yields correct results where it should. The result is due in two slightly different forms to Euler and Maclaurin. Euler first discovered the result in 1732 and published it in 1738[76], but the most accessible version of his proof is probably his Calculus book, *Institutiones calculi differentialis cum ejus usu in analysi finitorum ac doctrina serierum*[77].

---

[75] Mark McKinzie and Curtis Tuckey, "Hidden lemmas in Euler's summation of the reciprocals of the squares", *Archive for History of Exact Sciences* 51 (1997), pp. 29 - 57.

[76] Leonhard Euler, "Methodus generalis summandi progressiones", *Commentarii academiæ scientiarum Petropolitanæ* 25 (1732 - 3), 1738, pp. 68 - 97. I take this citation from Chabert, *op. cit.*, p. 439. Chabert also includes an extract from from Euler's "Inventio summæ cuiusque seriei ex dato termino generali", *Commentarii academiæ scientiarum Petropolitanæ* 47 (1736), published 1741, in which Euler derives the result and applies it to the problem of the Euler-Mascheroni constant,

$$C = \lim_{n \to \infty} \left(\sum_{k=1}^{\infty} \frac{1}{k} - \ln n \right).$$

Reiff, *op. cit.*, p. 95 also cites this work.

[77] A German translation appeared in the years 1790 - 1793. At the time I'm writing this, only the first of several volumes of an English translation by John D. Blanton has appeared: *Foundations of Differential Calculus*, Springer-Verlag, New York, 2000. However, Euler's treatment of the summation formula is described in some detail in Herman Goldstine, *A History of Numerical Analysis fom the 16th through the 19th Century*, Springer-Verlag, New York, 1977, pp. 126 - 137. A less detailed, but also illuminating, discussion also appears in Reiff, *op. cit.*, pp. 109 - 118.

Maclaurin published the formula in his *Treatise on Fluxions* in 1742, having written this book in 1737.[78]

Euler's proof is simpler than Maclaurin's and he gives a good determination of the sequence of coefficients, a determination which paved the way for the modern derivation of the Euler-Maclaurin Summation Formula first given by Lagrange. All these proofs— Maclaurin's, Euler's, and Lagrange's— are formal and non-rigourous. A rigorous treatment would be given first by Siméon-Denis Poisson and more definitively in 1834 by Carl Gustav Jacobi. It is Lagrange's proof that I wish to present.

But, "proof of what?", one might ask. I have not yet stated the result. For this we need some background material.

Recall the operators

$$\Delta_h f(x) = f(x+h) - f(x), \quad E_h f(x) = f(x+h),$$

from the preceding section. We shall only be interested in the present section in the case $h = 1$ and will thus write Δ for Δ_1 and E for E_1. As remarked back in Chapter I (cf. page 71), Δ is a discrete analogue to the differentiation operator. One has, for example, the formulæ

$$\Delta c = 0, \text{ for any constant } c \in \mathbb{R}$$
$$\Delta x^{(n)} = nx^{(n-1)}, \text{ for positive integral } n$$
$$\Delta a^x = (a-1)a^x, \text{ for } a \text{ positive}$$
$$\Delta(f+g) = \Delta f + \Delta g$$
$$\Delta r f(x) = r\Delta f(x), \text{ for } r \in \mathbb{R}.$$

If one defines
$$x^{|n|} = x(x+1)\cdots(x+n-1),$$

for positive integral n in analogy to the definition of $x^{(n)}$, then one also has

$$\Delta \frac{1}{x^{|n|}} = \frac{-n}{x^{|n+1|}}.$$

The analogy goes further. Just as integration relates to anti-differentiation, summation is a sort of inverse to differencing:

4.2 Theorem (Discrete Fundamental Theorem of Calculus). *Let f be given and $m, n \in \mathbb{N}$. Fix m and assume $n > m$.*
i. Suppose

$$F(n) = \sum_{k=m}^{n} f(k).$$

Then: $\Delta F(n) = f(n+1)$.
ii. Suppose $\Delta F(x) = f(x)$. Then

---

[78] Excerpts from Maclaurin can be found in Chabert's source book, *op. cit.*, pp. 434 - 439. His work is also discussed in Goldstine, *op. cit.*, pp. 84 - 94, and Reiff, *op. cit.*, pp. 84 - 87.

$$\sum_{k=m}^{n} f(k) = F(n+1) - F(m).$$

Both assertions are easy to prove. For the first, one simply calculates:

$$F(n+1) - F(n) = \sum_{k=m}^{n+1} f(k) - \sum_{k=m}^{n} f(k) = f(n+1).$$

The second assertion is just the familiar business of telescoping sums:

$$\sum_{k=m}^{n} f(k) = f(m) + f(m+1) + \ldots + f(n-1) + f(n)$$
$$= f(n) + f(n-1) + \ldots + f(m+1) + f(m)$$
$$= F(n+1) - F(n) + F(n) - F(n-1) + \ldots$$
$$+ F(m+2) - F(m+1) + F(m+1) - F(m)$$
$$= F(n+1) - F(m).$$

The crucial difference between the Discrete and Continuous Fundamental Theorems is the shift in the upper index of summation in part ii: The sum from m to n is not $F(n) - F(m)$, but $F(n+1) - F(m)$.

Taylor's Theorem has discrete analogues. For example,

$$f(x) \sim \sum_{k=0}^{n} \frac{\Delta^k f(0)}{k!} x^{(k)}, \qquad (46)$$

with "$\sim$" replaced by equality when $x = 0, 1, \ldots, n$ or when f is a polynomial of degree at most n. For other functions f, the sum on the right-hand-side of (46) is an approximation to f and is used to interpolate values for f on $[0, n]$.

It is summation and anti-differencing that interest us here. There are any number of notations that could be used to denote the anti-difference of a function. We could write "Σ" without specifying any limits of summation. Euler essentially did this, but using the latin "S". It is, however, customary (and suggestive) to use Δ^{-1}. As with the indefinite integral, which is unique up to an additive constant, the anti-difference is unique up to a constant (or, more correctly, up to a periodic function of period 1— but we are usually interested in discrete arguments of the function and write a constant). Thus we have for example

$$\Delta^{-1} c = cx + C, \text{ any } c \in \mathbb{R}$$

$$\Delta^{-1} x^{(n)} = \frac{x^{(n+1)}}{n+1} + C, \quad n \text{ a positive integer}$$

$$\Delta^{-1} \frac{1}{x^{|n|}} = \frac{-1}{(n-1)x^{|n-1|}} + C, \quad n > 1 \text{ integral}$$

$$\Delta^{-1} a^x = \frac{a^x}{a-1} + C, \quad a \neq 1, a > 0.$$

It can be kind of fun to play with these formulæ and see what kinds of sums one can express in closed form by means of them. For example, if $r \in \mathbb{R}$,

$$\sum_{k=0}^{n} r^k = \Delta^{-1} r^x \Big|_{x=0}^{n+1} = \frac{r^x}{r-1} \Big|_0^{n+1} = \frac{r^{n+1} - 1}{r - 1},$$

a result we applied back in Chapter I in discussing geometric progressions. Another sum familiar from Calculus is

$$\sum_{k=0}^{n} k = \Delta^{-1} x \Big|_{x=0}^{n+1} = \frac{x^{(2)}}{2} \Big|_0^{n+1}$$
$$= \frac{(n+1)n}{2} - \frac{0(-1)}{2} = \frac{n(n+1)}{2}.$$

Finding $\sum k^2$ is not quite as easy, but can be done by noting that $k^2 = k^{(2)} + k$:

$$\sum_{k=0}^{n} k^2 = \sum_{k=0}^{n} (k^{(2)} + k) = \frac{x^{(3)}}{3} + \frac{x^{(2)}}{2} \Big|_0^{n+1}$$
$$= \frac{(n+1)n(n-1)}{3} + \frac{(n+1)n}{2}$$
$$= \frac{2(n+1)n(n-1) + 3(n+1)n}{6}$$
$$= \frac{(n+1)n(2n-2+3)}{6} = \frac{(n+1)n(2n+1)}{6},$$

which should also be familiar from the Calculus.

More generally, for each m there is a polynomial P_m of degree $m + 1$ such that, for all $n \in \mathbb{N}$

$$\sum_{k=0}^{n} k^m = P_m(n).$$

One can find P_m by writing

$$k^m = \sum_{i=0}^{m} a_i k^{(i)}$$

and summing the right-hand-side after first calculating the coefficients $a_0, \ldots, a_m$. Alternatively, one could do it inductively using Summation by Parts[79], as the more ambitious reader will be asked to do in Exercise 6.10 at the end of this Chapter. But these methods are later developments. In his *Ars conjectandi*[80], Jacob Bernoulli gave the general formula

---

[79] Due to Abel, *op. cit.* Counting the reference in Chapter I, this is the third mention I've made of this fundamental paper. The reader with a knowledge of German ought to take a look at it.

[80] Posthumously published in 1713. An English translation of the relevant passages by Jekuthiel Ginsburg appears in Smith's source book (*op. cit.*, pp. 85 - 90) under the title "On the 'Bernoulli Numbers'" and is reprinted in Struik's source book (*op. cit.*, pp. 316 - 320) as the first half of a section titled "Jacob Bernoulli. Sequences and Series". Goldstine, *op. cit.*, discusses Bernoulli's work on pp. 94 - 97.

$$\sum_{k=0}^{n} k^m = \frac{n^{m+1}}{m+1} + \frac{n^m}{2} + \frac{m}{2}An^{m-1} + \frac{m(m-1)(m-2)}{4!}Bn^{m-3}$$

$$+ \frac{m(m-1)\cdots(m-4)}{6!}Cn^{m-5}$$

$$+ \frac{m(m-1)\cdots(m-6)}{6!}Dn^{m-7} + \cdots, \tag{47}$$

where $A, B, C, D, \ldots$ are the coefficients of the first degree terms of the sums $\sum k^2, \sum k^4, \sum k^6, \ldots$:

$$A = \frac{1}{6}, \quad B = \frac{-1}{30}, \quad C = \frac{1}{42}, \quad D = \frac{-1}{30}, \quad \ldots$$

The Euler-Maclaurin Summation Formula generalises this:

4.3 Theorem (Euler-Maclaurin Summation Formula). *Let f be given.*

$$\sum_{k=0}^{n} f(k) = \left[\int f(x)dx + \frac{1}{2}f(x) + \frac{A}{2!}f'(x) - \frac{B}{4!}f^{(3)}(x) + \frac{C}{6!}f^{(5)}(x) - \cdots\right]_{x=0}^{n}.$$
$$\tag{48}$$

Note that this is not quite an expression of the identity,

$$\sum_{k=0}^{n} f(k) = \Delta^{-1}f(k)\Big|_{0}^{n+1}. \tag{49}$$

Δ^{-1}, in fact, has a very similar form as one can see by subtracting $f(n)$ from each side of (48):

$$\sum_{k=0}^{n-1} f(k) = \left[\int f(x)dx - \frac{1}{2}f(x) + \frac{A}{2!}f'(x) - \frac{B}{4!}f^{(3)}(x) + \frac{C}{6!}f^{(5)}(x) - \cdots\right]_{x=0}^{n}.$$
$$\tag{50}$$

Thus

$$\Delta^{-1}f(k) = \left[\int_{0}^{k} f(x)dx - \frac{1}{2}f(k) + \frac{A}{2!}f'(k) - \frac{B}{4!}f^{(3)}(k) + \frac{C}{6!}f^{(5)}(k) - \cdots\right]$$

and $\sum_{k=0}^{n} f(k)$, when evaluated according to (49) becomes

$$\left[\int f(x)dx - \frac{1}{2}f(x) + \frac{A}{2!}f'(x) - \frac{B}{4!}f^{(3)}(x) + \frac{C}{6!}f^{(5)}(x) - \cdots\right]_{x=0}^{n+1}. \tag{51}$$

The close similarities of (48), (50), and (51) can lead to some confusion.

Euler's derivation of the Euler-Maclaurin Summation Formula is fairly straightforward, but computational. It is formal in that he doesn't worry about questions of convergence, but it lacks in entertainment value: it does not shock the senses in any way like a good Type I formal argument is supposed to. He simply starts from a Taylor expansion,

$$f(x-1) = f(x) - f'(x) + \frac{f^{(2)}(x)}{2!} - \frac{f^{(3)}(x)}{3!} + \dots$$

and forms the indefinite sums

$$\sum f(x-1) = \sum f(x) - \sum f'(x) + \sum \frac{f^{(2)}(x)}{2!} - \sum \frac{f^{(3)}(x)}{3!} + \dots$$

Transposing a couple of terms he concludes

$$\sum f'(x) = f(x) + C + \sum \frac{f^{(2)}(x)}{2!} - \sum \frac{f^{(3)}(x)}{3!} + \dots \tag{52}$$

for some constant C. He now replaces f by its antiderivative $\int f(x)dx$ and allows it to absorb C:

$$\sum f(x) = \int f(x)dx + \sum \frac{f'(x)}{2!} - \sum \frac{f^{(2)}(x)}{3!} + \dots \tag{53}$$

and differentiates this:

$$\sum f'(x) = f(x) + \sum \frac{f^{(2)}(x)}{2!} - \sum \frac{f^{(3)}(x)}{3!} + \dots \tag{54}$$

This is the same as (52), but with no C. (*Exercise.* Where did it go?) He plugs this back into (53) and simplifies:

$$\sum f(x) = \int f(x)dx + \frac{1}{2}\left(f(x) + \sum \frac{f^{(2)}(x)}{2!} - \sum \frac{f^{(3)}(x)}{3!} + \dots\right)$$
$$- \sum \frac{f^{(2)}(x)}{3!} + \sum \frac{f^{(3)}(x)}{4!} - \sum \frac{f^{(4)}(x)}{5!} + \dots$$
$$= \int f(x)dx + \frac{1}{2}f(x) + \left(\frac{1}{4} - \frac{1}{6}\right)\sum f^{(2)}(x)$$
$$+ \left(-\frac{1}{12} + \frac{1}{24}\right)\sum f^{(3)}(x) + \left(\frac{1}{2 \cdot 24} - \frac{1}{120}\right)\sum f^{(4)}(x) + \dots$$
$$= \int f(x)dx + \frac{1}{2}f(x) + \frac{1}{12}\sum f^{(2)}(x)$$
$$- \frac{1}{24}\sum f^{(3)}(x) + \frac{1}{80}\sum f^{(4)}(x) + \dots \tag{55}$$

He next differentiates (54) to get

$$\sum f^{(2)}(x) = f'(x) + \sum \frac{f^{(3)}(x)}{2!} - \sum \frac{f^{(4)}(x)}{3!} + \dots$$

and substitutes this into (55) to obtain

$$\sum f(x) = \int f(x)dx + \frac{1}{2}f(x) + \frac{1}{12}\left(f'(x) + \sum \frac{f^{(3)}(x)}{2!} - \sum \frac{f^{(4)}(x)}{3!} + \dots\right)$$

$$-\frac{1}{24}\sum f^{(3)}(x) + \frac{1}{80}f^{(4)}(x) + \dots$$

$$= \int f(x)dx + \frac{1}{2}f(x) + \frac{1}{12}f'(x)+$$

$$\left(\frac{1}{24} - \frac{1}{24}\right)\sum f^{(3)}(x) + \left(-\frac{1}{72} + \frac{1}{80}\right)\sum f^{(4)}(x) + \dots$$

$$= \int f(x)dx + \frac{1}{2}f(x) + \frac{1}{12}f'(x) - \frac{1}{720}\sum f^{(4)}(x) + \dots$$

One more iteration would yield

$$\sum f(x) = \int f(x)dx + \frac{1}{2}f(x) + \frac{1}{12}f'(x) - \frac{1}{720}f^{(3)}(x) + \dots . \qquad (56)$$

At some point one has to stop carrying out the above iteration and analyse what one is doing. Generally this means determining the recurrence rule for generating the sequence $a_0, a_1, \dots$ of coefficients in the development

$$\sum f(x) = \int f(x)dx + a_0 f(x) + a_1 f'(x) + a_2 f^{(2)}(x) + \dots$$

Euler determined the coefficients $A, B, C, \dots$ to be what he called the *Bernoulli numbers* in honour of Jacob Bernoulli. There are several definitions of the Bernoulli numbers. The most convenient choice is to declare them to be the sequence of coefficients $B_0, B_1, \dots$ of the Taylor expansion

$$\frac{t}{e^t - 1} = \sum_{i=0}^{\infty} \frac{B_i}{i!} t^i. \qquad (57)$$

Under this choice, Euler's definition gives the sequence $|B_2|, |B_4|, |B_6|, \dots$ of absolute values of Bernoulli numbers with positive even indices. From

$$e^t = \sum_{i=0}^{\infty} \frac{t^i}{i!}$$

it follows that

$$\frac{e^t - 1}{t} = \sum_{i=1}^{\infty} \frac{t^{i-1}}{i!} = 1 + \frac{t}{2} + \frac{t^2}{3!} + \frac{t^4}{4!} + \dots$$

A simple long division will determine the coefficients in (57). (See the box.) We quickly recognise the coefficients of the quotient to be those of (51).

If one actually does the arithmetic one sees that the same basic arithmetic steps are being taken as in the derivation proceeding from (52) to (56), only in a more familiar and less notationally opaque manner. One may also calculate the coefficients via a recurrence relation, and this is very transparently provided by multiplication. From

$$
\begin{array}{r|l}
 & 1 - t/2 + t^2/12 + \quad 0 \quad - t^4/720 + \ldots \\ \hline
1 + t/2 + t^2/3! + t^3/4! + \ldots & 1 \\
\end{array}
$$

$$1 + t/2 + t^2/3! + t^3/4! + t^4/5! + \ldots$$
$$- t/2 - t^2/2 - t^3/4! - t^4/5! + \ldots$$
$$- t/2 - t^2/4 - t^3/12 - t^4/48 + \ldots$$
$$t^2/12 + t^3/24 + t^4/80 + \ldots$$
$$t^2/12 + t^3/24 + t^4/72 + \ldots$$
$$- t^4/720 + \ldots$$
$$\vdots$$

$$
1 = \left(\frac{1}{1!} + \frac{t}{2!} + \frac{t^2}{3!} + \ldots \right) \left(\frac{B_0}{0!} + \frac{B_1 t}{1!} + \frac{B_2 t^2}{2!} + \ldots \right)
$$

one obtains on collecting terms and comparing coefficients

1: $1 = B_0$

t: $0 = \frac{B_0}{0!} \cdot \frac{1}{2!} + \frac{B_1}{1!} \cdot \frac{1}{1!}$

t^2: $0 = \frac{B_0}{0!} \cdot \frac{1}{3!} + \frac{B_1}{1!} \cdot \frac{1}{2!} + \frac{B_2}{2!} \cdot \frac{1}{1!}$

t^3: $0 = \frac{B_0}{0!} \cdot \frac{1}{4!} + \frac{B_1}{1!} \cdot \frac{1}{3!} + \frac{B_2}{2!} \cdot \frac{1}{2!} + \frac{B_3}{3!} \cdot \frac{1}{1!}$

and so on. Multiplying the coefficient of t^k by $(k+1)!$, this yields the fairly memorable recurrence formula

$$
\binom{k+1}{0} B_0 + \binom{k+1}{1} B_1 + \ldots + \binom{k+1}{k} B_k = 0. \tag{58}
$$

Indeed, if one now writes B^n instead of B_n, this reads

$$
(1+B)^{k+1} - B^{k+1} = 0,
$$

i.e. $B^{k+1} = (1+B)^{k+1}$, which, of course, does not allow one to determine B_{k+1} from $B_0, \ldots, B_k$ but rather to find B_k from $B_0, \ldots, B_{k-1}$.

The Bernoulli numbers have several interesting properties. I mention only one as it explains the missing terms in (48): For any $n > 0, B_{2n+1} = 0$. To see this, one sets

$$
f(t) = \frac{t}{e^t - 1}
$$

and observes

$$
f(t) - f(-t) = \frac{t}{e^t - 1} - \frac{-t}{e^{-t} - 1} = t \left(\frac{1}{e^t - 1} + \frac{1}{e^{-t} - 1} \right)
$$

$$= t \left(\frac{e^{-t/2}}{e^{t/2} - e^{-t/2}} + \frac{e^{t/2}}{e^{-t/2} - e^{t/2}} \right)$$

$$= t \left(\frac{e^{-t/2} - e^{t/2}}{e^{t/2} - e^{-t/2}} \right) = -t.$$

But

$$f(t) - f(-t) = \sum_{k=0}^{\infty} \frac{B_k}{k!} \left(t^k - (-t)^k \right)$$

and, comparing coefficients with $-t$, we see

$$2B_1 = -1, B_3 = B_5 = B_7 = \ldots = 0.$$

Aside from pointing to the first few coefficients of (56), I haven't offered anything remotely resembling a proof— rigorous or purely formal— that these coefficients depend on the Bernoulli numbers. This is most readily done by Lagrange's proof of the Euler-Maclaurin Summation Formula— or, better stated, his proof of an Euler-Maclaurin formula: his formula is slightly different from Euler's.

Lagrange's proof, or, at least, the modern presentation of it, is quite simple. It begins again by appealing to Taylor's Theorem,

$$f(x + 1) = \sum_{i=0}^{\infty} \frac{f^{(i)}(x)}{i!} = \sum_{i=0}^{\infty} \frac{D^i f(x)}{i!} = \left(\sum_{i=0}^{\infty} \frac{D^i}{i!} \right) f(x) = \left(e^D \right) f(x),$$

where $D = d/dx$ is the differentiation operator. Thus

$$\Delta f(x) = f(x + 1) - f(x) = e^D f(x) - f(x) = \left(e^D - 1 \right) f(x),$$

i.e.

$$\Delta = e^D - 1.$$

But then

$$\Delta^{-1} = \frac{1}{e^D - 1}$$

and

$$\Delta^{-1} = \frac{1}{D} \cdot \frac{D}{e^D - 1} = \frac{1}{D} \sum_{i=0}^{\infty} \frac{B_i}{i!} D^i = \sum_{i=0}^{\infty} \frac{B_i}{i!} D^{i-1},$$

and we have

$$\Delta^{-1} f(x) = B_0 D^{-1} f(x) + \frac{B_1}{1!} f(x) + \frac{B_2}{2!} f'(x) + \ldots$$

$$= \int f(x) dx - \frac{1}{2} f(x) + \sum_{i=2}^{\infty} \frac{B_i}{i!} f^{(i-1)}(x), \tag{59}$$

since $D^{-1} = \int$.

I suppose, after the tremendous build-up I've given this proof, it may disappoint somewhat. But it always amazes me that one can use a meaningless expression like e^D to obtain a real result like (59).

To illustrate the use of the Euler-Maclaurin Summation Formula, let us find $\sum_{k=0}^{n} k^3$. Using (50) we have

$$\sum_{k=0}^{n} k^3 = \frac{x^4}{4} + \frac{x^3}{2} + \frac{3x^2}{12} - \frac{6}{720}\Big|_{k=0}^{n} = \frac{x^4}{4} + \frac{x^3}{2} + \frac{x^2}{4} - \frac{1}{120}\Big|_{k=0}^{n}$$

$$= \frac{x^4}{4} + \frac{x^3}{2} + \frac{x^2}{4} - \frac{1}{120}\Big|_{0}^{n}$$

$$= \left(\frac{n^4}{4} + \frac{n^3}{2} + \frac{n^2}{4} - \frac{1}{120}\right) - \left(-\frac{1}{120}\right)$$

$$= \frac{n^4}{4} + \frac{n^3}{2} + \frac{n^2}{4} = \frac{n^2}{4}(n^2 + 2n + 1)$$

$$= \frac{n^2(n+1)^2}{4} = \left(\frac{n(n+1)}{2}\right)^2,$$

Or, using (51) we have

$$\sum_{k=0}^{n} k^3 = \frac{x^4}{4} - \frac{x^3}{2} + \frac{3x^2}{12} + 0 \cdot 6x - \frac{6}{720}\Big|_{0}^{n+1}$$

$$= \left(\frac{(n+1)^4}{4} - \frac{(n+1)^3}{2} + \frac{(n+1)^2}{4} - \frac{1}{120}\right) - \left(\frac{-1}{120}\right)$$

$$= \frac{(n+1)^2}{4}\left((n+1)^2 - 2(n+1) + 1\right)$$

$$= \frac{(n+1)^2}{4}((n+1)-1)^2 = \frac{(n+1)^2 n^2}{4} = \left(\frac{n(n+1)}{2}\right)^2.$$

5 The Dirac Delta Function

5.1 Basics

The Dirac delta function did not receive its name because P.A.M. Dirac first introduced it. It had been around in one form or another for just over a century when Dirac introduced it into quantum theory in his paper, "The physical interpretation of quantum mechanics"[81]. What he did do was to define the function formally and derive some of its properties. This was discussed more fully in his book *The Principles of Quantum Mechanics* (1930) and the list of properties extended in later editions of that book. The Dirac delta function proved useful in physics and mathematicians had to take note.

Dirac begins with the definition:

---

[81] *Proceedings of the Royal Society of London*, A, 113 (1926), pp. 621 - 641; the discussion of the delta function is on pp. 625 - 627.

One cannot go far in the development... without needing a notation for that function of a... number x that is equal to zero except when x is very small, and whose integral through a range that contains the point $x = 0$ is equal to unity. We shall use the symbol $\delta(x)$ to denote this function, *i.e.* $\delta(x)$ is defined by

$$\delta(x) = 0 \quad \text{when } x \neq 0, \tag{60}$$

and

$$\int_{-\infty}^{\infty} x\delta(x) = 1.$$

This last equation was a typographical error and should have read

$$\int_{-\infty}^{\infty} \delta(x)dx = 1. \tag{61}$$

If one now takes any interval $[a, b]$ containing the point $x = 0$, one has

$$1 = \int_{-\infty}^{\infty} \delta(x)dx = \int_{-\infty}^{a} \delta(x)dx + \int_{a}^{b} \delta(x)dx + \int_{b}^{\infty} \delta(x)dx$$
$$= \int_{-\infty}^{a} 0dx + \int_{a}^{b} \delta(x)dx + \int_{b}^{\infty} 0dx = \int_{a}^{b} \delta(x)dx,$$

and his claim for the integral's equalling 1 over any interval containing 0 is seen to follow from the assumption for the improper integral. Of course, no such function can exist as no value of $\delta(0)$ can make such an integral non-zero so long as $\delta(x) = 0$ for all $x \neq 0$. Dirac himself acknowledges this:

> Strictly, of course, $\delta(x)$ is not a proper function of x, but can be regarded only as a limit of a certain sequence of functions. All the same one can use $\delta(x)$ as though it were a proper function for practically all the purposes of quantum mechanics without getting incorrect results. One can also use the differential coefficients of $\delta(x)$, namely $\delta'(x), \delta''(x), \ldots$, which are even more discontinuous and less "proper" than $\delta(x)$ itself.

Dirac then cites a few elementary properties of $\delta(x)$. In his paper he introduces these saying, "A few elementary properties of these functions will now be given so as not to interrupt the argument later". In his informative discussion of delta functions in his history of distribution theory[82], Jesper Lützen quotes from Dirac's book that this is a list of "certain elementary properties of the δ-function which are deducible from, or at least not inconsistent with the definition". Those in his paper are:

$$\delta(-x) = \delta(x) \tag{62}$$
$$\delta'(-x) = -\delta'(x) \tag{63}$$

---

[82] Jesper Lützen, *The Prehistory of the Theory of Distributions*, Springer-Verlag, New York, 1982. The discussion of the δ-function covers pp. 110 - 143, with pp. 123 - 126 devoted to Dirac's work.

$$x\delta(x) = 0 \tag{64}$$

$$\int_{-\infty}^{\infty} f(x)\delta(a-x)dx = f(a), \text{ for any regular function } f \tag{65}$$

$$\int_{-\infty}^{\infty} f(x)\delta^{(n)}(a-x)dx = f^{(n)}(a) \tag{66}$$

$$\int_{-\infty}^{\infty} \delta(a-x)\delta(x-b)dx = \delta(a-b) \tag{67}$$

$$\int_{-\infty}^{\infty} \delta'(a-x)\delta(x-b)dx = \delta'(a-b) \tag{68}$$

$$\int_{-\infty}^{\infty} \delta'(a-x)\delta^{(n)}(x-b)dx = \delta^{(n+1)}(a-b) \tag{69}$$

$$\int_{-\infty}^{\infty} -x\delta'(x)dx = 1 \tag{70}$$

$$-x\delta'(x) = \delta(x). \tag{71}$$

Dirac's justification of (62) - (64) goes as follows:

We can obviously take $\delta(-x) = \delta(x), \delta'(-x) = -\delta'(x)$, etc. The condition $\delta(x) = 0$ except when $x = 0$ may be expressed by the algebraic equation $x\delta(x) = 0$.

Now the first line sounds like he is proposing (62) and (63) as axioms. However, we can almost derive them formally: for $x \neq 0, \delta(-x) = 0 = \delta(x)$ and for $x = 0$, whatever $\delta(0)$ should be, because $-0 = 0$, we must have $\delta(-0) = \delta(0)$. We can now establish (63) by differentiating (62) using the Chain Rule,

$$\delta'(-x)(-1) = \delta'(x),$$

and multiplying by -1: $\delta'(x) = -\delta'(-x)$.

Equation (64) is a little more problematic. Dirac says merely, "The condition $\delta(x) = 0$ except when $x = 0$ may be expressed by the algebraic equation $x\delta(x) = 0$". This isn't quite sufficient as the expression $x\delta(x)$ is of the indeterminate form $0 \cdot \infty$ at $x = 0$. And $0 \cdot \infty$ can take on any real value. Thus, it appears (64) is another condition imposed on δ. This is not the case: one can formally derive (64) using a lemma Dirac applies a couple of times without proof. This lemma requires (65).

Proving (65) is fairly easy. First note that $\delta(a-x) = 0$ for $x \neq a$, whence

$$\int_{-\infty}^{\infty} f(x)\delta(a-x)dx = \int_{a-h}^{a+h} f(x)\delta(a-x)dx$$

for any positive h. Choose h and let m_h and M_h be the minimum and maximum values, respectively, of $f(x)$ on $[a-h, a+h]$. Then

$$m_h = \int_{a-h}^{a+h} m_h\delta(a-x)dx \leq \int_{a-h}^{a+h} f(x)\delta(a-x)dx \leq \int_{a-h}^{a+h} M_h\delta(a-x)dx = M_h,$$

the two equalities holding because

$$\int_{a-h}^{a+h} \delta(a-x)dx = \int_{h}^{-h} \delta(u)(-du), \text{ for } u = a - x$$

$$= \int_{-h}^{h} \delta(u)du = 1.$$

Thus we see that

$$m_h \le \int_{-\infty}^{\infty} f(x)\delta(a-x)dx \le M_h.$$

Letting h go to 0, m_h and M_h both tend to $f(a)$, i.e.

$$f(a) \le \int_{-\infty}^{\infty} f(x)\delta(a-x)dx \le f(a).$$

We are now in a position to prove the promised lemma:

5.1 Lemma. *Let δ_1, δ_2 be "functions" satisfying*

$$\delta_i(x) = 0 \text{ for } x \ne 0 \qquad and \qquad \int_{-\infty}^{\infty} \delta_i(x)dx = 1.$$

Then: $\delta_1(x) = \delta_2(x)$.

Proof. Observe, for any a,

$$\delta_1(a) = \int_{-\infty}^{\infty} \delta_1(x)\delta_2(a-x)dx, \text{ by (65)}$$

$$= \int_{\infty}^{-\infty} \delta_1(a-u)\delta_2(u)(-du), \text{ for } u = a - x$$

$$= \int_{-\infty}^{\infty} \delta_2(u)\delta_1(a-u)dx$$

$$= \delta_2(a), \text{ by (65).} \qquad \square$$

The application of (65) is, of course, only heuristic as the derivation of (65) assumed the continuity of f and does not apply to δ.

We can use the Lemma to derive (64) by considering $\delta_1(x) = x\delta(x) + \delta(x)$. For $x \ne 0$,

$$\delta_1(x) = x \cdot 0 + 0 = 0,$$

and (60) is satisfied. To see that (61) holds, integrate:

$$\int_{-\infty}^{\infty} \delta_1(x)dx = \int_{-\infty}^{\infty} (x\delta(x) + \delta(x))\, dx$$

$$= \int_{-\infty}^{\infty} x\delta(x)dx + \int_{-\infty}^{\infty} \delta(x)dx$$

$$= \int_{-\infty}^{\infty} x\delta(x)dx + 1.$$

But

$$\int_{-\infty}^{\infty} x\delta(x)dx = \int_{\infty}^{-\infty} -y\delta(-y)(-dy), \text{ for } y = -x$$

$$= \int_{\infty}^{-\infty} y\delta(0-y)dy$$

$$= -\int_{-\infty}^{\infty} y\delta(0-y)dy$$

$$= -0, \text{ by (65)}.$$

Thus

$$\int_{-\infty}^{\infty} \delta_1(x)dx = 0 + 1 = 1$$

and $\delta_1(x) = \delta(x)$, i.e. $x\delta(x) + \delta(x) = \delta(x)$. Subtracting $\delta(x)$ yields (64).

Dirac establishes (66) by integration-by-parts:

$$\int_{-\infty}^{\infty} f(x)\delta'(a-x)dx =$$

$$= \left[-f(x)\delta(a-x) \right]_{-\infty}^{\infty} + \int_{-\infty}^{\infty} f'(x)\delta(a-x)dx = f'(a), \quad (72)$$

"since the integrated term vanishes at both limits", that is

$$\left[-f(x)\delta(a-x) \right]_{-\infty}^{\infty} = 0 - 0 = 0, \tag{73}$$

and

$$\int_{-\infty}^{\infty} f'(x)\delta(a-x)dx = f'(a),$$

by (65) applied to f'. I suppose the most questionable step here is the equation (73), which we might try justifying as follows:

$$\left[-f(x)\delta(a-x) \right]_{-\infty}^{\infty} = \lim_{k\to\infty} \left[-f(x)\delta(a-x) \right]_{-k}^{k}$$

$$= \lim_{k\to\infty} \left[-f(k)\cdot 0 + f(-k)\cdot 0 \right]$$

$$= \lim_{k\to\infty} 0 = 0.$$

Thus we have the special case of (66) for $n = 1$. One obtains the general case by induction:

$$\int_{-\infty}^{\infty} f(x)\delta^{(n+1)}(a-x)dx = \left[-f(x)\delta^{(n)}(a-x) \right]_{-\infty}^{\infty} + \int_{-\infty}^{\infty} f'(x)\delta^{(n)}(a-x)dx$$

$$= \int_{-\infty}^{\infty} f'(x)\delta^{(n)}(a-x)dx$$

$$= (f')^{(n)}(a) = f^{(n+1)}(a).$$

Dirac offers two justifications for (67). The first is by appeal to Lemma 5.1:
Define

$$\phi(b) = \int_{-\infty}^{\infty} \delta(a - x)\delta(x - b)dx.$$

For $b \neq a$, let h be a positive number smaller than half the distance between a
and b: $0 < h < |b-a|/2$. Now, $\delta(a-x) = 0$ outside $[a-h, a+h]$ and $\delta(b-x) = 0$
outside $[b - h, b + h]$ and these two intervals do not overlap. Thus

$$\phi(b) = \int_{a-h}^{a+h} \delta(a - x)\delta(x - b)dx + \int_{b-h}^{b+h} \delta(a - x)\delta(x - b)dx$$

$$= \int_{a-h}^{a+h} \delta(a - x)0dx + \int_{b-h}^{b+h} 0\delta(x - b)dx = 0.$$

whence the first defining condition for a delta function holds. Now consider

$$\int_{-\infty}^{\infty} \phi(b)db = \int_{-\infty}^{\infty} \int_{-\infty}^{\infty} \delta(a - x)\delta(x - b)dx\, db$$

$$= \int_{-\infty}^{\infty} \int_{-\infty}^{\infty} \delta(a - x)\delta(x - b)db\, dx$$

$$= \int_{-\infty}^{\infty} \delta(a - x)\left(\int_{-\infty}^{\infty} \delta(x - b)db \right) dx$$

$$= \int_{-\infty}^{\infty} \delta(a - x) \cdot 1dx, \text{ by (65) for } f(x) = 1$$

$$= \int_{\infty}^{-\infty} \delta(u)(-du), \text{ for } u = a - x$$

$$= \int_{-\infty}^{\infty} \delta(u)du = 1.$$

"Hence $\phi(b)$ has all the properties of $\delta(a-b)$ and may be put equal to $\delta(a-b)$."
I rather like his second proof: Set $f(x) = \delta(x - b)$ and apply (65). He says

This is thus a case in which we can use $\delta(x - b)$ as though it were a
regular function of x without getting a wrong result. Another such case
is the putting of $f(x)$ equal to $\delta(x - b)$, or, more generally, $\delta^{(n)}(x - b)$,
in (66), leading to equations

(68) and (69). Similarly, setting $f(x) = x$ and $a = 0$ in (66), he obtains (70).
To derive (71) Dirac says

But $-x\delta'(x)$, considered as a function of x, vanishes when $|x|$ is very
small. Hence $-x\delta'(x)$ has all the properties of $\delta(x)$, and we can write

$$-x\delta'(x) = \delta(x).$$

In other words, he says that Lemma 5.1 applies. Thus, let $\delta_1(x) = -x\delta'(x)$. By
(70), δ_1 satisfies the integral property of a delta function. For $x \neq 0$,

$$\frac{\delta(x+h) - \delta(x)}{h} = \frac{0-0}{h} = 0,$$

for h sufficiently small. Thus

$$\delta'(x) = \lim_{h \to 0} 0 = 0,$$

and $-x\delta'(x) = 0$. This establishes the first defining property(60) of the delta function and the Lemma applies: $\delta_1(x) = \delta(x)$.

As mentioned earlier, in later editions of his subsequent book, Dirac extended the list of elementary properties. Lützen cites, among others, the following two from the third edition:

$$\delta(ax) = a^{-1}\delta(x) \tag{74}$$

$$\delta(x^2 - a^2) = \frac{1}{2}a^{-1}\{\delta(x-a) + \delta(x+a)\}, \text{ for } a > 0, \tag{75}$$

remarking that "All of this shows Dirac as a skilful manipulator of the δ-function. Some of the above theorems, especially (75), are not even obvious in distribution theory, since the changes of variables are hard to perform in $\mathcal{D}'$ ".[83]

I won't attempt to derive (75). The equation (74), however, is another easy application of Lemma 5.1, provided we correct Lützen's statement by adding the condition that a be positive. Let $\delta_1(x) = a\delta(ax)$. For $x \neq 0$, we have $ax \neq 0$, whence

$$\delta_1(x) = a\delta(ax) = 0.$$

Thus the first defining property of δ holds for δ_1. The second:

$$\int_{-\infty}^{\infty} \delta_1(x)dx = \int_{-\infty}^{\infty} a\delta(ax)dx$$

$$= \int_{-\infty}^{\infty} a\delta(u)\frac{du}{a}, \text{ for } u = ax$$

$$= \int_{-\infty}^{\infty} \delta(u)du = 1.$$

Thus, $\delta_1 = \delta$, i.e. $a\delta(ax) = \delta(x)$, whence (74) follows by division.

Everything we have done so far is, of course, purely formal. No function δ satisfying the defining properties (60) and (61) exists. Assuming we had a generalised notion of function under which δ existed, there is still no guarantee that it satisfies all the properties used in deriving (62) - (71) or (74) and (75). The most questionable assumptions above were 1) the application of (65), which was proven for continuous f, to δ in the proof of Lemma 5.1, and 2) Dirac's application of integration-by-parts (equation (72), above). Assumption 1) speaks for itself, and as concerns 2), I am willing to accept the equality of the anti-derivatives,

---

[83] *Op. cit.*, pp. 125 - 126. Distribution theory is a branch of analysis which allows a rigorous treatment of δ and similar *generalised functions*. $\mathcal{D}'$ is the set of these functions.

$$\int f(x)\delta'(a-x)dx = -f(x)\delta(a-x) + \int f'(x)\delta(a-x)dx,$$

but the step from here to (72), i.e., to

$$\int_{-\infty}^{\infty} f(x)\delta'(a-x)dx = \left[-f(x)\delta(a-x) + \int f'(x)\delta(a-x)dx\right]_{-\infty}^{\infty}$$

requires the Fundamental Theorem of Calculus: if $F' = f$,

$$\int_{\alpha}^{\beta} f(x)dx = F(\beta) - F(\alpha),$$

which requires f to be continuous in $[\alpha, \beta]$. But consider $f(x) = x^{-2}, F(x) = -x^{-1}$ and note that

$$\int_{-1}^{1} \frac{dx}{x^2} = \infty, \text{ while } F(1) - F(-1) = 0.$$

Perhaps δ is as badly behaved.

Along these same lines, one might ask for the value of

$$\int_{0}^{\infty} \delta(x)dx.$$

Formally applying the rules of the Calculus, we have

$$1 = \int_{-\infty}^{\infty} \delta(x)dx = \int_{-\infty}^{0} \delta(x)dx + \int_{0}^{\infty} \delta(x)dx$$

$$= \int_{\infty}^{0} \delta(-y)(-dy) + \int_{0}^{\infty} \delta(x)dx, \text{ for } y = -x$$

$$= \int_{0}^{\infty} \delta(-y)dy + \int_{0}^{\infty} \delta(x)dx$$

$$= \int_{0}^{\infty} \delta(y)dy + \int_{0}^{\infty} \delta(x)dx, \text{ by (62)}$$

$$= 2\int_{0}^{\infty} \delta(x)dx.$$

Thus

$$\int_{0}^{\infty} \delta(x)dx = \frac{1}{2}. \tag{76}$$

That was simple enough. But consider: all the effect of δ is concentrated at the single point $x = 0$. Thus we should have

$$\int_{-\infty}^{\infty} \delta(x)dx = \int_{0}^{0} \delta(x)dx = \int_{0}^{\infty} \delta(x)dx,$$

and thus also

$$\int_0^\infty \delta(x)dx = 1. \tag{77}$$

The situation can be clarified by considering some examples of δ-functions, something not easy to do considering the fact that the delta function, which is unique, doesn't actually exist. But we shouldn't let that stop us. There are actually several possible approaches. We shall see one in Chapter III, section 3, below. That construction, however, will be based on the interval $[0, \infty)$ and will not serve to help us choose between the alternatives (76) and (77). It simply assumes the latter.

5.2 Nonstandard δ-Functions

Allowing infinitesimals and infinite numbers it is easy to give several examples of δ-functions. This can be done rigorously in Nonstandard Analysis, or purely formally by just assuming their existence and dealing with them. As this chapter is devoted to Type I Formalism, I follow the latter approach.

To clarify the extent to which the examples qualify as δ-functions, i.e. the sense in which (60) and (61) are true, let me introduce here a few basic distinctions and a little notation. We think of the infinitesimal and infinite numbers as residing in some extension $\mathbb{H}$ of $\mathbb{R}$. The elements of $\mathbb{H}$ that are not in $\mathbb{R}$ are called *hyperreals*, and come in three flavours. A hyperreal ρ is *infinite* if $|\rho| > r$ for every real number r. If $|\rho| < r$ for some real r, then ρ is *finite*. And, if $|\rho| < r$ for every positive real number r, then ρ is an *infinitesimal*. The practice in the 18th century was to calculate with infinitesimals as if they were distinct from 0, and in the end when two quantities differed by an infinitesimal, to ignore the difference. As this can be confusing and occasionally objectionable, modern nonstandard analysts introduce a second equality relation, namely, equality up to an infinitesimal: for $\rho, \sigma \in \mathbb{H}$, we write

$$\rho \approx \sigma \quad \text{iff} \quad \rho - \sigma \text{ is infinitesimal.}$$

Note that two real numbers that are equal up to an infinitesimal are, in fact, equal. This does not hold for hyperreals.

5.2 Definition. *A function $\delta : \mathbb{H} \to \mathbb{H}$ is a delta function iff it satisfies the following two conditions:*
i. for all $x \in \mathbb{R}$, if $x \neq 0$, then $\delta(x) \approx 0$
ii. $\int_{-\infty}^\infty \delta(\alpha)d\alpha = 1$.

5.3 Example. Let $\eta > 0$ be infinitesimal and define, for $\alpha \in \mathbb{H}$,

$$\delta_\eta(\alpha) = \begin{cases} 0, & |\alpha| > \eta \\ \frac{1}{2\eta}, & |\alpha| \leq \eta. \end{cases}$$

For any real number $x \neq 0$, either $x < \eta$ or $x > \eta$ (since otherwise $x \approx 0$ and x would then equal 0), and $\delta_\eta(x) = 0$. Thus condition i of Definition 5.2 is satisfied. Moreover, δ_η is a simple step-function, whence

$$\int_{-\infty}^{\infty} \delta_\eta(\alpha)d\alpha = \int_{-\infty}^{-\eta} 0 d\alpha + \int_{-\eta}^{\eta} \frac{d\alpha}{2\eta} + \int_{\eta}^{\infty} 0 d\alpha$$

$$= 0 + \left[\frac{1}{2\eta} \cdot \alpha\right]_{-\eta}^{\eta} + 0 = \frac{\eta}{2\eta} - \frac{-\eta}{2\eta} = \frac{2\eta}{2\eta} = 1.$$

Thus 5.2.ii is also satisfied and δ_η is a δ-function. For this choice of δ, clearly (76) holds:

$$\int_0^{\infty} \delta_\eta(\alpha)d\alpha = \int_0^{\eta} \frac{d\alpha}{2\eta} + \int_{\eta}^{\infty} 0 d\alpha = \left[\frac{1}{2\eta} \cdot \alpha\right]_0^{\eta} + 0 = \frac{1}{2\eta} \cdot \eta - 0 + 0 = \frac{1}{2}.$$

As for Dirac's properties (62) - (63), it is obvious that (62) holds for all real x, and indeed for all $\alpha \in \mathbb{H}$.

If we differentiate δ_η, we find

$$\delta_\eta'(\alpha) = \begin{cases} 0, & |\alpha| \neq \eta \\ undefined, & |\alpha| = \eta, \end{cases} \tag{78}$$

for the graph of δ_η is flat everywhere except at the points $\alpha = \pm\eta$ where there are jumps. So, except for these two points at which δ_η is undefined,

$$\delta_\eta'(-\alpha) = -\delta_\eta'(\alpha).$$

In particular, (63) holds for all $x \in \mathbb{R}$.

Equation (64) does not hold for all α: if $|\alpha| < \eta$ and $\alpha \neq 0$,

$$\alpha\delta_\eta(\alpha) = \alpha \cdot \frac{1}{2\eta} \neq \frac{1}{2\eta} = \delta_\eta(\alpha).$$

Depending on how much smaller α is than η, $\alpha\delta_\eta(\alpha)$ could be infinitesimal (e.g., if $\alpha = \eta^2$) or finite (e.g., if $\alpha = \eta/2$). But for real x we have

$$x \cdot \delta_\eta(x) = x \cdot 0 = 0$$

for $x \neq 0$; and

$$0\delta_\eta(0) = 0 \cdot \frac{1}{2\eta} = 0$$

if $x = 0$. Thus (64) does hold for $x \in \mathbb{R}$.

The all important (65) which was used in the proof of Lemma 5.1 and on which consequently the proofs of several later properties depended holds for all continuous $f : \mathbb{R} \to \mathbb{R}$. For, if f is continuous and $a \in \mathbb{R}$,

$$\int_{-\infty}^{\infty} f(\alpha)\delta_\eta(a-\alpha)d\alpha = \int_{\infty}^{-\infty} f(a-\beta)\delta_\eta(\beta)(-d\beta), \text{ for } \beta = a - \alpha$$

$$= \int_{-\infty}^{\infty} f(a-\beta)\delta_\eta(\beta)d\beta$$

$$= \int_{-\eta}^{\eta} f(a-\beta)\delta_\eta(\beta)d\beta$$

$$= \int_{-\eta}^{\eta} f(a - \beta) \frac{1}{2\eta} d\beta$$

$$= \frac{1}{2\eta} \int_{-\eta}^{\eta} f(a - \beta) d\beta$$

$$= \frac{1}{2\eta} f(a - \beta_0)(\eta - (-\eta)),$$

for some $\beta_0 \in (-\eta, \eta)$ by the Mean-Value Theorem for Integrals,

$$= f(a - \beta_0)$$
$$\approx f(a),$$

by continuity: a function $g : \mathbb{R} \to \mathbb{R}$ is continuous at a just in case $g(a + \epsilon) \approx g(a)$ for any infinitesimal ϵ. (And here, of course, since $0 < \beta_0 < \eta$, β_0 is infinitesimal.)

We have already seen that (64) holds for δ_η. Dirac's proof is suspect because it depended on Lemma 5.1, which in turn depended on the application of (65) to two delta functions,— and we have only proven (65) for continuous functions. Indeed, the Lemma is false, as is seen by considering δ_ϵ for an infinitesimal $\epsilon \neq \eta$:

$$\delta_\eta(0) = \frac{1}{2\eta} \not\approx \frac{1}{2\epsilon} = \delta_\epsilon(0).$$

Indeed, the difference

$$\delta_\eta(0) - \delta_\epsilon(0) = \frac{1}{2\eta} - \frac{1}{2\epsilon} = \frac{\epsilon - \eta}{2\eta\epsilon}$$

can be infinite: choose $\epsilon = 2\eta$ and observe

$$\frac{\epsilon - \eta}{2\eta\epsilon} = \frac{\eta}{4\eta^2} = \frac{1}{4\eta}.$$

This brings into question the validity of (67) and (71). The former had two derivations, one appealing to Lemma 5.1 and one applying (65) to a δ-function. The proof of (71) was a simple application of the Lemma, as was the proof of (74) which we might as well also consider at this point. Both (71) and (74) fail for δ_η.

To see the failure of (71), recall from (78) that $\delta'_\eta(0) = 0$, whence

$$0\delta'_\eta(0) = 0 \neq \frac{1}{2\eta} = \delta_\eta(0).$$

And for (74), let $a > 0$ and observe

$$\delta_\eta(a \cdot 0) = \frac{1}{2\eta} \not\approx \frac{1}{a} \cdot \frac{1}{2\eta} = \frac{1}{a} \cdot \delta_\eta(0).$$

Equation (67), however, does hold. For all $a, b \in \mathbb{R}$,

$$\int_{-\infty}^{\infty} \delta_\eta(a - \alpha)\delta_\eta(\alpha - b)d\alpha = \delta_\eta(a - b)$$

The proof splits into two cases, according as to whether or not $a = b$. If $a = b$,

$$\int_{-\infty}^{\infty} \delta_\eta(a - \alpha)\delta_\eta(\alpha - b)d\alpha = \int_{-\infty}^{\infty} \delta_\eta(a - \alpha)\delta_\eta(\alpha - a)d\alpha, \text{ for } a = b$$

$$= \int_{-\infty}^{\infty} \delta_\eta(a - \alpha)^2 d\alpha, \text{ by (62)}$$

$$= \int_{\infty}^{-\infty} \delta_\eta(\beta)^2(-d\beta), \text{ for } \beta = a - \alpha$$

$$= \int_{-\infty}^{\infty} \delta_\eta(\beta)^2 d\beta$$

$$= \int_{-\eta}^{\eta} \left(\frac{1}{2\eta}\right)^2 d\beta = \left[\left(\frac{1}{2\eta}\right)^2 \beta\right]_{-\eta}^{\eta}$$

$$= \left(\frac{1}{2\eta}\right)^2 (\eta - (-\eta)) = \frac{2\eta}{(2\eta)^2} = \frac{1}{2\eta}$$

$$= \delta_\eta(0) = \delta_\eta(a - b).$$

If $a \neq b$,

$$\int_{-\infty}^{\infty} \delta_\eta(a - \alpha)\delta_\eta(\alpha - b)d\alpha = \int_{-\infty}^{\infty} 0\, d\alpha = 0 = \delta_\eta(a - b),$$

because, for any $c \in \mathbb{R}$, $\delta_\eta(\alpha - c)$ is nonzero only if $-\eta < \alpha - c < \eta$ and, for $a \neq b$, a and b are too far apart for both $\delta_\eta(a - \alpha) = \delta_\eta(\alpha - a)$ and $\delta_\eta(\alpha - b)$ to be non-zero.

Dirac's derivations of (68) - (70) all refer back to (66), which we haven't considered yet. I can prove that this equation fails: Let f have a continuous derivative. Now

$$f(\alpha)\delta_\eta'(a - \alpha) = \begin{cases} 0, & |a - \alpha| \neq \eta \\ undefined, & |a - \alpha| = \eta. \end{cases}$$

This differs from the constant 0 function at only two points $\alpha = a \pm \eta$. It follows that

$$\int_{-\infty}^{\infty} f(\alpha)\delta_\eta'(a - \alpha)d\alpha = \int_{-\infty}^{\infty} 0\, d\alpha = 0,$$

which fails to equal $f'(a)$ for any number of choices of f, a.

Lützen quotes the third edition of *The Principles of Quantum Mechanics* for an "alternative way of defining the δ-function" as the derivative of the Heaviside function[84]:

---

[84] Lützen, *op. cit.*, p. 126. Indeed, Lützen points out (p. 130) that Heaviside favoured this definition of the delta function.

$$H(x) = \begin{cases} 0, & x < 0 \\ 1, & 0 \le x. \end{cases}$$

Note that $H'(x) = 0$ for any non-zero real number x. $H'(0)$ is undefined, but if one ignores this fact and applies the Fundamental Theorem of Calculus,

$$\int_{-\infty}^{\infty} H'(x)dx = H(x)|_{-\infty}^{\infty} = \lim_{k \to \infty} H(x)|_{-k}^{k} = 1 - 0 = 1.$$

So Lemma 5.1 tells us $H' = \delta$.

Now δ_η is not the derivative of H. In $\mathbb{H}$ it has an antiderivative, namely

$$h_\eta(\beta) = \int_{-\infty}^{\beta} \delta_\eta(\alpha)d\alpha$$

$$= \begin{cases} 0, & \beta \le -\eta \\ \dfrac{\beta + \eta}{2\eta}, & -\eta \le \beta \le \eta \\ 1, & \eta \le \beta. \end{cases} \tag{79}$$

Thus, for $x \in \mathbb{R}$,

$$h_\eta(x) = \begin{cases} 0, & x < 0 \\ 1/2, & x = 0 \\ 1, & 0 < x. \end{cases}$$

We can change this by shifting δ_η slightly. For ε being 0 or infnitesimal, define

$$\delta_{\eta,\varepsilon}(\alpha) = \delta_\eta(\alpha - \varepsilon).$$

This is still a δ-function under our definition, as the reader may easily verify. Define

$$h_{\eta,\varepsilon}(\beta) = \int_{-\infty}^{\beta} \delta_{\eta,\varepsilon}(\alpha)d\alpha$$

$$= \begin{cases} 0, & \beta < \eta + \varepsilon \\ \dfrac{\beta - \varepsilon + \eta}{2\eta}, & -\eta + \varepsilon \le \beta \le \eta + \varepsilon \\ 1, & \eta + \varepsilon < \beta. \end{cases}$$

Now $h_{\eta,\varepsilon}(0)$ need not equal a real number, but it is infinitesimally close to one (namely, the least upper bound of the set of $r \in \mathbb{R}$ for which $r \le h_{\eta,\varepsilon}(0)$). This real number can be taken to be any number in the interval $[0, 1]$ by careful choice of ε. For example, if we choose $\varepsilon = -\eta + \varepsilon^*$, where ε^* is infinitesimally small compared to η (e.g. $\varepsilon^* = \eta^2$), we will have $h_{\eta,\varepsilon}(0) \approx 1$. Choosing $\varepsilon^* = 0$ (i.e. $\varepsilon = -\eta$) will in fact make $h_{\eta,\varepsilon}(0) = 1$. If $\varepsilon^* < 0$, the entire non-zero portion of the graph of $\delta_{\eta,\varepsilon}$ lies in some infinitesimal neighbourhood to the left of the origin, i.e. $\delta_{\eta,\varepsilon}(0) = 0$, i.e. $\delta_{\eta,\varepsilon}(x) = 0$ for all real x (but, of course, not for all hyperreal α).

Detlef Laugwitz, one of the pioneers of Nonstandard Analysis and one of the more interesting historians of mathematics, gave a thoughtful treatment of δ-functions in his book on nonstandard analysis[85]. He says

> Occasionally one says, for $x \neq 0$ δ is the derivative of the Heaviside-function $x \mapsto 1/2$ for $x \gtrless 0$, and the second derivative of $|x|/2$. This mode of expression describes the intuitive background, but can mislead.[86]

He takes this intuition as basic however, and gives a (Type II) formal definition of a δ-function as follows:

5.4 Definition. *A function $\delta : \mathbb{H} \to \mathbb{H}$ is called an* (L)-delta-function *if there is a function $Q : \mathbb{H} \to \mathbb{H}$ such that $Q(\alpha) \approx |\alpha|/2$ for all finite $\alpha \in \mathbb{H}$ and $Q''(\alpha) = \delta(\alpha)$ for all $\alpha \in \mathbb{H}$.*

I have introduced the "(L)" prefix because this definition is not equivalent to our earlier Definition 5.2. It is not even at first sight evident that δ_η is an (L)-delta-function. To see that it is indeed such, rewrite (79) as

$$h_\eta(\beta) = \begin{cases} 0, & \beta < -\eta \\ 1/2 + \beta/(2\eta), & -\eta \leq \beta \leq \eta \\ 1, & \eta < \beta. \end{cases}$$

Subtracting $1/2$ yields another anti-derivative:

$$P_\eta(\beta) = \begin{cases} -1/2, & \beta < -\eta \\ \beta/(2\eta), & -\eta \leq \beta \leq \eta \\ 1/2, & \eta < \beta. \end{cases}$$

An antiderivative of this is

$$Q_{\eta,C_1,C_2}(\beta) = \begin{cases} -\beta/2 + C_1, & \beta < -\eta \\ \beta^2/(4\eta), & -\eta \leq \beta \leq \eta \\ \beta/2 + C_2, & \eta < \beta, \end{cases}$$

where C_1, C_2 are arbitrary constants. We choose C_1, C_2 so as to make Q_{η,C_1,C_2} continuous at $\beta = \pm\eta$: From

---

[85] Detlef Laugwitz, *Infinitesimalkalkül: Kontinuum und Zahlen— Eine elementare Einführung in die Nichtstandard Analysis*, Bibliographisches Institut, Mannheim, 1978, pp. 159 - 163. Presumably this dates back to his paper, "Eine Einführung der Deltafunktionen" (*Sitzungsberichte Bayer. Akademie Wiss.*, 1959, pp. 41 - 49), which he cites in the list of references to his discussion of δ-functions and nonstandard distribution theory. The discussion in his initial joint paper with Curt Schmieden ("Eine Erweiterung der Infinitesimalrechnung", *Mathematische Zeitschrift* 69 (1958), pp. 1 - 39) was certainly not as complete.

[86] *Ibid.*, p. 159.

$$\frac{\beta^2}{4\eta} = \pm\frac{\beta}{2} + C_i$$

at $\beta = \pm\eta$, it follows that

$$C_1 = \frac{(-\eta)^2}{4\eta} + \frac{-\eta}{2} = \frac{\eta^2}{4\eta} - \frac{\eta}{2} = \frac{-\eta}{4}$$

$$C_2 = \frac{\eta^2}{4\eta} - \frac{-\eta}{2} = \frac{-\eta}{4}.$$

This gives us

$$Q_\eta(\beta) = \begin{cases} -\beta/2 - \eta/4, & \beta \leq -\eta \\ \beta^2/(4\eta), & -\eta \leq \beta \leq \eta \\ \beta/2 - \eta/4, & \eta \leq \beta. \end{cases}$$

In particular, for $|\beta| \geq \eta$,

$$Q_\eta(\beta) = \frac{|\beta|}{2} - \frac{\eta}{4} \approx \frac{|\beta|}{2},$$

while for $|\beta| \leq \eta$,

$$Q_\eta(\beta) = \frac{\beta^2}{4\eta} \leq \frac{\eta^2}{4\eta} \leq \frac{\eta}{4},$$

whence $Q_\eta(\beta) \approx 0 \approx |\beta|/2$ for such β. Thus we see that δ_η is an (L)-delta-function.

Laugwitz provides a Theorem which allows the construction of (L)-delta-functions and thus allows us to see where his definition is more general than Definition 5.2.

5.5 Theorem. *Let* $g : \mathbb{R} \to \mathbb{R}$ *be a continuous function for which*

$$\int_0^\infty g(x)dx = \frac{1}{2} = \int_{-\infty}^0 g(x)dx.$$

Then: for any positive infinite number Ω,

$$\delta_{g,\Omega}(\alpha) = \Omega \cdot g(\Omega\alpha)$$

is an (L)-delta-function.

I shall not prove this here. Instead I prove the easier variant[87]:

5.6 Theorem. *Let* $g : \mathbb{R} \to \mathbb{R}$ *be a continuous function for which*

$$\int_0^\infty g(x)dx = \frac{1}{2} = \int_{-\infty}^0 g(x)dx. \quad and \quad \lim_{|x|\to\infty} xg(x) = 0$$

Then: for any positive infinite number Ω,

$$\delta_{g,\Omega}(\alpha) = \Omega \cdot g(\Omega\alpha)$$

is a delta function.

---

[87] The result, stated under the more stringent requirement that g be monotone decreasing as $|x| \to \infty$ was stated already in Schmieden and Laugwitz, *op. cit.*

Proof. First observe that

$$\int_{-\infty}^{\infty} \delta_{g,\Omega}(\alpha)d\alpha = \int_{-\infty}^{\infty} \Omega \cdot g(\Omega\alpha)d\alpha$$

$$= \int_{-\infty}^{\infty} g(u)du, \text{ where } u = \Omega\alpha$$

$$= 1, \text{ by assumption.}$$

The other requirement that $\delta_{g,\Omega}$ be a delta function is that $\delta_{g,\Omega}(x) \approx 0$ for all real $x \neq 0$. To see that this is the case, let $\epsilon > 0$ be any positive real number. Because $\lim_{|x| \to \infty} xg(x) = 0$, there is some x_0 such that

$$|xg(x)| = |xg(x) - 0| < \epsilon$$

whenever $|x| > x_0$. But, if $x \neq 0$ is finite and Ω is infinite, we have $|\Omega x| > x_0$, whence $|\Omega \cdot g(\Omega x)| < \epsilon$. As this holds for all real $\epsilon > 0$, it follows that $\Omega \cdot g(\Omega x) \approx 0$. □

That the two integrals from $-\infty$ to 0 and from 0 to ∞ agree is not necessary for this Theorem, but underscores the natural symmetry assumption embodied in (62). Thus, we can use the Theorem to construct delta functions by starting with any continuous function g which goes to 0 rapidly as $|x| \to \infty$ and for which $\int_{-\infty}^{\infty} g(x)dx$ exists, normalising the function to make the integral 1, and then applying the Theorem. For this, it is useful to have a table of definite integrals handy.[88] The boxed table below lists the results of my search.

Of these, (2) and (5) are cited explicitly by Laugwitz and certainly occur in various guises in the standard literature. The others are just cited as additional examples. (1), (4), (6), and (7) have graphs with cute cusps, but are otherwise of no special interest.

As for the evaluation of the integrals, (1), (2), and (4) are readily done by standard techniques of the Calculus. Function (5) is a cornerstone of probability theory and its integration is sometimes presented in Calculus textbooks because it reduces via a trick to a simple integral involving polar coordinates. Its corresponding δ-function goes back to Kirchhoff[89]. Number (3) is probably best handled by the methods of Complex Analysis.[90] I confess I haven't checked the integration of functions (6) and (7).

As for the crucial limit of $xg(x)$ as $|x| \to \infty$, all of these functions can be handled by standard methods of the Calculus. Only number (7) is a little tricky: Assume x positive to avoid having to write absolute value signs and observe:

$$\lim_{x \to \infty} x \ln(1 + e^{-x}) = \lim_{x \to \infty} \frac{\ln(1 + e^{-x})}{1/x}$$

---

[88] I used Herbert Bristol Dwight, *Tables of Integrals and Other Mathematical Data*, 4th ed., The Macmillan Company, New York, 1961. In doing so, I only took the simplest looking examples.

[89] Cf. Lützen, *op. cit.*, p. 130.

[90] Cf., e.g., Joseph Bak and Donald Newman, *Complex Analysis*, Springer-Verlag, New York, 1982, p. 134 (Exercise 2).

Table 1. Some Delta Functions

	$g(x)$	$\int_{-\infty}^{\infty} g(x)dx$	$\delta_{g,\Omega}(\alpha)$
(1)	$\dfrac{1}{(1+|x|)^2}$	2	$\dfrac{\Omega}{2(1+\Omega|\alpha|)^2}$
(2)	$\dfrac{1}{1+x^2}$	π	$\dfrac{\Omega}{\pi(1+\Omega^2\alpha^2)}$
(3)	$\dfrac{\sin^2 x}{x^2}$	π	$\dfrac{\sin^2(\Omega\alpha)}{\pi\Omega\alpha^2}$
(4)	$e^{-|x|}$	2	$\dfrac{\Omega}{2}e^{-|\Omega\alpha|}$
(5)	e^{-x^2}	$\sqrt{\pi}$	$\dfrac{\Omega e^{-\Omega^2\alpha^2}}{\sqrt{\pi}}$
(6)	$\dfrac{|x|}{e^{|x|}-1}$	$\dfrac{\pi^2}{3}$	$\dfrac{3|\Omega\alpha|}{e^{|\Omega\alpha|}-1}$
(7)	$\ln(1+e^{-|x|})$	$\dfrac{\pi^2}{6}$	$\dfrac{6\Omega\ln(1+e^{-\Omega|\alpha|})}{\pi^2}$

$$= \lim_{x\to\infty} \frac{\dfrac{-e^{-x}}{1+e^{-x}}}{\dfrac{-1}{x^2}}, \text{ by L'Hôpital's Rule}$$

$$= \lim_{x\to\infty} \frac{x^2 e^{-x}}{1+e^{-x}} = \lim_{x\to\infty} x^2 e^{-x}$$

$$= \lim_{x\to\infty} \frac{x^2}{e^x} = 0,$$

after 2 further applications of L'Hôpital's Rule.

If we don't worry about the limit of $xg(x)$, Theorem 5.5 can similarly be applied to construct (L)-delta-functions. I found 3 interesting such functions in the tables:

$$g_1(x) = \begin{cases} \dfrac{\sin x}{x}, & x \neq 0 \\ 1, & x = 0 \end{cases}, \quad g_2(x) = \sin\frac{\pi x^2}{2}, \quad g_3(x) = \cos\frac{\pi x^2}{2}.$$

I don't know about g_2 and g_3, but g_1, which is cited by Laugwitz, is the function from which a particularly famous delta function is derived, namely the one used by Joseph Fourier in 1822 in dealing with Fourier integrals[91]. Appealing to Complex Analysis, one finds

$$\int_{-\infty}^{\infty} \frac{\sin x}{x}dx = \pi,$$

whence

---

[91] Cf. Lützen, *op. cit.*, pp. 112 - 115.

$$\delta_{g_1,\Omega}(\alpha) = \frac{\sin(\Omega\alpha)}{\pi\alpha}$$

is an (L)-delta-function. The limit crucial to application of Theorem 5.6 does not exist, however, as

$$x g_1(x) = x\frac{\sin x}{x} = \sin x$$

oscillates and does not approach 0 as a limit as $|x|$ grows without bound. Thus we cannot conclude that $\delta_{g_1,\Omega}$ is a delta function according to Definition 5.2. In fact, if we assume Ω is an infinite odd integer,

$$\delta_{g_1,\Omega}(\pi/2) = \frac{\Omega}{\pi}\frac{\sin(\Omega\pi/2)}{\Omega\pi/2} = \frac{\pm 2}{\pi^2} \not\approx 0,$$

whence $\delta_{g_1,\Omega}$ is definitely not a delta function under that definition.

We have strayed a bit from the path I intended to follow, which was to consider the Dirac delta function and its applications as an example of Type I Formalism. Instead I have been using infinitely large and infinitely small numbers in a Type I manner to give what aspires to be a Type II Formal definition of a delta function and thus a Type II replacement for Dirac's purely formal fiction. (Indeed, to those familiar with Nonstandard Analysis and the properties of $\mathbb{H}$ and its elements, I have been giving a rigorous Type II formal treatment of the delta function.) I want to get back to δ itself and its formal use in applications. So, forget now that there is a legitimate meaning to δ, or pretend that our longish discussion of δ-functions on $\mathbb{H}$ was mere heuristic and let us return to the days when all we had were (60) - (71) and see how they are formally applied in mathematics.

5.3 Differential Equations

The simplest application of the δ-function is to differential equations. One uses δ to represent a pulse function or an explosion— any sudden imparting of a force that has a negligibly short duration. This is a contribution to acceleration. One integration yields velocity and a step function given by some translate of the Heaviside function. One more integration gives distance and a continuous function.

Now, δ is a normalised function, chosen to have overall effect 1. Different instantaneous overall forces would be represented by multiplying by different constants. A typical problem would involve some moving object with initial position $y(0)$ and initial velocity $y'(0)$. At some time $t = a$, a fairly instantaneous acceleration is imparted, perhaps through a collision or an explosion.

A simple example using only Calculus would be the following:

5.7 Example. An object 5 meters away is initially moving away at 2 m/sec, with a constant acceleration of 1 m/sec^2. Three seconds from now it will receive an impulse of 2 m/sec^2. How far will it be in 10 seconds?

Setting up the equation, we have

$$y''(t) = 1 + 2\delta(t-3), \quad y(0) = 5, \quad y'(0) = 2.$$

Now, for the purposes of such problems we take the Heaviside function as the antiderivative of δ. There is, as we saw in dealing with δ_n, some question about what value to give to H at 0, but once we've eliminated δ, we will be dealing with step functions, which we integrate normally, and the value at one point will not matter— so long as it is not infinite. Also, whether one takes $H(x)$ or $H(x) - \frac{1}{2}$, as Laugwitz prefers, doesn't matter because of the constant of integration. Following all this verbiage, we simply integrate and get

$$y'(t) = t + 2H(t-3) + C,$$

where C is determined by the initial value $y'(0) = 2$:

$$2 = 0 + 2 \cdot 0 + C.$$

Thus we have

$$y'(t) = t + 2H(t-3) + 2.$$

Now, $H(t-3)$ is a step function:

$$H(t-3) = \begin{cases} 0, & t < 3 \\ 1, & 3 \le t. \end{cases}$$

So an antiderivative on $(-\infty, 3)$ is 0 and an antiderivative on $[3, \infty)$ is t. Thus H has the antiderivatives

$$f(t) = \begin{cases} 0 + C_1, & 0 < 3 \\ t + C_2, & 3 \le t, \end{cases}$$

with C_1, C_2 constants of integration. If we choose $C_1 = 0$, we must choose $C_2 = -3$ to make f continuous. If we do this, we have

$$f(t) = (t-3)H(t-3).$$

Putting this together we have

$$y(t) = \frac{t^2}{2} + 2(t-3)H(t-3) + 2t + C,$$

where C again is determined by an initial condition: $Y(0) = 5$.

$$5 = 0 + 2 \cdot \quad 3 \cdot 0 + 2 \cdot 0 + C,$$

and $C = 5$. Thus our final distance function is

$$y(t) = \frac{t^2}{2} + 2(t-3)H(t-3) + 2t + 5$$

and the answer to our little exercise is

$$y(10) = \frac{100}{2} + 2 \cdot 7 \cdot 1 + 20 + 5 = 50 + 14 + 20 + 5 = 89.$$

In "real life" one is never given anything so simple as an equation $y''(t) = f(t)$. The acceleration may depend on $y(t)$ (as with gravity) or $y'(t)$ (resistance) and a typical equation may be $y''(t) = ay'(t) + by(t) + f(t)$. When that happens, one introduces new techniques, especially the use of the *Laplace Transform*. The Laplace transform of a function $f : [0, \infty) \to \mathbb{R}$ is defined by

$$\mathcal{L}\{f(t)\} = \int_0^\infty e^{-st} f(t) dt.$$

$\mathcal{L}\{f(t)\}$ is a function of the variable s and is defined for sufficiently large s if f is bounded by some exponential function. Insofar as we are working formally anyway, I won't bother worrying about convergence in the ensuing discussion.

In applying the Laplace transform, one assumes

$$\int_0^\infty \delta(x) dx = 1.$$

Then one readily calculates

$$\mathcal{L}\{\delta(t)\} = \int_0^\infty e^{-st} \delta(t) dt = e^{-s0} = 1,$$

just as we previously concluded (65). More generally,

$$\mathcal{L}\{\delta(t-a)\} = \int_0^\infty e^{-st} \delta(t-a) dt = e^{-as}.$$

The application of the Laplace transform in differential equations is as follows: First, one calculates the transform of some basic functions and creates a small table, as in Table 2, below:

Table 2. Some Laplace Transforms

$f(t)$	1	e^{at}	t^n	$\delta(t)$	$\delta(t-a)$	$H(t-a)$
$\mathcal{L}\{f(t)\}$	$\dfrac{1}{s}$	$\dfrac{1}{s-a}$	$\dfrac{n!}{s^{n+1}}$	1	e^{-as}	$\dfrac{e^{-as}}{s}$

One also proves a few useful lemmas, two of which I cite here:

5.8 Lemma. *Let f have a continuous derivative. If $\mathcal{L}\{f(t)\}$ and $\mathcal{L}\{f'(t)\}$ exist, then*

$$\mathcal{L}\{f'(t)\} = s\mathcal{L}\{f(t)\} - f(0).$$

Proof. Observe

$$\mathcal{L}\{f'(t)\} = \int_0^\infty e^{-st} f'(t)dt$$

$$= e^{-st} f(t)\Big|_0^\infty - \int_0^\infty -se^{-st} f(t)dt$$

$$= 0 - f(0) + s\int_0^\infty e^{-st} f(t)dt$$

$$= -f(0) + s\mathcal{L}\{f(t)\}. \qquad \square$$

5.9 Lemma. *If $\mathcal{L}\{(f(t)\}$ exists, then for $a > 0$*

$$\mathcal{L}\{H(t-a)f(t-a)\} = e^{-as}\mathcal{L}\{f(t)\}.$$

I leave the proof as an exercise to the more ambitious reader. With these one would solve an equation like

$$y'(t) = y(t) + \delta(t-1), \quad y(0) = 1$$

by applying Lemma 5.8 to the transform

$$\mathcal{L}\{y'(t)\} = \mathcal{L}\{y(t)\} + \mathcal{L}\{\delta(t-1)\}$$

to get

$$s\mathcal{L}\{y(t)\} - y(0) = \mathcal{L}\{y(t)\} + e^{-s},$$

i.e.

$$(s-1)\mathcal{L}\{y(t)\} = 1 + e^{-s},$$

whence we conclude

$$\mathcal{L}\{y(t)\} = \frac{1 + e^{-s}}{s-1} = \frac{1}{s-1} + \frac{e^{-s}}{s-1}.$$

One would now determine $y(t)$ by finding the inverse to the transform. In the present case, we can almost use the table we have: $1/(s-1)$ is clearly the transform of e^t. But $e^s/(s-1)$ does not match any entry. Fortunately Lemma 5.9 applies to tell us that this is the transform of $H(t-1)e^{t-1}$. Hence we have

$$y(t) = e^t + H(t-1)e^{t-1}.$$

5.4 Closing Remarks

I have been rather brief in discussing the Laplace transform as my purpose in bringing the subject up is not to discuss the transform itself, but to give some of the flavour of the formal use of the delta function in mathematics. For the curious reader who has no prior familiarity with this transform, which is not a topic universally covered in Calculus courses, I have added a few exercises on them at the end of the Chapter. A quick check at Amazon.com reveals a number of books devoted to the Laplace transform. A slim volume that was popular when I was a student (but is now apparently out of print) is Earl D.Rainville's *The Laplace Transform: An Introduction*[92]. Curiously, although Rainville deals explicitly with step functions, he nowhere introduces the δ-function. This is done in most modern textbooks on differential equations. Martin Braun's book, *Differential Equations and Their Applications*[93], has a good introductory treatment of the Laplace transform, including a section on the delta function that offers intuitive motivation, a physical justification of the use of the δ-function in differential equations, and a pat history of the function that, I suppose, is the standard presentation:

> In the early 1930's the Nobel Prize winning physicist[94] P.A.M. Dirac developed a very controversial method for dealing with impulsive functions...
>
> Now, most mathematicians, of course, usually ridiculed this method. "How can you make believe that $\delta(x)$ is an ordinary function if it is obviously not," they asked. However, they never laughed too loud since Dirac and his followers always obtained the right answers. In the late 1940's, in one of the great success stories of mathematics, the French mathematician Laurent Schwartz succeeded in placing the delta function on a firm mathematical foundation. He accomplished this by enlarging the class of all functions so as to include the delta function.[95]

A great deal more historical information is to be had in Lützen's *The Prehistory of the Theory of Distributions*[96]. I quote Lützen:

> Dirac's book on quantum mechanics became a most influential classic, published in four English editions and in a French and a German translation; with it the δ-function became a tool generally applied by physicists. In 1926, when Dirac's first article on the new treatment of quantum mechanics was published similar ideas were being explored by a group of mathematicians and physicists in Göttingen, including D. Hilbert, J. von Neumann, and L. Nordheim. Inspired by Hilbert's work on integral equations, they assumed as Dirac did... that operators T associated with physical observables were integral operators

---

[92] The Macmillan Company, New York, 1963.

[93] 3rd edition, Springer-Verlag, New York, 1983. The first edition came out in 1973, and a 4th edition appeared in 1993.

[94] Dirac shared the 1933 Nobel Prize in physics with Erwin Schrödinger "for the discovery of new productive forms of atomic theory". This was after he had introduced the delta function into physics and its use had spread. The description of Dirac as "the Nobel Prize winning physicist" is correct as a description of the man, but not as a description of the man at the time he introduced the δ-function.

[95] Braun, *op. cit.*, pp. 242 - 243.

[96] *Op. cit.*, pp. 110 - 143. Pp. 110 - 123 discuss precursors to Dirac; pp. 124 - 126 discuss Dirac's work; and pp. 126 - 143 discuss attempts to bypass or rigorise the use of the delta-function.

$$T(x) = \int \varphi(x,y) f(y) dy. \tag{80}$$

For this reason their quantum theory also depended heavily on the δ-function. In this way the δ-function came into the hands of mathematicians. However, during the same year that the joint work at Göttingen was published..., one of its authors, von Neumann, dissociated himself from it and formulated a new and revolutionary basis for quantum mechanics... His ideas were printed in book form in 1932 as *Mathematische Grundlagen der Quantenmechanik*, another highlight in the history of quantum mechanics. In his introductory remarks von Neumann proved that the representation of the identity operator by (80) would require that φ had the properties (60) and (61) which were incompatible with both the Riemann and Lebesgue integrals.[97]

Lützen then offers a couple of choice quotations from von Neumann's book:

Despite this Dirac fabricates the existence of such a function.[98]

and

We wish here to follow no further these trains of thought which were provided by Dirac and Jordan to a standard theory of the quantum events. The "improper" objects (like $\delta(x), \delta'(x), \dots$) play a decisive rôle in them— they lie outside the frame of generally customary mathematical methods, and we wish to describe quantum mechanics with the help of the latter.[99]

That Johann von Neumann, who would go on to become the 20th century's most outstanding applied mathematician (and a leading pure mathematician as well), was indeed motivated by his disapproval of the use of the δ-function is borne out by reading the opening paragraphs of the book:

The object of this book is the standard, and, so far as possible and reasonable, mathematically objection-free presentation of modern quantum mechanics, which in the course of recent years has acquired in its essential parts an expectedly definitive form: the so-called "theory of transformations"...

In several discourses, such as in his recently published book, Dirac has given a presentation of quantum mechanics, which... is scarcely to be outdone in brevity and elegance. Thus it is perhaps appropriate to supply here a few arguments for our methodology, which is essentially different from that named.

The methodology of Dirac referred to, which because of its transparency and elegance has inundated a large portion of the quantum mechanical literature, in no way comes up to the requirement of mathematical rigour— also not, if moreover this is to be reduced naturally and fairly,

---

[97] *Ibid.*, pp. 126 - 127.

[98] *Ibid.*, p. 127.

[99] *Ibid.*, p. 127.

to the usual norm of theoretical physics. So, for example, in consequence of holding on to the fiction that every self-adjoint operator can be brought into diagonal form, which in fact is not the case for these operators, the introduction of "improper" functions with self-contradictory properties is necessitated. Such an inclusion of mathematical "fictions" is inevitable under the circumstances, if it is only a matter of numerically calculating the result of an intuitively defined experiment. This would be no objection, if these concepts which are unsuitable in the framework of analysis were really essential for the new physical theory. Just as Newtonian mechanics first gave rise to the development of an infinitesimal calculus unquestionably self-contradictory in its form at the time, the quantum mechanics would suggest a new construction of our "analysis of infinitely many variables"— i.e. the mathematical apparatus would have to be changed, not the physical theory. That is, however, in no way the case; rather it should be shown, that the theory of transformations can also be mathematically unobjectionably founded in just as clear and standard a manner. Thereby it is required that the correct construction not consist of a mathematical precisioning and explication of Dirac's methods, but rather that it makes necessary from the outset a different approach, namely the dependence on Hilbert's spectral theory of operators.[100]

Historical remarks made by mathematicians are often vaguely remembered repetitions of what they have heard and not what they have checked. But von Neumann supports the gist of Braun's comments, if not the drama. There is criticism and the unambiguous declaration that Dirac's methods are incorrect, but there is no ridicule. Such may have been reserved for private conversation or correspondence; or it may even be found in a careful search of the literature, which I admit I haven't made. In any event, it shows a serious attempt by a serious mathematician to bypass the delta function. Lützen's book documents a number of such attempts by mathematicians to bypass or justify the use of this function before the latter half of the 1940s when Laurent Schwartz worked out his theory of distributions under which δ was a legitimate object. And, after the publication of a rigorous treatment by Schwartz in the early 1950s, such foundational work continued. We will see an example of this in section 3 of the next chapter.

6 Exercises

6.1 Exercise. (Cavalieri's Paradox) A common statement of Cavalieri's Principle is the following:

---

[100] Johann von Neumann, *Mathematische Grundlagen der Quantenmechanik*, Springer-Verlag, Berlin, 1932, pp. 1 - 2.

> If two solids have equal altitudes, and if sections made by planes parallel
> to the bases at equal distances from them are always in a given ratio,
> then the volumes of the solids are also in this ratio.[101]

It is important that the sections be parallel. Consider a right circular cone
inscribed in a right circular cylinder. If one passes a vertical plane through the
central axis, the cross sections of the cone and cylinder will be a triangle and
rectangle, respectively, of equal height and base. The triangle will have half the
area of the rectangle, whence on summing the areas the volume of the cone will
be half that of the cylinder. This result is incorrect. Why?

The arithmetic of infinite cardinal numbers is fairly simple after a few
grubby results have been proven. The next exercise is simple enough and the
reader may choose to skip it. The other messy beginnings are the basic laws
of calculation. I refer the reader to any elementary exposition of set theory for
the proofs. Our purpose here is simply to calculate the cardinal numbers of a
few simple sets. Thus, before beginning the exercises on cardinality I list a few
simple facts.

Every set X has a cardinal number we denote by $\overline{\overline{X}}$. If $\mathfrak{m}, \mathfrak{n}$ are cardinal
numbers and X, Y are sets of cardinalities $\mathfrak{m}, \mathfrak{n}$, respectively, then we define

$$\mathfrak{m} + \mathfrak{n} = \overline{\overline{X \cup Y}}, \text{ provided } X, Y \text{ are disjoint}$$

$$\mathfrak{m} \cdot \mathfrak{n} = \overline{\overline{X \times Y}}, \text{where } X \times Y = \{(x, y) | x \in X \ \& \ y \in Y\}$$

$$\mathfrak{m}^{\mathfrak{n}} = \overline{\overline{X^Y}}, \text{ where } X^Y = \{f \mid f : Y \to X\}.$$

There are two definitions of weak inequality of cardinal numbers:

$$\mathfrak{m} \leq_1 \mathfrak{n} \text{ iff there is a function } f \text{ mapping } X \text{ one-to-one into } Y$$

$$\mathfrak{m} \leq_2 \mathfrak{n} \text{ iff there is a function } g \text{ mapping } Y \text{ onto } X.$$

The two definitions are equivalent and we can dispense with the subscripts and
simply write $\mathfrak{m} \leq \mathfrak{n}$. We also define strict inequality by

$$\mathfrak{m} < \mathfrak{n} \text{ iff } \mathfrak{m} \leq \mathfrak{n} \ \& \ \mathfrak{m} \neq \mathfrak{n}.$$

Addition and multiplication are commutative and associative, while exponentiation satisfies the familiar laws of exponents, e.g.

$$(\mathfrak{m}^{\mathfrak{n}})^{\ell} = \mathfrak{m}^{\mathfrak{n}\ell}, \quad \mathfrak{m}^{\mathfrak{n}+\ell} = \mathfrak{m}^{\mathfrak{n}} \cdot \mathfrak{m}^{\ell}.$$

In addition, if one of $\mathfrak{m}, \mathfrak{n}$ is infinite,

$$\mathfrak{m} + \mathfrak{n} = \max\{\mathfrak{m}, \mathfrak{n}\},$$

and if neither set is empty,

---

[101] C.H. Edwards, Jr., *The Historical Development of the Calculus*, Springer-Verlag,
New York, 1979, p. 104.

$$\mathfrak{m} \cdot \mathfrak{n} = \max\{\mathfrak{m}, \mathfrak{n}\},$$

so that, in particular, if $\mathfrak{m}$ is infinite,

$$\mathfrak{m} + \mathfrak{m} = \mathfrak{m} \cdot \mathfrak{m} = \mathfrak{m}.$$

If $\mathfrak{m} > 1$, then

$$\mathfrak{n} < \mathfrak{m}^{\mathfrak{n}}.$$

A final remark is that $\leq$ is a linear ordering:

Dichotomy. $\mathfrak{m} \leq \mathfrak{m}$ or $\mathfrak{n} \leq \mathfrak{m}$

Transitivity. $\mathfrak{m} \leq \mathfrak{n}$ & $\mathfrak{n} \leq \mathfrak{k}$ $\Rightarrow$ $\mathfrak{m} \leq \mathfrak{k}$

Antisymmetry. $\mathfrak{m} \leq \mathfrak{n}$ & $\mathfrak{n} \leq \mathfrak{m}$ $\Rightarrow$ $\mathfrak{m} = \mathfrak{n}.$

Of these, only transitivity is trivial. Dichotomy and Antisymmetry are not even intuitively obvious. Dichotomy is, in fact, equivalent to the Axiom of Choice[102] (as are most assertions about infinite cardinal arithmetic), which has been accepted by some as a "Law of Thought" and harshly criticised by some, the most vocal among whom often used it themselves without realising it. Antisymmetry is one of the few facts about cardinal numbers that does not rely on the Axiom of Choice. The result is not that difficult to prove, but is sufficiently non-obvious to have three authors and is called the Cantor-Schröder-Bernstein Theorem after Georg Cantor, Ernst Schröder, and Felix Bernstein.

6.2 Exercise. (Well-Definedness) Let A_1, A_2, B_1, B_2 be given sets and assume $\overline{\overline{A_1}} = \overline{\overline{A_2}}, \overline{\overline{B_1}} = \overline{\overline{B_2}}$ (i.e., A_1 and A_2 have the same cardinality, as do B_1 and B_2). Show:

i.   if A_1, B_1 are disjoint and A_2, B_2 are disjoint, then

$$\overline{\overline{A_1 \cup B_1}} = \overline{\overline{A_2 \cup B_2}};$$

ii. $\overline{\overline{A_1 \times B_1}} = \overline{\overline{A_2 \times B_2}};$

iii. $\overline{\overline{A_1^{B_1}}} = \overline{\overline{A_2^{B_2}}};$

iv. $\overline{\overline{A_1}} \leq \overline{\overline{B_1}}$ iff $\overline{\overline{A_2}} \leq \overline{\overline{B_2}}.$

6.3 Exercise. ($\aleph_0$) The cardinal number of $\mathbb{N}$ is usually denoted $\aleph_0$ (pronounced "aleph-nought" or "aleph-null", $\aleph$ being the Hebrew letter aleph).

i.   Show the following sets to have cardinality $\aleph_0$:

    a. $\mathbb{N}^+$

    b. $\mathbb{Z}$

---

[102] There are many formulations of the Axiom of Choice. A simple one is the equivalence of the relations $\leq_1$ and $\leq_2$ above. If one tries to prove that, if there is a function $g : Y \to X$ which is onto, then there is a one-to-one function f going in the opposite direction, one might start by defining $F(x) = \{y \in Y | g(y) = x\}$ for $x \in X$ and then attempt to *choose* an element of $F(x)$ as the value of $f(x)$. It is the Axiom of Choice that says such choices can all be made at once and combined into a single function.

c. $\mathbb{N} \times \mathbb{N}$

d. $\mathbb{Q}^{\geq 0}$

e. $\mathbb{Q}$.

[Most of these are trivial if one uses the Cantor-Schröder-Bernstein Theorem. So I suggest using it as a last resort and trying to give an explicit one-to-one correspondence with $\mathbb{N}$ where possible.]

ii. Show that the set of finite sequences of natural numbers has cardinality $\aleph_0$.

iii. Show that the set of polynomials with integral coefficients has cardinality $\aleph_0$.

iv. Show that the set of algebraic real numbers has cardinality $\aleph_0$. Conclude the existence of transcendental numbers.

6.4 Exercise. ($\mathfrak{c}$) The cardinal number of the set $\mathbb{R}$ of real numbers is usually denoted $\mathfrak{c}$ ("c" for "continuum").

i.   Show by constructing an explicit one-to-one correspondence between the open interval $(0,1)$ and $\mathbb{R}$ that $\overline{\overline{(0,1)}} = \mathfrak{c}$. Conclude that $\overline{\overline{[0,1]}} = \mathfrak{c}$.

ii. Construct one-to-one functions $f : [0,1] \to \mathbb{R}$ and $g : \mathbb{R} \to [0,1]$ to conclude $\overline{\overline{[0,1]}} = \mathfrak{c}$ via the Cantor-Schröder-Bernstein Theorem.

iii. Show: $\mathfrak{c} = 2^{\aleph_0}$. [Hint: Use the dyadic expansion of a real number in $[0,1]$. The expression of a number $r \in [0,1]$ in base 2 is not unique because $(.1111\ldots)_2$ is the same as $1.000\ldots$ However, picking a representation without repeating 1's will show
$$\overline{\overline{[0,1]}} \leq 2^{\aleph_0}.$$
For the converse inequality, assign each to sequence $f \in \{0,1\}^{\mathbb{N}}$ an element of $[0,1]$ with no repeating 1's in its dyadic expansion.]

iv. Show: $\mathfrak{c}^{\aleph_0} = \mathfrak{c}$.

v. Show: $\overline{\overline{C[0,1]}} = \mathfrak{c}$. [Hint: A continuous function is completely determined by its behaviour on the rationals. Hence $\overline{\overline{C[0,1]}} \leq \overline{\overline{\mathbb{R}^{\mathbb{Q}}}} \leq \mathfrak{c}^{\aleph_0}$.]

By part v of the preceding exercise, there are only $\mathfrak{c}$ continuous functions defined on the unit interval. But there are $\mathfrak{c}^{\mathfrak{c}}$ functions from $[0,1]$ into the reals and $\mathfrak{c} < \mathfrak{c}^{\mathfrak{c}}$. Thus, most functions are not continuous. The following exercise covers the main tool for proving non-equality of cardinal numbers.

6.5 Exercise. (Cardinal Exponentiation) Let X be a set. Define the *power set* of X, written $\mathcal{P}(X)$ by
$$\mathcal{P}(X) = \{Y \,|\, Y \subseteq X\}.$$

Also, define
$$2^X = \{f \,|\, f : X \to \{0,1\}\}.$$

i.   Construct a one-to-one correspondence between $\mathcal{P}(X)$ and 2^X by identifying each subset $Y \subseteq X$ with its *characteristic function* χ_Y:

$$\chi_Y(x) = \begin{cases} 1, & x \in Y \\ 0, & x \notin Y. \end{cases}$$

Conclude: $\overline{\overline{\mathcal{P}(X)}} = \overline{\overline{2^X}} = 2^{\overline{\overline{X}}}$.

ii. Show, without appeal to the inequality $\mathfrak{n} < \mathfrak{m}^{\mathfrak{n}}$ cited earlier, that $\overline{\overline{X}} < \overline{\overline{\mathcal{P}(X)}}$. Conclude that $\mathfrak{n} < 2^{\mathfrak{n}}$ for any cardinal number $\mathfrak{n}$. [Hint: Assume f maps X onto $\mathcal{P}(X)$ and consider $\{x \in X \mid x \notin f(x)\}$.]

iii. Show: for $\mathfrak{n} > 1$, and any $\mathfrak{m}, \mathfrak{k}$,

$$\mathfrak{m} < \mathfrak{k} \;\Rightarrow\; \mathfrak{m}^{\mathfrak{n}} \leq \mathfrak{k}^{\mathfrak{n}}$$

directly using the definition of $\leq_1$. Conclude $\mathfrak{m} < \mathfrak{n}^{\mathfrak{m}}$.

iv. Can $\mathfrak{m} = \mathfrak{m}^{\mathfrak{n}}$ for $\mathfrak{n}$ infinite?

6.6 Exercise. (Cauchy's Counterexample) Define

$$f(x) = \begin{cases} e^{-1/x^2}, & x \neq 0 \\ 0, & x = 0. \end{cases}$$

i. Show: $f(x)$ is continuous at $x = 0$, i.e.

$$\lim_{x \to 0} e^{-1/x^2} = 0.$$

$f(x)$ clearly has derivatives of all orders at $x \neq 0$. Taking one or two differentiations should convince the reader that the successive derivatives of $f(x)$ are all of the form $P(1/x)e^{-1/x^2}$ for some polynomial $P(X)$. The proof that this is the case is an easy induction based on the following:

ii. Let $P(X)$ be a polynomial. Show:

$$\frac{d}{dx}\left(P\left(\frac{1}{x}\right) e^{-1/x^2} \right) = Q\left(\frac{1}{x}\right) e^{-1/x^2}$$

for some polynomial $Q(X)$.

As we wish to show $f^{(k)}$ is continuous and $f^{(k)}(0) = 0$, we will have to show

$$\lim_{x \to 0} Q\left(\frac{1}{x}\right) e^{-1/x^2} = 0$$

for whatever $Q(X)$ defines the polynomial factor of $f^{(k)}$. To do this, it suffices to prove

$$\lim_{x \to 0} \frac{1}{x^n} e^{-1/x^2} = 0$$

for all $n \in \mathbb{N}$. Thus, define

$$f_n(x) = \begin{cases} \frac{1}{x^n} e^{-1/x^2}, & x \neq 0 \\ 0, & x = 0. \end{cases}$$

iii. Let $n \in \mathbb{N}^+$. Show: For $x \neq 0$,

$$f_n'(x) = \frac{d}{dx}\left(\frac{1}{x^n} e^{-1/x^2} \right) = \frac{1}{x^{n+3}} e^{-1/x^2} (2 - nx^2).$$

Successively conclude:

    a. for $0 < x < \sqrt{2/n}$, $f_n'(x) > 0$, whence $f_n(x)$ is strictly increasing on $(0, \sqrt{2/n})$;

    b. $\lim_{x \to 0^+} f_n(x)$ exists; [Hint: Use the Greatest Lower Bound Principle.]

    c. $\lim_{x \to 0^-} f_n(x) = (-1)^n \lim_{x \to 0^+} f_n(x)$ and thus also exists;

    d. if the limit of $f_n(x)$ as x goes to 0 exists, then the limit must be 0.

iv. Show: $\lim_{x \to 0} f_n(x) = 0$. [Hint: If this limit is not 0, find

$$\lim_{x \to 0^+} \frac{f_{n+1}(x)}{f_n(x)}.$$

               ]

v. Let $P(X)$ be any polynomial. Show:

$$\lim_{x \to 0} P\left(\frac{1}{x}\right) e^{-1/x^2} = 0.$$

vi. Prove by induction on k that $f^{(k)}(0)$ exists and equals 0. Conclude that Maclaurin's expansion for $f(x)$ agrees with $f(x)$ only at $x = 0$.

6.7 Exercise. (Binomial Theorem) A *commutative ring* with unit element is a structure $\mathcal{R} = (R, +, \cdot, 0, 1)$ consisting of a set R, special elements $0, 1 \in R$, and two operations $+, \cdot$ satisfying the familiar algebraic laws satisfied by the integers, rational and real numbers, namely

Commutativity. for all $x, y \in R$,

$$x + y = y + x, \quad x \cdot y = y \cdot x$$

Associativity. for all $x, y, z \in R$,

$$x + (y + z) = (x + y) + z, \quad x \cdot (y \cdot z) = (x \cdot y) \cdot z$$

Distributivity. for all $x, y, z \in R$,

$$x \cdot (y + z) = (x \cdot y) + (x \cdot z), \quad (x + y) \cdot z = (x \cdot z) + (y \cdot z)$$

Identity Elements. for all $x \in R$,

$$x + 0 = 0 + x = x, \quad 1 \cdot x = x \cdot 1 = x$$

Additive Inverse. for all $x \in R$ there is a $y \in R$ such that

$$x + y = 0.$$

A structure $\mathcal{R}$ satisfying all of the above other than the commutativity of multiplication is called a *noncommutative* ring.

i.   Show that the set of linear transformations $T : \mathbb{R}^2 \to \mathbb{R}^2$ forms a noncommutative ring when addition is defined pointwise,

$$(T + U)(v) = T(v) + U(v)$$

for all $v \in \mathbb{R}^2$; and multiplication is composition,

$$(T \cdot U)(v) = T(U(v))$$

for all $v \in \mathbb{R}^2$. That is, show that the set of such linear transformations satisfies all the properties listed other than the commutativity of multiplication. Give explicit examples of T, U for which $T \cdot U \neq U \cdot T$.

ii. Show: If R is a noncommutative ring, then the Binomial Theorem fails: there are $x, y \in R$ such that

$$(x + y)^2 = x^2 + 2xy + y^2 \tag{81}$$

fails. [Hint: Derive commutativity from the assumed identity (81).]

iii. Prove the Binomial Theorem for any commutative ring: for all $n \in \mathbb{N}^+$ and for all $x, y \in R$,

$$(x + y)^n = \sum_{k=0}^{n} \binom{n}{k} x^k y^{n-k}.$$

iv. Show that the set of linear transformations of the form

$$F = \sum_{k=0}^{n} a_i \Delta^i$$

on the vector space $\mathbb{R}^{\mathbb{R}}$ forms a commutative ring. Justify the the use of the Binomial Theorem in section 3, above.

6.8 Exercise. (Vandermonde's Theorem I) For any polynomial $P(X)$ we set $\Delta P(x) = P(x + 1) - P(x)$.

i.   Show: If $\Delta P(X)$ is identically 0, then P is constant. [Hint: A nonconstant polynomial of degree $n > 0$ has at most n roots.]

ii. Let q be an arbitrary rational number and k a positive integer. Define

$$P_k(X) = \binom{X + q}{k}.$$

Show: $\Delta P_{k+1} = P_k$.

iii. For each $i \in \mathbb{N}$, show

$$\Delta \binom{X}{i} = \begin{cases} \binom{X}{i - 1}, & i > 0 \\ 0, & i = 0. \end{cases}$$

iv. Let q, k be as in part ii and define

$$Q_k(X) = \binom{X + q}{k} - \sum_{i=0}^{k} \binom{X}{i} \binom{q}{k - i}.$$

Show: $\Delta Q_{k+1} = Q_k$.

v.   Show by induction on k that each polynomial Q_k is identically 0.

vi. Show: for all $p, q \in \mathbb{Q}$, and all positive integral k,

$$\binom{p+q}{k} = \sum_{i=0}^{k} \binom{p}{i}\binom{q}{k-i}.$$

6.9 Exercise. (Vandermonde's Theorem II) Write

$$Q_k(X,Y) = \binom{X+Y}{k} - \sum_{i=0}^{k}\binom{X}{i}\binom{Y}{k-i}.$$

Note that each Q_k is a polynomial in the variables X, Y.

i. Show by considering the coefficient of X^k in the polynomial

$$(1+X)^{m+n} = (1+X)^m(1+X)^n$$

that $Q_k(m,n) = 0$ for all positive integral m, n for which $m+n > k$.

ii. Use a combinatorial argument to establish the conclusion of part i. [Hint: If $m+n > k$, how many ways can one form a committee of k people from a group of m men and n women?]

iii. Show: For each fixed positive integer m, the polynomial $Q(m, Y)$ is identically 0. Conclude, for $q \in \mathbb{Q}$, the polynomial $Q_k(X, q)$ is 0 for all positive integral values m of X.

iv. Prove Vandermonde's Identity.

6.10 Exercise. (Summation by Parts) Let $f(x), g(x)$ be functions on $\mathbb{Z}$.

i. Show: $\Delta(f(x)g(x)) = g(x+1)\Delta f(x) + f(x)\Delta g(x)$.

ii. Show: $\Delta^{-1}(f(x)\Delta g(x)) = f(x)g(x) - \Delta^{-1}(g(x+1)\Delta f(x)))$.

Recall that

$$\sum_{k=0}^{n} h(k) = \Delta^{-1}h(x)\Big|_0^{n+1}.$$

iii. Let $f(x) = x, \Delta g(x) = x$ and apply ii to obtain

$$\sum_{k=0}^{n} k^2 = \frac{(n+1)n(2n+1)}{6}.$$

iv. Let $f(x) = x, \Delta g(x) = x^2$ and show

$$\sum_{k=0}^{n} k^3 = \left(\frac{(n+1)n}{2}\right)^2.$$

6.11 Exercise. (Nonstandard δ-Functions) Examine two or three of the δ-functions of *Table 1* with an eye to seeing which properties (62) - (71) hold.

6.12 Exercise. (Formal Calculation of Laplace Transforms) Use only the formula

$$\mathcal{L}\{f'(t)\} = s\mathcal{L}\{f(t)\} - f(0)$$

to derive the following transforms formally without integrating:

i. $\mathcal{L}\{e^{at}\} = 1/(s-a)$;

ii. $\mathcal{L}\{t^{n+1}\} = \frac{n+1}{s}\mathcal{L}\{t^n\}$;
iii. $\mathcal{L}\{t^{n+1}\} = (n+1)!/s^{n+2}$; [Hint: Use $a = 0$ in i for $n = 0$.]
iv. $\mathcal{L}\{\delta(t)\} = 0$. [Hint: $H(t) = 1$ on $[0, \infty)$. This value does not agree with the value in *Table 2*. Explain.]

6.13 Exercise. (More Laplace Transforms) By integration prove Lemma 5.9: For $a > 0$

$$\mathcal{L}\{H(t-a)g(t-a)\} = e^{-as}\mathcal{L}\{g(t)\}.$$

Use this to conclude
i. $\mathcal{L}\{\delta(t-a)\} = e^{-as}$. [Use the value of $\mathcal{L}\{\delta(t)\}$ from *Table 2*.]
ii. $\mathcal{L}\{H(t-1)e^{t-1}\} = e^{-s}/(s-1)$.

6.14 Exercise. (Linearity of Laplace Transform) i. Show that $\mathcal{L}$ is linear:

$$\mathcal{L}\{c_1 f_1(t) + c_2 f_2(t)\} = c_1\mathcal{L}\{f_1(t)\} + c_2\mathcal{L}\{f_2(t)\}.$$

ii. Find $\mathcal{L}\{\sin t\}$ using Euler's equation,

$$\sin t = \frac{e^{it} - e^{-it}}{2i}.$$

iii. Similarly, find $\mathcal{L}\{\cos t\}$.
iv. Find $\mathcal{L}\{\cos(t - \frac{\pi}{2})H(t - \frac{\pi}{2})\}$.

6.15 Exercise. (A Differential Equation) Use the Laplace transform to solve the initial value problem

$$y'' - y' = 1, \quad y(0) = y'(0) = 1.$$

When you are finished, read Chapter III, sections 1 - 3 and compare your work with the solution given in Example III.3.6.

6.16 Exercise. ($\mathcal{L}\{\delta'(t)\}$) Note that

$$\mathcal{L}\{f(t)\} = \int_0^\infty e^{-st} f(t)dt = \int_{-\infty}^\infty e^{-st} H(t)f(t)dt. \tag{82}$$

Do the following formally (i.e. without worrying about questions of convergence):
i. Find $\mathcal{L}\{\delta'(t)\}$ via (82) and (66).
ii. Find $\mathcal{L}\{\delta'(t)\}$ via Lemma 5.8.
iii. Find $\mathcal{L}\{t\delta'(t)\}$ via (82) and (66).
iv. Prove

$$\mathcal{L}\{-tf(t)\} = \frac{d}{ds}\mathcal{L}\{f(t)\}. \tag{83}$$

Use this and iii to find $\mathcal{L}\{\delta'(t)\}$.
v. Find $\mathcal{L}\{-t\delta'(t)\}$ by appeal to (71). Use (83) to determine $\mathcal{L}\{\delta'(t)\}$ from this. Compare the values of iv and v with those of ii and iii.

6.17 Exercise. (Another Differential Equation) Solve the initial value problem,
$$y''(t) = 1 + 2\delta(t - 3), \quad y(0) = 5, \quad y'(0) = 2$$
of Example 5.7 by using the Laplace Transform.

6.18 Exercise. ($\mathcal{L}\{\delta_\eta(t)\}$) Calculate $\mathcal{L}\{\delta_\eta(t)\}$.

III

New Numbers From Old

1 Negative Numbers

1.1 Background

A famous quote by Leopold Kronecker is often loosely rendered as "God created the integers; all the rest is man's work".[1] Anthropologists have reported on tribal societies so primitive that their counting goes, "One, two, many." Thus Kronecker may be attributing too much to God. Let us assume, nonetheless, that God gave man the positive integers and man had to do the rest. The positive integers work fine so long as one is counting sheep or any discrete commodity. However, civilisation reaches a point where one has to distribute divisible quantities like bushels of grain or vessels of wine, and the number of bushels or vessels is not evenly divided by the number of people among whom the quantities are to be distributed. Rational numbers are thus born. Eventually computation grows more complex and 0 is introduced, initially as a place-holder, and eventually as a number in its own right. Irrational numbers like $\sqrt{2}$ or π will slip in, more-or-less unnoticed and, aside from the annoyance of one's never quite having the exact value in the form of a fraction, they would not be noticed as being different until mathematics reached the point of theorising about numbers. This was done by the Greeks, who discovered the irrationality of numbers like $\sqrt{2}$. The Greeks, who had based mathematics on number, now switched their foundations to geometry. Arithmetically they could not explain $\sqrt{2}$ in terms of number, but they could point to the diagonal of a square of side 1 and see that it had a magnitude. Euclid, in devoting a book of the *Elements* to treating quadratic surds (i.e. expressions involving square roots), can be viewed abstractly as extending the rational number system to include certain irrationals. The extent to which the rationals can be extended wasn't really thought about until the 19th century when the demands of the

---

[1] Heinrich Weber reports in his obituary, "Leopold Kronecker" (*Jahresbericht der Deutschen Mathematiker-Vereinigung* 2 (1892), pp. 5 - 31 (here: p. 19); reprinted in *Mathematische Annalen* 43 (1893), pp. 1 - 25 (here: p. 15)) that Kronecker made this remark in a lecture before the Berlin natural scientists meeting in 1886.

Calculus required more powerful closure properties than encompassed by the arithmetic operations. The new closure properties turned out to reduce to a single one— the completeness of the number line as given by the Least Upper Bound Principle or the convergence of Cauchy sequences. Oddly enough, this was also the century that negative numbers were finally universally accepted.

Negative numbers go back a long way. The earliest extant work of Chinese mathematics is the *Jiǔzhāng suànshù* [*Nine Chapters on the Mathematical Art*], which covers the developments in Chinese mathematics from approximately the 11th century BC to 220 AD, and it treats positive and negative numbers, giving the usual rules for addition and subtraction. The Chinese seem not, however, to have divined the rules for multiplication and division of positive and negative numbers until the 13th century AD.[2]

Negative numbers make their first European appearance in the *Arithmetic* of Diophantus of Alexandria (fl. c. 250 AD). How much later this appearance was than the Chinese debut depends on which point in the interval stretching from the 11th century BC to 220 AD that such numbers arrived in China, and on the accuracy in the dating of Diophantus. Diophantus broke with the Euclidean tradition whereby rational numbers aren't really numbers but only relations and referred to them as numbers (arithmos=$\alpha\rho\iota\vartheta\mu o\varsigma$). Positive rational numbers were "existent" (iparxis=$\upsilon\pi\alpha\rho\xi\iota\varsigma$) and negative ones "lacking" or "defective" (leipsis=$\lambda\varepsilon\iota\psi\iota\varsigma$). He gave correct rules for the arithmetic operations, including the operations of multiplication and division. That said, he did not completely elevate negative rational numbers to the same full status as positive rationals. His interest was in the solution of certain algebraic equations, and he accepted only positive rational numbers as solutions.[3]

In India, Brahmagupta, writing in the 7th century A.D., gave the correct rules for all the arithmetic operations in dealing with negative numbers. And sometime in the 12 century, Bhaskara definitely accepted negative numbers as solutions to problems.[4]

By the middle of the 13th century AD, the Chinese were able to find the roots of polynomials to any desired degree of accuracy through a procedure now known as Horner's Method after the British mathematician William George Horner. While the coefficients could be positive or negative and the Chinese did not restrict their search to rational solutions, they nevertheless likewise ignored negative solutions.

---

[2] Lǐ Yan and Dù Shíràn, *Chinese Mathematics; A Concise History*, (translated by John N. Crossley and Anthony W.-C. Lun) Oxford University Press, Oxford, 1987, pp. 48 - 50.

[3] Isabella Gregor'evna Bašmakova, *Diophant und diophantische Gleichungen*, Birkhäuser Verlag, Basel, 1974, pp. 17 - 18. Bašmakova's book offers an ideal alternative to reading Diophantus. It selects examples from Diophantus and analyses his solutions in modern terms in clear, intelligible symbolic notation. It also traces the Diophantine development into more modern times and the rudiments of algebraic geometry. I believe an English translation now exists.

[4] John N. Crossley, *The Emergence of Number*, World Scientific Publishing Co., Singapore, 1987, pp. 64 - 65.

Cardano, in his *Ars Magna sive de Regulis Algebraicis*[5] (1545) reverses things. Negative coefficients do not occur, but negative solutions are allowed. While positive solutions are termed *true*, negative ones are *false* or *fictitious*. This seems to be a description of them as numbers, not as solutions.

Cardano's book was written at the height of European reverence for Euclid and Greek geometry. He explains why he does not go into great detail beyond the cubic equation:

> Although a long series of rules might be added and a long discourse given about them, we conclude our detailed consideration with the cubic, others being merely mentioned, even if generally, in passing. For as *positio* [the first power] refers to a line, *quadratum* [the square] to a surface, and *cubum* [the cube] to a solid body, it would be very foolish for us to go beyond this point. Nature does not permit it.[6]

The geometric interpretation of multiplication of positive real numbers and devotion to Euclid meant that the demonstrations offered for the rules for solving equations were represented geometrically, with rectangles of given areas. Negative numbers do not occur as lengths and thus did not appear in the diagrams or in the equations represented by the diagrams. If one wanted to consider an equation,

$$X^2 - 3X + 2 = 0,$$

one would write it as

$$X^2 + 2 = 3X,$$

and consider it to be entirely different in character from

$$X^2 + 3X = 2.$$

In a lecture to the Mathematical Association of America in 1932, G.A. Miller pointed out that the history books were not in complete agreement on when Europeans had a "complete conception" of negative numbers. Some dated this with the appearance of *La Gèomètrie* by Rene Descartes in 1637. Others said Cardano's conception was "sufficient" and that Albert Girard already had a complete conception before Descartes.[7] Such completeness refers, however, only to computation with negative quantities and not with all aspects of negativity.

In 1655, the paradoxes that arise from Type I formalism reared their ugly heads with the publication by John Wallis of his book *Arithmetica infinitorum*. One has in general

$$\frac{1}{m} < \frac{1}{m-1},$$

whence

---

[5] Girolamo Cardano, *Ars Magna, or Rules of Algebra*, Dover Publications, Inc., New York, 1993; translated by T. Richard Witmer. Originally published by MIT Press, 1968.

[6] *Ibid.*, p. 9.

[7] G.A. Miller, "Historical note on negative numbers", *The American Mathematical Monthly* 40 (1933), pp. 4 - 5.

$$\frac{1}{0} < \frac{1}{-1},$$

i.e.

$$\infty < \frac{1}{-1},$$

and the ratio of a positive to a negative number is infinite.[8] It follows that $-1 = \frac{1}{-1}$ is infinite.[9]

About half a century after Wallis made his observation, Antoine Arnauld, a colleague and close friend of Blaise Pascal, noted a related difficulty: Mathematicians held the proportion,

$$1 : -1 = -1 : 1$$

to be correct. But this meant that the ratio of a larger to a smaller was the same as that of a smaller to a larger.[10]

Leonhard Euler would find fault with the argument of Wallis in his book on the Differential Calculus[11]: for $m \geq 1$,

$$\frac{1}{m^2} > \frac{1}{(m+1)^2},$$

but one cannot continue the inequality and derive

---

[8] Richard Reiff, *Geschichte der unendlichen Reihen*, Verlag der Laupp'schen Buchhandlung, 1889, p. 8. Another source, dealing extensively with the history of negative numbers and with many source materials quoted, is: Abraham Arcavi and Maxim Bruckheimer, *The Negative Numbers; A Source-work Collection for In-Service and Pre-Service Teacher Courses*, Weizman Institute, Rehovot, 1983.

[9] Moritz Cantor, *Vorlesungen über Geschichte der Mathematik, III*, ? ed.,Teubner, 1901, p. 367. Cantor refers to p. 902 of the second volume (which I haven't seen) of his work for a fuller discussion of Wallis on this point. As neither Reiff nor Arcavi and Bruckheimer mention it, I consulted Wallis, whose book is now available in English translation: Jacqueline A. Stedall, *The Arithmetic of Infinitesimals, John Wallis 1656*, Springer-Verlag, New York, 2004. If Wallis took this additional step, it wasn't in this work. On Wallis and the infinite, Stedall says

> The strangest of Wallis's concepts concerning infinity is that the ratio of a positive number to a negative number might be somehow 'greater than infinite'. He was led to this conclusion by the fact that $1/a$ grows infinitely large as a moves toward zero. If, therefore, a decreases *through* zero, the quantity $1/a$ must become both negative and 'greater than infinite'. At other times, however, Wallis used the usual rules of division for negative numbers, thus $\frac{1}{-2} = -\frac{1}{2} = \frac{-1}{2}$, so had no reason to consider the reciprocal of a negative quantity as 'greater than infinite', and his assertion has to be read in the specific geometric context to which it pertains, the quadrature of curves whose equations contain negative indices.

(Here, "negative indices" means negative powers.)

[10] Cantor, *op.cit.*, p. 367.

[11] Euler is quoted in Arcavi and Bruckheimer, *op.cit.*

$$\ldots > \frac{1}{9} > \frac{1}{4} > \frac{1}{1} > \frac{1}{0} > \frac{1}{1} > \frac{1}{4} > \frac{1}{9} > \ldots$$

However, he did point to the sum,

$$1 + 2 + 4 + 8 + \ldots = \frac{1}{1-2} = -1,$$

as proof that -1 was infinite.[12] And Nikolaus Bernoulli, in discussing his disagreement with the usual summation of the Grandi series, pointed out that because -1 is an infinite quantity, -1 was infinitely greater than 1 and thus the sum $1 - 1 + 1 - 1 + \ldots$ had to be infinitely greater than $1 + 1 + 1 + 1 + \ldots$, which is already infinite.[13] [14]

To further complicate matters, in his algebra textbook, Euler pointed out that $m + 1 > m$, whence one has

$$0 < 1 < 2 < \ldots$$

Continuing in the opposite direction yields

$$\ldots < -2 < -1 < 0 < 1 < 2 < \ldots,$$

and the negative numbers are "less than nothing".[15]

Euler's *Elements of Algebra* is an amazing document. It is an elementary textbook in which one finds such basics as the handling of negative numbers, simplifying algebraic expressions, and solving equations, alongside the introduction of imaginary numbers, infinity— defined by $\infty = \frac{1}{0}$, and infinite power series— introduced via long division and reference to Newton's Binomial Theorem. He does mention "greater" and "lesser", but one looks in vain for a list of properties of "greater" and "lesser" and one concludes the confusion to be about order and size, not negativity *per se*. Indeed, a number of paradoxes in mathematics were resolved by replacing intuitive notions of size by pairs of Type II formal definitions: measure and cardinality (Galileo's paradoxes), absolute value and position (for negative numbers), and cardinality and ordinality (Georg Cantor). But even after one separated the notion of order ($<=$ "to the left of") from that of size (smaller = "lesser in absolute value"), there was still a major obstacle to the full acceptance of negative numbers.

The problem was the nature of the numbers themselves.

---

[12] Reiff, *op. cit.*, p. 129.

[13] *Ibid.*, p. 69.

[14] Here, I cannot resist proving that God does not exist. Since -1 is infinite and 1 was proven infinite by Galileo, it follows that $0 = 1 + -1$ is infinite, i.e. nothing is infinite. But, God is infinite, whence God is nothing, i.e. God does not exist. If one is too pious to accept this conclusion, one can point to the fact that nothing is infinite and say that the creation of the universe out of nothing is not something to be doubted on arithmetic grounds; for, God had an infinite amount of material to work with!

[15] Leonhard Euler, *Elements of Algebra*, Springer-Verlag, New York, nd, p. 5.

In the 18th century mathematicians defined mathematics as the science of quantity, and negative numbers variously as "quantities less than nothing" and "quantities obtained by the subtraction of a greater quantity from a lesser"... In lieu of an adequate definition, some mathematicians had tried to justify the negatives through analogy to debts, lines drawn in certain directions, and the like.[16]

Standard textbooks of the eighteenth century defined the negatives as "quantities less than nothing" and routinely justified them through analogy with debts, times past, and the like. From around mid-century, however, three British mathematicians, Robert Simson, Francis Maseres, and William Frend, raised the problem of the status of the negative numbers. This triumvirate declared the concept of a "quantity less than nothing" to be nonsense and called for the elimination from algebra of the ill-defined negatives and the unrestricted subtraction operation which permitted the taking of a greater from a lesser.[17]

Simson, Maseres, and Frend were pretty much alone in their opposition to negative numbers, but they were not alone in seeing a problem:

Greenfield urged algebraists, such as Maseres (whom he explicitly mentioned) to "exert... industry and ingenuity rather to confirm than to destroy; rather to demonstrate, how far we might rely on the method of negative quantities, than to overturn at once so great a part of the labours of the modern algebraists". In a paper read before the Royal Society of London in 1801, Woodhouse also took a pragmatic approach to the problem. Acknowledging that "an abstract negative quantity is indeed unintelligible", he yet argued for retention of negative and imaginary numbers. "If operations with any characters or signs lead to just conclusions," he maintained, "such operations must be true by virtue of some principle or other".[18]

In France at about this time Lazare Carnot published a couple of works, *De la corrélation des figures de géométrie* (1801) and *Géométrie de position* (1803), in which he attempted a theory of negative numbers. Writing on Carnot, Charles Gillispie says

---

[16] Helena M. Pycior, "George Peacock and the British origins of symbolical algebra", *Historia Mathematica* 8 (1981), pp. 23 - 44; here p. 28.

[17] Helena M. Pycior, "Benjamin Peirce's *Linear Associative Algebra*", *Isis* 70 (1979), pp. 537 - 551; here p. 538.

[18] Pycior, "George Peacock...", *op. cit.*, p. 30. William Greenfield, a professor of rhetoric in Edinburgh, was an amateur mathematician. His remark cited here is from an address given in 1784 to the Royal Society of Edinburgh. Robert Woodhouse was a Cambridge man, first a fellow, then Lucasian Professor of mathematics, and later Plumian Professor of astronomy and experimental philosophy. Other examples cited by Pycior are John Playfair, famous for his formulation of Euclid's parallel postulate, and Adrien-Quentin Buée, one of the discoverers of the geometric representation of complex numbers.

As for negative quantity, however, two schools had developed in the course of the many discussions the problem had evoked in the eighteenth century. One interpretation accepted the notion of quantities less than zero. The other considered that a minus sign meant that a quantity was to be evaluated in the direction opposite to that of a positive quantity. Carnot found both explanations unconvincing. As for the former, he asserted roundly that the notion of something being less than nothing was absurd. It was here that he introduced his distinction between a quantity properly speaking and an algebraic value, the latter being a merely fictitious entity introduced for purposes of calculating. If it was permissible in a calculation to neglect quantities of no magnitude (as it was), then it ought surely to be justifiable to ignore those less than zero if such there were. But everyone knows that negative terms cannot be neglected. Thus, whatever they signify, it cannot be quantity.[19]

According to Gillispie, "Carnot's contemporaries did not find it easy to form a clear idea of what his theory of negative quantity was" and I am sure, not being a linguist, I would have even greater difficulty than they if I had his book before me. I don't and I rely on the brief description and simple example given by Gillispie. The essence of it seems to be this: Suppose we drop a perpendicular CD from one vertex of a triangle ABC to the opposite side AB, as in *Figure 1*, and we wish to calculate how much larger AB is

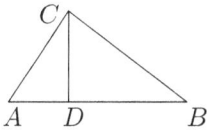

Figure 1

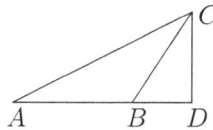

Figure 2

than AD. After we carry out the calculation we discover the length of BD to be -5. This means we have drawn the wrong picture, that *Figure 1* should be replaced by something like *Figure 2*, and that the actual length of BD is 5. The answer -5 was an error we got by drawing the wrong picture and we get the correct answer by erasing the minus sign. This is not unlike an assertion earlier made by Maseres that the root -5 of the equation $x^2 + 2x = 15$ actually pointed to 5 as the root of $x^2 - 2x = 15$, i.e. $x^2 = 2x + 15$. In Carnot's case, however, the statement seems less arbitrary. And Carnot did not ban negative numbers, but sought to explain, if not them, why their use was unproblematic. However ineffective, it seems Carnot's theory was in the spirit of Greenfield and Woodhouse.

Back in England, where negative numbers were more hotly debated, two solutions to the problem of negative numbers were proposed. The first was George Peacock's *symbolical algebra*, an extreme Type III formal axiomatisation that denied meaning to negative numbers and gave a name to the principle Woodhouse was asking for, but did not offer a demonstration of the validity of

---

[19] Charles Coulston Gillispie, *Lazare Carnot, Savant*, Princeton University Press, Princeton, 1971, pp. 125 - 126.

the principle. I postpone discussion of Peacock to Chapter IV, which is devoted
to the Type III formal approach.

The second solution came from William Rowan Hamilton, whose paper[20]
containing this solution is best known for its construction of complex numbers
as ordered pairs of real numbers, but also included a justification of negative
numbers. I find Hamilton's treatment of negatives to be an all but unreadable
mass of verbiage that does not look to be worthy of repetition. Thus, at this
point I desert the historical discussion and present instead a more modern
variation on the construction[21] of (structures to stand in for) the full systems
of integers, rational numbers, and real numbers, given the systems of positive
or nonnegative such numbers.

1.2 The Construction

The construction is quite general, allowing one to start with the positive or
non-negative integers and to finish with all the integers, or to start with the
positive or non-negative rational or real numbers and finish with all the rational
or real numbers, respectively. As the constructions of the real numbers from
the rationals are occasionally simplified if there are no negative numbers to
deal with, it is probably best to present the construction in such a way as to
make clear its generality. To this end, we offer the following definitions.

1.1 Definitions. *A* semi-ring *is an algebraic structure* $(R, +, \cdot)$, *where R is a
nonempty set and $+, \cdot$ are binary operations on R satisfying the following laws:*
Commutativity. *for all $a, b \in R$,*

$$a + b = b + a, \qquad a \cdot b = b \cdot a$$

Associativity. *for all $a, b, c \in R$,*

$$a + (b + c) = (a + b) + c, \qquad a \cdot (b \cdot c) = (a \cdot b) \cdot c$$

Distributivity. *for all $a, b, c \in R$,*

$$a \cdot (b + c) = (a \cdot b) + (a \cdot c), \qquad (a + b) \cdot c = (a \cdot c) + (b \cdot c)$$

Additive Cancellation. *for all $a, b, c \in R$,*

$$\text{if } a + c = b + c, \text{ then } a = b.$$

If there is an element $0 \in R$ such that

---

[20] "Theory of conjugate functions, or algebraic couples; with a preliminary essay on
algebra as the science of pure time", *Transactions of the Royal Irish Academy* 17
(1837), pp. 293 - 422.

[21] This may originate with Richard Dedekind. On page 490 of the third volume of his
collected works there is a note saying that at age 82 Dedekind wrote to a former
student of his named Lachmann outlining a construction of the set of integers from
the set of natural numbers using ordered pairs of natural numbers.

$$\text{for all } a \in R, \ a + 0 = a,$$

then $(R, +, \cdot, 0)$ *is called a* semi-ring with zero. *If, further, for every* $a \in R$ *there is an additive inverse, i.e. an element* b *such that* $a + b = 0$, *then* $(R, +, \cdot, 0)$ *is called a* ring.

Actually, in abstract algebra the multiplication operations in semi-rings and rings are not required to be commutative and I have defined *commutative* semi-rings and *commutative* rings. As we shall have no reason to consider non-commutative semi-rings or rings from this point on in this book, I have dropped the adjective. Note that, since we assume commutativity, we could have done with only one of the two listed Distributivity Laws.

As in ordinary arithmetic we often drop the raised dot when indicating multiplication, thus writing ab for $a \cdot b$.

Using the ordinary arithmetic operations, $\mathbb{Z}$ (the set of all integers), $\mathbb{Q}$, and $\mathbb{R}$ are all rings, while the subsets $\mathbb{N} = \mathbb{Z}^{\geq 0} = \{n \in \mathbb{Z} | n \geq 0\}, \mathbb{Q}^{\geq 0} = \{q \in \mathbb{Q} | q \geq 0\}$, and $\mathbb{R}^{\geq 0} = \{r \in \mathbb{R} | r \geq 0\}$ are all semi-rings with zero, and $\mathbb{Z}^+ = \{n \in \mathbb{Z} | n > 0\}, \mathbb{Q}^+ = \{q \in \mathbb{Q} | q > 0\}$, and $\mathbb{R}^+ = \{r \in \mathbb{R} | r > 0\}$ are semi-rings. The reader should pick one of the last six semi-rings and keep it in mind while reading through the details of the construction for the first time.

Some semi-rings have nicer properties than others. Two properties not required, but having special significance for the construction are i) the existence of a zero element, and ii) the ability, for any pair of distinct $a, b \in R$, to perform one of the subtractions $a - b$ or $b - a$: for any $a, b \in R$ there is an element $c \in R$ such that $b + c = a$ or $a + c = b$. All the semi-rings cited above have this property. An example that doesn't have this property would be given by choosing

$$\mathcal{R}_1 = \{3m + 5n \,|\, m, n \in \mathbb{N}\}.$$

For this set, $3 \in \mathcal{R}_1$ and $5 \in \mathcal{R}_1$, but neither $3 - 5$ nor $5 - 3$ exists in $\mathcal{R}_1$. Another example, slightly more complex, but perhaps a bit more natural, is

$$\mathcal{R}_2 = \{m - n\sqrt{2} \,|\, m, n \in \mathbb{N}\}.$$

Neither of the numbers $2 - \sqrt{2}$ and $1 - 2\sqrt{2}$ can be subtracted from the other in $\mathcal{R}_2$.

The reader who has not previously been exposed to modern algebra will find the following construction something of a *tour de force*. To soften the blow, let me outline in brief how the construction proceeds before getting down to the details.

We will be given a semi-ring $\mathcal{R}$, let us say $\mathbb{N}$, and attempt to construct a copy of the integers from scratch. Our first step will be to construct a set $\mathcal{R}^{\pm}$ of representations $a - b$ of differences of elements of $\mathcal{R}$. Just choosing expressions $a - b$ themselves will not do because, e.g. $2 - 7$ and $1 - 6$ are different representations of the same difference. Thus we define when two expressions are supposed to be "equal" and "identify" equal expressions by lumping them together into what are called equivalence classes. We will also show that each element of $\mathcal{R}$ can be associated with an equivalence class, whence $\mathcal{R}$ can be identified with a subset of the structure being defined. Definition 1.2 to Lemma 1.5 handle this.

The next step is to define the ring operations of addition and multiplication. This is done by implicitly appealing to Hankel's Principle of the Permanence of Formal Laws: we want

$$(a - b) + (c - d) = (a + c) - (b + d),$$

so we define addition this way. In doing so, we must verify that the sum does not depend on the particular representation of the differences: $2 - 7$ added to any difference must equal the result of adding $1 - 6$ to the given difference. In algebraic lingo, this is expressed by saying we show the new addition to be *well-defined*. This is handled by Lemma 1.6.

We are not yet finished with addition. Having shown it to be well-defined, we must then show it to satisfy all the addition axioms postulated for a ring. Lemma 1.8 will do this.

We then repeat the process for multiplication (Lemma 1.9 to Lemma 1.11, below).

The result of all this is that we have a ring that includes $\mathcal{R}$ (under the identification referred to above (Theorem 1.12). A final step is to show that the addition and multiplication of elements of $\mathcal{R}$ agrees with the new operations under the identification (Lemma 1.13).

That will complete the basic construction. Of course, we do not stop there. One of the reasons for introducing negative numbers is to be able to perform arbitrary subtractions. So we will define subtraction and use it to examine the close relation between $\mathcal{R}$ and the ring constructed.

The construction is based on the observation that every element of, say, $\mathbb{Z}$ is the difference of two elements of $\mathbb{N}$ or $\mathbb{Z}^+$. Given a semi-ring $\mathcal{R}$, we will construct a ring extending $\mathcal{R}$ essentially by taking all such differences. Until the ring has been constructed, we cannot actually take a difference $a - b$, but we can represent it by the ordered pair (a, b). To avoid typographical confusion with all the other parentheses, we shall denote ordered pairs by angled brackets, thus: $\langle a, b \rangle$.

We begin by defining

$$\mathcal{R}^\pm = \{\langle a, b \rangle | a \in \mathcal{R} \ \& \ b \in \mathcal{R}\}.$$

What we have to do next is to define when two elements of $\mathcal{R}^\pm$ are *equal* and to define addition and multiplication operations on this set.

Equality is nowadays taken to mean identity. So we are not defining when $\langle a, b \rangle$ and $\langle c, d \rangle$ are identical pairs, but when they represent the same difference, i.e. when $a - b = c - d$. Now this will happen when $a + d = b + c$. Thus we define

$$\langle a, b \rangle \cong \langle c, d \rangle \quad \text{iff} \quad a + d = b + c,$$

using the symbol "$\cong$" for congruence in place of the equality symbol to avoid confusion.

1.2 Definition. *A relation* $\sim$ *on a nonempty set* X *is an* equivalence relation *if the following hold for all* $a, b, c \in X$:
i.   *(Reflexivity)* $a \sim a$
ii. *(Symmetry)* $a \sim b \Rightarrow b \sim a$
iii. *(Transitivity)* $a \sim b \ \& \ b \sim c \Rightarrow a \sim c.$

1.3 Lemma. $\cong$ *is an equivalence relation on* $\mathcal{R}^{\pm}$.

Proof. This is fairly easy.
i. Notice
$$\langle a, b \rangle \cong \langle a, b \rangle \quad \Leftrightarrow \quad a + b = b + a.$$

ii. Again,
$$
\begin{aligned}
\langle a, b \rangle \cong \langle c, d \rangle \quad &\Leftrightarrow \quad a + d = b + c \\
&\Leftrightarrow \quad b + c = a + d \\
&\Leftrightarrow \quad c + b = d + a \\
&\Leftrightarrow \quad \langle c, d \rangle \cong \langle a, b \rangle.
\end{aligned}
$$

iii. This one is a little more fun: From $\langle a, b \rangle \cong \langle c, d \rangle$ and $\langle c, d \rangle \cong \langle e, f \rangle$ we conclude
$$a + d = b + c, \qquad c + f = d + e, \tag{1}$$
respectively. Now, by Associativity,
$$(a + d) + e = a + (d + e).$$

Combining this with (1), we have
$$(b + c) + e = a + (c + f).$$

Associativity and Commutativity yield
$$(b + e) + c = (a + f) + c,$$

and Cancellation yields
$$b + e = a + f,$$
i.e. $a + f = b + e$, i.e. $\langle a, b \rangle \cong \langle e, f \rangle$. $\square$

If one prefers dealing with objects and equality to dealing with representations and equivalences, one replaces $\mathcal{R}^{\pm}$ by the set of equivalence classes of its elements and its equivalence relation by identity.

1.4 Definition. *Let* $\sim$ *be an equivalence relation on a nonempty set* X. *For each* $a \in X$ *define its* equivalence class *by*
$$[a] = \{ b \in X \,|\, a \sim b \}.$$

Let
$$[X] = \{ [a] \,|\, a \in X \}$$
be the set of all equivalence classes of elements of X.

In the case at hand, we can think of $[\langle a,b\rangle]$ as denoting the actual difference $a-b$ and $\langle a,b\rangle$ the particular expression for the difference. For example, if we start with $\mathcal{R}=\mathbb{N}$, $\langle 2,7\rangle$ is the representation $2-7$ of the difference $2-7=-5$. $\langle 1,6\rangle$ is another representation of $-5=1-6$. The classes $[\langle 2,7\rangle]$ and $[\langle 1,6\rangle]$ are both -5. $\mathbb{N}^{\pm}$ will be the set of representations of the integers, and $[\mathbb{N}^{\pm}]$ will be the set $\mathbb{Z}$ of integers constructed as differences of elements of $\mathbb{N}$. Notice that the elements of $\mathbb{N}$ are numbers, not equivalence classes of pairs of numbers and thus not actual elements of $\mathbb{Z}$. However, we can identify the elements of $\mathbb{N}$ with specific elements of $\mathbb{Z}$, i.e. of $[\mathbb{N}^{\pm}]$. In general, we have:

1.5 Lemma. *The map*

$$a \mapsto [\langle a+a,a\rangle]$$

maps $\mathcal{R}$ one-to-one into $[\mathcal{R}^{\pm}]$.

Proof. The idea is that since $\langle a,b\rangle$ represents $a-b$, one can identify a with $a-0$, i.e. $[\langle a,0\rangle]$. Because we do not assume the existence of a zero element in $\mathcal{R}$, we identify a with $a+a-a$ instead.

To see that the function $a \mapsto [\langle a+a,a\rangle]$ is one-to-one, notice

$$
\begin{aligned}
[\langle a+a,a\rangle]=[\langle b+b,b\rangle] \quad &\Rightarrow \quad \langle a+a,a\rangle \cong \langle b+b,b\rangle \\
&\Rightarrow \quad (a+a)+b=a+(b+b) \\
&\Rightarrow \quad (a+a)+b=(a+b)+b \\
&\Rightarrow \quad a+a=a+b, \text{ by Cancellation} \\
&\Rightarrow \quad a+a=b+a \\
&\Rightarrow \quad a=b, \text{ by Cancellation.} \qquad \square
\end{aligned}
$$

If $0 \in \mathcal{R}$, then $[\langle a+a,a\rangle]=[\langle a,0\rangle]$ by the definition of $\cong$:

$$\langle a+a,a\rangle \cong \langle a,0\rangle \quad \text{iff} \quad (a+a)+0=a+a.$$

Thus the map of the Lemma agrees with the more obvious choice when a zero element exists in $\mathcal{R}$. The reader might notice that, for any fixed b_0, the assignment $a \mapsto [\langle a,b_0\rangle]$ also maps $\mathcal{R}$ one-to-one into $[\mathcal{R}^{\pm}]$. The choice of $b_0=0$, however, is best: It preserves addition and multiplication and thus embeds $\mathcal{R}$ into $[\mathcal{R}^{\pm}]$ as a sub-semi-ring and not just as a subset. Before proving this, of course, we must first define addition and multiplication on $[\mathcal{R}^{\pm}]$, a task we now turn to.

We define addition and multiplication on $[\mathcal{R}^{\pm}]$ by first defining auxiliary addition and multiplication functions on $\mathcal{R}^{\pm}$ and showing they are compatible with the equivalence relation $\cong$. This allows us then to lift the definitions up to $[\mathcal{R}^{\pm}]$.

For convenience and to avoid confusion we let $+,\cdot$ denote the addition and multiplication operations of $\mathcal{R}$ and $\oplus,\odot$ the defined operations on $\mathcal{R}^{\pm}$.

The definition of $\oplus$ is straightforward. Recalling that $\langle a,b\rangle,\langle c,d\rangle$ are supposed to represent $a-b,c-d$, respectively, their sum should represent $(a-b)+(c-d)=(a+c)-(b+d)$. This leads to the definition:

$$\langle a, b \rangle \oplus \langle c, d \rangle = \langle a + c, b + d \rangle.$$

In order to lift this definition to the set $[\mathcal{R}^\pm]$ of equivalence classes by taking

$$[\langle a, b \rangle] \oplus [\langle c, d \rangle] = [\langle a, b \rangle \oplus \langle c, d \rangle]$$
$$= [\langle a + c, b + d \rangle],$$

we have to show that the class $[\langle a + c, b + d \rangle]$ does not depend on the choices $\langle a, b \rangle, \langle c, d \rangle$ of elements of $[\langle a, b \rangle], [\langle c, d \rangle]$:

1.6 Lemma. *For all $a_1, b_1, a_2, b_2, c_1, d_1, c_2, d_2 \in \mathcal{R}$,*

$$\langle a_1, b_1 \rangle \cong \langle a_2, b_2 \rangle \ \& \ \langle c_1, d_1 \rangle \cong \langle c_2, d_2 \rangle \ \Rightarrow \ \langle a_1, b_1 \rangle \oplus \langle c_1, d_1 \rangle \cong \langle a_2, b_2 \rangle \oplus \langle c_2, d_2 \rangle.$$

Proof. From $\langle a_1, b_1 \rangle \cong \langle a_2, b_2 \rangle$ and $\langle c_1, d_1 \rangle \cong \langle c_2, d_2 \rangle$, we conclude

$$a_1 + b_2 = a_2 + b_1, \qquad c_1 + d_2 = c_2 + d_1.$$

Thus

$$a_1 + b_2 + c_1 + d_2 = a_2 + b_1 + c_2 + d_1$$
$$(a_1 + c_1) + (b_2 + d_2) = (a_2 + c_2) + (b_1 + d_1),$$

i.e.

$$\langle a_1 + c_1, b_1 + d_1 \rangle = \langle a_2 + c_2, b_2 + d_2 \rangle,$$

i.e.

$$\langle a_1, b_1 \rangle \oplus \langle c_1, d_1 \rangle = \langle a_2, b_2 \rangle \oplus \langle c_2, d_2 \rangle \qquad \square$$

1.7 Corollary. *We can define a function $\oplus$ on $[\mathcal{R}^\pm]$ by*

$$[\langle a, b \rangle] \oplus [\langle c, d \rangle] = [\langle a, b \rangle \oplus \langle c, d \rangle].$$

That is, the equivalence class on the right does not depend on the choice of representations $\langle a, b \rangle, \langle c, d \rangle$ of the classes on the left.

Defining $\oplus$ is, of course, not enough. One must show that the defined function behaves like addition.

1.8 Lemma. *The following hold:*
i.   for all $x, y \in [\mathcal{R}^\pm]$, $x \oplus y = y \oplus x$;
ii.   for all $x, y, z \in [\mathcal{R}^\pm]$, $x \oplus (y \oplus z) = (x \oplus y) \oplus z$;
iii. $[\mathcal{R}^\pm]$ has a zero element $[0]$;
iv.   for all $x \in [\mathcal{R}^\pm]$, there is an element $\ominus x \in [\mathcal{R}^\pm]$ such that

$$x \oplus (\ominus x) = [0].$$

Proof. i. Let $x, y \in [\mathcal{R}^\pm]$, choose $a, b, c, d \in \mathcal{R}$ such that $x = [\langle a, b \rangle], y = [\langle c, d \rangle]$, and observe

$$\langle a, b \rangle \oplus \langle c, d \rangle = \langle a + c, b + d \rangle = \langle c + a, d + b \rangle = \langle c, d \rangle \oplus \langle a, b \rangle,$$

whence $x \oplus y = y \oplus x$.

ii. Let $x, y, z \in [\mathcal{R}^{\pm}]$, choose $a, b, c, d, e, f \in \mathcal{R}$ such that $x = [\langle a, b \rangle], y = [\langle c, d \rangle], z = [\langle e, f \rangle]$, and observe

$$
\begin{aligned}
\langle a, b \rangle \oplus (\langle c, d \rangle \oplus \langle e, f \rangle) &= \langle a, b \rangle \oplus \langle c + e, d + f \rangle \\
&= \langle a + (c + e), b + (d + f) \rangle \\
&= \langle (a + c) + e, (b + d) + f \rangle \\
&= \langle a + c, b + d \rangle \oplus \langle e, f \rangle \\
&= (\langle a, b \rangle \oplus \langle c, d \rangle) \oplus \langle e, f \rangle.
\end{aligned}
$$

iii. The zero element is just $[\langle a, a \rangle]$ for any $a \in \mathcal{R}$.[22] For, let $x \in [\mathcal{R}^{\pm}]$ and $c, d \in \mathcal{R}$ be such that $x = [\langle c, d \rangle]$. Note

$$
\begin{aligned}
\langle c, d \rangle \oplus \langle a, a \rangle &= \langle c + a, d + a \rangle \\
&\cong \langle c, d \rangle,
\end{aligned}
$$

since $(c + a) + d = (d + a) + c$.

iv. Let $x \in [\mathcal{R}^{\pm}]$, say $x = [\langle a, b \rangle]$. Choose

$$
\ominus x = [\langle b, a \rangle].
$$

First, notice that

$$
\begin{aligned}
\langle a, b \rangle \cong \langle c, d \rangle \;\Rightarrow\; & a + d = b + c \\
\Rightarrow\; & b + c = a + d \\
\Rightarrow\; & \langle b, a \rangle \cong \langle d, c \rangle,
\end{aligned}
$$

and the operation $\ominus$ is well-defined. Then observe

$$
x \oplus \ominus x = \langle a, b \rangle \oplus \langle b, a \rangle = \langle a + b, a + b \rangle = [0]. \qquad \square
$$

To define multiplication, consider the product

$$
(a - b)(c - d) = ac - ad - bc + bd = (ac + bd) - (ad + bc).
$$

Thus we define

$$
\langle a, b \rangle \odot \langle c, d \rangle = \langle ac + bd, ad + bc \rangle.
$$

The first thing we need to do is to prove the multiplicative analogue to Lemma 1.6

1.9 Lemma. *For all* $a_1, b_1, a_2, b_2, c_1, d_1, c_2, d_2 \in \mathcal{R}$,

$$
\langle a_1, b_1 \rangle \cong \langle a_2, b_2 \rangle \;\&\; \langle c_1, d_1 \rangle \cong \langle c_2, d_2 \rangle \;\Rightarrow\; \langle a_1, b_1 \rangle \odot \langle c_1, d_1 \rangle \cong \langle a_2, b_2 \rangle \odot \langle c_2, d_2 \rangle.
$$

---

[22] Note that $\langle a, a \rangle \cong \langle b, b \rangle$ for any $a, b \in \mathcal{R}$.

Proof. This is a little tricky.

From $\langle a_1, b_1 \rangle \cong \langle a_2, b_2 \rangle$ and $\langle c_1, d_1 \rangle \cong \langle c_2, d_2 \rangle$, we conclude

$$a_1 + b_2 = a_2 + b_1, \qquad c_1 + d_2 = c_2 + d_1. \tag{2}$$

We have to show

$$\langle a_1, b_1 \rangle \odot \langle c_1, d_1 \rangle = \langle a_1 c_1 + b_1 d_1, a_1 d_1 + b_1 c_1 \rangle \tag{3}$$

equal to

$$\langle a_2, b_2 \rangle \odot \langle c_2, d_2 \rangle = \langle a_2 c_2 + b_2 d_2, a_2 d_2 + b_2 c_2 \rangle. \tag{4}$$

We do this by first showing (3) to equal

$$\langle a_2, b_2 \rangle \odot \langle c_1, d_1 \rangle = \langle a_2 c_1 + b_2 d_1, a_2 d_1 + b_2 c_1 \rangle \tag{5}$$

and then showing (5) to equal (4). This means we must show $\alpha = \beta$ and then $\gamma = \delta$, where

$$\alpha : \ a_1 c_1 + b_1 d_1 + a_2 d_1 + b_2 c_1$$
$$\beta : \ a_2 c_1 + b_2 d_1 + a_1 d_1 + b_1 c_1$$
$$\gamma : \ a_2 c_1 + b_2 d_1 + a_2 d_2 + b_2 c_2$$
$$\delta : \ a_2 c_2 + b_2 d_2 + a_2 d_1 + b_2 c_1.$$

To see that $\alpha = \beta$, rewrite α:

$$\alpha = a_1 c_1 + b_2 c_1 + b_1 d_1 + a_2 d_1$$
$$= (a_1 + b_2) c_1 + (b_1 + a_2) d_1$$
$$= (a_2 + b_1) c_1 + (b_2 + a_1) d_1, \text{by (2)}$$
$$= a_2 c_1 + b_1 c_1 + b_2 d_1 + a_1 d_1$$
$$= \beta.$$

The proof that $\gamma = \delta$, by factoring the a_2's and b_2's, is similar. $\qquad\square$

1.10 Corollary. *We can define a function $\odot$ on $[\mathcal{R}^{\pm}]$ by*

$$[\langle a, b \rangle] \odot [\langle c, d \rangle] = [\langle a, b \rangle \odot \langle c, d \rangle].$$

Defining $\odot$ is, of course, not enough. One must show that the defined function behaves like multiplication.

1.11 Lemma. *The following hold:*
i. for all $x, y \in [\mathcal{R}^{\pm}]$, $x \odot y = y \odot x$;
ii. for all $x, y, z \in [\mathcal{R}^{\pm}]$, $x \odot (y \odot z) = (x \odot y) \odot z$;
iii. for all $x, y, z \in [\mathcal{R}^{\pm}]$, $x \odot (y \oplus z) = (x \odot y) \oplus (x \odot z)$.

Proof. i. Let $x, y \in [\mathcal{R}^{\pm}]$, choose $a, b, c, d \in \mathcal{R}$ such that $x = [\langle a, b \rangle], y = [\langle c, d \rangle]$, and observe

$$\langle a,b\rangle \odot \langle c,d\rangle = \langle ac+bd, ad+bc\rangle$$
$$= \langle ca+db, da+cb\rangle$$
$$= \langle c,d\rangle \odot \langle a,b\rangle.$$

ii. Let $x,y,z \in [\mathcal{R}^{\pm}]$, choose $a,b,c,d,e,f \in \mathcal{R}$ such that $x = [\langle a,b\rangle], y = [\langle c,d\rangle], z = [\langle e,f\rangle]$, and observe

$$\langle a,b\rangle \odot (\langle c,d\rangle \odot \langle e,f\rangle) = \langle a,b\rangle \odot \langle ce+df, cf+de\rangle$$
$$= \langle ace+adf+bcf+bde, acf+ade+bce+bdf\rangle$$
$$= \langle (ac+bd)e+(ad+bc)f, (ac+bd)f+(ad+bc)e\rangle$$
$$= \langle ac+bd, ad+bc\rangle \odot \langle e,f\rangle$$
$$= (\langle a,b\rangle \odot \langle c,d\rangle) \odot \langle e,f\rangle.$$

iii. Let $x,y,z \in [\mathcal{R}^{\pm}]$, choose $a,b,c,d,e,f \in \mathcal{R}$ such that $x = [\langle a,b\rangle], y = [\langle c,d\rangle], z = [\langle e,f\rangle]$, and observe

$$\langle a,b\rangle \odot (\langle c,d\rangle \oplus \langle e,f\rangle) = \langle a,b\rangle \odot \langle c+e, d+f\rangle$$
$$= \langle a(c+e)+b(d+f), a(d+f)+b(c+e)\rangle$$
$$= \langle ac+ae+bd+bf, ad+af+bc+be\rangle$$
$$= \langle ac+bd, ad+bc\rangle \oplus \langle ae+bf, af+be\rangle$$
$$= (\langle a,b\rangle \odot \langle c,d\rangle) \oplus (\langle a,b\rangle \odot \langle e,f\rangle). \qquad \square$$

Putting it all together, we have the following Theorem.

1.12 Theorem. $[\mathcal{R}^{\pm}]$ *is a ring.*

This isn't quite everything. For we have the embedding of $\mathcal{R}$ into $[\mathcal{R}^{\pm}]$ of Lemma 1.5 to consider. We promised to show that it preserved $+,\cdot$. And we should also consider the close relationship between $\mathcal{R}$ and $[\mathcal{R}^{\pm}]$.

From Lemma 1.5 we have the existence of a one-to-one map $a \mapsto [\langle a+a,a\rangle]$ embedding $\mathcal{R}$ into $[\mathcal{R}^{\pm}]$. For notational convenience, let us denote the image of a under the map, namely $[\langle a+a,a\rangle]$, by $[a]$. Thus

$$[a] : \ [\langle a+a,a\rangle].$$

1.13 Lemma. *For all $a,b \in \mathcal{R}$,*
i. $[a+b] = [a] \oplus [b]$;
ii. $[a \cdot b] = [a] \odot [b].$

Proof. i. Observe

$$[a+b] = [\langle (a+b)+(a+b), a+b\rangle]$$
$$= [\langle a+a,a\rangle \oplus \langle b+b,b\rangle]$$
$$= [\langle a+a,a\rangle] \oplus [\langle b+b,b\rangle]$$
$$= [a] \oplus [b].$$

ii. Observe

$$[a] \odot [b] = [\langle a+a,a\rangle] \odot [\langle b+b,b\rangle]$$
$$= [\langle a+a,a\rangle \odot \langle b+b,b\rangle]$$
$$= [\langle (a+a)(b+b)+ab,(a+a)b+a(b+b)\rangle]$$
$$= [\langle ab+ab+ab+ab+ab,ab+ab+ab+ab\rangle]$$
$$= [\langle ab+ab,ab\rangle],$$

because

$$(ab+ab+ab+ab+ab)+ab = (ab+ab+ab+ab)+ab+ab.$$

But $[\langle ab+ab,ab\rangle] = [ab]$.   □

I note that if one assumes $\mathcal{R}$ to contain a zero element 0, the proof simplifies a bit. For, then the map is

$$a \mapsto [\langle a,0\rangle],$$

and

$$[a+b] = [\langle a+b,0\rangle] = [\langle a,0\rangle \oplus \langle b,0\rangle]$$
$$= [\langle a,0\rangle] \oplus [\langle b,0\rangle] = [a] \oplus [b]$$
$$[a] \odot [b] = [\langle a,0\rangle] \odot [\langle b,0\rangle] = [\langle a,0\rangle \odot \langle b,0\rangle]$$
$$= [\langle ab+0\cdot 0,a\cdot 0+b\cdot 0\rangle] = [\langle ab,0\rangle] = [ab].$$

Thus $\mathcal{R}$ embeds in an operation-preserving manner into the ring $[\mathcal{R}^{\pm}]$. Such functions have a special name— or, *names*, as there are varying degrees of strength associated with them.

1.14 Definitions. *A function* $f : R_1 \to R_2$ *of semi-rings* $(R_1,+,\cdot),(R_2,+,\cdot)$ *is a* homomorphism *iff for all* $x,y \in R_1$ *the following hold:*

$$f(x+y) = f(x)+f(y), \quad f(x\cdot y) = f(x)\cdot f(y).$$

If a homomorphism f *is one-to-one, it is called a* homomorphic embedding *or an* isomorphic embedding. *An isomorphic embedding of* R_1 *into* R_2 *that maps onto all of* R_2 *is an* isomorphism.

Thus, what we have proven can be stated succinctly as:

1.15 Theorem. *There is an isomorphic embedding of* $\mathcal{R}$ *into* $[\mathcal{R}^{\pm}]$.

1.16 Corollary. *Every semi-ring can be isomorphically embedded into a ring.*

The Corollary is a less specific restatement of the Theorem that does not name names. Neither is really a good summary of what has actually been achieved by the construction, because neither statement takes into account the closeness of the relation between $\mathcal{R}$ and $[\mathcal{R}^{\pm}]$.

1.17 Theorem. *Let $\mathcal{R}$ be a semi-ring and $[\mathcal{R}^{\pm}]$ the ring constructed therefrom. Let $f : \mathcal{R} \rightarrow R_1$ be any homomorphism of $\mathcal{R}$ into a ring R_1. There is a unique homomorphic extension $\overline{f} : [\mathcal{R}^{\pm}] \rightarrow R_1$ of f, i.e. there is a unique homomorphism $\overline{f} : [\mathcal{R}^{\pm}] \rightarrow R_1$ such that $\overline{f}([a]) = f(a)$ for all $a \in \mathcal{R}$. Moreover, if f is an embedding, then so is $\overline{f}$.*

This is how one says in abstract algebraic terms that any way one extends $\mathcal{R}$ to allow subtraction, the set of differences $a - b$ for $a, b \in \mathcal{R}$ will look exactly like $[\mathcal{R}^{\pm}]$. Of course, we haven't even defined subtraction yet.

1.18 Definition. *Let $(R_1, +, \cdot, 0)$ be a ring. For each $a \in R_1$, we let $-a$ denote the additive inverse of a:*

$$a + (-a) = 0.$$

We define

$$a - b = a + (-b),$$

and call the operation subtraction.

I leave to the reader the verification that the additive inverse is unique, and thus that inversion and subtraction are indeed functions.

In $[\mathcal{R}^{\pm}]$ we have used $\ominus x$ to denote the additive inverse of an element x, so we might as well also use $\ominus$ to denote subtraction in that ring.

The first thing to notice is that every element of $[\mathcal{R}^{\pm}]$ is a difference of (images of) elements of $\mathcal{R}$ as promised:

$$[\langle a, b \rangle] = [a] \ominus [b]. \tag{6}$$

For,

$$\begin{aligned}
[a] \ominus [b] &= [\langle a + a, a \rangle)] \oplus (\ominus [\langle b + b, b \rangle]) \\
&= [\langle a + a, a \rangle \oplus \langle b, b + b \rangle] \\
&= [\langle a + a + b, a + b + b \rangle] \\
&= [\langle a, b \rangle];
\end{aligned}$$

since

$$(a + a + b) + b = (a + b + b) + a$$

yields $\langle a + a + b, a + b + b \rangle \cong \langle a, b \rangle$.

Before getting down to considering Theorem 1.17, let us pause to reflect on the two conditions cited on page 199 above. First, if $0 \in \mathcal{R}$, the embedding can be taken to be $a \mapsto [\langle a, 0 \rangle]$ and the verification just given simplifies:

$$\begin{aligned}
[a] \ominus [b] &= [\langle a, 0 \rangle] \oplus [\ominus \langle b, 0 \rangle] \\
&= [\langle a, 0 \rangle] \oplus [\langle 0, b \rangle] \\
&= [\langle a, b \rangle].
\end{aligned}$$

Second, if we assume for any distinct $a, b \in \mathcal{R}$ the existence of $c \in \mathcal{R}$ such that

$$a + c = b \quad \text{or} \quad b + c = a,$$

then every nonzero element of $[\mathcal{R}^{\pm}]$ is of the form $[c]$ or $\ominus[c]$ for $c \in \mathcal{R}$. For, if $b + c = a$, then $[b] \oplus [c] = [a]$, whence $[a] \ominus [b] = [c]$. And, if $a + c = b$, then $[a] \oplus [c] = [b]$ and $[c] = [b] \ominus [a]$, whence $[a] \ominus [b] = \ominus[c]$.

Getting back to Theorem 1.17, we prove a simple lemma.

1.19 Lemma. *Let $g : R_1 \to R_2$ be a homomorphism of rings. For all $x, y \in R_1$,*
i. *$g(0) = 0$;*
ii. *$g(-x) = -g(x)$;*
iii. *$g(x - y) = g(x) - g(y)$.*

Proof. i. Observe,

$$g(0) + g(0) = g(0 + 0) = g(0) = g(0) + 0,$$

whence additive cancellation yields $g(0) = 0$.

ii. $g(x) + g(-x) = g(x + (-x)) = g(0) = 0$, whence $g(-x)$ is an additive inverse to $g(x)$, whence $g(-x) = -g(x)$.

iii. $g(x - y) = g(x + (-y)) = g(x) + g(-y) = g(x) + (-g(y)) = g(x) - g(y)$.
□

Proof of Theorem 1.17. The uniqueness of a homomorphism $\overline{f} : [\mathcal{R}^{\pm}] \to R_1$ extending a given $f : \mathcal{R} \to R_1$ is easy to verify. If $x \in [\mathcal{R}^{\pm}]$, write $x = [a] \ominus [b]$ for $a, b \in \mathcal{R}$ and observe

$$\overline{f}(x) = \overline{f}([a] \ominus [b]) = \overline{f}([a]) - \overline{f}([b]) = f(a) - f(b). \tag{7}$$

This last equation tells us that $\overline{f}$ is unique because it must take on the value given. It also tells us how to define $\overline{f}$. We have to show, however, that the value given does not depend on the choice of representation of x, i.e. that $\overline{f}$ actually exists. Thus, let

$$x = [a] \ominus [b] = [c] \ominus [d].$$

Then

$$[a] \oplus [d] = [b] \oplus [c],$$

whence

$$[a + d] = [b + c]$$

by Lemma 1.13. But the map from $\mathcal{R}$ to $[\mathcal{R}^{\pm}]$ is one-to-one, whence this yields $a + d = b + c$. But f maps $\mathcal{R}$ homomorphically into R_1, whence

$$f(a) + f(d) = f(a + d) = f(b + c) = f(b) + f(c),$$

and

$$f(a) - f(b) = f(c) - f(d),$$

as was to be shown.

Thus (7) defines a function. We have yet to show that it is a homomorphism (i.e. it preserves $+,\cdot$) and that, if f is one-to-one, then $\overline{f}$ is one-to-one. Let $x, y \in [\mathcal{R}^{\pm}]$ with $x = [a] \ominus [b], y = [c] \ominus [d]$ and observe

$$\begin{aligned}
\overline{f}(x) + \overline{f}(y) &= f(a) - f(b) + f(c) - f(d) \\
&= f(a) + f(c) - (f(b) + f(d)) \\
&= f(a + c) - f(b + d) \\
&= \overline{f}([a + c] \ominus [b + d]), \text{ by definition} \\
&= \overline{f}([a] \oplus [c] \ominus ([b] \oplus [d])) \\
&= \overline{f}(([a] \ominus [b]) \oplus ([c] \ominus [d])) \\
&= \overline{f}(x \oplus y) \\
\overline{f}(x) \cdot \overline{f}(y) &= (f(a) - f(b)) \cdot (f(c) - f(d)) \\
&= f(a)f(c) + f(b)f(d) - (f(a)f(d) + f(b)f(c)) \\
&= f(ac + bd) - f(ad + bc)
\end{aligned}$$

while

$$\begin{aligned}
\overline{f}(x \odot y) &= \overline{f}(([a] \ominus [b]) \odot ([c] \ominus [d])) \\
&= \overline{f}([a] \odot [c] \oplus [b] \odot [d] \ominus ([a] \odot [d] \oplus [b] \odot [c])) \\
&= \overline{f}([ac + bd] \ominus [ad + bc]) \\
&= f(ac + bd) - f(ad + bc).
\end{aligned}$$

Thus $\overline{f}$ is a homomorphism.

Finally, suppose f is one-to-one and $\overline{f}(x) = \overline{f}(y)$ for some $x, y \in [\mathcal{R}^{\pm}]$, with $x = [a] \ominus [b], y = [c] \ominus [d]$. Then

$$f(a) - f(b) = \overline{f}(x) = \overline{f}(y) = f(c) - f(d)$$

and

$$f(a) + f(d) = f(b) + f(c),$$

whence $f(a + d) = f(b + c)$. But f is one-to-one, whence $a + d = b + c$, i.e. $\langle a, b \rangle \cong \langle c, d \rangle$, i.e. $[\langle a, b \rangle] = [\langle c, d \rangle]$. But

$$\begin{aligned}
[a] \ominus [b] &= [\langle a, b \rangle], \text{ by (6)} \\
&= [\langle c, d \rangle], \text{ as just shown} \\
&= [c] \ominus [d], \text{ by (6)},
\end{aligned}$$

and we conclude $\overline{f}$ is one-to-one. $\square$

1.3 Variants of the Construction

The construction of negative numbers given is fairly standard in discussions of the foundations of the real number system. It is not the only one. There are two others, both a bit narrow in applicability but general enough to cover the constructions of all the integers from the positive or nonnegative integers, and the first of these also handles the rationals and reals as well.

The simpler method consists of simply taking the nonzero elements of one's given ring, copying them and tacking a minus sign in front, then taking all these numbers and 0 and putting them together and defining operations of addition and multiplication on them: Given $\mathcal{R} = \mathbb{N}, \mathbb{Z}^+$, etc., define

$$\mathcal{R}' = \{r|r \in \mathcal{R} \,\&\, r \neq 0\} \cup \{0\} \cup \{\ominus r|r \in \mathcal{R} \,\&\, r \neq 0\}.$$

For these particular semi-rings no element other than 0 has an additive inverse, whence there will be no equivalence of elements from different components of $\mathcal{R}'$ and thus no need to introduce an equivalence relation and its equivalence classes.

The operation of addition is defined for $r, s \in \mathcal{R}, x \in \mathcal{R}'$ by

$$r \oplus s = r + s$$

$$r \oplus \ominus s = \begin{cases} r - s, & \text{if } r > s \\ 0, & \text{if } r = s \\ \ominus(s - r), & \text{if } r < s \end{cases}$$

$$\ominus r \oplus s = \begin{cases} \ominus(r - s), & \text{if } r > s \\ 0, & \text{if } r = s \\ s - r, & \text{if } r < s \end{cases}$$

$$\ominus r \oplus \ominus s = \ominus(r + s)$$

$$x \oplus 0 = x$$

$$0 \oplus x = x.$$

Because there is no overlap in the components of $\mathcal{R}'$, each pair $x, y \in \mathcal{R}'$ falls into one of these cases and the operation is well-defined.

The operation of multiplication is similarly defined by an enumeration of cases. When this is done, the construction is complete. One now has to verify the commutative, associative, and distributive laws for the operations so defined. This turns out to be a rather tedious enumeration of trivialities that sheds no light on anything.

This simpler construction is limited in its applicability. The semi-ring,

$$\mathcal{R}_3 = \{m_0 + m_1\pi + \ldots + m_k\pi^k | k, m_0 \in \mathbb{N} \,\&\, m_1, \ldots, m_k \in \mathbb{Z}\},$$

of polynomials with integral coefficients and nonnegative constant term evaluated at π is an example of a semi-ring for which the construction does not apply. For, every polynomial with constant term equal to 0 will yield an element of $\mathcal{R}_3$ with additive inverse when evaluated at π. One can tinker with the construction to get it to work here, but one must tinker. And what if one restricts $\mathcal{R}_3$ further by disallowing, say, $m_0 = 1$ in the definition?

And, finally, there is no guarantee that the construction so described will be convincing to one who has objections to negative numbers. For, putting the minus sign in front of a number is the same as calling it a debt and the operations defined are just the rules for calculating with debts. Let us quote William Frend:

> Numbers are there[23] divided into two sorts, positive and negative; and an attempt is made to explain the nature of negative numbers, by allusions to book-debts and other arts. Now, when a person cannot explain the principles of a science without reference to metaphor, the probability is, that he has never thought accurately upon the subject. A number may be greater or less than another number; it may be added to, taken from, multiplied into, and divided by another number; but in other respects it is very untractable: though the whole world should be destroyed, one will be one, and three will be three; and no art whatever can change their nature. You may put a mark before one, which it will obey: it submits to be taken away from another number greater than itself, but to attempt to take it away from a number less than itself is ridiculous. Yet this is attempted by algebraists, who talk of a number less than nothing, of multiplying a negative number into a negative number and thus producing a positive number...[24]

Perhaps the construction would be more acceptable philosophically if one used ordered pairs and defined $\mathcal{R}'$ as

$$\mathcal{R}' = \{\langle 0, r\rangle | r \in \mathcal{R} \,\&\, r \neq 0\} \cup \{0\} \cup \{\langle 1, r\rangle | r \in \mathcal{R} \,\&\, r \neq 0\}$$

and defined operations $\oplus, \odot$ on $\mathcal{R}'$ as follows

$$\langle 0, r\rangle \oplus \langle 0, s\rangle = \langle 0, r + s\rangle$$

$$\langle 0, r\rangle \oplus \langle 1, s\rangle = \begin{cases} \langle 0, r - s\rangle, & r > s \\ 0, & r = s \\ \langle 1, s - r\rangle & r < s \end{cases}$$

etc. One could then proceed to show $(\mathcal{R}', \oplus, \odot, 0)$ to have all the ring properties, identify the nonzero elements of $\mathcal{R}$ with the elements $\langle 0, r\rangle \in \mathcal{R}'$ and declare a meaning for negative numbers to have been found in the elements $\langle 1, r\rangle$. I don't know if this would have convinced Frend, but Hamilton found that such a trick made the complex numbers more acceptable to him. He objected adding real and imaginary numbers because they were different things. However, he made complex numbers homogeneous by defining them to be ordered pairs of real numbers with their own defined addition and multiplication operations, and this comforted him.

Speaking of Hamilton, his paper, as I said earlier, is not very readable. The treatment is reported to be incomplete, and I haven't the patience to read it carefully. That said, I can say that a casual glance at his treatment of negative

---

[23] The reference is to Colin Maclaurin's algebra textbook.

[24] William Frend, *The Principles of Algebra*, Lodon, 1796; quoted by A. Arcavi and M. Bruckheimer, *op. cit.*

numbers suggests that, if we cut out the interpretation of algebra as a science of pure time, extract and complete the mathematics, his construction of the negative integers from the positive ones is the same as the one given here by means of ordered pairs representing differences— except in his treatment of multiplication. What he did is far more abstract, but somewhat pleasing because the derivation of some of the algebraic laws is rather slick. The point is that, starting with $\mathcal{R} = \mathbb{Z}^+$, and obtaining $\mathbb{Z} = [\mathcal{R}^\pm]$ and defining addition on it, he has contructed $(\mathbb{Z}, \oplus, 0, 1)$ which is a special structure called a *group* (Cf. Exercise 9.1 at the end of this chapter.). Now, multiplication in $\mathbb{Z}$ by any nonzero element is a *group homomorphism*. That is, for fixed $a \neq 0, a \in \mathbb{Z}$, the map

$$f_a(x) = ax$$

satisfies

$$f_a(0) = 0 \qquad f_a(x + y) = f_a(x) + f_a(y).$$

Moreover, every group homomorphism is determined by its action on 1: if $f(1) = a$, then $f = f_a$. He thus defines f_a to be the homomorphism mapping 1 to a (which can be defined for the negative integers without reference to multiplication). The product $a \odot b$ for integers a, b is then defined to be $a \odot b = f_a(b)$. Commutativity requires some work, but distributivity follows from the fact that f_a is a homomorphism:

$$a \odot (b \oplus c) = f_a(b \oplus c) = f_a(b) \oplus f_a(c) = (a \odot b) \oplus (a \odot c).$$

And associativity reduces to the associativity of composition:

$$a \odot (b \odot c) = f_a(b \odot c) = f_a(f_b(c)) = (f_a \circ f_b)(c) = f_{a \odot b}(c) = (a \odot b) \odot c,$$

where we conclude $f_a \circ f_b = f_{a \odot b}$ by applying both functions to 1:

$$(f_a \circ f_b)(1) = f_a(f_b(1)) = f_a(b) = f_{a \odot b}(1).$$

Because the composition of homomorphisms is a homomorphism, and because homomorphisms on $(\mathbb{Z}, \oplus, 0, 1)$ are determined by their action on 1, $f_a \circ f_b = f_{a \odot b}$.[25]

2 Rational Numbers

A standard algebraic fallacy proceeds as follows[26]: Assume $x = y$. Multiply by x:

---

[25] I have been quite sketchy on this construction and refer the reader to Günter Pickert and Lilly Görke, "Constructions of the system of real numbers", in: H. Behnke, F. Bachmann, K. Fladt, and W. Süss (eds.), *Foundations of Mathematics, vol. I: The Real Number System and Algebra*, MIT Press, Cambridge (Mass.), 1974; translated from *Grundzüge der Mathematik, Band I, Grundlagen der Mathematik, Arithmetik und Algebra*, 2nd edition, Vandenhoeck & Ruprecht, Göttingen, 1962, by S.H. Gould.

[26] Eugene P. Northrop, *Riddles in Mathematics; A Book of Paradoxes*, D. van Nostrand Company, Inc., New York, 1944, has a ton of these. I have copied the present one from pp. 82 - 83.

$$x^2 = xy.$$

Subtract y^2:

$$x^2 - y^2 = xy - y^2.$$

Factor:

$$(x + y)(x - y) = y(x - y).$$

Divide by $x - y$:

$$x + y = y.$$

Taking $x = y = 1$, this yields

$$1 + 1 = 1,$$

i.e. $2 = 1$. It doesn't take long to find where the computation went astray. Since $x = y$, $x - y = 0$ and the final division cannot be carried out. In short, this is not a paradox of Type I formal manipulation, but a simple error. There do not seem to be any genuine paradoxes involved with rationality, no objections that a smaller cannot be divided by a larger number, or that nothing can be smaller than unity. That is to say, there is no philosophical necessity for reducing the rational numbers to the integers. It is a purely (Type II) formal exercise. This is not to say that it is a pointless exercise. It is the natural next step after the construction of the integers in the arithmetisation of analysis, it is necessary for translating mathematics into the language of set theory, and the construction applies to other structures.

Once again, because the final step of defining the real numbers in terms of rational numbers works more smoothly under some approaches when negative numbers aren't around, we want to present the construction in such a way that it will apply to $\mathbb{N}, \mathbb{Z}^+, \mathbb{Z}$ to yield $\mathbb{Q}^{\geq 0}, \mathbb{Q}^+, \mathbb{Q}$, respectively. Thus we shall present the construction in some generality. The construction is analogous to that of the previous section, but treats multiplication instead of addition. And, where we previously had to assume the cancellation property for addition, we must now assume a multiplicative cancellation property.

2.1 Definitions. *A semi-ring R is called* nontrivial *if it contains a nonzero element. A nontrivial ring $(R, +, \cdot, 0)$ is called an* integral domain *if it satisfies the* multiplicative cancellation property,

for all $a, b, c \in R$, if $a \neq 0$, then $ac = bc \Rightarrow a = b$.

A nontrivial semi-ring that satisfies the multiplicative cancellation property will be called a semi-domain. *A ring $(R, +, \cdot, 0)$ is called a* field *if R has a multiplicative identity 1, and all nonzero elements have* multiplicative inverses. *That is, R is a field if there is an element $1 \in R$ such that*
i. for all $a \in R$, $1a = a$,
ii. for all $a \in R$, if $a \neq 0$, there is an element $b \in R$ such that $ab = 1$.
A semi-ring possessing a multiplicative identity and multiplicative inverses for its nonzero elements will be called a semi-field.

The terms "integral domain" and "field" are standard designations and except for our implicit rather than stated assumption of the commutativity of multiplication (recall page 199, above), so are their definitions. The terms "semi-domain" and "semi-field" are more-or-less obvious choices for the generalisations and may or may not be common in more specialised works of Abstract Algebra. These latter two concepts are certainly not as central to mathematics as are the former two.

The standard example of an integral domain is $\mathbb{Z}$ and the standard examples of fields are $\mathbb{Q}, \mathbb{R}$ and the field $\mathbb{C}$ of complex numbers. $\mathbb{N}, \mathbb{Z}^+$ are semi-domains, and $\mathbb{Q}^{\geq 0}, \mathbb{Q}^+, \mathbb{R}^{\geq 0}, \mathbb{R}^+$ are semi-fields.

Most Abstract Algebra textbooks will present a proof that every integral domain can be embedded into a field called the *field of fractions* or *quotient field* of the domain. The construction does not require the presence of negative numbers.

2.2 Theorem. *Every semi-domain can be embedded into a semi-field.*

This is a weak statement of the result. There is a certain canonicality to the construction, resulting in a theorem analogous to Theorem 1.17 concerning the construction of the previous section. I will leave the statement and proof of the result to the more enthusiastic reader.

We start by assuming given a semi-domain $\mathcal{R}$ and using ordered pairs to represent quotients:

$$Q(\mathcal{R}) = \{\langle a, b \rangle \,|\, a, b \in \mathcal{R} \,\&\, b \text{ is not a zero element}\}.$$

(I have used the awkward "b is not a zero element" in place of "$b \neq 0$" in case $\mathcal{R}$ has no zero element to say b is not equal to. In the sequel I shall not be so pedantic.) Equivalence of elements mirrors equality of fractions:

$$\langle a, b \rangle \cong \langle c, d \rangle \quad \text{iff} \quad ad = bc.$$

And again one defines equivalence classes and the set of such classes:

$$[\langle a, b \rangle] = \{\langle c, d \rangle \,|\, \langle c, d \rangle \in Q(\mathcal{R}) \,\&\, \langle a, b \rangle \cong \langle c, d \rangle\}$$
$$[Q(\mathcal{R})] = \{[\langle a, b \rangle] \,|\, \langle a, b \rangle \in Q(\mathcal{R})\}.$$

The proof that $\cong$ is an equivalence relation proceeds exactly as the proof of Lemma 1.3 with $+$ replaced by $\cdot$.

$\mathcal{R}$ naturally embeds into $[Q(\mathcal{R})]$ by mapping

$$a \mapsto [\langle aa, a \rangle], \text{ if } a \neq 0, \tag{8}$$

and, if $0 \in \mathcal{R}$,

$$0 \mapsto [\langle 0, b \rangle], \tag{9}$$

for some $b \neq 0$. Note that the class 0 maps to is not dependent on the choice of the particular nonzero $b \in \mathcal{R}$ chosen in defining the image: If $b, c \in \mathcal{R}$ are both nonzero, then $\langle 0, b \rangle$ and $\langle 0, c \rangle$ are defined and $0 \cdot c = 0 \cdot b$, whence $\langle 0, b \rangle = \langle 0, c \rangle$, i.e. $[\langle 0, b \rangle] = [\langle 0, c \rangle]$. Thus the value assigned to 0 is unique and the function is well-defined.

The fact that 0 has to be treated differently from the nonzero elements of $\mathcal{R}$ in defining the function also slightly complicates the proof that the embedding is one-to-one. First, assume $a, b \neq 0$ and note that

$$
\begin{aligned}
[\langle aa, a \rangle] = [\langle bb, b \rangle] \ &\Rightarrow \ \langle aa, a \rangle \cong \langle bb, b \rangle \\
&\Rightarrow \ (aa)b = a(bb) \\
&\Rightarrow \ a = b,
\end{aligned}
$$

after two cancellations. One must also rule out the possibility that $a \neq 0$ and 0 might map to the same element:

$$
\begin{aligned}
[\langle aa, a \rangle] = [\langle 0, a \rangle] \ &\Rightarrow \ \langle aa, a \rangle \cong \langle 0, a \rangle \\
&\Rightarrow \ (aa)a = a \cdot 0 \\
&\Rightarrow \ aa = 0, \ \text{by cancellation} \\
&\Rightarrow \ aa = a \cdot 0 \\
&\Rightarrow \ a = 0, \ \text{by cancellation.}
\end{aligned}
$$

But we assumed $a \neq 0$, a contradiction. Hence the map is indeed one-to-one.

Note that the above simplifies if $1 \in \mathcal{R}$. We can replace (8) and (9) by the single assignment,

$$
a \mapsto \langle a, 1 \rangle,
$$

for all $a \in \mathcal{R}$. There is no question of the mapping not being well-defined and the proof that it is one-to-one goes simply:

$$
\begin{aligned}
\langle a, 1 \rangle \cong \langle b, 1 \rangle \ &\Rightarrow \ a \cdot 1 = 1 \cdot b \\
&\Rightarrow \ a = b.
\end{aligned}
$$

We will have a reason in the next section for not having made this simplifying assumption.

Our next step is to define the operations $\oplus, \odot$ on $[Q(\mathcal{R})]$. Because the present construction is based on multiplication, it is the better operation to start with. For $\langle a, b \rangle, \langle c, d \rangle \in Q(\mathcal{R})$, we set

$$
\langle a, b \rangle \odot \langle c, d \rangle = \langle ac, bd \rangle.
$$

The proof that $\odot$ respects $\cong$ parallels that of Lemma 1.6 showing the addition operation on $R^{\pm}$ preserved $\cong$ and I leave the details to the reader. Having settled this, we define $\odot$ on $[Q(\mathcal{R})]$ by

$$
[\langle a, b \rangle] \odot [\langle c, d \rangle] = [\langle a, b \rangle \odot \langle c, d \rangle].
$$

2.3 Lemma. *The following hold:*

i.   for all $x, y \in [Q(\mathcal{R})]$, $x \odot y = y \odot x$;
ii.   for all $x, y, z \in [Q(\mathcal{R})]$, $x \odot (y \odot z) = (x \odot y) \odot z$;
iii. $[Q(\mathcal{R})]$ has an identity element 1 such that, for all $x \in [Q(\mathcal{R})]$, $1 \odot x = x$;
iv.   for all $x \in [Q(\mathcal{R})]$, if $x \neq 0$, there is an element $y \in [Q(\mathcal{R})]$ such that $x \odot y = 1$.

Proof. The proofs of i and ii are the same as those of the corresponding claims of Lemma 1.8.i - 1.8.ii and I omit them.

iii. This is also the same as the proof of Lemma 1.8.iii, only now we write 1 for $[\langle a, a \rangle]$ for some nonzero $a \in \mathcal{R}$. Again, 1 does not depend on the choice of a:

$$ab = ab \;\Rightarrow\; \langle a, a \rangle \cong \langle b, b \rangle.$$

And it is an identity element because

$$\langle a, a \rangle \odot \langle c, d \rangle = \langle ac, ad \rangle \cong \langle c, d \rangle$$

the congruence holding since $(ac)d = (ad)c$.

iv. Again, the inverse of $[\langle a, b \rangle]$ for $a, b \neq 0$ is given by $[\langle b, a \rangle]$:

$$\begin{aligned}
[\langle a, b \rangle] \odot [\langle b, a \rangle] &= [\langle a, b \rangle \odot \langle b, a \rangle] \\
&= [\langle ab, ba \rangle] = [\langle ab, ab \rangle] \\
&= 1,
\end{aligned}$$

by what was shown in part iii. $\square$

In the present construction it is addition that is slightly complicated, but not as complicated as was multiplication in the construction of $\mathcal{R}^{\pm}$. The definition is obvious. Because we want to have

$$\frac{a}{b} + \frac{c}{d} = \frac{ad + bc}{bd},$$

we must define

$$\langle a, b \rangle \oplus \langle c, d \rangle = \langle ad + bc, bd \rangle.$$

The well-definedness of the operation is routinely established. Assume $\langle a_1, b_1 \rangle \cong \langle a_2, b_2 \rangle$ and $\langle c_1, d_1 \rangle \cong \langle c_2, d_2 \rangle$, i.e.

$$a_1 b_2 = a_2 b_1 \quad \text{and} \quad c_1 d_2 = c_2 d_1. \tag{10}$$

To show $\langle a_1, b_1 \rangle \oplus \langle c_1, d_1 \rangle \cong \langle a_2, b_2 \rangle \oplus \langle c_2, d_2 \rangle$, i.e.

$$\langle a_1 d_1 + b_1 c_1, b_1 d_1 \rangle \cong \langle a_2 d_2 + b_2 c_2, b_2 d_2 \rangle,$$

we must show

$$(a_1 d_1 + b_1 c_1)(b_2 d_2) = (a_2 d_2 + b_2 c_2)(b_1 d_1).$$

But

$$\begin{aligned}
(a_1 d_1 + b_1 c_1)(b_2 d_2) &= a_1 d_1 b_2 d_2 + b_1 c_1 b_2 d_2 \\
&= a_1 b_2 d_1 d_2 + c_1 d_2 b_1 b_2 \\
&= a_2 b_1 d_1 d_2 + c_2 d_1 b_1 b_2, \text{ by (10)} \\
&= (a_2 d_2 + b_2 c_2)(b_1 d_1).
\end{aligned}$$

Thus we can define

$$[\langle a, b \rangle] \oplus [\langle c, d \rangle] = [\langle a, b \rangle \oplus \langle c, d \rangle].$$

Again, the important properties of addition hold:

2.4 Lemma. *The following hold:*

i.   for all $x, y \in [Q(\mathcal{R})]$, $x \oplus y = y \oplus x$;
ii. for all $x, y, z \in [Q(\mathcal{R})]$, $x \oplus (y \oplus z) = (x \oplus y) \oplus z$;
iii. for all $x, y, z \in [Q(\mathcal{R})]$, $x \odot (y \oplus z) = (x \odot y) \oplus (x \odot z)$.

Proof. i. Let $x, y \in [Q(\mathcal{R})]$ and choose $a, b, c, d \in \mathcal{R}$ such that $x = [\langle a, b \rangle], y = [\langle c, d \rangle]$, and observe

$$
\begin{aligned}
\langle a, b \rangle \oplus \langle c, d \rangle &= \langle ad + bc, bd \rangle \\
&= \langle cb + da, db \rangle \\
&= \langle c, d \rangle \oplus \langle a, b \rangle.
\end{aligned}
$$

ii. Let $x, y, z \in [Q(\mathcal{R})]$ and choose $a, b, c, d, e, f \in \mathcal{R}$ such that $x = [\langle a, b \rangle], y = [\langle c, d \rangle], z = [\langle e, f \rangle]$, and observe

$$
\begin{aligned}
\langle a, b \rangle \oplus (\langle c, d \rangle \oplus \langle e, f \rangle) &= \langle a, b \rangle \oplus \langle cf + de, df \rangle \\
&= \langle a(df) + b(cf + de), b(df) \rangle \\
&= \langle (ad)f + (bc)f + (bd)e, (bd)f \rangle \\
&= \langle ad + bc, bd \rangle \oplus \langle e, f \rangle \\
&= (\langle a, b \rangle \oplus \langle c, d \rangle) \oplus \langle e, f \rangle.
\end{aligned}
$$

iii. Let $x, y, z \in [Q(\mathcal{R})]$ and choose $a, b, c, d, e, f \in \mathcal{R}$ such that $x = [\langle a, b \rangle], y = [\langle c, d \rangle], z = [\langle e, f \rangle]$, and observe

$$
\begin{aligned}
\langle a, b \rangle \odot (\langle c, d \rangle \oplus \langle e, f \rangle) &= \langle a, b \rangle \odot \langle cf + de, df \rangle \\
&= \langle a(cf) + a(de), b(df) \rangle. \quad (11)
\end{aligned}
$$

But

$$
\begin{aligned}
(\langle a, b \rangle \odot \langle c, d \rangle) \oplus (\langle a, b \rangle \odot \langle e, f \rangle) &= \langle ac, bd \rangle \oplus \langle ae, bf \rangle \\
&= \langle acbf + bdae, bdbf \rangle. \quad (12)
\end{aligned}
$$

To check the equivalence of (11) and (12), we must compare

$$
\begin{aligned}
(acf + ade)(bdbf) &= abbcdff + abbddef \\
(bdf)(acbf + bdae) &= abbcdff + abbddef.
\end{aligned}
$$

These being equal, we have (11) $\cong$ (12).   □

We have gone through most of the steps we went through in the last section in constructing $\mathcal{R}^{\pm}$: We defined $Q(\mathcal{R}), \cong, [Q(\mathcal{R})], \oplus, \odot$, verified that $(Q(\mathcal{R}), \oplus, \odot)$ is a semi-field, and embedded $\mathcal{R}$ into $[Q(\mathcal{R})]$. We haven't yet verified that the embedding is a homomorphism. To this end, for $a \in \mathcal{R}$ let $[a]$ be the image of a in $[Q(\mathcal{R})]$ under the embedding given by (8) and, if $0 \in \mathcal{R}$, (9):

$$
[a] = \begin{cases} [\langle aa, a \rangle], & a \neq 0 \\ [\langle 0, b \rangle], & a = 0 \text{ and some } b \neq 0. \end{cases}
$$

2.5 Lemma. *For all $a, b \in \mathcal{R}$,*

i. $[a + b] = [a] \oplus [b]$;

ii. $[ab] = [a] \odot [b]$.

Moreover,

iii. *if $0 \in \mathcal{R}$, then $[0]$ is the zero element of $[Q(\mathcal{R})]$;*

iv.   *if $1 \in \mathcal{R}$, then $[1]$ is the identity element of $[Q(\mathcal{R})]$.*

Proof. The proof for multiplication is slightly easier than that for addition, so I prove ii first.

ii. Assume neither a nor b equals 0 and observe

$$\begin{aligned}
[a] \odot [b] &= [\langle aa, a \rangle \odot \langle bb, b \rangle] \\
&= [\langle aabb, ab \rangle] \\
&= [\langle (ab)(ab), ab \rangle] \\
&= [ab].
\end{aligned}$$

If, say, $a = 0$, then $ab = 0$ regardless of what b is. Thus $[ab] = [0]$. But, for any $c, d \in \mathcal{R}$ with $d \neq 0$,

$$\begin{aligned}
[a] \cdot [\langle c, d \rangle] &= [\langle 0, d \rangle] \cdot [\langle c, d \rangle] \\
&= [\langle 0c, dd \rangle] \\
&= \langle 0, dd \rangle] \\
&= [0].
\end{aligned}$$

Thus $[a] \odot [b] = [ab]$ for any $b \in \mathcal{R}$. The case $b = 0$ follows from that for $a = 0$ by commutativity.

i. For addition, again consider first the case where neither a nor b is 0 and calculate

$$\begin{aligned}
[a] \oplus [b] &= [\langle aa, a \rangle \oplus \langle bb, b \rangle] \\
&= [\langle aab + abb, ab \rangle] \\
&= [\langle (a + b)(ab), ab \rangle]. \tag{13}
\end{aligned}$$

If $a + b \neq 0$,

$$[a + b] = [\langle (a + b)(a + b), a + b \rangle], \tag{14}$$

and since

$$((a + b)(a + b))(ab) = ((a + b)(ab))(a + b),$$

the right-hand-sides of (13) and (14) are equal, i.e. $[a] \oplus [b] = [a + b]$.

If $a + b = 0$, the right-hand-side of (13) reads $[\langle 0, ab \rangle]$, which is $[0] = [a + b]$.

To handle the case where $a = 0$ or $b = 0$, let us reduce this to assertion iii. Suppose, say, $a = 0$. Then for any $b \in \mathcal{R}$,

$$[a] \oplus [b] = [0] \oplus [b] = [b] = [a + b].$$

iii. Suppose $0 \in \mathcal{R}$ and $x = [\langle a, b \rangle] \in [Q(\mathcal{R})]$. Then $b \neq 0$ and we may write $[0] = [\langle 0, b \rangle]$. Observe

$$[\langle a, b\rangle] \oplus [0] = [\langle a, b\rangle \oplus \langle 0, b\rangle]$$
$$= [\langle ab + 0b, bb\rangle]$$
$$= [\langle ab, bb\rangle]$$
$$= [\langle a, b\rangle],$$

because $(ab)b = (bb)a$ entails $\langle ab, bb\rangle \cong \langle a, b\rangle$.

iv. Let $1 \in \mathcal{R}$ and observe

$$[1] \odot [\langle a, b\rangle] = [\langle 1 \cdot 1, 1\rangle \odot \langle a, b\rangle] = [\langle 1, 1\rangle \odot \langle a, b\rangle] = [\langle a, b\rangle]. \qquad \square$$

The proof of the Lemma is simplified if $1 \in \mathcal{R}$. For then the definition of the embedding can be taken to be

$$[a] = [\langle a, 1\rangle]$$

for all $a \in \mathcal{R}$ and one needn't consider the various cases in the proofs of parts i and ii. For example, the treatment of multiplication is:

$$[a] \odot [b] = [\langle a, 1\rangle \odot \langle b, 1\rangle] = [\langle ab, 1 \cdot 1\rangle] = [\langle ab, 1\rangle] = [ab],$$

regardless of whether or not one of a, b is 0.

By Lemma 2.5, if $\mathcal{R}$ is a ring and thus has a 0 element, then $[Q(\mathcal{R})]$ has a 0 element as well. Also, the additive inverses of elements of $\mathcal{R}$ are, by the homomorphism, mapped to additive inverses in $[Q(\mathcal{R})]$. This doesn't quite establish that every element in $[Q(\mathcal{R})]$ has an additive inverse, but it is the essential ingredient in doing so.

2.6 Theorem. *If $\mathcal{R}$ is an integral domain, then $[Q(\mathcal{R})]$ is a field.*

Proof. Suppose every element $a \in \mathcal{R}$ has an additive inverse[27] $-a$. Let $\langle a, b\rangle \in Q(\mathcal{R})$ and consider

$$\langle a, b\rangle \oplus \langle -a, b\rangle = \langle ab + (-a)b, bb\rangle = \langle (a + -a)b, bb\rangle = \langle 0, bb\rangle,$$

whence

$$[\langle a, b\rangle] \oplus [\langle -a, b\rangle] = [\langle 0, bb\rangle] = [0]. \qquad \square$$

As in the preceding section, we can appeal to the uniqueness of the multiplicative inverse (*Exercise!*) to introduce the reciprocal function,

$$[\langle a, b\rangle] \mapsto 1/[\langle a, b\rangle] = [\langle b, a\rangle],$$

and fractions,

$$x/y = x \odot (1/y).$$

One can then show that

$$[\langle a, b\rangle] = [a]/[b],$$

and derive all the usual rules for dealing with fractions.

---

[27] The existence of a multiplicative inverse was established in Lemma 2.3.

3 Fields of Fractions

It is sometimes hard to see that anything has been gained by the constructions of the preceding two sections. If we are familiar with fractions, why on Earth would we want to define them to be equivalence classes of ordered pairs of integers? The very definitions of the operations presupposed familiarity with the actual operations on actual fractions. One way to see that something has been gained is to apply the construction to other rings.

The rings that pop up most naturally in mathematics are rings of functions.

3.1 Examples. A few simple examples of function rings are:

i.   The set $\mathbb{R}[X]$ of polynomials in one variable over the reals, considered as functions, is a ring under the operations

$$(P_1 + P_2)(X) = P_1(X) + P_2(X)$$
$$(P_1 \cdot P_2)(X) = P_1(X) \cdot P_2(X).$$

The constant polynomials $0^P(X) \equiv 0$ and $1^P(X) \equiv 1$ are the zero and unit elements of $\mathbb{R}[X]$. The additive inverse of P is just $-P$, obtained by replacing all the coefficients of P by their negations.

ii.   The set $C[0,1]$ of continuous real-valued functions on the unit interval $[0,1] = \{x \in \mathbb{R} \mid 0 \le x \le 1\}$ forms a ring when the operations are defined pointwise:

$$(f + g)(x) = f(x) + g(x)$$
$$(f \cdot g)(x) = f(x) \cdot g(x).$$

The zero and unit elements are, once again, the constant functions, and the additive inverse of f is just $-f$ defined by $(-f)(x) = -(f(x))$.

iii.   The set $\mathbb{R}^{\mathbb{N}}$ of all infinite sequences of real numbers forms a ring when operations are defined pointwise:

$$(a + b)_n = a_n + b_n$$
$$(ab)_n = a_n \cdot b_n.$$

The constant sequences again yield the zero and identity elements and the additive inverse of a sequence a is $-a$ defined by $(-a)_n = -a_n$.

iv.   Another ring $\mathcal{R}$ may be formed from $\mathbb{R}^{\mathbb{N}}$ by taking a different multiplication. The *Cauchy product* of two sequences a, b is defined by

$$(a * b)_n = \sum_{i=0}^{n} a_i b_{n-i}. \tag{15}$$

This last example requires some explanation. If one thinks of sequences a and b as the sequences of coefficients of power series,

$$a_0 + a_1 x + a_2 x^2 + \ldots$$

$$b_0 + b_1 x + b_2 x^2 + \ldots,$$

then, ignoring questions of convergence and proceeding formally, one finds

$$(a_0 + \ldots)(b_0 + \ldots) = c_0 + c_1 x + c_2 x^2 + \ldots,$$

where $c_n = (a * b)_n$. If one assumes the uniqueness of the coefficients of the power series representing a given function to have been established, then for convergent power series one can reduce the commutativity, associativity and distributivity of the Cauchy product to the validity of these laws for multiplication of real numbers very simply. For example, to establish commutativity, one notes

$$(a_0 + \ldots)(b_0 + \ldots) = (b_0 + \ldots)(a_0 + \ldots).$$

When the power series do not converge, one has to work a little harder.

Commutativity of the Cauchy product is easy to demonstrate:

$$
\begin{aligned}
(a * b)_n &= \sum_{i=0}^{n} a_i b_{n-i} \\
&= \sum_{j=0}^{n} a_{n-j} b_j, \text{ substituting } j = n - i \\
&= \sum_{j=0}^{n} b_j a_{n-j} \\
&= (b * a)_n.
\end{aligned}
$$

Distributivity is also an easy matter:

$$
\begin{aligned}
(a * (b + c))_n &= \sum_{i=0}^{n} a_i (b + c)_{n-i} \\
&= \sum_{i=0}^{n} (a_i b_{n-i} + a_i c_{n-i}) \\
&= \sum_{i=0}^{n} a_i b_{n-i} + \sum_{i=0}^{n} a_i c_{n-i} \\
&= (a * b)_n + (a * c)_n.
\end{aligned}
$$

It is also easy to see that the sequence

$$\mathbf{0} = (0, 0, 0, \ldots)$$

is a zero element and $\mathbf{1}$, defined by

$$
\mathbf{1}_n = \begin{cases} 1, & n = 0 \\ 0, & n > 0, \end{cases}
$$

is a multiplicative identity element. Again, $-a$ defined by $(-a)_n = -(a_n)$ is an additive inverse.

Associativity is a little tricky. To this end, note that a product $a_i b_j$ occurs as a summand of (15) just in case $i + j = n$. Thus we can write

$$(a * b)_n = \sum_{i+j=n} a_i b_j,$$

it being assumed $i, j \in \mathbb{N}$. [Note that this allows us to prove commutativity thus:

$$(a * b)_n = \sum_{i+j=n} a_i b_j = \sum_{i+j=n} b_j a_i = (b * a)_n.]$$

Now,

$$(a * (b * c))_n = \sum_{i+m=n} a_i (b * c)_m$$

$$= \sum_{i+m=n} a_i \left(\sum_{j+k=m} b_j c_k \right)$$

$$= \sum_{i+m=n} \sum_{j+k=m} (a_i b_j c_k)$$

$$= \sum_{i+j+k=n} (a_i b_j c_k).$$

Similarly,

$$((a * b) * c)_n = \sum_{m+k=n} (a * b)_m c_k$$

$$= \sum_{m+k=n} \left(\sum_{i+j=m} a_i b_j \right) c_k$$

$$= \sum_{m+k=n} \sum_{i+j=m} (a_i b_j c_k)$$

$$= \sum_{i+j+k=n} (a_i b_j c_k),$$

and we see $a * (b * c) = (a * b) * c$.

Finally, we prove that $\mathcal{R} = (\mathbb{R}^{\mathbb{N}}, +, *, \mathbf{0})$ is an integral domain by showing it to have no zero divisors:

$$\text{for all } a, b \in \mathbb{R}^{\mathbb{N}}, \ a * b = \mathbf{0} \Rightarrow a = \mathbf{0} \text{ or } b = \mathbf{0}.$$

(The reader is asked to prove this implies the multiplicative cancellation property in a ring in Exercise 9.4 at the end of the Chapter.) To prove this, assume $a \neq \mathbf{0}$ and $b \neq \mathbf{0}$. Then there are m, n such that $a_m \neq 0$ and $b_n \neq 0$, respectively. Choose m_0, n_0 minimal with these properties and observe:

$$(a * b)_{m_0+n_0} = \sum_{i+j=m_0+n_0} a_i b_j = a_{m_0} b_{n_0} \neq 0$$

since, for $i + j = m_0 + n_0$, one of $i < m_0, i = m_0, i > m_0$ holds. In the first case, $a_i = 0$, whence so does $a_i b_j$, and in the last case $j = m_0 + n_0 - i < m_0 + n_0 - m_0 = n_0$ and $b_j = 0$, whence $a_i b_j = 0$.

It follows from the results of section 2 that $\mathcal{R}$ can be embedded into a field. [$\mathcal{R}$ is not itself a field because no sequence a with $a_0 = 0$ is invertible.]

That $\mathcal{R}$ can be embedded into a field is moderately interesting. Of greater interest, however, is a continuous analogue which, as shown by Jan Mikusiński from the late 1940's through the 1950's, has applications.[28]

3.2 Definitions. *We define* $\mathcal{C} = (C[0, \infty), +, *, \mathbf{0})$, *where*

i. $C[0, \infty)$ *is the set of continuous functions* $f : [0, \infty) \to \mathbb{R}$, *where* $[0, \infty) = \{x \in \mathbb{R} \mid 0 \leq x < \infty\} = \mathbb{R}^{\geq 0}$;

ii. $+$ *is defined pointwise:* $(f + g)(x) = f(x) + g(x)$;

iii. $*$ *is convolution:*

$$(f * g)(x) = \int_0^x f(t)g(x - t)dt; \text{ and}$$

iv. $\mathbf{0}$ *is the constant function* $\mathbf{0}(x) = 0$.

3.3 Theorem. $\mathcal{C}$ *is an integral domain.*

The proofs of the basic additive properties are fairly trivial and the proofs of the basic multiplicative ring properties are similar to those given in the discrete case. Commutativity and distributivity are quite simple:

$$(f * g)(x) = \int_0^x f(t)g(x - t)dt = \int_x^0 f(x - u)g(u)(-du)$$
$$= \int_0^x g(u)f(x - u)du = (g * f)(x),$$

using the transformation $u = x - t$; and

$$(f * (g + h))(x) = \int_0^x f(t)\big(g(x - t) + h(x - t)\big)dt$$
$$= \int_0^x f(t)g(x - t)dt + \int_0^x f(t)h(x - t)dt$$
$$= (f * g)(x) + (f * h)(x).$$

Associativity is established by appeal to commutativity and a change in the order of integration:

$$(f * (g * h))(x) = (f * (h * g))(x)$$
$$= \int_0^x f(t)(h * g)(x - t)dt$$

---

[28]   A good exposition of Mikusiński's theory, only a small taste of which I can present here, is: Arthur Erdélyi *Operational Calculus and Generalized Functions*, Holt, Rinehart and Winston, New York, 1962.

$$= \int_0^x f(t) \left(\int_0^{x-t} h(u)g(x-t-u)du \right) dt$$

$$= \int_0^x \int_0^{x-t} f(t)h(u)g(x-t-u)\, du\, dt$$

$$= \int_0^x \int_0^{x-u} f(t)g(x-u-t)h(u)\, dt\, du,$$

since one is integrating over the region bounded by the t- and u-axes and the line $u + t = x$. Continuing,

$$(f * (g * h))(x) = \int_0^x \left(\int_0^{x-u} f(t)g(x-u-t)\, dt \right) h(u)\, du$$

$$= \int_0^x (f * g)(x-u)\, h(u)\, du$$

$$= (h * (f * g))(x)$$

$$= ((f * g) * h)(x),$$

by a second application of commutativity.

That $\mathcal{C}$ is an integral domain, i.e. that it has no zero divisors, is a much deeper result than the corresponding result for the discrete structure $\mathcal{R}$. It is a special case of a more general result due to E.C. Titchmarsh and is usually proven by appeal to the theory of functions of a complex variable. An elementary proof by Jan Mikusiński and Czesław Ryll-Nardzewski exists, but it relies on some intricate calculations involving functions that seemingly come from nowhere and the proof would take us far off course. Thus, for our purposes we shall accept the non-existence of zero divisors as given and explore the consequences of the result.[29]

3.4 Lemma. $\mathcal{C}$ *has no multiplicative identity.*

This is the case for a rather trivial reason: for any f, g,

$$(f * g)(0) = \int_0^0 f(t)g(x-t)dt = 0.$$

Hence $(\mathbf{1} * f)(0) = 0$ and $\mathbf{1} * f$ cannot equal f for any f for which $f(0) \neq 0$.

We can perhaps get more insight into why $\mathcal{C}$ has no multiplicative identity by assuming for the moment it had one and considering its properties. Let i denote this supposed identity and notice that for the constant function[30],

---

[29] The Mikusiński-Ryll-Nardzewski proof can be found in Erdélyi, *op. cit.*, pp. 16 - 20.

[30] The letter "h" stands for Heaviside, in honour of Oliver Heaviside, who made extensive use of the step functions,

$$h_a(x) = \begin{cases} 0, & x < a \\ 1, & x \geq a. \end{cases}$$

On $[0, \infty)$, the function h is Heaviside's h_0.

$$h(x) = 1,$$

we have for all x,

$$\int_0^x i(t)dt = \int_0^x i(t)h(x-t)\,dt$$
$$= (i * h)(x) = h(x) = 1.$$

Now, assuming i continuous on $(0, \infty)$, it follows that $i(x) = 0$ for all such x. For otherwise, $i(x)$ would assume some value $m \neq 0$ at some point $x_0 > 0$. Then one can choose $\delta < x_0$ such that for all x,

$$|x - x_0| < \delta \Rightarrow |i(x) - m| < \frac{|m|}{2}.$$

This would make

$$\left| \int_{x_0-\delta}^{x_0+\delta} i(t)dt \right| \geq \int_{x_0-\delta}^{x_0+\delta} \frac{|m|}{2}\,dt \geq 2\delta\frac{|m|}{2} = |m|\delta > 0.$$

However,

$$\int_{x_0-\delta}^{x_0+\delta} i(t)dt = \int_0^{x_0+\delta} i(t)dt - \int_0^{x_0-\delta} i(t)dt = 1 - 1 = 0.$$

In short, we have described the Dirac delta function δ discussed in Chapter II. As such a function doesn't exist, we see again that $\mathcal{C}$ can have no multiplicative identity. But its field of fractions does have such an identity, which we may denote either by δ or $\mathbf{1}$ as occasion demands.

The elements of the field $[Q(\mathcal{C})]$ of fractions of functions in $C[0, \infty)$ are called *generalised functions* and some interesting operations on $C[0, \infty)$ are represented by elements of $[Q(\mathcal{C})]$ and operate on $[Q(\mathcal{C})]$ by multiplication. For example, the aforementioned constant function $h(x) \equiv 1$ satisfies

$$(h * f)(x) = (f * h)(x)$$
$$= \int_0^x f(t)\,h(x-t)\,dt$$
$$= \int_0^x f(t)dt,$$

whence multiplication by h yields the definite integral. If f is continuous,

$$[h]^{-1} * \left[\int_0^x f(t)dt \right] = [h]^{-1} * ([h] * [f])$$
$$= ([h]^{-1} * [h]) * [f]$$
$$= \delta * [f] = [f],$$

whence $[h]^{-1}$ behaves like differentiation

Now $[h]^{-1}$ is not quite differentiation. To explain this, first observe that $C[0,\infty)$ is a vector space over $\mathbb{R}$. In particular, it has a scalar multiplication: for $r \in \mathbb{R}, f \in C[0,\infty)$,

$$(rf)(x) = r \cdot f(x).$$

By the Linearity of the Integral, scalar multiplication commutes and associates with convolution:

$$r(f * g) = (rf) * g = f * (rg).$$

For,

$$r \int_0^x f(t) \, g(x-t) \, dt = \int_0^x \big(rf(t)\big) \, g(x-t) \, dt = \int_0^x f(t) \, \big(rg(x-t)\big) \, dt.$$

We can extend the scalar multiplication to $[Q(\mathcal{C})]$ by defining

$$r\langle f, g \rangle = \langle rf, g \rangle$$

and noting that

$$\langle f_1, g_1 \rangle \cong \langle f_2, g_2 \rangle \Rightarrow f_1 * g_2 = f_2 * g_1$$
$$\Rightarrow r(f_1 * g_2) = r(f_2 * g_1)$$
$$\Rightarrow (rf_1) * g_2 = (rf_2) * g_1$$
$$\Rightarrow \langle rf_1, g_1 \rangle \cong \langle rf_2, g_2 \rangle,$$

whence we can also define

$$r[\langle f, g \rangle] = [r\langle f, g \rangle].$$

I leave to the reader the verification of the usual scalar properties:

$$r(x + y) = rx + ry$$
$$(r + p)x = rx + px,$$

for any $r, p \in \mathbb{R}, x, y \in [Q(\mathcal{C})]$.

Now, to see just how $[h]^{-1}$ differs from differentiation, let $f \in C[0,\infty)$ have a continuous derivative $f' \in C[0,\infty)$ and let $F(x) = f(x) - f(0)$. By the Fundamental Theorem of Calculus,

$$F(x) = F(x) - F(0) = \int_0^x F'(t)dt = \int_0^x f'(t)dt = (h * f')(x),$$

whence

$$f(x) - f(0) = (h * f')(x),$$

i.e.

$$f(x) = (h * f')(x) + f(0),$$

i.e.

$$f = h * f' + f(0)h. \tag{16}$$

3.5 Notational Convention. We write s for $[h]^{-1}$ and, when there is no danger of confusion, we also write f, h etc. for $[f], [h]$ etc.

Multiplying (16) by s yields

$$\begin{aligned}
s * f &= s * (h * f') + s * (f(0)h) \\
&= (s * h) * f' + f(0)(s * h) \\
&= \delta * f' + f(0)\delta \\
&= f' + f(0)\delta.
\end{aligned} \tag{17}$$

Replacing f by f' and assuming $f' \in C[0, \infty)$,

$$\begin{aligned}
s * f' &= f'' + f'(0)\delta \\
s * (s * f - f(0)\delta) &= f'' + f'(0)\delta \\
s^2 * f - s * (f(0)\delta) &= f'' + f'(0)\delta \\
s^2 * f - f(0)s &= f'' + f'(0)\delta \\
s^2 * f &= f'' + f'(0)\delta + f(0)s
\end{aligned} \tag{18}$$

More generally,

$$s^n * f = f^{(n)} + f^{(n-1)}(0)\delta + f^{(n-2)}(0)s + \ldots + f(0)s^{n-1}, \tag{19}$$

provided all the derivatives exist. Turning this around we have

$$f^{(n)} = s^n * f - f^{(n-1)}(0)\delta - f^{(n-2)}(0)s - \ldots - f(0)s^{n-1}, \tag{20}$$

familiar from our discussion of the Laplace Transform in Chapter II.

Using (17), we can make a small table:

f	h	x	x^2	$\sin x$	$\cos x$	e^x	e^{ax}
$s * f$	δ	h	$2x$	$\cos x$	$-\sin x + \delta$	$e^x + \delta$	$ae^{ax} + \delta$

The exponential case is particularly interesting. Notice that for $f(x) = e^{ax}$,

$$s * f = af + \delta,$$

or, using more algebraic notation,

$$sf = af + 1,$$

whence

$$(s - a\mathbf{1})f = 1,$$

i.e.

$$f = \frac{1}{s - \mathbf{a}},$$

where we write $\mathbf{a}$ for $a\mathbf{1}$, i.e. $a\delta$. Recalling what f is, this reads

$$(s - a\delta)^{-1} = e^{ax}. \tag{21}$$

Using this, we can solve some differential equations.

3.6 Example. Solve the equation,

$$f'' - f' = 1, \tag{22}$$

satisfying the initial conditions $f(0) = f'(0) = 1$.

Notice that

$$s^2 f = f'' + \delta + s$$

by (18), whence

$$f'' = s^2 f - \delta - s.$$

Similarly, $f' = sf - \delta$. Thus,

$$\begin{aligned} f'' - f' &= s^2 f - \delta - s - (sf - \delta) \\ &= s^2 f - sf - s, \end{aligned}$$

and the equation (22) is rendered

$$s^2 f - sf - s = h.$$

Multiplying by h,

$$\begin{aligned} sf - 1f - 1 &= h^2 \\ (s - 1)f &= h^2 + 1 \\ f &= \frac{h^2}{s - 1} + \frac{1}{s - 1}, \end{aligned}$$

But

$$\frac{1}{s - 1} = e^x,$$

whence

$$f = h^2 * e^x + e^x. \tag{23}$$

We evaluate the first term on the right in two steps. First,

$$h * e^x = e^x * h = \int_0^x e^t \cdot 1 \, dt = e^x - e^0 = e^x - 1$$

and second

$$\begin{aligned} h * (h * e^x) = h * (e^x - 1) = (e^x - 1) * h &= \int_0^x (e^t - 1) \cdot 1 \, dt \\ = e^t - t \Big|_0^x = e^x - x - e^0 &= e^x - x - 1. \end{aligned}$$

Plugging this into (23), we conclude

$$f(x) = e^x - x - 1 + e^x = 2e^x - x - 1.$$

The reader can easily verify that f does indeed satisfy (22) and the stated initial value conditions.

The marvellous thing about this Example is that it is *not* Type I formal heuristic, but a rigorously correct derivation. The reader who is up on his integration by parts may like to try his hand on the following:

Exercise. Solve the equation $f' - f = -x$ subject to the initial condition $f(0) = 1$.

The δ function usually shows up not in solving simple differential equations, but in setting them up.

3.7 Example. Solve the equation

$$f'(x) - f(x) = \delta(x - 1), \tag{24}$$

under the initial condition $f(0) = 1$.

Now the first problem is to interpret $\delta(x - 1)$. Although $\delta \in [Q(\mathcal{C})]$, it is not an actual function itself and thus it does not take on values at individual arguments. We can, nonetheless, assign a meaning to expressions of the form $\delta(x - a)$. To this end, one introduces the Heaviside step functions,

$$h_a(x) = \begin{cases} 0, & x < a \\ 1, & x \geq a. \end{cases}$$

Note that $h_0 = h$. The fact that $\delta = [h]^{-1} \cdot [h] = [h]^{-1} \cdot [h_0]$ is something like the derivative of h_0 suggests defining

$$\delta_a = [h]^{-1} \cdot [h_a], \tag{25}$$

i.e. $\delta_a = sh_a$, the derivative of h_a.[31] Indeed, examination of the graph of h_a shows

$$h_a'(x) = \begin{cases} 0, & x \neq a \\ undefined, & x = a. \end{cases}$$

Thus, we interpret (24) to mean

$$f' - f = \delta_a$$

with δ_a defined by (25) for $a = 1$. To solve this, note that

$$f' = sf - f(0)\delta = sf - \delta,$$

whence

$$f' - f = sf - f - \delta = \delta_a$$
$$(s - \delta)f = \delta + \delta_a$$

---

[31] Strictly speaking, h_a is not in $C[0, 1)$ as it is not continuous. However, it is integrable and its convolution with a continuous function is continuous. It is thus represented in $[Q(\mathcal{C})]$ by, e.g., $[\langle h * h_a, h \rangle]$.

$$f = \frac{\delta}{s - \delta} + \frac{\delta_a}{s - \delta} = \frac{\delta}{s - \delta} + \frac{\delta \delta_a}{s - \delta}$$
$$= e^x + e^x * \delta_a$$
$$= e^x + \delta_a * e^x.$$

To continue from here, it is probably best to determine the product $\delta_a * g = s * h_a * g$ for general $g \in C[0, \infty)$. We may assume $a > 0$. First we compute

$$h_a * g(x) = \int_0^x g(t) \, h_a(x - t) \, dt$$
$$= \begin{cases} \int_0^{x-a} g(t) \, dt, & a \le x \\ 0, & x < a, \end{cases}$$

since $x - t \ge a$ exactly when $x - a \ge t$, and $h_a(x - t) = 0$ for $t > x - a$. But $sk = k' + k(0)\delta$ for any $k \in C[0, \infty)$, whence

$$sh_a g = (h_a g)' + (h_a g)(0)\delta$$
$$= (h_a g)', \text{ since } a > 0 \text{ yields } (h_a g)(0) = 0$$
$$= \begin{cases} g(x - a) & \text{if } a \le x, \text{ by the Fundamental Theorem of Calculus} \\ 0, & \text{if } x < a \end{cases}$$
$$= h_a(x)g(x - a).$$

Thus, for the function at hand with $a = 1$, we conclude

$$f(x) = e^x + h_1(x)e^{x-1}$$
$$= \begin{cases} e^x, & x < 1 \\ e^x + e^{x-1}, & 1 \le x. \end{cases}$$

One can easily verify that f satisfies

$$f'(x) - f(x) = \begin{cases} 0, & x \ne 1 \\ undefined, & x = 1 \end{cases}$$
$$= \delta_1(x)$$

under the intuitive definition

$$\delta_a(x) = \begin{cases} 0, & x \ne a \\ undefined, & x = a. \end{cases}$$

This will end our discussion of Mikusiński's Type II definition of the δ-function. For more on Mikusiński's applications of the field of fractions of the ring $C[0, \infty)$, I refer the reader to Erdélyi's book cited in footnote 28. For more on the δ-function itself I suggest a good book on differential equations[32].

---

[32] I am rather fond of Martin Braun, *Differential Equations and Their Applications*, 3rd ed., Springer-Verlag, New York, 1983.

Other Type II treatments of the δ-function can be found. I refer to Braun's text on differential equations[33] for a brief but well-motivated description, and to Beals's textbook[34] for an exposition of the theory of distributions, which forms an alternative justification for the use of generalised functions like δ. The historical background to these theories is the subject of a book by Jesper Lützen[35].

4 The Real Numbers, I; Bolzano

4.1 Background

Among those who discovered the geometric interpretation of complex numbers as points in the plane was Carl Friedrich Gauss, one of the greatest mathematicians of all time. He considered opposition to negative and imaginary numbers to be linguistic:

> If one has previously considered this situation from a false perspective and found thereby a mysterious darkness, then this is in large part to be ascribed to the barely proper terminology. Had one not called $+1, -1, \sqrt{-1}$ positive, negative, imaginary (or even impossible) units, but rather something like direct, inverse, lateral units, then there could hardly have been talk of such darkness.[36]

Gauss neglected to defend the negative numbers against being "fictitious" or "less than nothing", but the point is made. With real numbers, however, there is a genuine foundational problem. The rise of the Calculus meant one had to deal with limits and, ultimately, prove those limits existed. This meant one had to have a good description of the real numbers and not an open-ended, ongoing, piecemeal accumulation of ever more numbers.

In the *Arithmetica infinitorum*, after deriving his infinite product

$$\frac{4}{\pi} = \frac{3}{2} \cdot \frac{3}{4} \cdot \frac{5}{4} \cdot \frac{5}{6} \cdot \frac{7}{6} \cdot \frac{7}{8} \cdots ,$$

John Wallis defends his introduction of such a product:

---

[33] *Ibid.*, pp. 246 - 248.

[34] Richard Beals, *Advanced Mathematical Analysis*, Springer-Verlag, New York, 1973.

[35] Jesper Lützen, *The Prehistory of the Theory of Distributions*, Springer-Verlag, New York, 1982.

[36] Quoted in: Reinhold Remmert, "Komplexe Zahlen", in: H.-D. Ebbinghaus, H. Hermes, F. Hirzebruch, M. Koecher, K. Mainzer, A. Prestel, R. Remmert, *Zahlen*, Springer-Verlag, Berlin, 1983; here, p. 50. This book, incidentally, is a collection of chapters by various authors on numbers and number systems with a great deal of historical background and a lot of interesting mathematics. Another fine reference from the historical point of view but with less mathematical detail is Helmuth Gericke, *Geschichte des Zahlbegriffs*, Bibliographisches Institut, Mannheim, 1970. An English source for some of the history is: Jerome H. Manheim, *The Genesis of Point Set Topology*, Pergamon Press, Oxford, and The Macmillan Company, New York, 1964.

And indeed I am inclined to believe (what from the beginning I suspected) that this ratio we seek is such that it cannot be forced out in numbers according to any method of notation so far accepted, not even by surds... so that it seems necessary to introduce another method of explaining a ratio of this kind, than by true numbers or even by the accepted means of surds...

And therefore what arithmeticians usually do in other work, must also be done here; that is, where some impossibility is arrived at, which indeed must be assumed to be done, but nevertheless cannot actually be done, they consider some method of representing what is assumed to be done, though it may not be done in reality.

And this indeed happens in all operations of arithmetic involving resolution[37], for example, in subtraction: if it is proposed that a larger number must be taken from a smaller, thus 3 from 2 or 2 from 1, since this can not be shown in reality, there are considered negative numbers by means of which a supposed subtraction of this kind may be expressed, thus $2 - 3$, or $1 - 2$, or -1.

In division, if it is proposed that a number must be divided by another which is not a divisor, thus 3 by 2, since this can not be shown in reality, there is invented a method of indicating a supposed division of this kind, in this form: $\frac{3}{2}$ or $1\frac{1}{2}$.

In the extraction of roots, if there is proposed a number that is not in its nature truly a power, for example, if there is sought the square root of 12, since that root cannot be expressed as any integer or fractional number, there is invented a method of indicating any supposed root of this kind in this form: $\sqrt{12}$ or $2\sqrt{3}$.[38]

Wallis followed these comments with a brief closer look at the ratio $4/\pi$, which he called $\square$ and said[39]

After this our description of that quantity $\square$, we may also add another, which I have received from that most noble person and very skilled geometer, Lord William Viscount and Baronet Brouncker.
... That Noble Gentleman, having thought it over himself, judged by a method of infinites of his own that the same quantity could be most conveniently described in this form:

$$\square = 1\cfrac{1}{2\cfrac{9}{2\cfrac{25}{2\cfrac{49}{2\cfrac{81}{2 \text{ etc.}}}}}}$$

He then proceeded to derive Brouncker's formula and explain continued fractions more generally.

---

[37] I.e. the inversion of operations, e.g. subtraction as opposed to addition.
[38] Stedall, *op. cit.*, pp. 161 - 162.
[39] *Ibid.*, p. 167.

Continued fractions were not a complete novelty. Unbeknownst to the Europeans, the Hindu mathematicians had earlier used them to solve linear equations. And in Europe Rafael Bombelli had represented square roots by continued fractions. While some infinitary operations were mere novelties— some examples from Jacob Bernoulli cited in Chapter I come to mind— infinitary operations— series, infinite products, and continued fractions in particular— had permanently settled on the mathematical landscape. They would have to be accounted for.

Nowadays we would describe the situation thus: The step from $\mathbb{N}$ to $\mathbb{Z}$ guarantees the existence of solutions to all equations of the form,

$$X + a = b.$$

The step from $\mathbb{Z}$ to $\mathbb{Q}$ adds solutions to all equations,

$$aX + b = 0.$$

The geometrical discovery of the incommensurability of certain line segments led Euclid to, in effect, extend the rationals by allowing closure under taking square roots, i.e. adding roots to equations of the form

$$X^2 = a, \text{ for positive } a.$$

Indeed, he developed to a small extent a geometric theory of simple quadratic surds. He probably intuited that this was insufficient because he also presented Eudoxus's theory of proportion, the last word on the structure of the real numbers before Wallis's day.

Of course, with the Greeks real numbers were not numbers, nor were rationals. But if we translate into modern terms, we can say that Eudoxus postulated the density of the rational numbers among the reals. This places an upper bound on the extent of the reals— there cannot be so many that they pile up so closely as not to be able to squeeze a rational between two of them. The only lower bound it supplies are those magnitudes that pop up from time to time that are not rational— $\sqrt{2}$ (the ratio of the diagonal to the side of a square), $\frac{\sqrt{5}+1}{2}$ (the *golden ratio* that pops up several times in the pentagram), and π.

The time was coming when a complete description of the real numbers would be needed. The series representations of functions— Taylor series or Fourier series— needed general convergence criteria. Even something as simple as the Comparison Test:

If a, b are sequences of non-negative numbers and, for all $n, a_n > b_n$, then if $\sum a_n$ converges so does $\sum b_n$;

requires the completeness property of the real numbers for a proof.

The situation with regard to the completeness of the real number system has turned out to be quite uncomplicated. In Chapter I we have already seen and remarked on two expressions of the completeness of the real numbers and their equivalence— namely, the *Least Upper Bound Principle* and the *Convergence of all Cauchy-Convergent Sequences*. To this we added the *Nested Interval Property* in Chapter II: The intersection of any nested sequence $I_0 \supseteq I_1 \supseteq I_2 \ldots$ of closed bounded intervals is nonempty. It could have been the case that there is an ever-growing, dazzling confusion of such manifestations of completeness, like the closure under finding roots of 2nd, 3rd, 4th... degree polynomials. But there isn't. The completeness properties are all equivalent and the postulation of any one of them provides a description of the real numbers.

Although such completeness properties were being stated, used, and derived from other implicit completeness results throughout the 19th century, no one took the axiomatic route until after various constructions of the real numbers had been given. With the exception of Bolzano's, these constructions were each based on one of these three geometric completeness properties: the real numbers were constructed as the completion of the rational numbers with respect to gaps in the order, non-convergent Cauchy sequences, or nested sequences of closed, bounded intervals with empty intersections. Thus the constructions were based on belief in the completeness axioms.

The various constructions were given by Bernard Bolzano, Karl Weierstrass, Richard Dedekind, Georg Cantor, Charles Méray, and Paul Bachmann. Weierstrass's and Dedekind's constructions are motivated by the Least Upper Bound Principle, Cantor's and Méray's by the Convergence of Cauchy Sequences, and Bachmann's by the Nested Interval Property. Bolzano's construction appears to be an infinitary algebraic completion.

The first of these to be carried out was Bolzano's in the early 1830's. It is not so widely known as some of the others for a variety of reasons. It should not be surprising from what we know from his treatment of sums to find that the work suffered from a lack of clarity. This was due in part to the fact that it was never completed. It was also not published until 1962, and then with not all the corrections taken into account.[40] And the construction is very closely

---

[40] The work of Bolzano referred to is a portion, generally called the *Reine Zahlenlehre* [*Pure Theory of Numbers*] or simply the *Zahlenlehre* [*Theory of Numbers*], of a larger work called the *Größenlehre* [*Theory of Magnitudes*] and was left behind unfinished. The first published version of that part of the work dealing with the construction of the real numbers was by Karel Rychlík (*Theorie der reellen Zahlen in Bolzanos handschriftlichem Nachlass*, Czech Academy of Science, Prague, 1962). Rychlík offered his own interpretations of some of the vaguely defined terms and pointed out that there were difficulties. This inspired Bob van Rootselaar to offer an alternative and refutations of some of Bolzano's assertions ("Bolzano's theory of real numbers", *Archive for History of Exact Sciences* 2 (1964), pp. 168 - 180). Detlef Laugwitz came to the rescue ("Bemerkungen zu Bolzanos Größenlehre", *Archive for History of Exact Sciences* 2 (1965), pp. 398 - 409), with a slightly alterred interpretation under which everything carried through. The full version of the *Reine Zahlenlehre* can be found in E. Winter, J. Berg, F. Kambartel, J. Loužil, and B. van Rootselaar (eds.), *Bolzano–Gesamtausgabe*, vol. 2A8, Friedrich Frommann Verlag, Stuttgart, 1976. A fairly thorough mathematical as well as historical treatment is given by Detlef Spalt, "Bolzanos Lehre von den meßbaren Zahlen 1830 - 1839", *Archive for History of Exact Sciences* 42 (1991), pp. 15 - 70. I base my discussion

tied to the rationals and does not generalise as readily as do some of the other constructions. Nonetheless, it is worth considering even if only for historical and philosophical reasons.

In point of fact, Bolzano was not offering a construction of the real numbers so much as a description of them as he saw them used in mathematical practice. If we recall his comment cited on page 26 in Chapter I, that to a mathematician the value of an infinite sum of a set of number expressions is "the simplest possible expression that is equivalent to the given set of number expressions", we can pretty much predict how his description is going to go: real numbers are certain infinite expressions, equality among which will be determined by equivalence according to algebraic rules. For a number of reasons his attempt failed. However, modulo some re-interpretation, it has been reworked into a rigorous construction of the real numbers. I wish here to discuss both the failed attempt and the ultimate successful re-interpretation.

Bolzano begins with a classical Greek approach starting with a unit. The unit generates the positive integers, which constitute *actual numbers* [wirkliche Zahlen]. When subtracting a larger from a smaller, he says, "if M is larger than S, then the representation of a number, which comes about if one subtracts M from S, is an objectless number representation. But the case becomes otherwise so soon as we set up a unit opposing that we had originally taken, i.e. one that is negative". Thus, he readily accepts negative numbers as well. Rational numbers, however, he regards as objectless number expressions built up from actual numbers by finitely many additions, subtractions, multiplications, and divisions other than by 0. An objectless number expression, of course, does not denote anything, but is merely a calculational device.

It is, of course, his explanation of real numbers that interests us here. These he takes to be *measurable* infinite number expressions, an *infinite number expression* being an expression built up from rational numbers by means of (possibly[41]) infinitely many additions, subtractions, multiplications and divisions. This already is rather vague and is the first element of Bolzano's theory to be "interpreted", i.e. replaced by a precisely defined formal concept. Bolzano does give some examples:

$$1 + 2 + 3 + \ldots$$
$$\tfrac{1}{2} - \tfrac{1}{4} + \tfrac{1}{8} - \tfrac{1}{16} + \ldots$$
$$(1 - \tfrac{1}{2})(1 - \tfrac{1}{4})(1 - \tfrac{1}{8})(1 - \tfrac{1}{16}) \cdots$$
$$a + \frac{b}{1+1+1+1+\ldots}, a, b \text{ integers}$$
$$1 - 1 + 1 - 1 + \ldots$$

---

on Gericke (*op. cit.*), Laugwitz, and Spalt. A portion of Bolzano's paper appears in English translation in Steve Russ (ed.), *The Mathematical Works of Bernard Bolzano*, Oxford University Press, Oxford, 2004.

[41] Note that by adding infinitely many 0's or multiplying by infinitely many 1's every expression is equivalent to an infinite number expression. Thus we can take the adjective "infinite" to be inclusive, allowing finite sums and products as well. In the sequel, we will in fact drop the adjective when the phrase including it would be unwieldy.

Examples like the first and last of these show that not all infinite number expressions are acceptable as real numbers. Thus he defines an infinite number expression S to be *measurable* if it can be placed to an arbitrary degree of accuracy among the rationals. Basically, this would mean that for any rational number $\epsilon > 0$, we can find another rational number p/q such that

$$\left| \frac{p}{q} - S \right| < \epsilon,$$

i.e.

$$\frac{p}{q} - \epsilon < S < \frac{p}{q} + \epsilon.$$

One of the difficulties with Bolzano's construction is that he tries to be too explicit here and defines S to be measurable if for any positive integer q there is an integer p and two infinite number expressions P_1, P_2 such that

$$S = \frac{p}{q} + P_1 \quad \text{and} \quad S = \frac{p+1}{q} - P_2, \tag{26}$$

where P_1 is a "purely positive" infinite number expression or 0 and P_2 is "purely positive". In other words,

$$\frac{p}{q} \leq S < \frac{p+1}{q}. \tag{27}$$

Bolzano calls the number p/q a *measuring fraction* of S.

   A quick example of a measurable number expression is given by

$$S = a + \frac{b}{1 + 1 + 1 + 1 + \ldots},$$

where a, b are positive integers. For any positive integer q, if we take $p = qa$, $P_1 = \frac{b}{1+1+1+1+\ldots}$, $P_2 = \frac{1}{q} - P_1$, we have

$$S = \frac{p}{q} + P_1 = \frac{p+1}{q} - P_2, \quad P_1 \geq 0, \quad P_2 > 0.$$

   This example brings up the second major problem with Bolzano's description of the real numbers. What does, or should, he mean by a *purely positive* [rein positiv] or *thoroughly* or *completely positive* [durchaus positiv] number expression? The simplest meaning that comes to mind is that a purely positive expression P is an expression built up from positive integers using (possibly) infinitely many additions, multiplications, and divisions, but no subtractions. This would make expressions like

$$\frac{b}{1 + 1 + 1 + \ldots}, \quad b > 0$$

purely positive, albeit infinitely small. Bolzano had no problem accepting infinitesimals. As a final step, however, he would identify number expressions that differed by an infinitesimal.

The expression P_2 of our quick example,

$$P_2 = \frac{1}{q} - \frac{b}{1+1+1+\ldots}, \qquad (28)$$

or, more generally, expressions

$$\frac{s(1+1+1+\ldots)-qb}{q(1+1+1+\ldots)}, \qquad (29)$$

with s, b, q positive integers, were also considered purely positive by Bolzano, even though minus signs appear in them. Now (28) is obtained by subtracting an infinitesimal from a finite number, and (29) essentially has as numerator a finite number subtracted from $+\infty$. These are the most obvious exceptions[42] needed in working out the details of Bolzano's construction.

Finally, two infinite number expressions would be deemed equal if one expression could be transformed into the other by means of elementary arithmetical and algebraic identities. This, of course, is the third source of ambiguity. As we saw in Chapter I, too liberal a use of algebra applied to the infinite expression,

$$S = 1 - 1 + 1 - 1 + \ldots$$

allows us to conclude $S = 1$ and $S = 0$. For infinite sums, we have some experience in choosing a "safe" list of transformation rules to be used in dealing with infinite expressions. Presumably we could analogously find rules for infinite products. But what about other infinite expressions, like continued fractions?

Thus, in brief, Bolzano constructs the real numbers as a collection of infinite number expressions. Not all expressions are to be used, just those which are measurable. In his exposition, he shows the measurable numbers to include the rationals, to be closed under addition, subtraction, multiplication, and division provided the divisor is not zero or infinitesimal. He also proves the Least Upper Bound Principle and that every measurable number equals an infinite sum of rational numbers.

4.2 Beginning Results

Bolzano's development is not unproblematic. There are ambiguities in the definitions of infinite number expression and purely positive infinite number expression; the necessary equality preserving transformations on infinite expressions needed to derive equations like (26) are nowhere delineated; and his definition of measurable is too narrow.

As I said, he first shows that all rational number expressions are measurable.

---

[42] According to Spalt, *op.cit.*, p. 29, this is not to be considered an exception, as $s(1+1+\ldots)-q$ can be written with no minus signs as qb is a fixed finite number and is less than the sum of he first few terms of $s(1+1+\ldots)$ and can be incorporated into them. Moreover, as (28) is easily rewritten in the form (29), it too is not an exception.

4.1 Lemma. *Every rational number is measurable.*

Proof. Let α/β be rational, $\beta > 0$. Let q be given and divide αq by β:

$$\alpha q = p\beta + r, \quad 0 \le r < \beta.$$

Then

$$\frac{\alpha}{\beta} = \frac{\alpha q}{\beta q} = \frac{p\beta + r}{\beta q} = \frac{p}{q} + \frac{r}{\beta q}.$$

But also

$$\frac{\alpha}{\beta} = \frac{p + 1 - 1}{q} + \frac{r}{\beta q} = \frac{p+1}{q} + \frac{r}{\beta q} - \frac{1}{q}$$

$$= \frac{p+1}{q} + \frac{r - \beta}{\beta q} = \frac{p+1}{q} - \frac{\beta - r}{\beta q},$$

and we can take

$$P_1 = \frac{r}{\beta q}, \quad P_2 = \frac{\beta - r}{\beta q}.$$

As written, there is a minus sign in P_2, but we can simply replace $\beta - r$ by its positive integral value. □

If we next try to show the measurable infinite number expressions to be closed under addition, we run into immediate trouble: Suppose A and B are infinite number expressions and

$$A = \frac{p_q(A)}{q} + P_1(A) = \frac{p_q(A) + 1}{q} - P_2(A)$$

$$B = \frac{p_q(B)}{q} + P_1(B) = \frac{p_q(B) + 1}{q} - P_2(B),$$

where $q, p_q(A), p_q(B)$ are integers, $q > 0$, $P_1(A), P_1(B)$ are purely positive or 0, and $P_2(A), P_2(B)$ are purely positive. When we add them we get

$$A + B = \frac{p_q(A) + p_q(B)}{q} + \big(P_1(A) + P_1(B)\big)$$

$$= \frac{p_q(A) + p_q(B) + 2}{q} - \big(P_2(A) + P_2(B)\big),$$

which is not what we wanted.

In terms of inequalities, we have only established

$$\frac{p}{q} \le A + B < \frac{p+2}{q},$$

which certainly shows $A + B$ can be placed very closely to, within $2/q$ of, a rational number. But it does not establish the measurability of $A + B$. In their reworkings of Bolzano's construction, van Rootselaar actually constructed counterexamples, measurable infinitesimal number expressions A, B for which

$A + B$ is not measurable[43]; Laugwitz proves the result after generalising the definition of measurable[44]; and Spalt points out that van Rootselaar's counterexample is a counterexample to the truth of the result under van Rootselaar's interpretation and is not a counterexample under a faithful reinterpretation of Bolzano's intent. However, he constructs a new counterexample and shows that if A and B are sufficiently closely related then the result holds.[45]

We consider Spalt's counterexample:

4.2 Example. Let S, T be the infinite number expressions,

$$S : \quad 2 + 1 + 2 + 1 + 2 + 1 + \ldots$$
$$T : \quad 1 + 2 + 1 + 2 + 1 + 2 + \ldots$$

Then

$$S - T = 1 - 1 + 1 - 1 + 1 - 1 + \ldots$$

is the Grandi series, which in Bolzano's words is "neither finite, nor infinitely small, nor infinitely large". However,

$$A = -\frac{1}{S} \quad \text{and} \quad B = \frac{1}{T}$$

are measurable numbers, and

$$A + B = -\frac{1}{S} + \frac{1}{T} = \frac{S - T}{S \cdot T}$$

is not measurable because of the numerator.

The validity of this as a counterexample seems to me a matter of interpretation. We have, for any positive integer q,

$$A + B = \frac{0}{q} + \frac{S - T}{S \cdot T} = \frac{1}{q} - \left(\frac{1}{q} - \frac{S - T}{S \cdot T} \right),$$

with

$$P_1 = \frac{S - T}{S \cdot T} \geq 0, \quad P_2 = \frac{1}{q} - \frac{S - T}{S \cdot T} > 0,$$

reasonable assumptions to make. However, P_1 is not purely positive in the strict sense discussed above.

More damaging to the reasonableness of Bolzano's choice of definition of measurability would seem to be the following:

4.3 Example. The infinite number expression,

$$S = \prod_{i=1}^{\infty} \left(\frac{-1}{i} \right),$$

is infinitesimal, but not measurable.

---

[43] Van Rootselaar, *op. cit.*, §3.3, pp. 175 - 177.
[44] Laugwitz, *op. cit.*, pp. 407 - 408.
[45] Spalt, *op.cit.*, pp. 39 - 42.

Of course, one is indulging in a little Type I formalism in concluding that S is infinitesimal. But I can say that there are valid mathematical models in which one has

$$-\frac{1}{q} < S < \frac{1}{q} \tag{30}$$

for all positive integers q, but for which neither

$$-\frac{1}{q} \leq S < \frac{0}{q} \quad \text{nor} \quad \frac{0}{q} \leq S < \frac{1}{q},$$

holds.

Digression. Before proceeding, we should examine this point a bit more carefully. How do we know that S is infinitesimal? Presumably, the order is determined by the positive elements:

$$a < b \quad \text{iff} \quad b = a + p \text{ for some positive } p.$$

To say, e.g., that $S < 1/q$ would thus mean that $1/q - S$ is positive. Now this follows from our assertion (made in discussing (28) and (29)) that subtracting an infinitesimal from a positive finite number yields a positive number. We can thus conclude S to be infinitesimal by assuming it is. This, of course, is circular. We can avoid the circularity by postulating the algebraic rules we would ordinarily use to conclude the infinitely small nature of S. First, we assume that the absolute value of a product is the product of the absolute values:

$$\left| \prod a_i \right| = \prod |a_i|.$$

Then we make the usual assumption, for b positive,

$$|a| < b \quad \text{iff} \quad -b < a < b.$$

Thus, to show (30) we need only show

$$\prod_{i=1}^{\infty} \frac{1}{i} < \frac{1}{q}.$$

To this end, we would postulate agreement with products of finite support and some form of order preservation:

$$\text{if } a_i = 1 \text{ for all } i > n, \text{ then } \prod_{i=0}^{\infty} a_i = a_0 \cdots a_n$$

$$\text{if } a_i > b_i > 0 \text{ for all } i, \text{ then } \prod a_i \geq \prod b_i.$$

[As the example,

$$a : 1, 1/2, 1/4, \ldots, \qquad b : 1/2, 1/4, 1/8, \ldots$$

shows, the products can be equal.] Given these, define a sequence c by

$$c_i = \begin{cases} \dfrac{1}{q+1}, & i = q+1, \\ 1, & i \neq q+1. \end{cases}$$

Then, agreement with finite products yields

$$\prod_{i=1}^{\infty} c_i = c_1 \cdots c_q \cdot c_{q+1} = 1 \cdots 1 \cdot \frac{1}{q+1} = \frac{1}{q+1} < \frac{1}{q},$$

and order preservation yields

$$\prod_{i=1}^{\infty} \frac{1}{i} \leq \prod_{i=1}^{\infty} c_i < \frac{1}{q}.$$

The general ordering of measurable number expressions is a complicated relation. It is not a true linear ordering and only becomes so when one identifies number expressions that differ only infinitesimally. That is, one defines two measurable number expressions to be equivalent if their difference is infinitesimal. The equivalence classes can then be taken to be the real numbers and will be linearly ordered. Until this step is taken, one has an order-like relation that satisfies some properties of order, but not all. What holds are the defining properties of asymmetry,

$$x \not< x$$

and transitivity,

$$x < y \ \& \ y < z \Rightarrow x < z,$$

but *not* trichotomy,

$$x < y \text{ or } x = y \text{ or } y < x.$$

In constructive mathematics one establishes a useful partial trichotomy,

$$x < y \Rightarrow x < z \text{ or } z < y.$$

In words, if x and y are separated, z cannot simultaneously be close to both of them. Thus, if $x < y$ and z is not greater than x, it is less than y. Even this can fail among measurable number expressions, as one sees by taking

$$x = -\frac{|S|}{2}, \quad y = \frac{|S|}{2}, \quad z = S.$$

If, however, the separation is not infinitesimal, i.e. if $y - x$ is greater than some positive rational number, then one of $x < z$ and $z < y$ will hold. At least, this should be the case when one finally gets down to properly defining the ordering of measurable number expressions, which is something Bolzano apparently never did. (End Digression.)

In a passage on the next to the last page of his manuscript, Bolzano writes

Should it not be possible to simplify the theory of measurable numbers if one sets up the definition thereof so that A is called measurable if one has 2 equations of the form

$$A = \frac{p}{q} + P = \frac{p+n}{q} - P,$$

wherein n, q can increase to infinity at the same time?[46]

Spalt points out that the two P's should represent different numbers here: P_1, P_2. What is not clear is how n, q are to increase to infinity at the same time. If one cannot fix n, there is no guarantee that the intervals

$$\frac{p_q(A)}{q} \leq A < \frac{p_q(A) + n}{q}$$

grow smaller as q grows larger. Thus, the best interpretation is that A is measurable if there is a number n_A such that for any $n > n_A$ and any positive integer q, numbers p, P_1, P_2 satisfying the necessary inequality can be found.

This passage was not included in Rychlík's 1962 edition of Bolzano's manuscript and did not appear until Jan Berg published the full, painstakingly decoded manuscript in 1976 in Bolzano's collected works. Unaware of Bolzano's suggestion, Laugwitz made a similar modification of the definition of measurability in his 1965 paper: S is a measurable infinite number expression iff for every positive integer q, there are an integer p and two purely positive infinite number expressions P_1, P_2 such that

$$S = \frac{p}{q} + P_1 = \frac{p+2}{q} - P_2, \tag{31}$$

i.e.,

$$\frac{p}{q} < S < \frac{p+2}{q}. \tag{32}$$

We do not lose the already established measurability of rational number expressions as they satisfy the even stronger condition (26). (If p_0 satisfies (26), then choose $p = p_0 - 1$ in (31).)

Bolzano's and Laugwitz's definitions are equivalent:

4.4 Lemma. *Let A be measurable under Bolzano's modified definition. Then A is measurable under Laugwitz's definition.*

Proof. Suppose for any $n > n_A$ and any positive integer q there is an integer p such that

$$\frac{p}{q} < A < \frac{p+n}{q}.$$

Let q be given, $n > n_A$, and let r be a positive integer to be determined. Find p such that

---

[46] Cited from Spalt, *op. cit.*, p. 65.

$$\frac{p}{qr} < A < \frac{p+n}{qr} = \frac{p}{qr} + \frac{n}{qr}.$$

Let $p_0 = [p/r]$ be the greatest integer in p/r, so that $p_0 \leq p/r < p_0 + 1$.[47] Then we have

$$\frac{p_0}{q} \leq \frac{p}{qr} < A.$$

But we want

$$\frac{p_0 + 2}{q} > \frac{p+n}{qr},$$

i.e.

$$p_0 + 2 > \frac{p}{r} + \frac{n}{r}.$$

Now $p_0 + 1 > p/r$, so it suffices that $1 > n/r$, i.e. $r > n$.   □

4.5 Lemma. *Let A, B be measurable number expressions. Then $A + B$ is a measurable number expression.*

Proof. Let A, B be measurable. Choose n_A such that, for any $n > n_A$, for all positive integers q there are an integer $p_n(A)$ and purely positive expressions $P_{1,q,n}(A), P_{2,q,n}(A)$ such that

$$A = \frac{p_n(A)}{q} + P_{1,q,n}(A) = \frac{p_n(A) + n}{q} - P_{2,q,n}(A),$$

and choose n_B similarly so that

$$B = \frac{p_n(B)}{q} + P_{1,q,n}(B) = \frac{p_n(B) + n}{q} - P_{2,q,n}(B),$$

can be solved for $n > n_B$. Then, for $n > \max\{n_A, n_B\}$,

$$A + B = \frac{p_n(A) + p_n(B)}{q} + P_{1,q,n}(A) + P_{1,q,n}(B)$$

$$= \frac{p_n(A) + p_n(B) + 2n}{q} - \left(P_{2,q,n}(A) + P_{2,q,n}(B)\right),$$

and we see that we can find the necessary $p_m(A+B), P_{1,q,m}(A+B), P_{2,q,m}(A+B)$ for $m > 2\max\{n_A, n_B\}$.   □

The modified concept of measurability also resolves the problem raised by Example 4.3:

4.6 Lemma. *Every infinitesimal is a measurable number expression.*

Proof. If S is a number expression representing an infinitesimal, then

$$-\frac{1}{q} < S < \frac{1}{q}$$

for all positive integral q, i.e. S is measurable with $p = -1, n \geq 2$.   □

Lemma 4.5 should be followed by a few related lemmas.

---

[47] I.e., choose p_0 by appeal to Lemma 4.1 for $q = 1$.

4.7 Lemma. *Let A be a measurable number expression. Then $-A$ is a measurable number expression.*

For, if

$$\frac{p}{q} < A < \frac{p+2}{q},$$

then obviously,

$$\frac{-p-2}{q} < -A < \frac{-p}{q}.$$

4.8 Corollary. *Let A, B be measurable number expressions. Then $A - B$ is a measurable number expression.*

For, $A - B = A + -B$ and Lemmas 4.5 and 4.7 apply.

4.9 Lemma. *Let A, B be measurable number expressions. Then $A \cdot B$ is a measurable number expression.*

4.10 Lemma. *Let A be a measurable number expression that is not zero or an infinitesimal. Then $1/A$ is a measurable number expression.*

4.11 Corollary. *Let A, B be measurable number expressions and suppose B is not zero or an infinitesimal. Then A/B is a measurable number expression.*

The Corollary of course follows from Lemmas 4.9 and 4.10. The proofs of the lemmas are fairly straightforward adaptations of the proofs that the limits of products and quotients are the products and quotients of the limits. I prove the former and leave the latter as an exercise to the reader.

Proof of Lemma 4.9. Let A, B be measurable number expressions. If one of A, B is infinitesimal, then the product AB is infinitesimal and by Lemma 4.6 is measurable. Thus we may assume neither A nor B is infinitesimal and they are bounded away from 0. Without loss of generality, we may assume them both to be positive.

For r, s to be chosen later find $p_r(A), p_s(B)$ such that

$$0 < \frac{p_r(A)}{r} < A < \frac{p_r(A) + 2}{r}$$

$$0 < \frac{p_s(B)}{s} < B < \frac{p_s(B) + 2}{s}.$$

Then

$$0 < \frac{p_r(A)p_s(B)}{rs} < AB < \frac{p_r(A) + 2}{r} \cdot \frac{p_s(B) + 2}{s}.$$

If we now set $s = 1$, the number $p_1(B)$ is just some rational constant c and we have

$$0 < \frac{c\,p_r(A)}{r} < AB < \frac{p_r(A) + 2}{r} \cdot (c + 2) = \frac{c\,p_r(A)}{r} + 2 \cdot \frac{c + 2}{r},$$

and we can make $2(c+2)/r$ as small as we please by choosing r sufficiently large.

Now, let q be given and choose r large enough so that

$$\frac{2(c+2)}{r} < \frac{1}{q},$$

i.e. choose $r > 2(c+2)q$. Choose p by Lemma 4.1 so that

$$\frac{p}{q} \leq \frac{c\,p_r(A)}{r} < \frac{p+1}{q}.$$

Then

$$\frac{p}{q} \leq \frac{c\,p_r(A)}{r} < AB < \frac{c\,p_r(A)}{r} + \frac{2(c+2)}{r} < \frac{p+1}{q} + \frac{1}{q} = \frac{p+2}{q}. \qquad \square$$

Bolzano proves a general representation theorem, a sort of error term for an infinite generalised decimal expansion, which we may express in the following way:

4.12 Lemma. *Let $b_1, b_2, \ldots, b_m$ be a sequence of integers greater than 1. For any measurable number expression A, there are integers $a_0, a_1, a_2, \ldots, a_m$, with $0 \leq a_i < b_i$ for $i = 1, 2, \ldots m$ such that*

$$a_0 + \frac{a_1}{b_1} + \frac{a_2}{b_1 b_2} + \ldots + \frac{a_{m-1}}{b_1 \cdots b_{m-1}} + \frac{a_m}{b_1 \cdots b_m} < A <$$

$$a_0 + \frac{a_1}{b_1} + \frac{a_2}{b_1 b_2} + \ldots + \frac{a_{m-1}}{b_1 \cdots b_{m-1}} + \frac{a_m + 2}{b_1 \cdots b_m}.$$

Proof. Because A is measurable, we have

$$\frac{p}{b_1 b_2 \cdots b_m} < A < \frac{p+2}{b_1 b_2 \cdots b_m}$$

for some integer p. By the Euclidean algorithm, there are a_0, p_1 such that

$$p = a_0 \cdot b_1 b_2 \cdots b_m + p_1, \quad 0 \leq p_1 < b_1 b_2 \cdots b_m.$$

Thus

$$\frac{p}{b_1 b_2 \cdots b_m} = a_0 + \frac{p_1}{b_1 b_2 \cdots b_m}$$

$$\frac{p+2}{b_1 b_2 \cdots b_m} = a_0 + \frac{p_1 + 2}{b_1 b_2 \cdots b_m}$$

and we have

$$a_0 + \frac{p_1}{b_1 b_2 \cdots b_m} < A < a_0 + \frac{p_1 + 2}{b_1 b_2 \cdots b_m}. \qquad (33)$$

Now divide p_1 by $b_2 \cdots b_m$:

$$p_1 = a_1 \cdot b_2 \cdots b_m + p_2, \quad 0 \le p_2 < b_2 \cdots b_m.$$

Then

$$\frac{p_1}{b_1 b_2 \cdots b_m} = \frac{a_1}{b_1} + \frac{p_2}{b_1 b_2 \cdots b_m}. \tag{34}$$

We have

$$\frac{a_1}{b_1} \le \frac{p_1}{b_1 b_2 \cdots b_m} < 1,$$

whence $a_1 < b_1$. Also, because $b_1, b_2, \ldots, b_m, p_1$ are non-negative we have $0 \le a_1$.

After plugging (34) into (33) we obtain

$$a_0 + \frac{a_1}{b_1} + \frac{p_2}{b_1 b_2 \cdots b_m} < A < a_0 + \frac{a_1}{b_1} + \frac{p_2 + 2}{b_1 b_2 \cdots b_m}.$$

Next, divide p_2 by $b_3 \cdots b_m$:

$$p_2 = a_2 \cdot b_3 \cdots b_m + p_3, \quad 0 \le p_3 < b_3 \cdots b_m,$$

so that

$$\frac{p_2}{b_1 b_2 \cdots b_m} = \frac{a_2}{b_1 b_2} + \frac{p_3}{b_1 b_2 \cdots b_m}.$$

Again

$$\frac{a_2}{b_1 b_2} < \frac{p_2}{b_1 b_2 \cdots b_m} < 1,$$

etc.

Iterating the procedure yields the Lemma.   □

Taking $b_1 = b_2 = \ldots = 10$, we see that every measurable number expression A has a decimal expansion in the weak sense that the finite chunks of the decimal approximate A very well:

$$0 < A - \sum_{i=0}^{m} \frac{a_i}{10^i} < \frac{2}{10^m}.$$

Now, *if* the full decimal,

$$B = \sum_{i=0}^{\infty} \frac{a_i}{10^i}$$

is measurable, then B is approximated equally well and A and B will differ infinitesimally.

4.3 Infinite Number Expressions Clarified

Before proceeding farther, it might be a good idea to make rigorous what has been done up till now. Thus far, we have a vaguely defined collection of "infinite number expressions", an undefined notion of equality presumably based on the assumption of the validity of common algebraic identities, and a partial ordering based on a not yet clearly defined notion of "pure positivity".

In modern terms, the definition of an "infinite number expression" can be given in several, unfortunately inequivalent, ways. First, one would offer an auxiliary definition:

4.13 Definition. *The class of* rational number expressions *is the smallest class of expressions satisfying the following:*
i. every natural number is a rational number expression;
ii. if A, B are rational number expressions, then so are A + B, A − B, A · B, and A/B.

These give the finite number expressions. One would use them to generate the infinite ones, perhaps as follows:

4.14 Definition. *The class of* infinite number expressions *(narrowly construed) is the smallest class of expressions satisfying the following:*
i. every rational number expression is an infinite number expression;
ii. if A, B are infinite number expressions, then so are A + B, A − B, A · B, and A/B;
iii. if $A_0, A_1, \ldots$ is a sequence of rational number expressions, then $\sum A_i, \prod A_i$ are infinite number expressions.

Note the restriction here on the sequence $A_0, A_1, \ldots$ in the third clause. This rules out any nesting of infinite sums or products in infinite number expressions. For our modern purpose of constructing the real numbers from the rational numbers, the infinite number expressions, narrowly construed, form a sufficiently broad class. They will yield all real numbers. And, every example from Bolzano's manuscript cited by van Rootselaar, Laugwitz and Spalt is of this form.[48] However, for the purpose of describing mathematical practice, as apparently Bolzano's goal was, they are insufficient and do not include all infinite expressions that had occurred in the literature by Bolzano's day. From Euler, whose elementary textbooks were widely read, one can cite examples. Plugging 1 into his formula for the sine we have the product

$$\sin 1 = \prod_{k=1}^{\infty} \left(1 - \frac{1}{k^2 \pi^2}\right),$$

which doesn't fit Definition 4.14.iii because of the presence of π^2: the factors are not rational. One could replace π^2 by Euler's sum,

$$\pi^2 = 6 \sum_{i=1}^{\infty} \frac{1}{i^2}$$

and obtain an infinite expression,

$$\sin 1 = \prod_{k=1}^{\infty} \left(1 - \frac{1}{\sum (6k^2/i^2)}\right),$$

with factors that again are not rational expressions, but are infinite number expressions themselves.

---

[48] In his much later work, *Paradoxien des Unendlichen* (Verlag von Felix Meiner, 2nd edition, Leipzig, 1921, p. 45), Bolzano does briefly consider an infinite sum of infinite sums.

If one wants to model real arithmetic in terms of that of the rationals, one must generalise the third clause in the definition to allow for infinitely many infinite number expressions as summands or factors:

4.15 Definition. *The class of* infinite number expressions *(broadly construed) is the smallest class of expressions satisfying the following:*
i.   every rational number expression is an infinite number expression;
ii. if A, B are infinite number expressions, then so are $A + B, A - B, A \cdot B$, and A/B;
iii. if $A_0, A_1, \ldots$ is a sequence of infinite number expressions, then $\sum A_i, \prod A_i$ are infinite number expressions.

Of the accounts of Bolzano's theory I've read, Spalt alone wants to hold on to a literal interpretation of "infinite number expression" as an expression as just described. He does not, however, attempt an exhaustive explanation of the algebraic rules permitted in transforming one infinite number expression into an "equivalent" or "equal" one. This may not be an insurmountable difficulty. One would start out readily enough with the familiar rules of calculating with rational numbers and declare two rational number expressions equivalent if they evaluate to the same rational number. For infinite number expressions, one would assume that substitution of equals yields equals:

$$A_0 = B_0, A_1 = B_1, \ldots \quad \Rightarrow \quad \Sigma A_i = \Sigma B_i \ \& \ \Pi A_i = \Pi B_i.$$

It is the choice of axioms for sums and products that is delicate. Back in Chapter I, we gave axioms for generalised sums and order-preserving generalised sums and these would seem to be the obvious choices. However, the situation is different here. In Chapter I, the mere act of writing down a sum did not guarantee its existence. In the present situation it does: The expression itself is the sum. The proof of Lemma 7.20.ii that $\Sigma 1$ does not exist no longer yields the same conclusion— because $\Sigma 1$ exists and is infinite. Instead, it proves $0 = 1$, from which we can conclude all infinite number expressions are equal.

The culprit is the Truncation/Prefixing Property, which simply does not hold for unbounded sums. However, without truncatability, we would be hard pressed to prove an inequality,

$$\frac{p}{q} \leq S.$$

If S is a sum, say $S = \Sigma a_i$, the natural proof of such an inequality would proceed by peeling off n terms $a_0, \ldots, a_{n-1}$ of the sum S so that

$$\frac{p}{q} \leq a_0 + \ldots + a_{n-1},$$

and there are no negative numbers among $a_n, a_{n+1}, \ldots$ Letting

$$P = \left(a_0 + \ldots + a_{n-1} - \frac{p}{q} \right) + \sum_{i=n}^{\infty} a_i,$$

we would have P "purely positive" and

$$\frac{p}{q} + P = S,$$

i.e. $p/q \leq S$. However, this requires truncatability.

And: how would we establish an inequality like

$$S < \frac{p+2}{q}?$$

The first test of a faithful and rigorous interpretation of Bolzano would seem to be to give a consistent set of algebraic transformation rules that apply to *all* number expressions (Since they must be used to prove measurability, we cannot restrict their use to measurable expressions.), and to prove the measurability of some nontrivial infinite sum, e.g. any convergent geometric progression.

As discussed in Chapter I, Bolzano gave what he called a rigorous proof of the formula for the sum of a geometric series in his book *Paradoxien des Unendlichen*. The reader will recall that Hahn found the proof not to be based on anything and I simply dismissed it as being incorrect. We all make mistakes, but, according to Laugwitz[49], Bolzano is not the one who did so this time. The great difficulties I have been bemoaning almost vanish once we realise that Agreement with Finite Sums in conjunction with the usual definition of the sum of two series yields a safe form of the Truncation/Prefixing Property: Given a sequence $a_0, a_1, \ldots$ of rational numbers, when peeling off the first n numbers to add separately, instead of truncating the sequence, we replace the elements by 0:

$$
\begin{aligned}
a_0 + a_1 + \ldots + a_{n-1} &+ a_n + a_{n+1} + \ldots = \\
&= (a_0 + a_1 + \ldots + a_{n-1} + 0 + 0 + \ldots) \\
&\quad + (0 + 0 + \ldots + 0 + a_n + a_{n+1} + \ldots) \\
&= (a_0 + a_1 + \ldots + a_{n-1}) + (0 + 0 + \ldots + 0 + a_n + a_{n+1} + \ldots).
\end{aligned}
$$

Let us re-examine Bolzano's proof in this light.

Let S be the infinite expression,

$$S = 1 + e + e^2 + \ldots,$$

where $0 < e < 1, e$ rational. Notice

$$
\begin{aligned}
S &= 1 + e + e^2 + \ldots + e^{n-1} + e^n + e^{n+1} + \ldots \\
&= (1 + e + e^2 + \ldots + e^{n-1}) + (0 + 0 + \ldots + 0 + e^n + e^{n+1} + \ldots) \\
&= \frac{1 - e^n}{1 - e} + (0 + 0 + \ldots + 0 + e^n + e^{n+1} + \ldots) \quad\quad (35) \\
&= \frac{1 - e^n}{1 - e} + P_1, \quad\quad (36)
\end{aligned}
$$

---

[49] Laugwitz, *op. cit.*, pp. 402 - 404.

with P_1 purely positive. This much we agreed to earlier. Now we can also write (35) as

$$S = \frac{1 - e^n}{1 - e} + e^n(0 + 0 + \ldots + 0 + 1 + e + \ldots).$$

Here Bolzano made the claim that the sum inside the parentheses, $A = 0 + 0 + \ldots + 0 + 1 + e + \ldots$, could be written as $S - P_2$, where P_2 was positive. Is there a sense in which P_2 is positive, i.e. in which S is greater than A? Comparison of terms doesn't work. The first n terms of S are positive, hence greater than the 0's forming the beginning of A. But after that, the terms of A dominate for $0 < e < 1$. However, A can be looked at as a late-starting version of S and its partial sums will thus always be less than those of S. Indeed, the partial sums of $P_2 = S - A$ are all positive, though in the limit they tend to 0. We may thus wish to consider P_2 to be a positive infinitesimal. If we do this, the rest of the proof carries through.

So imagine

$$S = \frac{1 - e^n}{1 - e} + e^n(S - P_2).$$

Then

$$S(1 - e^n) = \frac{1 - e^n}{1 - e} - e^n P_2$$

$$S = \frac{1}{1 - e} - \frac{e^n}{1 - e^n} P_2. \tag{37}$$

Combining this with (36), we have

$$\frac{1 - e^n}{1 - e} + P_1 = \frac{1}{1 - e} - \frac{e^n}{1 - e^n} P_2$$

$$\frac{-e^n}{1 - e} + P_1 = \frac{-e^n}{1 - e^n} P_2$$

$$P_1 + \frac{e^n}{1 - e^n} P_2 = \frac{e^n}{1 - e}.$$

Because $0 < e < 1$, we can make the right-hand side of this last as small as we choose by picking n large enough. So let q be a positive integer and pick n so large that

$$\frac{e^n}{1 - e} < \frac{1}{q}.$$

We thus also have

$$0 < \frac{e^n}{1 - e^n} P_2 < \frac{1}{q}. \tag{38}$$

By (37), we have

$$\frac{1}{1 - e} - S = \frac{e^n}{1 - e^n} P_2,$$

whence (38) yields

$$\frac{0}{q} = 0 < \frac{1}{1 - e} - S = \frac{e^n}{1 - e^n} P_2 < \frac{1}{q}. \tag{39}$$

Now (39) tells us two things. First, the difference $\frac{1}{1-e} - S$ is a measurable number expression, and thus $S = \frac{1}{1-e} - (\frac{1}{1-e} - S)$, being a difference of measurable number expressions is measurable. The second thing it tells us is that the difference is infinitesimal or 0 and, thus, that when we identify measurable number expressions which differ only infinitesimally, S and $\frac{1}{1-e}$ will be considered equal.

From the modern viewpoint of constructing the reals, it is simply easier to avoid all of the difficulties inherent in dealing with infinite expressions and to replace them entirely by infinite sequences. This is the approach of Rychlík, van Rootselaar, and Laugwitz. Now, I've not seen Rychlík's publications and van Rootselaar doesn't offer much by way of justifying the change other than the lack of clarity in Bolzano, but Laugwitz seems to suggest that all the infinite number expressions cited by Bolzano in his manuscript on real numbers are infinite number expressions narrowly construed and that we nowadays regard all of these as limits of sequences and do calculations with them by calculating with the sequences themselves. Further, such an identification seems to lie implicitly behind both the acceptance of the modified Truncation/Prefixing Property and our heuristic justification for assuming P_2 to be positive in the proof just given. In any event, this seems to be the simplest Type II formal replacement of Bolzano's rather vague conception and we shall adopt it here.

4.16 Definition. *An* infinite number expression A *is a sequence of rational numbers:* $A \in \mathbb{Q}^{\mathbb{N}}$.

For the equality relation, van Rootselaar uses identity:

$$A = B \quad \text{iff} \quad \text{for all } n, A_n = B_n.$$

Laugwitz used an equivalence relation, namely eventual agreement:

$$A \equiv B \quad \text{iff} \quad \text{there is some } n_0 \text{ such that for all } n > n_0, A_n = B_n.$$

This ought to look familiar, as we used something like this before in Chapter II, Section 1, in discussing Laugwitz's interpretation of the Horn Angle. Indeed, before the initial publication of Bolzano's theory of real numbers by Rychlík, Laugwitz had already used $\mathbb{Q}^{\mathbb{N}}$ to construct a model of nonstandard analysis, as will be discussed in section 6.

Van Rootselaar defined an expression $P \in \mathbb{Q}^{\mathbb{N}}$ to be purely positive if eventually P consisted only of positive terms:

$$P \text{ is purely positive} \quad \text{iff} \quad \text{there is some } n_0 \text{ such that for all } n > n_0, P_n > 0.$$

For van Rootselaar, the arithmetic operations are defined pointwise:

$$(A + B)_n = A_n + B_n, \quad (A - B)_n = A_n - B_n, \quad (A \cdot B)_n = A_n \cdot B_n$$

$$(A/B)_n = A_n/B_n, \text{ provided } B_n \text{ is never 0}.$$

With $+$, he defines inequality by

$$A < B \quad \text{iff} \quad \text{for some purely positive } P \in \mathbb{Q}^{\mathbb{N}}, B = A + P.$$

Laugwitz reverses the order of definition:

4.17 Definition. *Let $A, B \in \mathbb{Q}^{\mathbb{N}}$.*

$$A < B \quad \textit{iff} \quad \textit{there is an } n_0 \textit{ such that for all } n > n_0, A_n < B_n.$$

$P \in \mathbb{Q}^{\mathbb{N}}$ is purely positive *iff $\mathbf{0} < P$, where $\mathbf{0}$ is the constant function $\mathbf{0}_n = 0$.*

Laugwitz also defines the arithmetic operations pointwise. That his equivalence relation respects the operations, i.e. that

$$A \equiv B \ \& \ C \equiv D \quad \Rightarrow \quad A + C \equiv B + D, \text{ etc.,}$$

is left unstated. It is a simple enough exercise, as is that of showing their respect of the order relation:

$$B < C \quad \Rightarrow \quad A + B < A + C$$
$$0 < C \ \& \ A < B \quad \Rightarrow \quad A \cdot C < B \cdot C.$$

Van Rootselaar's and Laugwitz's structures,

$$(\mathbb{Q}^{\mathbb{N}}, +, -, \cdot, /, =, <) \quad \text{and} \quad (\mathbb{Q}^{\mathbb{N}}, +, -, \cdot, /, \equiv, <),$$

are both rings, but not fields as they have 0-divisors, e.g.

$$(1, 0, 1, 0, \ldots) \cdot (0, 1, 0, 1, \ldots) = (0, 0, 0, 0, \ldots).$$

The field $\mathbb{Q}$ embeds naturally into both structures via the constant functions: for $q \in \mathbb{Q}$, map q to $\mathbf{q} = (q, q, q, \ldots)$. Because of this identification, we cease writing $\mathbf{q}$ and use q in both settings.

Both rings are only partially ordered, with incomparable elements:

$$(1, 0, 1, 0, \ldots) \nless (0, 1, 0, 1, \ldots) \quad \& \quad (0, 1, 0, 1, \ldots) \nless (1, 0, 1, 0, \ldots).$$

Moreover, they contain infinitely large elements,

$$\mathbf{q} = (q, q, q, \ldots) < (1, 2, 3, \ldots), \text{ for all } q \in \mathbb{Q}$$

and infinitely small elements,

$$(1, 1/2, 1/3, \ldots) < (q, q, q, \ldots), \text{ for all } q \in \mathbb{Q}^{+}.$$

With respect to a construction of the real numbers, the idea should now be clear: Single out those sequences that ought to correspond to real numbers, i.e. those which converge in $\mathbb{R}$, and identify those with the same limit by introducing a new equivalence relation. The notion of measurability performs the first task.

4.18 Definition. *Define $\mathcal{M} \subseteq \mathbb{Q}^{\mathbb{N}}$ to be the set of* measurable infinite number expressions, *i.e. the set of sequences A for which, for any positive integer q an integer p exists such that*

$$\frac{p}{q} < A < \frac{p+2}{q}.$$

With respect to Laugwitz's structure, note that A is measurable just in case we have

$$\frac{p}{q} < A_n < \frac{p+2}{q}$$

for all sufficiently large n. Now, if $A \equiv B$, i.e. if $A_n = B_n$ for all sufficiently large n, then from the measurability of A follows

$$\frac{p}{q} < B_n < \frac{p+2}{q}$$

for all sufficiently large n, i.e. B is also measurable. Thus, measurability is preserved under Laugwitz's equivalence relation. (For van Rootselaar this, of course, is not an issue.)

The proofs we gave a little earlier of the measurability of the sums, differences, etc. of measurable numbers are clearly valid under these interpretations and we may thus consider the structures,

$$(\mathcal{M}, +, -, \cdot, /, =, <) \quad \text{and} \quad (\mathcal{M}, +, -, \cdot, /, \equiv, <).$$

The next step is to replace $=$ and $\equiv$, respectively by an equivalence relation $\approx$ identifying two sequences which converge to the same limit. Bolzano, of course, was dealing with actual expressions and not with sequences, so the obvious step of identifying sequences whose difference is a null sequence, i.e. of defining

$$A \approx B \quad \text{iff} \quad \lim_{n \to \infty} (A_n - B_n) = 0,$$

was not available to him. Today we can see that he should have done the equivalent and defined

$$A \approx B \quad \text{iff} \quad A - B \text{ is infinitesimal.}$$

Instead he made a hopeless mess of things trying to use the sequences $p_1(A), p_2(A), p_3(A)$ of numerators of measuring fractions of expressions A to determine A. This led to a relation,

$$A \cong B \quad \text{iff} \quad A = B \text{ or one of } A - B, B - A \text{ is a positive infinitesimal,}$$

which required further work. I refer the curious reader to Spalt's exposition for elaboration.

4.19 Definitions. *For $A, B \in \mathcal{M}$, we define*
i. $A \approx B$ *iff* $A - B$ *is infinitesimal*
ii. $[A] = \{B \in \mathcal{M} | A \approx B\}$
iii. $\mathbb{R} = [\mathcal{M}] = \{[A] \,|\, A \in \mathcal{M}\}$.

To justify identifying $\mathbb{R}$ with $[\mathcal{M}]$, we must show $[\mathcal{M}]$ to have the usual properties expected. First, of course, we must show that the operations on $\mathcal{M}$ define operations on $[\mathcal{M}]$:

4.20 Lemma. *For $A, B, C, D \in \mathcal{M}$,*

i. $A \approx C \;\&\; B \approx D \;\;\Rightarrow\;\; A + B \approx C + D$

ii. $A \approx C \;\&\; B \approx D \;\;\Rightarrow\;\; A - B \approx C - D$

iii. $A \approx C \;\&\; B \approx D \;\;\Rightarrow\;\; A \cdot B \approx C \cdot D$

iv. $A \approx C \;\&\; B \approx D \;\&\; B \not\approx 0 \;\;\Rightarrow\;\; A/B \approx C/D$.

Proof. As with the proofs that sums, differences, etc. of measurable numbers are measurable, this is essentially a version of the proof that the limit of a sum, difference, etc. is the sum, difference, etc., respectively, of the limits. I shall prove i and iii and leave the rest to the reader.

i. Let $A \approx C, B \approx D$ and let a positive integer q be given, and suppose

$$\frac{-1}{2q} < A - C < \frac{1}{2q}, \quad \frac{-1}{2q} < B - D < \frac{1}{2q}.$$

Then

$$\frac{-1}{q} = \frac{-1}{2q} + \frac{-1}{2q} < (A - C) + (B - D) < \frac{1}{2q} + \frac{1}{2q},$$

i.e.

$$\frac{-1}{q} < (A + B) - (C + D) < \frac{1}{q}$$

and we conclude $A + B \approx C + D$.

iii. Let $A \approx C, B \approx D$ and let a positive integer q be given. By measurability, A, B, C, D are bounded. Let r be a large positive integer bounding all of them:

$$-r < A, B, C, D < r.$$

Now

$$AB - CD = AB - AD + AD - CD = A(B - D) + (A - C)D$$

and we have a sum of a bounded number times an infinitesimal and an infinitesimal times a bounded number, i.e. an infinitesimal. More formally, since $A - C, B - D$ are infinitesimal, they are smaller than any rational number we choose, like $1/(2rq)$:

$$\frac{-1}{2rq} < A - C < \frac{1}{2rq} \quad \text{and} \quad \frac{-1}{2rq} < B - D < \frac{1}{2rq}.$$

Thus,

$$r\left(\frac{-1}{2rq}\right) + \left(\frac{-1}{2rq}\right)r < AB - CD < r\left(\frac{1}{2rq}\right) + \left(\frac{1}{2rq}\right)r,$$

i.e.

$$\frac{-1}{q} < AB - CD < \frac{1}{q}. \qquad \square$$

With the proof that the arithmetic operations are well-defined on $[\mathcal{M}]$, we have finished constructing the real numbers; or, more exactly, we have completed the construction of a field $\mathbb{R} = ([\mathcal{M}], +, \cdot, [0], [1])$ to fill the rôle of the real numbers. It remains to be shown that it is a viable candidate for the position. This is usually done by defining the notion of a complete Archimedean ordered field, proving the constructed structure is a complete Archimedean ordered field, and then proving the isomorphism of any two complete Archimedean ordered fields.

4.4 Complete Archimedean Ordered Fields

We begin with a few definitions.

4.21 Definitions. *A* structure $\mathcal{F} = (F, +, \cdot, 0, 1, <)$ *is an* ordered field *if it satisfies the following conditions:*
i. $(F, +, \cdot, 0, 1)$ *is a field according to Definition 2.1;*
ii. $<$ *is an ordering respected by the operations* $+, \cdot$:
    a. for all $x, y, z \in F$, $x < y$ & $y < z$ $\Rightarrow$ $x < z$
    b. for all $x \in F$, $x \not< x$
    c. for all $x, y \in F$, $x < y$ *or* $x = y$ *or* $y < x$
    d. for all $x, y, z \in F$, $y < z$ $\Rightarrow$ $x + y < x + z$
    e. for all $x, y, z \in F$, $0 < x$ & $y < z$ $\Rightarrow$ $xy < xz$.
An ordered field $\mathcal{F} = (F, +, \cdot, 0, 1, <)$ *is an* Archimedean ordered field *if the ordering is Archimedean: for any positive* $x, y \in F$, *some sum* $y + \ldots + y$ *is greater than* x.

It is not hard to see or even prove that $\mathbb{Q}$ can be homomorphically embedded into any ordered field by mapping m/n to the ratio of the $|m|$-fold sum of 1's to the $|n|$-fold sum of 1's, with the proper choice of sign affixed. The proof consists of verifying a lot of grubby little details and I do not propose to do this here. We can still be rigorous in our treatment by simply making the inclusion of $\mathbb{Q} = (\mathbb{Q}, +, \cdot, 0, 1, <)$ an additional clause in the definition. In the case at hand, namely $\mathbb{R} = [\mathcal{M}]$, we have the explicit embedding,

$$q \mapsto [(q, q, q, \ldots)],$$

and the obvious truths of

$$[p + q] = [(p, p, p, \ldots) + (q, q, q, \ldots)]$$
$$= [(p, p, p, \ldots)] + [(q, q, q, \ldots)] = [p] + [q],$$

$$[p \cdot q] = [(p, p, p, \ldots) \cdot (q, q, q, \ldots)]$$
$$= [(p, p, p, \ldots)] \cdot [(q, q, q, \ldots)] = [p] \cdot [q],$$

$$[p] = [q] \quad \text{iff} \quad (p, p, p, \ldots)_n = (q, q, q, \ldots)_n \text{ for sufficiently large } n$$
$$\text{iff} \quad p = q \text{ for sufficiently large } n$$
$$\text{iff} \quad p = q.$$

I haven't yet defined the ordering on $\mathbb{R}$, so we have to do this before proving that the order on $\mathbb{R}$ agrees with that of $\mathbb{Q}$: for $p, q \in \mathbb{Q}$, $[p] < [q]$ iff $p < q$. The definition is simple enough:

$$[A] < [B] \quad \text{iff} \quad [A] \neq [B] \ \& \ A < B$$
$$\text{iff} \quad 0 < B - A \ \& \ B - A \text{ is not infinitesimal}$$
$$\text{iff} \quad \text{for some positive integer } q, \ \frac{1}{q} < B - A.$$

To see that this is well-defined, i.e. that the truth of the statement $[A] < [B]$ does not depend on the choice of representatives $A \in [A], B \in [B]$, we must show:

$$A \approx C \ \& \ B \approx D \quad \Rightarrow \quad B - A \text{ is "big" iff } D - C \text{ is "big"},$$

where "big" means greater than $1/q$ for some positive integer q. To this end, assume $A \approx C, B \approx D$, and $A < B$. Then there is a positive integer q such that

$$\frac{1}{q} < B - A,$$

i.e.

$$\frac{1}{q} < B_n - A_n,$$

for sufficiently large n, say $n > n_1$. But $A \approx C$ and $B \approx D$, whence

$$\frac{-1}{4q} < A_n - C_n < \frac{1}{4q},$$

for sufficiently large n, say $n > n_2$, and

$$\frac{-1}{4q} < D_n - B_n < \frac{1}{4q},$$

for sufficiently large n, say $n > n_3$. But

$$D_n - C_n = D_n - B_n + B_n - A_n + A_n - C_n,$$

whence for $n > n_0 = \max\{n_1, n_2, n_3\}$,

$$\frac{-1}{4q} + \frac{1}{q} + \frac{-1}{4q} < D_n - B_n + B_n - A_n + A_n - C_n,$$

i.e.

$$\frac{1}{2q} < D_n - C_n,$$

and thus

$$\frac{1}{2q} < D - C.$$

The converse argument is identical.

We can now quickly verify that the embedding of $\mathbb{Q}$ into $\mathbb{R}$ preserves order. For $p, q \in \mathbb{Q}$:

$$[p] < [q] \quad \Rightarrow \quad (p, p, p, \ldots) < (q, q, q, \ldots)$$
$$\Rightarrow \quad p < q.$$

Conversely, if $p < q$, then for any positive integer $r > 1/(q - p)$ we have

$$\frac{1}{r} < q - p.$$

Thus $[p] < [q]$.

We have yet to define the notion of a complete Archimedean ordered field and prove that $\mathbb{R}$ is such. First, let us show that $\mathbb{R}$ is an ordered field:

4.22 Lemma. $\mathbb{R} = ([\mathcal{M}], +, \cdot, 0, 1)$ *is a field.*

Proof. The field axioms come in two types— identities, which are universal equations, and existential assertions (existence of the inverse). The validity of the first sort of axiom is preserved under each of the two steps performed in the passage from $\mathbb{Q}$ to $[\mathcal{M}]$. Consider, for example, commutativity of addition. By the pointwise nature of the definition, $A + B = B + A$ holds for $A, B \in \mathcal{M}$ just in case $A_n + B_n = B_n + A_n$ for all $n \in \mathbb{N}$ (in van Rootselaar's case and all but finitely many in Laugwitz's case). But this holds by commutativity in $\mathbb{Q}$. Validity in $[\mathcal{M}]$ is equally trivial:

$$[A] + [B] = [A + B] = [B + A] = [B] + [A].$$

The same argument applies to commutativity of multiplication, both associative laws, and the distributive law, as well as the identity laws governing 0 and 1.

Existential assertions that are universally valid, such as the existence of an additive inverse for every element, will also be preserved in the two steps of the construction. Conditional existential assertions, like the existence of a multiplicative inverse for non-zero elements, are not preserved under the passage from $\mathbb{Q}$ to $\mathcal{M}$. Indeed, we exhibited zero divisors in $\mathcal{M}$. The second passage to $[\mathcal{M}]$, however, corrects for this. To see this, let $[A] \in [\mathcal{M}]$ be non-zero. That is, A is not an infinitesimal: There is a natural number n_0, and a positive integer q such that for $n > n_0$,

$$|A_n| > \frac{1}{q}, \text{ i.e. } A_n < \frac{-1}{q} \text{ or } \frac{1}{q} < A_n.$$

Define B by

$$B_n = \begin{cases} 1, & n \leq n_0 \\ A_n, & n > n_0. \end{cases}$$

Then $A \equiv B$ and $1/B$ exist, and

$$A_n \cdot 1/B_n = \begin{cases} A_n, & n \le n_0 \\ 1, & n > n_0, \end{cases}$$

whence $A \cdot 1/B \equiv 1$, i.e.

$$[A] \cdot [1/B] = [A \cdot 1/B] = [1] = 1. \qquad \square$$

4.23 Lemma. $\mathbb{R} = ([\mathcal{M}], +, \cdot, 0, 1, <)$ *is an ordered field.*

Proof. We must prove assertions ii.a-e of Definitions 4.21. Most of these verifications are trivial, e.g. transitivity (ii.a): Let $[A] < [B]$ and $[B] < [C]$. Then there are positive integers p, q such that $\frac{1}{p} < B - A$ and $\frac{1}{q} < C - B$. But

$$\frac{1}{q} < \frac{1}{p} + \frac{1}{q} < C - B + B - A = C - A,$$

whence $[A] < [C]$.

The only nontrivial case is trichotomy (ii.c). Let $[A], [B]$ be given and suppose $[A] \neq [B]$, i.e. $A - B$ is not infinitesimal. This means that there is some positive integer q such that for infinitely many n either

$$\frac{1}{q} < A_n - B_n \quad \text{or} \quad \frac{1}{q} < B_n - A_n.$$

One of these two alternatives must hold for infinitely many n. For the sake of definiteness, assume it is the former.

Now $A - B$ is measurable, whence for such q there is an integer p such that

$$\frac{p}{3q} < A - B < \frac{p+2}{3q},$$

i.e. for all but finitely many n, say $n > n_0$,

$$\frac{p}{3q} < A_n - B_n < \frac{p+2}{3q}.$$

Now choose $n > n_0$ so that

$$\frac{1}{q} < A_n - B_n,$$

and thus

$$\frac{1}{q} < A_n - B_n < \frac{p+2}{3q}.$$

Then

$$\frac{p+2}{3q} > \frac{1}{q},$$

whence $p + 2 > 3$, i.e. $p > 1$. Thus, for all but finitely many n,

$$\frac{1}{3q} < A_n - B_n,$$

i.e.

$$\frac{1}{3q} < A - B,$$

i.e. $[B] < [A]$. $\qquad \square$

4.24 Lemma. $\mathbb{R} = ([\mathcal{M}], +, \cdot, 0, 1, <)$ *is an Archimedean ordered field.*

Proof. This is quite easy. Let $[A], [B] \in \mathcal{M}$ be positive. By the positivity of $[A]$, we have

$$\frac{1}{q} < A$$

for some positive integer q. By the positivity and measurability of $[B]$, we have

$$\frac{p}{q} < B < \frac{p+2}{q},$$

with $p + 2$ positive. But

$$B < (p+2) \cdot \frac{1}{q} < (p+2)A,$$

i.e. $[B] < (p+2)[A] = [A] + \ldots + [A]$. □

We wish, of course, to show $\mathbb{R}$ to be a *complete* Archimedean ordered field. For this we must settle on a definition of completeness. As I mentioned earlier, all the various completness properties are equivalent; so it doesn't matter much which definition we choose. For our purposes, the most convenient definition is probably given in terms of Cauchy convergence, which we adapt to the present context as follows:

4.25 Definition. *Let $\mathcal{F}$ be an ordered field. A sequence $b \in F^{\mathbb{N}}$ is a* Cauchy sequence *if for any positive integer q there is an integer n_0 such that for all integers $m, n > n_0$ one has*

$$\frac{-1}{q} < b_m - b_n < \frac{1}{q}.$$

The differences between this definition and that of Definition 5.18 of Chapter I are i) the elements of the sequence come from F, and ii) the general real ϵ has been restricted to being rational.

4.26 Definition. *An ordered field $\mathcal{F}$ is* complete *if every Cauchy sequence has a limit. That is, if $b \in F^{\mathbb{N}}$ is a Cauchy sequence, there is an element $L \in F$ such that, for any positive integer q there is an integer n_0 such that for any $n > n_0$ one has*

$$\frac{-1}{q} < b_n - L < \frac{1}{q}.$$

Bolzano's first attempt[50] to prove completeness was inadequate. He attempted to show a Cauchy sequence converges by showing it was consistent

---

"Rein analytischer Beweis des Lehrsatzes, daß zwischen je zwei Werthen, die ein entgegegesetztes Resultat gewähren, wenigstens eine reelle Wurzel der Gleichung liege", *Abhandlungen der königlichen Gesellschaft der Wissenschaften*, Prague, 1817. The work was reprinted in 1894 and most importantly again in 1905 when it was accompanied by a paper by Hermann Hankel as volume 153 of Ostwald's series of reprints of scientific classics. This last version has traditionally been the most

to assume it had a limit, that such a limit could be approximated arbitrarily closely, and that it was unique. This criticism, which goes back at least to the annotated remarks of the 1905 reprint of Bolzano's paper, is stated rather definitively by Manheim:

> Bolzano did not, of course, prove this theorem. It is equivalent to establishing the sufficiency criterion of Cauchy's theorem[51] of convergence of a sequence and cannot be proved without an arithmetical construction of the real number system. What he did prove was that the existence of a limit is not self-contradictory and that this limit, when it exists, can be approximated as close as is desired. The absence of self-contradiction and the possibility of approximation once existence is known are not, however, enough to establish existence, a fact he overlooked.[52]

Manheim softens this a bit by continuing,

> Bolzano announced Cauchy's convergence criterion before Cauchy. But, what is far more important, Bolzano recognized, as Cauchy did not, that the criterion had to be proved without reference to geometric intuition. This view was not at all evident in 1817.[53]

Bolzano returned to the problem in his *Zahlenlehre*. I do not have Bolzano's work at hand and at present find even Spalt's clarification confusing and hard going. Plus it is rather lengthy and, as the reader may have noticed, this section is getting rather lengthy itself. Thus, instead of trying to expound upon Bolzano at this point, I shall simply prove completeness directly.

4.27 Theorem. $\mathbb{R} = ([\mathcal{M}], +, \cdot, 0, 1, <)$ *is a complete Archimedean ordered field.*

The proof will be a little more intelligible if we first present the following special case.

4.28 Lemma. *Let* $a : a_0, a_1, \ldots$ *be a Cauchy sequence of rational numbers. Then: a is measurable and, considered as a sequence of constant sequences of $\mathcal{M}$, is its own limit.*

A small notational remark before beginning the proof: It is convenient to use the absolute value notation. Now we can legitimise this in at least three ways. We can define absolute value in $\mathcal{M}$ by defining, for $a \in \mathbb{Q}^{\mathbb{N}}$,

$$|a|_n = |a_n|,$$

---

accessible and, indeed, I found it available online. The paper was again reprinted in his collected works. An excerpt including this proof attempt in English translation appeared in Garrett Birkhoff, *A Source Book in Classical Analysis*, Harvard University Press, Cambridge (Mass.), 1973. A full English translation of the paper appears in Russ, *op.cit.*

[51] In point of fact, Bolzano dealt with Cauchy convergent series and not sequences.
[52] Manheim, *op. cit.*, p. 68.
[53] *Ibid.*

and then defining
$$\big|[a]\big| = \big[\,|a|\,\big]$$

and prove the usual properties of absolute values:

$$|x| \geq 0 \quad \text{and} \quad |x| = 0 \text{ iff } x = 0 \tag{40}$$
$$|xy| = |x| \cdot |y| \tag{41}$$
$$|x + y| \leq |x| + |y|, \tag{42}$$

for $x, y \in [\mathcal{M}]$. More generally, we can define absolute value in any ordered field by first proving that multiplication by -1 reverses order, and thus that, for any nonzero element x of the field in question, exactly one of x and $-x$ is greater than 0, and then defining

$$|x| = \begin{cases} x, & x = 0 \text{ or } 0 < x \\ -x, & x < 0. \end{cases}$$

One then derives (40) to (42). Finally, one can view our uses of absolute value,

$$|x| + \ldots + |z| < w,$$

as mere abbreviations for statements,

$$-w < x \pm \ldots \pm z < w.$$

Proof of Lemma 4.28. Let q be a positive integer and choose n_0 so large that

$$|a_m - a_n| < \frac{1}{q}$$

whenever $m, n > n_0$. For $m = n_0 + 1$ find an integer p such that

$$\frac{p}{q} < a_m < \frac{p+2}{q}.$$

It quickly follows that

$$\frac{p-1}{q} < a_n < \frac{p+3}{q}$$

for all $n > n_0$. Thus

$$\frac{p-1}{q} < a < \frac{p+3}{q}$$

by the definition of order on $\mathcal{M} = \mathbb{Q}^{\mathbb{N}}$. But this establishes the measurability of a under Bolzano's modified definition. Hence $a \in \mathcal{M}$.

Note too that

$$\frac{p-1}{q} - \frac{p+3}{q} < a - a_n < \frac{p+3}{q} - \frac{p-1}{q},$$

i.e.

$$\frac{-4}{q} < a - a_n < \frac{4}{q}$$

for all $n > n_0$, whence in $[\mathcal{M}]$

$$\lim_{n \to \infty} [a_n] = [a]. \qquad \square$$

Proof of Theorem 4.27. Let $x : x_0, x_1, \ldots$ be a sequence in $[\mathcal{M}]$. Each x_i is an equivalence class of sequences of rational numbers, and we might as well identify x_i with one of these sequences. Thus:

$$x_i : x_{i0}, x_{i1}, x_{i2}, \ldots,$$

each x_{ij} being a rational number. We will define a new sequence

$$a : a_0, a_1, a_2, \ldots$$

of rational numbers to serve as the limit. We will in fact guarantee through the construction that a is a Cauchy sequence, whence measurable, and that a is the limit of x.

We construct a and an auxiliary sequence $n_0, n_1, \ldots$ in stages in such a way that at stage q, for all $n > n_q$,

$$\frac{-1}{q+1} < x_n - a_q < \frac{1}{q+1}. \qquad (43)$$

At stage 0, we appeal to the Cauchy convergence of $x_0, x_1, \ldots$ to find n_0 such that for all $m, n > n_0$,

$$|x_n - x_m| < \frac{1}{2}.$$

For $m = n_0 + 1$, find an integer p by the measurability of x_m so that

$$\frac{p}{2} < x_m < \frac{p+2}{2},$$

and choose

$$a_0 = \frac{p+1}{2}.$$

Then

$$\frac{-1}{2} < x_m - a_0 < \frac{1}{2},$$

and for any $n > n_0$,

$$|x_n - a_0| = |x_n - x_m + x_m - a_0| \le |x_n - x_m| + |x_m - a_0| < \frac{1}{2} + \frac{1}{2} = 1.$$

At stage q, for q a positive integer, we assume given $n_0, \ldots, n_{q-1}, a_0, \ldots, a_{q-1}$. Choose n' so large that

$$|x_n - x_m| < \frac{1}{2(q+1)}$$

for $m, n > n'$. Let $n_q = \max\{n_{q-1}, n'\}$ and $m = n_q + 1$, and find an integer p such that

$$\frac{p}{2(q+1)} < x_m < \frac{p+2}{2(q+1)}.$$

Choose

$$a_q = \frac{p+1}{2(q+1)}$$

and observe that

$$|x_n - a_q| < \frac{1}{2(q+1)} + \frac{1}{2(q+1)} = \frac{1}{q+1}$$

for $n > n_q$ as before.

The sequence $a_0, a_1, \ldots$ is a Cauchy sequence: For any positive integer q, choose $q' = 2q$ and let $r, s > q'$. For $n > \max\{n_q, n_r\}$, we have

$$|a_q - a_r| \leq |a_q - x_n| + |x_n - a_r| < \frac{1}{s+1} + \frac{1}{r+1}$$
$$< \frac{1}{q'+1} + \frac{1}{q'+1} = \frac{2}{q'+1} < \frac{1}{q}.$$

By Lemma 4.28, a is measurable.

Finally, to see that $a = \lim x_n$, let a positive integer q be given, and let $r > q$ be another positive integer. For $n > q$ we have, as just shown,

$$|a_n - a_r| < \frac{1}{q}.$$

It follows that

$$|a - a_r| < \frac{1}{q}. \tag{44}$$

Choose n_1 so large that

$$|x_n - a_r| < \frac{1}{r} \tag{45}$$

for $n > n_1$. If we now choose $n > n_0 = \max\{q, n_1\}$, (44) and (45) yield

$$|x_n - a| \leq |x_n - a_r| + |a_r - a| < \frac{1}{r} + \frac{1}{q} < \frac{1}{q} + \frac{1}{q} = \frac{2}{q},$$

i.e. $\lim x_n = a$. □

We have thus finished the construction of a complete Archimedean ordered field extending the rationals. Moreover, we call it $\mathbb{R}$ and claim it represents the real numbers. It certainly possesses the algebraic properties we expect of the real numbers, and by completeness it has all the limits we could hope the real numbers to have. There is, moreover, an invariant significance to the construction:

4.29 Theorem. *Any two complete Archimedean ordered fields are isomorphic.*

4.30 Theorem. *Let $\mathcal{F}$ be a complete Archimedean ordered field. No proper extension of $\mathcal{F}$ is a complete Archimedean ordered field.*

We can prove these by appealing to the following lemma, the proof of which is given in the Exercises at the end of the Chapter.

4.31 Lemma. *Let $\mathcal{F} = (F, +, \cdot, 0, 1, <)$ be an Archimedean ordered field. Every element $x \in F$ is the limit of a sequence of rational numbers.*

From the Lemma, the truth of the theorems is fairly immediate. For Theorem 4.29, suppose $\mathcal{F}_1 = (F_1, +, \cdot, 0, 1, <)$ and $\mathcal{F}_2 = (F_2, +, \cdot, 0, 1, <)$ are complete Archimedean ordered fields. Now $\mathbb{Q}$ can be assumed to be a subset of each, so we begin defining the isomorphism $f : F_1 \to F_2$ by defining $f(q) = q$ for all $q \in \mathbb{Q}$. If $x \in F_1$ is not rational, then Lemma 4.31 tells us that there is a sequence $x_0, x_1, \ldots$ of elements of F_1 which converges to x:

$$x = \lim_{n \to \infty} x_n.$$

We simply define

$$f(x) = \lim_{n \to \infty} f(x_n).$$

There are, of course, some formal details to attend to. One must show that the sequence $f(x_0), f(x_1), f(x_2), \ldots$ is a Cauchy sequence and hence has a limit as implicitly defined in Definition 4.26. To see that f is a homomorphism, one must show that the limit of a sum or product of two sequences is the sum or product of their respective limits. One must show that f preserves order. And, finally, one must show that f is one-to-one and onto. I shall leave these details to the more ambitious reader.

Theorem 4.30 is a simpler matter. If $\mathcal{F}_1 \subseteq \mathcal{F}_2$ are complete Archimedean ordered fields, and $x \in F_2$, find a sequence $x_0, x_1, \ldots$ of rationals which converges in $\mathcal{F}_2$ to x. But $\mathcal{F}_1$ is also complete, whence the sequence has a limit y in $\mathcal{F}_1$. But $y \in F_2$ because of the inclusion, whence $x = y$ by the uniqueness of the limit in $\mathcal{F}_2$.

This ends our discussion of Bolzano's theory of real numbers and the construction of the real numbers based on it. It does not, of course, end our discussion of the real numbers. There are several separate directions this discussion can branch into at this point. We will cover these in the next three sections. In the immediately following section, we consider as alternatives to the present section the constructions of Weierstrass and Dedekind. Following that, in section 6, we consider the next stage in the evolution of numbers, namely the construction of hyperreals, i.e. an extension of the real numbers including infinite and infinitesimal numbers. Unsurprisingly, this was first given by the construction used here— minus the restriction to measurable number expressions. Finally, there is the construction of a system of real numbers as Cauchy sequences of rational numbers. By Lemma 4.28 every Cauchy sequence is measurable. The converse is also true[54] and we have thus already presented a version of this construction. However, it generalises well beyond the construction of the reals and we will consider these generalisations briefly in section 7.

---

[54] If A is measurable and

5 The Real Numbers, II; Weierstrass and Dedekind

5.1 Weierstrass

As mentioned in the previous section, the mid-to-late 19th century saw a number of such constructions of the real number system, i.e. infinitary constructions that begin with the rational numbers and finish with a complete Archimedean ordered field. Thus, up to isomorphism, they all result in the same structure. The two most cited are the constructions of Dedekind and that done independently by Cantor and Méray. The earliest and least satisfying, is the construction by Weierstrass.

It is interesting to note that the constructions by Weierstrass and Dedekind were pædagogically motivated. Weierstrass never published his construction, but lectured on it frequently and his students published various accounts of it. And Dedekind reports,

> My attention was first directed toward the considerations which form the subject of this pamphlet in the autumn of 1858. As professor in the Polytechnic School in Zürich I found myself for the first time obliged to lecture upon the elements of the differential calculus and felt more keenly than ever before the lack of a really scientific foundation for arithmetic. In discussing this notion of the approach of a variable magnitude to a fixed limiting value, and especially in proving the theorem that every magnitude which grows continually, but not beyond all limits, must certainly approach a limiting value, I had recourse to geometric evidences. Even now such resort to geometric intuition in a first presentation of the differential calculus, I regard as exceedingly useful, from the didactic standpoint, and indeed indispensable, if one does not wish to lose too much time. But that this form of introduction into the differential calculus can make no claim to being scientific, no one will deny. For myself this feeling of dissatisfaction was so overpowering that I made the fixed resolve to keep meditating on the question till I should find a purely arithmetic and perfectly rigorous foundation for the principles of infinitesimal analysis. The statement is so frequently made that the differential calculus deals with continuous magnitude, and yet an explanation of this continuity is nowhere given; even the

$$\frac{p}{q} < A < \frac{p+2}{q}$$

then for large enough n, say $n > n_0$,

$$\frac{p}{q} < A_n < \frac{p+2}{q}.$$

Hence for $m, n > n_0$,

$$\frac{-2}{q} = \frac{p}{q} - \frac{p+2}{q} < A_n - A_m < \frac{p+2}{q} - \frac{p}{q} = \frac{2}{q}.$$

most rigorous expositions of the differential calculus do not base their proofs upon continuity but, with more or less consciousness of the fact, they either appeal to geometric notions or those suggested by geometry, or depend upon theorems which are never established in a purely arithmetic manner. Among these, for example, belongs the above-mentioned theorem, and a more careful investigation convinced me that this theorem, or any one equivalent to it, can be regarded in some way as a sufficient basis for infinitesimal analysis. It then only remained to discover its true origin in the elements of arithmetic and thus at the same time to secure a real definition of the essence of continuity. I succeeded Nov. 24, 1858, and a few days afterward I communicated the results of my meditations to my dear friend Durège with whom I had a long and lively discussion. Later I explained these views of a scientific basis of arithmetic to a few of my pupils, and here in Braunschweig read a paper upon the subject before the scientific club of professors, but I could not make up my mind to its publication, because in the first place, the presentation did not seem altogether simple, and further, the theory itself had little promise. Nevertheless I had already half determined to select this theme as subject for this occasion, when a few days ago, March 14, by the kindness of the author, the paper *Die Elemente der Funktionenlehre* by E. Heine (*Crelle's Journal*[55], Vol. 74) came into my hands and confirmed me in my decision. In the main I fully agree with the substance of this memoir, and indeed I could hardly do otherwise, but I will frankly acknowledge that my own presentation seems to me to be simpler in form and to bring out the vital point more clearly. While writing this preface (March 20, 1872), I am just in receipt of the interesting paper *Ueber die Ausdehnung eines Satzes aus der Theorie der trigonometrischen Reihen*, by G. Cantor (*Math. Annalen*, Vol. 5)[56], for which I owe the ingenious author my hearty thanks. As I find on a hasty perusal, the axiom given in Section II, of that paper, aside from the form of presentation, agrees with what I designate in Section III, as the essence of continuity. But what advantage will be gained by even a purely abstract definition of real numbers of a higher type, I am as yet unable to see, conceiving as I do of the domain of real numbers as complete in itself.[57]

---

[55] The formal title of the journal edited by August Leopold Crelle is *Journal für die reine und angewandte Mathematik*. Heine's paper, "Die Elemente der Functionenlehre", appeared in volume 74 in 1872 (pp. 172 - 188). The construction given therein was an elaboration of ideas put forth by Cantor.

[56] Cantor's paper appeared in volume 5 of *Mathematische Annalen* in 1872 (pp. 123 - 132). It incorporated Heine's modifications of his original ideas.

[57] Richard Dedekind, *Essays on the Theory of Numbers*, Dover Publications, New York, 1963, pp. 1 - 3. This volume, which was originally published by Open Court in 1901, consists of translations from the German by Wooster Woodruff Beman of two essays by Dedekind, "Stetigkeit und irrationale Zahlen" and "Was sind und was sollen die Zahlen?". The first of these was originally published in 1872 and

Weierstrass had "certainly conceived his theory already in 1841/42" according to Gösta Mittag-Leffler[58] and regularly began his lectures on the theory of analytic functions with the construction probably as early as 1859/60. Publication of it began in 1872 with Ernst Kossak's *Die Elemente der Arithmetik.* Salvatore Pincherle published a description in 1880[59] based on Weierstrass's lectures of 1877/78 and Otto Biermann followed Pincherle's exposition in his own book[60] in 1887. And, in 1920, Mittag-Leffler published a description in a Japanese mathematics journal.

The most accessible of these expositions is probably Biermann's. It is accessible in the material sense of being available online: I was able to consult it without leaving home. It is also accessible in being simply presented. By choosing to construct the positive reals from the positive rationals, Biermann avoids all the complications inherent in dealing with negative numbers— complications we have just encountered in discussing Bolzano's theory of real numbers, and which we earlier ran into back in Chapter I in discussing Bolzano's set-theoretic, accumulative notion of summation.

I do not mention Bolzano's notion of sum introduced in Chapter I merely in passing. Weierstrass chose to construct the real numbers as sums of sets of rational numbers, which sums coming prior to real numbers are not conceived of as limits. They are in fact sets behaving exactly as Bolzano described sums to behave in his *Wissenschaftslehre.* Let me quote Biermann:

> We bring into consideration completely general magnitudes, which in contrast to the rational magnitudes are built through the assembling of an unbounded number of (for the present only positive) elements. Abstractly, such a magnitude is completely defined through a "series"
>
> $$\varepsilon_{n_1} + \varepsilon_{n_2} + \ldots + \varepsilon_{n_m} + \ldots,$$
>
> if one can state which elements enter into the series and how often they do.
> We imagine an object, a magnitude produced via the series, taking the series itself, in contrast to the elements and the sums of a finite number of elements, as its own magnitude.
> It must first be shown how one can compare the new magnitudes among themselves and with rational number magnitudes.
> The definition of the equality of rational numbers, which rests on the

concerns his construction of the real numbers; the latter originally appeared in 1888 and presented a set theoretic construction of the natural numbers.

[58] Gericke, *op.cit.*, p. 109. Most of my information on Weierstrass's construction comes from Gericke's exposition (pp. 109 - 113) and Biermann's (cf. footnote 60, below). Gericke's description of the construction itself is based on notes taken by Adolf Kneser of lectures given by Weierstrass in 1880/81.

[59] S. Pincherle, "Saggio di una introduzione alla Teoria della funzioni analitiche secondo i principii del Prof. C. Weierstrass", *Giornale de Matematica* 18 (1880), pp. 178 - 254 and 317 - 357.

[60] O. Biermann, *Theorie der analytischen Functionen*, B.G. Teubner, Leipzig, 1887.

transformations into identical elements ε_ν, is not applicable here, because an unbounded number of transformations is not performable. One cannot for example demonstrate the equality of

$$\frac{1}{3} \quad \text{and} \quad \frac{3}{10} + \frac{3}{10^2} + \frac{3}{10^3} + \dots,$$

... in the earlier manner, and one must find a new definition of equality. To this end introduce the following definition: If one takes out of a magnitude a of the new kind an arbitrary but bounded number of elements, one says that one has taken out a *component*. Accordingly b is called a component of a if a bounded number of elements in a can be so transformed that in a together with b other elements can be found.[61]

We can rephrase Biermann more formally as follows: One is given the positive rational numbers, or perhaps better, expressions built up from them (say, finite sums). One determines the equality of such expressions by the usual methods of simplification (the "transformations" referred to). The real numbers, his "completely general magnitudes", are (possibly) infinite sums of rational numbers. These can be taken to be infinite number expressions *à la* Bolzano's theory, sets of rational numbers *à la* Weierstrass, or as sequences of rational numbers considered as sums.[62] The expression, set, or sequence is itself taken to be the value of the sum. Biermann points out that we cannot determine if two such infinite sums are equal by means of the same sort of simplifications used to determine if two finite sums are equal, as infinitely many simplifications would be required. So he introduces a way of comparing rational numbers to the new magnitudes: We say that a positive rational number q is a *component* of a sum-magnitude a, and write $q < a$, just in case, if we write

$$a = \sum_{i=0}^{\infty} a_i,$$

we have for some $q' \in \mathbb{Q}^+$ and $n \in \mathbb{N}$,

$$q + q' = \sum_{i=0}^{n} a_i, \text{ i.e. } q < \sum_{i=0}^{n} a_i,$$

where this last inequality, being between rational numbers, is taken to be the usual order.

At this point, Biermann defines a to be *finite* if there is some rational number that is not a component of a and says we are only interested in the finite magnitudes and can ignore the rest. He then proceeds to define "two magnitudes

---

[61] *Ibid.*, p. 19.

[62] Biermann writes his sums sequentially. The difference is not entirely cosmetic. For technical reasons, should one opt for dealing with sequences, one should either allow 0 (i.e. deal with sequences of nonnegative rationals instead of positive ones) or allow finite— even empty— sequences.

of the new kind [to be] equal, if every component of the one magnitude is contained in the other as a component", i.e. he defines

$$a = a' \quad \text{iff} \quad \text{for all } q \in \mathbb{Q}^+, q < a \Leftrightarrow q < a'. \tag{46}$$

Ordering is also readily defined:

$$a < a' \quad \text{iff} \quad \text{there is some } q \in \mathbb{Q}^+ \text{ such that } q < a' \text{ but } q \not< a.$$

As for the algebraic operations, addition and multiplication are fairly obvious. Thinking of the sums as sets, one simply takes the sum to be the union, adding multiplicities of occurrence. Thinking in terms of sequences, one uses the usual definition,

$$\sum_{i=0}^{\infty} a_i + \sum_{i=0}^{\infty} b_i = \sum_{i=0}^{\infty} (a_i + b_i),$$

with the obvious modifications should one or both of the sums be finite. Thinking in terms of sets, the product of two sums would be the set consisting of products of elements of the sets:

$$\sum_{q \in X} q \cdot \sum_{p \in Y} p = \sum_{q \in X, p \in Y} qp,$$

i.e. $X \cdot Y = \{qp | q \in X \ \& \ p \in Y\}$ and

$$\sum_{q \in X} q \cdot \sum_{p \in Y} p = \sum_{r \in X \cdot Y} r.$$

If one prefers sums as sequences, one uses the Cauchy product.

Subtraction and division are defined algebraically at the end of the construction process and are nontrivial matters. I refer the reader to Biermann's exposition for the details.

As for completeness, not surprisingly a bounded infinite sum turns out to be its own limit as a series of rational numbers. Biermann gives the proof, but does not prove completeness directly. Instead he introduces the Cantor-Méray construction and proves completeness for that. However, he does show that every bounded monotonically increasing sequence of positive *rational* numbers converges. Indeed, if $a_0 < a_1 < \ldots < b$ are rational, one has

$$\lim_{n \to \infty} a_n = \lim_{n \to \infty} \left(a_0 + \sum_{i=0}^{n-1} (a_{i+1} - a_i) \right),$$

the right-hand limit existing because it is a bounded infinite sum of positive rationals. If $b_0 < b_1 < \ldots < b$ is a bounded monotonically increasing sequence of *real* numbers, there are *rational* numbers $a_0, a_1, \ldots, b'$ satisfying

$$b_0 < a_0 < b_1 < a_1 < \ldots < b < b'.$$

Evidently
$$\lim_{n \to \infty} b_n = \lim_{n \to \infty} a_n.$$
This actually suffices to establish the existence of limits to all Cauchy sequences. The conscientious reader has, in fact, given the proof of this already back in Chapter I, Exercise 9.9.

Gericke, who based his exposition of Weierstrass's construction on Kneser's notes, gives a different proof. We can illustrate this proof by applying its reasoning to sketch a proof of the Least Upper Bound Principle. Let X be a bounded nonempty set of positive real numbers. Suppose q, d are rational numbers such that there is some element of X greater than q and every element of X is less than $q + d$. If some element of X is greater than $q + \frac{d}{2}$, replace q by $q + \frac{d}{2}$ and d by $\frac{d}{2}$. Otherwise replace q by $q + \frac{0}{2}$ and d by $\frac{d}{2}$. Either way, we have a number $q + \varepsilon_1 \cdot \frac{d}{2}$ with $\varepsilon = 0$ or 1 such that some element of X lies between $q + \varepsilon_1 \cdot \frac{d}{2}$ and $q + \varepsilon_1 \cdot \frac{d}{2} + \frac{d}{2}$ and no element is greater than $q + \varepsilon_1 \cdot \frac{d}{2} + \frac{d}{2}$. Thus the procedure can be iterated and we generate an infinite sum
$$q + \varepsilon_1 \cdot \frac{d}{2} + \varepsilon_2 \cdot \frac{d}{4} + \varepsilon_3 \cdot \frac{d}{2^3} + \cdots$$

This turns out to be the least upper bound of X. As a mathematician I feel I am committing some kind of moral offence in not proving this to be the case, but I think the chief point here is that the bound is given as a bounded sum, i.e. it is a real in Weierstrass's sense and hence belongs to the structure he defines. Hence his reals satisfy the Least Upper Bound Principle and, ultimately, the completeness property as defined in Definition 4.26.

Gericke remarks that

> The theory of Weierstrass distinguishes itself from that of Bolzano through this, that arbitrary computation-operations are not permitted, rather only the "gathering together", thus the addition. By restriction to positive elements the definition of finite number magnitudes as "smaller than a finite number" is made possible. This characterisation is genuinely easier to apply than Bolzano's, e.g. in the proof that the sum of two real numbers is a real number.[63]

If I haven't given sufficient detail on the construction by Weierstrass, it is because an even further simplification is possible. Recall (46),
$$a = a' \quad \text{iff} \quad \text{for all } q \in \mathbb{Q}^+, q < a \Leftrightarrow q < a'. \tag{46}$$

Assuming we have established that the rationals can be embedded in the reals under construction in some obvious fashion and that the ordering of rationals as rationals agrees with their ordering as reals, we see that (46) tells us that a positive real number r is determined by the positive rationals less than it:
$$r \mapsto X_r = \{q \in \mathbb{Q}^+ | q < r\}.$$

In essence, Dedekind can be said to have taken this as the starting point of his simpler construction of the reals.

---

[63] Gericke, *op. cit.*, p, 113.

5.2 Dedekind

In actual fact, Dedekind approached the problem from a different direction. His inspiration was geometrical rather than arithmetical. He sought to capture the continuity of the real number line and constructed the real numbers as fillers of gaps in the ordering of the rational numbers. The arithmetic operations are secondary and are carried along by continuity. Indeed, in his own exposition, Dedekind was not very explicit on how multiplication can be defined directly in his structure.

The gaps Dedekind filled he called *cuts*. These may be defined in a number of ways. The basic version is as follows:

5.1 Definition. *Let* $\mathcal{L} = (L, <)$ *be a linearly ordered set. A* cut *in* $\mathcal{L}$ *is a pair* $\langle A, B \rangle$ *of subsets of L satisfying*
i . every element of L is in one of the two sets A, B: $L = A \cup B$;
ii. every element of A is less than every element of B.
A cut $\langle A, B \rangle$ is trivial *if one of A, B is empty; it is* nontrivial *otherwise.*

If $\langle A, B \rangle$ is a nontrivial cut in the real numbers, then A has a least upper bound b, which either belongs to A and

$$A = \{x \in \mathbb{R} | x \le b\}, \quad B = \{x \in \mathbb{R} | b < x\},$$

or belongs to B and

$$A = \{x \in \mathbb{R} | x < b\}, \quad B = \{x \in \mathbb{R} | b \le x\}.$$

Note that b is uniquely determined by the cut $\langle A, B \rangle$ and, up to the choice of whether to place b in A or B, the cut is uniquely determined by b.

The cuts in the rationals are just the restrictions to the rationals of cuts in the reals. In particular, if $\langle A, B \rangle$ is a cut in the rationals, there is a real number b such that
$$A = \{x \in \mathbb{Q} | x \le b\}, \quad B = \{x \in \mathbb{Q} | b < x\},$$

or

$$A = \{x \in \mathbb{Q} | x < b\}, \quad B = \{x \in \mathbb{Q} | b \le x\}.$$

However, b need not be in $\mathbb{Q}$ and the cut need not determine a rational number.

As his goal is the construction of the real numbers, Dedekind takes great pains in describing the situation without reference to the reals. Thus, he illustrates this point by explicitly constructing cuts $\langle A, B \rangle$ which do not correspond to any rational number and proving they do not without relying on the existence of irrational numbers. A special case of his construction is the following.

5.2 Example. Define $A = \{x \in \mathbb{Q} \mid x < 0 \text{ or } 0 \le x^2 < 2\}$ and $B = \{x \in \mathbb{Q} \mid 0 < x \text{ \& } 2 < x^2\}$. Then: $\langle A, B \rangle$ is a cut for which A has no greatest and B no least element.

Clearly every element of A is less than every element of B. Moreover, by the irrationality of $\sqrt{2}$, i.e. the nonexistence of a rational number x such that $x^2 = 2$, every rational number belongs to one of the sets A, B. Thus $\langle A, B \rangle$ is a cut in $\mathbb{Q}$.

Referring to the reals, one can see that A has no greatest element by noting that for any $x \in A$, $x < \sqrt{2}$ and there must be a rational number y between x and $\sqrt{2}$ by the density of the rationals in the reals, whence $x < y \in A$. Similarly, one concludes B to have no least element. Not presupposing the existence of the real numbers, Dedekind explicitly constructs y:

$$y = \frac{x(x^2 + 6)}{3x^2 + 2}.$$

Then

$$y - x = \frac{x(x^2 + 6)}{3x^2 + 2} - \frac{x(3x^2 + 2)}{3x^2 + 2} = \frac{4x - 2x^3}{3x^2 + 2} = \frac{2x(2 - x^2)}{3x^2 + 2}$$

and we see, for $x > 0$,

$$y > x \ \text{ iff } \ y - x > 0 \ \text{ iff } \ x^2 - 2 < 0 \ \text{ iff } \ x \in A$$
$$y < x \ \text{ iff } \ y - x < 0 \ \text{ iff } \ x^2 - 2 > 0 \ \text{ iff } \ x \in B.$$

But also

$$y^2 - 2 = \frac{x^2(x^2 + 6)^2}{(3x^2 + 2)^2} - \frac{2(3x^2 + 2)^2}{(3x^2 + 2)^2} = \frac{x^6 + 12x^4 + 36x^2 - 18x^4 - 24x^2 - 8}{(3x^2 + 2)^2}$$
$$= \frac{x^6 - 6x^4 + 12x^2 - 8}{(3x^2 + 2)^2} = \frac{(x^2 - 2)^3}{(3x^2 + 2)^2},$$

whence

$$y^2 < 2 \ \text{ iff } \ y^2 - 2 < 0 \ \text{ iff } \ x^2 - 2 < 0 \ \text{ iff } \ x \in A$$
$$y^2 > 2 \ \text{ iff } \ y^2 - 2 > 0 \ \text{ iff } \ x^2 - 2 > 0 \ \text{ iff } \ x \in B.$$

Thus, whenever $x > 0$, if $x \in A$ we can find a greater element $y \in A$ and if $x \in B$ we can find a lesser element $y \in B$; i.e. A has no greatest and B no least element.

5.3 Remark. We should not proceed without remarking on this example. Dedekind seemingly pulls the function

$$f(x) = \frac{x(x^2 + 6)}{3x^2 + 2} \tag{47}$$

out of the air and verifies that it works via a miraculous computation. As he wished to establish the existence of such cuts in the rationals, he had to produce some such function. But how did he do it? A little familiarity with the Calculus might suggest some possibilities. Newton's Method applied to approximating solutions to the equation $X^2 - 2 = 0$ yields the rational function

$$g(x) = x - \frac{x^2 - 2}{2x} = \frac{2x^2 - x^2 + 2}{2x} = \frac{x^2 + 2}{2x}.$$

For $x > 0$, one has

$$y - x = g(x) - x = \frac{2 - x^2}{2x}$$

and

$$y > x \ \text{ iff } \ y - x > 0 \ \text{ iff } \ x^2 - 2 < 0 \ \text{ iff } \ x \in A$$
$$y < x \ \text{ iff } \ y - x < 0 \ \text{ iff } \ x^2 - 2 > 0 \ \text{ iff } \ x \in B.$$

Unfortunately, the next step only partly works:

$$y^2 - 2 = \left(\frac{x^2 + 2}{2x}\right)^2 - 2 = \frac{x^4 + 4x^2 + 4 - 8x^2}{4x^2}$$
$$= \frac{x^4 - 4x^2 + 4}{4x^2} = \frac{(x^2 - 2)^2}{4x^2},$$

and $y \in B$ for all rational values of x. Thus Newton's method only allows us to conclude B to have no least element, but does not rule out a greatest element for A. So, how did Dedekind arrive at his function (47)? Exercises 9.11 - 9.12 at the end of the Chapter offer a couple of possible explanations.

Anyway, the above comments digress from the digression I was about to make with this remark. Type II Formalism seems to encourage this sort of non-intuitive proof whereby one presents the final result as a *fait accompli* with no hint of where it came from, how an auxiliary function was constructed, etc. This was the standard complaint of the Renaissance mathematicians about the ritual formal proofs of the Greeks, a dissatisfaction that resulted in several Type I formal approaches to the subject— as we saw, for example, with Cavalieri's Principle in Chapter II. In the present case, the result of the Example becomes conceptually obvious once the reals have been constructed, the density of the rationals established, and the irrationality of $\sqrt{2}$ demonstrated. The point of the Example was, for the sake of exposition, to establish without reference to $\sqrt{2}$ as a number that the cut was not determined by a rational number. He could assume that any of his contemporaries knew how to construct better and better rational approximations to $\sqrt{2}$ from above and below, and there was no need in the present case to explain the construction. He had, for the sake of completeness, just to give it and verify it worked. For the construction given, the verification is particularly simple— like the use of $\epsilon/2$ or $\epsilon/|r|$ in proving theorems about limits (cf. Chapter I, section 3, where I slipped a few of these in without mention)— and one might forget to ask where his rational function came from.

My digressive remark aside, we can say that through Example 5.2, Dedekind showed the rational number line to be incomplete, to have gaps. By the Least Upper Bound Principle, the real number line has no gaps. The reals may thus be viewed as the order completion of the rational numbers, and Dedekind chose to construct the reals numbers by the simple process of filling in these gaps.

The basic idea behind Dedekind's construction is very simple: Every cut that is not determined by a rational number tells us where a new real number ought to be, so we simply put one there. In doing so, there are basically three parameters to choose from in presenting the proof. The first is the choice of ordering to start with. If we construct the real numbers from the rational numbers, the definition of multiplication is slightly more complicated than that used if we choose to start with the positive rational numbers and construct the positive real numbers. In this latter case, however, the discussion of completeness is ever so slightly more subtle. The crucial relations between the initial and final orders are, however, the same and we can treat that part of the construction uniformly by presenting the construction for an arbitrary dense linear ordering.

5.4 Definition. *Let $\mathcal{L} = (L, <)$ be a linearly ordered set. $\mathcal{L}$ is a* dense linear ordering *if between any two elements of L a third element exists.*

5.5 Definition. *Let $\mathcal{L} = (L, <)$ be a dense linear ordering. $\mathcal{L}$ is* order-complete *if every nontrivial cut is determined by an element of L: If $\langle A, B \rangle$ is a nontrivial cut in $\mathcal{L}$, then there is an element $q \in L$ such that either $A = \{x \in L | x \leq q\}$ or $B = \{x \in L | q \leq x\}$.*

The order-theoretic kernal of Dedekind's construction is the following:

5.6 Theorem. *Every dense linear ordering embeds into an order-complete dense linear ordering.*

As with our initial statement of the main result of section 1, this is a weak statement of the result. In fact, the original ordering is dense in the new ordering and, with this added condition, the order completion is unique up to isomorphism.

The second parameter of Dedekind's construction is the presentation of the cuts themselves. Virtually every possible variation can be found in the literature: The cut $\langle A, B \rangle$ is determined by the *lower cut A* or its *upper cut B*. Hence one can formally define a cut to be a lower cut, an upper cut, or a pair of complementary lower and upper cuts. The choice here seems to be largely a matter of taste. My preference is to stick with the original concepts so long as doing so does not complicate the proofs to any noticeable degree. Thus, I shall present the construction using the full cut consisting of both the lower and upper elements.

The decision of what to do with those cuts determined by elements of L is, however, a different matter. The number of cases to consider increases and definitions become slightly more complicated if one allows both cuts determined by a given element of L to be used in one's construction. Therefore, I shall allow only those cuts in which the second component, i.e. the upper cut, has no minimum element. For later reference I introduce some terminology:

5.7 Definitions. *Let* $\mathcal{L} = (L, <)$ *be a dense linear ordering. A cut* $\langle A, B \rangle$ *in* $\mathcal{L}$ *is called a* principal cut *if there is an element* $q \in L$ *such that* q *is the greatest element of* A *or the least element of* B. *For such a cut, we call* q *the* principal generator *of the cut. A principal cut* $\langle A, B \rangle$ *is the* primary *principal cut generated by* q *if* q *is the maximum element of* A, *and it is the* secondary *principal cut generated by* q *if* q *is the minimum element of* B. *A cut that is not principal is called* nonprincipal.

I suppose I should warn the reader that this terminology may be highly local. It is standard mathematical practice to call an object *principal* if it is generated by a single element, so the designations "principal cut" and "nonprincipal cut" are likely fairly common if not universal. Probably more common, however, is the term "rational cut", which is used because when $\mathcal{L}$ is $\mathbb{Q}$ these are the cuts determined by rational numbers. The terms "primary" and "secondary" are, in all likelihood, not in general use in this context and I introduce them here merely to avoid clumsy phrasing later.

The third parameter in the construction is the exact manner in which we use the cuts to obtain the elements of the order completion of a given dense linear ordering, in particular, to determine the reals from the rationals. After the earlier constructions of sections 1, 2, and 4, it should be obvious that the new ordering will consist of cuts or equivalence classes of cuts. This was not always so obvious. In olden times one was not too explicit on this point. Dedekind wrote:

> Whenever, then, we have to do with a cut (A_1, A_2) produced by no rational number, we create a new, an *irrational* number a, which we regard as completely defined by this cut (A_1, A_2); we shall say that the number a corresponds to this cut, or that it produces this cut. From now on, therefore, to every definite cut there corresponds a definite rational or irrational number, and we regard two numbers, as *different* or *unequal* always and only when they correspond to essentially different cuts.[64]

Today this sounds a bit vague and unfinished. We would not "create" new irrational numbers, but would define them to be the cuts themselves, or equivalence classes thereof obtained by identifying two cuts that are essentially the same. Heinrich Weber did just this in the introductory section of his *Lehrbuch der Algebra*[65]. In a letter to Weber of 24 January 1888, Dedekind explained his position:

> We are of divine lineage and possess without any doubt creative power not merely in material things (railroads, telegraphs), but rather most especially in mental things. This is wholly the same question, of which you speak on the close of your letter regarding my theory of irrationals, where you declare the irrational number to be nothing other than the cut itself, while I prefer to create something *new* (distinct from cuts)

[64] Dedekind, *op. cit.*, p. 15.
[65] Heinrich Weber, *Lehrbuch der Algebra I*, Vieweg, Braunschweig, 1895, p. 6.

which corresponds to the cut and from which I say that it produces or generates the cut. We have the right to indulge in such a creative power and besides it is on account of the similarity of all numbers much more expedient to proceed thus. The rational numbers surely also generate cuts, but I certainly do not pass the rational number off as identical with the cut generated by it; and too after the introduction of the irrational numbers one will often speak of cut-phenomena with such expressions to bestow upon them such attributes which applied to the corresponding numbers themselves would indeed sound peculiar.[66]

Dedekind's point is pædagogical, not philosophical. By creating new numbers to be placed where the cuts tell us to place them, only the essential arithmetical properties are introduced. One will not worry that the reals you construct using, say, lower cuts to represent cuts and the ones I construct using, say, upper cuts to represent them are not the same collections of cuts. They determine the same positions in which to place reals and will result in the same structures. Today we like to know where these reals come from and we more-or-less follow Weber. We do not "bestow upon real numbers such attributes" as would apply to cuts and be meaningless applied to numbers because we have other constructions of the reals to which these attributes do not apply. The multiplicity of structures is no problem, for by Theorem 4.29, they are unique up to essential (i.e., arithmetical) properties, and non-uniqueness, where it occurs, concerns only accidental features of the given constructions.

But let us get down to the construction itself. Let $\mathcal{L} = (L, <)$ be a dense linear ordering, which we will shortly take to be $\mathbb{Q}$ or $\mathbb{Q}^+$.

5.8 Definition. *Let $\mathcal{L} = (L, <)$ be any dense linear ordering. Define the structure $\mathcal{L}^C = (L^C, <_C, \equiv_C)$ as follows. L^C is the collection of all nontrivial cuts in $\mathcal{L}$ that are nonprincipal or are primary principal cuts, i.e. L^C consists of all nontrivial cuts that are not secondary principal cuts. We further define two relations $\equiv_C, <_C$ on L^C as follows:*

$$\langle A, B \rangle \equiv_C \langle C, D \rangle \quad \textit{iff} \quad A = C \ \& \ B = D$$
$$\langle A, B \rangle <_C \langle C, D \rangle \quad \textit{iff} \quad A \subsetneq C,$$

where $X \subsetneq Y$ means X is a proper subset of Y: $X \subseteq Y$, but $X \neq Y$.

Notice that $\equiv_C$ is just the relation of equality. Had we allowed both primary and secondary principal cuts, we would have had to alter the definition slightly to allow for the equivalence of primary and secondary cuts with the same generator. We would then have to prove that the relation is an equivalence relation.

5.9 Lemma. *$(L^C, <_C)$ is a linear ordering:*
i.   *(Irreflexivity) $\langle A, B \rangle \not<_C \langle A, B \rangle$;*

---

[66] Dedekind, *Gesammelte mathematische Werke, III*, (ed. by Robert Fricke, Emmy Noether, and Öystein Ore), Vieweg, Braunschweig, 1932, pp. 489 - 490.

ii. (Transitivity) if $\langle A_1, B_1 \rangle <_C \langle A_2, B_2 \rangle$ *and* $\langle A_2, B_2 \rangle <_C \langle A_3, B_3 \rangle$, *then* $\langle A_1, B_1 \rangle <_C \langle A_3, B_3 \rangle$;

iii. (Trichotomy) one of the following holds:

$$\langle A_1, B_1 \rangle <_C \langle A_2, B_2 \rangle, \langle A_1, B_1 \rangle = \langle A_2, B_2 \rangle, \langle A_2, B_2 \rangle <_C \langle A_1, B_1 \rangle.$$

Again, had we allowed both types of principal cuts, $\equiv_C$ would only be an equivalence relation, not equality, and we would have to prove a 4th condition, namely: iv. if $\langle A_1, B_1 \rangle \equiv_C \langle C_1, D_1 \rangle$ and $\langle A_2, B_2 \rangle \equiv_C \langle C_2, D_2 \rangle$, then

$$\langle A_1, B_1 \rangle <_C \langle A_2, B_2 \rangle \quad \text{iff} \quad \langle C_1, D_1 \rangle <_C \langle C_2, D_2 \rangle.$$

Proof of the Lemma. These assertions are fairly trivial.

i. $\langle A, B \rangle <_C \langle A, B \rangle$ would entail $A \subsetneq A$, which is not the case.

ii. Observe

$$\langle A_1, B_1 \rangle <_C \langle A_2, B_2 \rangle \ \& \ \langle A_2, B_2 \rangle <_C \langle A_3, B_3 \rangle \Rightarrow A_1 \subsetneq A_2 \ \& \ A_2 \subsetneq A_3$$
$$\Rightarrow A_1 \subsetneq A_3$$
$$\Rightarrow \langle A_1, B_1 \rangle <_C \langle A_3, B_3 \rangle.$$

iii. Let $\langle A, B \rangle$ and $\langle C, D \rangle$ be given and suppose $A \neq C$. Then there is some $q \in L$ such that either $q \in A$ and $q \notin C$ or $q \in C$ and $q \notin A$.

Suppose $q \in A$ and $q \notin C$. Because C is downward closed, i.e. $x < y \in C$ implies $x \in C$, q is not less than any element of C, i.e.

$$C \subseteq \{x \in L | x \leq q\}. \tag{48}$$

But A is also downward closed and it contains q, whence it contains the superset of inclusion (48). Hence $C \subseteq A$. The inclusion being proper because of q, we have $\langle C, D \rangle <_C \langle A, B \rangle$.

If $q \in C$ and $q \notin A$, the symmetric argument yields $\langle A, B \rangle <_C \langle C, D \rangle$.   □

5.10 Definition. *For* $q \in L$, *we write*

$$\langle q \rangle = \langle \{x \in L | x \leq q\}, \{x \in L | q < x\} \rangle$$

for the primary cut generated by q.

5.11 Lemma. *The map* $q \mapsto \langle q \rangle$ *is an order preserving embedding of* $\mathcal{L} = (L, <)$ *into* $\mathcal{L}^C = (L^C, <_C)$.

The proof is trivial.

As with the embeddings of previous constructions, we identify L with its image in L^C and $<$ with the restriction of $<_C$ to L. Eventually, we will write q for $\langle q \rangle$ and $<$ for $<_C$. For the time being, however, it may be less confusing to keep the various rôles of q separate, so we will continue to distinguish q and $\langle q \rangle$.

5.12 Lemma. *Under the embedding/identification,* L *is dense in* L^C, *i.e. if* $\langle A, B \rangle <_C \langle C, D \rangle$ *are cuts in* L, *there is an element* $q \in L$ *such that* $\langle A, B \rangle <_C \langle q \rangle <_C \langle C, D \rangle$.

Proof. By definition, $\langle A, B \rangle <_C \langle C, D \rangle$ just in case $A \subsetneq C$, i.e. just in case there is an element $q \in L$ such that $q \in C$ but $q \notin A$.

As in the proof of Lemma 5.9.iii, because $q \notin A$, and A is downward closed, it follows that $A \subsetneq \{x \in L | x \leq q\}$, i.e. $\langle A, B \rangle <_C \langle q \rangle$. We also want $\langle q \rangle <_C \langle C, D \rangle$. There are two cases to consider.

If q is not the principal generator of $\langle C, D \rangle$, then q is not the greatest element of C and there is $q' \in L$ with $q < q' \in C$. Thus

$$\{x \in L | x \leq q\} \subsetneq \{x \in L | x \leq q'\} \subseteq C,$$

and $\langle q \rangle <_C \langle C, D \rangle$.

If, however, q is the principal generator of $\langle C, D \rangle$, then $\langle q \rangle = \langle C, D \rangle$. I claim that there must be some $q' < q$ which is not in A. For, otherwise $A = \{x \in L | x < q\}$, which would make $B = \{x \in L | q \leq x\}$. But then $\langle A, B \rangle$ would be the secondary cut generated by q. This cannot be the case because we disallowed secondary principal cuts in constructing $\mathcal{L}^C$. Thus there is some $q' < Q$ with $q' \notin A$. But $q' \in C$ and q' is not the principal generator of $\langle C, D \rangle$, whence the argument already given shows $\langle A, B \rangle <_C \langle q' \rangle <_C \langle C, D \rangle$.   □

A quick corollary is the following.

5.13 Theorem. *Let $\mathcal{L}$ be a dense linear ordering. Then $\mathcal{L}^C$ is order-complete.*

Proof. Assume to the contrary that some nonprincipal cut $\langle \mathfrak{A}, \mathfrak{B} \rangle$ of $\mathcal{L}^C$ exists. Define

$$A = \{q \in L | \langle q \rangle \in \mathfrak{A}\}$$
$$B = \{q \in L | q \notin A\} = \{q \in L | \langle q \rangle \in \mathfrak{B}\}.$$

We will arrive at a contradiction by showing $\langle A, B \rangle$ to be a principal generator of $\langle \mathfrak{A}, \mathfrak{B} \rangle$.

Let $\langle C, D \rangle \in L^C$ be a cut other than $\langle A, B \rangle$. By trichotomy, $\langle C, D \rangle <_C \langle A, B \rangle$ or $\langle A, B \rangle <_C \langle C, D \rangle$. In the former case, there is some $q \in L$ such that

$$\langle C, D \rangle <_C \langle q \rangle <_C \langle A, B \rangle.$$

But $\langle q \rangle <_C \langle A, B \rangle$ implies $q \in A$, i.e. $\langle q \rangle \in \mathfrak{A}$. But $\langle \mathfrak{A}, \mathfrak{B} \rangle$ is a cut, whence

$$\langle C, D \rangle <_C \langle q \rangle \in \mathfrak{A} \quad \Rightarrow \quad \langle C, D \rangle \in \mathfrak{A}.$$

Likewise, if $\langle A, B \rangle <_C \langle C, D \rangle$, there is some $q \in L$ such that

$$\langle A, B \rangle <_C \langle q \rangle <_C \langle C, D \rangle.$$

This puts $q \in B$, whence $\langle q \rangle \in \mathfrak{B}$, and thus $\langle C, D \rangle \in \mathfrak{B}$.

Hence, for any $x \in L^C$, one has

$$x <_C \langle A, B \rangle \quad \Rightarrow \quad x \in \mathfrak{A}$$
$$\langle A, B \rangle <_C x \quad \Rightarrow \quad x \in \mathfrak{B}$$

and $\langle \mathfrak{A}, \mathfrak{B} \rangle$ has $\langle A, B \rangle$ as a principal generator, contrary to assumption.   □

5.14 Corollary. *The Least Upper Bound Principle holds in $\mathcal{L}^C$.*

Proof. The proof holds for any order-complete dense linear ordering. Let $X \subseteq L^C$ be bounded above and nonempty. Define

$$\mathfrak{A} = \{y \in L^C \,|\, \text{for some } x \in X, y \leq_C x\}$$
$$\mathfrak{B} = \{y \in L^C \,|\, \text{for all } x \in X, x <_C y\}.$$

Then $\langle \mathfrak{A}, \mathfrak{B} \rangle$ is a cut in $\mathcal{L}^C$. Let a be the principal generator of $\langle \mathfrak{A}, \mathfrak{B} \rangle$. Then $X \subseteq \{y \in L^C \,|\, y \leq_C a\}$, and a is an upper bound on X.

Let b be any upper bound on X. If $b \in X$, then b must be the greatest element of X and it generates the cut $\langle \mathfrak{A}, \mathfrak{B} \rangle$, whence $b = a$. If $b \notin X$, then $b \in \mathfrak{B}$ and, since a generates the cut, $a <_C b$. Either way, $a \leq_C b$ and we conclude a to be the least upper bound of X.   □

This completes the order theoretic part of Dedekind's construction. It really is this simple. The uniqueness of the order completion up to isomorphism is equally simple: If $(L_1, <_1)$ and $(L_2, <_2)$ are order-complete linear orderings in which $(L, <)$ is dense, one defines an isomorphism of $(L_1, <_1)$ and $(L_2, <_2)$ by first mapping each element of L to itself. Then, given $a \in L_1$, find the cut in L determined by a. This is also a cut in $(L_2, <_2)$, which has a principal generator b. Map a to b. I leave to the reader the verification that the map so defined is one-to-one, onto, and order-preserving.

There is still more work to do. Indeed the main part of the work lies before us. The rationals come equipped with arithmetic operations making them a field and the reals must inherit this structure. Without this, the construction is not as complete as that of the preceding section.

5.3 Algebraic Operations on Cuts

We now assume $\mathcal{L}$ to be one of $\mathbb{Q}^+$ and $\mathbb{Q}$ and show how to lift the arithmetic from $\mathcal{L}$ to $\mathcal{L}^C$, i.e. from $\mathbb{Q}^+$ and $\mathbb{Q}$ to $\mathbb{R}^+$ and $\mathbb{R}$, respectively.

The definition of addition ought to be fairly straightforward. One would expect to define $\langle A, B \rangle + \langle C, D \rangle = \langle E, F \rangle$, where

$$E = \{a + c \,|\, a \in A \,\&\, c \in C\}, \quad F = \{b + d \,|\, b \in B \,\&\, d \in D\}.$$

The proofs that E is downward closed, F is upward closed, and E and F have no overlap are simple enough, but that $E \cup F = L$ requires some thought. One could get around this by defining

$$E = \{a + c \,|\, a \in A \,\&\, c \in C\}, \quad F = \{x \,|\, x \notin E\} \tag{49}$$

or

$$F = \{b + d \,|\, b \in B \,\&\, d \in D\}, \quad E = \{x \,|\, x \notin F\}. \tag{50}$$

Then all one has to deal with is the downward closure of E and the upward closure of F in these respective cases. Dedekind himself defined

$$E = \{z \in L| \text{ for some } a \in A \ \& \ c \in C, z \le a + c\}, \tag{51}$$

thus getting the downward closure for free. The proof of downward closure of E for the definition given by (49) is simple enough and the simpler definition makes it preferable to (51). However, my earlier arbitrary decision not to allow secondary principal cuts actually makes working with (50) slightly easier. The whole business of formalisation can introduce delicate subtleties that don't always seem to belong. (49) is a correct definition, but results do not fall into place using it as easily as they do when one uses (50).

5.15 Definition. *For* $\langle A, B\rangle, \langle C, D\rangle \in L^C$, *define* $\langle A, B\rangle + \langle C, D\rangle = \langle E, F\rangle$, *where*

$$F = \{b + d | b \in B \ \& \ d \in D\}, \quad E = \{x | x \notin F\}.$$

5.16 Lemma. *The following hold:*
i. (Well-Definedness of +) $\langle A, B\rangle + \langle C, D\rangle \in L^C$, *i.e. it is a nontrivial cut, but not a secondary principal cut;*
ii. (Commutativity) $\langle A, B\rangle + \langle C, D\rangle = \langle C, D\rangle + \langle A, B\rangle$;
iii. (Associativity) $\langle A_1, B_1\rangle + (\langle A_2, B_2\rangle + \langle A_3, B_3\rangle) = (\langle A_1, B_1\rangle + \langle A_2, B_2\rangle) + \langle A_3, B_3\rangle$;
iv. (Cancellation) $\langle A_1, B_1\rangle + \langle C, D\rangle = \langle A_2, B_2\rangle + \langle C, D\rangle \Rightarrow \langle A_1, B_1\rangle = \langle A_2, B_2\rangle$.

Proof. i. Let $\langle A, B\rangle + \langle C, D\rangle = \langle E, F\rangle$. Note that, for $b \in B, d \in D$,

$$b + d \in F \ \& \ b + d < z \ \Rightarrow \ \text{for some } w \in \mathbb{Q}^+, (b + d) + w = z$$
$$\Rightarrow \ z = b + (d + w).$$

But $d < d + w$ and $d \in D$, whence $d + w \in D$ and we see $z \in F$. Thus F is upward closed, whence its complement E is downward closed, It follows that $\langle E, F\rangle$ is a cut.

The cut is nontrivial since, for any $a \in A, c \in C, b \in B$, and $d \in D$, $a + c < b + d$, whence $a + c \notin F$, i.e. $a + c \in E$.

To place $\langle E, F\rangle$ in L^C, we must verify it is not a secondary principal cut, i.e. F does not have a least element: If $b_0 + d_0$ were a minimum element of F, with $b_0 \in B, d_0 \in D$, then since B has no minimum element there is some $b \in B$ with $b < b_0$. But then $b + d_0 \in F$ and $b + d_0 < b_0 + d_0$.

ii. $\langle A, B\rangle + \langle C, D\rangle = \langle E_1, F_1\rangle$ and $\langle C, D\rangle + \langle A, B\rangle = \langle E_2, F_2\rangle$, where

$$F_1 = \{b + d | b \in B \ \& \ d \in D\} = \{d + b | d \in D \& \ b \in B\} = F_2$$

by the commutativity of addition in the rationals.

iii. Again, for B_1, B_2, B_3,

$$\{x + (y + z) | x \in B_1 \ \& \ y \in B_2 \ \& \ z \in B_3\}$$
$$= \{(x + y) + z | x \in B_1 \ \& \ y \in B_2 \ \& \ z \in B_3\}$$

by the associativity of addition in the rationals.

iv. This requires some work. Let

$$\langle A_1, B_1 \rangle + \langle C, D \rangle = \langle E, F \rangle = \langle A_2, B_2 \rangle + \langle C, D \rangle,$$

but assume $\langle A_1, B_1 \rangle \neq \langle A_2, B_2 \rangle$. By trichotomy, either $\langle A_1, B_1 \rangle <_C \langle A_2, B_2 \rangle$ or $\langle A_2, B_2 \rangle <_C \langle A_1, B_1 \rangle$. Assume the former, i.e. $B_2 \subsetneq B_1$. Now B_1 has no minimum element, so given any element of B_1 not in B_2 there is a smaller element of B_1 which necessarily is also not in B_2. Thus we can choose $q_1 < q_2$ with each $q_i \in B_1, q_i \notin B_2$. Let $r = q_2 - q_1$.

Because C, D are not empty, there are some $c \in C$ and $d \in D$. There is some $n \in \mathbb{N}^+$ (by the Archimedean Property of the rationals) so that $c + nr > d$. Choose n_0 to be the largest $n \in \mathbb{N}$ such that $c + nr \in C$. Let $c_0 = c + n_0 r$ and, since D has no minimum element, choose $d_0 \in D$ with $d_0 < c + (n_0 + 1)r$. Thus $0 < d_0 - c_0 < r$.

We are almost done. $q_1 + d_0 \in F$, whence there are $y \in B_2, d \in D$ such that $q_1 + d_0 = y + d$. Thus

$$d_0 - d = y - q_1 > q_2 - q_1 = r$$

and $d < d_0 - r < c_0$, whence $d \in C$, a contradiction. $\square$

A quick corollary to the cancellation property is that addition preserves order:

5.17 Corollary. *If* $\langle A_1, B_1 \rangle <_C \langle A_2, B_2 \rangle$, *then*

$$\langle A_1, B_1 \rangle + \langle C, D \rangle <_C \langle A_2, B_2 \rangle + \langle C, D \rangle.$$

Proof. Writing $\langle A_i, B_i \rangle + \langle C, D \rangle = \langle E_i, F_i \rangle$, we clearly have

$$F_2 = \{b + d | b \in B_2 \ \& \ d \in D\} \subseteq \{b + d | b \in B_1 \ \& \ d \in D\} = F_1.$$

The inclusion must be proper, else the sums are equal and by cancellation $\langle A_1, B_1 \rangle = \langle A_2, B_2 \rangle$, contrary to assumption. $\square$

We have one more fact about addition to prove before specialising the construction to $\mathbb{Q}$.

5.18 Lemma. *The embedding* $q \mapsto \langle q \rangle$ *preserves addition:*

$$\langle q_1 + q_2 \rangle = \langle q_1 \rangle + \langle q_2 \rangle.$$

Proof. Write $\langle q_1 \rangle + \langle q_2 \rangle = \langle E, F \rangle$. Note

$$F = \{x + y | q_1 < x \ \& \ q_2 < y\} \subseteq \{z | q_1 + q_2 < z\}.$$

The converse inclusion is quickly proven: Let $z > q_1 + q_2$ and $r = z - (q_1 + q_2)$. Then

$$z = q_1 + q_2 + r = \left(q_1 + \frac{r}{2} \right) + \left(q_2 + \frac{r}{2} \right) \in \langle q_1 \rangle + \langle q_2 \rangle.   \square$$

Addition obeys a couple of laws in $\mathbb{Q}$ that it does not obey in $\mathbb{Q}^+$:

5.19 Lemma. *For $\mathcal{L} = \mathbb{Q}$, we have, for any $\langle A, B \rangle \in L^C$,*
i. $\langle A, B \rangle + \langle 0 \rangle = \langle A, B \rangle$;
ii. there is a cut $\langle C, D \rangle \in L^C$ such that $\langle A, B \rangle + \langle C, D \rangle = \langle 0 \rangle$.

Proof. i. Recall that $\langle 0 \rangle = \langle \{x \in \mathbb{Q} | x \leq 0\}, \{x \in \mathbb{Q} | 0 < x\} \rangle$. For any $\langle A, B \rangle$, if we write $\langle A, B \rangle + \langle 0 \rangle = \langle E, F \rangle$, we have

$$F = \{b + y | b \in B \ \& \ y > 0\} \subseteq B.$$

To see that $F = B$, observe that B has no least element, whence for any $b \in B$ there is some $b' \in B$ with $b' < b$. But

$$b = b' + (b - b') \in F,$$

since $b - b' > 0$.

ii. If $\langle A, B \rangle$ is principal, then $\langle A, B \rangle = \langle q \rangle$ for some $q \in \mathbb{Q}$ and we simply choose $\langle C, D \rangle = \langle -q \rangle$ and appeal to the homomorphism of Lemma 5.18: $\langle q \rangle + \langle -q \rangle = \langle q + -q \rangle = \langle 0 \rangle$.

If $\langle A, B \rangle$ is nonprincipal, we choose

$$\langle C, D \rangle = \langle -B, -A \rangle = \langle \{-b | b \in B\}, \{-a | a \in A\} \rangle.$$

Writing $\langle A, B \rangle + \langle C, D \rangle = \langle E, F \rangle$, we have

$$F = \{b + a | b \in B \ \& \ a \in -A\} = \{b - a | b \in B \ \& \ a \in A\} \subseteq \{z | z > 0\}$$

since every element of B is greater than every element of A. To prove the converse inclusion, let $r > 0$ and, as in the proof of Lemma 5.16.iv, find $b \in B, a \in A$ such that $0 < b - a < r$. By the upward closure of F, $r \in F$.   □

There are two essentially different approaches to defining the multiplication of Dedekind cuts. The most direct approach is to determine the cut that would be formed by the product of two numbers, write down its description as a definition, and then verify step-by-step that the operation thereby defined satisfies all the laws multiplication is supposed to satisfy. The second, more abstract approach is to take the reals as so far defined— a complete ordering with an addition operation defined upon it and in which the rationals are dense— and extending multiplication from $\mathbb{Q}$ to $\mathbb{R}$ by continuity.

Dedekind announced both approaches, but did not himself give any details.

The former, direct approach is favoured by most authors. The proofs are grubby, but the concreteness of the approach makes good pædagogical sense. The peculiar difficulty, if it can be called that, arising in the treatment of multiplication is caused by the fact that multiplication by negative numbers reverses order. This difficulty, of course, does not arise when there are no negative numbers around. For this reason one may choose to construct $\mathbb{R}^+$ from $\mathbb{Q}^+$ and then adjoin the negative numbers as in section 1.

So let us begin by choosing $\mathcal{L} = \mathbb{Q}^+$.

5.20 Definition. *Let $\langle A, B \rangle, \langle C, D \rangle$ be nontrivial cuts in $\mathbb{Q}^+$. The product $\langle A, B \rangle \cdot \langle C, D \rangle$ is the pair $\langle E, F \rangle$ given by*

$$F = \{bd | b \in B \ \& \ d \in D\}, \quad E = \{x \in \mathbb{Q}^+ | x \notin F\}.$$

5.21 Lemma. *For $L = \mathbb{Q}^+$, the following hold:*
i.   (Well-Definedness of $\cdot$) $\langle A, B \rangle \cdot \langle C, D \rangle \in L^C$, i.e. it is a nontrivial cut, but not a secondary principal cut;
ii. (Commutativity) $\langle A, B \rangle \cdot \langle C, D \rangle = \langle C, D \rangle \cdot \langle A, B \rangle$;
iii. (Associativity)

$$\langle A_1, B_1 \rangle \cdot (\langle A_2, B_2 \rangle \cdot \langle A_3, B_3 \rangle) = (\langle A_1, B_1 \rangle \cdot \langle A_2, B_2 \rangle) \cdot \langle A_3, B_3 \rangle;$$

iv. (Distributivity)

$$\langle A, B \rangle \cdot (\langle C_1, D_1 \rangle + \langle C_2, D_2 \rangle) = (\langle A, B \rangle \cdot \langle C_1, D_1 \rangle) + (\langle A, B \rangle \cdot \langle C_2, D_2 \rangle);$$

v.   (Unit Element) $\langle 1 \rangle \cdot \langle A, B \rangle = \langle A, B \rangle$;
vi. (Multiplicative Inverse) for any $\langle A, B \rangle \in L^C$, there is a $\langle C, D \rangle \in L^C$ such that $\langle A, B \rangle \cdot \langle C, D \rangle = \langle 1 \rangle$;
vii. (Order Preservation)

$$\langle C_1, D_1 \rangle <_C \langle C_2, D_2 \rangle \;\Rightarrow\; \langle A, B \rangle \cdot \langle C_1, D_1 \rangle <_C \langle A, B \rangle \cdot \langle C_2, D_2 \rangle.$$

The proof is routine and I leave most of the details to the reader, commenting only on part vi. Because we use $\mathcal{L} = \mathbb{Q}^+$, there is no zero in L^C and thus we do not need to qualify the statement to exclude zero from the cuts for which we claim the existence of a multiplicative inverse. The construction of the inverse is simple: If $\langle A, B \rangle = \langle q \rangle$ is principal, we take

$$\langle A, B \rangle^{-1} = \langle q^{-1} \rangle.$$

If $\langle A, B \rangle$ is not principal, we take

$$\langle A, B \rangle^{-1} = \langle \{q^{-1} | q \in B\}, \{q^{-1} | q \in A\} \rangle.$$

If one chooses to construct $\mathbb{R}$ from $\mathcal{L} = \mathbb{Q}$, Definition 5.20 just doesn't work. There is an obvious workaround: Apply the definition only to positive cuts and then extend the definition to other cuts algebraically.

5.22 Definition. *Let $\langle A, B \rangle, \langle C, D \rangle$ be nontrivial cuts in $\mathbb{Q}$ for which B, D contain only positive rationals. The product $\langle A, B \rangle \cdot \langle C, D \rangle$ is the pair $\langle E, F \rangle$ given by*

$$F = \{bd | b \in B \ \& \ d \in D\}, \quad E = \{x \in \mathbb{Q} | x \notin F\}.$$

The proof of Lemma 5.21 carries over *mutatis mutandis* for positive cuts. For more general cuts, one defines

$$x \cdot y = \begin{cases} \langle 0 \rangle, & x = \langle 0 \rangle \text{ or } y = \langle 0 \rangle \\ -((-x) \cdot y), & x <_C \langle 0 \rangle \ \& \ y >_C \langle 0 \rangle \\ -(x \cdot (-y)), & x >_C \langle 0 \rangle \ \& \ y <_C \langle 0 \rangle \\ (-x) \cdot (-y), & x <_C \langle 0 \rangle \ \& \ y <_C \langle 0 \rangle. \end{cases}$$

One then has to verify that Lemma 5.21.i-vi is still valid, but for the nonexistence of an inverse for $\langle 0 \rangle$. I leave all of this to the reader to prove or to accept without proof.

5.4 Multiplication of Cuts; An Algebraic Approach

These remaining two subsections are intended to be independent of each other and can thus be read in either order, or, indeed, given their more specialised nature, the reader can skip either or both of them. They offer alternative approaches to defining multiplication on cuts by appeal to higher theory. These approaches are more abstract than that of determining what the cuts should be and defining them accordingly. Moreover, they both involve as much or more work than the direct definition. The proofs, however, are interesting in their own rights and I make no apologies for including them here. One nice thing about the proofs is that neither requires us to treat the reals as cuts in the rationals. They start from the assumption that the reals are given as a structure $(\mathbb{R}, +, <, 0, 1)$ in which $(\mathbb{Q}, +, <, 0, 1)$ has been densely embedded. The goal is then to use some abstract principle to extend the multiplication on $\mathbb{Q}$ in some unique way to a multiplication operation on $\mathbb{R}$ so as to make $(\mathbb{R}, +, \cdot, <, 0, 1)$ a complete archimedean ordered field.

The proof given in the present subsection is based on the observation, going back at least as far as Cauchy's *Cours d'Analyse* (1821), that the only continuous functions $f : \mathbb{R} \to \mathbb{R}$ satisfying the *functional equation*

$$f(x + y) = f(x) + f(y), \tag{52}$$

are those given by multiplication by a fixed constant: $f(x) = ax$ for some a depending on f. Continuity does come into play here, but for the most part the crucial property of a continuous function satisfying (52) is that it either preserves or reverses order, and the proof is thus largely algebraic. It is a cousin to Hamilton's definition of multiplication in constructing the integers (cited in section 1, above), and again it is the commutativity of multiplication that is hardest to prove.

The proof in the next subsection proceeds by extending multiplication on the rationals to a function on the reals by continuity and was announced by Dedekind himself. I do not know who deserves credit for the algebraic proof given here.[67]

We begin by considering equation (52). It is a special equation: it is the defining condition for a *homomorphism* of *abelian groups*. We mentioned groups and their homomorphisms in passing in section 1. It is now time to take a closer look at them.

5.23 Definition. *A structure $(G, +, 0)$ is an* abelian group *if it satisfies*
i.   *for all $x, y \in G, x + y = y + x$;*
ii.   *for all $x, y, z \in G, x + (y + z) = (x + y) + z$;*
iii. *for all $x \in G, x + 0 = x$;*
iv.   *for all $x \in G$, there is a $y \in G$ such that $x + y = 0$.*

The structures $(\mathbb{Z}, +, 0), (\mathbb{Q}, +, 0)$, and $(\mathbb{R}, +, 0)$ are abelian groups. Another pair of abelian groups is given by $(\mathbb{Q}^+, \cdot, 1)$ and (the not yet formally defined) $(\mathbb{R}^+, \cdot, 1)$. One can even take $(\mathbb{Q}^{\neq 0}, \cdot, 1)$ and $(\mathbb{R}^{\neq 0}, \cdot, 1)$, where

---

[67] I learned of such a proof from G. Pickert and L. Görke, *op. cit.*

$$\mathbb{Q}^{\neq 0} = \{q \in \mathbb{Q} \mid q \neq 0\}, \qquad \mathbb{R}^{\neq 0} = \{r \in \mathbb{R} \mid r \neq 0\}.$$

These last examples are less likely to be thought of immediately because of the notation: The operation of an abelian group is usually written additively; but what is important is the structure— a commutative, associative binary operation with an identity element and inverses.

Abelian groups can also be obtained from vector spaces. If we take a vector space over the reals and strip it of its scalar multiplication we are left with an abelian group.

5.24 Definition. *A function* $f: G_1 \to G_2$, *where* $(G_1, +_1, 0_1), (G_2, +_2, 0_2)$ *are abelian groups, is called a* group homomorphism *(or, homomorphism for short) if, for all* $x, y \in G_1, f(x +_1 y) = f(x) +_2 f(y)$. *A homomorphism is an* isomorphism *if it is one-to-one and onto.*

There are lots of examples of group homomorphisms one can cite. Every linear transformation from one vector space to another is a group homomorphism when the scalar multiplication is ignored. The map $x \mapsto -x$ taking an element of an abelian group to its additive inverse is a homomorphism; in fact, it is an isomorphism. If $(R, +, \cdot, 0, 1)$ is a ring and $a \in R$, then multiplication by a, i.e. the map $x \mapsto ax$, is a homomorphism of $(R, +, 0)$ into itself considered as an additive group. Our comment about the continuous solutions to (52) asserts that for $R = \mathbb{R}$, these are the only additive homomorphisms that are continuous.

Any discussion of the groups $(\mathbb{Q}, +, 0)$ and $(\mathbb{R}, +, 0)$ involving Dedekind cuts had better acknowledge that they are ordered sets:

5.25 Definition. *A structure* $(G, +, <, 0)$ *is an* ordered abelian group *if* $(G, +, 0)$ *is an abelian group,* $(G, <)$ *is a linear ordering, and* $+$ *preserves* $<$*: for all* $x, y, z \in G$,

$$x < y \Rightarrow x + z < y + z. \tag{53}$$

Readily cited examples of ordered abelian groups include $(\mathbb{Z}, +, <, 0)$, $(\mathbb{Q}, +, <, 0), (\mathbb{R}, +, <, 0), (\mathbb{Q}^+, \cdot, <, 1)$, and $(\mathbb{R}^+, \cdot, <, 1)$. The abelian groups $(\mathbb{Q}^{\neq 0}, \cdot, 1)$ and $(\mathbb{R}^{\neq 0}, \cdot, 1)$ cannot be ordered. (*Exercise.*)

For ordered abelian groups, the proper notion of homomorphism would be a group homomorphism that preserves order. We shall also be interested in group homomorphisms that reverse order.

5.26 Definitions. *Let* $(G_1, +_1, <_1, 0_1)$ *and* $(G_2, +_2, <_2, 0_2)$ *be ordered abelian groups and let* $f : G_1 \to G_2$ *be a group homomorphism of* $(G_1, +_1, 0_1)$ *into* $(G_2, +_2, 0_2)$. *We say* f *is* order preserving, *or* f preserves order *if* f *is strictly increasing: for all* $x, y \in G_1$,

$$x <_1 y \Rightarrow f(x) <_2 f(y).$$

We say f *is* order reversing, *or* f reverses order, *if* f *is strictly decreasing: for all* $x, y \in G_1$,

$$x <_1 y \implies f(y) <_2 f(x).$$

If f is order preserving, order reversing, or the constant function $f(x) = 0_2$, we say f is order aware.

The terminology of this last definition is not standard.

Familiar examples of order preserving homomorphisms are the embeddings of $(\mathbb{Z}, +, <, 0)$ into $(\mathbb{Q}, +, <, 0)$ and of $(\mathbb{Q}, +, <, 0)$ into $(\mathbb{R}, +, <, 0)$. The map $x \mapsto \ln x$ is an order preserving isomorphism of $(\mathbb{R}^+, \cdot, <, 1)$ onto $(\mathbb{R}, +, <, 0)$.

Mapping to the additive inverse, $x \mapsto -x$, yields an order reversing group isomorphism of any ordered abelian group onto itself. Given any ordered abelian group $(G, +, <, 0)$, the identity map, $x \mapsto x$ is an order reversing group isomorphism of $(G, +, <, 0)$ to $(G, +, >, 0)$, the ordered group obtained by reversing the order.

Another special type of abelian group needs to be introduced:

5.27 Definition. *A structure $(G, +, 0)$ is a* divisible abelian group *if for every element $x \in G$ and every positive integer n there is an element $y \in G$ such that*

$$ny = \underbrace{y + \ldots + y}_{n} = x.$$

The obvious examples of divisible abelian groups are $(\mathbb{Q}, +, 0)$ and $(\mathbb{R}, +, 0)$. $(\mathbb{R}^+, \cdot, 1)$ is another example. In such a multiplicative group, divisibility means that n-th roots exist. Thus, $(\mathbb{Q}^+, \cdot, 1), (\mathbb{Q}^{\neq 0}, \cdot, 1)$, and $(\mathbb{R}^{\neq 0}, \cdot, 1)$ are not divisible. Also, of course, $(\mathbb{Z}, +, 0)$ is not a divisible abelian group.

The element y postulated by the definition of divisibility is not necessarily unique. The simplest example of a divisible abelian group in which these divisors are not unique is probably $(\mathbb{C}^{\neq 0}, \cdot, 1)$, where $\mathbb{C}^{\neq 0}$ denotes the set of nonzero complex numbers. If, however, the divisible group is also ordered, then uniqueness holds.

5.28 Lemma. *Let $(G, +, <, 0)$ be an ordered abelian group. Then, for any positive integer n and any $x \in G$ there is at most one $y \in G$ such that $ny = x$.*

Proof. Let y, z be given and suppose $y \neq z$. Then $y < z$ or $z < y$. Suppose the former case obtains: $y < z$. But then $y + y < y + z < z + z, y + y + y < z + z + z$, etc. Thus,

$$ny = \underbrace{y + \ldots + y}_{n} < \underbrace{z + \ldots + z}_{n} = nz,$$

and $ny \neq nz$. $\qquad\square$

5.29 Corollary. *Let $(G, +, <, 0)$ be an ordered divisible abelian group, i.e let $(G, +, <, 0)$ be an ordered abelian group such that $(G, +, 0)$ is divisible. Then, for any positive integer n and any $x \in G$ there is a unique $y \in G$ such that $ny = x$.*

We may, as in the Corollary, make a few more combinations of definitions to arrive at the notions of Archimedean ordered abelian groups, Archimedean ordered divisible abelian groups, and even order complete Archimedean ordered divisible abelian groups. I leave it to the reader to provide formal definitions of these structures.

The introduction of all these definitions serves two immediate purposes. They allow succinct statements of the results we prove along the way to defining multiplication of cuts. Indeed, before we prove these auxiliary results they even suggest the results to be proven. We already know that the reals as so far developed, i.e. $(\mathbb{R}, +, <, 0)$, form an ordered abelian group. Is there any doubt in the reader's mind that we are going to prove it to be an ordered divisible abelian group?

The second service provided by these definitions is that, through the list of properties they postulate they codify those properties that will be used in the proofs. Once we have proven $(\mathbb{R}, +, <, 0)$ to be divisible, we will know that it is an order complete Archimedean ordered divisible abelian group and we will only need to use these properties and not the fact that $\mathbb{R}$ consists of cuts, as opposed to lower cuts or upper cuts or equivalence classes of "measurable infinite number expressions" or even Weierstrassian sums.

Our first goal is the proof that $(\mathbb{R}, +, 0)$ is a divisible abelian group. To get there we must first take a quick look at the n-fold sum $nx = x + \ldots + x$. This will be the seed from which the definition of multiplication on $\mathbb{R}$ will grow. It will be convenient to give it its own proprietary notation and establish a few of its properties.

5.30 Definition. *Let $(G, +, 0)$ be an abelian group, $n \in \mathbb{N}$, and $x \in G$. We define*

$$n \odot x = \underbrace{x + \ldots + x}_{n}.$$

In particular, $0 \odot x = 0$.

5.31 Lemma. *Let $n \in \mathbb{N}$.*
i. in any abelian group $(G, +, 0)$: for all $x, y \in G$,

$$n \odot (x + y) = (n \odot x) + (n \odot y);$$

ii. in any ordered abelian group $(G, +, <, 0)$: for $n \neq 0$ and all $x, y \in G$,

$$x < y \Rightarrow n \odot x < n \odot y;$$

iii. in $(\mathbb{R}, +, 0)$: for all $q \in \mathbb{Q}$, $n \odot q = nq$.

Proof. i. Observe

$$n \odot x + n \odot y = \underbrace{x + \ldots + x}_{n} + \underbrace{y + \ldots + y}_{n}$$

$$= \underbrace{(x + y) + \ldots + (x + y)}_{n} = n \odot (x + y).$$

ii. Again, if $n \neq 0$ and $x < y$

$$n \odot x = x + \ldots + x < y + \ldots + y = n \odot y.$$

iii. Trivial. □

5.32 Lemma. *For any* $n \in \mathbb{N}^+$ *and every* $x \in \mathbb{R}$ *there is a* $y \in \mathbb{R}$ *such that* $n \odot y = x$.

Proof. Let $x \in \mathbb{R}$. There are three cases: $x = 0, x > 0, x < 0$. Obviously, if $x = 0$, one can take $y = 0$. The other two cases are symmetric, so we need only consider $x > 0$.

Let $X = \{z \in \mathbb{R} \mid nz \leq x\}$. X is not empty since $n0 = 0 + \ldots + 0 < x$ and thus $0 \in X$. Moreover, X is bounded since x itself is an upper bound on X: Let $z \in X$. If $z \leq 0$, then Lemma 5.31 yields $nz \leq n0 = 0 < x$, and if $z > 0$, then

$$z \leq \underbrace{z + z + \ldots + z}_{n-1} = nz \leq x.$$

Choose y to be the least upper bound of X. We will show $ny = x$. If $ny < x$, there is some rational number q such that $ny < q < x$. Now rational numbers can be divided by positive integers, whence there is some rational number q' such that $nq' = q$. Observe:

$$q < x \Rightarrow nq' = q \leq x \Rightarrow q' \in X$$
$$\Rightarrow q' \leq y, \text{ since } y \text{ bounds } X$$
$$\Rightarrow q = nq' \leq ny, \text{ by 5.31.ii,}$$

but this last contradicts the choice of $q > ny$. Thus we cannot have $ny < x$.

If $ny > x$, there is some rational q satisfying $ny > q > x$. Again, let $q = nq'$. By 5.31.ii,

$$q' \geq y \Rightarrow q = nq' \geq ny,$$

which is not the case, whence $q' < y$. But, for $z \in X$,

$$q' \leq z \Rightarrow q = nq' \leq nz \leq x,$$

which is also not the case and we see that q' is an upper bound on X smaller than y, contrary to the choice of y. Hence the assumptions $ny < x$ and $ny > x$ both lead to contradictions and it follows that ny must equal x. □

5.33 Definition. *Let* $(G, +, <, 0)$ *be an ordered divisible abelian group. For* $n \in \mathbb{N}^+, x \in G$, *we define* $\frac{1}{n} \odot x$ *to be the unique* $y \in G$ *such that* $n \odot y = x$.

The analogue to Lemma 5.31 holds:

5.34 Lemma. *Let* $n \in \mathbb{N}^+$.

i. *in any ordered divisible abelian group* $(G, +, <, 0)$: *for all* $x, y \in G$,

$$\frac{1}{n} \odot (x + y) = \frac{1}{n} \odot x + \frac{1}{n} \odot y;$$

ii. in any ordered divisible abelian group $(G, +, <, 0)$: for all $x, y \in G$,

$$x < y \quad \Rightarrow \quad \frac{1}{n} \odot x < \frac{1}{n} \odot y;$$

iii. in $(\mathbb{R}, +, 0)$: for all $q \in \mathbb{Q}$,

$$\frac{1}{n} \odot q = \frac{1}{n} q.$$

Proof. i. By Lemma 5.31,

$$n \odot \left(\frac{1}{n} \odot x + \frac{1}{n} \odot y \right) = n \odot \left(\frac{1}{n} \odot x \right) + n \odot \left(\frac{1}{n} \odot y \right) = x + y,$$

whence $\frac{1}{n} \odot x + \frac{1}{n} \odot y$ is the unique z such that $n \odot z = x + y$, i.e. it is $\frac{1}{n} \odot (x + y)$.

ii. Let $x < y$. Then

$$\frac{1}{n} \odot x = \frac{1}{n} \odot y \;\Rightarrow\; n \odot \left(\frac{1}{n} \odot x \right) = n \odot \left(\frac{1}{n} \odot y \right) \;\Rightarrow\; x = y,$$

contrary to assumption. Similarly,

$$\frac{1}{n} \odot y < \frac{1}{n} \odot x \;\Rightarrow\; n \odot \left(\frac{1}{n} \odot y \right) < n \odot \left(\frac{1}{n} \odot x \right) \;\Rightarrow\; y < x,$$

also contrary to assumption.

iii. Trivial.   □

The next step is to define $q \odot x$ for any nonnegative rational number q. The definition is very simple: If we write $q = m/n$, we should define $q \odot x = m \odot (\frac{1}{n} \odot x)$. However, the representation of q as the ratio of two integers is not unique, so we must show that we get the same result if we write $q = m_1/n_1$ and $q = m_2/n_2$.

5.35 Lemma. *Let $(G, +, <, 0)$ be an ordered divisible abelian group, and let $m \in \mathbb{N}$ and $n, k \in \mathbb{N}^+$. For any $x \in G$,*

$$m \odot \left(\frac{1}{n} \odot x \right) = (mk) \odot \left(\frac{1}{nk} \odot x \right).$$

Proof. Note that

$$x = (nk) \odot \left(\frac{1}{nk} \odot x \right) = \underbrace{\frac{1}{nk} \odot x + \ldots + \frac{1}{nk} \odot x}_{nk}$$

$$= \underbrace{\Big(\underbrace{\frac{1}{nk} \odot x + \ldots + \frac{1}{nk} \odot x}_{k} \Big) + \ldots + \Big(\underbrace{\frac{1}{nk} \odot x + \ldots + \frac{1}{nk} \odot x}_{k} \Big)}_{n},$$

whence

$$k \odot \left(\frac{1}{nk} \odot x \right) = \frac{1}{n} \odot x.$$

The same argument shows $m \odot (k \odot y) = (mk) \odot y$. Thus,

$$(mk) \odot \left(\frac{1}{nk} \odot x \right) = m \odot \left(k \odot \left(\frac{1}{nk} \odot x \right) \right) = m \odot \left(\frac{1}{n} \odot x \right). \qquad \square$$

5.36 Corollary. *Let $(G, +, <, 0)$ be an ordered divisible abelian group. For all $m_1, m_2 \in \mathbb{N}, n_1, n_2 \in \mathbb{N}^+$, and $x \in G$,*

$$\frac{m_1}{n_1} = \frac{m_2}{n_2} \quad \Rightarrow \quad m_1 \odot \left(\frac{1}{n_1} \odot x \right) = m_2 \odot \left(\frac{1}{n_2} \odot x \right).$$

Proof. From $m_1/n_1 = m_2/n_2$ we conclude $m_1 n_2 = m_2 n_1$. For any $x \in G$,

$$\frac{1}{m_1 n_2} \odot x = \frac{1}{m_2 n_1} \odot x,$$

whence

$$(m_1 m_2) \odot \left(\frac{1}{m_1 n_2} \odot x \right) = (m_1 m_2) \odot \left(\frac{1}{m_2 n_1} \odot x \right).$$

But Lemma 5.35 yields

$$m_2 \odot \left(\frac{1}{n_2} \odot x \right) = (m_1 m_2) \odot \left(\frac{1}{m_1 n_2} \odot x \right)$$

$$= (m_1 m_2) \odot \left(\frac{1}{m_2 n_1} \odot x \right) = m_1 \odot \left(\frac{1}{n_1} \odot x \right). \qquad \square$$

We may now extend our definition to all nonnegative rational numbers q:

5.37 Definition. *Let $(G, +, <, 0)$ be an ordered divisible abelian group. For nonnegative rational q and $x \in G$ we define*

$$q \odot x = m \odot \left(\frac{1}{n} \odot x \right),$$

where $q = m/n$, with $m \in \mathbb{N}, n \in \mathbb{N}^+$.

And we have the usual lemma:

5.38 Lemma. *Let $q \in \mathbb{Q}^{\geq 0}$.*

i.   in any ordered divisible abelian group $(G, +, <, 0)$: for all $x, y \in G$,

$$q \odot (x + y) = q \odot x + q \odot y;$$

ii. in any ordered divisible abelian group $(G, +, <, 0)$: for all $x, y \in G$, if $q \neq 0$, then

$$x < y \Rightarrow q \odot x < q \odot y;$$

iii. in $(\mathbb{R}, +, <, 0)$: for all $q' \in \mathbb{Q}, q \odot q' = qq'$.

Proof. Write $q = m/n$.

i. Observe

$$q \odot (x + y) = m \odot \left(\frac{1}{n} \odot (x + y) \right) = m \odot \left(\frac{1}{n} \odot x + \frac{1}{n} \odot y \right), \text{ by Lemma 5.34}$$

$$= m \odot \left(\frac{1}{n} \odot x \right) + m \odot \left(\frac{1}{n} \odot y \right), \text{ by Lemma 5.31}$$

$$= q \odot x + q \odot y, \text{ by definition.}$$

ii. This also follows from Lemmas 5.31 and 5.34:

$$x < y \implies \frac{1}{n} \odot x < \frac{1}{n} \odot y \implies m \odot \frac{1}{n} \odot x < m \odot \frac{1}{n} \odot y \implies q \odot x < q \odot y.$$

iii. Trivial.   □

The final step in defining multiplication by rationals is the definition of multiplication by a negative rational number.

5.39 Definition. *Let $(G, +, <, 0)$ be an ordered divisible abelian group. Let q be a negative rational number and $x \in G$. We define $q \odot x = -(|q| \odot x)$.*

Here I use the notation $|q|$ for the absolute value of q, which is supposedly already known to us for rational numbers.

5.40 Lemma. *Let q be a negative rational number.*

i.   in any ordered divisible abelian group $(G, +, <, 0)$: for all $x, y \in G$,

$$q \odot (x + y) = q \odot x + q \odot y;$$

ii. in any ordered divisible abelian group $(G, +, <, 0)$: for all $x, y \in G$, if $q \neq 0$, then

$$x < y \implies q \odot y < q \odot x;$$

iii. in $(\mathbb{R}, +, <, 0)$: for all $q' \in \mathbb{Q}$, $q \odot q' = qq'$.

Proof. i. Observe

$$q \odot (x + y) = -(|q| \odot (x + y)) = -(|q| \odot x + |q| \odot y)$$
$$= -(|q| \odot x) + -(|q| \odot y) = q \odot x + q \odot y.$$

ii. Again,

$$x < y \implies |q| \odot x < |q| \odot y$$
$$\implies -(|q| \odot y) < -(|q| \odot x)$$
$$\implies q \odot y < q \odot x.$$

iii. Trivial.   □

There are three interpretations of what we have accomplished thus far in defining $q \odot x$. Taking $\mathbb{R}$ for the group G, we have partially defined multiplication on $\mathbb{R}$ and shown it to agree with multiplication in $\mathbb{Q}$ when q, x are both rational. We could now complete the definition by appeal to continuity and the Least Upper Bound Principle. Let $y \in \mathbb{R}$, if $y \geq 0$, define

$$x \odot y = \text{ least upper bound of } \{q \odot y | q \leq x\},$$

and, if $y < 0$,

$$x \odot y = \text{ greatest lower bound of } \{q \odot y | q \leq x\}.$$

Of course, one has to show that the sets indicated are bounded, i.e. that the products are indeed defined. And, too, there is the task of proving commutativity, associativity, etc.

The second interpretation is that when $G = \mathbb{R}$ we have shown that for each fixed $q \in \mathbb{Q}$, the map f_q taking x to $q \odot x$ is an order aware homomorphism $\mathbb{R} \to \mathbb{R}$. As hinted in the beginning of this subsection, we want to find appropriate homomorphisms indexed by reals and use them to define multiplication. The maps f_q don't suggest any method of doing this. However, if we fix x, we have a map $q \mapsto q \odot x$ of $\mathbb{Q} \to \mathbb{R}$, which we can extend to a map $\mathbb{R} \to \mathbb{R}$ by appeal to continuity and the Least Upper Bound Principle. What we haven't done yet, however, is show that the map $q \mapsto q \odot x$ is a homomorphism.

This brings us to the third interpretation: $\odot$ is a scalar multiplication when we consider $(G, +, 0)$ as a vector space over the field, not of real numbers, but of rational numbers. The reader will recall that we defined vector spaces back in Chapter I as spaces of objects which could be added to each other and be multiplied by real numbers. Algebraically, we get an equally useful notion if we replace the real numbers by rational numbers. Indeed, one can define something like a vector space when the scalar multipliers come from any ring, but the best behaviour occurs when the ring is a field and the name *vector space* is reserved for vector spaces over fields, such as $\mathbb{Q}, \mathbb{R}, \mathbb{C}$. If we peek back at the definition of a vector space given in Chapter I, but replace the real scalar multipliers by rational ones, we will see that there are still two properties of scalar multiplication we haven't yet proven. These are collected in the following lemma.

5.41 Lemma. *Let $(G, +, <, 0)$ be an ordered divisible abelian group. For all $q, r \in \mathbb{Q}$ and all $x \in G$,*
i. $(q + r) \odot x = q \odot x + r \odot x$;
ii. $q \odot (r \odot x) = (qr) \odot x$.

Proof. i. For $q = m \in \mathbb{N}$ and $r = n \in \mathbb{N}$, we have

$$(q + r) \odot x = (m + n) \odot x = \underbrace{x + \ldots + x}_{m+n}$$

$$= \underbrace{x + \ldots + x}_{m} + \underbrace{x + \ldots + x}_{n} = m \odot x + n \odot x = q \odot x + r \odot x.$$

For $q, r \in \mathbb{Z}$, say $q = m, r = n$, note first that for any $k \in \mathbb{Z}$, if k is negative we have $k \odot x = -(|k| \odot x)$ by definition. But

$$0 = |k| \odot (x + -x) = |k| \odot x + |k| \odot (-x) \tag{54}$$

by Lemma 5.31. Thus $|k| \odot (-x)$ is the additive inverse of $|k| \odot x$: $|k| \odot (-x) = -(|k| \odot x) = k \odot x$.

Now consider

$$m \odot x + n \odot x = \pm(|m| \odot x) + \pm(|n| \odot x) = |m| \odot (\pm x) + |n| \odot (\pm x).$$

Except when m, n have the same sign, we haven't yet reduced this to the case already proven. If, say, m is nonnegative and n is negative, we have

$$m \odot x + n \odot x = m \odot x + |n| \odot (-x)$$
$$= \underbrace{x + \ldots + x}_{m} + \underbrace{-x + \ldots + -x}_{|n|},$$

which, for $m \geq |n|$ equals

$$\underbrace{x + \ldots + x}_{m-|n|} = \underbrace{x + \ldots + x}_{m+n} = (m+n) \odot x,$$

and for $|n| > m$ equals

$$\underbrace{-x + \ldots + -x}_{|n|-m} = \underbrace{-x + \ldots + -x}_{|n|-m} = (|n| - m) \odot (-x)$$
$$= -((|n| - m) \odot x), \text{ using } k = m - |n| \text{ in } (54)$$
$$= (-|n| + m) \odot x, \text{ by definition}$$
$$= (n + m) \odot x = (m + n) \odot x.$$

The case where m is negative and n nonnegative is similar.

The general case reduces to the integral case as follows: Let $q = \pm|m_1|/n_1$, $r = \pm|m_2|/n_2$, with $m_1, m_2 \in \mathbb{N}, n_1, n_2 \in \mathbb{N}^+$. Then

$$q \odot x + r \odot x = \pm\frac{|m_1|}{n_1} \odot x + \pm\frac{|m_2|}{n_2} \odot x$$
$$= \pm\frac{|m_1|n_2}{n_1 n_2} \odot x + \pm\frac{|m_2|n_1}{n_1 n_2} \odot x, \text{ by Lemma 5.35}$$
$$= \pm(|m_1|n_2) \odot \left(\frac{1}{n_1 n_2} \odot x\right) + \pm(|m_2|n_1) \odot \left(\frac{1}{n_1 n_2} \odot x\right),$$

by definition,

$$= \pm(|m_1|n_2) \odot y + \pm(|m_2|n_1) \odot y,$$

where $y = (1/(n_1 n_2)) \odot x$. By the integral case,

$$q \odot x + r \odot x = (\pm|m_1|n_2 + \pm|m_2|n_1) \odot y$$

$$= \frac{\pm|m_1|n_2 + \pm|m_2|n_1}{n_1 n_2} \odot x, \text{ by definition of } y$$

$$= (q + r) \odot x.$$

ii. The general structure of the proof is the same as that of part i. We begin with the case $q = m \in \mathbb{N}, r = n \in \mathbb{N}$ and observe

$$q \odot (r \odot x) = m \odot (n \odot x) = \underbrace{\underbrace{x + \ldots + x}_{n} + \ldots + \underbrace{x + \ldots + x}_{n}}_{m}$$

$$= \underbrace{x + \ldots + x}_{mn} = (mn) \odot x = (qr) \odot x.$$

For $q, r \in \mathbb{Q}^{\geq 0}$, i.e. $q = m_1/n_1, r = m_2/n_2$ with $m_1, m_2 \in \mathbb{N}$, $n_1, n_2 \in \mathbb{N}^+$, we have

$$(qr) \odot x = \left(\frac{m_1}{n_1} \cdot \frac{m_2}{n_2} \right) \odot x = \frac{m_1 m_2}{n_1 n_2} \odot x = (m_1 m_2) \odot \left(\frac{1}{n_1 n_2} \odot x \right). \quad (55)$$

But also

$$q \odot (r \odot x) = \frac{m_1}{n_1} \odot \left(\frac{m_2}{n_2} \odot x \right) = \frac{m_1}{n_1} \odot \left(m_2 \odot \left(\frac{1}{n_2} \odot x \right) \right)$$

$$= \frac{m_1}{n_1} \odot \left(\frac{1}{n_2} \odot x + \ldots + \frac{1}{n_2} \odot x \right)$$

$$= m_1 \odot \left(\frac{1}{n_1} \odot \left(\frac{1}{n_2} \odot x + \ldots + \frac{1}{n_2} \odot x \right) \right), \text{ by definition}$$

$$= m_1 \odot \left(\frac{1}{n_1} \odot \left(\frac{1}{n_2} \odot x \right) + \ldots + \frac{1}{n_1} \odot \left(\frac{1}{n_2} \odot x \right) \right),$$

by Lemma 5.38,

$$= m_1 \odot \left(m_2 \odot \left(\frac{1}{n_1} \odot \left(\frac{1}{n_2} \odot x \right) \right) \right)$$

$$= (m_1 m_2) \odot \left(\frac{1}{n_1} \odot \left(\frac{1}{n_2} \odot x \right) \right), \quad (56)$$

by the nonnegative integral case. By (55) and (56), to prove $(qr) \odot x = q \odot (r \odot x)$, it suffices to show

$$\frac{1}{n_1 n_2} \odot x = \frac{1}{n_1} \odot \left(\frac{1}{n_2} \odot x \right).$$

But this is easy:

$$(n_1 n_2) \odot \left(\frac{1}{n_1} \odot \left(\frac{1}{n_2} \odot x \right) \right) = (n_2 n_1) \odot \left(\frac{1}{n_1} \odot \left(\frac{1}{n_2} \odot x \right) \right)$$

$$= n_2 \odot \left(n_1 \odot \left(\frac{1}{n_1} \odot \left(\frac{1}{n_2} \odot x \right) \right) \right),$$

by the integral case,

$$= n_2 \odot \left(\frac{n_1}{n_1} \odot \left(\frac{1}{n_2} \odot x \right) \right), \text{ by definition}$$

$$= n_2 \odot \left(1 \odot \left(\frac{1}{n_2} \odot x \right) \right), \text{ by Corollary 5.36}$$

$$= n_2 \odot \left(\frac{1}{n_2} \odot x \right), \text{ since } 1 \odot y = y$$

$$= \frac{n_2}{n_2} \odot x = 1 \odot x = x.$$

Now we have only to consider how a minus sign affects things. A little algebra shows that for any $q, r \in \mathbb{Q}$,

$$(q(-r)) \odot x = -((qr) \odot x)$$
$$q \odot ((-r) \odot x) = -(q \odot (r \odot x))$$
$$((-q)r) \odot x = -((qr) \odot x)$$
$$(-q) \odot (r \odot x) = -(q \odot (r \odot x)).$$

To establish the first of these, for example, note that

$$(q(-r)) \odot x + (qr) \odot x = (q(-r) + q(r)) \odot x, \text{ by part i}$$
$$= 0 \odot x = 0,$$

whence $(q(-r)) \odot x = -((qr) \odot x)$. The other proofs are similar.

Thus for any $q, r \in \mathbb{Q}$ we have

$$(qr) \odot x = \pm((\, |q| \cdot |r|) \odot x)$$
$$= \pm(\, |q| \odot (\, |r| \odot x)), \text{ by the nonnegative case}$$

and

$$q \odot (r \odot x) = \pm(\, |q| \odot (\, |r| \odot x)),$$

and we have $(qr) \odot x = q \odot (r \odot x)$ as in each case the $\pm$ sign is $+$ if 0 or 2 of q, r are negative and $-$ if exactly one of them is negative. $\square$

This Lemma establishes that $(\mathbb{R}, +, 0)$ is a vector space over $\mathbb{Q}$ when we use $\odot$ as the scalar multiplication. Of greater immediate interest, however, is the first part, by which it follows that for each $r \in \mathbb{R}$ the map

$$g_r(q) = q \odot r \tag{57}$$

is a homomorphism of $(\mathbb{Q}, +, 0)$ into $(\mathbb{R}, +, 0)$. It is, in fact, order aware:

5.42 Lemma. *Let* $(G, +, <, 0)$ *be an ordered divisible abelian group, and let* $r \in G$. *The homomorphism* g_r *defined by* (57) *is*

i.   *an order preserving isomorphism if* $0 < r$
ii. *an order reversing isomorphism if* $r < 0$
iii. *the constant map* $g_r(q) = 0$ *if* $r = 0$.

Proof. i. Let $r \in G$ be positive. Suppose $q_1, q_2 \in \mathbb{Q}$ with $q_1 < q_2$. We have

$$g_r(q_2) = g_r(q_1 + (q_2 - q_1)) = g_r(q_1) + g_r(q_2 - q_1)$$

and we see that to show $g_r(q_1) < g_r(q_2)$ it suffices to show $g_r(q) > 0$ for any positive q. But

$$g_r(q) = q \odot r = f_q(r),$$

where f_q is order preserving, whence

$$0 < r \;\Rightarrow\; 0 = f_q(0) < f_q(r) = g_r(q).$$

Assertion ii is proven similarly and iii is trivial.   □

The rest of the abstract treatment of multiplication of real numbers consists of showing
i.   every order aware homomorphism of $(\mathbb{Q}, +, <, 0)$ into $(\mathbb{R}, +, <, 0)$ is of this form;
ii. every order aware homomorphism $g : \mathbb{Q} \to \mathbb{R}$ extends uniquely to an order aware homomorphism $\bar{g} : \mathbb{R} \to \mathbb{R}$;
iii. the operation $x, y \mapsto x \otimes y$ defined by

$$x \otimes y = \bar{g}_y(x)$$

defines multiplication on $\mathbb{R}$.

The first of these tasks is easy:

5.43 Theorem. *Let* $(G, +, <, 0)$ *be an ordered divisible abelian group and let* $r \in G$. g_r *is the unique homomorphism* $f : \mathbb{Q} \to G$ *such that* $f(1) = r$.

Proof. First note that
$$g_r(1) = 1 \odot r = r,$$
by definition.

Assume $f : \mathbb{Q} \to G$ is a homomorphism mapping 1 to r.
Let $m \in \mathbb{N}, n \in \mathbb{N}^+$ and observe

$$f\left(\frac{m}{n}\right) = f\left(\frac{1}{n} + \ldots + \frac{1}{n}\right) = f\left(\frac{1}{n}\right) + \ldots + f\left(\frac{1}{n}\right).$$

If $m = n$, this yields

$$f(1) = n \odot f\left(\frac{1}{n}\right),$$

i.e.

$$f\left(\frac{1}{n}\right) = \frac{1}{n} \odot f(1) = \frac{1}{n} \odot r = g_r\left(\frac{1}{n}\right).$$

For other m, it yields

$$f\left(\frac{m}{n}\right) = m \odot f\left(\frac{1}{n}\right) = m \odot \left(\frac{1}{n} \odot f(1)\right) = \frac{m}{n} \odot f(1) = \frac{m}{n} \odot r = g_r\left(\frac{m}{n}\right).$$

Finally, if $q = \frac{-m}{n}$, the proof of Lemma 1.19 shows that group homomorphisms preserve $-$:

$$f\left(\frac{-m}{n}\right) = -f\left(\frac{m}{n}\right) = -\left(\frac{m}{n} \odot r\right) = \left(\frac{-m}{n}\right) \odot r = g_r\left(\frac{-m}{n}\right). \qquad \square$$

I remark that we did not have to call on the order awareness of f or g_r in the proof of the Theorem. This property is needed in performing the second task, where it deputises for continuity. This is also where we use the Archimedean Property and the order completeness of $\mathbb{R}$.

The second task requires a couple of small lemmas. First:

5.44 Lemma. *Let $(G, +, <, 0)$ be an Archimedean ordered divisible abelian group. Let $f : \mathbb{Q} \to G$ be an order preserving or order reversing homomorphism of $(\mathbb{Q}, +, 0)$ into $(G, +, 0)$. Then: $f(\mathbb{Q}) = \{f(q) | q \in \mathbb{Q}\}$ is dense in G.*

Proof. The cases of an order preserving and an order reversing homomorphism are symmetric, whence we need only prove, say, the former case. Thus, assume f is order preserving and $f(1) = a > 0$. Let $b, c \in G$ be given with $b < c$. We wish to find $q \in \mathbb{Q}$ such that $b < f(q) < c$.

Assume first that $0 \le b < c$. By the Archimedean order of G, there is some $n \in \mathbb{N}^+$ such that $n \odot (c - b) > a > 0$. By divisibility, $\frac{1}{n} \odot a$ exists and we have

$$c - b > \frac{1}{n} \odot a > 0,$$

i.e.

$$b + \frac{1}{n} \odot a < c. \tag{58}$$

One can also choose $m \in \mathbb{N}$ such that

$$\frac{m}{n} \odot a = m \odot \left(\frac{1}{n} \odot a\right) > b.$$

Choose m minimum with this property and observe

$$b < \frac{m}{n} \odot a = \frac{m-1}{n} \odot a + \frac{1}{n} \odot a \le b + \frac{1}{n} \odot a,$$

by choice of m. Thus

$$b < \frac{m}{n} \odot a < c,$$

by (58). But

$$\frac{m}{n} \odot a = \frac{m}{n} \odot f(1) = f\left(\frac{m}{n}\right),$$

by Lemma 5.43.

If $b < c < 0$, find $\frac{m}{n}$ such that

$$-c < f\left(\frac{m}{n}\right) < -b,$$

and observe

$$b < f\left(\frac{-m}{n}\right) < c.$$

And if $b < 0 < c$, then $b < f(0) < c$.   □

The second lemma is a simple property of least upper bounds and is very simply stated if we introduce two pieces of notation. First, in any linear ordering, we write $\mathrm{lub}(X)$ for the least upper bound and $\mathrm{glb}(X)$ for the greatest lower bound of X when it exists.[68] And second, if X, Y are two subsets of an abelian group, we write

$$X + Y = \{x + y \mid x \in X \ \& \ y \in Y\}.$$

5.45 Lemma. *Let X, Y be nonempty bounded subsets of an order complete ordered divisible abelian group. Then*

$$\mathrm{lub}(X + Y) = \mathrm{lub}(X) + \mathrm{lub}(Y)$$
$$\mathrm{glb}(X + Y) = \mathrm{glb}(X) + \mathrm{glb}(Y).$$

Proof. I prove the first of these assertions and leave the second to the reader. Let $a = \mathrm{lub}(X), b = \mathrm{lub}(Y)$. Note that for any $z \in X + Y$, we have, for some $x \in X, y \in Y$,

$$z = x + y \leq a + b,$$

whence $a + b$ is an upper bound on $X + Y$.

Let, by way of contradiction, $c < a + b$ be a smaller upper bound on $X + Y$ and let $\epsilon = a + b - c > 0$. Because a, b are least upper bounds on X, Y, respectively, there are $x_0 \in X, y_0 \in Y$ such that

$$a - \frac{\epsilon}{2} < x_0 \quad \text{and} \quad b - \frac{\epsilon}{2} < y_0.$$

But then

$$c = a + b - \epsilon = a - \frac{\epsilon}{2} + b - \frac{\epsilon}{2} < x_0 + y_0$$

and c was not an upper bound after all.   □

We are now ready to carry out the second task:

---

[68] Some authors prefer to call the least upper bound the *supremum* of X, and thus write $\sup(X)$ for $\mathrm{lub}(X)$. Similarly, they prefer to call the greatest lower bound the *infimum* and write $\inf(X)$ for $\mathrm{glb}(X)$.

5.46 Theorem. *Let $(G, +, <, 0)$ be an order complete Archimedean ordered divisible abelian group. Let $f : \mathbb{Q} \to G$ be an order aware homomorphism. There is a unique order aware homomorphism $\overline{f} : \mathbb{R} \to G$ extending f: for all $q \in \mathbb{Q}$, $\overline{f}(q) = f(q)$.*

Proof. I consider the case of an order preserving homomorphism and leave the rest to the reader.

Suppose first that some order preserving extension $\overline{f}$ of f exists. Let $x \in \mathbb{R}$ and define

$$a = \mathrm{lub}(\{f(q)|q \in \mathbb{Q} \ \& \ q \le x\})$$
$$b = \mathrm{glb}(\{f(q)|q \in \mathbb{Q} \ \& \ x < q\}).$$

By order preservation we have

$$a \le \overline{f}(x) \le b.$$

But we cannot have $a < b$ because $f(\mathbb{Q})$ is dense in G and no $f(q)$ lies strictly between a and b. Thus $a = b = \overline{f}(x)$ and $\overline{f}$ is unique.

As for existence, we have just seen that we must define

$$\overline{f}(x) = \mathrm{lub}(\{f(q)|q \in \mathbb{Q} \ \& \ q \le x\}).$$

This is a legitimate definition of a function. We must show that i. it extends f, ii. it is order preserving, and iii. it is a homomorphism.

i. Let $q_0 \in \mathbb{Q}$. Then

$$\{f(q)|q \in \mathbb{Q} \ \& \ f(q) \le f(q_0)\} = \{f(q)|q \in \mathbb{Q} \ \& \ q \le q_0\},$$

by order preservation, whence

$$\overline{f}(q_0) = \mathrm{lub}(\{f(q)|q \in \mathbb{Q} \ \& \ q \le q_0\})$$
$$= \mathrm{lub}(\{f(q)|q \in \mathbb{Q} \ \& \ f(q) \le f(q_0)\}) = f(q_0).$$

ii. If $x < y$,

$$\{f(q)|q \in \mathbb{Q} \ \& \ q \le x\} \subseteq \{f(q)|q \in \mathbb{Q} \ \& \ q \le y\},$$

whence

$$\overline{f}(x) = \mathrm{lub}(\{f(q)|q \in \mathbb{Q} \ \& \ q \le x\})$$
$$\le \mathrm{lub}(\{f(q)|q \in \mathbb{Q} \ \& \ q \le y\}) = \overline{f}(y).$$

Thus $\overline{f}(x) \le \overline{f}(y)$. To see that the inequality is strict, we appeal to the density of $\mathbb{Q}$ in $\mathbb{R}$: There are $q_0, q_1 \in \mathbb{Q}$ with $x < q_0 < q_1 < y$. We have just shown that

$$\overline{f}(x) \le \overline{f}(q_0) \le \overline{f}(q_1) \le \overline{f}(y).$$

But $\overline{f}(q_i) = f(q_i)$ and f is order preserving, whence the central inequality is strict:

$$\overline{f}(x) \leq \overline{f}(q_0) < \overline{f}(q_1) \leq \overline{f}(y),$$

and we have $\overline{f}(x) < \overline{f}(y)$.

iii. Let $x, y \in \mathbb{R}$ and consider

$$\begin{aligned}
\overline{f}(x) + \overline{f}(y) &= \text{lub}(\{f(q_0)|q_0 \leq x\}) + \text{lub}(\{f(q_1)|q_1 \leq y\}) \\
&= \text{lub}(\{f(q_0) + f(q_1)|q_0 \leq x \ \& \ q_1 \leq y\}), \text{ by Lemma 5.45} \\
&= \text{lub}(\{f(q_0 + q_1)|q_0 \leq x \ \& \ q_1 \leq y\}), \quad (59)
\end{aligned}$$

where q_0, q_1 range over $\mathbb{Q}$ and we use the fact that f is a homomorphism in concluding (59). Now

$$x = \text{lub}(\{q_0|q_0 \leq x\}) \quad \text{and} \quad y = \text{lub}(\{q_1|q_1 \leq y\})$$

whence Lemma 5.45 yields

$$\begin{aligned}
x + y &= \text{lub}(\{q_0|q_0 \leq x\}) + \text{lub}(\{q_1|q_1 \leq y\}) \\
&= \text{lub}(\{q_0 + q_1|q_0 \leq x \ \& \ q_1 \leq y\}).
\end{aligned}$$

Thus, for any $q \in \mathbb{Q}$,

$$q \leq x + y \text{ iff } q \leq q_0 + q_1 \text{ for some } q_0 \leq x \ \& \ q_1 \leq y. \quad (60)$$

But, if $q \leq q_0 + q_1$ then $q = q_0' + q_1'$, where

$$q_i' = q_i - \frac{d}{2}, \qquad d = q_0 + q_1 - q \geq 0.$$

This means the right-hand inequality of (60) can be replaced by an equality and

$$\{f(q_0 + q_1)|q_0 \leq x \ \& \ q_1 \leq y\} = \{f(q)|q \leq x + y\}$$

and (59) yields

$$\overline{f}(x) + \overline{f}(y) = \text{lub}(\{f(q)|q \leq x + y\}) = \overline{f}(x + y). \qquad \square$$

5.47 Corollary. *Let $(G, +, <, 0)$ be an order complete Archimedean ordered divisible abelian group and let $r \in G$. $\overline{g}_r$ is the unique order aware homomorphism $f : \mathbb{R} \to G$ for which $f(1) = r$.*

Our final task is to define multiplication on $\mathbb{R}$, and prove that it extends $\odot$ (and hence multiplication on $\mathbb{Q}$) and that it satisfies all the field axioms.

5.48 Definition. *Let $x, y \in \mathbb{R}$. We define*

$$x \otimes y = \overline{g}_y(x).$$

5.49 Lemma. *$\otimes$ extends $\odot$: For all $q \in \mathbb{Q}$ and $y \in \mathbb{R}, q \otimes y = q \odot y$.*

Proof. By definition

$$q \otimes y = \overline{g}_y(q) = g_y(q) = q \odot y. \qquad \square$$

Our sole remaining task is to prove the following theorem.

5.50 Theorem. $(\mathbb{R}, +, \otimes, 0, 1)$ *is a field.*

Proof. In alphabetical order, we must prove Associativity, Commutativity, Distributivity, and the Existence of an Inverse. The most difficult of these is Commutativity, which we save for last.

Associativity is probably the easiest to prove. By Corollary 5.47, for any $w \in \mathbb{R}$, $\overline{g}_w$ is the unique order aware homomorphism that maps 1 to w. Thus

$$\overline{g}_{x \otimes y}(1) = x \otimes y.$$

But

$$x \otimes y = \overline{g}_y(x) = \overline{g}_y(\overline{g}_x(1)),$$

whence the composition $\overline{g}_y \circ \overline{g}_x$ must equal $\overline{g}_{x \otimes y}$. By the associativity of the composition of functions, we thus have

$$\overline{g}_{x \otimes (y \otimes z)} = \overline{g}_{y \otimes z} \circ \overline{g}_x = (\overline{g}_z \circ \overline{g}_y) \circ \overline{g}_x$$
$$= \overline{g}_z \circ (\overline{g}_y \circ \overline{g}_x) = \overline{g}_z \circ \overline{g}_{x \otimes y}$$
$$= \overline{g}_{(x \otimes y) \otimes z},$$

and,

$$x \otimes (y \otimes z) = \overline{g}_{x \otimes (y \otimes z)}(1) = \overline{g}_{(x \otimes y) \otimes z}(1) = (x \otimes y) \otimes z.$$

The existence of an inverse is handled similarly. If $x \neq 0$, then $\overline{g}_x$ is an isomorphism of $(\mathbb{R}, +, 0)$ onto itself. By Corollary 5.47, its inverse is of the form $\overline{g}_y$ for some $y \in \mathbb{R}$. But

$$x \otimes y = \overline{g}_y(x) = \overline{g}_y(\overline{g}_x(1)) = 1.$$

There are left- and right-distributive laws, which reduce to one another by commutativity. Without appeal to commutativity, we can only easily prove right-distributivity:

$$(x + y) \otimes z = \overline{g}_z(x + y)$$
$$= \overline{g}_z(x) + \overline{g}_z(y), \text{ since } \overline{g}_z \text{ is a homomorphism,}$$
$$= (x \otimes z) + (y \otimes z).$$

To establish commutativity, we first assume both x, y to be positive and observe that

$$x \otimes y = \overline{g}_y(x) = \text{lub}\{g_y(q_1) \mid q_1 \leq x \ \& \ q_1 \in \mathbb{Q}\}$$
$$= \text{lub}\{q_1 \odot y \mid q_1 \leq x \ \& \ q_1 \in \mathbb{Q}\}$$
$$= \text{lub}\{q_1 \odot y \mid 0 < q_1 \leq x \ \& \ q_1 \in \mathbb{Q}\}, \tag{61}$$

since $q_1 \odot y \leq 0$ for $q_1 \leq 0$ by Lemma 5.40.ii. Now, for fixed $q_1 > 0$, the map $y \mapsto q_1 \odot y$ is the unique order preserving homomorphism extending the map $q \mapsto q_1 q$, whence by the proof of Theorem 5.46,

$$q_1 \otimes y = \text{lub}\{q_1 q_2 \mid q_2 \leq y \ \& \ q_2 \in \mathbb{Q}\}$$

$$= \text{lub}\{q_1 q_2 \mid 0 < q_2 \leq y \ \& \ q_2 \in \mathbb{Q}\}. \tag{62}$$

Combining (61) and (62),

$$x \otimes y = \text{lub}\{\text{lub}\{q_1 q_2 \mid 0 < q_2 \leq y \ \& \ q_2 \in \mathbb{Q}\} \mid 0 < q_1 \leq x \ \& \ q_1 \in \mathbb{Q}\}.$$

I claim that

$$x \otimes y = \text{lub}\{q_1 q_2 \mid 0 < q_1 \leq x \ \& \ 0 < q_2 \leq y \ \& \ q_1, q_2 \in \mathbb{Q}\}. \tag{63}$$

Interchanging x and y will then yield

$$y \otimes x = \text{lub}\{q_2 q_1 \mid 0 < q_1 \leq x \ \& \ 0 < q_2 \leq y \ \& \ q_1, q_2 \in \mathbb{Q}\},$$

and commutativity in $\mathbb{Q}$ will yield $x \otimes y = y \otimes x$, thereby yielding the Theorem in the case when x, y are both positive.

To prove (63), first note that $x \otimes y$ is an upper bound on

$$\{q_2 q_1 \mid 0 < q_1 \leq x \ \& \ 0 < q_2 \leq y \ \& \ q_1, q_2 \in \mathbb{Q}\}.$$

For, if $0 < q_1 \leq x, 0 < q_2 \leq y$, then

$$q_1 q_2 \leq \text{lub}\{q_1 q_2' \mid 0 < q_2' \leq y \ \& \ q_2' \in \mathbb{Q}\}$$
$$\leq \text{lub}\{\text{lub}\{q_1' q_2' \mid 0 < q_2' \leq y \ \& \ q_2' \in \mathbb{Q}\} \mid 0 < q_1' \leq x \ \& \ q_1' \in \mathbb{Q}\}.$$

If $B < x \otimes y$ is another upper bound, there is some q_1 in the appropriate interval such that

$$B < \text{lub}\{q_1 q_2 \mid 0 < q_2 \leq y \ \& \ q_2 \in \mathbb{Q}\}.$$

But then there is some q_2 such that $B < q_1 q_2$ and B was not an upper bound on the set in question.

The cases in which one of x and y is negative are handled by the usual tricks:

$$(-x) \otimes y = -(x \otimes y)$$

since $x \otimes y = \overline{g}_y(x)$ is a homomorphism for fixed y.

$$x \otimes (-y) = -(x \otimes y)$$

since $x \otimes y = \overline{g}_y(x)$, $x \otimes (-y) = \overline{g}_{-y}(x)$, and $\overline{g}_{-y}$ is the unique order aware homomorphism mapping 1 to $-y$. But $-\overline{g}_y$ is also an order aware homomorphsim mapping 1 to $-y = -\overline{g}_y(1)$. Similarly,

$$x \otimes 0 = 0, \quad 0 \otimes x = 0. \qquad \square$$

Digression. I am quite pleased with the above treatment up to and including the proof of Lemma 5.44. There is some excess of detail, but the progression of results is natural and reproducible, and the individual proofs follow obvious lines. After that point, a better approach might have been to have introduced the formal definition of continuity, proven that a homomorphism from $\mathbb{Q}$ or

$\mathbb{R}$ into $\mathbb{R}$ is order aware iff it is continuous, and then proven directly that the continuous g_r had a unique continuous homormorphic extension $\overline{g}_r$. In the proof of Theorem 5.46, the existence of the limit would still be proven by appeal to lub's and glb's. One could then proceed in the proof of Theorem 5.50 as follows: For rational q,

$$g_{y+z}(q) = q \odot (y + z) = q \odot y + q \odot z = g_y(q) + g_z(q)$$

Now $g_y + g_z$ is the sum of two continuous homomorphisms, whence it too is a continuous homomorphism and, by Theorem 5.46 has a unique continuous homomorphic extension, namely $\overline{g}_y + \overline{g}_z$. But $\overline{g}_{y+z}$ is also such, whence for all x,

$$x \otimes (y + z) = \overline{g}_{y+z}(x) = \overline{g}_y(x) + \overline{g}_z(x) = (x \otimes y) + (x \otimes z).$$

We then get commutativity by noting that this shows

$$h_x(y) = x \otimes y$$

to be a homomorphism and considering the continuous homormorphism, for each fixed x,

$$f(y) = \overline{g}_x(y) - h_x(y).$$

By continuity, f is order aware. But $f(x) = \overline{g}_x(x) - h_x(x) = x \otimes x - x \otimes x = 0$. Thus f is the constant homomorphism

$$f(y) = y \otimes x - x \otimes y = 0,$$

and $x \otimes y = y \otimes x$. I refer the reader to Exercises 9.15 and 9.16 at the end of this chapter for some details.

5.5 Multiplication of Cuts; Appeal to Continuity

[The material in this subsection is more specialised and one may wish to skip it on a first reading.]

As mentioned earlier, Dedekind did not give a detailed treatment of multiplication. It is probably worth quoting him at some length:

> Just as addition is defined, so can the other operations of the so-called elementary arithmetic be defined, viz., the formation of differences, products, quotients, powers, roots, logarithms, and in this way we arrive at real proofs of theorems (as, e.g. $\sqrt{2}\sqrt{3} = \sqrt{6}$), which to the best of my knowledge have never been established before.[69] The excessive length that is to be feared in the definitions of the more complicated operations is partly inherent in the nature of the subject but can for the

---

[69] It goes without saying that Dedekind would be challenged on this claim. He defended it in a letter to Rudolf Lipschitz of 10 June 1876 saying that $\sqrt{2}\sqrt{3} = \sqrt{6}$ could not have been proven previously because the very definitions of the terms involved had been lacking. Cf. his *Gesammelte mathematische Werke*, vol. 3, pp. 471 - 474.

most part be avoided. Very useful in this connection is the notion of an *interval*, i.e., a system A of rational numbers possessing the following characteristic property: if a and a' are numbers of the system A, then are all rational numbers lying between a and a' contained in A. The system R of all rational numbers, and also the two classes of any cut are intervals. If there exist a rational number a_1 which is less and a rational number a_2 which is greater than every number of the interval A, then A is called a finite interval; there then exist infinitely many numbers in the same condition as a_1 and infinitely many in the same condition as a_2; the whole domain R breaks up into three parts A_1, A, A_2 and there enter two perfectly definite rational or irrational numbers α_1, α_2 which may be called respectively the lower and upper (or the less and greater) *limits* [i.e., endpoints] of the interval; the lower limit α_1 is determined by the cut for which the system A_1 forms the first class and the upper α_2 by the cut for which the system A_2 forms the second class. Of every rational or irrational number α lying between α_1 and α_2 it may be said that it lies *within* the interval A. If all numbers of an interval A are also numbers of an interval B, then A is called a portion [i.e., subset] of B.

Still lengthier considerations seem to loom up when we attempt to adapt the numerous theorems of the arithmetic of rational numbers (as, e.g., the theorem $(a + b)c = ac + bc$) to any real numbers. This, however, is not the case. It is easy to see that it all reduces to showing that the arithmetic operations possess a certain continuity. What I mean by this statement may be expressed in the form of a general theorem:

"If the number λ is the result of an operation performed on the numbers $\alpha, \beta, \gamma, \ldots$ and λ lies within the interval L, then intervals $A, B, C, \ldots$ can be taken within which lie the numbers $\alpha, \beta, \gamma, \ldots$ such that the result of the same operation in which the numbers $\alpha, \beta, \gamma, \ldots$ are replaced by arbitrary numbers of the intervals $A, B, C, \ldots$ is always a number lying within the interval L." The forbidding clumsiness, however, which marks the statement of such a theorem convinces us that something must be brought in as an aid to expression; this is, in fact, attained in the most satisfactory way by introducing the ideas of *variable magnitudes, functions, limiting values*, and it would be best to base the definitions of even the simplest arithmetic operations upon these ideas, a matter which, however, cannot be carried further here.[70]

I consider these remarks cryptic. If one doesn't already know what Dedekind is talking about, they don't really point the way.

Today, one would say that the arithmetic operations on $\mathbb{Q}$ are to be extended to $\mathbb{R}$ "by continuity" and every professional mathematician and graduate student, as well as the more advanced undergraduate would know how to proceed and recognise such a procedure in Dedekind's remarks. The intervals he would

---

[70] Dedekind, *Essays...*, *op. cit.*, pp. 22 - 24.

find "useful" are ϵ- and δ-intervals; and the condition that all numbers of some intervals $A, B, C \ldots$ containing points $\alpha, \beta, \gamma, \ldots$ that map to λ must themselve map into the interval L is a variant of continuity: Think of L as the interval $(\lambda - \epsilon, \lambda + \epsilon)$ and $A, B, C, \ldots$ as $(\alpha - \delta_0, \alpha + \delta_0), (\beta - \delta_1, \beta + \delta_1), (\gamma - \delta_2, \gamma + \delta_2), \ldots$

The idea, thus, is to extend any continuous function on $\mathbb{Q}$ or $\mathbb{Q}^2$ (in particular, $+, \cdot$) or more generally on $\mathbb{Q}^n$ by taking limits. This doesn't work for all continuous functions, but it does work for *uniformly* continuous functions. Let us first recall the definition of a continuous function.

5.51 Definition. *Let $n > 1$, let X be one of $\mathbb{Q}, \mathbb{R}$, and let $I \subseteq X^n$. A function $f : I \to \mathbb{R}$ is* continuous *at a point $\langle x_0, \ldots, x_{n-1} \rangle \in I$ if, for any $\epsilon \in \mathbb{Q}^+$ there is a $\delta \in \mathbb{Q}^+$ such that for all $\langle y_0, \ldots, y_{n-1} \rangle \in I$,*

$$|x_0 - y_0| < \delta \ \& \ \ldots \ \& \ |x_{n-1} - y_{n-1}| < \delta$$
$$\Rightarrow |f(x_0, \ldots, x_{n-1}) - f(y_0, \ldots, y_{n-1})| < \epsilon.$$

f is continuous *on I if f is continuous at all $\langle x_0, \ldots, x_{n-1} \rangle \in I$.*

[A small remark: In the definitions of limit, continuity, etc., it makes no difference if the ϵ's and δ's are restricted to being rational or not. Insofar as we will be operating with them arithmetically, and we supposedly have not yet defined multiplication for real numbers, it is convenient to restrict ourselves to rational numbers. Hence, even when I forget to state it explicitly, the letters ϵ and δ are assumed throughout the rest of this subsection to be positive rational numbers.]

Not every continuous function on $\mathbb{Q}^n$ has a continuous extension to $\mathbb{R}^n$. Indeed, there are easy counterexamples for $n = 1$:

5.52 Counterexample. The function

$$f(x) = \frac{1}{x^2 - 2}$$

is continuous for all rational values of x, but there is no continuous extension of f to all of $\mathbb{R}$: no value of f at $x = \sqrt{2}$ will make the function continuous.

To be able to extend a function f continuously, we need to know that the limit of a sequence $f(\mathbf{x}_0), f(\mathbf{x}_1), f(\mathbf{x}_2), \ldots$ exists whenever $\mathbf{x}_0, \mathbf{x}_1, \mathbf{x}_2, \ldots$ converges to a point of I. It turns out that this is guaranteed by a strong form of continuity:

5.53 Definition. *Let $n > 1$, let X be one of $\mathbb{Q}, \mathbb{R}$, and let $I \subseteq X^n$. A function $f : I \to \mathbb{R}$ is* uniformly continuous *on I if, for any $\epsilon \in \mathbb{Q}^+$ there is a $\delta \in \mathbb{Q}^+$ such that for all $\langle x_0, \ldots, x_{n-1} \rangle, \langle y_0, \ldots, y_{n-1} \rangle \in I$,*

$$|x_0 - y_0| < \delta \ \& \ \ldots \ \& \ |x_{n-1} - y_{n-1}| < \delta$$
$$\Rightarrow |f(x_0, \ldots, x_{n-1}) - f(y_0, \ldots, y_{n-1})| < \epsilon.$$

Uniform continuity is an important concept that makes its first appearance in an Advanced Calculus or Real Analysis course in the discussion of the definite integral: It is the uniform continuity of a continuous function f on a closed bounded interval $[a, b]$ that is appealed to in the proof of the existence of the definite integral of f on $[a, b]$. The notion, however, is not always stressed in the introductory Calculus course. In a small, non-representative sampling of Calculus textbooks, I find the concept mentioned briefly in George B. Thomas, Jr., *Calculus*[71], the book I learned my Calculus from. Thomas mentions that a function that is continuous on a closed, bounded interval is uniformly continuous there. He does not mention later that this result is needed in proving the existence of the definite integral, which is left unproven. The first edition of Thomas was pre-Sputnik, having appeared in 1953. After the Soviet launch of the first artificial satellite in the late 1950's, greater emphasis was given on science and mathematics in American schools and this manifested itself in the production of more rigorous textbooks. Two examples of this are Albert G. Fadell, *Calculus with Analytic Geometry*[72], and R.E. Johnson and F.L. Kiokemeister, *Calculus with Analytic Geometry*[73]. Each of these books, reflecting the rigour of the period, defines uniform continuity, proves the uniform continuity of a continuous function on a closed, bounded interval, and applies it in a rigorous proof of the integrability of continuous functions. The days of that level of rigour in an introductory Calculus course are probably gone and I assume James Stewart, *Single Variable Calculus; Concepts & Contexts*[74] is representative. In this book the notion of uniform continuity is not even mentioned and the existence of the integral of a continuous function is simply stated as a fact without any hint of a proof.

It follows that the intended reader may well never have encountered uniform continuity before and I should digress to say a few words about it.

Uniform continuity on I certainly implies continuity at all points in I, but it is more than that. Continuity at some n-tuple $\mathbf{x} = \langle x_0, \ldots, x_{n-1} \rangle$ means that for any $\epsilon \in \mathbb{Q}^+$ there is a $\delta \in \mathbb{Q}^+$ satisfying the familiar condition. But for different values of $\mathbf{x}$, one may need different δ's. Uniform continuity says that a single value of δ works uniformly for all values of $\mathbf{x}$ in I.

5.54 Example. Addition is uniformly continuous in $\mathbb{Q}^2$. To see this, let ϵ be any positive rational number and $\delta = \epsilon/2$. Observe, for $|x_0 - y_0| < \delta$ and $|x_1 - y_1| < \delta$,

$$\left| (x_0 + x_1) - (y_0 + y_1) \right| \leq |x_0 - y_0| + |x_1 - y_1| < \delta + \delta = \frac{\epsilon}{2} + \frac{\epsilon}{2} = \epsilon.$$

5.55 Example. Multiplication is continuous on $\mathbb{Q}^2$. It is not uniformly continuous on all of $\mathbb{Q}^2$, but it is uniformly continuous on

$$I = \{ \langle x_0, x_1 \rangle \in \mathbb{Q}^2 \mid |x_0| \leq M \ \& \ |x_1| \leq M \} \text{ for any } M \in \mathbb{Q}^+.$$

---

[71] 2nd ed., Addison-Wesley Publishing Company, Reading (Mass), 1961.
[72] D. van Nostrand Company, Inc., Princeton, 1964.
[73] 3rd ed., Allyn and Bacon, Inc., Boston, 1964.
[74] 3rd ed., Thomson, Brooks/Cole, Belmont (Cal), 2005.

Given ϵ, let $\delta = \epsilon/(2M)$ and observe that for $\langle x_0, x_1 \rangle, \langle y_0, y_1 \rangle \in I$, if we have $|x_0 - y_0| < \delta$ and $|x_1 - y_1| < \delta$, then

$$
\begin{aligned}
|x_0 x_1 - y_0 y_1| &= |x_0 x_1 - x_0 y_1 + x_0 y_1 - y_0 y_1| \\
&\leq |x_0| \cdot |x_1 - y_1| + |x_0 - y_0| \cdot |y_1| \\
&< M \cdot \delta + \delta \cdot M = M\frac{\epsilon}{2M} + M\frac{\epsilon}{2M} = \epsilon.
\end{aligned}
$$

Thus multiplication is uniformly continuous on I. That it is not uniformly continuous on all of $\mathbb{Q}^2$ can be seen as follows. Let $\epsilon, \delta \in \mathbb{Q}^+$ be given, choose $\langle x_0, x_1 \rangle = \langle M, M \rangle, \langle y_0, y_1 \rangle = \langle M, M + \delta/2 \rangle$, and observe

$$
|x_0 x_1 - y_0 y_1| = \left| M^2 - M\left(M + \frac{\delta}{2} \right) \right| = \left| -\frac{M\delta}{2} \right| = \frac{M\delta}{2} > \epsilon,
$$

if we choose $M > 2\epsilon/\delta$.

I leave to the reader the verification that the function f of Counterexample 5.52 is continuous on all of $\mathbb{Q}$ but not uniformly so on any interval

$$
[a, b]_Q = \{ q \in \mathbb{Q} \mid a \leq q \leq b \},
$$

where $a < \sqrt{2} < b$, i.e., $a^2 < 2 < b^2$, b positive.

The modern definition of continuity is generally credited to Bolzano (1817) and Cauchy (1821), as we will discuss in a bit more detail in the next section. On the surface, their definitions appear to define continuity for a function on a closed interval $[a, b]$ to mean continuity at all points of the interval. However, their intervals contained additional points at infinitesimal distances from our modern reals and we can now recognise that their definitions of continuity are equivalent to our definitions of uniform continuity, a fact not always realised in historical accounts.[75] It was Weierstrass (1860's) who banished infinitesimals from analysis and gave us our familiar ϵ-δ definition of continuity at a point. "Continuity" now meant continuity in our sense and Eduard Heine isolated and named the notion of uniform continuity in a paper of 1870 and two years later proved that any function defined and continuous on a closed, bounded interval is uniformly continuous there.[76]

---

[75] Cf., e.g., C.H. Edwards, Jr., *The Historical Development of the Calculus* (Springer-Verlag, New York, 1979), p. 309, where he says, "it may be noted that Cauchy occasionally stumbled conspicuously as in failing to distinguish between continuity and uniform continuity or between convergence and uniform convergence", or again on p. 319 where, in discussing Cauchy's proof of the existence of the integral, he says, "This is where he overlooks the need to prove that the continuous function f is *uniformly* continuous on $[x_0, X]$". Cauchy was, in fact, assuming something equivalent to uniform continuity when he used the word "continuity".

[76] The two papers appeared in Crelle's Journal, the *Journal für die reine und angewandte Mathematik*, in 1870 (volume 71, pp. 353 - 365) and 1872 (volume 74, pp. 172 - 188). Both papers have additional tangential historical interest. The first, "Ueber trigonometrische Reihen", was strongly objected to by the arch-constructivist Leopold Kronecker, who tried to get Heine to withdraw the paper; the second, "Die Elemente der Functionenlehre", is the one cited by Dedekind as that in which Heine presented his working out of Cantor's construction of the real numbers.

5.56 Theorem. *Let $a, b \in \mathbb{R}$ and f a continuous function on $[a, b]$. f is uniformly continuous on $[a, b]$.*

Proof. If f were not uniformly continuous, there would be some $\epsilon > 0$ such that for all $n \in \mathbb{N}$, one could find $x_n, y_n \in [a, b]$ with

$$|x_n - y_n| < \frac{1}{n} \quad \text{but} \quad |f(x_n) - f(y_n)| \geq \epsilon.$$

By the Bolzano-Weierstrass Theorem (Theorem I.5.20), there is a convergent subsequence $x_{i_0}, x_{i_1}, x_{i_2}, \ldots$ of $x_0, x_1, x_2, \ldots$ Let $x = \lim_{n \to \infty} x_{i_n}$ and note that $x \in [a, b]$. Now, by the continuity of f at x we can find $n_0 \in \mathbb{N}$ and $\delta > 0$ such that for all $n > n_0$,

$$|x_{i_n} - x| < \delta,$$

and for all $y \in [a, b]$,

$$|x - y| < \delta \Rightarrow |f(x) - f(y)| < \frac{\epsilon}{2}.$$

But we can also choose n_1 so large that for all $n > n_1$,

$$|x - y_{i_n}| \leq |x - x_{i_n}| + |x_{i_n} - y_{i_n}| < \frac{\delta}{2} + \frac{\delta}{2} = \delta,$$

whence

$$\left| f(x_{i_n}) - f(y_{i_n}) \right| \leq \left| f(x_{i_n}) - f(x) \right| + \left| f(x) - f(y_{i_n}) \right| < \frac{\epsilon}{2} + \frac{\epsilon}{2} = \epsilon,$$

a contradiction. □

5.57 Remark. The same argument establishes the existence of a maximum value of a continuous function on a closed, bounded interval $[a, b] \subseteq \mathbb{R}$. For, either the range is bounded and has a least upper bound B and there is a sequence $f(x_0), f(x_1), \ldots$ converging to B and thus $x_0, x_1, \ldots$ has a subsequence converging to a point x at which the maximum is attained, or there is no bound and one constructs a sequence $f(x_0), f(x_1), \ldots$ that grows without bound. But this latter cannot hold because some subsequence of $x_0, x_1, \ldots$ converges to a limit x and $f(x)$ is defined and finite. One similarly proves the Intermediate Value Theorem.

Our present concern is not in applying the uniform continuity of a function of a real variable, but to apply the uniform continuity of a function of a rational variable to construct a continuous extension.

5.58 Theorem. *Let $a_0 < b_0, a_1 < b_1, \ldots, a_{n-1} < b_{n-1}$ and let I be the intersection of $\mathbb{Q}^n$ with*

$$[a_0, b_0] \times \ldots \times [a_{n-1}, b_{n-1}]$$
$$= \{\langle x_0, \ldots, x_{n-1} \rangle \mid a_0 \leq x_0 \leq b_0 \, \& \, \ldots \, \& \, a_{n-1} \leq x_{n-1} \leq b_{n-1}\}.$$

Let $f : I \to \mathbb{R}$ be uniformly continuous. There is a unique continuous function $\overline{f} : [a_0, b_0] \times \ldots \times [a_{n-1}, b_{n-1}] \to \mathbb{R}$ such that

$$\overline{f}(x_0, \ldots, x_{n-1}) = f(x_0, \ldots, x_{n-1})$$

for all $\langle x_0, \ldots, x_{n-1} \rangle \in I$.

Proof sketch. The idea behind the proof is very simple. If $\mathbf{x} \in [a_0, b_0] \times \ldots \times [a_{n-1}, b_{n-1}]$ is not rational, one can find a sequence $\mathbf{x}_0, \mathbf{x}_1, \mathbf{x}_2, \ldots \in I$ which converges to $\mathbf{x}$. The sequence $f(\mathbf{x}_0), f(\mathbf{x}_1), f(\mathbf{x}_2), \ldots$ will be Cauchy convergent (by the uniform continuity of f) and hence will have a limit L. For $\overline{f}$ to be continuous at $\mathbf{x}$, we must define $\overline{f}(\mathbf{x}) = L$ (whence unicity will follow) and, if we do so, the function $\overline{f}$ so defined will be continuous at $\mathbf{x}$.   □

A rigorous proof of the Theorem introduces no ideas not mentioned in the sketch, but merely sets up all the necessary notation and fills in some detail. It is formal both in the sense of using a formalism (all the notation) and being ritualistic. I am torn between accepting the sketch as sufficient and, this being a book about formalism after all, carrying out the ritual. A reasonable compromise is to give the details, but advise the reader to skip them, pausing only to count how many lines of text the formal proof requires.

The first order of business in performing our formal ritual is to clarify notation and present a few definitions. As regards notation, we formally declare the use of boldface letters $\mathbf{a}, \mathbf{x}, \mathbf{x}_i$, etc., to denote n-tuples and the corresponding subscripted italic letters to denote the elements of the n-tuples, thus, e.g.,

$$\mathbf{x} = \langle x_0, x_1, \ldots, x_{n-1} \rangle$$
$$\mathbf{x}_i = \langle x_{i,0}, x_{i,1}, \ldots, x_{i,n-1} \rangle.$$

The definitions needed concern limits of sequences in $\mathbb{R}^n$ and of functions from $X = \mathbb{Q}^n$ or $\mathbb{R}^n$ into $\mathbb{R}$. For such definitions we need a decent notion of size or distance in $\mathbb{R}^n$. We will discuss the notions of size and distance in greater detail in section 7, below; for now it suffices merely to pick one. For size, the following is probably the simplest.

5.59 Definition. *Let $\mathbf{x} \in \mathbb{R}^n$. We define $|\mathbf{x}| = \max\{|x_0|, \ldots, |x_{n-1}|\}$.*

The definition of the limit of a sequence in $\mathbb{R}^n$ is a straightforward adaptation of the usual definition.

5.60 Definition. *Let $\mathbf{x}_0, \mathbf{x}_1, \ldots$ be a sequence of elements of $\mathbb{R}^n$. An element $\mathbf{x} \in \mathbb{R}^n$ is the limit of the sequence $\mathbf{x}_0, \mathbf{x}_1, \ldots$, written*

$$\lim_{m \to \infty} \mathbf{x}_m = \mathbf{x},$$

if, for all $\epsilon > 0$ there is some $m_0 \in \mathbb{N}$ such that, for all $m \in \mathbb{N}$, if $m > m_0$, $|\mathbf{x}_m - \mathbf{x}| < \epsilon$.

It must be emphasized that these definitions presuppose an addition operation on $\mathbb{R}$. For, $|x - y|$ refers directly to subtraction $x - y = x + (-y)$, and even the definition of absolute value is usually given in terms of the additive inverse:

$$|x| = \begin{cases} x, & 0 \le x \\ -x, & x < 0. \end{cases}$$

We have an addition function already defined on cuts. But we also have a multiplication function so defined and if we are going to use the one, why not the other? A sense of elegance demands that, if possible, we treat addition and multiplication similarly. We can do so by defining the distance function $|x - y|$ directly as

$$d(x, y) = \begin{cases} \mathrm{lub}\{q_1 - q_2 \mid x \le q_2 \le q_1 \le y \ \& \ q_1 \in \mathbb{Q} \ \& \ q_2 \in \mathbb{Q}\}, & x < y \\ 0, & x = y \\ \mathrm{lub}\{q_1 - q_2 \mid y \le q_2 \le q_1 \le x \ \& \ q_1 \in \mathbb{Q} \ \& \ q_2 \in \mathbb{Q}\}, & y < x, \end{cases}$$

where lub denotes the least upper bound, and proving the necessary properties a distance function should have. As discussed in greater detail in section 7, below, the crucial properties of a distance function are these:
i. $d(x, y) \ge 0$ and $d(x, y) = 0$ iff $x = y$;
ii. $d(x, y) = d(y, x)$;
iii. $d(x, z) \le d(x, y) + d(y, z)$.
Of these, ii is trivial and i is an easy consequence of the density of the rationals in the reals: if, say, $x < y$ find $x < q_1 < y$ and then $x < q_2 < q_1$ and observe $d(x, y) \ge q_1 - q_2 > 0$. Property iii is, of course, problematic. The right-hand side of the inequality adds real numbers. However, in establishing continuity, one needs ϵ's and δ's— which may be assumed rational— for upper bounds on inequalities and iii can be replaced in this context by
iii'. if $d(x, y) < q_1$ and $d(y, z) < q_2$, then $d(x, z) < q_1 + q_2$,
where q_1, q_2 are assumed rational and their addition is already defined. Property iii' is an easy exercise the reader might want to work out. Defining

$$d(\mathbf{x}, \mathbf{y}) = \max\{d(x_0, y_0), \ldots, d(x_{n-1}, y_{n-1})\},$$

the more industrious reader can also verify i-ii, iii' remain valid for distances between n-tuples. I am placing the responsibility on the shoulders of the reader because I am going to take the easy way out and assume addition is already given (with the usual algebraic properties— commutativity, associativity, 0 as identity, the existence of an additive inverse, and the preservation of order), and thus also subtraction, absolute value, the distance $d(x, y) = |x - y|$ and properties i-iii.

So we are given the archimedian ordered field $(\mathbb{Q}, +, \cdot, <, 0, 1)$ of rational numbers and an order- and addition-preserving embedding of $(\mathbb{Q}, +, <, 0, 1)$ into $(\mathbb{R}, +, <, 0, 1)$ under which $\mathbb{Q}$ is dense in $\mathbb{R}$, and $(\mathbb{R}, <)$ is a complete ordering. Our task is to extend the multiplication operation from $\mathbb{Q}$ to $\mathbb{R}$ and derive the usual algebraic laws. We wish to do so by appeal to continuity and limits.

For the definitions of continuity and uniform continuity we take Definitions 5.51 and 5.53, noting that the premise,

$$|x_0 - y_0| < \delta \ \& \ \dots \ \& \ |x_{n-1} - y_{n-1}| < \delta,$$

of the underlying implication can now be abbreviated as

$$|\mathbf{x} - \mathbf{y}| < \delta.$$

I leave it to the reader to sort out the trivial consequences of these definitions, such as the uniqueness of the limit where it exists and the continuity of the composition of continuous functions.

In mentioning continuity I am getting a little ahead of myself. We have first a couple of lemmas about sequences and their limits to prove.

5.61 Lemma. *Let* $\mathbf{x}_0, \mathbf{x}_1, \dots$ *be a sequence of elements of* $\mathbb{R}^n$. *The sequence* $\mathbf{x}_0, \mathbf{x}_1, \dots$ *converges to a limit in* $\mathbb{R}^n$ *iff each of the sequences,*

$$x_{00}, x_{10}, x_{20}, \dots$$
$$x_{01}, x_{11}, x_{21}, \dots$$
$$\vdots$$
$$x_{0,n-1}, x_{1,n-1}, x_{2,n-1}, \dots,$$

converges in $\mathbb{R}$. *Moreover, when these sequences converge,*

$$\lim_{m \to \infty} \mathbf{x}_m = \left\langle \lim_{m \to \infty} x_{m0}, \lim_{m \to \infty} x_{m1}, \dots, \lim_{m \to \infty} x_{m,n-1} \right\rangle.$$

The proof of this can safely be left to the reader.

Our next lemma is a manifestation of the density of $\mathbb{Q}^n$ in $\mathbb{R}^n$. It will be convenient for the typesetter if I give a name to the n-dimensional cube mentioned in the statement of Theorem 5.58. We shall call it C:

$$C = [a_0, b_0] \times \dots \times [a_{n-1}, b_{n-1}]. \tag{64}$$

Theorem 5.58 thus says that any uniformly continuous function mapping $I = \mathbb{Q}^n \cap C$ to $\mathbb{R}$ can be lifted uniquely to such a function from C to $\mathbb{R}$. This depends, of course, on the density of I in C.

5.62 Lemma. *For all* $\mathbf{x} \in C$, *there is a sequence* $\mathbf{x}_0, \mathbf{x}_1, \dots$ *in* I *such that* $\lim_{m \to \infty} \mathbf{x}_m = \mathbf{x}$.

Proof. We have already seen that $\mathbb{Q}$ is dense in $\mathbb{R}$. Thus, for each $m \in \mathbb{N}$ and $i = 0, \dots, n - 1$, there is a rational number $x_{mi} \in [a_i, b_i]$ such that

$$|x_{mi} - x_i| < \frac{1}{m + 1}.$$

Thus we have, for $i = 0, \dots, n - 1$,

$$\lim_{m \to \infty} x_{mi} = x_i.$$

Lemma 5.61 thus yields

$$\lim_{m \to \infty} \mathbf{x}_m = \mathbf{x}.$$

Moreover, each $\mathbf{x}_m$ is an element of I.   □

Proof of Theorem 5.58. The proof goes pretty much as sketched. Let C, I, f be given with $f : I \to \mathbb{R}$ uniformly continuous. Choose

$$\mathbf{x} \in C = [a_0, b_0] \times \ldots \times [a_{n-1}, b_{n-1}].$$

By Lemma 5.62, there is a sequence $\mathbf{x}_0, \mathbf{x}_1, \ldots$ in I such that

$$\lim_{m \to \infty} \mathbf{x}_m = \mathbf{x}.$$

Towards proving there to be at most one continuous extension of f to all of C, let $g : C \to \mathbb{R}$ be any continuous such extension. Because g is continuous at $\mathbf{x}$, for any $\epsilon > 0$ there is some $\delta > 0$ such that, for all $\mathbf{y} \in C$,

$$|\mathbf{y} - \mathbf{x}| < \delta \implies |g(\mathbf{y}) - g(\mathbf{x})| < \epsilon.$$

But, because $\mathbf{x}$ is the limit of the sequence $\mathbf{x}_0, \mathbf{x}_1, \ldots$, there is a number m_0 such that, for all $m > m_0$, $|\mathbf{x}_m - \mathbf{x}| < \delta$, whence

$$m > m_0 \implies |g(\mathbf{x}_m) - g(\mathbf{x})| < \epsilon,$$

and we see

$$g(\mathbf{x}) = \lim_{m \to \infty} g(\mathbf{x}_m) = \lim_{m \to \infty} f(\mathbf{x}_m),$$

since g agrees with f on I. The uniqueness of g follows.

The argument for the uniqueness of any continuous extension of f tells us how to define $\overline{f}$: for any $\mathbf{x} \in C$, find $\mathbf{x}_0, \mathbf{x}_1, \ldots$ in I such that $\lim_{m \to \infty} \mathbf{x}_m = \mathbf{x}$ and set

$$\overline{f}(\mathbf{x}) = \lim_{m \to \infty} f(\mathbf{x}_m). \tag{65}$$

We have three things to prove:
i.   the limit (65) exists;
ii. the limit does not depend on the choice of sequence converging to $\mathbf{x}$; and
iii. the function $\overline{f}$ so defined is uniformly continuous.

[A small remark: The uniqueness part of the proof required only continuity; and it should be no surprise to read that, if one is only interested in proving the continuity of $\overline{f}$, uniform continuity is not needed in proving assertion iii. The necessity of assuming the uniform continuity of f on I comes in proving claims i and ii.]

i. The function f is uniformly continuous on I, whence, for any $\epsilon > 0$ there is a $\delta > 0$ such that for all $\mathbf{c}, \mathbf{d} \in I$,

$$|\mathbf{c} - \mathbf{d}| < \delta \implies |f(\mathbf{c}) - f(\mathbf{d})| < \epsilon. \tag{66}$$

But the sequence $\mathbf{x}_0, \mathbf{x}_1, \ldots$ converges to $\mathbf{x}$, whence, for any given δ, there is an m_0 such that for all $m > m_0$,

$$|\mathbf{x}_m - \mathbf{x}| < \frac{\delta}{2}.$$

But then, for any $m, k > m_0$,

$$|\mathbf{x}_m - \mathbf{x}_k| \le |\mathbf{x}_m - \mathbf{x}| + |\mathbf{x} - \mathbf{x}_k| < \frac{\delta}{2} + \frac{\delta}{2} = \delta,$$

whence (66) yields

$$m, k > m_0 \;\Rightarrow\; |f(\mathbf{x}_m) - f(\mathbf{x}_k)| < \epsilon.$$

Thus the sequence $f(\mathbf{x}_0), f(\mathbf{x}_1), \ldots$ is Cauchy convergent and has a limit L.

[A not so small remark: The various formulations of completeness discussed in this Chapter are equivalent, and we have proven the Least Upper Bound Principle for $(\mathbb{R}, <)$. But how much algebraic structure is needed to derive the other completeness properties from it? If we look back at Chapter I, where we derived Cauchy completeness from the Least Upper Bound Principle, we see that only addition and not multiplication of reals was needed.]

ii. Suppose $\mathbf{x}_0, \mathbf{x}_1, \ldots$ and $\mathbf{y}_0, \mathbf{y}_1, \ldots$ are sequences in I converging to $\mathbf{x}$. By i, the sequences

$$f(\mathbf{x}_0), f(\mathbf{x}_1), \ldots \quad \text{and} \quad f(\mathbf{y}_0), f(\mathbf{y}_1), \ldots$$

have limits L_1 and L_2, respectively. Let $\epsilon > 0$ and choose m_1 so large that for $m > m_1$,

$$|f(\mathbf{x}_m) - L_1| < \frac{\epsilon}{3},$$

and choose m_2 so large that for $m > m_2$,

$$|f(\mathbf{y}_m) - L_2| < \frac{\epsilon}{3}.$$

Now,

$$|L_1 - L_2| \le |L_1 - f(\mathbf{x}_m)| + |f(\mathbf{x}_m) - f(\mathbf{y}_m)| + |f(\mathbf{y}_m) - L_2|$$
$$< \frac{\epsilon}{3} + |f(\mathbf{x}_m) - f(\mathbf{y}_m)| + \frac{\epsilon}{3}$$

for $m > m_1, m_2$. It only remains to appeal to the uniform continuity of f again to bound the centre term by $\epsilon/3$. To this end, choose δ such that, for all $\mathbf{c}, \mathbf{d} \in I$,

$$|\mathbf{c} - \mathbf{d}| < \delta \;\Rightarrow\; |f(\mathbf{c}) - f(\mathbf{d})| < \frac{\epsilon}{3},$$

and choose m_3, m_4 so that, for $m > m_3$,

$$|\mathbf{x}_m - \mathbf{x}| < \frac{\delta}{2},$$

and for $m > m_4$,

$$|\mathbf{y}_m - \mathbf{x}| < \frac{\delta}{2}.$$

Then, for any $m > m_3, m_4$,

$$|\mathbf{x}_m - \mathbf{y}_m| \le |\mathbf{x}_m - \mathbf{x}| + |\mathbf{x} - \mathbf{y}_m| < \frac{\delta}{2} + \frac{\delta}{2} = \delta.$$

Choosing $m > m_0 = \max\{m_1, m_2, m_3, m_4\}$, we see

$$|L_1 - L_2| < \frac{\epsilon}{3} + \frac{\epsilon}{3} + \frac{\epsilon}{3} = \epsilon.$$

As $\epsilon > 0$ was arbitrary, $|L_1 - L_2| = 0$, i.e. $L_1 = L_2$.

It follows that (65) does indeed define a function $\overline{f} : C \to \mathbb{R}$.

iii. To see that $\overline{f}$ is uniformly continuous, let $\epsilon > 0$ be given and choose δ so that for all $\mathbf{c}, \mathbf{d} \in I$,

$$|\mathbf{c} - \mathbf{d}| < \delta \;\Rightarrow\; |f(\mathbf{c}) - f(\mathbf{d})| < \frac{\epsilon}{3}.$$

Let $\mathbf{x}, \mathbf{y} \in C$ and suppose $|\mathbf{x} - \mathbf{y}| < \delta/3$. Now, by Lemma 5.62, $\mathbf{x}, \mathbf{y}$ are limits of sequences $\mathbf{x}_0, \mathbf{x}_1, \ldots$ and $\mathbf{y}_0, \mathbf{y}_1, \ldots$, respectively, in I and $\overline{f}(\mathbf{x}) = \lim f(\mathbf{x}_m), \overline{f}(\mathbf{y}) = \lim f(\mathbf{y}_m)$, whence, for sufficiently large m,

$$|\mathbf{x} - \mathbf{x}_m| < \frac{\delta}{3}, \; |\mathbf{y} - \mathbf{y}_m| < \frac{\delta}{3}, \; |\overline{f}(\mathbf{x}) - f(\mathbf{x}_m)| < \frac{\epsilon}{3}, \; |\overline{f}(\mathbf{y}) - f(\mathbf{y}_m)| < \frac{\epsilon}{3}.$$

But

$$|\mathbf{x}_m - \mathbf{y}_m| \le |\mathbf{x}_m - \mathbf{x}| + |\mathbf{x} - \mathbf{y}| + |\mathbf{y} - \mathbf{y}_m| < \frac{\delta}{3} + \frac{\delta}{3} + \frac{\delta}{3} = \delta,$$

whence

$$|\overline{f}(\mathbf{x}) - \overline{f}(\mathbf{y})| \le |\overline{f}(\mathbf{x}) - f(\mathbf{x}_m)| + |f(\mathbf{x}_m) - f(\mathbf{y}_m)| + |f(\mathbf{y}_m) - \overline{f}(\mathbf{y})|$$
$$< \frac{\epsilon}{3} + \frac{\epsilon}{3} + \frac{\epsilon}{3} = \epsilon. \qquad \square$$

The main task is over. What remains are the tasks of applying Theorem 5.58 and drawing such conclusions as we might.

The simplest application should be to addition[77], which is uniformly continuous on all of $\mathbb{Q}^2$. However, I have stated Theorem 5.58 with multiplication in mind. Multiplication is not uniformly continuous on all of $\mathbb{Q}^2$, but only on ever

---

[77] Why addition if we are assuming it already given? There are several reasons. First, it ought to be the simplest example. Second, the distributive laws involve both addition and multiplication and will be quickly established if addition is defined in this manner. Third, it is nice to show that the addition operation so defined agrees with the old one. And, of course, the reader who may have chosen not to presuppose addition in following this treatment does not yet have an addition operation on the reals and needs to define one.

larger boxes. Thus, we have to apply Theorem 5.58 to construct ever larger chunks of multiplication to be pieced together into a single function on $\mathbb{R}^2$. We shouldn't have to do this for addition. Indeed, we don't have to if we go back and replace all references to C and I in the proof of Theorem 5.58 by corresponding references to $\mathbb{R}^n$ and $\mathbb{Q}^n$ to obtain a proof of the following:

5.63 Theorem. *Let $f : \mathbb{Q}^n \to \mathbb{R}^n$ be uniformly continuous. There is a unique continuous function $\overline{f} : \mathbb{R}^n \to \mathbb{R}^n$ such that*

$$\overline{f}(x_0, \ldots, x_{n-1}) = f(x_0, \ldots, x_{n-1})$$

for all $\langle x_0, \ldots, x_{n-1} \rangle \in \mathbb{Q}^n$.

I leave the verification that the proof of Theorem 5.58, modified as suggested, yields the result. And, of course, our new Theorem 5.63 yields a uniformly continuous extension of addition to $\mathbb{R}^2$.

Addition, as hinted, can also be treated in the same manner as multiplication by piecing together fragments defined on boxes C. Proving the validity of the axioms asserting the existence of inverses will most simply be done by extending the inverse functions from the rationals to the reals, which will mean more piecing together. We can do all these unifications as applications of one master construction if we introduce a couple of simple concepts.

5.64 Definition. *Let C be the closed box defined by (64),*

$$C = [a_0, b_0] \times \ldots \times [a_{n-1}, b_{n-1}]$$

for $a_0 < b_0, a_1 < b_1, \ldots, a_{n-1} < b_{n-1}$. The interior *of C, written C^o is the open box,*

$$C^o = (a_0, b_0) \times \ldots \times (a_{n-1}, b_{n-1}).$$

5.65 Definition. *A subset $E \subseteq \mathbb{R}^n$ is* open *if, for every element $\mathbf{x} \in E$ there is a box $C \subseteq E$ of the form (64) such that $\mathbf{x} \in C^o$.*

The notion of an open set is a multi-dimensional generalisation of that of an open interval and would more normally be defined by only demanding $C^o \subseteq E$. As we will not be dealing with open sets elsewhere in this book, I have taken the liberty of giving the definition that best fits the use we are going to put it to in the proof of the following corollary.

5.66 Corollary. *Let $E \subseteq \mathbb{R}^n$ be open, $I = E \cap \mathbb{Q}^n$, and let $f : I \to \mathbb{R}$ be uniformly continuous on every intersection $I_C = E \cap \mathbb{Q}^n = I \cap C$, C being a closed box. There is a unique continuous function $\overline{f} : E \to \mathbb{R}$ which agrees with f on I.*

Proof. The definition of $\overline{f}$ is obvious: for $\mathbf{x} \in E$, choose $C \subseteq E$ such that $\mathbf{x} \in C^o$, let $\overline{f}_C$ be given by Theorem 5.58 so that $\overline{f}_C$ is continuous and $\overline{f}_C(\mathbf{y}) = f(\mathbf{y})$ for all $\mathbf{y} \in I_C$, and set

$$\overline{f}(\mathbf{x}) = \overline{f}_C(\mathbf{x}).$$

We must, of course, show that $\overline{f}$ is well-defined, i.e. $\overline{f}(\mathbf{x})$ does not depend on the choice of C: If $\mathbf{x} \in C_1^o$ and $\mathbf{x} \in C_2^o$ with each $C_i \subseteq E$, then

$$\overline{f}_{C_1}(\mathbf{x}) = \overline{f}_{C_2}(\mathbf{x}).$$

But this is quite easy. Let $\mathbf{x}$ be in the interiors of both boxes. By Lemma 5.62, $\mathbf{x}$ is the limit of a sequence $\mathbf{x}_0, \mathbf{x}_1, \ldots$ in I_{C_1}. Because $\mathbf{x}$ is in the interior of C_2, from some m_0 on all the terms $\mathbf{x}_i$ of the sequence are also in C_2. (*Exercise.*) But

$$\overline{f}_{C_1}(\mathbf{x}) = \lim_{m \to \infty} f(\mathbf{x}_m) = \overline{f}_{C_2}(\mathbf{x}).$$

The continuity of $\overline{f}$ at $\mathbf{x}$ follows from the continuity of $\overline{f}_C$ at $\mathbf{x}$ for any C for which $\mathbf{x} \in C^o \subseteq C \subseteq E$. And unicity follows as in Theorem 5.58 from continuity and the density of $\mathbb{Q}^n \cap E$ in E.   □

By Examples 5.54 and 5.55, taking $E = \mathbb{R}^n$, we quickly conclude the existence of unique continuous extensions of $+$ and $\cdot$ from $\mathbb{Q}^2$ to $\mathbb{R}^2$. Following Dedekind, we can even apply the Corollary to verify that the "numerous theorems of the arithmetic of the rational numbers" carry over. The crucial "numerous theorems" are, of course, the axioms of a complete archimedean ordered field. We have already verified the completeness of the order in constructing $(\mathbb{R}, <)$. The rest of the axioms divide into 4 groups:

i.   identities: commutativity, associativity, distributivity, 0 and 1 as identity elements;

ii. existence of inverse elements;

iii. order respecting properties of addition and multiplication; and

iv. the archimedean axiom.

The first batch of these axioms are all of the form

$$f(\mathbf{x}) = g(\mathbf{x}) \tag{67}$$

for continuous functions f, g for which $f(\mathbf{x}) = g(\mathbf{x})$ holds for all $\mathbf{x} \in \mathbb{Q}^n$. The uniqueness portion of the proof of Corollary 5.66, depending only on continuity, now extends the identity from $\mathbb{Q}^n$ to $\mathbb{R}^n$. Thus, for example, consider the functions

$$f(x_0, x_1, x_2) = x_0(x_1 + x_2)$$
$$g(x_0, x_1, x_2) = (x_0 \cdot x_1) + (x_0 \cdot x_2).$$

Being compositions of continuous functions, they are themselves continuous. By the distributive law in $\mathbb{Q}$, if we denote the restrictions of f, g to $\mathbb{Q}^3$ by f_Q, g_Q, respectively, we have

$$f_Q(\mathbf{x}) = g_Q(\mathbf{x}),$$

for all $\mathbf{x} \in \mathbb{Q}^3$. Thus f, g are both continuous extensions of f_Q to $\mathbb{R}^3$ and the uniqueness of the extension yields (67).[78]

---

[78] It is convenient, but not essential, here that uniqueness of the extension among continuous functions held under the weaker assumption of continuity of the original

The existence of inverse elements is proven in almost the same manner. For addition, one observes that negation, $f(x) = -x$, is uniformly continuous on $\mathbb{Q}$ and thus has a continuous extension to $\mathbb{R}$. The identity

$$x + (-x) = 0$$

on $\mathbb{Q}$ thus extends by Corollary 5.66 to $\mathbb{R}$.

The multiplicative inverse, $f(x) = 1/x$, is not uniformly continuous on $\mathbb{Q}$, or even defined on all of $\mathbb{Q}$. But f is uniformly continuous on the intersection of $\mathbb{Q}$ with any closed interval $C = [a, b]$ which does not contain 0:

$$|x - y| < \epsilon(\min\{|a|, |b|\})^2 \ \Rightarrow \ \left|\frac{1}{x} - \frac{1}{y}\right| = \left|\frac{y - x}{xy}\right| < \epsilon\frac{(\min\{|a|, |b|\})^2}{|x| \cdot |y|} \leq \epsilon.$$

Thus, applying Corollary 5.66 to f and $E = \{x \in \mathbb{R} \mid x \neq 0\}$, f has a continuous extension $\overline{f}$ to E and the identity

$$x \cdot f(x) = 1$$

on $\{x \in \mathbb{Q} \mid x \neq 0\}$ extends to an identity

$$x \cdot \overline{f}(x) = 1$$

on E.

That order is preserved under addition we almost get for free. We have assumed the addition operation on cuts given and, for this operation we have already proven (Corollary 5.17) for all $x, y, z \in \mathbb{R}$,

$$x < y \ \Rightarrow \ x + z < y + z.$$

Now the proof of the uniform continuity of addition on $\mathbb{Q}$ applies to the real case for this addition function as well. Hence the uniqueness assertion of Theorem 5.63 or Corollary 5.66 tells us that this addition function agrees with the new one constructed by appeal to these powerful results. The reader who chose not to follow me in taking the easy way out and assuming addition given at the outset has his work cut out for him.

The multiplicative order properties are:

$$x < y \ \& \ 0 < z \ \Rightarrow \ zx < zy$$
$$x < y \ \& \ z < 0 \ \Rightarrow \ zy < zx.$$

---

function on $\mathbb{Q}^n$ and did not require uniform continuity. Had uniqueness required uniform continuity, we would have had to prove $f(x_0, x_1, x_2) = x_0(x_1 + x_2)$ uniformly continuous on boxes C, which would have required proving that the image of a box C under addition is contained in such a box. That is, the functions f, g of (67) are compositions of functions uniformly continuous on boxes and, to conclude their uniform continuity, we would have had to generalise Remark 5.57 to show that the uniformly continuous image of a box C is a bounded set.

The idea behind the proofs of these is very simple. We know that they hold for rational values of x, y and z. So choose $p, q, r \in \mathbb{Q}$ so close to x, y, z, respectively, that

$$|(zy - zx) - (rp - rq)| < |zy - zx| \quad \text{and} \quad |z - r| < |z|.$$

These inequalities guarantee that r has the same sign as z and $rp - rq$ the same sign as $zy - zx$. I leave the detailed verification that this argument works to the reader.

And, finally, the validity of the Archimedean Axiom in $\mathbb{R}$ follows quickly from its validity in $\mathbb{Q}$: Let $\epsilon, M > 0$ and find rational numbers q_1, q_2 by the density of $\mathbb{Q}$ in $\mathbb{R}$ such that $q_1 < \epsilon$ and $M < q_2$. By the archimedean property of $(\mathbb{Q}, <)$ there is some $n \in \mathbb{N}$ such that $nq_1 > q_2$. By what we have just proven,

$$n\epsilon > nq_1 > q_2 > M.$$

With this last we have established anew that $(\mathbb{R}, +, \cdot, <, 0, 1)$ is a complete archimedean ordered field as promised, and we come to the end of our discussion of Dedekind cuts.

6 Hyperreals and Nonstandard Analysis

6.1 Background

In the last chapter we made free use of infinitesimals and infinite numbers. These were, of course, common in mathematics before the Bolzano-Cauchy-Weierstrass rigorisation of mathematics banished them from pure, if not entirely from applied, mathematics. Infinitesimals and infinite numbers, however, are not essentially different from any other numbers introduced by man after God created the positive integers for Kronecker. The history of the expansion of the number systems is one of dealing with numbers that did not exist and expanding the then current systems to include them. If this did not officially happen with infinitesimal and infinite numbers, it was only because there was no need for them after Bolzano-Cauchy-Weierstrass. The actual construction of a system of hyperreal numbers extending the reals to allow for the infinitely large and infinitely small is not that difficult, or even that different from the construction of the reals from the rationals using Cauchy sequences. There is a bit more leeway in performing the construction, and greater sophistication and the use of purely 20th century mathematics yield "better" nonstandard extensions.

The banished infinitesimals made their initial reappearance in geometry in the late 1800's when Giuseppe Veronese demonstrated the independence of the Archimedean axiom in geometry[79]. I quote Felix Klein:

---

[79] Veronese's book, *Fundamenti di geometria* was published in Padua in 1891 and translated into German as *Grundzüge der Geometrie* in 1894.

In the most recent mathematics, "actually" infinitely small quantities have come to the front again, but in entirely different connection, namely in the geometric investigations of Veronese and also in Hilbert's *Grundlagen der Geometrie*[80]. The guiding thought of these investigations can be stated briefly as follows: A geometry is considered in which $x = a$ (a an ordinary real number) determines not only *one* point on the x-axis, but infinitely many points, whose abscissas differ by finite multiples of infinitely small quantities of different orders $\eta, \zeta, \ldots$ A point is thus determined only when one assigns

$$x = a + b\eta + c\zeta + \cdots ,$$

where a, b, c are ordinary real numbers, and the $\eta, \zeta, \ldots$ actually infinitely small quantities of decreasing orders. Hilbert uses this guiding idea by subjecting these new quantities $\eta, \zeta, \ldots$ to such axiomatic assumptions as will make it evident that one can operate with them consistently. To this end it is of chief importance to determine appropriately the relation as to size between x and a second quantity $x_1 = a_1 + b_1\eta + c_1\zeta + \cdots$. The first assumption is that $x >$ or $< x_1$ if $a >$ or $< a_1$; but if $a = a_1$, the determination as to size rests with the second coefficient, so that $x \gtreqless x_1$ according as $b \gtreqless b_1$; and if, in addition, $b = b_1$, the decision lies with the c, etc. These assumptions will be clearer to you if you refrain from attempting to associate with the letters any sort of concrete representation.[81]

The coordinates and their ordering referred to by Klein have a simple description. One thinks of the quantities x, x_1 as ordinary sequences,

$$x : \ (a, b, c, \ldots)$$
$$x_1 : (a_1, b_1, c_1, \ldots)$$

of real numbers, and orders them *lexicographically*: $x < x_1$ if the coordinate of x is less than that of x_1 at the first place they differ. Klein has not said how

[80] David Hilbert, *Grundlagen der Geometrie*, B.G. Teubner, Leipzig, 1899. This book was immediately translated into French with some additions. The English translation by E.J. Townsend, *The Foundations of Geometry*, Open Court Publishing Company, LaSalle (Illinois), 1902, appeared shortly after. The book has gone through a number of editions with varying collections of appendices. The 7th edition (1939) included 5 papers by Hilbert on geometry, as well as edited versions of his most important papers on the foundations of mathematics. From the 8th edition on only the geometric papers were retained, but in the 10th edition (1968) supplementary appendices written by his former assistant Paul Bernays were added. The book was a major influence and is still in print.

[81] Felix Klein, *Elementary Mathematics from an Advanced Standpoint. Arithmetic. Algebra. Analysis.*, MacMillan & Co., London, 1932, p. 218. This is a translation by E. R. Hedrick and C.A. Noble of the 3rd edition of lecture notes of a course given by Klein in Göttingen in 1907/1908. The translation is still in print as one of two volumes by Dover Publishing Company in New York.

one is to add such sequences, but the componentwise operation would seem to be called for as one would expect

$$x + x_1 = (a + a_1) + (b + b_1)\eta + (c + c_1)\zeta + \cdots .$$

Multiplication would not have a clear geometric interpretation, but algebraically it would make sense if one used $1, \eta, \eta^2, \ldots$ in place of $1, \eta, \zeta, \ldots$ Then one would imagine the sequences as coefficients of a power series,

$$a + bx + cx^2 + dx^3 + \cdots$$

evaluated at an infinitesimal η. Multiplication would thus be convolution as described in §3, above.

6.1 Example. $\mathcal{R}_K = (\mathbb{R}^{\mathbb{N}}, +, *, <, \mathbf{0}, \mathbf{1})$ is a non-Archimedean extension of $\mathbb{R}$, where

$$\mathbf{0} = (0, 0, \ldots), \quad \mathbf{1} = (1, 0, 0, \ldots),$$

$a + b$ is defined componentwise, $*$ is convolution, $<$ is the lexicographical ordering, and $r \in \mathbb{R}$ is identified with $r \cdot \mathbf{1} = (r, 0, 0, \ldots)$. The elements

$$\eta = (0, 1, 0, 0, \ldots)$$
$$\eta^2 = (0, 0, 1, 0, \ldots)$$

etc. are all infinitesimal: $\eta < r$ for every positive real r (for: $(0, 1, 0, 0, \ldots)$ precedes $(r, 0, 0, \ldots)$ in "alphabetical" order; $(0, 0, 1, 0, \ldots)$ precedes $(0, 1, 0, \ldots)$; etc.).

Before continuing, I should pause to define carefully the term "non-Archimedean" [or "nonarchimedean"]. I do this by continuing the quotation from Klein:

> Now it turns out that, after imposing upon these new quantities this rule, together with certain others, it is possible to operate with them as with finite numbers. One essential theorem, however, which holds in the system of ordinary real numbers, now loses its validity, namely the theorem: *Given two positive numbers e, a, it is always possible to find a finite integer n such that $n \cdot e > a$, no matter how small e is nor how large a may be.* In fact, it follows immediately from the above definition that an arbitrary finite multiple $n \cdot \eta$ of η is smaller than any positive finite number a, and it is precisely this property that characterizes the η as an infinitely small quantity. In the same way $n \cdot \zeta < \eta$, that is, ζ is an infinitely small quantity of higher order than η.[82]
> This number system is called non-Archimedean. The above theorem concerning finite numbers is called, namely, the *axiom of Archimedes*, because he emphasized it as an unprovable assumption, or as a fundamental one which did not need proof, in connection with the numbers

---

[82] Is it Klein or the translator? In the previous quote $\eta, \zeta, \ldots$ were of "decreasing orders". Generally, an infinitesimal ζ is said to be a higher order infinitesimal than η if ζ/η is infinitesimal.

which he used. The denial of this axiom characterizes the possibility of actually infinitely small quantities. The name *Archimedean axiom*, however, like most personal designations, is historically inexact. Euclid gave prominence to this axiom more than half a century before Archimedes; and it is said not to have been invented by Euclid, either, but, like so many of his theorems, to have been taken over from Eudoxus of Knidos. The study of non-Archimedean quantities, which have been used especially as coordinates in setting up a non-Archimedean geometry, aims at deeper knowledge of the nature of continuity and belongs to the large group of investigations concerning the logical dependence of different axioms of ordinary geometry and arithmetic. For this purpose, the method is always to set up artificial number systems for which only a part of the axioms hold, and to infer the logical independence of the remaining axioms from these.

The question naturally arises whether, starting from such number systems, it would be possible to modify the traditional foundations of infinitesimal calculus, so as to include actually infinitely small quantities in a way that would satisfy modern demands as to rigor; in other words, to construct a non-Archimedean analysis. The first and chief problem of this analysis would be to prove the mean-value theorem

$$f(x + h) - f(x) = h \cdot f'(x + \vartheta h)$$

from the assumed axioms. I will not say that progress in this direction is impossible, but it is true that none of the investigators who have busied themselves with actually infinitely small quantities have achieved anything positive.[83]

Klein's quotation not only formally defines "non-Archimedean" for us, but it also states succinctly the problem facing us. As a foundation for the Calculus it is not enough just to construct a non-Archimedean extension of $\mathbb{R}$. One must construct a non-Archimedean extension that inherits a lot of structure from $\mathbb{R}$.

One has but to consider some of the uses infinitesimals and infinite numbers were put to to see what sort of structure needs to be inherited. For example, if f is a function defined in a neighbourhood of a, one might try to find the limit of $f(x)$ as x approaches a by calculating $f(a + \eta)$ for infinitesimal η, performing some simplifications, and then observing that the simplified number is infinitesimally close to a real value which one concludes to be the limit. But what is $f(a + b\eta + c\eta^2 + \cdots)$? If f is expandable into a power series, say

$$f(x) = \sum_{n=0}^{\infty} b_n x^n,$$

and we write $x \in \mathcal{R}_K$ as $x = \sum_{i=0}^{\infty} a_i \eta^i$, we can evaluate $f(x)$ algebraically by making the substitution,

---

[83] Klein, *op. cit.*, pp. 218 - 219.

$$f(x) = \sum_{n=0}^{\infty} b_n \left(\sum_{i=0}^{\infty} a_i \eta^i \right)^n,$$

and rearranging the terms,

$$f(x) = \sum_{i=0}^{\infty} c_i \eta^i,$$

where

$$c_0 = \sum_{n=0}^{\infty} b_n a_0^n$$

$$c_1 = \sum_{n=0}^{\infty} b_n \binom{n}{1} a_0^{n-1} a_1$$

$$c_2 = \sum_{n=0}^{\infty} b_n \left(\binom{n}{1} a_0^{n-2} a_2 + \binom{n}{2} a_0^{n-2} a_1^2 \right),$$

etc. But what do we do if f is only continuous? Some functions, like $f(x) = |x|$, offer no difficulty, but generally we haven't a clue as to how to define $f(a + b\eta + c\eta^2 + \cdots)$.

And what about the infinitely large? $\mathcal{R}_K$ has no infinitely large elements. This is not an insurmountable obstacle, as $\mathcal{R}_K$ is an integral domain (it is the ring of Example 3.1.iv of §3, above) and thus embedds into its field of fractions. It doesn't take much imagination to see that $\eta^{-1} = (0, 1, 0, \ldots)^{-1}$ is infinite. Is it an integer, i.e. can one consider an Eulerian product,

$$\left(1 + \frac{1}{\eta^{-1}} \right)^{\eta^{-1}},$$

as an approximation to e? Assuming the product exists (i.e., assuming we can assign a meaning to it) and it is infinitesimally close to e, is it in fact in $\mathcal{R}_K$?

As Klein said, infinitesimal elements were re-introduced into mathematics in a geometric context; non-Archimedean geometries were introduced to study the necessity, or lack thereof, of assuming the Archimedean axiom in proving geometric propositions. Now, Euclidean geometry does not make as many existential demands on our ontology as does the Calculus: If one has a point corresponding to the real number a on the geometric line, one doesn't require points corresponding to $f(a)$ for all that many continuous functions f. One needs only closure under the four arithmetic operations $(+, -, \cdot, /)$ and taking square roots.

This brings us to a curious digression. Klein was certainly aware that one did not need much to construct a model of geometry. He had presented the construction himself in 1895 in his lectures on the famous construction problems in geometry and their impossibility.[84] In *Grundlagen der Geometrie*, Hilbert

---

[84] Felix Klein, *Famous Problems of Elementary Geometry*, Dover Publishing Company, New York.

alluded to the construction[85] in his proof of the consistency of the axioms, and generalised it in his construction of a model of non-Archimedean geometry. I quote Hilbert's description:

> We construct a domain $\Omega(t)$ of all those algebraic functions of t which may be obtained from t by means of the four arithmetical operations of addition, subtraction, multiplication, division, and the fifth operation $\sqrt{1+\omega^2}$, where ω represents any function arising from the application of these five operations... These five operations may all be performed without introducing imaginaries, and that in only one way. The domain $\Omega(t)$ contains, therefore only single valued functions of t.[86]

The quotation comes from the 1902 English translation of the first edition as augmented in the French translation and the same construction appears in each of the 7th and 12th editions.[87] It in no way resembles Klein's $\mathcal{R}_K$. Perhaps Klein was describing Veronese's construction or something Hilbert had done in one of his other lecture courses on geometry. In any event, Hilbert's non-Archimedean structure is even less suited for a foundation for the Calculus than $\mathcal{R}_K$. Imagine trying to prove

$$\lim_{x \to 0} \frac{\sin x}{x} = 1$$

by calculating $(\sin \eta)/\eta$ for some infinitesimal η. One can choose the function $1/t$ for η[88], but what value does one assign to $\sin \eta$? The obvious definition of $\sin f$ for a function f is the composition, but the sine of an element of $\Omega(t)$, in particular $\sin(1/t)$, is not generally an element of $\Omega(t)$.

It would take almost half a century for the infinitely large and infinitely small to be convincingly re-introduced into the Calculus. And I think this is not because it was so difficult— it wasn't— but because attention was diverted away from the problem. Structures like $\mathcal{R}_K$ and $\Omega(t)$ were studied, but they were studied for their own sakes. Where techniques of the Calculus were introduced, it was not calculus on $\mathbb{R}$ to be established by appeal to the new structures, but Calculus on the new function spaces established by generalising the standard proofs. This changed with Curt Schmieden and Detlef Laugwitz, the unsung heroes of the forthcoming Nonstandard Analysis.

6.2 Schmieden and Laugwitz

The origins of nonstandard analysis are nicely summed up in an abstract by Detlef Laugwitz:

---

[85] Hilbert, *op. cit.*, §9, pp. 27 - 30 of the English translation.

[86] *Ibid.*, p. 34.

[87] I checked only those editions I have on my own bookshelf.

[88] The functions in $\Omega(t)$ are linearly ordered by eventual dominance:

$$f < g \quad \text{iff} \quad \text{for sufficiently large } x, f(x) < g(x).$$

So there is no difficulty in seeing that η is an infinitesimal.

In the early 1950's C. Schmieden developed the idea to "adjoin" an infinitely large natural number Ω to real analysis. A formula $A(\Omega)$ was true if $A(n)$ was true for almost all finite natural n. In a paper of 1958 we used rational (and later real) sequences to establish a model in which Ω was represented by the sequence of finite natural numbers. Clearly, this approach was less powerful with respect to new applications than Robinson's (1961) was. Yet it turned out that the Omega-calculus could serve as a means to clarify earlier uses of infinitesimals and infinitely large numbers. The sequential approach was successful in re-considering the foundational aspects of Cauchy's textbooks and also early Fourier analysis and the use of delta functions around 1820. Adjoining an ideal element was more in the spirit of Euler (and, in some sense of Leibniz and Bolzano). Like most of pre-Cantorian mathematics the two versions of Schmieden's approach were basically constructive.[89]

In another abstract[90], Detlef Spalt adds to this:

> The "Darmstadt Version" of nonstandard analysis first saw the light of day through an article published in 1958 in the *Mathematische Zeitschrift*, which [article] would be jointly authored by Curt Schmieden and Detlef Laugwitz. In the period following, this approach would further be put into concrete form by Detlef Laugwitz.
> Manuscript finds from Curt Schmieden's Nachlass show that the mathematical substance of this publication of the year 1958 was formulated by him already in the year 1952.

Schmieden's initial approach was not Type II formalism, but Type III. He assumed as we did in Chapter II in discussing delta functions, that there was some extension $\mathbb{H}$ of $\mathbb{R}$ which contained an infinite element Ω. This new element was assumed in fact to be an infinite integer and he postulated for it the truth of any formula $A(x)$ for which there was a natural number n_0 such that, for all $n > n_0$, $A(n)$ is true. In accordance with the Peacock-Hankel Principle of the Permanence of Formal Laws, this is the minimum requirement one ought to impose on Ω. For, if the universal law

$$\text{for all } n \in \mathbb{N}, \text{ if } n > n_0 \text{ then } A(n)$$

is to remain true, then, since Ω is assumed to be in $\mathbb{N}$ and $\Omega > n_0$, it must follow that $A(\Omega)$ is true.

---

[89] Abstract to: Detlef Laugwitz, "Curt Schmieden's approach to infinitesimals— an eye-opener to the historiography of analysis", in: P. Schuster, U. Berger, and H. Osswald (eds.), *Reuniting the Antipodes— Constructive and Nonstandard Views of the Continuum*, Kluwer Academic Publishers, Dordrecht, 2001.

[90] The title is "Die Nichtstandardanalysis Curt Schmiedens aus dem Jahr 1952". I found this abstract online and do not know the publication history. I did, however, find a paper: Detleft Spalt, "Curt Schmieden's non-standard analysis— a method of dissolving the standard paradoxes of analysis", *Centaurus* 43 (2001), pp. 137 - 175.

Now, the Principle of the Permanence of Formal Laws is not an axiom that yields existence, but is, rather, a desideratum. Its validity in any case depends on the domain to be extended and the choice of formulæ $A(x)$ used to express the laws. Now, Peacock and Hankel were quite modest in this respect, limiting their permanent laws to equational identities. There is, however, a general theorem of mathematical logic that tells us any infinite domain can be properly extended in such a way as to ensure the validity of all laws "for all x, $A(x)$" for an extensive class of formulæ $A(x)$. Abraham Robinson would later apply this result to give a much more powerful version of nonstandard analysis. Such a result, however, will not apply to all situations. For example, in the constructions of $\mathbb{Z}$ from $\mathbb{N}$ or $\mathbb{Q}$ from $\mathbb{Z}$, we did not just add new elements to the given semi-ring or ring to get a larger semi-ring or ring, but we added solutions to equations that were not initially solvable in order to construct a ring or field, respectively. Thus, the validity of simple "laws",

$$\text{for all } x, f(x) \neq g(x),$$

for certain functions f, g was not preserved: for

$$f(x) = x + 2, \quad g(x) = 1$$

in the passage from $\mathbb{N}$ to $\mathbb{Z}$, and

$$f(x) = 2x, \quad g(x) = 1,$$

in the passage from $\mathbb{Z}$ to $\mathbb{Q}$.

Schmieden, however, made a fortunate choice. The general result of mathematical logic covers the sort of extension he was looking for and, although he did not realise it[91], he was in no danger of running into a contradiction and concluding $0 = 1$ on the basis of his assumption. Schmieden used the infinitely large and infinitely small to analyse what he took to be almost contradictory paradoxes of modern analysis[92]. Unlike Robinson, who would go straight for nonstandard proofs of the "big theorems", Schmieden devoted a lot of his attention to the small, but interesting asides. For example, early on he reproduced some Eulerian derivations, such as the summation of the alternating harmonic series[93]:

$$\ln 2 = 1 - \frac{1}{2} + \frac{1}{3} - \frac{1}{4} + \dots$$

From the existence of the Euler-Mascheroni constant C,

---

[91] Laugwitz and Spalt report that Schmieden's work began in 1948. The logical result— the Compactness Theorem to be discussed below— was first published in applicable form in Russian in 1941, and was only being full appreciated by specialists in Mathematical Logic in the early 1950's, most notably in Robinson's dissertation published in 1951.

[92] Cf. especially, Spalt, "Curt Schmieden's non-standard analysis...", *op. cit.*, pp. 141 - 154.

[93] Laugwitz, "Curt Schmieden's approach...", *op.cit.*, p. 129; Spalt, "Curt Schmieden's non-standard analysis...", *op. cit.*, pp. 144 - 145.

$$C = \lim_{n \to \infty} \left(\sum_{i=1}^{n} \frac{1}{i} - \ln n \right) = .577\dots,$$

it follows that the error C_N of the N-th partial sum of the *harmonic series* as an approximation to $\ln N$,

$$H(n) = 1 + \frac{1}{2} + \dots + \frac{1}{n} = \ln n + C_n,$$

is infinitesimally close to C for infinite N. Thus

$$1 - \frac{1}{2} + \frac{1}{3} - \frac{1}{4} + \dots + \frac{1}{2N-1} - \frac{1}{2N} = H(2N) - 2 \cdot \frac{1}{2} H(N)$$
$$= \ln 2N + C_{2N} - (\ln N + C_N)$$
$$\approx \ln 2N - \ln N = \ln 2;$$

and

$$1 + \frac{1}{3} - \frac{1}{2} + \frac{1}{5} + \frac{1}{7} - \frac{1}{4} + \dots + \frac{1}{4N-3} + \frac{1}{4N-1} - \frac{1}{2N} =$$
$$= H(4N) - \frac{1}{2} H(2N) - \frac{1}{2} H(N)$$
$$\approx \ln 4N - \frac{1}{2} \ln 2N - \frac{1}{2} \ln N$$
$$\approx \ln 4 + \ln N - \frac{1}{2} \ln 2 - \frac{1}{2} \ln N - \frac{1}{2} \ln N$$
$$\approx \frac{3}{2} \ln 2;$$

etc.

Laugwitz joined Schmieden in 1954 and reports:

> After years of debates among the two of us, and with the referees (one of them was Paul Lorenzen[94]), a joint paper was eventually accepted.[95]

He adds,

> The referees of our paper SL58[96] did … not accept the approach. We rewrote the paper and defined the new numbers in terms of equivalence classes of rational (and later real) sequences, in other words, we supplied a model.[97]

---

[94] As a logician, Lorenzen should have been able to steer Schmieden and Laugwitz to the Compactness Theorem. Lorenzen, however, was a staunch constructivist and the application of such a nonconstructive, abstract existence theorem would not have appealed to him.

[95] Laugwitz, "Curt Schmieden's approach…", *op. cit.*, p. 128.

[96] I.e., Schmieden and Laugwitz, *op. cit.*

[97] Laugwitz, "Curt Schmieden's approach…", *op. cit.*, p. 131. Laugwitz does not separate his contributions from Schmieden's, but we have on Spalt's authority ("Curt Schmieden's non-standard analysis…", *op. cit.*, pp. 159 - 160) the attribution of the construction of the model to Laugwitz.

Thus, when Schmieden did publish his results in a joint paper with his student Detlef Laugwitz in 1958[98], they did not present the work as a calculus carried out in some hypothetical axiomatically determined extension $\mathbb{H}$ of $\mathbb{R}$, but, rather, based it on a concrete construction of such an $\mathbb{H}$.

Strictly speaking, this last sentence is not correct. In their 1958 paper, Schmieden and Laugwitz did not develop $\mathbb{H}$ as an extension of $\mathbb{R}$, but as an alternative completion of $\mathbb{Q}$ to be studied in its own right. They took for $\mathbb{H}$ the structure $(\mathbb{Q}^{\mathbb{N}}, +, \cdot, <, =, 0, 1)$, defined the limit of a sequence of rational numbers to be the sequence itself, and Ω to be the identity sequence $a_n = n$. The set of real numbers then coincides with certain equivalence classes of elements of $\mathbb{Q}^{\mathbb{N}}$ in the now familiar way. Subsequently, Laugwitz simplified matters by assuming the real numbers to be given at the outset and considering a structure based on $\mathbb{R}^{\mathbb{N}}$, which Laugwitz eventually denoted $^{\Omega}\mathbb{R}$.

6.2 Definition. *The structure* $\mathbb{R}^{\mathbb{N}} = (\mathbb{R}^{\mathbb{N}}, +, \cdot, <, =, 0, 1)$ *is given by taking as domain the set* $\mathbb{R}^{\mathbb{N}}$ *of all sequences* $a : a_0, a_1, a_2 \ldots$ *of real numbers, with operations and relations defined componentwise,*

$$(a+b)_n = a_n + b_n, \quad (a \cdot b)_n = a_n \cdot b_n$$
$$a = b \quad \text{iff} \quad \text{for all } n, a_n = b_n$$
$$a < b \quad \text{iff} \quad \text{for all } n, a_n < b_n,$$

and the identity elements being the constant sequences,

$$\mathbf{0}_n = 0, \quad \mathbf{1}_n = 1.$$

In general every real number $r \in \mathbb{R}$ is identified with its constant sequence,

$$\mathbf{r}_n = r,$$

and both objects are ambiguously denoted by r. In particular, we write $0, 1$ for $\mathbf{0}, \mathbf{1}$, respectively.

As we saw in discussing van Rootselaar's interpretation of Bolzano's theory of real numbers, $\mathbb{R}^{\mathbb{N}}$ itself constitutes a viable candidate for $\mathbb{H}$. For the development of nonstandard analysis, however, it is convenient to replace the structure by a "collapsed" version consisting of equivalence classes under an appropriate equivalence relation. After having seen Laugwitz's treatment of the Horn angle in Chapter II and his treatment of Bolzano's theory, his choice of equivalence relation will come as no surprise.

6.3 Definitions. *We define, for* $a, b \in \mathbb{R}^{\mathbb{N}}$,

$$a \equiv b \quad \text{iff} \quad a_n = b_n \text{ for all but finitely many } n \in \mathbb{N}$$
$$a < b \quad \text{iff} \quad a_n < b_n \text{ for all but finitely many } n \in \mathbb{N},$$

i.e. $a \equiv b$ *iff for some* $n_0 \in \mathbb{N}$, *for all* $n \in \mathbb{N}$, *if* $n > n_0$, *then* $a_n = b_n$; *and similarly for* $<$.

---

[98] Curt Schmieden and Detlef Laugwitz, "Eine Erweiterung des Infinitesimalrechnung", *Mathematische Zeitschrift* 69 (1958), pp. 1 - 39.

The reader may easily verify that $\equiv$ is an equivalence relation and that $\equiv, <$ respect the operations:

$$a \equiv c \ \& \ b \equiv d \ \Rightarrow \ a + b \equiv c + d$$
$$a \equiv c \ \& \ b \equiv d \ \Rightarrow \ ab \equiv cd$$
$$a \equiv c \ \& \ b \equiv d \ \& \ a < b \ \Rightarrow \ c < d.$$

6.4 Definitions. *For $a \in \mathbb{R}^{\mathbb{N}}$, we define*

$$[a] = \{b \in \mathbb{R}^{\mathbb{N}} | a \equiv b\}.$$

For $a, b \in \mathbb{R}^{\mathbb{N}}$, we define

$$[a] + [b] = [a + b], \quad [a] \cdot [b] = [a \cdot b]$$
$$[a] < [b] \quad iff \quad a < b.$$

Finally, we define

$$^{\Omega}\mathbb{R} = (\{[a] \, | \, a \in \mathbb{R}^{\mathbb{N}}\}, +, \cdot, =, <, 0, 1),$$

where $0, 1$ denote the classes $[0], [1]$, i.e. $[(0, 0, 0, \ldots)], [(1, 1, 1, \ldots)]$, respectively.

As announced, $^{\Omega}\mathbb{R}$ is the Schmieden-Laugwitz candidate for $\mathbb{H}$. One of the things that particularly recommends this choice is that the choice of the underlying structure $\mathbb{R}^{\mathbb{N}}$, unlike that of the rings cited earlier in this section, guarantees the closure of $^{\Omega}\mathbb{R}$ under *all* operations defined on $\mathbb{R}$ and not just the simple algebraic ones: If $f : \mathbb{R} \to \mathbb{R}$ is any function whatsoever, and $a = (a_0, a_1, a_2, \ldots) \in \mathbb{R}$, we can define

$$\hat{f}(a) = (f(a_0), f(a_1), f(a_2), \ldots) \in \mathbb{R}^{\mathbb{N}}.$$

Notice that if $a \equiv b$, then for some $n_0 \in \mathbb{N}$ we have $a_n = b_n$ for all $n > n_0$. But then $f(a_n) = f(b_n)$ for all $n > n_0$, whence $\hat{f}(a) \equiv \hat{f}(b)$ and we can thus define:

$$^{\Omega}f(a) = [\hat{f}(a)].$$

There is, of course, no need to restrict ourselves to unary functions: every $f : \mathbb{R}^n \to \mathbb{R}$ has a similar extension $^{\Omega}f : \mathbb{R}^n \to \mathbb{R}$. In a like manner, every relation R on the reals lifts to a relation $^{\Omega}R$ on $^{\Omega}\mathbb{R}$:

$$^{\Omega}R(a, b, c, \ldots) \quad \text{iff} \quad \text{for all but finitely many } k, R(a_k, b_k, c_k, \ldots).$$

It can also easily be seen that some form of the Principle of the Permanence of Formal Laws holds. For example, if f, g are functions of several variables, then an identity,

$$\text{for all } x_0, \ldots, x_{n-1} \in \mathbb{R}, \ f(x_0, \ldots, x_{n-1}) = g(x_0, \ldots, x_{n-1}), \tag{68}$$

entails the corresponding identity,

for all $x_0, \ldots, x_{n-1} \in^\Omega \mathbb{R}$, $^\Omega f(x_0, \ldots, x_{n-1}) = {}^\Omega g(x_0, \ldots, x_{n-1})$.   (69)

I leave it to the reader to verify this. It is a good exercise in not getting confused by one's own notation.

Now the inference of the truth of (69) from that of (68) is a weak formulation of the Principle of the Permanence of Formal Laws. Even at the equational level it does not express the full extent of the validity of the Principle. There is no evident way, for example, of establishing from this inference the further validity of inferences like: from the truth of

$$\text{for all } x, y \in \mathbb{R}, \ f(f_1(x,y), f_2(x,y)) = g(g_1(x,y), g_2(x,y)),$$

conclude the truth of

$$\text{for all } x, y \in {}^\Omega\mathbb{R}, \ {}^\Omega f({}^\Omega f_1(x,y), \ {}^\Omega f_2(x,y)) = {}^\Omega g({}^\Omega g_1(x,y), \ {}^\Omega g_2(x,y)).$$

That is, the inference of (69) from (68) does not obviously cover the reduction of the validity of, say,

$$\text{for all } x, y, z, \ x + (y + z) = (x + y) + z$$

in $^\Omega\mathbb{R}$ from the validity of the associative law in $\mathbb{R}$.

Concomitant with this question of the proper expression of the equational validity of the Principle of the Permanence of Formal Laws is the question of the extent of the validity of the Principle in a non-equational setting. For, note that $^\Omega\mathbb{R}$ does not inherit a number of properties of $\mathbb{R}$. It has zero divisors:

$$[(0, 1, 0, 1, \ldots)] \cdot [(1, 0, 1, 0, \ldots)] = [(0, 0, 0, 0, \ldots)] = 0.$$

The ordering is not linear: $[(0, 1, 0, 1, \ldots)] \neq [(1, 0, 1, 0, \ldots)]$ and yet

$$[(0, 1, 0, 1, \ldots)] \not< [(1, 0, 1, 0, \ldots)] \quad \text{and} \quad [(1, 0, 1, 0, \ldots)] \not< [(0, 1, 0, 1, \ldots)].$$

And, of course, the Archimedean property fails.

In order to delineate properly a large class of formal laws to which the permanence principle applies, as well as to state later the contribution of Abraham Robinson to the field, we must now go ultraformal and introduce the language of mathematical logic.

Mathematical Logic is Type II Formalism on steroids. In it one does not just replace this or that particular fuzzy concept by a formally defined one; one replaces language itself by a formal concept. The notions of proof and truth are formalised. When one does this for a given theory, the notion of validity within the theory can be studied and the possibility of establishing transfer principles like the Principle of the Permanence of Formal Laws arises. However, when one actually attempts not to study the theory, but to carry out the development formally, the theory generally becomes unintelligible and students have been known to rebel against it. It might also be added that the elementary syntactical parts of logic can be a bit boring as every concept is defined inductively and every elementary fact then proven by induction on the complexity (number of symbols, number of generating steps, etc.) of a term, formula, or derivation. I do not intend to subject the reader to all of this, but merely to give as informally as possible a brief introduction to the notions of a formal language and its interpretation within a given structure.

A *structure* is just a set (e.g. $\mathbb{Q}, \mathbb{R}$) together with some designated functions of one or more arguments defined on it (e.g., $+, \cdot$), some designated relations— possibly unary— defined on the set (e.g., $<, =, \equiv$, and even $\mathbb{N}$ considered as a unary relation on $\mathbb{Q}$ or $\mathbb{R}$), and some designated *individuals* of the set (e.g., $0, 1$). In mathematical logic, one introduces *names* for all the designated functions, relations, and individuals of the structure and uses them to generate the *terms* and *formulæ* of a language called the *language of the structure* in which to state properties of the structure. The generation is inductive in nature.

The key structure in nonstandard analysis is that of the real numbers. Thus we take $\mathbb{R}$ to be the underlying set of the structure. If our interest were only algebra, we could take $+, \cdot$ as designated functions, $<, =$ as designated relations, and $0, 1$ as designated individuals. In nonstandard analysis, however, we wish to do Calculus, which deals with more general functions. Rather than trying to pin down those functions necessary for the task in advance, we might as well consider every function $f : \mathbb{R}^n \to \mathbb{R}$ for some $n \geq 1$ to be designated, and every number $r \in \mathbb{R}$ to be designated. The language $L_{\mathbb{R}}$ of the reals will thus be generated from the following *primitives*:

function symbols. $\overline{f}$ for every function $f : \mathbb{R}^n \to \mathbb{R}$ for some $n \geq 1$;
relation symbols. $\overline{R}$ for every relation $R \subseteq \mathbb{R}^n$ for some $n \geq 1$;
constant symbols. $\overline{r}$ for every $r \in \mathbb{R}$.

The inductive generation of the language begins with the generation of the terms of the language.

6.5 Definition. *The class of terms of $L_{\mathbb{R}}$ is the smallest class of expressions satisfying the following:*
i.   every constant $\overline{r}$ for $r \in \mathbb{R}$ is a term;
ii. every variable v is a term;
iii. if $\overline{f}$ is a function symbol naming an n-ary function f, and $t_0, \ldots, t_{n-1}$ are terms, then $\overline{f}(t_0, \ldots, t_{n-1})$ is a term.

In mathematical logic, one explains where all the names and variables come from and that $\overline{f}(t_0, \ldots, t_{n-1})$ is a sequence of symbols. I don't think we need to go into this here, except perhaps to emphasise that the names and variables are considered syntactic objects. The names, i.e. function symbols and constant symbols, refer to specific objects and in that sense have meanings. The variables do not refer to anything in particular and have no meanings of their own. In terms they are mere place-holders, waiting to be replaced by constants so that the terms acquire meanings.

6.6 Definition. *An atomic formula of $L_{\mathbb{R}}$ is an expression of one of the forms,*

$$t_0 = t_1, \quad \overline{R}(t_0, \ldots, t_{n-1}),$$

where $t_0, \ldots, t_{n-1}$ are real terms and $\overline{R}$ is an n-ary relation symbol.

6.7 Definition. *The class of formulæ of $L_\mathbb{R}$ is the smallest class of expressions satisfying the following:*

i.   every atomic formula φ is a formula;
ii. if φ, ψ are formulæ, then so are $(\varphi \wedge \psi), (\varphi \vee \psi), (\varphi \to \psi)$ and $(\neg\varphi)$;
iii. if φ is a formula and v a variable, then $(\forall v \varphi)$ and $(\exists v \varphi)$ are formulæ.

As in algebra, the parentheses are necessary for the unique readability of a formula. Also as in algebra one omits them by creating an order of precedence. Negation ($\neg$) and quantification ($\forall$ and $\exists$) have the smallest possible scopes, like the first minus sign in the expression,

$$-7 + 6 - 3 \cdot 5.$$

Then come conjunction ($\wedge$), disjunction ($\vee$), and implication ($\to$) in that order. These conventions pretty much follow informal usage. One place where informal usage is not followed is the placement of the quantifiers. Informally, one may write something like

for all $\epsilon > 0$, there is an n_0 such that $|a_n - L| < \epsilon$ for all $n > n_0$

to express that L is the limit of a sequence $a_0, a_1, a_2, \ldots$ In formal logic, the quantifier always precedes its *scope*. Thus

$$\forall \epsilon > \overline{0} \, \exists n_0 \, |a_n - \overline{L}| < \epsilon \, \forall n > n_0$$

is not a formula, but

$$\forall \epsilon > \overline{0} \, \exists n_0 \forall n > n_0 (|a_n - \overline{L}| < \epsilon)$$

is. A bit more formally, we would write

$$\forall \epsilon ((\epsilon > \overline{0}) \to (\exists n_0 ((n_0 \in \mathbb{N}) \wedge (\forall n (((n \in \mathbb{N}) \wedge (n > n_0)) \to (|a_n - \overline{L}| < \epsilon)))))).$$

One can be even more formal than this by i) writing the inequalities in the form $\overline{R}(t_0, t_1)$ where $R \subseteq \mathbb{R} \times \mathbb{R}$ is the ordering relation, ii) writing the assertions $n \in \mathbb{N}$ as $\overline{\mathbb{N}}(n)$, and iii) replacing $|a_n - L|$ by a compound term built up from function symbols for absolute value, subtraction, and the sequence $a_0, a_1, a_2, \ldots$ viewed as a function, say,

$$a(x) = \begin{cases} a_x, & x \in \mathbb{N} \\ 0, & x \notin \mathbb{N}. \end{cases}$$

Fortunately, we needn't approach such an extreme level of formality here.

We have defined the *syntax* of the language $L_\mathbb{R}$ of the reals needed for any sort of nonstandard analysis. Standing side-by-side with syntax is *semantics*. The language itself does not tell us the meaning of an expression or the truth of a formula. The term $\overline{f}(x, y)$ could as easily denote the multiplication function as the addition one. For the language and structure defined, the intended meanings are clear. $\overline{f}$ denotes the function f, $\overline{R}$ the relation R, etc. But we also have the structure $^\Omega R$ in which to interpret the real terms and formulæ.

6.8 Definition. *Let* $\mathfrak{M}$ *be a structure and suppose each constant* $\bar{c}$ *of the language names an individual* $c^{\mathfrak{M}}$ *of the domain of* $\mathfrak{M}$ *and each n-ary function symbol* $\bar{f}$ *is assigned an n-ary function* $f^{\mathfrak{M}}$ *of the domain of* $\mathfrak{M}$. *We define the value* $t^{\mathfrak{M}}$ *of any variable-free term* t *inductively by:*

i. $\bar{c}^{\mathfrak{M}}$ *is* $c^{\mathfrak{M}}$;

ii. $\bar{f}(t_0, \ldots, t_{n-1})^{\mathfrak{M}}$ *is* $f^{\mathfrak{M}}(t_0^{\mathfrak{M}}, \ldots, t_{n-1}^{\mathfrak{M}})$.

Thus, for example, for the language $L_{\mathbb{R}}$ and the choices $\mathbb{R}$ and $^{\Omega}\mathbb{R}$ for $\mathfrak{M}$, where $\bar{r}^{\mathfrak{M}}$ is r and $\bar{f}^{\mathfrak{M}}$ is f and $^{\Omega}f$, respectively, we have

$$\bar{f}(\bar{2}, \bar{3})^{\mathfrak{M}} \text{ is } f(2,3)$$

for $\mathfrak{M} = \mathbb{R}$, while

$$\bar{f}(\bar{2}, \bar{3})^{\mathfrak{M}} \text{ is } {}^{\Omega}f(\,(2,2,\ldots),(3,3,\ldots)\,)$$

for $\mathfrak{M} = {}^{\Omega}\mathbb{R}$. But this last is just $(f(2,3), f(2,3), \ldots)$, which we identify with $f(2,3)$ itself. Under the identification of reals with the equivalence classes of their constant sequences, every term of the language $L_{\mathbb{R}}$ will have the same value in $^{\Omega}\mathbb{R}$ as in $\mathbb{R}$.

Notice that we do not define a value for any term t that contains a variable. This is because the variable has no value of its own. In logic, one can define the value of a term under an *assignment* of values to its variables, but we shall not need this here.

The language $L_{\mathbb{R}}$ is perfect for $\mathbb{R}$, but it lacks names for elements of $^{\Omega}\mathbb{R}$ that are not in $\mathbb{R}$. To this end, we simply enlarge the language.

6.9 Definition. *Let* $\mathfrak{M}$ *be a structure for a language* L, *i.e.* $\mathfrak{M}$ *is given by a set* M, *functions* $\bar{f}^{\mathfrak{M}}$, *relations* $\bar{R}^{\mathfrak{M}}$, *and individuals* $\bar{c}^{\mathfrak{M}}$ *for all function symbols* $\bar{f}$, *relation symbols* $\bar{R}$, *and constants* $\bar{c}$ *of* L. *We define the* augmented language $L(\mathfrak{M})$ *to be the language obtained by adding to* L *new constants* $\overline{m}$ *for all* $m \in \mathfrak{M}$. *The value of the new constant* $\overline{m}$ *is the element* m *itself.*

Thus, $L_{\mathbb{R}}(^{\Omega}\mathbb{R})$ is obtained by adding to $L_{\mathbb{R}}$ names for all the equivalence classes of sequences of real numbers. For example, the element $\Omega = [(0, 1, 2, \ldots)]$ itself now has a constant $\overline{\Omega}$ naming it.

The reason for introducing the augmented language is not just to accommodate the previously unnamed elements of $^{\Omega}\mathbb{R}$, but to allow us to define the semantic notion of truth within a structure.

6.10 Definition. *A sentence* in a language L *is a formula of* L *which contains no* free variables, *i.e. unquantified variables— variables lying outside the scopes of any quantifiers binding them.*

6.11 Definition. *Let* $\mathfrak{M}$ *be a structure for a language* L. *We inductively define what it means for a sentence* φ *of* $L(\mathfrak{M})$ *to be* true in $\mathfrak{M}$, *written* $\mathfrak{M} \vDash \varphi$, *as follows*

i.   *if* t_0, t_1 *are terms with no variables,*

$$\mathfrak{M} \models t_0 = t_1 \quad \textit{iff} \quad t_0^{\mathfrak{M}} = t_1^{\mathfrak{M}};$$

ii.   *if $\overline{R}$ is an n-ary relation symbol and $t_0, \dots, t_{n-1}$ are terms with no variables, then*

$$\mathfrak{M} \models \overline{R}(t_0, \dots, t_{n-1}) \quad \textit{iff} \quad \overline{R}^{\mathfrak{M}}(t_0^{\mathfrak{M}}, \dots, t_{n-1}^{\mathfrak{M}});$$

iii. *if φ is a sentence of $L(\mathfrak{M})$,*

$$\mathfrak{M} \models \neg\varphi \quad \textit{iff} \quad \mathfrak{M} \not\models \varphi \ \ (\textit{i.e. } \varphi \text{ is not true in } \mathfrak{M});$$

iv.   *if φ, ψ are sentences of $L(\mathfrak{M})$,*

$$\mathfrak{M} \models \varphi \wedge \psi \quad \textit{iff} \quad \mathfrak{M} \models \varphi \ \& \ \mathfrak{M} \models \psi;$$

v.   *if φ, ψ are sentences of $L(\mathfrak{M})$,*

$$\mathfrak{M} \models \varphi \vee \psi \quad \textit{iff} \quad \mathfrak{M} \models \varphi \ \textit{or} \ \mathfrak{M} \models \psi;$$

vi.   *if φ, ψ are sentences of $L(\mathfrak{M})$,*

$$\mathfrak{M} \models \varphi \rightarrow \psi \quad \textit{iff} \quad \mathfrak{M} \not\models \varphi \ \textit{or} \ \mathfrak{M} \models \psi;$$

vii. *if $\forall v\varphi(v)$ is a sentence of $L(\mathfrak{M})$,*

$$\mathfrak{M} \models \forall v\varphi(v) \quad \textit{iff} \quad \textit{for all } m \in M, \ \mathfrak{M} \models \varphi(\overline{m});$$

viii. *if $\exists v\varphi(v)$ is a sentence of $L(\mathfrak{M})$,*

$$\mathfrak{M} \models \exists v\varphi(v) \quad \textit{iff} \quad \textit{for some } m \in M, \ \mathfrak{M} \models \varphi(\overline{m}).$$

The formal definition is fairly intuitive. Indeed, before Alfred Tarski first gave a formal definition of truth in the 1930's[99], mathematical logicians had no difficulty with the concept of truth in a structure. Just as terms with free variables had no definite values, formulæ with free variables have no truth values assigned to them for the intuitively obvious reason that varying assignments of values to the unquantified variables will result in varying truth values of the resulting sentences. Just as one can modify the interpretation of terms to determine this value under assignments of values to the variables, one can— and indeed this is what Tarski initially did— define what it means for an assignment of values to the free variables to *satisfy* the given formula. The definition of satisfaction is a lot more daunting on first exposure than is Definition 6.11, but it does more readily impress one that something definite has been achieved, and it is of greater use in more advanced logic. We do not need it here. What one can say is whether or not a formula is *valid* in a structure: If φ has $v_0, \dots, v_{n-1}$

---

[99] Alfred Tarski, "Der Wahrheitsbegriff in den formalisierten Sprachen", *Studia Philosophica* 1 (1936), pp.. 261 - 405. An English translation by J.H. Woodger appears in: Alfred Tarski, *Logic, Semantics, Metamathematics; Papers from 1923 to 1938*, Oxford University Press, Oxford, 1956.

as its only free variables, we say φ is valid in a structure $\mathfrak{M}$ and write $\mathfrak{M} \vDash \varphi$ just in case $\forall v_0 \ldots \forall v_{n-1}\varphi$ is true in $\mathfrak{M}$, i.e. in case $\varphi(\overline{m}_0, \ldots, \overline{m}_{n-1})$ is true in $\mathfrak{M}$ for every choice of $m_0, \ldots, m_{n-1}$.[100]

We can now ask, with respect to the Principle of the Permanence of Formal Laws, which "formal laws" carry over from $\mathbb{R}$ to $^{\Omega}\mathbb{R}$. It turns out that a related question had already been asked and partially answered in 1951 by Alfred Horn who proved that a certain class of formulæ now called Horn formulæ in his honour had their validity preserved in taking direct products of structures and, in particular, in taking direct powers, i.e. passages like that from $\mathbb{R}$ to $\mathbb{R}^{\mathbb{N}}$. Around 1962, Chen Chung Chang proved that Horn formulæ were also preserved under *reduced direct products*, special types of homomorphic images of direct products that included the passage from $\mathbb{R}$ to $^{\Omega}\mathbb{R}$.[101] [102] These formulæ are readily defined:

6.12 Definition. *A formula φ is a* Horn formula *if it is of the form*

---

[100] I have cheated a tiny bit here and in stating clauses vii and viii of Definition 6.11 in writing $\varphi(\overline{m})$ without explanation. The intention, of course, is to denote the formula obtained from φ by replacing every *free* occurrence of the variable v by the constant $\overline{m}$. The situation is familiar from the Calculus, where dx *binds* a variable much as a quantifier does. If $f(x)$ is

$$x^2 + \int_0^5 x\,dx,$$

then $f(2)$ is

$$2^2 + \int_0^5 x\,dx,$$

and not

$$2^2 + \int_0^5 2\,dx \quad \text{or} \quad 2^2 + \int_0^5 2\,d2.$$

If one wishes to be absolutely precise, one can give a formal inductive definition of substitution of a term for the free occurrences of a variable in a formula. I refer the reader to any textbook on logic for such.

[101] I suppose I should mention that introducing Horn formulæ into the account is not historically accurate. As late as 1983 Laugwitz ("Ω-calculus as a generalization of field extension; an alternative approach to nonstandard analysis", in: A.E. Hurd (ed.), *Nonstandard Analysis— Recent Developments*, Springer-Verlag, Heidelberg, 1983, p. 124) stated that the principle that $A(\Omega)$ held when $A(n)$ was almost always true had only been assumed for formulæ A with no logical symbols. Reference to Horn formulæ allows us to streamline our exposition ever so slightly.

[102] A converse also holds: If a formula is preserved under passage to reduced direct products, then it is *logically equivalent* to a Horn formula. This result, first proven in 1965 by H. Jerome Keisler under the assumption of the *continuum hypothesis* (i.e., the hypothesis that there is no cardinal strictly between $\aleph_0$ and $\mathfrak{c}$), is rather deep and only to be found in more advanced logic books. We will not need it here. Incidentally, Horn formulæ have proven to be of some interest in computer science. *Cf.* Jean H. Gallier, *Logic for Computer Science; Foundations of Automatic Theorem Proving*, Harper & Row, New York, 1986.

$$Q_0 v_0 Q_1 v_1 \ldots Q_{n-1} v_{n-1} \bigwedge_{i=0}^{k} (\psi_{i,0} \wedge \ldots \wedge \psi_{i,m_i-1} \to \psi_i), \qquad (70)$$

where each Q_i is a quantifier $\forall$ or $\exists$, and each ψ_i and each $\psi_{i,j}$ is atomic, and $\bigwedge_{i=0}^{k}$ denotes the $(k+1)$-fold conjunction.

In this definition, the choice $n = 0$ is intended to mean that there are no quantifiers, and the choice $m = 0$ means the *matrix* of the formula, $\psi_0 \wedge \ldots \wedge \psi_{m-1} \to \psi$, consists solely of ψ. Notice that ψ is equivalent to $x = x \to \psi$ and the case $m = 0$ is not really different from any other case. If one is dealing with structures in which some atomic formula, like $\bar{0} = \bar{1}$, is false, then, using it for ψ, the matrix $\psi_0 \wedge \ldots \wedge \psi_{m-1} \to \bar{0} = \bar{1}$ can also be written in the form $\neg(\psi_0 \wedge \ldots \wedge \psi_{m-1})$, i.e. $\neg\psi_0 \vee \ldots \neg \vee \psi_{m-1}$. If there is no such common false atomic sentence, one must either broaden the definition to allow matrices of this form, or extend the language by the addition of an atomic sentence *falsum*, usually written $\curlywedge$, $\bot$, or even $\mathbf{f}$, allowing these disjunctions of negated atomic formulæ to be written in the Horn form:

$$\psi_0 \wedge \ldots \wedge \psi_{m-1} \to \curlywedge.$$

6.13 Theorem. *For any Horn formula φ,*

$$\mathbb{R} \vDash \varphi \quad \Rightarrow \quad {}^{\Omega}\mathbb{R} \vDash \varphi.$$

I leave it to the reader to verify this for himself, to look up the proof, or to accept the result without proof. I shall simply list a few Horn formulæ to which the Theorem applies and then a few non-Horn formulæ which are not preserved in the passage from $\mathbb{R}$ to ${}^{\Omega}\mathbb{R}$.

6.14 Examples. The following are expressed by Horn formulæ
i.   Commutativity.

$$\forall x \forall y (x + y = y + x), \quad \forall x \forall y (xy = yx);$$

ii. Associativity.

$$\forall x \forall y \forall z (x + (y + z) = (x + y) + z), \quad \forall x \forall y \forall z (x(yz) = (xy)z);$$

iii. Distributivity.

$$\forall x \forall y \forall z (x(y + z) = (xy) + (xz));$$

iv. Identity Elements.

$$\forall x (x + \bar{0} = x), \quad \forall x (\bar{1}x = x);$$

v.   Existence of Additive Inverse.

$$\forall x \exists y (x + y = \bar{0});$$

vi. Partial Existence of Multiplicative Inverse.

$$\forall x \exists y (\overline{0} < x \to xy = \overline{1})$$
$$\forall x \exists y (x < \overline{0} \to xy = \overline{1});$$

vii. Transitivity of Order.

$$\forall x \forall y \forall z (x < y \wedge y < z \to x < z);$$

viii. Antisymmetry.

$$\forall x \forall y (x < y \to \neg y < x),$$

i.e. $\forall x \forall y (x < y \wedge y < x \to \overline{0} = \overline{1});$
ix. The Limit of a Sequence $\overline{f}$ is L.

$$\forall \epsilon \, \exists n_0 \forall n \big(n > n_0 \wedge \overline{\mathbb{N}}(n) \wedge \epsilon > \overline{0} \to |\overline{f}(n) - L| < \epsilon\big)$$

(This will require a little explanation, which I will give after I've listed the rest
of the examples and non-examples.);
x. The Limit of a Function $\overline{f}$ at a is L.

$$\forall \epsilon \, \exists \delta \, \forall x \Big(\big(\delta > \overline{0}\big) \wedge \big(\overline{0} < |x - a| \wedge |x - a| < \delta \wedge \epsilon > \overline{0} \to |\overline{f}(x) - L| < \epsilon\big)\Big);$$

xi. $\overline{f}$ is continuous at a.

$$\forall \epsilon \, \exists \delta \, \forall x \Big(\big(\delta > \overline{0}\big) \wedge \big(|x - a| < \delta \wedge \epsilon > \overline{0} \to |\overline{f}(x) - \overline{f}(a)| < \epsilon\big)\Big).$$

6.15 Examples. The following are not definable by Horn formulæ.
i.   Full Existence of a Multiplicative Inverse.

$$\forall x \exists y (\neg x = \overline{0} \to xy = \overline{1});$$

ii. Non-Existence of Zero-Divisors.

$$\forall x \forall y (xy = \overline{0} \to x = \overline{0} \vee y = \overline{0});$$

iii. Trichotomy.

$$\forall x \forall y (x < y \vee x = y \vee y < x);$$

iv. Alternate Definition of Limit of a Function.

$$\forall \epsilon \, \exists \delta \, \forall x \Big(\big(\delta > \overline{0}\big) \wedge \big(\neg |x - a| = \overline{0} \wedge |x - a| < \delta \wedge \epsilon > \overline{0} \to |\overline{f}(x) - L| < \epsilon\big)\Big);$$

v. Induction.

$$\varphi(\overline{0}) \wedge \forall x \big(\overline{\mathbb{N}}(x) \wedge \varphi(x) \to \varphi(x + \overline{1})\big) \to \forall x \big(\overline{\mathbb{N}}(x) \to \varphi(x)\big).$$

The sentences of Examples 6.15.i-iii are valid in $\mathbb{R}$, but not in $^{\Omega}\mathbb{R}$; the
formula 6.15.iv, which correctly defines the notion of limit in $\mathbb{R}$, does not work
in $^{\Omega}\mathbb{R}$; and induction fails in $^{\Omega}\mathbb{N}$ for formulæ of $L_{\mathbb{R}}$. The reader is asked to
work through these as Exercises 9.19 to 9.21 at the end of the Chapter.

I mentioned in the middle of Examples 6.14 that the definition of the limit of a sequence required some comment. It is not the definition of limit, but the use of the function symbol $\overline{f}$ for a sequence that requires comment. A sequence $a : a_0, a_1, a_2, \ldots$ is a function from $\mathbb{N}$ into $\mathbb{R}$ and we have only introduced function symbols $\overline{f}$ for functions from $\mathbb{R}^n$ into $\mathbb{R}$ for some $n \in \mathbb{N}$. So the question is what to do with sequences. There are two solutions. One is to take any sequence a and extend it arbitrarily to a function from $\mathbb{R}$ into $\mathbb{R}$. For example, we could define

$$f_a(x) = \begin{cases} a_x, & x \in \mathbb{N} \\ 0, & x \notin \mathbb{N}. \end{cases}$$

Notice that, if f, g are any such extensions,

$$\mathbb{R} \vDash \forall x \big(\overline{\mathbb{N}}(x) \to \overline{f}(x) = \overline{g}(x) \big),$$

and, this being a Horn sentence,

$$^{\Omega}\mathbb{R} \vDash \forall x \big(\overline{\mathbb{N}}(x) \to \overline{f}(x) = \overline{g}(x) \big),$$

i.e. the restrictions of $^{\Omega}f, {}^{\Omega}g$ to $^{\Omega}\mathbb{N}$ agree and thus define the same "sequence" mapping $^{\Omega}\mathbb{N}$ into $^{\Omega}\mathbb{R}$.

An alternate approach is to consider the *graph* of the sequence viewed as a function:

$$R_a(x, y) : \ x \in \mathbb{N} \ \& \ y = a_x.$$

The functional nature of the sequence is expressed by Horn formulæ:

$$\forall x \exists y \big(\overline{\mathbb{N}}(x) \to \overline{R}_a(x, y) \big), \quad \forall x \forall y \big(\overline{R}_a(x, y) \to \overline{\mathbb{N}}(x) \big)$$
$$\forall x \forall y \forall z \big(\overline{\mathbb{N}}(x) \wedge \overline{R}_a(x, y) \wedge \overline{R}_a(x, z) \to y = z \big),$$

whence $^{\Omega}\mathbb{R}$ is the graph of a function with domain $^{\Omega}\mathbb{N}$ mapping into $^{\Omega}\mathbb{R}$, i.e. it is the graph of a "sequence" in $^{\Omega}\mathbb{R}$.

The same results apply *mutatis mutandis* to other partially defined functions of a real variable, such as functions defined on an interval $[\alpha, \beta]$. If D is the domain of a function f, we can consider, for example, the graph R_f and observe that the validity of the Horn formulæ,

$$\forall x \exists y \big(\overline{D}(x) \to \overline{R}_f(x, y) \big), \quad \forall x \forall y \big(\overline{R}_f(x, y) \to \overline{D}(x) \big)$$
$$\forall x \forall y \forall z \big(\overline{D}(x) \wedge \overline{R}_f(x, y) \wedge \overline{R}_f(x, z) \to y = z \big),$$

is preserved in the passage from $\mathbb{R}$ to $^{\Omega}\mathbb{R}$.

Note too that intervals are lifted to intervals: If D is $\{x \in \mathbb{R} \,|\, \alpha \leq x \leq \beta\}$, then the Horn formulæ

$$\forall x \big(\overline{D}(x) \to \overline{\alpha} \leq x \big), \quad \forall x \big(\overline{D}(x) \to x \leq \overline{\beta} \big), \quad \forall x \big(\overline{\alpha} \leq x \wedge x \leq \overline{\beta} \to \overline{D}(x) \big)$$

are preserved: $^{\Omega}[\alpha, \beta] = \{x \in {}^{\Omega}\mathbb{R} \,|\, \alpha \leq x \leq \beta\}$. This works for open and half-open intervals as well, but not for punctured intervals:

$$\Omega\big([\alpha,\beta)\cup(\beta,\gamma]\big)\neq {}^{\Omega}[\alpha,\beta)\cup {}^{\Omega}(\beta,\gamma].$$

For, the equivalence class of, e.g., the sequence $(\alpha,\gamma,\alpha,\gamma,\alpha,\gamma,\dots)$ lies in the first set but not in the second.

The point here is that the framework is broad enough to accommodate sequences and functions defined on intervals, but one may have to be ever so slightly careful in considering the domains of definition of such functions when lifting them to ${}^{\Omega}\mathbb{R}$.

The fact is that ${}^{\Omega}\mathbb{R}$ is a mess, but it doesn't matter because it is not ${}^{\Omega}\mathbb{R}$ we are interested in. The *raison d'être* of nonstandard analysis is, at least initially, to replace the cumbersome ϵ-δ-machinery of the theory of limits by the more intuitive appeal to the infinitely large and infinitely small, which ${}^{\Omega}\mathbb{R}$ has in abundance. It also has the infinitely wobbly, which we simply don't need to use.

The reader who has been going through this book in a linear fashion has already been introduced to infinitesimals three times. To him, the following definition is a mere formality.

6.16 Definition. *An element $x\in {}^{\Omega}\mathbb{R}$ is an* infinitesimal *if $|x|<\epsilon$ for every positive real number ϵ, i.e. if $-\epsilon<x<\epsilon$ for all $\epsilon\in\mathbb{R}^{+}$.*

Concomitant with this definition one can define x to be *infinite* if $|x|>r$ for every $r\in\mathbb{R}$, *finite* or *bounded* if $|x|<r$ for some $r\in\mathbb{R}$, *positive* if $x>0$, etc. Note that because of the oscillation of some sequences, an element $x\in {}^{\Omega}\mathbb{R}$ might be neither finite nor infinite, positive nor negative. This is an inconvenience, but not an insurmountable one.

The infinitesimals have some nice closure properties.

6.17 Lemma. *i. The sum of two infinitesimals is infinitesimal.*
ii. The product of an infinitesimal and a finite number is infinitesimal.

Proof. i. Let x,y be infinitesimal and let $\epsilon\in\mathbb{R}^{+}$ be given. Because x,y are infinitesimal, we have

$$|x|<\frac{\epsilon}{2},\quad |y|<\frac{\epsilon}{2}.$$

Thus,

$$|x+y|\leq|x|+|y|<\frac{\epsilon}{2}+\frac{\epsilon}{2}=\epsilon,$$

and $x+y$ is infinitesimal. [*N.B.* We are using basic properties of $\mathbb{R}$ in ${}^{\Omega}\mathbb{R}$. We can do this because the properties are expressible by Horn formulæ. For example,

$$\forall v\forall w\Big(|v|<\frac{\epsilon}{2}\wedge|w|<\frac{\epsilon}{2}\rightarrow|v+w|<\epsilon\Big).$$

For very elementary arguments like this, I leave it to the reader to provide proper justification.]

ii. Let x be infinitesimal and y finite, say $|y|<r\in\mathbb{R}$. Choose $\epsilon\in\mathbb{R}^{+}$ and appeal to the infinitesimality of x to conclude

$$|x| < \frac{\epsilon}{r}.$$

But

$$|xy| = |x| \cdot |y| < \frac{\epsilon}{r} \cdot r = \epsilon. \qquad \square$$

One can similarly prove the finite numbers to be closed under addition and multiplication, the multiplicative inverse of an infinite element to exist and to be infinitesimal, and the multiplicative inverse of an invertible infinitesimal to be infinite. [*N.B.* Some infinitesimals are not invertible.] I leave these considerations to the reader.

What I wish to do is to illustrate the use of infinitesimals in discussing limits. To this end, we formally introduce another familiar equivalence relation.

6.18 Definition. *Two elements* $x, y \in {}^{\Omega}\mathbb{R}$ *are* infinitesimally close, *written* $x \approx y$, *if their difference* $x - y$ *is infinitesimal.*

6.19 Theorem. *Let* $f : \mathbb{R} \to \mathbb{R}$ *be given and let* $a \in \mathbb{R}$. *The following are equivalent:*
i. f is continuous at a;
ii. ${}^{\Omega} f(a + \eta) \approx {}^{\Omega} f(a)$ for every infinitesimal η of ${}^{\Omega}\mathbb{R}$.

Proof. i $\Rightarrow$ ii. Suppose f is continuous at a and ϵ is any positive real number. By continuity, there is a positive real number δ such that

$$\mathbb{R} \vDash \forall x \big(|x - \overline{a}| < \overline{\delta} \to |\overline{f}(x) - \overline{f}(\overline{a})| < \overline{\epsilon} \big),$$

and thus

$${}^{\Omega}\mathbb{R} \vDash \forall x \big(|x - \overline{a}| < \overline{\delta} \to |\overline{f}(x) - \overline{f}(\overline{a})| < \overline{\epsilon} \big).$$

For η infinitesimal,

$$|a + \eta - a| = |\eta| < \delta,$$

whence

$$\big| {}^{\Omega} f(a + \eta) - {}^{\Omega} f(a) \big| < \epsilon.$$

As this is true for all positive real ϵ, it follows that ${}^{\Omega} f(a + \eta) \approx {}^{\Omega} f(a)$.

ii $\Rightarrow$ i. Assume f is not continuous at a: for some real $\epsilon > 0$ and any real $\delta > 0$ there is a real number x such that

$$|x - a| < \delta, \quad \text{but} \quad |f(x) - f(a)| > \epsilon.$$

For each $n \in \mathbb{N}$, choose a number $x_n \in \mathbb{R}$ satisfying

$$|x_n - a| < \frac{1}{n + 1}, \quad \text{but} \quad |f(x_n) - f(a)| > \epsilon.$$

Let x be the equivalence class in ${}^{\Omega}\mathbb{R}$ of $(x_0, x_1, x_2, \ldots)$. By definition, ${}^{\Omega} f(x)$ is the equivalence class of the sequence $(f(x_0), f(x_1), f(x_2), \ldots)$ and

$$\big| {}^{\Omega} f(x) - {}^{\Omega} f(a) \big| > \epsilon$$

since $|f(x_n) - f(a)| > \epsilon$ for all $n \in \mathbb{N}$. But, for all $n \in \mathbb{N}$,

$$|x - a| < \frac{1}{n+1},$$

since $|x_m - a| < 1/(n+1)$ for all $m > n+1$. This makes $x - a$ infinitesimal, say η, and $^{\Omega}f(a+\eta) = {}^{\Omega}f(x) \not\approx {}^{\Omega}f(a)$. $\square$

The corresponding nonstandard characterisation of the limit of a function at a point of possible discontinuity requires a bit of care, as suggested by the inequivalence of 6.14.x and 6.15.iv. To this end, another formal definition is in order.

6.20 Definition. *Two elements x, y are* separated *or* apart *from one another, written $x \# y$, if the absolute value of their difference is greater than 0: $|x - y| > 0$.*

The *apartness relation* and its notation are borrowed from constructive mathematics, where a more positive notion of being different than mere negated equality is needed. Note that there are lots of elements of $^{\Omega}\mathbb{R}$ that are not equal to 0, but are not separated from 0. Every 0-divisor is such a number: Such a sequence equals 0 infinitely often and thus is not a positive distance from 0, but does not equal 0 often enough for the sequence itself to be 0.

6.21 Theorem. *Let $f : \mathbb{R} \to \mathbb{R}$ be a function and let $a, L \in \mathbb{R}$. The following are equivalent:*
i. $\lim_{x \to a} f(x) = L$;
ii. *for all infinitesimal $\eta \# 0$, $^{\Omega}f(a+\eta) \approx L$.*

I leave the proof as an exercise to the reader.

Limits of sequences can similarly be characterised. First, however, another definition.

6.22 Definition. *The Ω-natural numbers are those elements x of $^{\Omega}\mathbb{R}$ that are in $^{\Omega}\mathbb{N}$. An element $N \in {}^{\Omega}\mathbb{N}$ is* infinite *if $N > n$ for all $n \in \mathbb{N}$.*

A word about notation: It is tempting to let "Ω" or "ω" denote an infinite natural number. However, "Ω" is reserved for the particular infinite natural number determined by the sequence $(1, 2, 3, \ldots)$ and Schmieden and Laugwitz use "ω" to denote its infinitesimal reciprocal, $1/\Omega$. The next obvious choice is "N".

An element $N \in {}^{\Omega}\mathbb{N}$ is an equivalence class of a sequence all but finitely many entries in which are natural numbers. Replacing these finitely many exceptions by natural numbers will not change the equivalence class, and we may thus choose a sequence $(i_0, i_1, i_2, \ldots)$ of natural numbers to represent N:

$$N = [(i_0, i_1, i_2, \ldots)].$$

Thinking of a sequence $a \in \mathbb{R}^{\mathbb{N}}$ as a function f,

$$f(n) = a_n,$$

we may consider its extension $^{\Omega}f : {}^{\Omega}\mathbb{R} \to {}^{\Omega}\mathbb{R}$:

$$a_N = {}^{\Omega}f(N) = [(f(i_0), f(i_1), f(i_2), \ldots)] = [(a_{i_0}, a_{i_1}, a_{i_2}, \ldots)].$$

In particular, $a_{\Omega} = [a]$. And, if $i_0 < i_1 < \ldots$, then a_N is the equivalence class of the subsequence determined by $(i_0, i_1, i_2, \ldots)$.

6.23 Theorem. *Let* $a : a_0, a_1, a_2, \ldots$ *be a sequence of real numbers and let* $L \in \mathbb{R}$. *The following are equivalent:*
i. $\lim_{n \to \infty} a_n = L$;
ii. for all infinite $N \in {}^{\Omega}\mathbb{N}$, $a_N \approx L$;
iii. $a_{\Omega} \approx L$.

Proof. i $\Rightarrow$ ii. Let L be the limit of the sequence and let a positive real number ϵ be given. There is some $n_0 \in \mathbb{N}$ such that

$$\mathbb{R} \vDash \forall n \big(\overline{\mathbb{N}}(n) \wedge n > \overline{n}_0 \to |a_n - \overline{L}| < \overline{\epsilon} \big).$$

This is a Horn sentence and thus true in $^{\Omega}\mathbb{R}$. For any infinite $N \in {}^{\Omega}\mathbb{N}$, one has

$$^{\Omega}\mathbb{R} \vDash \overline{\mathbb{N}}(\overline{N}) \wedge \overline{N} > \overline{n}_0,$$

whence $|a_N - L| < \epsilon$. This being true for all positive real ϵ, it follows that $a_N \approx L$.

ii $\Rightarrow$ iii. Ω is an infinite element of $^{\Omega}\mathbb{N}$.

iii $\Rightarrow$ i. Suppose $a_{\Omega} \approx L$, i.e. $a \approx L$. Then, for every $\epsilon \in \mathbb{R}^+$, $|a - L| < \epsilon$, i.e. for all but finitely many $n \in \mathbb{N}$, $|a_n - L| < \epsilon$. But this means there is an $n_0 \in \mathbb{N}$ such that for all $n > n_0, |a_n - L| < \epsilon$. As ϵ was arbitrary, this means $\lim_{n \to \infty} a_n = L$. $\square$

One can now use infinitesimals to prove basic properties of limits of sequences. For example, if

$$\lim_{n \to \infty} a_n = L_1 \quad \text{and} \quad \lim_{n \to \infty} b_n = L_2,$$

then $\lim_{n \to \infty}(a_n + b_n)$ exists and

$$\lim_{n \to \infty} (a_n + b_n) = L_1 + L_2.$$

For, from the assumptions, we have $a_{\Omega} \approx L_1, b_{\Omega} \approx L_2$, and thus $a_{\Omega} + b_{\Omega} \approx L_1 + L_2$. [Actually, we haven't proven that $\approx$ respects $+$. But this is easy:

$$a_{\Omega} + b_{\Omega} - (L_1 + L_2) = (a_{\Omega} - L_1) + (b_{\Omega} - L_2)$$

is the sum of two infinitesimals, whence by Lemma 6.17 is infinitesimal.]

Alongside the notion of the limit of a sequence is that of a limit point of the sequence.

6.24 Definition. *Let* $a : a_0, a_1, a_2, \ldots$ *be a sequence of real numbers. A number* L *is a limit point of* a *if for every* $\epsilon > 0$ *and any* $n_0 \in \mathbb{N}$ *there is an* $n \in \mathbb{N}$ *with* $n > n_0$ *such that* $|a_n - L| < \epsilon$.

In words, a limit point of a sequence a is a number L which the sequence infinitely often approaches, but, unlike a limit, the sequence can also infinitely often veer away from L. The simplest example is perhaps the sequence $0, 1, 0, 1, 0, 1, \ldots$, which has both 0 and 1 as limit points. A sequence can fail to have any limit points (as does $0, 1, 2, 3, \ldots$), or it can have infinitely many limit points. Indeed, if $q : q_0, q_1, q_2, \ldots$ is some enumeration of the rational numbers, then every real number is a limit point of q. (*Exercise.*)

6.25 Theorem. *Let* $a : a_0, a_1, a_2, \ldots$ *be a sequence of real numbers and let* $L \in \mathbb{R}$. *The following are equivalent:*
i. L is a limit point of a;
ii. for some infinite $N \in {}^{\Omega}\mathbb{N}$, $a_N \approx L$.

Proof. i $\Rightarrow$ ii. Assume L is a limit point of a:

$$\mathbb{R} \vDash \forall \epsilon \forall n_0 \exists n \big(\epsilon > \overline{0} \wedge \overline{\mathbb{N}}(n_0) \rightarrow \overline{\mathbb{N}}(n) \wedge n > n_0 \wedge |a_n - \overline{L}| < \epsilon \big).$$

This is a Horn sentence, whence it is true in ${}^{\Omega}\mathbb{R}$. Let ϵ be infinitesimal and $N_0 \in {}^{\Omega}\mathbb{N}$ infinite. There is an $N \in {}^{\Omega}\mathbb{N}$, with $N > N_0$ and $|a_N - L| < \epsilon$; i.e. there is an infinite N for which $a_N \approx L$.

[A less slick, but more constructive proof based on the structure of ${}^{\Omega}\mathbb{R}$ instead of the Horn nature of the formula asserting L to be a limit point of a proceeds as follows: Let L be a limit point of a and define a sequence $i_0 < i_1 < i_2 < \ldots$ of natural numbers as follows: i_0 is any natural number such that $|a_{i_0} - L| < 1$. Given $i_0 < i_1 < \ldots < i_n$, choose i_{n+1} to be any natural number greater than i_n such that $|a_{i_{n+1}} - L| < 1/(n + 2)$. (For the sake of definiteness, we can choose each i_n successively to be the least number satisfying the stated conditions.)

[Now, for each n, $|a_{i_n} - L| < 1/(n + 1)$, whence $|a_{i_m} - L| < 1/(n + 1)$ for $m > n$. Thus

$$\big|[(a_{i_0}, a_{i_1}, \ldots)] - L\big| < \frac{1}{n + 1},$$

i.e., if $N = [(i_0, i_1, i_2, \ldots)]$, we have $N \in {}^{\Omega}\mathbb{N}$ and

$$|a_N - L| < \frac{1}{n + 1}.$$

As this holds for all $n \in \mathbb{N}$, $a_N \approx L$.]

ii $\Rightarrow$ i. The converse is easy: From $a_N \approx L$ we have, for all real $\epsilon > 0$ and any $n_0 \in \mathbb{N}$,

$${}^{\Omega}\mathbb{R} \vDash \exists n \big(\overline{\mathbb{N}}(n) \wedge |a_n - \overline{L}| < \overline{\epsilon} \wedge n > \overline{n}_0 \big).$$

If $\mathbb{R}$ did not also satisfy this, we would have

$$\mathbb{R} \vDash \forall n \neg \big(\overline{\mathbb{N}}(n) \wedge |a_n - \overline{L}| < \overline{\epsilon} \wedge n > \overline{n}_0 \big)$$
$$\vDash \forall n \big(\overline{\mathbb{N}}(n) \wedge |a_n - \overline{L}| < \overline{\epsilon} \wedge n > \overline{n}_0 \rightarrow \overline{0} = \overline{1} \big).$$

But this is a Horn sentence and is also true in ${}^{\Omega}\mathbb{R}$, a contradiction. Thus, for any $\epsilon > 0$ and $n_0 \in \mathbb{N}$, there is some $n \in \mathbb{N}$ with $n > n_0$ such that $|a_n - L| < \epsilon$. I.e., L is a limit point of a. $\qquad\qquad\qquad\qquad\qquad\qquad\qquad\qquad\qquad\quad\square$

If, as in Schmieden's and Laugwitz's original paper, we had based our construction on $\mathbb{Q}^{\mathbb{N}}$, we would at some stage in our discussion have to introduce the reals and prove their completeness property. Because we have chosen to assume the reals as given, we may also assume their completeness. As we have already remarked, there are several equivalent formulations of completeness. In the early years of nonstandard analysis, expositions assumed Dedekind completeness: every cut is determined by a real number. If we follow tradition and assume this, we can easily derive the following result.

6.26 Theorem. *Let $a : a_0, a_1, a_2, \ldots$ be a sequence of real numbers and assume a is bounded, i.e. for some real number M, we have $|a_n| < M$ for every $n \in \mathbb{N}$. Then: a has a limit point in $\mathbb{R}$.*

Proof. Define sets A, B by

$$A = \{x \in \mathbb{R} \mid \text{for infinitely many } n \in \mathbb{N}, x \le a_n\}$$
$$B = \{x \in \mathbb{R} \mid x \notin A\}.$$

I claim $\langle A, B \rangle$ is a Dedekind cut. Assuming M to be a bound on the sequence:

$$\text{for all } n \in \mathbb{N}, \ |a_n| < M,$$

it follows that $-M \in A$ and $M \in B$, whence both sets are nonempty. They are clearly complementary and it is obvious that every element of A is less than every element of B.

Let $r \in \mathbb{R}$ determine the cut, so

$$A = \{x \in \mathbb{R} \mid x < r\} \quad \text{or} \quad A = \{x \in \mathbb{R} \mid x \le r\}$$

(exactly which is the case will vary with the choice of the sequence a). Let $\epsilon > 0$. Because $r - \epsilon/2 \in A$, there are infinitely many $n \in \mathbb{N}$ for which $a_n \ge r - \epsilon/2$. But only finitely many of these are greater than $r + \epsilon/2$, because this latter number is not in A. Hence there are infinitely many $n \in \mathbb{N}$ for which $r - \epsilon < a_n < r + \epsilon$, i.e. $|r - a_n| < \epsilon$. Thus r is a limit point of $a_0, a_1, \ldots$ □

6.27 Corollary (Bolzano-Weierstrass Theorem). *Every bounded sequence of real numbers has a convergent subsequence.*

Proof. I want to simply apply Theorems 6.25 and 6.26: If $a : a_0, a_1, a_2, \ldots$ is a bounded sequence of reals, it has a limit point L. There is then an infinite $N \in {}^{\Omega}\mathbb{N}$ such that $a_N \approx L$. Now N is represented in ${}^{\Omega}\mathbb{R}$ by a sequence of natural numbers, $N = [(i_0, i_1, i_2, \ldots)]$. So take the sequence b given by $b_n = a_{i_n}$. Then

$$b_{\Omega} = a_N \approx L$$

and $(b_0, b_1, b_2, \ldots) = (a_{i_0}, a_{i_1}, a_{i_2}, \ldots)$ is a sequence converging to L. Unfortunately, $a_{i_0}, a_{i_1}, a_{i_2}, \ldots$ need not be a subsequence of $a_0, a_1, a_2, \ldots$, as there may be some repetitions and reversals of order among the i_j's. Because N is infinite, one can pick a subsequence $i_{j_0}, i_{j_1}, i_{j_2}, \ldots$ that is strictly increasing:

$$i_{j_0} < i_{j_1} < i_{j_2} < \ldots$$

and the sequence b' defined by

$$b'_n = a_{i_{j_n}}$$

will do the trick. I shall not go into details on this because the problem will not arise if one applies, not the implication 6.25.i $\Rightarrow$ ii, but the alternate parenthetical proof thereof in which N was constructed as a monotone sequence thereby rendering b a subsequence of a. □

A first application of this is a nonstandard proof of the convergence of Cauchy sequences.

6.28 Lemma. *Let* $a : a_0, a_1, a_2, \ldots$ *be a sequence of real numbers. Then: a is Cauchy convergent iff for all infinite* $M, N \in {}^{\Omega}\mathbb{N}$, $a_M \approx a_N$.

Proof. Supppose a is a Cauchy sequence. Let $\epsilon > 0$ and choose $n_0 \in \mathbb{N}$ such that

$$\mathbb{R} \vDash \forall m \forall n \big(\overline{\mathbb{N}}(m) \wedge \overline{\mathbb{N}}(n) \wedge m > \overline{n}_0 \wedge n > \overline{n}_0 \rightarrow |a_m - a_n| < \overline{\epsilon} \big).$$

This must also be true in ${}^{\Omega}\mathbb{R}$. But for infinite $M, N \in {}^{\Omega}\mathbb{N}$, one has

$$\Omega\mathbb{R} \vDash \overline{\mathbb{N}}(\overline{M}) \wedge \overline{\mathbb{N}}(\overline{N}) \wedge \overline{M} > \overline{n}_0 \wedge \overline{N} > \overline{n}_0,$$

whence

$$\Omega\mathbb{R} \vDash |a_{\overline{M}} - a_{\overline{N}}| < \overline{\epsilon}, \text{ i.e. } |a_M - a_N| < \epsilon.$$

As ϵ was an arbitrary positive real number, it follows that $a_M \approx a_N$.

For the converse, assume a is not a Cauchy sequence. Then there is some $\epsilon > 0$ such that for all $n_0 \in \mathbb{N}$ there are $m, n \in \mathbb{N}$ with $m, n > n_0$ such that $|a_m - a_n| \geq \epsilon$:

$$\mathbb{R} \vDash \forall n_0 \exists m \exists n$$

$$\left(\overline{\mathbb{N}}(n) \wedge (n > n_0) \wedge \overline{\mathbb{N}}(m) \wedge (m > n_0) \wedge \big(\overline{\mathbb{N}}(n_0) \rightarrow |a_m - a_n| \geq \overline{\epsilon} \big) \right).$$

This is a Horn sentence, whence true in ${}^{\Omega}\mathbb{R}$. In particular,

$$\Omega\mathbb{R} \vDash |a_{\overline{M}} - a_{\overline{N}}| \geq \overline{\epsilon}, \text{ i.e. } |a_M - a_N| \geq \epsilon,$$

for some $M, N \in {}^{\Omega}\mathbb{N}$ with $M, N > \Omega$. But this contradicts the assumption that $a_M \approx a_N$. □

6.29 Corollary. *Every Cauchy sequence of real numbers converges in* $\mathbb{R}$.

Proof. Let a be a Cauchy sequence and choose n_0 so large that $|a_m - a_n| < 1$ for all $m, n > n_0$. Then, for any $n \in \mathbb{N}$,

$$|a_n| < \max\{|a_0|, |a_1|, \ldots, |a_{n_0}|, |a_{n_0+1}| + 1\}.$$

Thus, a is a bounded sequence and, by Theorem 6.26 it has a limit point L. By Theorem 6.25, $L \approx a_N$ for some infinite $N \in {}^{\Omega}\mathbb{N}$. By the Lemma, $a_M \approx L$ for all infinite $M \in {}^{\Omega}\mathbb{N}$. But then Theorem 6.23 tells us $L = \lim_{n \to \infty} a_n$.   □

Recall Klein's remark that the first task of any nonarchimedean analysis would be a proof of the Mean Value Theorem. The key analytic step in the proof thereof is a proof of the following Theorem.

6.30 Theorem (Extreme Value Theorem). *A continuous function on a closed, bounded interval attains a maximum and a minimum value on the interval.*

The idea behind the proof is very simple. Given a continuous function f defined on an interval $[a, b]$, and any positive integer n, the finite set

$$\left\{ f\left(a + \frac{k}{n}(b - a)\right) \;\middle|\; k \in \mathbb{N} \; \& \; 0 \le k \le n \right\} \tag{71}$$

has a maximum element, say

$$f\left(a + \frac{k_n}{n}(b - a)\right).$$

Assuming this carries over to the infinite integers, for $0 < N \in {}^{\Omega}\mathbb{N}$ there is some $K_N \in {}^{\Omega}\mathbb{N}$ such that

$$f\left(a + \frac{K_N}{N}(b - a)\right)$$

is the maximum element of the set of numbers of the form $f(a + (K/N)(b-a))$ for $K \le N$. If $c \approx a + (K_N/N)(b-a)$, then $f(c)$ ought to be the maximum value of $f(x)$ on $[a, b]$. This works with Robinson's formulation of nonstandard analysis, but not without major modification in the Schmieden-Laugwitz theory. The big block is that in the Schmieden-Laugwitz theory, K_N may arise from an oscillating sequence and thus $a + (K_N/N)(b - a)$ might not be infinitesimally close to any real number: c might not exist.

Proof of Theorem 6.30. Let $a < b$ be real numbers, f a real-valued function defined and continuous on $[a, b]$. For any finite $n \in \mathbb{N}^+$ the set (71) has a maximum element. Define a sequence $c_0, c_1, c_2, \ldots$ by choosing $c_0 = a$, say, and, when $n > 0$,

$$c_n = a + \frac{k_n}{n}(b - a),$$

the point at which $f(a + (k/n)(b - a))$ is maximised in (71). The sequence $c_0, c_1, c_2, \ldots$ is bounded and hence has, by the Bolzano-Weierstrass Theorem, a convergent subsequence $c_{i_0}, c_{i_1}, c_{i_2}, \ldots$ Let N be the sequence $(i_0, i_1, i_2, \ldots)$ and $c \approx c_N$ the limit thereof. Then $f(c)$ will be the maximum value of $f(x)$ for $x \in [a, b]$. To see this, let $d \in [a, b]$ be any other real number.

The following sentence is true in $\mathbb{R}$:

$$\forall n \forall x \exists k \left(\overline{\mathbb{N}}(n) \wedge n > \overline{0} \wedge \overline{a} \leq x \wedge x \leq \overline{b} \rightarrow \right.$$

$$\left. \overline{\mathbb{N}}(k) \wedge \overline{0} \leq k \wedge k \leq n \wedge \left| x - \left(a + \frac{k}{n}(\overline{b} - \overline{a}) \right) \right| \leq \frac{1}{n} \right).$$

This isn't quite written in the proper Horn fashion, but it is clearly equivalent to the less immediately intelligible

$$\forall n \forall x \exists k \left(\left(\overline{\mathbb{N}}(n) \wedge n > \overline{0} \wedge \overline{a} \leq x \wedge x \leq \overline{b} \rightarrow \overline{\mathbb{N}}(k) \right) \wedge \right.$$

$$\left(\overline{\mathbb{N}}(n) \wedge n > \overline{0} \wedge \overline{a} \leq x \wedge x \leq \overline{b} \rightarrow \overline{0} \leq k \right) \wedge$$

$$\left(\overline{\mathbb{N}}(n) \wedge n > \overline{0} \wedge \overline{a} \leq x \wedge x \leq \overline{b} \rightarrow k \leq n \right) \wedge$$

$$\left. \left(\overline{\mathbb{N}}(n) \wedge n > \overline{0} \wedge \overline{a} \leq x \wedge x \leq \overline{b} \rightarrow \left| x - \left(a + \frac{k}{n}(\overline{b} - \overline{a}) \right) \right| \leq \frac{1}{n} \right) \right),$$

which is written as a Horn sentence. Hence it is true in $^\Omega\mathbb{R}$ and, in particular, there is some $K \in {}^\Omega\mathbb{N}$ with $0 \leq K \leq N$ such that

$$\left| d - \left(a + \frac{K}{N}(b - a) \right) \right| < \frac{1}{N},$$

i.e. $d \approx a + (K/N)(b-a)$. But, thinking of K, N as sequences, $k_0, k_1, k_2 \dots,$ $n_0, n_1, n_2 \dots$, we have for all but finitely many j for which $n_j = 0$ (there are only finitely many such since N is infinite),

$$f\left(a + \frac{k_j}{n_j}(b - a) \right) \leq f(c_{n_j}),$$

and thus

$$f\left(a + \frac{K}{N}(b - a) \right) \leq f(c_N).$$

Now, by the continuity of f at c, d,

$$f(d) \approx f\left(a + \frac{K}{N}(b - a) \right) \leq f(c_N) \approx f(c),$$

and we cannot have $f(d) > f(c)$ as $f(d) - f(c)$ differs infinitesimally from $f(a + (K/N)(b-a)) - f(c_N)$, which is not positive. Thus $f(c)$ is the maximum value of f on $[a, b]$.

Reversing a few inequalities in the proof just given shows f also to possess a minimum. [Alternatively, note that a minimum of f must exist at the point at which $-f$ is maximum.] □

The rest of the proof of the Mean Value Theorem now follows the classical path. I shall complete it later. For now, I note that Laugwitz himself criticised the proof and a few related ones:

The proof suffers from the following lack in elegance [*Schönheitsfehler*]. We had to apply conventional theorems on real sequences, e.g. on the existence of a convergent subsequence of a bounded sequence. This was necessary in order to get x_{N_Ω} [c_N in the present exposition] infinitesimally close to a real number.[103]

Because the Schmieden-Laugwitz approach to nonstandard analysis provides a nonstandard proof of the Bolzano-Weierstrass Theorem, I don't regard the appeal to this Theorem to be a committed appeal to conventionality. What I do not like about the proof is the necessity of appealing to something to construct a specific N for which an appropriate K_N exists. I learned Robinson's theory before becoming aware of the Schmieden-Laugwitz one and, Robinson's theory being simpler and more elegant, I am too taken aback by what strikes me immediately as roundabout and unnecessary. It took me a while to appreciate the ingenuity of the argument, an ingenuity that seems wasted when one imagines it is not needed under Robinson's approach. However, the argument is there behind the scenes in related diagonal constructions. This is most evident in Thoralf Skolem's constructions of nonstandard arithmetics, the constructions which inspired Robinson.[104] We turn our attention now to Robinson's approach.

6.3 Robinson

Although he was aware of the work of Schmieden and Laugwitz by the time he first announced his version of nonstandard analysis in 1960, Robinson is not reported to have been aware of their work when the possibility of establishing a rigorous foundation for the use of infinitesimals suddenly occurred to him. He had previously toyed with nonstandard models of arithmetic, which had been around since the 1930's when the Norwegian number theorist and logician Thoralf Skolem first introduced them. Briefly, a nonstandard model of arithmetic is a structure $\mathfrak{N}$ for the language $L_\mathbb{N}$ of arithmetic (defined analogously to $L_\mathbb{R}$) for which all the sentences of $L_\mathbb{N}$ true in $\mathbb{N}$ are true in $\mathfrak{N}$.[105] Skolem had proven the existence of such for any countable sublanguage of $L_\mathbb{N}$ by a construction, but by Robinson's day it was easier to establish this existence by a simple appeal to some powerful general results in logic. This is essentially what Robinson did— in his own words:

> In the fall of 1960 it occurred to me that the concepts and methods of contemporary Mathematical Logic are capable of providing a suitable

---

[103] Detlef Laugwitz, *Infinitesimalkalkül: Kontinuum und Zahlen—Eine elementare Einführung in die Nichtstandard Analysis*, Bibliographisches Institut, Mannheim, 1978, p. 58.

[104] *Cf.* footnote 126 on page 539, below, for a list of Skolem's publications on the construction.

[105] Actually, this defines what is called a *strong* nonstandard model of arithmetic. Today we refer to any structure $\mathfrak{N}$ satisfying certain familiar axioms as a nonstandard model of arithmetic.

framework for the development of the Differential and Integral Calculus by means of infinitely small and infinitely large numbers. I first
reported my ideas in a seminar talk at Princeton University (November
1960) and, later, in an address at the annual meeting of the Association for Symbolic Logic (January 1961) and in a paper published in the
Proceedings of the Royal Academy of Sciences of Amsterdam... The
resulting subject was called by me Non-standard Analysis since it involves and was, in part, inspired by the so-called Non-standard models
of Arithmetic whose existence was first pointed out by T. Skolem.[106]

Whereas nonstandard models of arithmetic were primarily mere curiosities
until the late 1970's, when a theory of them finally began to take shape and a
few applications began to emerge, nonstandard models of analysis saw immediate application. Indeed, by 1964, together with Allen R. Bernstein, Robinson
had used it to solve an open problem in analysis.

Robinson's first published announcement of his version of nonstandard analysis was basically an extended abstract[107] that provided some orientation and
a catalogue of developments, but no proofs. In this abstract, he recited the
prehistory of the subject:

It is our main purpose to show that these models provide a natural
approach to the age old problem of producing a calculus involving infinitesimal (infinitely small) and infinitely large quantities. As is well
known, the use of infinitesimals, strongly advocated by Leibnitz [*sic*]
and unhesitatingly accepted by Euler fell into disrepute after the advent of Cauchy's methods which put Mathematical Analysis on a firm
foundation. Accepting Cauchy's standards of rigor, later workers in the
domain of non-archimedean quantities concerned themselves only with
fragments of the edifice of Mathematical Analysis... Finally, a recent
and rather successful effort at developing a calculus of infinitesimals is
due to SCHMIEDEN and LAUGWITZ [8] whose number system consists
of infinite sequences of rational numbers. The drawback of this system
is that it includes zero-divisors and that it is only partially ordered.
In consequence, many classical results of the Differential and Integral
calculus have to be modified to meet the changed circumstances.[108]

Robinson's first detailed introduction to the subject appeared in the final chapter on "Selected Topics" of his book *Introduction to Model Theory and to the*

---

[106] Abraham Robinson, *Non-Standard Analysis*, North-Holland Publishing Company,
Amsterdam, 1966, p. VII.

[107] Abraham Robinson, "Non-standard analysis", *Nederl. Akad. Wetensch. Proc. Ser.
A* 64, and *Indagationes Math.* 23 (1961), pp. 432 - 440. Reprinted in: H.J. Keisler,
S. Körner, W.A.J. Luxemburg, and A.D. Young (eds.), *Selected Papers of Abraham
Robinson, vol. 2, Nonstandard Analysis and Philosophy*, North-Holland Publishing
Company, Amsterdam, 1979, pp. 3 - 11.

[108] Robinson, *Selected Papers vol. 2*, p. 4. Reference [8] is Schmieden and Laugwitz, *op.
cit.* The curious mixture of lower and upper cases in the designation "Differential
and Integral calculus" is copied exactly from Robinson.

Metamathematics of Algebra[109]. He ends the chapter with a lot of references, concluding with the minimalist statement on Schmieden and Laugwitz:

> In this connection, the work of Schmieden and Laugwitz, see Schmieden-Laugwitz 1958, Laugwitz 1961, 1961a, deserves special mention.[110]

In 1966, Robinson's book *Non-Standard Analysis* appeared. The final chapter, "Concerning the History of the Calculus", briefly discussed the history of the Calculus with respect to the use of infinitesimals. Regarding attempts to reintroduce them rigorously, he had this to say about Schmieden and Laugwitz:

> We may complete the picture of recent developments by mentioning that during the period under consideration attempts were still made to define or justify the use of infinitesimals in Analysis... The most successful among these is the theory of Schmieden and Laugwitz (SCHMIEDEN and LAUGWITZ [1958], LAUGWITZ [1961, 1961a]). The starting point of the Schmieden-Laugwitz theory is, once again, the identification of infinitely small or infinitely large numbers with *functions*, with particular reference to their asymptotic behavior, except that these functions now are mappings from the natural into the rational or real numbers, i.e., sequences. The resulting system is not an ordered field but a ring with zero divisors. Nevertheless, the reader who consults the original papers will find that the theory is of considerable interest. Among other things, it includes a substitute for the theory of distributions.[111]

At this point, Robinson decides to quote Adolf Frænkel's introduction to set theory, first commenting that, "Frænkel mentions, with approval, the opinion that the test for the efficacity of infinitely small numbers is their applicability to the Differential and Integral Calculus, and then goes on"[112]:

> *However the infinitely small has completely failed at this test.* The sorts of infinitely small magnitudes that have been considered up till now and sometimes carefully grounded have proven themselves to be completely unusable for dealing with even only the simplest and most fundamental

---

[109] North-Holland Publishing Company, Amsterdam, 1963. This work, completed in 1961, appeared in 1963, with a second printing labelled a "second edition" in 1965. According to his collected works, it was translated into Russian in 1967 and into Italian in 1974, the year the official Second Edition appeared.

[110] *Ibid.*, p. 271. The references are to Schmieden and Laugwitz, *op. cit.*, and to two papers of Laugwitz on the δ-function that appeared in Crelle's Journal: "Anwendungen unendlich kleiner Zahlen, I. Zur Theorie der Distributionen" and "Anwendungen unendlich kleiner Zahlen, II. Ein Zugang zur Operatorenrechnung von Mikusiński", *Journal für die reine und angewandte Mathematik* 207 (1961), pp. 53 - 60, and 208 (1961), pp. 22 - 34, respectively.

[111] Abraham Robinson, *Non-Standard Analysis*, North-Holland Publishing Company, Amsterdam, 1966, p. 278.

[112] Robinson, *Ibid.*, p. 279. Frænkel was one of Robinson's teachers. Hence Frænkel's implied approval became, in effect, a public endorsement of Robinson and his work.

problems of the Infinitesimal Calculus (for example for the proof of the Mean Value Theorem or for the definition of the definite integral) (Cf., for example F. BERNSTEIN [1][113]). They had to leave the field undiminished [by the use] of the previous (relatively involved) foundational methods by means of the limit concept[114]; there is too no ground for the expectation, that anything herein would change in the future. Certainly it were *conceivable* in itself (if also extremely improbable on good grounds and, in any case, from the present stand of the science lying at an unreachable distance), that a second CANTOR will one day give an objection-free arithmetical foundation of new infinitely small numbers, which will prove themselves to be mathematically usable and for their part perhaps open a new entrance to the Infinitesimal Calculus. So long, however, as this is not the case, one may place neither the (in many respects interesting) VERONESEan nor any other infinitely small numbers in parallel to CANTOR's[115], but rather must take the position, that one can in no way speak of the mathematical and thereby logical existence of the *Infinitely Small* in the same or a similar sense as by the Infinitely Large.[116]

Robinson's biographer says that "Robinson was apparently prepared to see himself as a second Cantor".[117] One can readily believe this from his recitation of Frænkel's comment, as well as his lukewarm praise of Schmieden and Laugwitz. Does the word "effort" place the phrase "a recent and rather successful effort" into the category of damning with faint praise? Does "Nevertheless" in "Nevertheless...the theory is of considerable interest" indicate only a grudging respect? Schmieden and Laugwitz had shown how the use of infinitesimals and infinitely large numbers could be justified and they had even achieved the Kleinian Holy Grail of nonarchimedean analysis— a proof of the Mean

---

[113] The reference is to Felix Bernstein, "Über die Begründung der Differentialrechnung mit Hilfe der unendlichkleiner Größen", *Jahresbericht der Deutschen Mathematiker-Vereinigung* 13 (1904), pp. 241 - 246. Robinson omits the reference to Bernstein's short paper, which covers no new ground.

[114] Robinson omits this clause. As Robinson was by far a better linguist than I, and as the original is a trifle tricky, I am less than 100% confident in my translation. But I take the meaning to be that the authors in question did not succeed in replacing limit arguments by the use of infinitesimals.

[115] I.e., Cantor's infinite cardinal numbers.

[116] Adolf Frænkel, *Einleitung in die Mengenlehre*, 3rd. revised and expanded edition, Springer-Verlag, Berlin, 1928, pp. 116 - 117. But for the deletions cited in the footnotes above, Robinson quoted Frænkel verbatim in German. I have translated this passage from the American *Raubdruck* reprint of 1946. Following World War II, the U.S. Attorney General assigned copyrights to German publications to American publishers as reparations for the war. The Germans refer to such an American edition as a Raubdruck, or pirate edition. I cannot help but notice that, in this case, the author was Jewish.

[117] Joseph Warren Dauben, *Abraham Robinson: The Creation of Nonstandard Analysis: A Personal and Mathematical Odyssey*, Princeton University Press, Princeton, 1995, p. 354.

Value Theorem. Whether or not they would have knocked Robinson out of the running for consideration as the second Cantor, it would have been easy to credit them for their accomplishment in a more positive manner and still fairly compare his own approach favourably to theirs. That his system of hyperreals formed an ordered field is a powerful advantage, though I cannot agree with his judgement that, under the Schmieden-Laugwitz approach, "In consequence [of the existence of zero-divisors and the non-totality of the ordering], many classical results of the Differential and Integral calculus have to be modified to meet the changed circumstances". I cannot agree because I simply do not know what he is referring to.[118]

Robinson's approach to nonstandard analysis was superior to that of Schmieden and Laugwitz in a number of ways. After the initial appeal to logic, the working out of the details is much simpler than under the Schmieden-Laugwitz approach, and this, as we shall soon see, is manifest already in the most elementary stages of the theory. Robinson's hyperreals behaved more like real numbers than did those of Schmieden and Laugwitz. His approach revealed greater flexibility and applicability to many structures other than the reals. And, finally, he attracted many other workers to the area.

Robinson, like Schmieden and Laugwitz, had a background in applied mathematics. In fact, like Schmieden, his applied background encompassed ærodynamics. A final parallel with Schmieden was his initial approach to nonstandard analysis. Recall that Schmieden's approach in 1952 had been to adjoin an infinite integer Ω to $\mathbb{R}$ and assume as an axiom any assertion $A(\Omega)$ for which $A(n)$ held for all sufficiently large natural numbers n. Schmieden, however, gave no justification for this adjunction until he and Laugwitz constructed a variant of $^\Omega\mathbb{R}$ as an extension $\mathbb{H}$ of $\mathbb{R}$ containing such numbers Ω. Robinson largely did the same, but he justified his approach immediately. Robinson had been interested in mathematical logic for some time, submitting his thesis on the subject in 1949, a book length version of which he published in 1951[119]. In this work he made extensive use of the Compactness Theorem and what he called the Extended Completeness Theorem, essentially an equivalent result which can be applied immediately to yield the existence of a structure $\mathbb{H}$ better suited to the purpose of nonstandard analysis than $^\Omega\mathbb{R}$.

The Compactness Theorem is very easy to state. Suppose Γ is a set of sentences in a *first-order language*, i.e. a language with individual constants, function symbols, and relation symbols from which one can build up complex formulæ exactly as we did in $L_\mathbb{R}$ from atomic formulæ using conjunction, disjunction, negation, implication, and quantification over individuals— quantifi-

---

[118] I thought briefly of the various explicit nonstandard δ-functions constructed by Laugwitz which do not satisfy all of Dirac's various axioms. However, these functions are in no way tied to the Schmieden-Laugwitz theory, and work just as well in Robinson's theory. In fact, I tacitly assumed back in Chapter II that the hyperreals used constituted an ordered field. Perhaps he is referring to things like the difference between Definitions 6.14.x and 6.15.iv.

[119] Abraham Robinson, *On the Metamathematics of Algebra*, North-Holland Publishing Company, Amsterdam, 1951.

cation over functions and relations is not allowed. A structure $\mathfrak{M}$ for the given language is called a *model* of Γ, written $\mathfrak{M} \vDash \Gamma$, if $\mathfrak{M} \vDash \varphi$ for every sentence φ in Γ.

6.31 Theorem (Compactness Theorem). *A set Γ of sentences has a model iff every finite subset $\Gamma_0 \subseteq \Gamma$ has a model.*

The proof of the Compactness Theorem must lie beyond the scope of this book.[120] It is accessible in the first semester of a course in Mathematical Logic, but logic itself is a bit abstract and so best tackled after one has already taken the first semester of an Abstract or Modern Algebra course. The result was first proven for countable languages (i.e. languages with only countably many constants, function symbols, and relation symbols) by Kurt Gödel in 1930 in his dissertation and generalised to uncountable languages (like $L_{\mathbb{R}}$) in 1941 by Anatoliĭ Ivanovič Mal'cev, who applied the Theorem to algebra. Robinson further pioneered the application to algebraic systems in his dissertation and one day in 1960 it struck him that he could apply it to construct an extension of the field $\mathbb{R}$ that contained infinitesimals as well as infinitely large numbers.

Armed with the Compactness Theorem, the existence proof is really quite trivial. One starts with the theory of the real numbers,

$$\mathcal{T}_{\mathbb{R}} = \{\varphi \mid \mathbb{R} \vDash \varphi \ \& \ \varphi \text{ is a sentence of } L_{\mathbb{R}}\},$$

adds a constant, say $\overline{\Omega}$, to the language $L_{\mathbb{R}}$ and defines Γ to be the set

$$\Gamma = \mathcal{T}_{\mathbb{R}} \cup \{\overline{\Omega} > \overline{r} \mid r \in \mathbb{R}\}.$$

Any finite subset of Γ consists of sentences true in $\mathbb{R}$ and finitely many sentences,

$$\overline{\Omega} > \overline{r}_0, \ \overline{\Omega} > \overline{r}_1, \ \ldots, \ \overline{\Omega} > \overline{r}_{n-1}.$$

Now these can be interpreted in $\mathbb{R}$ itself by choosing $\overline{\Omega}^{\mathbb{R}} = r$ for any $r > \max\{r_0, r_1, \ldots, r_{n-1}\}$.

6.32 Theorem. $\mathbb{R}$ *has a proper extension*[121] $^*\mathbb{R}$ *which has an infinite element Ω and which satisfies: for any sentence φ of $L_{\mathbb{R}}$,*

$$\mathbb{R} \vDash \varphi \Rightarrow {}^*\mathbb{R} \vDash \varphi. \tag{72}$$

---

[120] Actually, I give a proof of it in the next chapter (p. 523, below) from the Extended Completeness Theorem (called the Completeness Theorem (Theorem 4.5) in that chapter), whose proof is only sketched.

[121] In his first few papers, Robinson used R^* to denote the structure $^*\mathbb{R}$. Eventually, he switched over to $^*R$, denoted here by $^*\mathbb{R}$. The interpretation of any constant, function symbol, or relation symbol of $L_{\mathbb{R}}$ is then denoted by pre-superscripting the symbol with a $*$ and dropping the overline. Thus, $\overline{r}$ is interpreted by $^*r$, $\overline{f}$ by $^*f$, and $\overline{R}$ by $^*R$. Since we identify $\mathbb{R}$ with its image in $^*\mathbb{R}$, we drop the $*$ from $^*r$ and do likewise for familiar functions and relations like $+, \cdot$ or $<$.

Proof. By the Compactness Theorem, Γ has a model $^*\mathbb{R}$. For any sentence φ of $L_\mathbb{R}$,

$$\mathbb{R} \vDash \varphi \;\Rightarrow\; \varphi \in \mathcal{T}_\mathbb{R} \;\Rightarrow\; \varphi \in \Gamma \;\Rightarrow\; {}^*\mathbb{R} \vDash \varphi.$$

The embedding is easy to describe. Given $r \in \mathbb{R}$, find the element $^*r$ interpreting the constant $\bar{r}$ in $^*\mathbb{R}$ and map r to it. The map is one-to-one since

$$\mathbb{R} \vDash \neg\,\bar{r}_1 = \bar{r}_2 \;\Rightarrow\; {}^*\mathbb{R} \vDash \neg\,\bar{r}_1 = \bar{r}_2.$$

Similarly we see that all functions and relations are preserved.

And, of course, since $^*\mathbb{R} \vDash \overline{\Omega} > \bar{r}$ for any $r \in \mathbb{R}$, the element Ω [perhaps I should write $^*\Omega$] interpreting $\overline{\Omega}$ is infinite.   □

Because (72) holds for all sentences of $L_\mathbb{R}$ and not just Horn sentences, extra properties are preserved:

6.33 Corollary. $^*\mathbb{R}$ *is an ordered field.*

Proof. The order axioms and field axioms are given by sentences of $L_\mathbb{R}$. For example, trichotomy: from

$$\mathbb{R} \vDash \forall x \forall y (x < y \lor x = y \lor y < x),$$

we can conclude

$$^*\mathbb{R} \vDash \forall x \forall y (x < y \lor x = y \lor y < x);$$

or the existence of a multiplicative inverse: from

$$\mathbb{R} \vDash \forall x \big(\neg x = \bar{0} \to \exists y (x \cdot y = \bar{1})\big),$$

we obtain

$$^*\mathbb{R} \vDash \forall x \big(\neg x = \bar{0} \to \exists y (x \cdot y = \bar{1})\big).   □$$

As usual, we distinguish between infinitesimals, finite numbers, and infinite numbers: An element $x \in {}^*\mathbb{R}$ is

infinitesimal iff for all $r \in \mathbb{R}^+$, $|x| < r$
finite iff for some $r \in \mathbb{R}^+$, $|x| < r$
infinite iff for all $r \in \mathbb{R}^+$, $r < |x|$.

In the present situation, there is nothing resembling oscillation and every finite element is close to an actual real:

6.34 Lemma. *Let* $x \in {}^*\mathbb{R}$ *be finite. There is a real number* r' *such that* $x \approx r'$.

Proof. x defines a cut in $\mathbb{R}$:

$$A_x = \{r \in \mathbb{R} \mid r \leq x\}, \quad B_x = \{r \mid x < r\}.$$

In $\mathbb{R}$ every cut is principal, whence $\langle A_x, B_x \rangle$ has a generator r'. For any real $\epsilon > 0$, $r' - \epsilon \in A_x$ and $r' + \epsilon \in B_x$, whence

$$r' - \epsilon < x < r' + \epsilon,$$

i.e.
$$-\epsilon < x - r' < \epsilon,$$
i.e. $|x - r'| < \epsilon$. It follows that $|x - r'|$ is infinitesimal, i.e. $x \approx r'$.   □

The number r' just constructed is called the *standard part* of x, and is usually written $st(x)$ in the literature.

The development of the theory of limits begins as before, with one small difference when it comes to sequences. First, however, we consider limits of functions of a real variable.

6.35 Theorem. *Let $f : \mathbb{R} \to \mathbb{R}$ be given and let $a \in \mathbb{R}$. The following are equivalent:*

i. f is continuous at a;

ii. ${}^*f(a + \eta) \approx {}^*f(a)$ for every infinitesimal $\eta \in {}^*\mathbb{R}$.

Proof. i $\Rightarrow$ ii. The proof is identical to that of Theorem 6.19.i $\Rightarrow$ ii and I omit it.

ii $\Rightarrow$ i. Assume ii. Then for all $x \in {}^*\mathbb{R}$,

$$x \approx a \Rightarrow {}^*f(x) \approx {}^*f(a).$$

Let ϵ be a positive real number, and η a positive infinitesimal:

$$\begin{aligned}{}^*\mathbb{R} &\vDash \forall x \big(|x - \overline{a}| < \overline{\eta} \to |\overline{f}(x) - \overline{f}(a)| < \overline{\epsilon} \big) \\ &\vDash \exists \delta \big(\delta > \overline{0} \wedge \forall x (|x - \overline{a}| < \delta \to |\overline{f}(x) - \overline{f}(a)| < \overline{\epsilon}) \big).\end{aligned}$$

Now this last sentence, call it φ, is in the language $L_\mathbb{R}$ and must be true in $\mathbb{R}$, since

$$\mathbb{R} \vDash \neg\varphi \Rightarrow {}^*\mathbb{R} \vDash \neg\varphi$$

by choice of ${}^*\mathbb{R}$. So we have:

$$\mathbb{R} \vDash \exists \delta \big(\delta > \overline{0} \wedge \forall x (|x - \overline{a}| < \delta \to |\overline{f}(x) - \overline{f}(a)| < \overline{\epsilon}) \big).$$

Thus for all $\epsilon > 0$ there is a $\delta > 0$ such that for all $x \in \mathbb{R}$,

$$|x - a| < \delta \implies |f(x) - f(a)| < \epsilon,$$

i.e. f is continuous at x.   □

For limits where f might be discontinuous, we have the following variant of Theorem 6.20:

6.36 Theorem. *Let $f : \mathbb{R} \to \mathbb{R}$ be a function and let $a, L \in \mathbb{R}$. The following are equivalent:*

i. $\lim_{x \to a} f(x) = L$;

ii. for all infinitesimal $\eta \neq 0$, ${}^*f(a + \eta) \approx L$.

Note that we assume in ii that $\eta \neq 0$ and not $\eta \# 0$. By the totality of the ordering in ${}^*\mathbb{R}$ the two notions coincide.

In ${}^\Omega\mathbb{R}$, Ω represented a very particular infinite integer. In ${}^*\mathbb{R}$, Ω is just some infinite real number. We could have made it an infinite integer by adding $\overline{\mathbb{N}}(\overline{\Omega})$ to Γ, but we would still not have pinned it down as much as we did in ${}^\Omega\mathbb{R}$. Theorem 6.23 thus loses its special clause concerning Ω and becomes

6.37 Theorem. *Let $a : a_0, a_1, a_2, \ldots$ be a sequence of real numbers and let $L \in \mathbb{R}$. The following are equivalent:*
i. $\lim_{n \to \infty} a_n = L$;
ii. for all infinite $N \in {}^*\mathbb{N}$, $a_N \approx L$.

Proof. i $\Rightarrow$ ii. The proof, minus the reference to Horn sentences is the same as before: Let L be the limit of the sequence and let a positive real number ϵ be given. There is some number $n_0 \in \mathbb{N}$ such that

$$\mathbb{R} \vDash \forall n \left(\overline{\mathbb{N}}(n) \land n > \overline{n}_0 \to |a_n - \overline{L}| < \overline{\epsilon} \right).$$

This is also true in ${}^*\mathbb{R}$. Now, for any infinite $N \in {}^*\mathbb{N}$, one has

$$ {}^*\mathbb{R} \vDash \overline{\mathbb{N}}(\overline{N}) \land \overline{N} > \overline{n}_0, $$

whence $|a_N - L| < \epsilon$. This being true for all positive real ϵ, it follows that $a_N \approx L$.

ii $\Rightarrow$ i. Suppose $a_N \approx L$ for every infinite $N \in {}^*\mathbb{N}$. Then, for any real $\epsilon > 0$, we have $|a_N - L| < \epsilon$ for all infinite $N \in {}^*\mathbb{N}$. Hence, for any infinite $N_0 \in {}^*\mathbb{N}$,

$$ {}^*\mathbb{R} \vDash \overline{\mathbb{N}}(\overline{N}_0) \land \forall n (n > \overline{N}_0 \land \overline{\mathbb{N}}(n) \to |a_n - \overline{L}| < \overline{\epsilon}) $$
$$ \vDash \exists n_0 \left(\overline{\mathbb{N}}(n_0) \land \forall n (n > n_0 \land \overline{\mathbb{N}}(n) \to |a_n - \overline{L}| < \overline{\epsilon}) \right). $$

Now this last sentence, call it φ, is in the language $L_{\mathbb{R}}$ and must be true in $\mathbb{R}$. As $\epsilon > 0$ was arbitrary, this means $\lim_{n \to \infty} a_n = L$. $\square$

Theorem 6.25 also carries over:

6.38 Theorem. *Let $a : a_0, a_1, a_2, \ldots$ be a sequence of real numbers and let $L \in \mathbb{R}$. The following are equivalent:*
i. L is a limit point of a;
ii. for some infinite $N \in {}^*\mathbb{N}$, $a_N \approx L$.

Proof. The proof is as before, minus reference to the Horn nature of the formulæ used.

i $\Rightarrow$ ii. Because L is a limit point of a, we have

$$ \mathbb{R} \vDash \forall \epsilon \forall n_0 \left(\epsilon > \overline{0} \land \overline{\mathbb{N}}(n_0) \to \exists n (\overline{\mathbb{N}}(n) \land n > n_0 \land |a_n - \overline{L}| < \epsilon) \right). $$

This is a sentence of $L_{\mathbb{R}}$, whence true in ${}^*\mathbb{R}$. Let ϵ be infinitesimal, N_0 an infinite natural number in ${}^*\mathbb{R}$, and $N \in {}^*\mathbb{N}$ larger than N_0 such that $|a_N - L| < \epsilon$. Obviously $a_N \approx L$.

ii $\Rightarrow$ i. From $a_N \approx L$, we have for any $n_0 \in \mathbb{N}$ and $\epsilon \in \mathbb{R}^+$,

$$ {}^*\mathbb{R} \vDash \exists n \left(\overline{\mathbb{N}}(n) \land n > \overline{n}_0 \land |a_n - \overline{L}| < \overline{\epsilon} \right), $$

whence

$$ \mathbb{R} \vDash \exists n \left(\overline{\mathbb{N}}(n) \land n > \overline{n}_0 \land |a_n - \overline{L}| < \overline{\epsilon} \right), $$

i.e. for any $\epsilon > 0$ and any $n_0 \in \mathbb{N}$ there is some $n > n_0$ such that $|a_n - L| < \epsilon$, i.e. L is a limit point of a. $\square$

6.39 Theorem. *Let $a : a_0, a_1, a_2, \ldots$ be a sequence of real numbers and assume a is bounded, i.e. for some real number M, we have $|a_n| < M$ for every $n \in \mathbb{N}$. Then: a has a limit point in $\mathbb{R}$.*

Proof. We have

$$\mathbb{R} \vDash \forall n(\overline{\mathbb{N}}(n) \to |a_n| < M),$$

whence $^*\mathbb{R} \vDash \forall n(\overline{\mathbb{N}}(n) \to |a_n| < M)$. Let $N \in {}^*\mathbb{N}$ be infinite. Because $|a_N| < M$, a_N is finite and Lemma 6.34 applies: $a_N \approx L$ for some $L \in \mathbb{R}$. This means L is a limit point of a. $\square$

Note that the proof does not really differ that much from the earlier proof of Theorem 6.26. There is still the appeal to Dedekind cuts; only now it comes in as a reference to Lemma 6.34.

6.40 Corollary (Bolzano-Weierstrass Theorem). *Every bounded sequence of real numbers has a convergent subsequence.*

Proof. Let $a : a_0, a_1, a_2, \ldots$ be a bounded sequence of real numbers and let L be a limit point of a. Let $i_0 = 0$ and define $0 < i_1 < i_2 < \ldots$ in succession so that, for $n > 0$, $|a_{i_n} - L| < 1/n$. The sequence $b_n = a_{i_n}$ has L as its limit since, for any infinite N,

$$|b_N - L| = |a_{i_N} - L| < \frac{1}{N},$$

i.e. $b_N \approx L$. $\square$

The proofs of the nonstandard characterisation of Cauchy convergence and the existence of limits of Cauchy convergent sequences are identical to the proofs given in the Schmieden-Laugwitz approach and I omit them. Other than in avoiding having to note the Horn nature of the formulæ whose validity is preserved from $\mathbb{R}$ to $^*\mathbb{R}$, the superiority of Robinson's approach has not yet been demonstrated. This changes with the proof of the Extreme Value Theorem.

6.41 Theorem (Extreme Value Theorem). *A continuous function f on a closed, bounded interval attains a maximum and a minimum value on the interval.*

Proof. Let f be continuous on $[a, b]$, with $a < b$ real numbers. Observe

$$\mathbb{R} \vDash \forall n \Big[\overline{\mathbb{N}}(n) \to \exists k_0 \Big(\overline{\mathbb{N}}(k_0) \wedge k_0 \leq n \wedge$$

$$\forall k \Big(\overline{\mathbb{N}}(k) \wedge k \leq n \to \overline{f}\Big(\overline{a} + \frac{k}{n}(\overline{b} - \overline{a})\Big) \leq \overline{f}\Big(\overline{a} + \frac{k_0}{n}(\overline{b} - \overline{a})\Big) \Big) \Big) \Big].$$

This sentence is also true in $^*\mathbb{R}$. Let $N \in {}^*\mathbb{N}$ be infinite and choose $K_0 \in {}^*\mathbb{N}$ for which $^*f(a + (K_0/N)(b - a))$ is maximum in $\{^*f(a + (K/N)(b - a)) \mid K \in {}^*\mathbb{N} \,\&\, 0 \leq K \leq N\}$. Now $a + (K_0/N)(b-a)$ is bounded between a and b, whence it has a standard part c, i.e. there is a number $c \in \mathbb{R}$ such that

$$a + \frac{K_0}{N}(b - a) \approx c.$$

c cannot be less than a nor greater than b, whence $c \in [a, b]$.

The proof that $f(c)$ is the maximum value of $f(d)$ for $d \in [a, b]$ proceeds as before: if $d \in [a, b]$, then, in $*\mathbb{R}$,

$$d \approx a + \frac{K}{N}(b - a)$$

for some $K \in *\mathbb{N}$ with $0 \leq K \leq N$. By continuity,

$$f(d) \approx {}^*f\left(a + \frac{K}{N}(b - a)\right) \leq {}^*f\left(a + \frac{K_0}{N}(b - a)\right),$$

whence $f(d) \leq f(c)$.

Again, the case for a minimum value is handled similarly or reduced to the case for a maximum. $\qquad\qquad\qquad\qquad\qquad\qquad\qquad\qquad\qquad\qquad\qquad\qquad\qquad\square$

Continuing from here to a proof of the Mean Value Theorem will be a bit disappointing as the proof is purely standard: Nonstandard analysis replaces the appeal to ϵ's and δ's by new arguments; it does not replace the algebraic or (as in the proof of Corollary 6.27) combinatorial parts of arguments.

The first step is to define the derivative of a function. One does this in the usual manner.

6.42 Definition. *A function* $f : \mathbb{R} \to \mathbb{R}$ *has* derivative d *at a point* $a \in \mathbb{R}$, *written* $f'(a) = d$, *if*

$$\lim_{h \to 0} \frac{f(a + h) - f(a)}{h} = d.$$

By the nonstandard characterisation of the limit of a function (Theorem 6.36), $f'(a) = d$ iff

$$\frac{{}^*f(a + \eta) - f(a)}{\eta} \approx d \qquad\qquad\qquad\qquad (73)$$

for all infinitesimal $\eta \neq 0$.

At this point, the reader may wish to test his skill in dealing with infinitesimals by using them to show that, if f, g are differentiable at a then

$$(f + g)'(a) = f'(a) + g'(a)$$
$$(rf)'(a) = r(f'(a)) \text{ for any constant } r \in \mathbb{R}$$
$$(fg)'(a) = f'(a)g(a) + f(a)g'(a)$$

and, if $f(a) \neq 0$,

$$(1/f)'(a) = \frac{-f'(a)}{f(a)^2}.$$

I prefer here to head straight for the Mean Value Theorem.

6.43 Lemma. *Let* $f : \mathbb{R} \to \mathbb{R}$ *have a local extremum at* a *and suppose* $f'(a)$ *exists. Then:* $f'(a) = 0$.

Proof. I shall handle the case in which $f(a)$ is a local maximum, leaving it to the reader to modify the proof to handle the case of a local minimum.

Let $f'(a) = d$ and suppose $\eta \neq 0$ is infinitesimal, so that (73) holds. Suppose $d > 0$. Then

$$\frac{{}^*f(a+\eta) - f(a)}{\eta} > \frac{d}{2} > 0,$$

which cannot be the case if $\eta > 0$ since ${}^*f(a+\eta) \leq f(a)$ makes the quotient negative or zero. Thus $d \not> 0$. Similarly, $d < 0$ cannot hold for $\eta < 0$. Thus $d = 0$. □

6.44 Theorem (Rolle's Theorem). *Let f be continuous on $[a, b]$ and differentiable on (a, b), and suppose $f(a) = f(b)$. There is a number $c \in (a, b)$ such that $f'(c) = 0$.*

Proof. If f is constant, the result is trivial. So suppose $f(d) \neq f(a)$ for some $d \in (a, b)$. If $f(d) > f(a)$, choose c so that $f(c)$ is the maximum of f on $[a, b]$ and apply the Lemma; and if $f(d) < f(a)$, choose c so that $f(c)$ is the minimum and apply the Lemma. □

Notice that the proof is exactly the same as the standard one: there is no limit argument in the reduction of Rolle's Theorem to the Extreme Value Theorem to be replaced by an appeal to infinitesimals. The same is true of the reduction of the Mean Value Theorem to Rolle's Theorem.

6.45 Theorem (Mean Value Theorem). *Let f be continuous on $[a, b]$ and differentiable on (a, b). There is a number $c \in (a, b)$ such that*

$$f'(c) = \frac{f(b) - f(a)}{b - a}.$$

Proof. One introduces the auxiliary function

$$g(x) = f(x) - f(a) - \frac{f(b) - f(a)}{b - a}(x - a),$$

and observes

$$g(a) = f(a) - f(a) - \frac{f(b) - f(a)}{b - a}(a - a) = 0 - 0 = 0$$

$$g(b) = f(b) - f(a) - \frac{f(b) - f(a)}{b - a}(b - a) = f(b) - f(a) - (f(b) - f(a)) = 0.$$

Thus $g(b) = g(a)$ and Rolle's Theorem applies: for some $c \in (a, b)$, $g'(c) = 0$. But the usual rules for calculating derivatives apply:

$$0 = g'(c) = f'(c) - \frac{f(b) - f(a)}{b - a},$$

whence

$$f'(c) = \frac{f(b) - f(a)}{b - a}. \qquad □$$

Digression. I should think that the proof of the Mean Value Theorem as just given is acceptible from an expository standpoint because, a course in Calculus being a prerequisite, the reader should have seen it before. It would not be appropriate in a textbook on the Calculus. It is simply bad exposition to introduce g at that level without some explanation where it came from. The matter has been discussed in the pages of the *American Mathematical Monthly* and *Mathematics Magazine* and some of the articles were collected in the Mathematical Association of America's *Selected Papers in Calculus*[122] Several authors criticise the unmotivated introduction of the function g. R.C. Yates writes

> In discussing the Law of the Mean, we consider the function
>
> $$\phi(x) = f(x) - f(a) - \frac{f(b) - f(a)}{b - a}(x - a).$$
>
> This is a formidable expression whose origin is puzzling until it is pointed out as the difference PQ of the ordinate of a point on the graph of $f(x)$ and the ordinate of the secant line for the same x.[123]

Louis C. Barrett and Richard A. Jacobson state

> The usual proofs of the first and extended mean value theorems involve the process of applying Rolle's theorem to functions happily designed to yield the desired conclusions. Frequently, no mention is made of how these functions are discovered.[124]

The "happy" appearance out of nowhere of the function, a sort of *deus ex machina*, ought to remind us of Dedekind's function used in Example 5.2 and the question of where it came from. In Exercises 9.29 and 9.30 the reader is given two answers to the question of the origin of the present function g, one algebraic and one geometric. But I encourage the reader, especially one whose Calculus text offered no explanation, to tackle the problem on his own before consulting the exercises.

With the proof of the Mean Value Theorem we have reached the all-important "first and chief" task of nonarchimedean analysis, and have thus reached a natural stopping point for an introductory essay such as this. Indeed, with respect to the further development of analysis along nonstandard lines, I shall stop here and leave it to the reader to continue on his own. I offer a couple of simple exercises at the end of the Chapter as a starting point, and shall here recommend a couple of references to proceed further.

There is a number of books dedicated to the development of nonstandard analysis and I should like to recommend two such in particular. James M.

---

[122] Tom M. Apostol, Herbert E. Chrestenson, C. Stanley Ogilvy, Donald E. Richmond, and N. James Schoonmaker (eds.), *Selected Papers in Calculus*, Dickenson Publishing Company, Belmont (California), 1969.

[123] *Ibid.*, p. 195.

[124] *Ibid.*, p. 198.

Henle and Eugene M. Kleinberg, *Infinitesimal Calculus*[125] offers a particularly elementary introduction to the subject that presupposes familiarity only with freshman Calculus. It is a nicely written, slim volume presented with a minimum amount of mathematical logic, and it reviews and proves all the major theorems of the first Calculus course. Martin Davis, *Applied Nonstandard Analysis*[126] has more logical and algebraic preliminaries, but once it gets through them the book takes off and covers a lot of space with chapters on real analysis, topology and metric spaces, normed linear spaces, and Hilbert space. The book assumes a stronger mathematical background than does that by Henle and Kleinberg— not knowledge of specific results, but greater familiarity and ease with abstraction.

A second direction to follow is the application of nonstandard analysis to understand and better approach the "nonrigorous" use of infinitesimals and the infinitely large in the past, as well as to clear up some "errors". We have seen this already in the present Chapter with Laugwitz's use of $^{\Omega}\mathbb{R}$ to interpret Bolzano's *Reine Zahlenlehre*, and I cited a paper of Mark McKinzie and Curtis Tuckey on the correctness of Euler's summation of the reciprocals of the squares back in Chapter II. I would like to add here a few words on Cauchy.

6.4 Cauchy

Robinson was the first to examine Cauchy's works from a nonstandard point of view in his 1966 book. His discussion was immediately criticised by the philosopher Imre Lakatos[127] and the number of authors who have contributed to the re-evaluation of Cauchy's rigour is not negligible. The most extensive discussion of Cauchy is Spalt's book[128], the twelfth chapter of which summarises the contributions of an even dozen contributors. It is also the interpretation of Cauchy's work most radically different from a reading based on the traditional Weierstrassian continuum. In a series of papers, Laugwitz, who was Spalt's teacher and who cited Spalt as an influence, gave a satisfying mathematical interpretation of Cauchy in nonstandard terms. I base the following discussion of Cauchy on some of Laugwitz's papers.

In his now legendary textbook *Cours d'analyse*[129], of 1821, Cauchy gave his definition of continuity. I suppose I ought to consult Cauchy and offer my

---

[125] MIT Press, Cambridge (Mass.), 1979. This book has been republished by Dover, New York, and is still in print.

[126] John Wiley & Sons, New York, 1977. This book has been republished by Dover, New York, and is still in print.

[127] Imre Lakatos, "Cauchy and the continuum: the significance of non-standard analysis for the history and philosophy of mathematics", in: Imre Lakatos, *Mathematics, science and epistemology: Philosophical Papers 2*, Cambridge University Press, Cambridge, 1978.

[128] Detlef Spalt, *Die Vernunft im Cauchy-Mythos: Synthetischer Aufbau einer Analysis: Herkunft, Mißverständnisse und Herkunft der Mißverständnisse*, Verlag Harri Deutsch, Frankfurt am Main, 1996.

[129] The full title is *Cours d'analyse de l'ecole royale polytechnique. 1re partie: analyse algébrique*. In text one usually abbreviates this by simple truncation to *Cours*

own translation, but I found several English translations of this definition at hand[130], all substantially agreeing with one another:

C.H. Edwards, Jr., *The Historical Development of the Calculus*, Springer-Verlag, New York, 1979, pp. 310 - 311;

Judith V. Grabiner, *The Origins of Cauchy's Rigorous Calculus*, MIT Press, Cambridge (Mass), 1981, p. 87;

Umberto Bottazzini, *The* HIGHER CALCULUS*: A History of Real and Complex Analysis from Euler to Weierstrass*, Springer-Verlag, New York, 1986, pp. 104 - 105. (English translation by Warren van Egmond. Original Italian edition: *Il Calcolo sublime: storia dell'analisi matematica da Euler a Weierstrass*, Editore Boringhieri società per azioni Torino, Torino, 1981.)

In a paper[131] on Cauchy, Laugwitz cites Cauchy in the original French and in the very literal translation of Edwards:

> If, starting from a value of x included between these limits[132], one assigns to the variable x an infinitely small increment α, the function itself will take on for an increment the difference $f(x+\alpha) - f(x)$, which will depend at the same time on the new variable α and on the value of x. This granted, the function $f(x)$ will be, between the two limits assigned to the variable x, a *continuous* function of the variable if, for each value of x intermediate between these limits, the numerical value of the difference $f(x + \alpha) - f(x)$ decreases indefinitely with that of α. In other words, *the function $f(x)$ will remain continuous with respect to x between the given limits if, between these limits, an infinitely small increment of the variable always produces an infinitely small increment of the function itself.*

The italics are Cauchy's and are reproduced by Edwards, Bottazzini, and Laugwitz, but not by Grabiner. Grabiner does, however, immediately follow her translated passage (which omits the first sentence given here) with a comparison with Bolzano's definition of 1817:

> Bolzano's definition was given in a slightly different and more precise language, but its meaning is the same:
>
> A function $f(x)$ for all values of x inside of, or outside of, certain

---

d'analyse, and the subtitle "analyse algébrique" is often omitted even in bibliographies. The 1828 German translation, however, is titled *Lehrbuch der algebraischen Analysis [Textbook on algebraic analysis]* and the 1885 translation is even more simply titled *Algebraische Analysis*. Cauchy's goal was to replace inadequate geometric intuition by algebraic rigour, hence the name "algebraic analysis".

[130] I suppose I should also have consulted the various source books, but that would have required me to get out of my chair and walk to the other end of the house. In view of the agreement of the three sources in easy reach, I have deemed such an expenditure of energy unnecessary.

[131] Detlef Laugwitz, "Infinitely small quantities in Cauchy's textbooks", *Historia Mathematica* 14 (1987), pp. 258 - 274; here: p. 261.

[132] I.e., x lies in some given closed interval $[a, b]$.

bounds, varies according to the law of continuity only insofar as, if x is any such value, the difference $f(x + \omega) - f(x)$ can be made less than any given magnitude, when ω is taken as small as desired.[133]

The first thing to notice is that both Cauchy and Bolzano define continuity on a closed interval, not continuity at a point. As Laugwitz notes[134], Cauchy subsequently defines f to be continuous at a point x_0 if f is continuous on some interval $[a, b]$, where $a < x_0 < b$. The second thing to notice is the italicised phrase. Laugwitz interprets "always" to include not only the real numbers $x \in [a, b]$, but the hyperreals as well.[135] To us, continuity on an interval, stated in terms of ϵ's and δ's, reads:

$$\forall x \in [a, b] \, \forall \epsilon > 0 \, \exists \delta > 0 \, \forall h \big(|h| < \delta \Rightarrow |f(x + h) - f(x)| < \epsilon \big),$$

which we earlier showed is equivalent to

for all $x \in [a, b]$ and all infinitesimal α, ${}^*f(x + \alpha) \approx f(x)$.

Laugwitz reads Cauchy as saying,

for all $x \in {}^*[a, b]$ and all infinitesimal α, ${}^*f(x + \alpha) \approx {}^*f(x)$.

This is a much stronger condition: in terms of ϵ's and δ's, it reads:

$$\forall \epsilon > 0 \, \exists \delta > 0 \, \forall x \in [a, b] \, \forall h \big(|h| < \delta \Rightarrow |f(x + h) - f(x)| < \epsilon \big).$$

I leave the proof of this equivalence as an exercise to the reader (Exercise 9.32). That Cauchy intends this stronger condition is unclear from the statement of the definition. However, it is the form in which he used continuity in the proof[136] of his first theorem on continuous functions:

6.46 Theorem. *Let $f : [a, b] \times [c, d] \to \mathbb{R}$ be "continuous" in each variable separately. Then f is a "continuous" function of both variables.*

Proof. Let $\langle x, y \rangle \in [a, b] \times [c, d]$, and let α, β be infinitesimals. Observe

$${}^*f(x+\alpha, y+\beta) - f(x, y) = {}^*f(x+\alpha, y+\beta) - {}^*f(x, y+\beta) + {}^*f(x, y+\beta) - f(x, y).$$

By the "continuity" of f in the first variable, with $y + \beta$ held fixed, the first difference on the right-hand side is infinitesimal. Likewise, fixing x, the second difference is infinitesimal by "continuity" in the second variable. Thus ${}^*f(x + \alpha, y + \beta) - f(x, y)$, being the sum of two infinitesimals, is infinitesimal. □

The proof is perfectly rigorous and only appears false because "continuous" does not here mean continuous in our modern sense.

To get a clearer picture of what's going on, consider one of the standard counterexamples to Theorem 6.46:

---

[133] Grabiner, *op. cit.*, p. 87.

[134] Laugwitz, "Infinitely small...", *op. cit.*, pp. 261 - 262.

[135] I am glossing over something here and will be more precise a little later.

[136] Grabiner adds in a footnote that "Cauchy and Bolzano did not seem to appreciate that they were assuming, in effect, that given an ε, their δ works for all x". (Grabiner, *op. cit.*, p. 204.)

6.47 Example. Define

$$f(x,y) = \begin{cases} 0, & \langle x,y \rangle = \langle 0,0 \rangle \\ \dfrac{xy}{x^2+y^2}, & \langle x,y \rangle \neq \langle 0,0 \rangle. \end{cases}$$

In the usual sense, f is everywhere continuous in each variable separately, but f is not continuous at $\langle 0,0 \rangle$ as a function of two variables. In Cauchy's sense f is not continuous near $\langle 0,0 \rangle$ in either x or y. To verify this latter assertion, let α be infinitesimal and consider

$$
\begin{aligned}
{}^*f(\alpha+\alpha,\alpha) - {}^*f(\alpha,\alpha) &= {}^*f(2\alpha,\alpha) - {}^*f(\alpha,\alpha) \\
&= \frac{2\alpha\alpha}{(2\alpha)^2+\alpha^2} - \frac{\alpha\alpha}{\alpha^2+\alpha^2} \\
&= \frac{2\alpha^2}{5\alpha^2} - \frac{\alpha^2}{2\alpha^2} \\
&= \frac{2}{5} - \frac{1}{2} = -\frac{1}{10},
\end{aligned}
$$

which is not infinitesimal. Thus ${}^*f$ is not continuous at $\langle \alpha,\alpha \rangle$ as a function of x. And in the same way it is seen not to be continuous as a function of y.

The most famous example of a "false" theorem by Cauchy, however, is his claim that the sum of a convergent series of continuous functions is continuous. Illustrating the general appraisal of Cauchy's result, Laugwitz cites Hans Freudenthal:

> For instance, it is well known that he [Cauchy] asserted the continuity of the sum of a convergent series of continuous functions; Abel gave a counterexample, and *it is clear that Cauchy himself knew scores of them.*[137]

Laugwitz does not go on to cite Freudenthal's next sentence. I'm not sure why.

> It is less known that later Cauchy correctly formulated and applied the uniform convergence that is needed here.[138]

The new concept here is *uniform convergence*. The industrious reader who went and worked through Exercise 9.32 when directed to it a page or so back and went on and worked the succeeding Exercise 9.33 as well is already familiar with the concept, its standard and nonstandard definitions, and the fact that the sum of a uniformly convergent series of continuous functions is continuous. The reader who hasn't already looked at this exercise is invited to do so now.

---

[137] Laugwitz, "Infinitely small...", *op. cit.*, p. 260. The emphasis was added by Laugwitz; the bracketed insertion is mine. Abel's example, cited in section 8, below, is from his oft-cited paper on the binomial series.

[138] Hans Freudenthal, "Cauchy, Augustin-Louis", in: Charles Coulston Gillispie (ed.), *Dictionary of Scientific Biography*, vol. 3, Charles Scribner's Sons, New York, 1971, p. 137.

The sum of a pointwise convergent series of continuous functions need not be continuous. Thus, whether or not Cauchy's "false theorem" is correct depends on what Cauchy meant by "continuous" and "convergent". We have already seen that what Cauchy means by "continuous" differs from what we mean today by the term. Is the same true of "convergent"? Cauchy gave an almost clear definition of continuity that is rightly celebrated and quoted by many authors including Bottazzini, Edwards, and Grabiner, as cited above. What Cauchy meant by the convergence of a series is not so easily determined, nor so readily discussed. Edwards merely mentions briefly that Cauchy gave an incorrect proof and tried to apply the result to give a rigorous proof of Newton's Binomial Theorem. With the broad scope of Edwards's book, the slightly more than half a page devoted to the topic therein is about right. Surprisingly, Grabiner's book on the *Origins of Cauchy's Rigorous Calculus* is not more expansive on this point. Perhaps this illustrates the difference between a mathematician and an historian in approaching the material. Grabiner explains the *origins* of Cauchy's rigorous calculus, even addressing the possibility raised by Ivor Grattan-Guinness that Cauchy had plagiarised Bolzano in developing a rigorous theory of the Calculus; a mathematician would probably not even mention the unsubstantiated charge of plagiarism, and would go straight for the mathematics. Bottazzini exemplifies this mathematical approach and is the only one of my three earlier sources on Cauchy's definition of continuity to discuss Cauchy's theorem on the sum of a convergent series of continuous functions in any detail. And, of course, there is Laugwitz's paper on "Infinitely small quantities in Cauchy's textbooks". And to this I should add the history of infinite series by Richard Reiff[139] referred to repeatedly in Chapter I.

At this point, the reader may be wondering why I don't simply consult Cauchy himself on the matter. Well, I have— but in German translation. French mathematical papers tend not to offer the English reader much difficulty. Modern French mathematicians tend not to be overly wordy, the grammar appears to be simple, and the number of irregular verbs seems to be small. However, when it comes to subtle points and ambiguous phrases, the usual supplementary practice of inferring the meaning of a passage from the mathematics does not apply here: for, the result as stated is false and Cauchy himself admitted his statement was poor[140]. Thus, I opted for Carl Itzigsohn's German translation of 1885[141], which is available online.

The material we are interested in is from Chapter VI, section 1 of *Cours d'analyse*. He begins by considering a sequence

$$u_0, u_1, u_2, \ldots, \tag{74}$$

and the partial sums,

$$s_n = u_0 + u_1 + \ldots + u_{n-1},$$

---

[139] Richard Reiff, *Geschichte der unendlichen Reihen*, Verlag der H. Laupp'schen Buchhandlung, 1889.

[140] Cf. Laugwitz, "Infinitely small...", *op. cit.*, p. 265.

[141] Augustin Louis Cauchy, *Algebraische Analysis*, Springer-Verlag, Berlin, 1885.

of the infinite series $\sum u_i$ and declares the series convergent to a sum s if the limit of the sequence is s. As a first example he cites the geometric progression,

$$1, x, x^2, \ldots,$$

and proves it has $1/(1-x)$ as limit for $|x| < 1$ and diverges for $|x| > 1$.

Cauchy then proves the necessity and states without proof the sufficiency of what we now call Cauchy convergence as a criterion for the convergence of a series and applies it to the geometric progression to conclude its divergence for $|x| = 1$.

He now considers a couple of *numerical series*, namely the harmonic series, which he proves divergent by the familiar Oresme-Bernoulli proof, which he declares a new proof, and the infinite series for e:

$$e = \sum_{k=0}^{\infty} \frac{1}{k!}.$$

The important thing here is that it causes us to realise that all along he has been talking about a series of functions.

The inadequacy of his definition of convergence of a series of functions becomes quite apparent when we read the various translations of his statement of the theorem on the sum of a convergent series of continuous functions. Whereas earlier I found three independent translations in substantial agreement on the definition of continuity, I now find three independent translations in substantial *dis*agreement. Itzigsohn reads (after translation into English):

> **Theorem.** *If the individual members of the series* (74) *are functions of the same variable x, and indeed* [*are*] *continuous with respect to this variable in the neighbourhood of a particular value, for which the series is convergent, so too the sum s of the series is a continuous function of x in the neighbourhood of this particular value.*[142]

Four years after Itzigsohn, Reiff offers a shorter, looser, paraphrased translation, but he gives it in quotation marks and I take this fact to indicate that the statement was intended to be faithful to the original meaning:

> **Theorem.** *If the u_n's are continuous functions of a variable x, and for all x in the neighbourhood of a certain x, the series u_n is convergent, then the sum of the series is likewise continuous for this value of x.*[143]

And van Egmond translates Bottazzini[144] as follows:

> **Theorem.** *When the different terms of the series* $\left[\sum_{n=1}^{\infty} u_n\right]$ *are functions of the same variable x, continuous with respect to this variable in*

---

[142] *Ibid.*, p. 90.

[143] Reiff, *op. cit.*, p. 168.

[144] Did Bottazzini quote Cauchy in French, or is this van Egmond's English translation of Bottazzini's Italian translation?

the neighbourhood of a particular value for which the series is convergent, the sum s of the series is also a continuous function of x in the neighbourhood of this particular value.[145]

Remember that continuity for Cauchy meant continuity in a neighbourhood. A point of commonality of these three definitions is their reference to the continuity of the various functions in *"the* neighbourhood". Thus they all seem to say we are dealing with a series of functions defined and continuous in some interval I. It is explicitly assumed in all cases that the series converges at some point x_0 in the interior of I. The definitions do not agree on where else in the interval the series is assumed to converge. My English translation of Itzigsohn is not as unambiguous as his German[146]: convergence is only assumed at x_0 and nowhere else. Reiff explicitly assumes convergence at all $x \in I$. And Bottazzini/van Egmond appears to be agnostic on this issue, though the closer proximity of "for which" to "value" than to "neighbourhood" suggests agreement with Itzigsohn. As I cannot completely rule out the reference's being to the neighbourhood for which the series is convergent, and in which, incidentally, the point x_0 lies, I am undecided on the proper interpretation here. Insofar as Itzigsohn and Reiff have interpreted the original French distinctly differently, I surmise the original French to have been ambiguous or Reiff to have quietly "corrected" an obvious error.

There are several ways of resolving this ambiguity. We can read Cauchy's proof and see if the convergence of the series s on all of I is assumed or proved. We can check ourselves whether or not this convergence can be concluded from the weak assumption of convergence at x_0. We can see how Cauchy treats related concepts. And, we can see what Cauchy explicitly assumes in his later statement of the result.

To discuss Cauchy's proof, we must first recall some notation. The series in question is a sum of functions u_n, the limit is s, and the partial sums are denoted s_n. To these Cauchy has added the remainders,

$$r_n = s - s_n.$$

With all of this, we can translate Itzigsohn's translation as follows:

If the terms of the series (74) contain only a single variable x, this series is convergent and its various terms are continuous functions in the neighbourhood of one of the ascribed special values of the variable, so too

$$s_n, r_n, \text{ and } s$$

will be functions of the variable x, of which the first is obviously continuous with respect to x in the neighbourhood of the particular value, here under consideration. Under this assumption we wish to investigate what increments these three functions experience, if one allows x to be

---

[145] Bottazzini, *op. cit.*, p. 110. The bracketed insertion is from the work cited.

[146] In the original German, the declension of "which" in "for which" is masculine, matching "value" and not the feminine "neighbourhood".

increased by an infinitely small numerical quantity α. For all possible values of x the increment of s_n is an infinitely small numerical quantity, and that of r_n is barely perceptible, if one assigns a very large value to n in r_n. The increment of the function s can thus be only an infinitely small numerical quantity.[147]

This is not much help. That r_n is continuous is concluded on the basis of— nothing. In effect, if not intent, it is an unmentioned assumption.

Can it be concluded from the continuity of all the functions $u_n(x)$ of the series in an interval I, and the convergence of the series at a specific value x_0 of x in I, that it converges throughout the interval I? The answer is no:

6.48 Example. Let u_n be defined on $[-1, 1]$ for $n \geq 1$ by

$$u_n(x) = \begin{cases} 1, & x < \frac{-1}{n} \\ |nx|, & \frac{-1}{n} \leq x \leq \frac{1}{n} \\ 1, & \frac{1}{n} < x. \end{cases}$$

Each function u_n is continuous on $[-1, 1]$ (whence continuous in Cauchy's sense). The series $\sum_{n=1}^{\infty} u_n(x)$ converges only at $x = 0$.

I leave the exploration of this example (sketching the graph and verifying the assertions) to the reader.

Two of our four cited methods of resolving ambiguity speak against Itzig-sohn's rendering of Cauchy's theorem. Our third method is to look at how he defines related concepts. We have already seen that he defines continuity on an interval; continuity at a point for him meaning continuity in an interval containing the point. Cauchy's definition of derivative is also not of a function differentiable at a point, but in the whole interval.[148] And his definition of the derivative in the complex case, whereby a function is assumed differentiable not at a point but throughout a neighbourhood of the point, is the standard defini-tion today. Thus, it is entirely consistent with his general practice for Cauchy to have implicitly assumed convergence of a series of functions at a point to mean convergence at all numbers in some interval containing the point. And, if we examine Cauchy's attempted proof cited above, we would take this to mean not just at all real numbers in the interval, say $[a, b]$, but also at all elements of $^*[a, b]$.

Finally, there is Cauchy's corrected statement of his theorem. Both Freuden-thal and Laugwitz refer to the same paper[149], and after defining

$$s_{n'} - s_n = u_n + u_{n+1} + \ldots + u_{n'-1} \tag{75}$$

---

[147] Itzigsohn/Cauchy, *op. cit.*, p. 90.

[148] For an English statement of his definition, cf. Edwards, *op. cit.*, p. 313. This is repeated by Laugwitz in "Infinitely small...", *op. cit.*, pp. 267 - 268.

[149] Augustin Louis Cauchy, "Note sur les séries convergentes dont les divers termes sont des fonctions continues d'une variable réelle ou imaginaire, entre des limites données". The paper is from 1853, but both refer to series 1, volume 12 (pp. 30 - 36) of Cauchy's collected works published in Paris in 1900 by Gauthier-Villars.

for $n' > n$, Laugwitz offers the following translation of the new statement of the theorem:

> If the different terms of the series (74)... are functions of the real variable x, which are continuous with respect to this variable between the given limits; and if, moreover, the sum (75)... becomes always infinitely small for infinitely large values of the integers n and $n' > n$, then the series will be convergent, and the sum of the series (74) will be, between the given limits, a continuous function of the variable x.[150]

The conditions on convergence given, expressed in terms of a limit instead of Cauchy convergence, are those of the nonstandard formulation of uniform convergence:

6.49 Definition. *A series of functions $\sum u_n$ converges in the Cauchy sense to a function s on an interval $[a, b]$ if for every infinite N and every $x \in {}^*[a, b]$, $s_N(x) \approx s(x)$.*

The concept so defined is indeed equivalent to uniform convergence, as the reader is asked to prove in Exercise 9.33 in the exercises at the end of the Chapter, where he is also asked to prove that the uniform limit of continuous functions is continuous. Thus, Freudenthal's remark that Cauchy correctly formulated and proved the continuity of the uniformly convergent sum of continuous functions is correct. The missing elements were the modern standard formulation of the notion and proof of equivalence.

Nonetheless, Laugwitz claims Cauchy's original proof is correct under an ostensibly weaker notion of convergence of functions on the hyperreals.

6.50 Definition. *A series of functions $\sum u_n$ converges in the Laugwitz-Cauchy sense to a function s on an interval $[a, b]$ if for every $x \in {}^*[a, b]$ and every real $\epsilon > 0$ there is an $n_0 \in \mathbb{N}$ such that, for all $n \in \mathbb{N}$, if $n > n_0$, then $|s_n(x) - s(x)| < \epsilon$.*

Laugwitz's explanation of this definition is that infinitesimals are used to explain continuity, but only finite integers are needed for limits of sequences, i.e. summation of series. His version of Cauchy's theorem is this:

6.51 Theorem. *Let the functions $u_0, u_1, \ldots$ be continuous on an interval $[a, b]$ and suppose the series $\sum u_n$ converges to s in the Laugwitz-Cauchy sense on $[a, b]$. Then: s is continuous on $[a, b]$.*

Proof. Let $x, y \in {}^*[a, b]$ with $x \approx y$. Observe

$$s(x) - s(y) = s(x) - s_n(x) + s_n(x) - s_n(y) + s_n(y) - s(y),$$

whence

The easily accessible online version appears to be a later edition and, as Laugwitz offers a convenient English translation of what I need, I have not checked which volume of this edition the paper is in.

[150] Laugwitz, "Infinitely small...", *op. cit.*, p. 265.

$$|s(x) - s(y)| \leq |s(x) - s_n(x)| + |s_n(x) - s_n(y)| + |s_n(y) - s(y)|. \qquad (76)$$

Now, by the continuity of s_n, the middle term of (76), $|s_n(x) - s_n(y)|$ is infinitesimal. By the Laugwitz-Cauchy convergence of the series, there are $n_1, n_2 \in \mathbb{N}$ such that for all $n \in \mathbb{N}$,

$$n > n_1 \;\Rightarrow\; |s(x) - s_n(x)| < \frac{\epsilon}{3}$$

$$n > n_2 \;\Rightarrow\; |s_n(y) - s(y)| < \frac{\epsilon}{3},$$

whence the right-hand side of (76) is less than $\epsilon/3 + \epsilon/3 + \epsilon/3 = \epsilon$ for $n > n_0 = \max\{n_1, n_2\}$ and we have

$$|s(x) - s(y)| < \epsilon.$$

This being true for all $\epsilon > 0$, $|s(x) - s(y)|$ is infinitesimal and s is continuous on $[a, b]$.   □

The interest in Theorem 6.51 is that it shows that Cauchy's original proof, when the details are filled in, is correct. The result itself is not a new one to be placed alongside the continuity of a uniformly convergent series of continuous functions. For, we have the following result:

6.52 Lemma. *Let the functions $u_0, u_1, \ldots, s$ be defined on an interval $[a, b]$. Then: $\sum u_n$ converges in the Laugwitz-Cauchy sense to s iff it converges to s in the Cauchy sense.*

Proof. The proof is essentially a variant of the solution to Exercise 9.33.i and the conscientious reader might not want to read on until after he has tried his hand at the Exercise.

Assume $\sum u_n$ is Laugwitz-Cauchy convergent to s on $[a, b]$ and let $x \in {}^{*}[a, b]$. Let $\epsilon \in \mathbb{R}^+$ and choose n_0 such that $|s_n(x) - s(x)| < \epsilon$ for all $n > n_0$. In particular, $|s_N(x) - s(x)| < \epsilon$ for any infinite N and, since ϵ was arbitrary, $s_N(x) \approx s(x)$, i.e. $\sum u_n$ converges in the Cauchy sense to s.

Conversely, assume $\sum u_n$ converges in the Cauchy sense to s. Let $\epsilon \in \mathbb{R}^+$ and N_0 be an infinite integer. By assumption,

$$^{*}\mathbb{R} \models \forall n \forall x \in [\overline{a}, \overline{b}] \big(\overline{\mathbb{N}}(n) \wedge n > \overline{N}_0 \to |\overline{s}_n(x) - \overline{s}(x)| < \overline{\epsilon} \big)$$

$$\models \exists n_0 \forall n \forall x \in [\overline{a}, \overline{b}] \big(\overline{\mathbb{N}}(n) \wedge n > n_0 \to |\overline{s}_n(x) - \overline{s}(x)| < \overline{\epsilon} \big).$$

This is a sentence of $L_{\mathbb{R}}$, whence it is true in $\mathbb{R}$:

$$\mathbb{R} \models \exists n_0 \forall n \forall x \in [\overline{a}, \overline{b}] \big(\overline{\mathbb{N}}(n) \wedge n > n_0 \to |\overline{s}_n(x) - \overline{s}(x)| < \overline{\epsilon} \big)$$

$$\models \forall n \forall x \in [\overline{a}, \overline{b}] \big(\overline{\mathbb{N}}(n) \wedge n > \overline{n}_0 \to |\overline{s}_n(x) - \overline{s}(x)| < \overline{\epsilon} \big),$$

for some $n_0 \in \mathbb{N}$. But then $^{*}\mathbb{R}$ satisfies this as well: for all $n > n_0$ and all $x \in {}^{*}[a, b]$, $|s_n(x) - s(x)| < \epsilon$. That is, $\sum u_n$ converges to s in the Laugwitz-Cauchy sense.   □

Because Laugwitz-Cauchy convergence entails convergence in the Cauchy sense, i.e. uniform convergence, Laugwitz's version of Cauchy's theorem offers nothing essentially new.

There is a caveat to this. While Laugwitz admits to the superiority of $^*\mathbb{R}$ to $^\Omega\mathbb{R}$ for the purpose of developing modern analysis, when it comes to interpreting history he prefers $^\Omega\mathbb{R}$, finding it closer in spirit to 19th century analysis. Let me quote him:

> When Robinson [Ro66] [151] had drawn attention to an interpretation of Cauchy by means of his Nonstandard Analysis this gave rise to a vivid debate, also on the question whether conceptions of the second half of the 20th century were permitted as a justification here. It seems to me that our approach is closer to the level of abstraction of the early 19th century. Also it permits an easy characterization of continuous functions which are central in Cauchy's texts. Let F be a real function which is fully defined, that is, has real values for all real x. The function is, in the obvious way, extended to all Ω-numbers. (The value of the number represented by the sequence x_n is represented by the sequence $y_n = F(x_n)$.) It is easy to see that F must be continuous if it maps the Cauchy continuum into itself. In Nonstandard Analysis this conclusion does not follow.[152]

The major drawback to $^\Omega\mathbb{R}$ is that not every hyperreal in it possesses a standard part. Laugwitz gets around this by defining the *Cauchy continuum* to be those hyperreals which do possess such and using them in place of the ordinary hyperreals. Formally, for $\mathbb{H}$ the space of hyperreals, the definition of the Cauchy continuum reads:

6.53 Definition. *The* Cauchy continuum, $^C\mathbb{R}$, *of a hyperreal extension* $\mathbb{H}$ *of* $\mathbb{R}$ *is defined by*

$$^C\mathbb{R} = \{x \in \mathbb{H} \mid x = y + \eta \text{ for some real } y \text{ and infinitesimal } \eta\}.$$

For $\mathbb{H} = {}^*\mathbb{R}$, the Cauchy continuum consists of the finite hyperreals; for $\mathbb{H} = {}^\Omega\mathbb{R}$, it consists of those finite hyperreals which do not oscillate too much. The two Cauchy continua behave quite differently, as Laugwitz notes in the paragraph cited. Formally, his claim is as follows:

6.54 Theorem. *(For* $\mathbb{H} = {}^\Omega\mathbb{R}$*) Let* $f : \mathbb{R} \to \mathbb{R}$. *The following are equivalent:*
i. *f is continuous;*
ii. $^\Omega f : {}^C\mathbb{R} \to {}^C\mathbb{R}$.

One probably gets a better feel for why this Theorem is true by considering first why the Cauchy continuum of $^\Omega\mathbb{R}$ is not closed under some particular discontinuous function, such as the Heaviside function,

$$H(x) = \begin{cases} 0, & x < 0 \\ 1, & 0 \le x. \end{cases}$$

---

[151] The reference is to Robinson's book on nonstandard analysis.
[152] Laugwitz, "Curt Schmieden's approach...", *op. cit.*, p. 134.

$^{\Omega}H(\eta) = 0$ for any negative infinitesimal, and 1 for any positive infinitesimal. But $H(\eta)$ is neither 0 nor 1 for many wobbly infinitesimals:

$$H\left(\left[\left(0, -1, \frac{1}{2}, -\frac{1}{3}, \frac{1}{4}, \ldots\right)\right]\right) = [(1, 0, 1, 0, 1, \ldots)] \notin {}^{C}\mathbb{R}.$$

The value of H here is bounded, but it is not in $^{C}\mathbb{R}$ because it oscillates too much to be infinitesimally close to a single real number. The very failure of trichotomy, which complicated the proof of the Extreme Value Theorem and which denied us the possibility of defining the standard part of an arbitrary finite element of $^{\Omega}\mathbb{R}$, here comes to the rescue in allowing us to prove Theorem 6.54: As Laugwitz notes, the result does not hold if one defines $^{C}\mathbb{R}$ in $^{*}\mathbb{R}$.

Proof of Theorem 6.54. i $\Rightarrow$ ii. Let $x \in {}^{C}\mathbb{R}$, say $x = y + \eta$ where $y \in \mathbb{R}$ and η is infinitesimal. Observe:

$$f \text{ continuous at } y \Rightarrow {}^{\Omega}f(x) = {}^{\Omega}f(y + \eta) \approx f(y)$$
$$\Rightarrow {}^{\Omega}f(x) \in {}^{C}\mathbb{R}.$$

ii $\Rightarrow$ i. If f is not continuous at some $y \in \mathbb{R}$, then there is an $\epsilon \in \mathbb{R}^{+}$ such that for all real $\delta > 0$ there is a $z \in \mathbb{R}$ with $|y - z| < \delta$, but $|f(y) - f(z)| > \epsilon$. Choose $z_0, z_1, z_2, \ldots \in \mathbb{R}$ such that for all $n \in \mathbb{N}$,

$$|y - z_n| < \frac{1}{n + 1}, \text{ but } |f(y) - f(z_n)| > \epsilon.$$

Define $x = [(z_0, y, z_2, y, z_4, y, \ldots)]$. The claim is that $x \approx y$, yet $^{\Omega}f(x) \notin {}^{C}\mathbb{R}$. That $x \approx y$ is clear:

$$|x_n - y_n| = \begin{cases} |z_n - y|, & n \text{ even} \\ 0, & n \text{ odd} \end{cases}$$
$$< \frac{1}{n + 1}.$$

Suppose $c = {}^{\Omega}f(x) = [(f(z_0), f(y), f(z_2), f(y), \ldots)] \in {}^{C}\mathbb{R}$. Writing $c = [(c_0, c_1, c_2, \ldots)]$, note that if $c \approx r \in \mathbb{R}$, we must have

$$|c_n - r| < \frac{\epsilon}{2}$$

for all sufficiently large n, say $n > n_0$. But for such n,

$$\epsilon < |f(z_{2n}) - f(y)| \leq |f(z_{2n}) - r| + |r - f(y)|$$
$$\leq |c_{2n} - r| + |r - c_{2n+1}|$$
$$\leq \frac{\epsilon}{2} + \frac{\epsilon}{2},$$

a contradiction. $\qquad\square$

Theorem 6.54 does not hold in $^{*}\mathbb{R}$, where there is no counterpart to the oscillation in $^{\Omega}\mathbb{R}$, and the extension $^{*}H$ of Heaviside's H indeed maps elements of the Cauchy continuum to the Cauchy continuum. As Cauchy only dealt with infinitesimals in a Type I manner, we cannot say if $^{\Omega}\mathbb{R}$ or $^{*}\mathbb{R}$ would have been closer to his vague conception of $\mathbb{H}$ and thus whether the Theorem is relevant to an understanding of Cauchy or not. However, it does illustrate the power of the assumption that one is working with the Cauchy continuum and not just with $\mathbb{R}$.

I am not sure what to make of Theorem 6.54 and the special status of the Cauchy continuum in $^{\Omega}\mathbb{R}$. Laugwitz might like it because it reminds one of constructive mathematics in which all constructive functions defined for all reals are continuous, or, perhaps because it offers an additional explanation of what is behind Theorem 6.46 or the continuity of the derivative of a differentiable function (Exercise 9.34, below)— a result he notes Cauchy used tacitly in some proofs. Or, it may simply be that some Robinsonian nonstandard proofs carry over to the Laugwitz-Schmieden theory if one partially replaces $^{\Omega}\mathbb{R}$ by $^{C}\mathbb{R}$. In any event I strongly recommend Laugwitz's paper, "Infinitely small quantities in Cauchy's textbooks" for some insights into some of Cauchy's more questionable proofs, and demonstrations that these proofs, in their proper settings, are indeed correct.

7 Metric Space Completions

I have put off discussing the Cantor-Méray construction of the real numbers for two reasons. One is that their construction of real numbers as equivalence classes of Cauchy convergent sequences of rational numbers was already given in the standard account of Bolzano's construction: the Cauchy convergent sequences of rationals are just Bolzano's measurable infinite number expressions when these latter are interpreted as sequences of rational numbers. A second reason is that, as first shown by Felix Hausdorff, the Cantor-Méray construction applies in a more general setting that I thought should be discussed only after finishing the discussion of the real numbers.

7.1 Definition. *Let X be a nonempty set. A function $d : X \times X \to \mathbb{R}$ is called a* distance function, *or* metric, *if it satisfies the following for all $x, y, z \in X$:*
i. $d(x, y) \geq 0$ and $d(x, y) = 0$ iff $x = y$;
ii. $d(x, y) = d(y, x)$;
iii. $d(x, z) \leq d(x, y) + d(y, z)$.
A pair (X, d) is called a metric space *if d is a distance function on X.*

(Some earlier authors preferred the Greek letter "ρ" for the distance function here denoted "d". Nowadays, "d" seems to be the standard, though I rather like "μ" for "metric". However, "μ" is generally used to denote a measure on the space when it has one.)

The familiar metric spaces from the Calculus are the Euclidean spaces $\mathbb{R}^n$, where

$$d_n\big((x_0, \ldots, x_{n-1}), (y_0, \ldots, y_{n-1})\big) = \sqrt{(x_0 - y_0)^2 + \ldots + (x_{n-1} - y_{n-1})^2}.$$

That d_n is indeed a distance function on $\mathbb{R}^n$ is not a trivial fact. Properties i and ii are quickly verified, but iii is sufficiently nonobvious as to have a name and a list of authors. It is called the *Triangle Inequality* because it asserts, in a triangle determined by three points A, B, C, the length of side AC is less than or equal to the sum of the lengths AB and BC. For $n = 2, 3$, the Pythagorean Theorem can be appealed to to show d_n to agree with our usual notion of distance. Such geometric reasoning, however, does not apply when $n > 3$ and, in any event, the general programme of arithmetising mathematics that the previous sections of this chapter represent and was carried out in the 19th century demands a more formal, arithmetic proof. This is supplied by the Cauchy-Schwarz Inequality, sometimes called the Cauchy-Schwarz-Bunyakovsky Inequality. The priority in this case is actually easy to sort out. Cauchy included it, along with other important inequalities, in an appendix to his *Cours d'analyse* in 1821.[153] According to *The Dictionary of Scientific Biography*[154], Viktor Yakovlevich Bunyakovsky published an analogue of the inequality for integrals in a monograph in 1859. His inequality was rediscovered by Hermann Amandus Schwarz in 1884 and became known as the Schwarz Inequality.

Cauchy's inequality is usually written in terms of the *dot product*, or *inner product*, of two vectors: Writing $\mathbf{x}$ for $(x_0, \ldots, x_{n-1})$, $\mathbf{y}$ for $(y_0, \ldots, y_{n-1})$, the dot product of the two vectors is the sum

$$\mathbf{x} \cdot \mathbf{y} = x_0 y_0 + \ldots + x_{n-1} y_{n-1}.$$

Writing

$$|\mathbf{x}| = \sqrt{x_0^2 + \ldots + x_{n-1}^2}, \quad |\mathbf{y}| = \sqrt{y_0^2 + \ldots + y_{n-1}^2},$$

the Cauchy inequality reads

$$|\mathbf{x} \cdot \mathbf{y}| \le |\mathbf{x}| \cdot |\mathbf{y}|, \tag{77}$$

the absolute value on the left and the dot on the right standing for the ordinary absolute value and multiplication of real numbers, respectively. Vectors weren't around in Cauchy's day, so he wrote the more instantly intelligible, but less compact

$$\left| \sum_{i=0}^{n-1} x_i y_i \right| \le \sqrt{\sum_{i=0}^{n-1} x_i^2} \cdot \sqrt{\sum_{i=0}^{n-1} y_i^2}. \tag{78}$$

(In fact, he used neither the $\sum$ nor indices, but wrote[155]

$$val.num.(a\alpha + a'\alpha' + a''\alpha'' + \ldots)$$
$$< \sqrt{a^2 + a'^2 + a''^2 + \ldots}\sqrt{\alpha^2 + \alpha'^2 + \alpha''^2 + \ldots},$$

---

[153] The importance of this inclusion is nicely appreciated by Judith V. Grabiner, *op. cit.*, pp. 75 - 76.

[154] Charles Scribner's Sons, New York, volume 15, 1980.

[155] I confess not to have consulted the original, but the German translation available online through the university in Göttingen.

which is quite ugly.) His proof was quite simple: One has the identity,

$$\left(\sum_{i=0}^{n-1} x_i y_i\right)^2 + \sum_{i<j} (x_i y_j - x_j y_i)^2 = \left(\sum_{i=0}^{n-1} x_i^2\right)\left(\sum_{i=0}^{n-1} y_i^2\right), \qquad (79)$$

from which follows

$$\left(\sum_{i=0}^{n-1} x_i y_i\right)^2 \leq \left(\sum_{i=0}^{n-1} x_i^2\right)\left(\sum_{i=0}^{n-1} y_i^2\right),$$

and thence follows (78).

Cauchy's proof is quite simple once one knows to prove (79). However, it is a bit unmotivated and many authors like to apply a simple trick. Letting $\mathbf{x}, \mathbf{y}$ be given, define

$$Q(t) = (t\mathbf{x} - \mathbf{y}) \cdot (t\mathbf{x} - \mathbf{y}) \qquad (80)$$
$$= t^2 \mathbf{x} \cdot \mathbf{x} - 2t\mathbf{x} \cdot \mathbf{y} + \mathbf{y} \cdot \mathbf{y}.$$

Now $\mathbf{x} \cdot \mathbf{x}, \mathbf{x} \cdot \mathbf{y}$ and $\mathbf{y} \cdot \mathbf{y}$ are real numbers and $Q(t)$ is a simple quadratic function with positive leading coefficient (We can ignore the case in which $\mathbf{x}$ consists of n 0's as the inequality is trivial in this case.) and thus has a minimum value, which can be found via the Calculus or by completing the square:

$$Q(t) = (\mathbf{x} \cdot \mathbf{x})\left(t - \frac{\mathbf{x} \cdot \mathbf{y}}{\mathbf{x} \cdot \mathbf{x}}\right)^2 + \mathbf{y} \cdot \mathbf{y} - (\mathbf{x} \cdot \mathbf{x})\left(\frac{\mathbf{x} \cdot \mathbf{y}}{\mathbf{x} \cdot \mathbf{x}}\right)^2.$$

The minimum value clearly occurs at $t = (\mathbf{x} \cdot \mathbf{y})/(\mathbf{x} \cdot \mathbf{x})$:

$$Q\left(\frac{\mathbf{x} \cdot \mathbf{y}}{\mathbf{x} \cdot \mathbf{x}}\right) = \mathbf{y} \cdot \mathbf{y} - \frac{(\mathbf{x} \cdot \mathbf{y})^2}{\mathbf{x} \cdot \mathbf{x}}. \qquad (81)$$

But, by (80), $Q(t)$ is a sum of squares, whence nonnegative. Equation (81) thus yields

$$\mathbf{y} \cdot \mathbf{y} - \frac{(\mathbf{x} \cdot \mathbf{y})^2}{\mathbf{x} \cdot \mathbf{x}} \geq 0,$$

i.e.

$$(\mathbf{x} \cdot \mathbf{x})(\mathbf{y} \cdot \mathbf{y}) \geq (\mathbf{x} \cdot \mathbf{y})^2.$$

Taking square roots yields (77).

A possibly more pleasing proof than either of these is given by Casper Goffman in his introductory essay on functional analysis[156]. There the result is proven by a clever reduction of the general case for $\mathbb{R}^n$ to the case for $n = 1$. The most straightforward proof, however, is probably inductive. (See Exercises 9.35 and 9.36 at the end of this Chapter.)

---

[156] Casper Goffman, "Preliminaries to functional analysis", in: R.C. Buck (ed.), *Studies in Mathematics, Volume 1: Studies in Modern Analysis*, Mathematical Association of America/Prentice Hall, Englewood Cliffs (New Jersey), 1962. The proof is on pp. 141 - 142.

To avoid a notational mess in deriving the Triangle Inequality for d_n, the standard thing to do is first to define a *norm* or *absolute value*, prove the Triangle Inequality for absolute values, and then derive the inequality for distances from this. To this end, we define

$$|\mathbf{a}| = \sqrt{\mathbf{a} \cdot \mathbf{a}} = \sqrt{\sum_{i=0}^{n-1} a_i^2}.$$

The triangle inequality for absolute value reads,

$$|\mathbf{a} + \mathbf{b}| \le |\mathbf{a}| + |\mathbf{b}|. \tag{82}$$

The proof is simple:

$$\begin{aligned}
|\mathbf{a} + \mathbf{b}|^2 &= \sum_{i=0}^{n-1} (a_i + b_i)^2 \\
&= \sum a_i^2 + 2 \sum a_i b_i + \sum b_i^2 \\
&= |\mathbf{a}|^2 + 2(\mathbf{a} \cdot \mathbf{b}) + |\mathbf{b}|^2 \\
&\le |\mathbf{a}|^2 + 2|\mathbf{a} \cdot \mathbf{b}| + |\mathbf{b}|^2 \\
&\le |\mathbf{a}|^2 + 2|\mathbf{a}| \cdot |\mathbf{b}| + |\mathbf{b}|^2, \text{ by (77)} \\
&\le (|\mathbf{a}| + |\mathbf{b}|)^2,
\end{aligned}$$

whence (82) follows. Letting $\mathbf{a} = \mathbf{y} - \mathbf{x}, \mathbf{b} = \mathbf{z} - \mathbf{y}$ (82) yields

$$d(x, z) = \sqrt{\sum (z_i - x_i)^2} = |\mathbf{z} - \mathbf{x}| \le |\mathbf{z} - \mathbf{y}| + |\mathbf{y} - \mathbf{x}| = d(y, z) + d(x, y).$$

Thus we have seen that $\mathbb{R}^n$ is a metric space under the usual "Euclidean metric", i.e. the usual measure of distance. There are other workable distance functions that turn $\mathbb{R}^n$ into a metric space:

$$m_n(\mathbf{x}) = \max\{|x_0|, |x_1|, \ldots, |x_{n-1}|\}$$

is called the Minkowski metric after Hermann Minkowski. Or one can take the sum of the absolute values of the components,

$$|x_0| + |x_2| + \ldots + |x_{n-1}|,$$

or their average:

$$\frac{|x_0| + |x_2| + \ldots + |x_{n-1}|}{n}.$$

While the spaces $\mathbb{R}^n$ are the most familiar metric spaces, they are far from being the only interesting and useful ones. Function spaces abound in the literature and have many applications. Particularly simple examples of such are afforded by supplying the set $C[0, 1]$ of continuous functions on the unit interval with various metrics. In analogy with the metrics defined for $\mathbb{R}^n$, there are the following

$$\|f - g\|_\infty = \max\{|f(x) - g(x)| \mid x \in [0,1]\}$$

$$\|f - g\|_1 = \int_0^1 |f(x) - g(x)| dx$$

$$\|f - g\|_2 = \left(\int_0^1 (f(x) - g(x))^2 dx \right)^{\frac{1}{2}}.$$

For an interval like $[0, 2\pi]$, it is common to take the average over the interval, e.g.

$$\|f - g\|_1 = \frac{1}{2\pi} \int_0^{2\pi} |f(x) - g(x)| dx.$$

In addition to varying the metric, one can strengthen or weaken the condition for inclusion of a function in the space, thus considering the continuously differentiable functions or the integrable functions, etc.[157]

So, as in Chapter I, where we noted that vector spaces pop up quite often in mathematics, we see that metric spaces do also. In fact, the most interesting metric spaces are often vector spaces themselves.

A central concern in a metric space is the convergence of a sequence of elements.

7.2 Definition. *Let (X, d) be a metric space. A sequence $x_0, x_1, \ldots$ of elements of X* converges *to an element $x \in X$, written*

$$\lim_{n \to \infty} x_n = x,$$

if for any real number $\epsilon > 0$ a number $n_0 \in \mathbb{N}$ exists such that, for all $n > n_0$, we have $d(x_n, x) < \epsilon$.

7.3 Definition. *Let (X, d) be a metric space. A sequence $x_0, x_1, \ldots$ of elements of X is a* Cauchy sequence, *or is said to be* Cauchy convergent, *if for any real number $\epsilon > 0$ a number $n_0 \in \mathbb{N}$ exists such that, for all $m, n > n_0$, we have $d(x_m, x_n) < \epsilon$.*

As in the reals, a necessary condition for convergence in a metric space is Cauchy convergence:

7.4 Lemma. *Let (X, d) be a metric space. If a sequence $x_0, x_1, \ldots$ is convergent, then it is Cauchy convergent.*

I leave the proof to the reader.

The converse can fail. The obvious example is given by the set of rational numbers with the distance function inherited from the reals: Any sequence of

---

[157] If one weakens this condition, however, condition i of Definition 7.1 fails and one must pass to equivalence classes using the equivalence relation,

$$f \equiv g : \quad d(f, g) = 0.$$

rationals converging to an irrational in $\mathbb{R}$ will be Cauchy convergent and not convergent in $\mathbb{Q}$. A more complex, but possibly instructive example is given by the following:

7.5 Example. On $C[0, 1]$, consider the sequence of piecewise linear continuous approximations to a step function,

$$f_n(x) = \begin{cases} 0, & 0 \le x \le \frac{n-1}{2n} \\ 2nx - n + 1, & \frac{n-1}{2n} < x < \frac{1}{2} \\ 1, & \frac{1}{2} \le x \le 1. \end{cases}$$

The sequence $f_0, f_1, \ldots$ is a Cauchy sequence if we use the distance function,

$$d(f, g) = \int_0^1 |f(x) - g(x)| dx;$$

for, given any n_0, if $m, n > n_0$, the graphs of f and g differ only inside the rectangle of height 1 based on the subinterval $[(n_0 - 1)/(2n), 1/2]$ and thus

$$d(f_m, f_n) < 1 \cdot \frac{1}{2n_0} = \frac{1}{2n_0}.$$

The ordinary pointwise limit of this sequence of functions is the discontinuous step function,

$$f(x) = \begin{cases} 0, & x < \frac{1}{2} \\ 1, & \frac{1}{2} \le x. \end{cases} \tag{83}$$

The fact that the pointwise limit is not continuous does not show the sequence to have no limit in $C[0, 1]$ under the given metric For, we have not proven any relation between the two notions of limit. Nor do we have to. The present example is simple enough to allow us to apply a simple *ad hoc* argument.

Let $x_0 \in [0, 1]$. Assume first that $x_0 < 1/2$. Choose n_1 large enough so that $x_0 < (n_1 - 1)/(2n_1)$ and pick a, b so that

$$x_0 \in [a, b] \subseteq \left[0, \frac{n_1 - 1}{2n_1}\right].$$

Let $\epsilon > 0$ be arbitrary and choose $n_0 > n_1$ so large that

$$\int_0^1 |f(x) - f_n(x)| dx < \epsilon,$$

whenever $n > n_0$, where we assume, by way of contradiction, that $\lim_{n \to \infty} f_n = f$ exists in $C[0, 1]$. Now:

$$\int_a^b |f(x) - f_n(x)| dx \le \int_0^1 |f(x) - f_n(x)| dx < \epsilon.$$

But, $f_n(x) = 0$ on $[a, b]$, whence

$$\int_a^b |f(x) - f_n(x)|dx = \int_a^b |f(x)|dx,$$

i.e.

$$\int_a^b |f(x)|dx < \epsilon.$$

As $\epsilon > 0$ was arbitrary, we in fact have

$$\int_a^b |f(x)|dx = 0.$$

If $f(x_0) \neq 0$, then we can choose $[a, b]$ small enough so that $f(x)$ is bounded away from 0 on $[a, b]$ and the integral would not be 0. Thus, $f(x_0) = 0$.
   If $x_0 \geq 1/2$, the same calculation for $[a, b] \subseteq [1/2, 1]$ gives

$$\int_a^b |f(x) - f_n(x)|dx = \int_a^b |f(x) - 1|dx = 0,$$

and $f(x_0) - 1 = 0$, i.e. $f(x_0) = 1$. Thus, f is the discontinuous step function (83), contrary to the assumption f was continuous and one can conclude the sequence $f_0, f_1, \ldots$ to have no limit in $C[0, 1]$.

   The step function (83) is the limit of this sequence in a larger space obtained by using all the Riemann integrable functions on $[0, 1]$ and the same distance function,

$$d(f, g) = \int_0^1 |f(x) - g(x)|dx,$$

after we identify functions f and g of distance 0 from one another. I.e., the sequence $f_0, f_1, \ldots$ has a limit in a larger space that consists, not of functions, but of equivalence classes of functions. This behaviour is quite general: Every sequence in a metric space can be given a limit in a larger space. Indeed, every metric space can be embedded into a complete metric space.

7.6 Definition. *A metric space (X, d) is a complete metric space if every Cauchy sequence in X has a limit in X.*

7.7 Definition. *A distance preserving embedding of a metric space (X, d_1) into a metric space (Y, d_2) is a function $f : X \to Y$ such that, for all $x, y \in X$,*

$$d_2(f(x), f(y)) = d_1(x, y).$$

   Note that, by Definition 7.1.i, a distance preserving function is one-to-one and we have not inadvertently expanded the notion of an embedding with this definition.
   Another adjective for "distance preserving" is "*isometric*", and distance preserving embeddings are often referred to as *isometries*.
   The result we wish to prove is the following

7.8 Theorem. *Every metric space has a distance preserving embedding into a complete metric space.*

This is a weak statement of what is actually proven. As in sections 1, 2, 4, and 5, there is a canonicality to the construction. We actually construct a minimum *completion* of the space (X, d).

7.9 Definition. *Let (X, d) be a metric space. A space $(\overline{X}, \overline{d})$ is a metric completion of (X, d) if*
i. there is a distance preserving embedding $\iota : X \to \overline{X}$; and
ii. for any distance preserving embedding $f : X \to Y$ of X into a complete metric space, there is a distance preserving embedding $\overline{f} : \overline{X} \to Y$ which agrees with f: for all $x \in X, \overline{f}(\iota(x)) = f(x)$.

7.10 Theorem. *Every metric space has a metric completion.*

The construction of $\overline{X}$ is fairly straightforward. $\overline{X}$ will consist of equivalence classes $[\mathbf{x}]$ of Cauchy sequences $\mathbf{x} = (x_0, x_1, \dots)$ of elements of X under the equivalence relation,

$$\mathbf{x} \equiv \mathbf{y} : \ \lim_{n \to \infty} d(x_n, y_n) = 0;$$

$\overline{d}$ is defined by

$$\overline{d}([\mathbf{x}], [\mathbf{y}]) = \lim_{n \to \infty} d(x_n, y_n); \tag{84}$$

and the embedding $\iota : X \to \overline{X}$ is defined by mapping $x \in X$ to its constant sequence:

$$\iota(x) = (x, x, x, \dots).$$

The proof that the construction works is a fairly routine matter.

Proof of Theorem 7.8. The first thing to do is to verify that $\equiv$ is an equivalence relation and that, as defined by (84), $\overline{d}$ is well-defined, i.e. its value does not depend on the choice of representatives $\mathbf{x}, \mathbf{y}$ in $[\mathbf{x}], [\mathbf{y}]$, respectively.

Verification that $\equiv$ is an equivalence relation consists of several trivial observations.
i. For any $\mathbf{x} \in \overline{X}, d(x_n, x_n) = 0$, whence $\lim_{n \to \infty} d(x_n, x_n) = 0$, i.e. $\mathbf{x} \equiv \mathbf{x}$.
ii. For any $\mathbf{x}, \mathbf{y} \in \overline{X}$,

$$\mathbf{x} \equiv \mathbf{y} \ \Rightarrow \ \lim_{n \to \infty} d(x_n, y_n) = 0$$
$$\Rightarrow \ \lim_{n \to \infty} d(y_n, x_n) = 0$$
$$\Rightarrow \ \mathbf{y} \equiv \mathbf{x}.$$

iii. Let $\mathbf{x}, \mathbf{y}, \mathbf{z} \in \overline{X}$ with $\mathbf{x} \equiv \mathbf{y}$ and $\mathbf{y} \equiv \mathbf{z}$. By the Triangle Inequality,

$$d(x_n, z_n) \leq d(x_n, y_n) + d(y_n, z_n),$$

whence

$$\lim_{n \to \infty} d(x_n, z_n) \leq \lim_{n \to \infty} d(x_n, y_n) + \lim_{n \to \infty} d(y_n, z_n) \leq 0 + 0 = 0.$$

Thus, $\mathbf{x} \equiv \mathbf{z}$.

That $\overline{d}$ is well-defined is established in a couple of stages. Note first that

$$
\begin{aligned}
|d(x_m, y_m) - d(x_n, y_n)| &= |d(x_m, y_m) - d(x_m, y_n) + d(x_m, y_n) - d(x_n, y_n)| \\
&\leq |d(x_m, y_m) - d(x_m, y_n)| + |d(x_m, y_n) - d(x_n, y_n)| \\
&\leq d(y_m, y_n) + d(x_m, x_n), \hspace{3cm} (85)
\end{aligned}
$$

by the Triangle Inequality:

$$
d(x_m, y_m) \leq d(x_m, y_n) + d(y_n, y_m), \quad d(x_m, y_n) \leq d(x_m, y_m) + d(y_m, y_n),
$$

whence

$$
|d(x_m, y_m) - d(x_m, y_n)| \leq d(y_n y_m).
$$

Likewise,

$$
|d(x_m, y_n) - d(x_n, y_n)| \leq d(x_m, x_n),
$$

and (85) holds. But $\mathbf{x}, \mathbf{y}$ are Cauchy sequences, whence each summand of (85) can be made less than, say $\epsilon/2$, by choosing $m, n > n_0$ for some $n_0 \in \mathbb{N}$. Thus

$$
|d(x_m, y_m) - d(x_n, y_n)| < \frac{\epsilon}{2} + \frac{\epsilon}{2} = \epsilon
$$

for $m, n > n_0$, and we see that the sequence

$$
d(x_0, y_0), d(x_1, y_1), \ldots
$$

is a Cauchy sequence of real numbers and thus has a limit[158]. We may thus define $\overline{d}$ as a function of Cauchy sequences,

$$
\overline{d}(\mathbf{x}, \mathbf{y}) = \lim_{n \to \infty} d(x_n, y_n).
$$

To define $\overline{d}$ as a function of the equivalence classes, i.e. to justify (84), we must show that equivalent sequences yield identical results. To this end, let $\mathbf{a} \equiv \mathbf{x}, \mathbf{b} \equiv \mathbf{y}$, and observe that

$$
\begin{aligned}
d(a_n, b_n) &\leq d(a_n, y_n) + d(y_n, b_n) \\
&\leq (d(a_n, x_n) + d(x_n, y_n)) + d(y_n, b_n) \\
&\leq d(x_n, y_n) + (d(a_n, x_n) + d(y_n, b_n))
\end{aligned}
$$

and, taking limits,

$$
\overline{d}(\mathbf{a}, \mathbf{b}) \leq \overline{d}(\mathbf{x}, \mathbf{y}) + (0 + 0),
$$

i.e. $\overline{d}(\mathbf{a}, \mathbf{b}) \leq \overline{d}(\mathbf{x}, \mathbf{y})$. Similarly, $\overline{d}(\mathbf{x}, \mathbf{y}) \leq \overline{d}(\mathbf{a}, \mathbf{b})$ and equality follows.

Thus far we have constructed $\overline{X}$ and $\overline{d}$, but we have not yet shown $(\overline{X}, \overline{d})$ to be a metric space, i.e. we have not shown $\overline{d}$ to be a distance function. This is very simple. To verify 7.1.i, observe

---

[158] Note that we are using the completeness of $\mathbb{R}$ in generalising the Cantor-Méray construction to arbitrary metric spaces. To subsume the construction of $\mathbb{R}$ from $\mathbb{Q}$ into the present construction, one has to make some modifications here.

$$d([\mathbf{x}], [\mathbf{y}]) = \lim_{n \to \infty} d(x_n, y_n) \geq 0,$$

since $d(x_n, y_n) \geq 0$ for all n. And:

$$\overline{d}([\mathbf{x}], [\mathbf{y}]) = 0 \;\; \text{iff} \;\; \lim_{n \to \infty} d(x_n, y_n) = 0$$
$$\text{iff} \;\; \mathbf{x} \equiv \mathbf{y}$$
$$\text{iff} \;\; [\mathbf{x}] = [\mathbf{y}].$$

To see that 7.1.ii holds, observe

$$\overline{d}([\mathbf{x}], [\mathbf{y}]) = \lim_{n \to \infty} d(x_n, y_n) = \lim_{n \to \infty} d(y_n, x_n) = \overline{d}([\mathbf{y}], [\mathbf{x}]).$$

And finally 7.1.iii:

$$\overline{d}([\mathbf{x}], [\mathbf{z}]) = \lim_{n \to \infty} d(x_n, z_n)$$
$$\leq \lim_{n \to \infty} (d(x_n, y_n) + d(y_n, z_n))$$
$$\leq \lim_{n \to \infty} d(x_n, y_n) + \lim_{n \to \infty} d(y_n, z_n)$$
$$\leq \overline{d}([\mathbf{x}], [\mathbf{y}]) + \overline{d}([\mathbf{y}], [\mathbf{z}]).$$

The embedding of X into $\overline{X}$ is clear: Map x to the constant sequence $(x, x, x, \dots)$ and observe

$$\overline{d}(\,(x, x, x, \dots), (y, y, y, \dots)\,) = \lim_{n \to \infty} d(x, y) = d(x, y).$$

Hence we have constructed a metric space $(\overline{X}, \overline{d})$ and a distance preserving embedding $\iota(x) = (x, x, x, \dots)$ of X into $\overline{X}$. By construction, every Cauchy convergent sequence in X has a limit in $\overline{X}$. To complete the proof of Theorem 7.8, one must still prove that every Cauchy convergent sequence of elements of $\overline{X}$ has a limit in $\overline{X}$.

I want to prove the completeness of $\overline{X}$ by starting with a Cauchy sequence $\mathbf{x}_0, \mathbf{x}_1, \dots$ of elements of $\overline{X}$,

$\mathbf{x}_0 : x_{00}, x_{01}, x_{02}, \dots$
$\mathbf{x}_1 : x_{10}, x_{11}, x_{12}, \dots$
$\vdots$

and diagonalising on it, choosing

$\mathbf{x} : x_{00}, x_{11}, x_{22}, \dots$

or at least some sequence $x_{0k_0}, x_{1k_1}, x_{2k_2}, \dots$ with $k_0 < k_1 < k_2 < \dots$ However, this is a delicate matter and instead we shall mimic the proof of Theorem 4.27 by defining the proper notion of the denseness of X in $\overline{X}$, proving this denseness, and then applying it.

7.11 Definition. *Let* (Y, ρ) *be a metric space. A subset* $X \subseteq Y$ *is* dense *in* Y *if, for every element* $y \in Y$ *and every positive real number* $\epsilon > 0$, *an element* $x \in X$ *can be found for which* $\rho(x, y) < \epsilon$.

When we speak of the density of the rationals in the reals, we usually refer to order: Between any two real numbers a, b, a rational number x exists: $a < x < b$. But, if we are given a real number y and a positive real $\epsilon > 0$, choosing $a = y - \epsilon, b = y + \epsilon$, we have $y - \epsilon < x < y + \epsilon$ for some x, i.e. $d(x, y) < \epsilon$. Thus, Definition 7.11 generalises this to general metric spaces.

7.12 Lemma. X *is dense in* $\overline{X}$.

Proof. Let $\mathbf{y} \in \overline{X}$ be given. $\mathbf{y} = (y_0, y_1, \ldots)$ is a Cauchy sequence, whence given $\epsilon > 0$ there is an $n_0 \in \mathbb{N}$ such that for all $m, n > n_0$, $d(y_m, y_n) < \epsilon$. Let $x = y_{n_0+1}$ and choose $\mathbf{x} = (x, x, x, \ldots)$. Observe

$$\overline{d}(\mathbf{x}, \mathbf{y}) = \lim_{n \to \infty} d(x_n, y_n) = \lim_{n \to \infty} d(x, y_n).$$

But, for $n > n_0$, $d(x, y_n) = d(y_{n_0+1}, y_n) < \epsilon$. Thus $\lim_{n \to \infty} d(x, y_n) < \epsilon$, i.e. $\overline{d}(\mathbf{x}, \mathbf{y}) < \epsilon$.   □

We can now finish the proof of the Theorem fairly quickly. Let $\mathbf{x}_0, \mathbf{x}_1, \ldots$ be a Cauchy sequence in $\overline{X}$ and let $\epsilon_0 > \epsilon_1 > \ldots$ be a sequence of positive real numbers converging to 0 (e.g., $\epsilon_n = 1/2^n$). For each n, choose $x_n \in X$ such that $\overline{d}(x_n, \mathbf{x}_n) < \epsilon_n$. Then $\mathbf{x} = (x_0, x_1, x_2, \ldots)$ is a Cauchy sequence and $\mathbf{x} = \lim_{n \to \infty} \mathbf{x}_n$.

To see that $\mathbf{x}$ is a Cauchy sequence, let $\epsilon > 0$ be given and choose n_1 so large that for all $n > n_1$, $\epsilon_n < \epsilon/3$ and n_2 so large that for all $m, n > n_2$, $\overline{d}(\mathbf{x}_m, \mathbf{x}_n) < \epsilon/3$. Let $n_0 = \max\{n_1, n_2\}$ and observe, for $m, n > n_0$,

$$d(x_m, x_n) \leq \overline{d}(x_m, \mathbf{x}_m) + \overline{d}(\mathbf{x}_m, \mathbf{x}_n) + \overline{d}(\mathbf{x}_n, x_n)$$
$$< \epsilon_m + \frac{\epsilon}{3} + \epsilon_n$$
$$< \frac{\epsilon}{3} + \frac{\epsilon}{3} + \frac{\epsilon}{3} = \epsilon.$$

Finally, to see that $\mathbf{x} = \lim_{n \to \infty} \mathbf{x}_n$ in $\overline{X}$, note that

$$\overline{d}(\mathbf{x}, x_n) = \overline{d}((x_0, x_1, \ldots), (x_n, x_n, \ldots)) = \lim_{m \to \infty} d(x_m, x_n),$$

which can be made as small as we please by choosing n large enough since $\mathbf{x} = (x_0, x_1, \ldots)$ is Cauchy convergent. Let $\epsilon > 0$ be given and choose n_1 so large that, for all $n > n_1$, $\overline{d}(\mathbf{x}, x_n) < \epsilon/2$. Choose n_2 so large that, for all $n > n_2$, $\epsilon_n < \epsilon/2$ and observe, for $n > n_0 = \max\{n_1, n_2\}$,

$$\overline{d}(\mathbf{x}, \mathbf{x}_n) \leq \overline{d}(\mathbf{x}, x_n) + \overline{d}(x_n, \mathbf{x}_n) < \frac{\epsilon}{2} + \epsilon_n < \frac{\epsilon}{2} + \frac{\epsilon}{2} = \epsilon.$$

Thus

$$\lim_{n \to \infty} \mathbf{x}_n = \mathbf{x}. \qquad □$$

I shall leave the details of the proof of Theorem 7.10 to the reader. We shall not need it. Its interest is more philosophical or doctrinaire in that it establishes the significance, via uniqueness up to isometry, of the metric completion of a metric space. Of greater utility is the more general result that any continuous function $f : X \to Y$ of a metric space into a complete metric space has a continuous extension $\overline{f} : \overline{X} \to Y$ of the completion into the second space. Again, we will not need this result and I leave it to the interested reader to consult any introductory topology book for a proof of the result (as well as for the definition of and basic facts about continuity of functions on metric spaces). What I do wish to do in the remainder of this section is to emulate section 3 above and apply the construction to a specific metric space to provide another Type II formal justification for some Type I formal manipulations we have already encountered. What shall it be? Will we see another approach to justifying the use of the Dirac delta function, a context in which the Lagrangian proof of the Euler-Maclaurin summation formula is rigorously correct, or at least a space in which e^D has a clear meaning as an infinite sum? These would be noble undertakings, but with all the heavy material we've had thus far in this chapter, I thought we should finish the section with something simple.

Number theoretic work of Kurt Hensel carried out from the 1890's on what he called g-adic numbers has been reformulated in terms of norms, metrics, and metric space completions, resulting in numerous different completions of the field of rational numbers. We shall take a small peek at this here.

What we are going to do is not difficult, but it is a little unmotivated. It has to do with congruences. To solve a polynomial equation $P(X) = 0$ over the integers, one might first try solving over the integers modulo some prime, say 5,

$$P(X) \equiv 0 \mod 5.$$

If the solution to the original equation is, say, 672, the

solution modulo 5^1 is $2 = 2 \cdot 5^0$

solution modulo 5^2 is $22 = 2 \cdot 5^0 + 4 \cdot 5^1$

solution modulo 5^3 is $47 = 2 \cdot 5^0 + 4 \cdot 5^1 + 1 \cdot 5^2$

solution modulo 5^4 is $47 = 2 \cdot 5^0 + 4 \cdot 5^1 + 1 \cdot 5^2 + 0 \cdot 5^3$

solution modulo 5^5 is $672 = 2 \cdot 5^0 + 4 \cdot 5^1 + 1 \cdot 5^2 + 0 \cdot 5^3 + 1 \cdot 5^4$.

Each successive step narrows down the possible range of solutions and thus better approximates the correct solution. The idea then is to use the powers of the prime 5 to measure the size of the range— larger powers denoting smaller ranges. We can, in fact, define a sort of absolute value by taking the reciprocal of the power of 5 dividing a number and defining the distance between two numbers to be this absolute value of their difference. Thus, letting $\| \cdot \|_5$ denote the absolute value and d_5 the distance associated with the number 5, we see that

$$d_5(2, 672) = \|4 \cdot 5^1 + 1 \cdot 5^2 + 0 \cdot 5^3 + 1 \cdot 5^4\|_5 = \frac{1}{5},$$

since 5^1 is the highest power of 5 dividing $672 - 2 = 670$. Similarly,

$$d_5(22, 672) = \|1 \cdot 5^2 + 0 \cdot 5^3 + 1 \cdot 5^4\|_5 = \frac{1}{5^2}.$$

$$d_5(47, 672) = \|1 \cdot 5^4\|_5 = \frac{1}{5^4}$$

$$d_5(672, 672) = 0.$$

Thus we see that the sequence of approximate solutions is indeed approaching the solution.

Now, the absolute value and distance functions can be defined for all rational numbers, not just integers, and, via the following rather strange looking lemma, we will see that we do not need to assume the base number (in the example just given: 5) is prime to obtain a good metric on $\mathbb{Q}$.

7.13 Lemma. *Let $p/q \in \mathbb{Q}$ and let g be an integer ≥ 2. There are unique integers k, r and a unique positive integer s such that*

$$g^k \cdot \frac{p}{q} = \frac{r}{s},$$

where the pairs r, s and g, s are relatively prime, and g does not divide r.

Given any p, q, g it is fairly easy to determine k, r, s. For example, if

$$\frac{p}{q} = \frac{32}{120}, \ g = 12,$$

first reduce p/q to lowest terms: $p/q = 4/15$. Since $g = 12$ does not divide 4, but has a factor in common with 15, multiply 4/15 by the smallest power of 12 it takes to cancel the common factor it has with 15:

$$12 \cdot \frac{p}{q} = 12 \cdot \frac{4}{15} = 4 \cdot \frac{4}{5} = \frac{16}{5}.$$

We have $k = 1, r = 16, s = 5$ and the conditions of the Lemma are satisfied. For

$$\frac{p}{q} = \frac{336}{55}, \ g = 12,$$

the fraction is already in reduced form, g divides p and has no common factor with q. Divide by the largest power of g that divides p:

$$\frac{1}{12} \cdot \frac{p}{q} = \frac{1}{12} \cdot \frac{336}{55} = \frac{28}{55},$$

and we have $k = -1, r = 28, s = 55$.

Proof of Lemma 7.13. There are two cases depending on whether g divides p or not, assuming p/q in lowest terms.

If g divides p, let g^h be the highest power of g dividing p and divide by g^h:

$$\frac{1}{g^h} \cdot \frac{p}{q} = \frac{r}{s},$$

where $s = q$, which is relatively prime to p and hence to g, and r is the quotient p/g. g does not divide r since otherwise g^{h+1} divides p, contrary to choice of h. Hence, in this case we have $k = -h, r = p/g^h, s = q$.

If g does not divide p, it may still have a common divisor with q. If it doesn't, then choose $k = 0, r = p, s = q$. If there is a common divisor of g and q, write

$$g \cdot \frac{p}{q} = \frac{p'}{q'}$$

in lowest terms. p' will be larger than p, but q' is smaller than q. We may thus inductively assume there to be some $h \in \mathbb{Z}$ such that

$$g^h \cdot \frac{p'}{q'} = \frac{r}{s},$$

with h, r, s satisfying the necessary condition. Thus

$$g^{h+1} \cdot \frac{p}{q} = \frac{r}{s}$$

also satisfies the necessary conditions $\qquad\qquad\qquad\qquad\qquad\qquad\square$

With this Lemma, we obtain our g-adic analogue to an absolute value as follows:

7.14 Definition. *We define the g-norm $\|x\|_g$ for $x \in \mathbb{Q}$ by*
i. $\|0\|_g = 0$;
ii. for $p/q \in \mathbb{Q}$ with $p/q \neq 0$, $\|p/q\|_g = g^k$ if $g^k(p/q) = r/s$, where g, s are relatively prime, r, s are relatively prime, and g does not divide r.

7.15 Lemma. *For all $x, y \in \mathbb{Q}$,*
i. $\|x\|_g \geq 0$ *and* $\|x\|_g = 0$ *iff* $x = 0$;
ii. $\|x + y\|_g \leq \max\{\|x\|_g, \|y\|_g\} \leq \|x\|_g + \|y\|_g$;
iii. $\|xy\|_g \leq \|x\|_g \cdot \|y\|_g$, *with equality always holding when g is prime.*

Proof. Part i is a trivial consequence of the definition of $\| \cdot \|_g$.
ii. Write

$$x = \frac{p_1}{q_1}, \quad y = \frac{p_2}{q_2}$$

and find $k_1, k_2, r_1, r_2, s_1, s_2$ by Lemma 7.13 with

$$g^{k_i} \frac{p_i}{q_i} = \frac{r_i}{s_i}.$$

Assume, without loss of generality, that $k_1 \leq k_2$ and consider

$$g^{k_2}(x + y) = g^{k_2}\left(\frac{p_1}{q_1} + \frac{p_2}{q_2}\right) = g^{k_2-k_1} \cdot \frac{r_1}{s_1} + \frac{r_2}{s_2} = \frac{g^{k_2-k_1}r_1 s_2 + r_2 s_1}{s_1 s_2}.$$

g has no common divisor with $s_1 s_2$ since it has no common divisor with either factor. Reduction to lowest terms will not change this. We will thus have

$$g^{k_2}(x + y) = \frac{r}{s},$$

with r, s and g, s relatively prime. It could happen, however, that g divides r. Let g^h be the highest power of g doing so (so that $h \geq 0$) and divide r by g^h to obtain

$$g^{k_2 - h}(x + y) = \frac{r'}{s}$$

satisfying the conditions of Lemma 7.13, whence

$$\|x + y\|_g = g^{k_2 - h} \leq g^{k_2} = \max\{\|x\|_g, \|y\|_g\}.$$

iii. Let $x, y, p_i, q_i, r_i, s_i, k_i$ be as in the proof of ii and observe

$$g^{k_1 + k_2} xy = g^{k_1} x \cdot g^{k_2} y = \frac{r_1}{s_1} \cdot \frac{r_2}{s_2} = \frac{r}{s},$$

after reducing $(r_1 r_2)/(s_1 s_2)$ to lowest terms. If g has no factor in common with s_1 or s_2, it has no common factor with $s_1 s_2$ and hence none with s. If g is prime and does not divide r_1, r_2, it does not divide $r_1 r_2$, hence not r. Thus, if g is prime,

$$\|xy\|_g = g^{k_1 + k_2} = g^{k_1} \cdot g^{k_2} = \|x\|_g \cdot \|y\|_g.$$

If g is not prime, g will still have no factor in common with s, but it could divide $r_1 r_2$. For example, for $g = 4, x = y = 2$,

$$x = y = 4^0 \cdot \frac{2}{1} \text{ but } xy = 2 \cdot 2 = 4 \text{ and } \frac{1}{4} \cdot 4 = \frac{1}{1},$$

whence

$$\|2\|_4 \cdot \|2\|_4 = 4^0 \cdot 4^0 = 1 \neq \frac{1}{4^1} = \|4\|_4;$$

and for $g = 6, x = 2, y = 3$,

$$x = 6^0 \cdot \frac{2}{1}, \ y = 6^0 \cdot \frac{3}{1}, \text{ but } x \cdot y = 2 \cdot 3 = 6,$$

and 6 divides $2 \cdot 3$ and we have

$$\|6\|_6 = \frac{1}{6} \neq 1 = 1 \cdot 1 = \|2\|_6 \cdot \|3\|_6.$$

For composite g, let g^k be the largest power of g dividing r and note that

$$g^{k_1 + k_2} xy = \frac{r}{s} = g^k \cdot \frac{r'}{s}, \text{ for some } r'$$

and

$$g^{k_1 + k_2 - k} xy = \frac{r'}{s}.$$

Thus

$$\|xy\|_g = g^{k_1 + k_2 - k} \leq g^{k_1 + k_2} = g^{k_1} g^{k_2} = \|x\|_g \cdot \|y\|_g. \qquad \square$$

7.16 Definition. *Let g be an integer greater than 1. The g-distance function d_g on $\mathbb{Q}$ is defined by*

$$d_g(x, y) = \|x - y\|_g.$$

7.17 Lemma. $(\mathbb{Q}, d_g)$ *is a metric space.*

Proof. There are three things to prove.

i. By definition, if $x = y$, $d_g(x, y) = \|x - y\|_g = \|0\|_g = 0$; and, if $x \neq y$, $d_g(x, y) = \|x - y\|_g = g^k$, for some k, but $g^k > 0$.

ii. Let $x - y = p/q$. If

$$g^k \cdot \frac{p}{q} = \frac{r}{s}$$

as in Lemma 7.13, then

$$g^k \frac{-p}{q} = \frac{-r}{s}$$

with the conditions of the Lemma still fulfilled. Thus

$$d_g(y, x) = \|y - x\|_g = \left\|\frac{-p}{q}\right\|_g = g^k = \left\|\frac{p}{q}\right\|_g = \|x - y\|_g = d_g(x, y).$$

iii. By Lemma 7.15.ii.   □

The completion of $\mathbb{Q}$ with respect to d_g is denoted $\mathbb{Q}_g$ and its elements are called the *g-adic numbers*.

As just constructed, $\mathbb{Q}_g$ is a complete metric space. We know no more about it. $\mathbb{Q}$ is a field, but we have not even defined addition or multiplication on $\mathbb{Q}_g$. Let us do so now. The elements $x \in \mathbb{Q}_g$ are equivalence classes of Cauchy sequences $a \in \mathbb{Q}^{\mathbb{N}}$ of rational numbers, two sequences a, b being equivalent if their difference is a *null sequence*, i.e. if the difference converges to 0:

$$a \equiv b \quad \text{iff} \quad \lim_{n \to \infty} \|a_n - b_n\|_g = 0.$$

The most obvious way to define addition and multiplication operations on such sequences is componentwise:

$$(a + b)_n = a_n + b_n, \quad (a \cdot b)_n = a_n \cdot b_n.$$

To define $x + y$, $x \cdot y$ for $x, y \in \mathbb{Q}_g$, we choose representatives $a, b \in \mathbb{Q}^{\mathbb{N}}$ such that $x = [a], y = [b]$ and define

$$x + y = [a + b], \quad x \cdot y = [a \cdot b].$$

To be able to do this, we must show first that, if a, b are Cauchy sequences then so are $a + b$ and $a \cdot b$ (so that $[a + b]$ and $[a \cdot b]$ are in $\mathbb{Q}_g$), and second that the sum and product do not depend on the choice a, b of representatives: for all Cauchy sequences a, b, c, d of rationals,

$$a \equiv c \ \& \ b \equiv d \quad \Rightarrow \quad a + b \equiv c + d \ \& \ a \cdot b \equiv c \cdot d.$$

7.18 Lemma. *i. The sum and product of Cauchy sequences are Cauchy sequences.*

ii. If a, b, c, d are Cauchy sequences with $a \equiv c$ and $b \equiv d$, then $a + b \equiv c + d$ and $a \cdot b \equiv c \cdot d$.

Proof. i. The case of addition is easy. Let a, b be Cauchy sequences, $\epsilon > 0$, and n_1, n_2 so large that for all $m, n \in \mathbb{N}$,

$$m, n > n_1 \ \Rightarrow \ \|a_m - a_n\|_g < \frac{\epsilon}{2}$$

$$m, n > n_2 \ \Rightarrow \ \|b_m - b_n\|_g < \frac{\epsilon}{2}.$$

Let $n_0 = \max\{n_1, n_2\}$ and observe, for $m, n > n_0$,

$$\begin{aligned}
\|(a_m + b_m) - (a_n + b_n)\|_g &= \|(a_m - a_n) + (b_m - b_n)\|_g \\
&\leq \|a_m - a_n\|_g + \|b_m - b_n\|_g \\
&< \frac{\epsilon}{2} + \frac{\epsilon}{2} = \epsilon.
\end{aligned}$$

For multiplication we have to work a little harder. First, one notes that every Cauchy convergent sequence is bounded in norm by some positive number. For, let n_0 be so large that for $m, n > n_0$,

$$\|a_m - a_n\|_g < \frac{1}{2}.$$

Then $A = \max\{\|a_0\|_g, \ldots, \|a_{n_0+1}\|_g\} + 1$ is larger than the g-norm of any element of the sequence $a_0, a_1, a_2, \ldots$

Given Cauchy sequences a, b, choose bounds A, B on the g-norms of all elements of a, b, respectively, and note

$$\begin{aligned}
\|a_m b_m - a_n b_n\|_g &= \|a_m b_m - a_n b_m + a_n b_m - a_n b_n\|_g \\
&\leq \|a_m b_m - a_n b_m\|_g + \|a_n b_m - a_n b_n\|_g \\
&\leq \|b_m\|_g \cdot \|a_m - a_n\|_g + \|a_n\|_g \cdot \|b_m - b_n\|_g \\
&< B \cdot \|a_m - a_n\|_g + A \cdot \|b_m - b_n\|_g,
\end{aligned}$$

and if we choose n_0 so large that for $m, n > n_0$ we have

$$\|a_m - a_n\|_g < \frac{\epsilon}{2B}, \quad \|b_m - b_n\|_g < \frac{\epsilon}{2A},$$

we have

$$\|a_m b_m - a_n b_n\|_g < \frac{\epsilon}{2} + \frac{\epsilon}{2} = \epsilon,$$

and conclude the product to be Cauchy convergent.

ii. I leave this as an exercise for the reader. □

7.19 Remark. It may not have escaped the reader's notice that in the proof of this Lemma I twice used the Triangle Inequality and bounds $\epsilon/2, \epsilon/(2A)$ and $\epsilon/(2B)$. Had I used the stronger conclusion of Lemma 7.15.ii, namely,

$$\|x + y\|_g \leq \max\{\|x\|_g, \|y\|_g\}, \tag{86}$$

I could have dispensed with the division by 2. Inequality (86) is called the *Ultrametric Inequaltiy* and has some interesting consequences. The exercises at the end of the chapter will offer a peek. We will not be needing any of this extra power here.

7.20 Theorem. $\mathbb{Q}_g$ *is a ring under the operations defined above.*

Other than the existence of an additive inverse, the ring axioms are identities of the form

$$\text{for all } x_0, \ldots, x_{n-1}, \ f(x_0, \ldots, x_{n-1}) = g(x_0, \ldots, x_{n-1}).$$

Because the operations are defined componentwise, their validity in $\mathbb{Q}$ is preserved in the passage to $\mathbb{Q}^\mathbb{N}$ and thence to $\mathbb{Q}_g$, which is the homomorphic image of a subring of $\mathbb{Q}^\mathbb{N}$. The existence of an additive inverse is also a simple matter: If a is the sequence $a_0, a_1, a_2, \ldots$, then $-a$ will be the sequence $-a_0, -a_1, -a_2, \ldots$ For,

$$a + (-a) = (a_0 - a_0, a_1 - a_1, a_2 - a_2, \ldots) = (0, 0, 0, \ldots).$$

And, of course, if a is a Cauchy sequence, so is $-a$:

$$\| -a_m - (-a_n) \|_g = \| (-1)(a_m - a_n) \|_g$$
$$\leq \| -1 \|_g \cdot \| a_m - a_n \|_g = \| a_m - a_n \|_g,$$

by Lemma 7.15.iii. I leave the details to the reader.

$\mathbb{Q}_g$ is not generally a field. In fact, $\mathbb{Q}_g$ is a field iff g is a prime number or a power of a prime. This has to do with the failure of the inequality of Lemma 7.15.iii to be an equality for nonprime g. Incidentally, when g is prime, it is customary to use the letter "p" for "g" and to refer to the field $\mathbb{Q}_p$ as the field of *p-adic numbers*. The fields $\mathbb{Q}_p$ are of great interest in mathematics and there are several good expositions devoted to them.[159]

One might imagine I have included this section on metric spaces to demonstrate the general nature of the construction of the real numbers as the metric completion of the rationals (under the usual metric), and that I have applied this to the construction of the rings $\mathbb{Q}_g$ to provide a simple class of examples of applications of this construction. This, indeed, is a laudable goal. But the truth is that I did all this work to prove the following theorem:

7.21 Theorem. *Let g be given and suppose $q \in \mathbb{Q}$ is such that $\|q\|_g < 1$. Then in $\mathbb{Q}_g$,*

---

[159] A very readable elementary introduction is George Bachman, *Introduction to p-Adic Numbers and Valuation Theory*, Academic Press, New York, 1964. Most of the other books I've seen on the subject are more advanced and less leisurely. An exception is Kurt Mahler, *Introduction to p-Adic Numbers and Their Functions*, Cambridge University Press, Cambridge, 1973. This slim volume includes a treatment of the *g*-adic numbers before specialising to the fields $\mathbb{Q}_p$ and developing a theory of continuous and differentiable functions of a *p*-adic variable. To achieve this depth in a mere 91 pages, Mahler's coverage is not as broad and wide-ranging, but he remains readable. Incidentally, although he recommended Bachman's book as "an excellent introductory presentation from the standpoint of valuation theory" (i.e., norms and metrics), Mahler considered Hensel's 1913 *Zahlentheorie* to be "still one of the best elementary books" on the subject, but "somewhat out of date".

$$\sum_{k=0}^{\infty} q^k = \lim_{n \to \infty} \sum_{k=0}^{n} q^k = \frac{-1}{q-1} = \frac{1}{1-q}.$$

In particular, in $\mathbb{Q}_2$,

$$1 + 2 + 4 + 8 + \ldots = -1$$
$$1 - 2 + 4 - 8 + \ldots = \frac{1}{3}.$$

Proof. Using the formula for the sum of a finite geometric progression, we have

$$\left\| \sum_{k=0}^{n} q^k - \frac{-1}{q-1} \right\|_g = \left\| \frac{q^{n+1}-1}{q-1} - \frac{-1}{q-1} \right\|_g$$
$$= \left\| \frac{q^{n+1}}{q-1} \right\|_g$$
$$\leq \left\| q^{n+1} \right\|_g \cdot \left\| \frac{1}{q-1} \right\|_g$$
$$\leq \left(\|q\|_g \right)^{n+1} \cdot \left\| \frac{1}{q-1} \right\|_g ,$$

and the right-hand side can be made as small as one pleases by choosing n sufficiently large.   □

7.22 Remark. Note that I have only stated this result for $q \in \mathbb{Q}$ with g-norm less than 1, and not for arbitrary $x \in \mathbb{Q}_g$ with equally restricted g-norm. The result is still valid in the more general case, but the proof, which can be found in the exercises at the end of the chapter, requires a new argument showing $1 - x$ to have an inverse when $\|x\|_g < 1$.

I like to think of Theorem 7.21 as the final nail in the coffin of him who would criticise Euler's summation of the powers of 2 or Christian Wolf's summations of the powers of -2 and -3. Back in Chapter I we saw that there were generalised sum functions yielding the formally obtained values for any geometric progression. But there was no definitively convincing sense in which such a sum function was a sum. Indeed, when Order Preservation was imposed as a condition, we saw that generalised sum functions no longer gave results for a divergent geometric progression. But now we have a situation in which the series $1 + 2 + 4 + 8 + \ldots$, interpreted as the limit of finite sums converges to -1 and one cannot deny it. There is, of course, a caveat to this. The evaluation of $1 + 2 + 4 + 8 + \ldots$ as -1 in $\mathbb{Q}_2$ is not an evaluation thereof in $\mathbb{R}$. Although the series consists of real numbers and the stated limit is a real number, the relation of the series to the limit is valid in $\mathbb{Q}_2$, but not in $\mathbb{R}$.

8 Monsters

8.1 Introduction

"And now for something completely different." In line with Monty Python, I wish here to make an abrupt change in direction. As promised in the preface, the present chapter has thus far devoted itself to Type II definitions and their use in making Type I formal mathematics rigorous by finding formal replacements for intuitive concepts. The 19th century program of the arithmetisation of analysis is the most outstanding example of this and we have been considering it in some detail, so much so that I have named the chapter after one phase in this development. The logically next phase, which chronologically partly preceded this phase, would be the introduction of formal definitions of limit, continuous function, derivative, integral, etc. and the formal derivation of the properties these objects are expected to have. But for the occasional historical snippet, the reader has already seen this in his basic Calculus course and nothing is served by my discussing it all here. The replacement of intuitive concepts by rigorously defined formal ones does have, however, unintended consequences as the new definitions either fail to capture the intuitive conceptions properly or they demonstrate the inadequacy of our intuition. The celebrated French mathematician, physicist, and philosopher Henri Poincaré, in writing for a general audience, described the situation thus:

> Logic sometimes makes monsters. Since half a century we have seen arise a crowd of bizarre functions which seem to try to resemble as little as possible the honest functions which serve some purpose. No longer continuity, or perhaps continuity, but no derivative, etc. Nay more, from the logical point of view, it is these strange functions which are the most general, those one meets without seeking no longer appear except as particular case. There remains for them only a very small corner.
>
> Heretofore when a new function was invented, it was for some practical end; to-day they are invented expressly to put at fault the reasonings of our fathers, and one never will get from them anything more than that.
>
> If logic were the sole guide of the teacher, it would be necessary to begin with the most general functions, that is to say with the most bizarre. It is the beginner that would have to be set grappling with this teratologic museum.[160]

---

[160] Henri Poincaré, *The Foundations of Science*, The Science Press, New York, 1913, pp. 435 - 436. I quote from the 1929 reprinting of George Bruce Halsted's omnibus translation of three popular works by Poincaré: *Science and Hypothesis*, *The Value of Science*, and *Science and Method*. The particular passage is from *Science and Method*, which was originally published in French in 1902. *Science and Hypothesis* and *Science and Method* are available in inexpensive paperback editions from Dover Publications.

While most mathematicians prefer to use milder language with expressions like "pathological function" or "counterexample" rather than terms like "monster", "bizarre", and "teratologic", the strength of even the toned down terminology is indicative of the reaction to the new functions. This reaction is best discussed in greater detail in Chapter V; the monstrous, teratological, bizarre, pathological counterexamples themselves will be discussed now.

On 18 July 1872 Carl Weierstrass read a paper before the Royal Academy of Sciences in Berlin presenting a result that would have profound consequences for mathematics. The result was not entirely new, but it was most convincingly presented and would be published in 1875 by his student Paul du Bois-Reymond and become widely known. The result he presented was the proof that a certain function defined by a relatively simple trigonometric series was everywhere continuous and yet did not possess a derivative at any point.

Bad behaviour in functions had been encountered since the earliest days of the Calculus. The rational functions,

$$f(x) = \frac{x^2 - 1}{x - 1}, \quad g(x) = \frac{1}{x^2}, \quad h(x) = \frac{1}{x},$$

all have gaps in their graphs. The gap in the graph of $y = f(x)$ can be filled in, while those in the graphs of $y = g(x)$ and $y = h(x)$ cannot unless one adds points at infinity, identifying $+\infty$ and $-\infty$ in the case of h. For rational functions, however, there are only finitely many such gaps. One can obtain infinitely many such using trigonometric functions. For example,

$$f(x) = \frac{1}{\sin x}$$

is undefined infinitely often.

Holes in a graph are only one kind of discontinuity. There are also jump discontinuities as exhibited by Heaviside's step function,

$$H(x) = \begin{cases} 0, & x < 0 \\ 1, & 0 \leq x. \end{cases}$$

In the days when "function" meant "analytic expression" and "continuity" meant "continuity of expression", this would have been deemed not one function but two. And, at what stage would the greatest integer function,

$$[x] = \text{ the greatest integer } \leq x,$$

have been considered an analytic expression? In 1783 Euler published another example, one which became famous when Abel used it again in his paper on the binomial theorem[161] as an example of a convergent sequence of continuous functions that converge to a discontinuous limit:

---

[161] N.H. Abel, "Untersuchungen über die Reihe: $1 + \frac{m}{1}x + \frac{m \cdot (m-1)}{1 \cdot 2} \cdot x^2 + \frac{m \cdot (m-1) \cdot (m-2)}{1 \cdot 2 \cdot 3} \cdot x^3 + \dots$ u.s.w.", *Journal für die reine und angewandte Mathematik* 1 (1826), pp. 311 - 339. For those keeping score, this is now the fifth time I've had occasion to mention Abel's paper in this book.

$$f(x) = \sin x - \frac{1}{2}\sin 2x + \frac{1}{3}\sin 3x - \ldots = \sum_{n=1}^{\infty}(-1)^{n+1}\frac{\sin nx}{n}. \qquad (87)$$

The limit of this sequence has the graph given in *Figure 3*, below. Determining

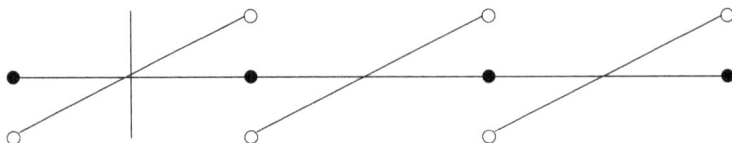

Figure 3

the graph from the analytic representation would seem to require an amazing intellect, which Euler certainly had. Some of the mystery vanishes, however, when one realises that $f(x)$ is the Fourier series of the function $g(x) = x/2$ on the interval $[-\pi, \pi]$ and Euler had calculated it. But the point here is not how one proves that f has the graph in question, but that f is given by an analytic expression and has infinitely many jump discontinuities.

The exceptional points, however, are discretely separated. Even that can fail, even with analytic expressions. In 1829, Johann Peter Gustav Lejeune Dirichlet introduced the function

$$D(x) = \begin{cases} 0, & x \text{ is irrational} \\ 1, & x \text{ is rational} \end{cases}$$

which is discontinuous at every real number. (*Exercise.*) It is also "analytically expressible"[162]:

$$f(x) = \lim_{m \to \infty}\left[\lim_{n \to \infty}(\cos(m!\pi x))^{2n}\right],$$

To see that this is correct note first that, if x is irrational, $m!\pi x$ is never of the form $k\pi$ for integral k, whence $|\cos(m!\pi x)| < 1$ and

$$\lim_{n \to \infty}(\cos(m!\pi x))^{2n} = 0,$$

whence $f(x) = 0$. If, on the other hand, x is rational, one still has $|\cos(m!\pi x)| \leq 1$ and

$$\lim_{n \to \infty}(\cos(m!\pi x))^{2n}$$

exists. But for m large, $m!\pi x$ is of the form $k\pi$ for integral k, and for such m,

$$\lim_{n \to \infty}(\cos(m!\pi x))^{2n} = \lim_{n \to \infty}(\pm 1)^{2n} = 1,$$

whence $f(x) = 1$.

---

[162] I am cheating. Dirichlet actually used arbitrary constants c, d in place of $0, 1$, respectively. Thus his actual function would be $c + d \cdot D(X)$ — and still analytically expressible. The analytic expression is not in Dirichlet's 1837 paper.

If one assumed continuity, a function could still fail to be differentiable at some isolated points. The absolute value function,

$$f(x) = |x| = \sqrt{x^2},$$

for example, has no derivative at 0, but it does have two one-sided derivatives,

$$\lim_{\Delta x \to 0^-} \frac{f(x + \Delta x) - f(x)}{\Delta x} = -1, \quad \lim_{\Delta x \to 0^+} \frac{f(x + \Delta x) - f(x)}{\Delta x} = 1.$$

Another example of bad behaviour is exhibited by the famously oscillating $\sin \frac{1}{x}$. The dampened version,

$$f(x) = x \sin \frac{1}{x},$$

can, as known to Cauchy, be made continuous by setting $f(0) = 0$, but it has no derivative, not even one-sided ones at $x = 0$.

Continuous functions with infinitely many points lacking derivatives were also known early on. One such function is the cycloid, a simple example of which is given by the parametric equations,

$$x(t) = t + \sin t, \qquad y(t) = 1 + \cos t.$$

The curve has vertical tangents at all points $x = k\pi$ for $k \in \mathbb{Z}$, whence the function $y = f(x)$ graphed by it has no derivatives at any of these points.

And, of course, one can readily imagine piecing together lots of copies of fragments of the graph of $y = x \sin \frac{1}{x}$ to create a curve with infinitely many points at which the derivative fails to exist or to be infinite.

But still, like the gaps in the curve of $y = 1/\sin x$, the bad points can all be separated from one another and there are only finitely many in any given interval. This was believed always to be the case.

Such points, where a function is discontinuous or nondifferentiable or it "misbehaves" in some other fashion, are called *singularities* because of their perceived *singular* behaviour. The word "singular" has several meanings. My copy of *The Concise Oxford Dictionary*[163] reads in part:

> **2.** *a.* Single, individual,...; unique; unusual, remarkable from rarity, much beyond the average in degree, extraordinary, surprising; eccentric, unconventional, strangely behaved; (Math.) possessing unique properties...

These points are not called singular because they are unique, but because the behaviour of the given function around them was considered out of the ordinary.[164] It was generally believed that a function defined on an interval was

---

[163] 6th edition, 5th printing, Oxford University Press, Oxford, 1978.

[164] Those familiar with the use of matrices in linear algebra will recall that singular matrices are precisely those matrices A for which the appropriate matrix equations $AX = B$ do *not* have unique solutions. Here too the word "singular" refers to what was considered extraordinary or surprising behaviour.

well-behaved at "most" points and the misbehaving singular points were isolated from one another. Indeed, André Marie Ampère made a famous attempt in 1806 to prove that every continuous function defined on a finite interval was differentiable except at a finite number of points. Writing in 1880, Salvatore Pincherle said

> It has been thought until recently that being continuous is enough for a function to be differentiable; many treatises of the differential calculus even give a demonstration of the theorem: "every continuous function is differentiable". But all these demonstrations implicitly admit certain properties that are not contained in the general concept of a function.[165]

Volkert cites textbooks by Sylvestre François Lacroix, J. Raabe, Augustus de Morgan and Joseph Louis François Bertrand as examples in which Ampère's proof is presented.[166]

8.2 Bolzano

The first person to recognise and prove that a continuous function could fail to be differentiable at a dense set of points was Bernard Bolzano who constructed an example around 1830. However, his manuscript went unpublished until it was rediscovered in 1921 by Martin Jašek. The following year, Karel Rychlík and Vojtěch Jarník independently proved that Bolzano's function was *nowhere differentiable*, i.e. it was not differentiable at any real argument.

Bolzano obtained his pathological function as the limit of a sequence of continuous, piecewise linear functions, $f_0, f_1, f_2, \ldots$ by doing three essential things at each stage of the construction:

i. he changed each function f_n by replacing each linear segment of its graph by a piecewise linear segment with corners[167];

ii. he guaranteed that the lengths of the segments on which f_n is linear went to 0 as n got large; and

iii. he made sure that each f_{n+1} did not deviate too much from f_n.

Basically, the first condition introduces singularities, the second guarantees the set of introduced singularities is dense, and the third condition guarantees the limit of the sequence is continuous. While one must exhibit some care, there is a great deal of flexibility to the construction.

Bolzano's own modification goes like this. He starts with a linear function,

$$g(x) = A + \frac{B - A}{b - a}(x - a),$$

---

[165] Umberto Bottazzini, *op. cit.*, pp. 282 - 288; translated from S. Pincherle, *op. cit.*, p. 247.

[166] Klaus Thomas Volkert, *Die Krise der Anschauung*, Vandenhoeck & Ruprecht, Göttingen, 1986, p. 112.

[167] Actually, it is not the corners but the oscillation that is important. This will be evident when we discuss Weierstrass's example.

over an interval $[a, b]$, then subdivides the interval into four subintervals of lengths $\frac{3}{8}(b-a), \frac{1}{8}(b-a), \frac{3}{8}(b-a), \frac{1}{8}(b-a)$ and assigns to their endpoints the values in *Table 1*.

Table 1

x	a	$a + \frac{3}{8}(b-a)$	$a + \frac{1}{2}(b-a)$	$a + \frac{7}{8}(b-a)$	b
y	A	$A + \frac{5}{8}(B-A)$	$A + \frac{1}{2}(B-A)$	$A + \frac{9}{8}(B-A)$	B

He then defined h to be the function with these values that is linear on each of the four subintervals. The overall construction then was to start with a function f_0 that is linear on a given interval, then define f_1 by applying the above process, f_2 by applying the process to to the restriction of f_1 to the constructed subintervals, f_3 by applying it to the pieces of f_2 on the constructed subintervals, and so on. The nowhere differentiable function constructed by Bolzano was the limit of this sequence:

$$B(x) = \lim_{n \to \infty} f_n(x).$$

One gets a feel for the function $B(x)$ by graphing the first few iterates f_0, f_1, f_2 as in *Figure 4*, based on taking $f_0(x) = x$ on $[0, 1]$.

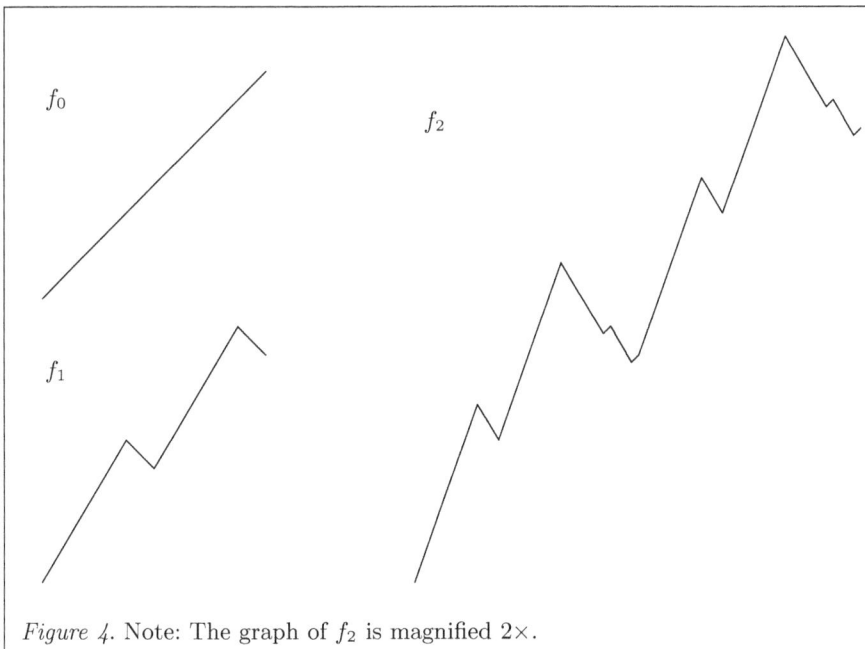

Figure 4. Note: The graph of f_2 is magnified 2×.

If x_0 is an endpoint of one of the intervals of linearity of f_n, then, by construction $f_{n+1}(x) = f_n(x)$. The graphs of successive f_n's exhibit an increasing number of corners, ever more tightly packed. Bolzano has built a little too much asymmetry into the construction of h from g in Table 1 to make his sequence easy to work with. Also, if $A = B$, i.e. if g is constant, h is not piecewise linear, but linear— the same constant function as g. Moreover, the varying sizes of the intervals is just another added complication.

One can simplify the construction as follows: On each interval on which g is linear, one obtains h by adding a *spike function*. If one adds two spikes which start at 0 go up to a fixed value and come down to 0, there will be, besides the endpoints of the original interval, a point in between where h will agree with g. If each successive spike function is only half as high as the preceding one, the total amount added to a given function will be finite and the limit will exist.

Now one simple way of describing all the spike functions needed is to start with a simple spike on the unit interval. On $[0, 1]$ we have the inverted spike

$$s(x) = 2\left|x - \frac{1}{2}\right|.$$

We can extend this by periodicity to the entire line:

$$s^*(x) = s(x - [x]),$$

where $[x] =$ the greatest integer less than or equal to x, i.e.

$$s^*(x) = 2\left|x - [x] - \frac{1}{2}\right|.$$

I promised a spike on the unit interval, but the graph of this tops off at 0 and 1. So we apply a shift:

$$s_0(x) = s^*\left(x - \frac{1}{2}\right) = 2\left|x - \left[x - \frac{1}{2}\right] - 1\right|.$$

Another transformation familiar from college algebra is compression. We get a k-fold horizontal compression, or in the case of a periodic function a multiplication by k of the number of periods of a function $y = g(x)$ by replacing it by $y = g(kx)$. And the vertical compression is achieved by dividing the function by the necessary factor. Now, at each step, each subinterval of linearity will be replaced by 4 subintervals (assuming 2 spikes, á la Bolzano, placed on each interval). And, we want the spikes to be, say, only half as high as the previous spikes. Thus, define

$$s_1(x) = \frac{1}{2}s_0(4x) = \frac{1}{2^1}s_0(4^1 x)$$

$$s_2(x) = \frac{1}{2}s_1(4x) = \frac{1}{4}s_0(16x) = \frac{1}{2^2}s_0\left(4^2 x\right)$$

$$s_3(x) = \frac{1}{2}s_2(4x) = \frac{1}{2^3}s_0\left(4^3 x\right),$$

and generally,

$$s_n(x) = \frac{1}{2^n}s_0\left(4^n x\right).$$

If we choose to start with $f_0(x) = s_0(x)$, we successively define

$$f_{n+1}(x) = f_n(x) + s_{n+1}(x),$$

i.e.

$$f_1(x) = f_0(x) + s_1(x) = s_0(x) + s_1(x)$$
$$f_2(x) = f_1(x) + s_2(x) = s_0(x) + s_1(x) + s_2(x)$$

$$\vdots$$

$$f_n(x) = \sum_{i=0}^{n} s_i(x) \tag{88}$$

and

$$f(x) = \lim_{n \to \infty} f_n(x) = \sum_{i=0}^{\infty} s_i(x)$$

$$= \sum_{n=0}^{\infty} s_n(x) = \sum_{n=0}^{\infty} \frac{1}{2^n} s_0 \left(4^n x\right). \tag{89}$$

Our main concern here is nondifferentiability, not continuity, so I leave the proof of continuity as an exercise to the reader. As to nondifferentiability, I suggest the reader graph the first few elements of the sequence on the computer or one's graphing calculator.[168] If one does so, the following facts stare one in the face.

8.1 Lemma. *Let $n \in \mathbb{N}$.*
i.   The endpoints of the maximal intervals on which f_n is linear are precisely the numbers of the form $k/2^{2n+1}$ for $k \in \mathbb{Z}$;
ii. the local minima of f_n are precisely the numbers of the form $k/2^{2n}$ for $k \in \mathbb{Z}$; they are local minima of f_m for all $m > n$ and, in fact, $f_m(k/2^{2n}) = f_n(k/2^{2n})$; and
iii. the local maxima of f_n are precisely the numbers of the form $(2k+1)/2^{2n+1}$ for $k \in \mathbb{Z}$.

Proof. Except for the constancy of the value of $f_m(k/2^{2n})$ for $m \geq n$, the proof is an induction on n. The constancy is a simple application of periodicity: if $j > n$,

---

[168] On my trusty *TI-83+*[TM], I entered

$$Y_1 = 2\text{abs}(X - \text{int}(X - 1/2) - 1)$$
$$Y_2 = Y_1(X) + 1/2Y_1(4X)$$
$$Y_3 = Y_2(X) + 1/4Y_1(16X).$$

to graph f_0, f_1, f_2. The window with X ranging from -1 to 2 and Y from $-.2$ to 2.5 gave an excellent overall view. To get a distinct view of Y_3, however, I had to zoom in.

$$s_j\left(\frac{k}{2^{2n}}\right) = \frac{1}{2^j}s_0\left(2^{2j}\frac{k}{2^{2n}}\right) = \frac{1}{2^j}s_0\left(k\cdot 2^{2(j-n)}\right) = 0$$

since s_0 vanishes on $\mathbb{Z}$. Thus

$$f_m\left(\frac{k}{2^{2n}}\right) = f_n\left(\frac{k}{2^{2n}}\right) + \sum_{j=n+1}^{m} s_j\left(\frac{k}{2^{2n}}\right) = f_n\left(\frac{k}{2^{2n}}\right).$$

As for the induction, the basis step, $n = 0$, is trivial.

For the induction step, one has to show that the transition from f_n to f_{n+1} does not result in a change in the graph similar to that in *Figure 5*, below (nor to the mirror image obtained when P is higher than Q)[169].

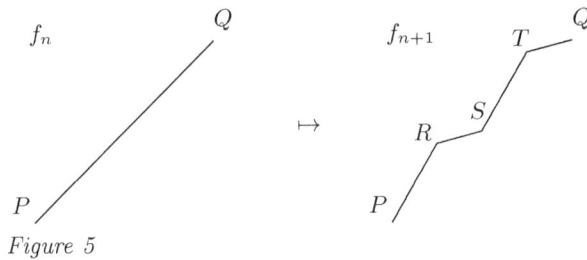

Figure 5

This is guaranteed by the following lemma:

8.2 Lemma. *For all $n \in \mathbb{N}$.*
i. The slope of a linear segment of the graph of s_n is $\pm 2^{n+1}$.
ii. The slope of a linear segment of the graph of f_n has absolute value at most $2 + 2^2 + \ldots + 2^{n+1} = 2^{n+2} - 2$.

Proof. i. By induction on n. For $n = 0$, each interval has length $1/2$ and the rise or fall is ± 1 yielding a slope of

$$\frac{\pm 1}{1/2} = \pm 2 = \pm 2^{0+1}.$$

In going from s_n to s_{n+1}, one applies two compressions: vertical by a factor of $1/2$ and horizontal by a factor of $1/4$. Thus

$$\text{slope of } s_{n+1}(x) = \frac{\frac{1}{2}\cdot(\text{slope of } s_n(x))}{\frac{1}{4}} = \frac{4}{2}\cdot(\pm 2^{n+1})$$

$$= \pm 2\cdot 2^{n+1} = \pm 2^{n+2}.$$

---

[169] If one examines the graph of Bolzano's second iterate in *Figure 4*, above, one will see two bad forms of behaviour. There are successive intervals (that between 21/64 and 3/8 and that between 3/8 and 27/64) where the segments of f_2 are collinear, making the candidate for an endpoint an interior point, as well as transitions (e.g. at 1/2) where the endpoint is neither a local maximum nor a local minimum.

ii. On the interior of each linear segment, $f_n(x) = \sum_{i=0}^{n} s_i(x)$ is the sum of differentiable functions, whence its derivative there is the sum of the derivatives, and the slope of f_n on the segment is

$$f'(x) = \sum_{i=0}^{n} s_i'(x) = \sum_{i=0}^{n} \pm 2^{i+1} \leq \sum_{i=0}^{n} 2^{i+1} = 2^{n+2} - 2. \qquad \square$$

Returning to the proof of Lemma 8.1, if *Figure 5* represents the graphs of f_n and f_{n+1}, then the slopes of RS and TQ are $-2^{n+2}+$ a sum less in absolute value than 2^{n+2} and thus are negative. Hence, in the graph of f_{n+1} the local minimum at P is followed by a local maximum at R, then a local minimum at S, a local maximum at T, and a value of Q less than that of T. The symmetric argument (with the individual graphs of *Figure 5* flipped horizontally) shows the value at Q is also lower than the values to its immediate right. $\qquad \square$

The nondifferentiability of f at each of the points $k/2^{2n}$ is intuitively obvious. From stage n on, $f_m(k/2^{2n}) = f(k/2^{2n})$, but the successive functions f_m are moving away from the point $\langle k/2^{2n}, f(k/2^{2n}) \rangle$ more and more rapidly, albeit in smaller and smaller neighbourhoods. The slope from the left is tending to $-\infty$ and that from the right to $+\infty$. Hence f cannot have a derivative at $k/2^{2n}$. More formally, let

$$x_0 = \frac{k}{2^{2n}}, \quad h = \frac{1}{2^{2m+1}},$$

for some $m > n$, and observe that

$$\frac{f(x_0 + h) - f(x_0)}{h} = \frac{f_{m+1}(x_0 + h) - f_{m+1}(x_0)}{h}$$
$$= \text{slope of } f_{m+1} \text{ on } [x_0, x_0 + h]$$
$$\geq \text{slope of } s_{m+1} \text{ on } [x_0, x_0 + h]$$
$$\geq 2^{m+2}.$$

Similarly,

$$\frac{f(x_0 - h) - f(x_0)}{-h} \leq -2^{m+2}.$$

As the set of rational numbers of the form $k/2^{2n}$ is dense in the real line, we have established that f is indeed much worse than had been believed possible. The function is, in fact, nowhere differentiable, but its nondifferentiability at points not of the form $k/2^{2n}$ requires a little more work. To this end, let x_0 be a point not of this form, let n be given, and choose k such that

$$\frac{k}{2^{2n+1}} < x_0 < \frac{k+1}{2^{2n+1}}.$$

Pick

$$h_n = \pm \frac{1}{2^{2n+2}}$$

so that $x_0 + h_n$ also lies in the open interval $((k-1)/2^{2n+1}, (k+1)/2^{2n+1})$. If $m = n + j > n$, then

$$s_m(x_0 + h_n) = \frac{1}{2^m} s_0 \left(4^m(x_0 + h_n)\right)$$

$$= \frac{1}{2^m} s_0 \left(4^m x_0 + 4^m h_n\right)$$

$$= \frac{1}{2^m} s_0 \left(4^m x_0\right) = s_m(x_0),$$

since

$$4^m h_n = \pm 2^{2n+2j} \cdot \frac{1}{2^{2n+2}} = \pm 2^{2(j-1)}$$

is an integer and s_0 has period 1. Thus

$$\frac{f(x_0 + h_n) - f(x_0)}{h_n} = \frac{1}{h_n} \sum_{i=0}^{\infty} (s_i(x_0 + h_n) - s_i(x_0))$$

$$= \frac{1}{h_n} \sum_{i=0}^{n} (s_i(x_0 + h_n) - s_i(x_0))$$

$$= \sum_{i=0}^{n} (\text{slope of } s_i \text{ at } x_0)$$

$$= \sum_{i=0}^{n} (-1)^{\varepsilon_i} 2^{i+1}, \tag{90}$$

where $\varepsilon_i \in \{0, 1\}$. This is an integer with the same sign as $(-1)^{\varepsilon_n}$, which is $+$ if k is even (since then $k/2^{2n+1}$ is a local minimum of f_n) and $-$ if k is odd (as then $k/2^{2n+1}$ is a local maximum of f_n). Thus, for f to be differentiable at x_0, from some n on, k must always be even or always odd.

Choose n_0 so large that for all $n > n_0$ the numerator k is always even or always odd, and so that h_n is small enough to make the difference quotient (90) differ from $f'(x_0)$ by at most 1. But observe, for $n > n_0$,

$$\frac{f(x_0 + h_{n+1}) - f(x_0)}{h_{n+1}} = \sum_{i=0}^{n+1} (\text{slope of } s_i \text{ at } x_0)$$

$$= \sum_{i=0}^{n} (\text{slope of } s_i \text{ at } x_0) + (\text{slope of } s_{n+1} \text{ at } x_0)$$

$$= \frac{f(x_0 + h_n) - f(x_0)}{h_n} \pm 2^{n+2},$$

and the two quotients

$$\frac{f(x_0 + h_{n+1}) - f(x_0)}{h_{n+1}}, \quad \frac{f(x_0 + h_n) - f(x_0)}{h_n}$$

cannot both be close to $f'(x_0)$, a contradiction. It follows that f is nowhere differentiable.

The exposition of this proof suffers a bit because I have tried to serve two masters. I wanted to present the motivation for the construction and at the same time prove the result with a fair amount of rigour. A more compact presentation could make the result plausible via some reference to obtaining a function with infinitely many oscillations in every neighbourhood by adding a dampened series of periodic oscillating functions with increasingly small periods. This is the common feature of most of the historical examples of nowhere differentiable functions. Alternatively, one could forego intuition entirely and simply present the function and a proof of its nowhere differentiability, leaving the reader to marvel that anyone could ever come up with such a thing. An especially short and sweet example of this was given as recently as 1953 by the computer scientist John McCarthy, who, in just under half a page, gave a fully rigorous formal proof, complete with references to the literature, but, of course, with no attempt to explain where his function came from.[170] The key difference between McCarthy's nowhere differentiable function and the modification of Bolzano's function cited above is the factor of horizontal compression. By using a factor of 2^{2^n} instead of $(2^2)^n$ in defining something like

$$M(x) = \sum_{n=0}^{\infty} \frac{1}{2^n} s_0 \left(2^{2^n} x \right),$$

he obtains a sequence $s_0, s_1, \ldots$ of spike functions with much steeper slopes, under which the difference quotient analogous to (90) is more easily seen to tend to infinity.

Bolzano's definition of a nowhere differentiable function yields a more irregular function than does (89) and the proof is correspondingly more difficult.[171]

Concerning Bolzano's function, it is worth quoting Manheim in full on the subject:

> In the period about 1830, Bolzano discovered a remarkable function. However, one must hasten to add that seemingly he did not appreciate the importance of his discovery nor understand fully its implications for analysis.
>
> Let PQ be a non-horizontal line with midpoint M. Divide PM in four parts equal in length and MQ in four parts equal in length. Let the points of division be $P_1, P_2, P_3, Q_1, Q_2, Q_3$, respectively. Let P_3' be the

---

[170] Cf. John McCarthy, "An everywhere continuous nowhere differentiable function", *American Mathematical Monthly* 60 (1953), p. 709. The proof is repeated in Johan Thim's master's thesis, "Continuous nowhere differentiable functions", Department of Mathematics, Luleå University of Technology, 2003. At the time of writing this book, Thim's thesis is available online and appears to be the most quoted reference online on the subject— justifiably so: Thim catalogues all the important nowhere differentiable functions, discusses some of the history, and presents either proofs or references to where proofs can be found for the individual results. Less technical historical treatments of nowhere differentiable functions are Manheim, *op. cit.*, and Volkert, *op.cit.*

[171] The obvious replacement for Lemma 8.1 does not hold, but one can still estimate the slopes of the linear segments of the iterates. Thim, *op. cit.*, presents the proof that it is not differentiable at a dense set of points and refers to the literature for full nowhere differentiability.

reflection of P_3 in the horizontal drawn through M and let Q_3' be the reflection of Q_3 in the horizontal drawn through Q. Now draw the broken line P_1, P_3', M, Q_3', Q. Continue the above process on each segment and repeat *ad infinitum*. Bolzano's manuscript demonstrates that there does not exist a derivative for a countable, everywhere dense set of points on the limit curve.

Bolzano's non-differentiable function would have pointed out the need to abandon reliance on pictures in proofs had he published it. Why he did not is unknown. Perhaps he treated it merely as a curiosity, for use in his lectures. Since the function was constructed about 1830, it preceded Dirichlet's extension of the notion of a function to include correspondences without equation representability. Bolzano did not give an equation for his nondifferentiable curve and, hence, most likely did not consider it a function. The manuscript was brought to light by Jâsek [*sic*] in 1921. In 1922 Rychlik [*sic*] showed that the function was nowhere differentiable.[172]

Manheim's comments raise a few questions. They also serve to make a point about exposition. In describing an historical work, a writer will emphasise what is important to the writer. In doing so he may overlook what was important to the original author or misrepresent what is important to another expositor. It is not always practical to consult the original. In the present case, Bolzano's major mathematical works are now published, even in English, so it is not necessary to travel to Prague, arrange permission to examine the manuscripts, perhaps to discover them undecipherable without the aid of a handwriting expert or stenographer or linguistic expert. But the books are expensive and only the wealthiest university libraries will have Bolzano's works.

One question raised by Manheim's comments, a question that doesn't interest me greatly, but which might have some bearing on the hypothetical issue of how convincing Bolzano's function might have been had it been published, is just how geometric Bolzano's own description of his function was. The description I gave earlier was algebraic and I lifted it from Thim's master's thesis[173]. Although he discusses history, Thim's primary purpose, like mine, is mathematical and translating geometric into algebraic language would have been legitimate. Manheim's non-algebraic description smacks either of authenticity or brevity. As he did not present a proof, he may have chosen the geometric formulation because it required less space to present than did the algebraic one[174]. Out of curiosity, I checked what Volkert had to say in his book on the crisis of intuition. He says, "To produce the function, Bolzano applied a geometric construction, the decisive step of which consists of replacing a line segment by four others". There then follows a diagram identical to the picture of Bolzano's

---

[172] Manheim, *op. cit.*, pp. 69 - 70.

[173] Thim, *op.cit.*, pp. 11- 17.

[174] Carl Boyer, *The History of the Calculus and Its Conceptual Development*, Dover, New York, 1959, pp. 269 - 270 gives the same geometric description. Both Manheim and Boyer refer to a paper of Gerhard Kowalewski for information on Bolzano's construction. So this explanation would be doubtful.

first iterate f_1 of *Figure 4*, above, and a concluding remark that the procedure when iterated results in a nowhere differentiable function in the limit.[175] His book on the history of analysis offers a more detailed picture in which the numerical constants $3/8, 5/8$, etc. are carefully labelled[176], but otherwise it sheds no light on the matter.

This is the point where I had originally left off in my exposition three years ago when I was too lazy to make the long drive to one of the two universities in the area whose library catalogues revealed copies of Bolzano's works. Recently, however, I found an affordable copy of the English translation of Bolzano's mathematical works online and can now report that Bolzano regarded geometrical intuition as not always reliable and gave a rigorous algebraic treatment.[177]

More interesting are the question why he didn't publish the result and the speculation that it would not have been considered a function. Bolzano was a professor of religion who was accused of heresy, dismissed from his post and forbidden to publish in 1819. The ban must have been lifted, for in 1837 he published his four volume work on the theory of science, which had occupied him from 1820 on. His nowhere differentiable function is to be found in his unfinished manuscript *Functionenlehre*, a companion to his *Reine Zahlenlehre* discussed earlier in this chapter. This work stems from the early to middle 1830's and was never completed. That he did not publish the example separately is probably best explained by the fact that Bolzano was a systematist and saw the result as part of a whole[178]. Had he finished his analytic work, he would have published the example therein. In any event, he certainly was not the last person to construct a nowhere differentiable function and not publish it.

As for the functional nature of Bolzano's function, let me also quote Morris Kline:

> In his *Funktionenlehre* [*sic*], which he wrote in 1834 but did not complete and publish, he gave an example of a continuous function which has no finite derivative at any point. Bolzano's example, like his other works, was not noticed. Even if it had been published in 1834 it probably would have made no impression because the curve did not have an analytic representation, and for mathematicians of that period functions were still entities given by analytical expressions.[179]

This is a good point to quote Bolzano himself:

---

[175] Volkert, *op. cit.*, p. 115.

[176] Klaus Volkert, *Geschichte der Analysis*, Bibliographisches Institut, Mannheim, 1988, pp. 209 - 210.

[177] Russ, *op. cit.*, pp. 487 - 489 and 507 - 508.

[178] Indeed, it is one among a number of counterexamples given by him.

[179] Morris Kline, *Mathematical Thought from Ancient to Modern Times*, Oxford University Press, New York, 1972, p. 955. I hasten to add that Kline points out in a footnote that Rychlík proved nowhere differentiability in 1922, thus correcting the tacit implication of the text that Bolzano knew his function to be nowhere differentiable. Not all authors are so careful: In Boyer, *op.cit.*, on page 284, one reads "Weierstrass in 1872 read a paper in which he showed what had been known

§136

Note. The last part of the previous theorem contradicts to a certain extent what *Lagrange* and many others sometimes explicitly claim, and sometimes just tacitly assume: that every function, with at most the exception of some isolated [*isolirt*] values of its variable, but in all other cases, has a *derivative*. But it is worth remarking (as I already mentioned in §39) that these scholars take the word *function* in a much narrower sense, because they understand by it only such numbers, dependent on another number x which can be expressed by one of the seven signs: $a + x, a - x, ax, \frac{a}{x}, x^n, a^x, \log x$, or by a combination of several of these. Now what they claim certainly holds of such [functions] especially as with some of these signs it is already in the *meaning* of them, that they should denote numbers that vary only by the law of continuity, or always have a derivative. But since I believe (§2) that a much wider concept must be associated with the word *function* then it will be necessary to allow of functions that they not only have no derivative, but they may even *break* the law of continuity not only for single values of their variables, but for all values lying within certain limits, and for all values *generally*.[180]

All general statements are oversimplifications and, despite our repeated reference to the fact that functions were given by analytic expressions, that is not quite true. There were "discontinuous" functions given by assigning different analytic expressions to different intervals and Dirichlet would, in 1837, incorporate them into his definition of a continuous function. That said, one can wonder how Bolzano's function would have been accepted. Volkert quotes Jarník:

> As mentioned by M. Jasek [*sic*], Bolzano presented the manuscript of his "Functionenlehre" to his favorite student A. Slivka from Slivice. Bolzano's estate includes Slivka's extensive critical answer. Slivka argues with Bolzano and claims among other things that he believes it possible to prove a theorem asserting every continuous function has a derivative everywhere except at some isolated points. Here we meet a wonderful proof how prejudice survives persistently in one's mind: Slivka saw with his own eyes the construction of Bolzano's function— and yet he did not believe in its existence because it contradicted (scientifically unjustifiable) ideas.[181]

It is commonly reported that the *Functionenlehre* has numerous errors. Slivka may have recognised this and simply been wary of an argument yielding such a nonintuitive result. Or he may have taken Bolzano's disclaimer at the beginning of §136 as trumping the example itself. Again, we would have to consult the

---

to Bolzano sometime before— that a function which is continuous throughout an interval need not have a derivative at any point in this interval".

[180] Russ, *op. cit.*, pp. 508 - 509.

[181] Volkert, *Krise, op.cit.*, p. 191.

original, but I am not aware if Slivka's extensive comments are published or otherwise readily available.

Today Bolzano's name is honoured in the Pantheon of mathematical heroes, but this was not the case in his own day. My speculation is that, had he published his result in 1834, he would have been ignored. In the latter half of the 20th century, Eduard Wette claimed to prove the inconsistency of analysis, then arithmetic, and ultimately the propositional calculus. I know of only one established logician who actually tried to read the proof. Everyone else simply assumed it was wrong. What could one expect for a relatively unknown author announcing so counter-intuitive a result as Bolzano's?[182] It would take the prestige of a Weierstrass (or, a Riemann) to have such a result taken seriously. And, of course, Riemann's and Weierstrass's non-differentiable functions, though a bit more complex than the finite expressions built up from the seven operations cited by Bolzano, were analytically expressible as infinite series.

8.3 Riemann and Weierstrass

Let me begin by quoting Weierstrass:

> Until most recently one had generally assumed that a definite and continuous function of a real variable also always had a first derivative, the values of which could only be undetermined or infinitely large at isolated points. Even in the writings of Gauss, Cauchy and Dirichlet there are to be found, to my knowledge, no statements from which it unambiguously follows that these mathematicians, who were generally accustomed to practise the most rigorous discrimination in their science, were of any other opinion. Riemann was first, as I have heard from some of his students, to assert with conviction (in 1861, or perhaps even earlier) that this assumption is not valid and, e.g., does not prove true for the function represented by the infinite series
>
> $$\sum_{n=0}^{\infty} \frac{\sin(n^2 x)}{n^2}.$$
>
> Unfortunately, Riemann's proof of this has not been published and appears not to have been preserved in his papers or through oral communication. This is all the more to be regretted as I have not once been able to hear with certainty how Riemann expressed himself to his students. Those mathematicians who have occupied themselves with the subject since Riemann's claim has become known in wider circles appear (at least the majority of them) to be of the opinion that it suffices to prove the existence of a function for which in every arbitrarily small

---

[182] When I was in graduate school, a paper purporting to solve a long outstanding open problem arrived in the mathematics department. One professor announced the proof incorrect because the papers listed in the references were all old. I myself will not read a paper that comes to me with the word "copyright" rubber-stamped on every other page.

interval in its domain points exist where it is not differentiable. That there are functions of this sort is extraordinarily easy to prove, and I therefore believe that Riemann had in mind only such functions which possessed a determinate derivative at no value of its argument. The proof of this, that the given trigonometric series represents a function of this sort, appears to me however to be somewhat difficult; one can however easily construct continuous functions of a real variable x, for which it can be proven with the simplest means that they possess a definite derivative for no value of x.

This can be done, for example, as follows.

Let x be a real variable, a an odd whole number, b a positive constant smaller than 1, and

$$f(x) = \sum_{n=0}^{\infty} b^n \cos(a^n x \pi);$$

so $f(x)$ is a continuous function, for which one can show that so long as the value of the product ab exceeds a certain bound the function nowhere possesses a determinate derivative.[183] [184]

Weierstrass was, of course, unaware of Bolzano's function which remained unpublished. But the rest of his history is still incomplete. In 1872 he was unaware of the work of Charles Cellérier who, around 1860, had also discovered a continuous, nowhere differentiable function,

$$C(x) = \sum_{n=0}^{\infty} \frac{1}{a^n} \sin(a^n x),$$

where $a > 1000$ was an even integer. Cellérier wrote a paper in which he proved the function to be nowhere differentiable, but it was only first posthumously published in 1890. Weierstrass's paper appeared in 1895 in his collected works and makes no mention of Cellérier.[185]

Let us first discuss Riemann.

Bernhard Riemann submitted his habilitation thesis in December 1853, but it went unpublished until Richard Dedekind published it in 1868 after Rie-

---

[183] Carl Weierstrass, "Über continuirliche Functionen eines reellen Arguments, die für keinen Werth des letzteren einen bestimmten Differentialquotienten besitzen", *Gesammelte Abhandlungen, II*, Mayer & Müller, Berlin, 1895.

[184] The conditions on a, b are that a be a positive odd integer, $b < 1$, and $ab > 1 + \frac{3}{2}\pi$. In lectures given in 1874 Weierstrass cites a relaxation of these conditions: a can be even or odd and ab needs only exceed 1. The extra conditions serve, however to simplify the proof. *Cf.* Georg Hettner, *Einleitung in die Theorien der analytischen Functionen von Prof. Dr. Weierstrass*, summer semester 1874, pp. 221, 222, 234.

[185] Volkert (*Krise, op. cit.*, p. 123.), citing P.L. Butzer, says "There is some reason to believe 'Weierstrass never mentioned Riemann in his talk (1872) but...this was added to the collected works of 1895'." In point of fact, Riemann and his example are cited in Weierstrass's lectures of 1874 (*Ibid.*, p. 220).

mann's death.[186] Dirichlet had proven the representability by Fourier series of any function with only finitely many discontinuities and finitely many maxima and minima on a closed, bounded interval. Riemann set himself the task of generalising this. This entailed a rigorous extension of the definition of the integral, whence our modern notion of the Riemann integral as the limit of Riemann sums. He not only showed that all bounded continuous functions were integrable[187], but characterised those more general functions that are integrable. In addition, he constructed a few pathological functions, most famously an integrable function with a dense set of discontinuities. Moreover, the integral of the function was continuous and failed to be differentiable on a dense set, a point not mentioned by Riemann in the thesis.

Riemann's function is easy to define. First, let $\langle x \rangle$ denote the *nearest integer* to x, if x is not of the form $(2k+1)/2$ for $k \in \mathbb{Z}$, and define

$$(x) = x - \langle x \rangle$$

to be the excess of x over the nearest integer. For x of the form $(2k+1)/2$, x is exactly halfway between two integers and has $\frac{1}{2}$ and $-\frac{1}{2}$ as the two excesses. For such x we average these values and consequently take $(x) = 0$. The graph of the function $y = (x)$ is very similar to that of the graph of the Euler-Abel series (87) pictured in *Figure 3*, above. The only difference is the scale: With (87), the points of discontinuity occur at numbers of the form $(2k+1)\pi$ for $k \in \mathbb{Z}$ and now they occur at numbers of the form $(2k+1)/2$. Thus, because I have not labelled these points, we can use *Figure 3* to visualise $y = (x)$.

Riemann's function is obtained, as was our modified Bolzano function, by summing a series of weighted compressions of the base function:

$$R(x) = \sum_{k=1}^{\infty} \frac{(nx)}{n^2}.$$

For any x one has $|(nx)| \leq \frac{1}{2}$, whence

$$\left| \frac{(nx)}{n^2} \right| \leq \frac{1/2}{n^2}$$

---

[186] *Abhandlungen der Königlichen Gesellschaft der Wissenschaften zu Göttingen* 13 (1868) and reprinted in his collected works, *Bernhard Riemanns Gesammelte Mathematische Werke und Wissenschaftlicher Nachlass*, Teubner, Leipzig, 1876, edited by Heinrich Weber with the cooperation of Richard Dedekind. An English translation of his collected works was published by Kendrick Press in 2004, and excerpts from his habilitation thesis in English translation earlier appeared in Garrett Birkhoff, *A Source Book in Classical Analysis*, Harvard University Press, Cambridge (Mass), 1973. An apparently full version, complete with Dedekind's or Weber's appendicial remarks, also appears in Stephen Hawking, *God Created the Integers: The Mathematical Breakthroughs that Changed History*, Running Press, Philadelphia, 2005. In the English language, this is probably the most accessible version, being commonly available in municipal libraries. Riemann's thesis is discussed in some detail in §6.3 (pp. 240 - 250) of Bottazzini, *op.cit.*

[187] By the Extreme Value Theorem, first proven later in 1870 by Eduard Heine, this means all continuous functions are integrable.

and $R(x)$ is absolutely convergent, whence convergent. Unlike the situation with our modified Bolzano function, where I cavalierly ignored the question of continuity, the key issue here is continuity and I must introduce the following definitions and prove a lemma.

8.3 Definitions. *Let $f_0, f_1, f_2, \ldots$ be a sequence of real-valued functions defined on some interval I (possibly the whole of $\mathbb{R}$). Let f be another function. We say that the sequence $f_0, f_1, f_2, \ldots$ converges pointwise to f, written,*

$$\lim_{n \to \infty} f_n = f,$$

if for each $x \in I$ the following holds: for all $\epsilon > 0$ there is an $n_0 \in \mathbb{N}$ such that for any $n > n_0$,

$$|f_n(x) - f(x)| < \epsilon. \tag{91}$$

We say that $f_0, f_1, f_2, \ldots$ converges uniformly to f if for every $\epsilon > 0$ there is an $n_0 \in \mathbb{N}$ such that for all $x \in I$ and all $n > n_0$ inequality (91) holds.

The difference between the two notions is that with pointwise convergence n_0 depends on both ϵ and x, while for uniform convergence, n_0 depends only on ϵ. $R(x)$ is an example of a uniformly convergent sequence, for

$$\left| R(x) - \sum_{k=1}^{n} \frac{(kx)}{k^2} \right| = \left| \sum_{k=n+1}^{\infty} \frac{(kx)}{k^2} \right|$$

$$\leq \sum_{k=n+1}^{\infty} \frac{|(kx)|}{k^2}$$

$$\leq \frac{1}{2} \sum_{k=n+1}^{\infty} \frac{1}{k^2}, \tag{92}$$

which, by the convergence of the series

$$\sum_{k=1}^{\infty} \frac{1}{k^2} = \frac{\pi^2}{6},$$

can be made smaller than any given ϵ by choosing n large enough, independent of the choice of x.

Examples of pointwise convergent but not uniformly convergent sequences abound. The function defined by the Euler-Abel trigonometric series (87) is such. In fact, whenever all the elements of the sequence f_n are continuous and their pointwise limit f is discontinuous somewhere, the convergence cannot be uniform:

8.4 Lemma. *Let $f_0, f_1, f_2 \ldots$ be a sequence of functions that converge uniformly to a function f on an interval I, and let $x_0 \in I$. If each f_n is continuous at x_0, then f is continuous at x_0.*

Proof. Let $\epsilon > 0$ be given. Choose n_0 so large that for all $n > n_0$,

$$|f_n(x) - f(x)| < \frac{\epsilon}{3}$$

for all $x \in I$. Let $n > n_0$ be given and choose δ so that for all $x \in I$

$$|x - x_0| < \delta \quad \Rightarrow \quad |f_n(x) - f_n(x_0)| < \frac{\epsilon}{3}.$$

For such x it follows that

$$
\begin{aligned}
|f(x) - f(x_0)| &= |f(x) - f_n(x) + f_n(x) - f_n(x_0) + f_n(x_0) - f(x_0)| \\
&\leq |f(x) - f_n(x)| + |f_n(x) - f_n(x_0)| + |f_n(x_0) - f(x_0)| \\
&< \frac{\epsilon}{3} + \frac{\epsilon}{3} + \frac{\epsilon}{3} = \epsilon.
\end{aligned}
$$
$\qquad\square$

Claim. Riemann's function R is discontinuous at all rational numbers x which in lowest terms are of the form $\frac{p}{2n}$ with p odd; R is continuous at all other values of x.

Proof. First note that any rational number of the form $(2m+1)/(2k)$ assumes the form $p/(2n)$ with p odd after reduction to lowest terms.
Define

$$s_k(x) = \frac{(kx)}{k^2}$$

for $k \in \mathbb{N}^+$. The points of discontinuity of s_k are precisely those values of x for which kx lies halfway between two integers, i.e. for which there is some $m \in \mathbb{Z}$ such that

$$kx = \frac{2m+1}{2},$$

i.e.

$$x = \frac{2m+1}{2k}.$$

It follows that for x_0 not of the forbidden form $p/(2n)$, every function s_k is continuous at x_0. Thus

$$f_k(x) = \sum_{i=1}^{k} s_i(x)$$

is continuous at x_0 and, by the Lemma, so is $R = \lim_{k \to \infty} f_k$.
Now let x_0 be of the form $p/(2n)$ with p, n relatively prime integers. We just saw that x_0 is a point of discontinuity of s_k just in case

$$x_0 = \frac{p}{2n} = \frac{2m+1}{2k}$$

for some $m \in \mathbb{Z}$, i.e. just in case

$$kp = (2m+1)n.$$

Because n, p are relatively prime, i.e. because they have no factor in common, it follows that n divides k. Moreover, if we set $j = k/n$, we have $jnp = (2m+1)n$, whence $jp = 2m + 1$ and j is odd. Thus, k is an odd multiple of n.

Conversely, if k is an odd multiple of n, $p/(2n)$ is a point of discontinuity of s_k. For, if $k = (2j + 1)n$,

$$\frac{p}{2n} = \frac{(2j+1)p}{2(2j+1)n} = \frac{(2j+1)p}{2k} = \frac{2m+1}{2k}$$

for some $m \in \mathbb{Z}$ because p is odd itself.

Now, for s_k we have

$$s_k\left(\frac{p}{2n}\right) = 0, \tag{93}$$

but

$$\lim_{h\to 0^+} s_k\left(\frac{p}{2n} + h\right) = -\frac{1/2}{k^2} \quad \text{and} \quad \lim_{h\to 0^-} s_k\left(\frac{p}{2n} + h\right) = \frac{1/2}{k^2}.$$

Thus

$$\lim_{h\to 0^+} R\left(\frac{p}{2n} + h\right) = \lim_{h\to 0^+} \sum_{k=1}^{\infty} s_k\left(\frac{p}{2n} + h\right)$$

$$= \lim_{h\to 0^+} \left(\sum_{k\in A} s_k\left(\frac{p}{2n} + h\right) + \sum_{k\in B} s_k\left(\frac{p}{2n} + h\right) \right)$$

where $A = \{k | n \text{ divides } k \text{ an odd number of times}\}$, $B = \{k \in \mathbb{N}^+ | k \notin A\}$,

$$= \frac{1}{2} \sum_{j=0}^{\infty} \frac{-1}{((2j+1)n)^2} + \sum_{k\in B} s_k\left(\frac{p}{2n}\right)$$

$$= \frac{-1}{2n^2} \sum_{j=0}^{\infty} \frac{1}{(2j+1)^2} + \sum_{k=0}^{\infty} s_k\left(\frac{p}{2n}\right), \text{ by (93)}$$

$$= \frac{-1}{2n^2} \sum_{j=0}^{\infty} \frac{1}{(2j+1)^2} + R\left(\frac{p}{2n}\right)$$

$$\neq R\left(\frac{p}{2n}\right),$$

and we see that R is not continuous at $p/(2n)$.   □

Here, *in toto*, is what Riemann has to say about his function:

Since these functions have not yet been looked at by anyone, it will be good to proceed on the basis of a specific example. For the sake of brevity, let us use (x) to indicate the excess of x over the next [i.e., nearest] whole number, or, if x is midway between two values and this definition becomes ambiguous, (x) indicates the average of both values $\frac{1}{2}$ and $-\frac{1}{2}$, i.e., zero. Furthermore let us use n to indicate a whole number and p an odd number, and after that let us construct the series

$$f(x) = \frac{(x)}{1} + \frac{(2x)}{4} + \frac{(3x)}{9} + \ldots = \sum_{1,\infty} \frac{(nx)}{nn};$$

as can easily be seen, this series converges steadily for every value of x. If the value supplied by the argument (both when steadily decreasing and increasing) becomes equal to x, the value of the series continually approaches a fixed limit, and in fact, if $x = \frac{p}{2n}$ (where p and n are relatively prime numbers)

$$f(x+0) = f(x) - \frac{1}{2n^2}\left(1 + \frac{1}{9} + \frac{1}{25} + \ldots\right) = f(x) - \frac{\pi\pi}{16nn},$$

$$f(x-0) = f(x) + \frac{1}{2n^2}\left(1 + \frac{1}{9} + \frac{1}{25} + \ldots\right) = f(x) + \frac{\pi\pi}{16nn},$$

but otherwise everywhere $f(x+0) = f(x)$, $f(x-0) = f(x)$.

This function is thus discontinuous for every rational value of x, which, reduced to its lowest terms, is a fraction with an even denominator. Thus the function has infinitely many discontinuities between any two limits however close, though in such a way that the number of jumps larger than a given value is always finite. It is completely integrable. Besides its finiteness, two properties are in fact sufficient for this purpose: that for every value of x it have a limit $f(x+0)$ and $f(x-0)$ on either side, and that the number of jumps [in any closed bounded interval] greater than or equal to a given value σ always be finite.[188]

Since Riemann's day, his integral has been much studied and many of its fundamental properties established. However, the treatment of discontinuous integrable functions is often rather subtle and not all textbooks in Advanced Calculus or Elementary Real Analysis state and prove the results I wish to appeal to. One exception is the textbook of Robert B. Bartle and Donald A. Sherbert[189] and I refer the reader to their book. The first result I wish to cite is their Theorem 7.2.4:

8.5 Theorem. *Let $f_0, f_1, f_2, \ldots$ be a sequence of functions that are integrable on $[a, b]$. Suppose the sequence converges uniformly to a function f. Then f is integrable and*

$$\int_a^b f(x)dx = \lim_{n\to\infty} \int_a^b f_n(x)dx.$$

If one assumes f integrable or that all the functions $f_0, f_1, f_2, \ldots$ are continuous, whence f is continuous and thus integrable, then the proof is a simple exercise in *epsilontics*. If such integrability is not assumed, one has to do some work[190].

Our interest in this Theorem is that it tells us that $R(x)$ is integrable. For, $R(x)$ is the uniform limit of integrable functions; in fact, it is the uniform

[188] Hawking, *op. cit.*, p. 838.
[189] *Introduction to Real Analysis*, John Wiley & Sons, New York, 1982.
[190] *Ibid.*, pp. 303 - 304.

limit of piecewise continuous functions, each having only finitely many (jump) discontinuities in any interval.

Because $R(x)$ is integrable, we can define

$$F(x) = \int_0^x R(t)dt.$$

The function F is continuous for all real x.[191] And, in fact, F is differentiable at all x at which R is continuous.[192] However, F is not differentiable everywhere: it fails to be differentiable at every point of discontinuity of R. The determination of where F is differentiable rests on the following variant of the theorem on the differentiability of the integral.

8.6 Theorem. *Let g be a bounded function integrable on an interval I, let $[a, b] \subseteq I$, and define*

$$G(x) = \int_a^x g(t)dt$$

for $x \in [a, b]$. Suppose

$$g(x + 0) = \lim_{t \to x^+} g(t)$$

exists. Then

$$\lim_{h \to 0^+} \frac{G(x + h) - G(x)}{h} = g(x + 0).$$

Similarly, if g has a limit from the left at x, then G has a left-sided derivative at x and it equals the limit from the left of g at x.

Applying the Theorem to F and R, we conclude immediately that, for x not of the form $p/(2n)$ with $p, 2n$ relatively prime, F is differentiable at x and $F'(x) = R(x)$, while for $x = p/(2n)$, with $p, 2n$ relatively prime,

$$\lim_{h \to 0^+} \frac{F(x + h) - F(x)}{h} = R(x + 0) = R(x) - \frac{\pi^2}{16n^2}$$

$$\lim_{h \to 0^-} \frac{F(x + h) - F(x)}{h} = R(x - 0) = R(x) + \frac{\pi^2}{16n^2}$$

and the derivative fails to exist.

The function F may have been the first widely known example of a function that fails to have a derivative at a dense set of points. It does not appear in Riemann's dissertation, but we have Weierstrass's testimony that Riemann made some unspecified claims in his lectures from 1861 on. We also have the function

$$R_2(x) = \sum_{n=0}^{\infty} \frac{\sin(n^2 x)}{n^2}$$

---

[191] *Ibid.*, pp. 257 - 258.
[192] *Ibid.*, pp. 258 - 259.

cited as an example of Riemann's of a function probably possessing no deriva-tives.[193] However, a search of Riemann's papers failed to find any trace of this function and attribution of the example to him is nowadays deemed in error.[194]

And this brings us chronologically to Weierstrass— not 1872, but 1861. The history books are a bit too definite on this point. Manheim states, without citing any reference:

> About 1861 Weierstrass gave, in his lectures at the University of Berlin, a function which was everywhere continuous but nowhere differen-tiable.[195]

Boyer states in a footnote to his remark[196] on Weierstrass's 1872 lecture on his nowhere differentiable function:

> This seems to have been presented by Weierstrass in his lectures as early as 1861. See Pringsheim, "Principes fondamentaux", p. 45, n.; Voss, "Calcul différential", pp. 260 - 261.[197]

David Burton, like Manheim, offers no source, but states things a little more cautiously:

> The mathematical world was profoundly surprised when Weierstrass read a paper to the Berlin Academy of Sciences in 1872, presenting an example of a continuous function that has a derivative at no point of the real line— or what is the same, a continuous curve having no tangent at any point. Weierstrass's example,
>
> $$\sum_{n=0}^{\infty} a^n \cos(b^n x), \ \ 0 < a < 1, \ ab > 1 + \frac{3}{2}$$
>
> is supposed to have been given in his classroom lectures as far back as 1861.[198]

---

[193] In 1970 it was proven that $R_2(x)$ possesses derivatives on a countable dense set of points and nowhere else.

[194] Cf. Volkert, *Krise*, p. 117 and 118 (footnote 120).

[195] Manheim, *op. cit.*, p. 72.

[196] Cited in footnote 179 above.

[197] Boyer, *op. cit.*, p. 284.

[198] David M. Burton, *The History of Mathematics: An Introduction*, 6th edition, McGraw-Hill, New York, 2007, p. 618. The attentive reader will no doubt have noticed that Burton has not accurately transcribed the definition of Weierstrass's function (cf. page 408, above). He has interchanged the letters "a" and "b", left out the condition that the new "b" be an odd integer, and omitted a factor of π next to the argument x. The curiously expressed upper bound $1 + \frac{3}{2}$ is also the result of the omission of a factor of π from Weierstrass's bound $1 + \frac{3}{2}\pi$. Mathe-matically, all of this is harmless enough as the conditions that b be an odd integer and Weierstrass's bound $1+\frac{3}{2}\pi$ could be weakened. *Cf.* footnote 184, above. More-over, writing $W(x)$ for Weierstrass's function as defined by Weierstrass, Burton's representation is thus

Perhaps the reason for Burton's cautious statement of a mere supposition that Weierstrass presented his function in lectures already in 1861 is that he had read Morris Kline's version:

> The example that attracted the most attention is due to Weierstrass. As far back as 1861 he had affirmed in his lectures that any attempt to prove that differentiability follows from continuity must fail.[199]

Kline lists both Manheim's and Boyer's books (the latter under an alternate title, *The Concepts of the Calculus*) as well as Weierstrass's collected works in the bibliography to the chapter from which this quote comes, but he does not state the specific source for his remark— something that obviously cannot be done for every remark in a massive 1200+ page volume.

The references to Alfred Pringsheim and Aurel Voss given by Boyer are to the French editions of the second volume of the German *Enzyklopädie der mathematischen Wissenschaften* [*Encyclopædia of mathematical sciences*] originally published in Leipzig in 1899. The German original can be consulted online, so I checked that version. The only relevant footnote I found in an admittedly perfunctory search of Pringsheim's article was a reference to Voss's immediately following contribution in the same volume. Section 4 of the latter's article is titled "Existence of derivatives" and reads in full:

> That a continuous function generally possessed a derivative with the possible exception of isolated points would probably first have been seen as self-evident; in any case it was not further investigated. First *A.M. Ampère* threw out the question why of all things the ratio of Δy to the *first* power of the increment Δx yielded a definite limit and attempted to prove[200], that a continuous function, apart from isolated points, must always be differentiable; yet he succeeded only with the verification, that for a nonconstant function the differential quotient cannot always be $+\infty$ and $-\infty$. The concept of the *Riemann* integral first led to knowledge of *continuous functions*, which are not differentiable at *infinitely many points of an interval* or at *any point* of the same, in particular *Riemann*'s function that is discontinuous at all rational points but is nevertheless integrable, the integral of which possesses no derivative at any rational point of the interval[201]; further *Hankel*'s

$$b(x) = W\left(\frac{1}{\pi}x\right)$$

> and is also continuous and nowhere differentiable for $0 < a < 1$ and $ab > 1$, and thus for the bounds given. I suppose it is too pedantic of me to add that a continuous curve need not be the graph of a function and thus "or what is the same" should read something like "or in particular".

[199] Kline, *op. cit.*, p. 956.

[200] In a footnote here Voss cites Ampère's paper and adds that the last attempt of this sort was published in Brussels by Philippe Gilbert in 1872.

[201] In a footnote Voss gives a reference to the page of Riemann's habilitation thesis on which R is described. R, of course, is not discontinuous at *all* rationals, but at a dense subset of them, which is the essential point.

Principle of Condensation of Singularities[202]. The full realisation however was probably given by the example rigorously worked out by *K.
Weierstrass*[203] under the application of a suitably chosen passage to
the limit for $\Delta(x)$ converging to 0, later clarified geometrically by *Ch.
Wiener*, [namely] the function

$$y = \sum_{0}^{\infty} a^n \cos(b^n \pi x),$$

which for $(0 < a < 1, b$ odd and integral) possesses a derivative at no
point, if $ab > 1 + \frac{3}{2}\pi$; many others soon followed this.

The statement in the footnote that Weierstrass had this example in 1861 is
unambiguous. The source for this assertion, however, is a bit less definite. In
1873 H.A. Schwarz published a paper[204] titled "Beispiel einer stetigen nicht dif-
ferentiirbaren function" [Example of a continuous not differentiable function].

Schwarz introduces his own nondifferentiable function "which on account of
its simplicity is perhaps not without interest" with some background remarks:

> The question, whether the existence of a derivative of a definite and
> continuous function of a real argument is already a necessary conse-
> quence of the presupposed continuity, or whether the demand of the
> existence of a derivative includes a new restricting condition imposed
> on the function, has for more than ten years ceased to be the subject
> of discussion in German mathematical circles.
> In his habilitation thesis presented to the Göttingen philosophical fac-
> ulty in 1854 *Riemann* reported an example of a function, which is dis-
> continuous for every rational value of the argument and yet permits

---

[202] Hermann Hankel, already familiar to us for his Principle of the Permanence of
Formal Laws, had attended Riemann's lectures and was inspired to write a paper
on oscillating functions in which he declared a general construction principle for
pathological functions to be to start with a function $\phi(x)$ with a singularity at 0
and define

$$f(x) = \sum_{n=0}^{\infty} a_n \phi(\sin n\pi x)$$

where $a_0, a_1, a_2, \ldots$ are chosen to guarantee uniform convergence. f will possess
singularities at every rational value of x. Hankel presented this Principle and a
number of applications of it in Tübingen in 1870, but died without publishing it.
His treatment was not without problems, but nonetheless (or, perhaps, therefore)
its posthumous publication in 1882 was influential in further developments. Voss's
subordinate clause referring to Hankel lacks a verb, but I surmise the overall intent
is that the subject started with Riemann's function and was furthered by Hankel's
contribution.

[203] The footnote to this begins, "In lectures since 1861 (cf. *H.A. Schwarz*, Ges. Abh.
2, p. 269)".

[204] According to the copy in his collected works (*Gesammelte Mathematische Abhandlungen II*, Berlin, 1890, pp. 269 - 274), this was published in two Swiss journals,
the *Archives des Sciences physiques et naturelles* (pp. 33 - 38) and *Verhandlungen
der Schweizerischen Naturforschenden Gesellschaft* (pp. 252 - 258).

integration throughout. Of the integral function belonging to this function one can thus assert, that it possesses a definite differential coefficient for no rational value of its argument.

In the year 1861 in the lectures, which he held in the Gewerbeinstitut in Berlin on the Differential and Integral Calculus, Weierstrass has as well given the correct laying out of the true facts of the case, in accordance with which *all attempts to prove the existence of a derivative for continuous functions of a real argument in general, must without exception be considered as failed.*[205]

Schwarz had studied at the Gewerbeinstitut, where Weierstrass lectured from 1854 till the winter semester of 1861/62[206], from 1860 on and may thus probably be regarded as an eyewitness in this matter. He offers no justification here for any claim that Weierstrass actually knew of the existence of functions which failed to be differentiable at infinitely many points in some closed finite interval, much less that Weierstrass was aware of functions lacking derivatives on a dense set of points, or that were nowhere differentiable. And certainly there is no mention of the 1872 function.

I know of only one other not very specific first hand account of what Weierstrass said with regard to nondifferentiability— Weierstrass himself confirms Schwarz's statement in a lecture of 1880:

In my lectures on the elements of the theory of functions I have stressed two propositions not concurring with the customary views, namely:

1) that one *cannot conclude from the continuity* of a function of a *real* argument that it also has at an individual point a definite differential quotient, let alone then— at least in an interval— likewise a continuous derivative;

2) that a function of a *complex* argument, which is defined for a bounded domain, *cannot always be extended* beyond the borders of this domain; and that the points for which the function is not differentiable can form not only isolated points but also lines and surfaces.[207]

---

[205] Cf. citation in footnote 203.

[206] The word "Gewerbe" means trade. The Gewerbeinstitut became the Gewerbeakademie in 1866 and merged with the Bauakademie in 1879 to become the Technische Hochschule Charlottenburg (Berlin). Weierstrass was professor there from 1856 until 1864. He was simultaneously extraordinary professor at the Berlin University until he became professor at the University in 1864. The dates cited are on the authority of Felix Klein and may be an error on his part or they may reveal something I don't know. I do know that Weierstrass was very ill before his appointment as full professor. Schwarz thus neither supports nor refutes Manheim's above declaration that Weierstrass lectured on the matter at the University in 1861. Cf. the biography of Weierstrass in the *Dictionary of Scientific Biography* and Winfried Scharlau, *Mathematische Institute in Deutschland 1800 - 1945*, Vieweg, Braunschweig, 1990, pp. 19, 38, and 39.

[207] Carl Weierstrass, "Zur Functionenlehre", *Gesammelte Abhandlungen, II*, pp. 201 - 233; here, p. 221. The second point refers to a common belief of the day that an analytic function defined by a power series could be extended beyond the radius

So sometime between Riemann's having written his habilitation thesis in 1853 and Weierstrass's giving his lecture in 1872, probably before 1861, either Riemann or Weierstrass or someone in one of their spheres recognised that functions could be constructed that were everywhere continuous but not differentiable at infinitely many points in any interval. As Schwarz stated in his 1873 paper cited above, no one in Germany had believed otherwise for over a decade, whence we can conclude knowledge of such functions was widely known in Germany. That Hankel does not mention nowhere differentiability in his 1870 paper in which he uses his Principle of Condensation of Singularities to create functions with various kinds of singularities at all rational numbers, and as Weierstrass emphasises in his 1872 lecture (cited pp. 407 - 408, above) that many were "of the opinion that it suffices to prove the existence of a function for which in every arbitrarily small interval in its domain points exist where it is not differentiable" and that such functions are "extraordinarily easy" to produce, we can conclude that the existence of his function was not widely known and that the probability he had been lecturing on it or some other nowhere differentiable function since 1861 is low.

Let us now consider Weierstrass's function. The citation on pages 407 - 408, above, is the beginning of his paper. Here follows the remainder:

Let x_0 be any definite value of x, and m an arbitrarily given positive whole number; so there is a certain whole number α_m, for which the difference

$$a^m x_0 - \alpha_m,$$

which will be denoted x_{m+1}, is $> -\frac{1}{2}$, but $\leq \frac{1}{2}$.
Then set

$$x' = \frac{\alpha_m - 1}{a^m}, \quad x'' = \frac{\alpha_m + 1}{a^m},$$

so one has

$$x' - x_0 = -\frac{1 + x_{m+1}}{a^m}, \quad x'' - x_0 = \frac{1 - x_{m+1}}{a^m};$$

one thus has

$$x' < x_0 < x''.$$

One can assume m so large, that both x', x'' come as close to the quantity x_0 as one wishes.
Now

$$\frac{f(x') - f(x_0)}{x' - x_0} = \sum_{n=0}^{\infty} \left(b^n \cdot \frac{\cos(a^n x' \pi) - \cos(a^n x_0 \pi)}{x' - x_0} \right)$$

of convergence of the series. On p. 223 of this work Weierstrass uses his nowhere differentiable function in this connexion. I must acknowledge that I did not find this reference myself. Kline's bibliography cites 7 volumes of Weierstrass's collected works, of which I found 5 online. For the point in question I have relied on the careful scholarship of Klaus Volkert (*Krise*, *op. cit.*, p. 124).

$$= \sum_{n=0}^{m-1} \left((ab)^n \cdot \frac{\cos(a^n x' \pi) - \cos(a^n x_0 \pi)}{a^n (x' - x_0)} \right) \tag{94}$$

$$+ \sum_{n=0}^{\infty} \left(b^{m+n} \cdot \frac{\cos(a^{m+n} x' \pi) - \cos(a^{m+n} x_0 \pi)}{x' - x_0} \right).$$

Because

$$\frac{\cos(a^n x' \pi) - \cos(a^n x_0 \pi)}{a^n (x' - x_0)} = -\pi \sin\left(a^n \frac{x' + x_0}{2} \pi \right) \cdot \frac{\sin\left(a^n \frac{x' - x_0}{2} \pi \right)}{a^n \frac{x' - x_0}{2} \pi}$$

and the value of

$$\frac{\sin\left(a^n \frac{x' - x_0}{2} \pi \right)}{a^n \frac{x' - x_0}{2} \pi}$$

always lies between -1 and 1, the summand (94) is smaller in absolute value than

$$\pi \sum_{n=0}^{m-1} (ab)^n,$$

and thus also smaller than

$$\frac{\pi}{ab - 1} (ab)^m.$$

Further, because a is an odd number, one has:

$$\cos(a^{m+n} x' \pi) = \cos(a^n (\alpha_m - 1)\pi) = -(-1)^{\alpha_m},$$
$$\cos(a^{m+n} x_0 \pi) = \cos(a^n \alpha_m \pi + a^n x_{m+1} \pi) = (-1)^{\alpha_m} \cos(a^n x_{m+1} \pi),$$

thus

$$\sum_{n=0}^{\infty} b^{m+n} \cdot \left(\frac{\cos(a^{m+n} x' \pi) - \cos(a^{m+n} x_0 \pi)}{x' - x_0} \right) =$$

$$(-1)^{\alpha_m} (ab)^m \sum_{n=0}^{\infty} \frac{1 + \cos(a^n x_{m+1} \pi)}{1 + x_{m+1}} b^n.$$

All members of the sum

$$\sum_{n=0}^{\infty} \frac{1 + \cos(a^n x_{m+1} \pi)}{1 + x_{m+1}} b^n$$

are positive, and the first, since $\cos(x_{m+1}\pi)$ is not negative and $1+x_{m+1}$ lies between $\frac{1}{2}$ and $\frac{3}{2}$, is not smaller than $\frac{2}{3}$.
From this one has

$$\frac{f(x') - f(x_0)}{x' - x_0} = (-1)^{\alpha_m} (ab)^m \cdot \eta \left(\frac{2}{3} + \varepsilon \frac{\pi}{ab - 1} \right),$$

where η denotes a positive quantity, which is > 1, while ε lies between -1 and 1.

Similarly, one has

$$\frac{f(x'') - f(x_0)}{x'' - x_0} = -(-1)^{\alpha_m} (ab)^m \cdot \eta_1 \left(\frac{2}{3} + \varepsilon_1 \frac{\pi}{ab - 1} \right),$$

where η_1 like η is positive and > 1, and ε_1 lies between -1 and 1.
Assuming now a, b are such that $ab > 1 + \frac{3}{2}\pi$, thus

$$\frac{2}{3} > \frac{\pi}{ab - 1},$$

then

$$\frac{f(x') - f(x_0)}{x' - x_0}, \quad \frac{f(x'') - f(x_0)}{x'' - x_0}$$

always have *opposite* signs, but *both*, when m grows without bound, become *infinitely large*.

From this it follows immediately, that $f(x)$ possesses neither a definite finite nor even a definite infinitely large differential quotient at the point $(x = x_0)$.[208]

Weierstrass's proof is not so much a proof as the skeleton of a proof to which the reader is expected to add the flesh. All motivation is lacking. The presentation is formal— not in the sense of my Types I, II, and III distinction, but in the familiar sense of being ritualistic symbol manipulation. A function is written down, a point x_0 assumed, further elements α_m, x_m, x', x'' are defined via formulæ with no motivating explanation, and a step-by-step derivation, almost a calculation, is carried out.

Once the work of finding the result has been done, one can explain the principles of the proof, i.e. the intuition behind it, and hope that the reader has the necessary technical skill to fill in the details; or, one can skip the motivation and simply give the details, leaving the reader to work out what is going on. Insofar as he offers no motivation for considering it in the first place, Riemann's discussion of the function I have labelled $R(x)$ in this chapter does not exactly fit the first approach, but his proof does: He says that if you calculate $R(x+0)$ and $R(x-0)$ you will get two different values at $x = p/(2n)$ with $p, 2n$ relatively prime, that the function is continuous everywhere else, and that it is integrable.

---

[208] Weierstrass, *Gesammelte Abhandlungen, II, op.cit.*, pp. 72 - 74. I have only taken a couple of minor liberties with the translation. Keeping the word order in German sentences can result in some rather awkward English. In correcting for this, the phrase "the first part of this expression" was separated from the expression in question by another. Hence I simply replaced it by numbering the line in question and referring directly to the number. Other than that I have copied the proof as Weierstrass presented it.

Students fresh out of a modern Calculus course would not be expected to be able to fill in the details, but advanced undergraduate and graduate students would. Weierstrass's proof can be followed line-by-line and verified by a student fresh out of Calculus (who hasn't forgotten his trigonometric identities) modulo his granting that the series converges to a continuous function (as uniform convergence is not a topic likely to be covered in any depth in the standard Calculus course); but Weierstrass's readers will have to work out on their own what is going on. This approach has been much maligned, but it is the most efficient way of giving a correct and rigorous proof. And the path to discovery is hidden, a fact that may have great pædagogical value as the stronger student tries to uncover that path and, in so doing, learns more than he would had the motivation been laid out for him.

If the purpose of the present work were to produce competent research mathematicians, I would have to end my discussion of Weierstrass's proof here with an exhortation to the reader to study the proof and explain the various steps, not only why the assertions are correct, but why the steps were taken and what would lead one to them. But this is not my purpose and my expository instincts compel me to proceed farther. Weierstrass's proof is indeed a calculation and one cannot replace it by some reference to intuition. Its various elements, however, can be explained and, I believe, in such a way as to make the proof more reproducible than mere memorisation could. First, as to the definition of the function, a quick nod to Hankel's Principle of Condensation of Singularities goes a long way. One wants something like

$$\sum_{n=0}^{\infty} c_n \cos(d_n x).$$

The choice of $c_n = b^n$ for some $0 < b < 1$ will guarantee uniform convergence of the series and thus continuity. This is also guaranteed by the sequence $c_n = 1/n^2$ of the function $R_2(x)$ attributed by Weierstrass to Riemann, which function is infinitely often differentiable. As we might expect from our experience with the modified Bolzano function, the exponential shrinking of b^n will contribute to the steepness of the individual contributing functions $c_n \cos(d_n x)$.

The choice $d_n = a^n \pi$, for an odd integer a, can also be motivated. The π simply changes the interval of periodicity from $[-\pi, \pi]$ or $[0, 2\pi]$ to $[-1, 1]$ or $[0, 2]$, respectively. The crucial choice is a^n. Each factor of a results in a periodic function with a times as many periods in a fixed interval. The spike function,

$$s_n(x) = b^n \cos(a^n \pi x),$$

will have a period of length $2/a^n$. What is more, $2/a^n$ will also be a period, albeit not a minimum one, of each $s_{n+k}(x)$ for $k \in \mathbb{N}^+$. This would also be the case for, say, $d_n = n! \pi$, but is not the case for the choice $d_n = n^2$ of $R_2(x)$. In fact, the choice of a being odd and integral reinforces the extremities of the oscillation: The maxima of $s_n(x)$ with value b^n occur precisely at the numbers of the form

$$x = \frac{2k}{a^n}, \quad k \in \mathbb{Z},$$

and the minima occur at those numbers of the form

$$x = \frac{2k+1}{a^n}, \quad k \in \mathbb{Z}.$$

Moreover, each number x retains this form as one proceeds from n to $n+1$:

$$\frac{2k}{a^n} = \frac{2ak}{a^{n+1}}, \quad \frac{2k+1}{a^n} = \frac{(2k+1)a}{a^{n+1}},$$

where $(2k+1)a$ is again odd.

For every integer α (getting back to Weierstrass's notation), the interval

$$\left[\frac{\alpha-1}{a^n}, \frac{\alpha+1}{a^n}\right]$$

is a minimum period of $s_n(x)$. If α is even, the graph of $s_n(x)$ starts with a value $s_n(\frac{\alpha-1}{a^n}) = -b^n$, grows to $s_n(\frac{\alpha}{a^n}) = b^n$, and drops down to $s_n(\frac{\alpha+1}{a^n}) = -b^n$; and, if α is odd, it drops from b^n to $-b^n$ before growing back to b^n. (Cf. *Figure 6*, bearing in mind that the x- and y- scales differ radically.)

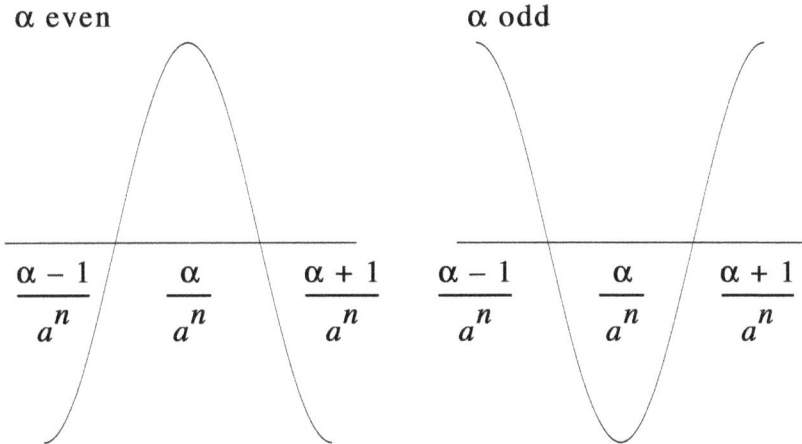

α even   **α odd**

$$\frac{\alpha-1}{a^n} \qquad \frac{\alpha}{a^n} \qquad \frac{\alpha+1}{a^n} \qquad \frac{\alpha-1}{a^n} \qquad \frac{\alpha}{a^n} \qquad \frac{\alpha+1}{a^n}$$

Figure 6

Weierstrass's proof proceeds as follows. Let x_0 be given and let α be the integer such that α/a^m is the peak closest to x_0 or for which x_0 lies exactly halfway between $(\alpha-1)/a^m$ and α/a^m. Weierstrass's x' and x'' are just the end points of one of the minimum periods of $s_m(x)$:

$$x' = \frac{\alpha-1}{a^m}, \quad x'' = \frac{\alpha+1}{a^m}.$$

Now, as m gets larger and larger, both x' and x'' tend to x_0. Hence, if Weierstrass's function— let us call it W— is differentiable, the difference quotients

$$\frac{W(x') - W(x_0)}{x' - x_0}, \quad \frac{W(x'') - W(x_0)}{x' - x_0}$$

must approach a common limit. Verifying that they do not, i.e. that, for m large the two quotients are in fact large and have opposite signs, is just a matter of calculation.

Now, the steps in a calculation are often forced upon us. In the present case, the choice of endpoints of a period of $s_m(x)$ for x', x'' gives us some information on the quotients

$$\frac{s_n(x') - s_n(x_0)}{x' - x_0}$$

for $n \geq m$: First, consider $n = m$. If, say, α is even, $s_m(\frac{\alpha-1}{a^m}) = -b^m$ and $s_m(x_0) \geq 0$, since x_0 is closest to α/a^m. Thus

$$\frac{s_m(x') - s_m(x_0)}{x' - x_0} > 0$$

since both numerator and denominator are negative. Moreover, because x_0 is in the middle half of the period,

$$\frac{1}{2a^m} \leq |x' - x_0| \leq \frac{3}{2a^m},$$

and, in fact,

$$\frac{s_m(x') - s_m(x_0)}{x' - x_0} \geq \frac{b^m}{\frac{3}{2a^m}} = \frac{2}{3}(ab)^m.$$

If we look at

$$\frac{s_m(x'') - s_m(x_0)}{x'' - x_0},$$

the same reasoning shows the quotient to be greater in absolute value than $\frac{2}{3}(ab)^m$, only it is now negative because the denominator has become negative.

If we now consider $n > m$ and α still even, $s_n(x') = -b^n$ is still a minimum in the interval $[x', x'']$. With all the new oscillations, $s_n(x_0)$ could be anywhere between $-b^n$ and b^n, but it cannot be less than $-b^n$ and

$$\frac{s_n(x') - s_n(x_0)}{x' - x_0}$$

is still nonnegative. Thus,

$$\sum_{n=m}^{\infty} \frac{s_n(x') - s_n(x_0)}{x' - x_0} \geq \frac{s_m(x') - s_m(x_0)}{x' - x_0} \geq \frac{2}{3}(ab)^m. \tag{95}$$

Similarly,

$$\sum_{n=m}^{\infty} \frac{s_n(x'') - s_n(x_0)}{x'' - x_0} \leq \frac{s_m(x'') - s_m(x_0)}{x'' - x_0} \leq -\frac{2}{3}(ab)^m. \tag{96}$$

If α is odd, as a glance at *Figure 6* shows, everything reverses:

$$\sum_{n=m}^{\infty} \frac{s_n(x') - s_n(x_0)}{x' - x_0} \leq -\frac{2}{3}(ab)^m \tag{97}$$

$$\sum_{n=m}^{\infty} \frac{s_n(x'') - s_n(x_0)}{x'' - x_0} \geq \frac{2}{3}(ab)^m. \tag{98}$$

Inequalities (95) - (98) can be written in equational form:

$$\sum_{n=m}^{\infty} \frac{s_n(x') - s_n(x_0)}{x' - x_0} = (-1)^{\alpha}\eta\frac{2}{3}(ab)^m$$

$$\sum_{n=m}^{\infty} \frac{s_n(x'') - s_n(x_0)}{x'' - x_0} = -(-1)^{\alpha}\eta_1\frac{2}{3}(ab)^m,$$

where $\eta, \eta_1 \geq 1$.

The use of a trigonometric identity and inequality in dealing with the finite sum, which use may seem clever or inspired, can be replaced by the more familiar and general appeal to the Mean Value Theorem:

$$\cos u_1 - \cos u_2 = (u_1 - u_2)(-\sin u),$$

for some u between u_1 and u_2. Thus

$$b^n \cos(a^n x'\pi) - b^n \cos(a^n x_0\pi) = (a^n x'\pi - a^n x_0\pi)(-b^n \sin u),$$

for some u between $a^n x'\pi$ and $a^n x_0\pi$,

$$= -(ab)^n \pi(x' - x_0)\sin u. \tag{99}$$

Thus

$$\left|\sum_{n=0}^{m-1} \frac{s_n(x') - s_n(x_0)}{x' - x_0}\right| \leq \sum_{n=0}^{m-1} \left|\frac{s_n(x') - s_n(x_0)}{x' - x_0}\right|$$

$$\leq \sum_{n=0}^{m-1} \frac{(ab)^n \pi|x' - x_0|}{|x' - x_0|}$$

by (99) and the inequality $|\sin u| \leq 1$,

$$\leq \sum_{n=0}^{m-1} (ab)^n \pi = \frac{(ab)^m - 1}{ab - 1}\pi$$

$$< (ab)^m \frac{\pi}{ab - 1} \leq \eta(ab)^m \frac{\pi}{ab - 1}.$$

The same argument applies to the sum with x' replaced by x'' and η by η_1. Moreover, introducing ε and ε_1 of absolute values less than 1, we can obtain, *á la* Weierstrass, equations in place of these inequalities:

$$\frac{W(x') - W(x_0)}{x' - x_0} = (-1)^\alpha \eta \frac{2}{3}(ab)^m + \eta\varepsilon \frac{\pi}{ab-1}(ab)^m$$

$$= (-1)^\alpha \eta (ab)^m \left(\frac{2}{3} \pm \varepsilon \frac{\pi}{ab-1}\right)$$

$$\frac{W(x'') - W(x_0)}{x'' - x_0} = -(-1)^\alpha \eta_1 (ab)^m \left(\frac{2}{3} \pm \varepsilon_1 \frac{\pi}{ab-1}\right).$$

All that remains to finish the proof of the nowhere differentiability of Weierstrass's function W is the application of a little elementary algebra to verify that

$$\frac{2}{3} > \frac{\pi}{ab-1}$$

when $ab > 1 + \frac{3}{2}\pi$.

I hope my leisurely repetition of Weierstrass's proof has been helpful and not merely redundant. His result was a milestone in the history of mathematics and we should stop to admire it and not merely quickly observe its existence, citing it as a counterexample to this or that (it shows not only that continuity does not imply differentiability, but that uniform convergence need not preserve differentiability anywhere). It was not a result that immediately led to a major mathematical development, such as a theory of nowhere differentiable functions, but was a result which changed attitudes; it changed the mathematical world view as the Germans would say. The rest of the 19th century saw the production of a variety of monsters. Cantor's study of cardinality soon revealed that the line and the plane had the same number of points, a fact that raised the question of dimensionality and its meaning. In 1890, Giuseppe Peano startled the mathematical world by producing a *space-filling curve*, that is, a continuous function defined on the unit interval that mapped onto the unit square. David Hilbert published another example the following year. Peano's and Hilbert's curves were both nowhere differentiable. And, of course, both strongly shook one's faith in the notion of dimensionality.

There is an asymmetry between positive results— general theorems— and negative results— the construction of counterexamples, pathological functions, monsters, or whatever one wishes to call them. Both can range from being easy (Dirichlet's everywhere discontinuous function) to being quite difficult (notice that I haven't proven the nowhere differentiability of the Bolzano function, the almost everywhere nondifferentiability of $R_2(x)$, nor the nowhere differentiability of $W(x)$ for a an arbitrary real number satisfying $ab > 1$). A positive result that requires a bit of work usually has lots of easy consequences. In the Calculus, for example, once one has found the derivative of $\sin x$ using the definition of the derivative as a limit, it is easy to differentiate any trigonometric function. Similarly, once one has proven the Fundamental Theorem of the Calculus, one can quickly find many areas, arc lengths, volumes, etc. that would be quite difficult to work out directly as the limits of sums. However, after proving a negative result, there often isn't much one can do with it. Hankel tried to give a general construction principle for a class of counterexamples. And others tried variations. G.H. Hardy (1916) relaxed the conditions on a, b in Weierstrass's function, and Konrad Knopp (1918) did the same for the variant of Bolzano's function considered above, showing the nowhere differentiability of the more general

$$K(x) = \sum_{n=0}^{\infty} \frac{1}{a^n} s_0(b^n x), \quad 0 < a < b.$$

Generally, however, there is not much one can do with a counterexample, other than to see if it, or some dressed up version of it, is a counterexample to something else. Books about counterexamples are not readable expositions of developments, but catalogues[209] comparable to a table of integrals or of logarithms, the difference being that such catalogues often show their work, while tables do not.

A glance at the eminent names cited in Thim's thesis[210] should convince anyone that my remark ought not to be taken as dismissive. It must, however, be stated that once the novelty has worn off, a counterexample ceases to please. And a surfeit of them can bring about a negative reaction. I have already cited Poincaré's reference to "a crowd of bizarre functions which seem to try to resemble as little as possible the honest functions which serve some purpose". In a letter to Thomas Stieltjes of 20 May 1893, Charles Hermite wrote

> I turn away with fear and horror from the lamentable plague of continuous functions which do not have derivatives...[211]

Hermite later softened his position on these functions, but general dissatisfaction continued. For the third International Congress of Mathematicians held in Heidelberg in 1904, a special booklet bearing the title *Dritter Jnternationaler Mathematiker-Kongress Heidelberg 1904*[212] and bearing the more cumbersome designation,

Liederbuch

den Teilnehmern am

Dritten Internationalen

Mathematiker-Kongress

in Heidelberg

als Andenken

an die Tage vom 8. bis 13. August 1904

überreicht von der

---

[209] Cf., for example, Bernard R. Gelbaum and John M.H. Olmsted, *Counterexamples in Analysis*, Holden Day Inc., San Francisco, 1964, a work that was popular in my schooldays among graduate students studying for their prelims.

[210] Thim, *op.cit.*, p. 6 has a table(!) of names and dates.

[211] I steal this quote from Thim, *op.cit.*, p. 4, who borrowed it from elsewhere. The French original is cited in Volkert, *Krise, op. cit.*, p. 132.

[212] The use of the same letter for "J" and "I" continued much longer in German than in English.

on the title page, was placed on sale. It gave the lyrics of drinking songs to be sung by the participants at the Friday evening gathering at the castle restaurant. Following 22 traditional German drinking songs (under the heading "General Songs") came 11 "Mathematical Songs". Hermann Schubert and Eugen Netto wrote songs 24, "Alte und neue Zeit" [Olden and modern times], and 29, "Die moderne Richtung" [The modern direction], respectively. Both are nostalgic remembrances of days when mathematics was simpler. Each mentions nondifferentiable functions, Schubert in the first stanza and Netto in the third. In nonlyrical translation, Netto's reference reads:

> And today? Three times woe,
> How everything now complicates itself!
> In view stand only those curves,
> Which infinitely often oscillate.
> Derivatives are out of fashion,
> Tangents are totally lacking,
> Singularities blossom,
> In the very wildest plurality.

This was, of course, intended humorously, and one should not read too much into it. But it should not be ignored either. The second stanza cites other pathologies, ending with

> And what "dimension" should be,
> That every 6th former understood.

Following Cantor, Peano, and Hilbert, it was no longer clear that the dimension of the space $\mathbb{R}^n$ was fixed. This would only first be settled by Luitzen Egbertus Jan Brouwer in 1911, in another triumph for Type II formalism, one, however, that lies beyond the scope of this book.

9 Exercises

9.1 Exercise. (Semigroups I) An *additive cancellation semigroup* is a structure $(G, +)$ where G is a nonempty set and $+$ a binary relation on G satisfying
i.   for all $x, y \in G$, $x + y = y + x$
ii. for all $x, y, z \in G$, $x + (y + z) = (x + y) + z$
iii. for all $x, y, z \in G$, $x + z = y + z \Rightarrow x = y$.
A structure $(G, +, 0)$ is called an *abelian group* if $(G, +)$ is an additive cancellation semi-group and
iv. for all $x \in G$, $x + 0 = x$

---

[213] Verlag von B.G. Teubner, Leipzig, 1904. The title translates to *Song Book for the Participants at the Third International Congress of Mathematicians in Heidelberg as a Memento of the Days from 8 to 13 August 1904 Presented by the German Mathematicians-Union*. The various lines of the title appeared in various font sizes, some in boldface.

v.   for all $x \in G$ there is an element $y \in G$ such that $x + y = 0$.
Show: Every additive cancellation semigroup can be embedded into an abelian group.

9.2 Exercise. (Semigroups II) i.   Let $(G, +)$ be an additive cancellation semigroup. Suppose for some $a, e \in G$, $a + e = a$. Show: e is a zero element: for all $x \in G$, $x + e = x$.
ii.   Show: The zero element of an additive cancellation semigroup (group, semi-ring, ring), when it exists, is unique.
iii.   Show: The multiplicative identity of a semi-ring, if it exists, is unique.
iv.   Let $(G, +, 0)$ be an abelian group. Show that additive inverses are unique: $a + b = 0 \,\&\, a + c = 0 \Rightarrow b = c$.

9.3 Exercise. (Negative Numbers and Order) A semi-ring $(R, +, \cdot, 0)$ with zero element 0 has a *positive ordering* $<$ if $<$ is a binary relation on R satisfying
i.   for all $a, b \in R$, $a < b$ or $b < a$ or $a = b$
ii.   for all $a, b, c \in R$, $a < b \,\&\, b < c \Rightarrow a < c$
iii.   for no element $a \in R$ do we have $a < a$
iv.   for all $a \in R$, $a \neq 0 \Rightarrow 0 < a$
v.   for all $a, b, c \in R$, $a < b \Rightarrow a + c < b + c$
vi.   for all $a, b, c \in R$, if $c \neq 0$ then

$$a < b \Rightarrow c \cdot a < c \cdot b.$$

If $(R, +, \cdot, 0)$ has a positive ordering, show that the relation $\bigcirc\!\!\!\!<$ on $[R^{\pm}]$ defined by
$$\langle a, b \rangle \bigcirc\!\!\!\!< \langle cd \rangle : \; a + d < b + c$$
satisfies the analogues of conditions i - ii and v. Further, show that for all $a, b, c \in [R^{\pm}]$,

$$a \bigcirc\!\!\!\!< b \,\&\, 0 \bigcirc\!\!\!\!< c \Rightarrow c \odot a \bigcirc\!\!\!\!< c \odot b$$
$$a \bigcirc\!\!\!\!< b \,\&\, c \bigcirc\!\!\!\!< 0 \Rightarrow c \odot b \bigcirc\!\!\!\!< c \odot a.$$

9.4 Exercise. (Multiplicative Cancellation in Semi-Rings) Consider the following variants of the multiplicative cancellation property on a semi-ring R:
No Zero Divisors   for all $a, b \in R$, $ab = 0 \Rightarrow a = 0$ or $b = 0$
Multiplicative Cancellation   for all $a, b, c \in R$, if $c \neq 0, ac = bc \Rightarrow a = b$
Strong Multiplicative Cancellation   for all $a, b, c, d \in R$, $ac + bd = ad + bc \Rightarrow a = b$ or $c = d$.
i.   Let R be a ring. Show: R has the multiplicative cancellation property iff R has no zero divisors.
ii.   Let
$$\mathcal{R}_4 = \{\langle m, n \rangle \mid m \in \mathbb{N} \,\&\, n \in \mathbb{Z}^+ \text{ or } m = n = 0\}$$
be given with addition and multiplication defined componentwise. Show:
a. $\langle 0, 0 \rangle$ is a zero element;
b. $\mathcal{R}_4$ has no zero divisors;

c. $\mathcal{R}_4$ does not satisfy the multiplicative cancellation property.

iii. Let
$$\mathcal{R}_5 = \{\langle m, n\rangle \,|\, m, n \in \mathbb{Z}^+ \text{ or } m = n = 0\}$$

Show: $\mathcal{R}_5$ is a semi-ring satisfying the multiplicative cancellation property, but $[\mathcal{R}_5^\pm]$ has zero divisors.

iv. Let R be a semi-ring. Show: $[R^\pm]$ satisfies the multiplicative cancellation property iff R satisfies the strong multiplicative cancellation property.

iv.   Let R be a semi-ring with no zero-divisors. Show: If for all $a, b \in R$ there is an element $c \in R$ such that $a + c = b$ or $b + c = a$, then R satisfies the strong multiplicative cancellation property.

9.5 Exercise. $([[Q(R)]^\pm])$ Define the partial subtraction property in a semiring R by:

for all distinct $a, b \in R$ there is a $c \in R$ such that $a + c = b$ or $b + c = a$.

i. Show: If R is a semi-field and R satisfies the partial subtraction property, then $[R^\pm]$ is a field.

ii. Show: If R is a semi-domain satisfying the partial subtraction property, then $[Q(R)]$ satisfies the partial subtraction property, whence $[[Q(R)]^\pm]$ is a field.

iii. (For more advanced readers) Show: If R is a semi-domain satisfying the partial subtraction property, then $[[Q(R)]^\pm]$ and $[Q([R^\pm])]$ are isomorphic.

9.6 Exercise. $(\mathcal{R} = (\mathbb{R}^\mathbb{N}, +, *, \mathbf{0}, \mathbf{1}))$ Let $\mathcal{R}$ be the ring of Example 3.1.iv. Define the special elements of $\mathbb{R}^\mathbb{N}$:

$$\mathbf{0} = (0, 0, 0, \ldots), \text{ i.e. } (\mathbf{0})_n = 0 \text{ for all } n,$$

$$\mathbf{1} = (1, 0, 0, \ldots), \text{ i.e. } (\mathbf{1})_n = \begin{cases} 1, & n = 0 \\ 0, & n > 0 \end{cases}$$

$$s = (0, 1, 0, 0, \ldots), \text{ i.e. } s_n = \begin{cases} 1, & n = 1 \\ 0, & n \neq 1 \end{cases}$$

$$\blacktriangle = (1, -1, 0, 0, \ldots), \text{ i.e. } \blacktriangle = \begin{cases} 1, & n = 0 \\ -1, & n = 1 \\ 0, & n > 1 \end{cases}$$

$$g_k = (1, k, k^2, \ldots), \text{ i.e. } (g_k)_n = k^n$$

$$\Sigma = g_1.$$

In Chapter I we met with the shift and truncation operators:
$$\sigma(a_0, a_1, a_2, \ldots) = (0, a_0, a_1, a_2, \ldots)$$
$$\tau(a_0, a_1, a_2, \ldots) = (a_1, a_2, a_3, \ldots).$$

From the Calculus of Finite Differences there are also the operators
$$Z(a) = \mathbf{0}$$
$$I(a) = a$$

and Δ, E defined by
$$(\Delta a)_n = a_{n+1} - a_n, \text{ i.e. } \Delta(a_0, a_1, \ldots) = (a_1 - a_0, a_2 - a_1, \ldots)$$
$$(E(a))_n = a_{n+1}, \text{ i.e. } E(a_0, a_1, \ldots) = (a_1, a_2, \ldots).$$

i.   Show: For all $a \in \mathbb{R}^\mathbb{N}$,

$$Z(a) = \mathbf{0} * a \qquad I(a) = \mathbf{1} * a \qquad \sigma(a) = s * a$$
$$\Sigma * a = (a_0, a_0 + a_1, a_0 + a_1 + a_2, \ldots), \text{ i.e. } (\Sigma * a)_n = \sum_{i=0}^{n} a_i$$
$$\blacktriangle * a = (a_0, a_1 - a_0, a_2 - a_1, \ldots)$$
$$= a_0 \mathbf{1} + \sigma(\Delta a),$$

whence $\Delta a = \tau(\blacktriangle * a)$.

ii.   Show: $E = \tau, \Sigma * \blacktriangle = \mathbf{1}, \blacktriangle = \mathbf{1} - s, g_0 = \mathbf{1}$.

iii.   Show: Δ, τ are not represented in $\mathcal{R}$ by multiplication: For no $d \in \mathbb{R}^{\mathbb{N}}$ do we have

$$\Delta a = d * a$$

for all $a \in \mathbb{R}^{\mathbb{N}}$; and, for no $t \in \mathbb{R}^{\mathbb{N}}$ do we have

$$\tau(a) = t * a$$

for all $a \in \mathbb{R}^{\mathbb{N}}$.

iv.   Show: $a \in \mathbb{R}^{\mathbb{N}}$ is invertible in $\mathcal{R}$ iff $a_0 \neq 0$, i.e. iff a is not in the range of the operator σ.

v.   Show: Every g_k is invertible, i.e. show $g_k - k\sigma(g_k) = \mathbf{1}$ and conclude

$$(g_k)^{-1} = \mathbf{1} - ks.$$

Show: For all $k, m \in \mathbb{N}$,

$$g_k * g_m = \begin{cases} \dfrac{\tau(g_m - g_k)}{m - k}, & m \neq k \\ (1, 2k, 3k^2, 4k^3, \ldots), & m = k. \end{cases}$$

(For graduate students and beyond.) In particular, for $k = -1$, g_{-1} is the Grandi sequence and its Cauchy product with itself is the Hölder sequence. The Grandi sequence sums to $\frac{1}{2}$ and the Hölder sequence $h = g_{-1} * g_{-1}$ sums to $\frac{1}{2} \cdot \frac{1}{2} = \frac{1}{4}$. Can you explain this?

9.7 Exercise. (Solving Difference Equations Via $[Q(\mathcal{R})]$) In $[Q(\mathcal{R})]$ define $t = s^{-1}$ (i.e., $t = [s]^{-1}$), $d = t - 1$. For $k \in \mathbb{R}$, write $\mathbf{k}$ for $k\mathbf{1}$.

i.   Show: For all $a \in \mathbb{R}^{\mathbb{N}}$, if $a_0 = 0$, then

$$t * a = \tau(a) \qquad d * a = \Delta a.$$

ii.   Show: for all $a \in \mathbb{R}^{\mathbb{N}}$,

$$\tau(a) = t * (a - a_0 \mathbf{1}) = t * a - a_0 t.$$

iii.   Show

$$t^{-1} = s$$
$$(t - 1)^{-1} = s * g_1$$
$$(t - \mathbf{k})^{-1} = s * g_k, \text{ for any } k \in \mathbb{R}$$
$$\frac{t}{t - \mathbf{k}} = g_k.$$

iv.   Recalling $E = \tau$, apply parts ii and iii to solve the difference equation

$$(E^2 - 3E + 2)a = \mathbf{0},$$

subject to the initial conditions $a_0 = a_1 = 1$.

v.   Repeat part iv for the equation

$$(E^2 - E - 1)f = 0,$$

using the initial values $f_0 = f_1 = 1$ to obtain a closed formula for the Fibonacci sequence $f_0, f_1, \ldots$.

9.8 Exercise. (Ordered Fields, I) Let $\mathcal{F} = (F, +, \cdot, 0, 1, <)$ be an ordered field.

i.   For $x, y \in F$, show $x < y$ iff $-y < -x$.

ii. For $x \in F$, show: if $x \neq 0$, then exactly one of $x, -x$ is greater than 0.

iii. Define

$$|x| = \begin{cases} x, & x = 0 \text{ or } 0 < x \\ -x, & x < 0. \end{cases}$$

Show:

   a. $|x| \geq 0$, and $|x| = 0$ iff $x = 0$

   b. $|x \cdot y| = |x| \cdot |y|$

   c. $|x + y| \leq |x| + |y|$,

where $a \leq b$ abbreviates "$a < b$ or $a = b$".

9.9 Exercise. (Ordered Fields, II) Let $\mathcal{F} = (F, +, \cdot, 0, 1, <)$ be an ordered field. Assume $\mathbb{Q}$ is given as a subset of F. Make the following definitions:

x is *finite* iff for some positive $q \in \mathbb{Q}, |x| < q$

x is *infinite* iff for all positive $q \in \mathbb{Q}, q < |x|$

x is *infinitesimal* iff for all positive $q \in \mathbb{Q}, |x| < q$

x is *measurable* iff for all positive $q \in \mathbb{Z}$ there is an integer $p \in \mathbb{Z}$ such that

$$\frac{p}{q} < x < \frac{p+2}{q}.$$

[Note that this definition is inclusive: 0 is infinitesimal and infinitesimals are finite.]

Show:

i.   The sum and product of finite numbers is finite. If x is finite and not infinitesimal, then $1/x$ is finite.

ii.   The sum and product of infinitesimals is infinitesimal. If $x \neq 0$ is infinitesimal, then $1/x$ is infinite.

iii. The sum of a finite number and an infinitesimal number is finite. The product of such is infinitesimal.

iv.   An element $x \in F$ is measurable iff it is finite.

9.10 Exercise. (Archimedean Ordered Fields) Let $\mathcal{F} = (F, +, \cdot, 0, 1, <)$ be an Archimedean ordered field. Assume $\mathbb{Q}$ is given as a subset of F. Let $b_0, b_1, \ldots$ be a sequence of integers greater than 1.

i.   Prove the analogue of Lemma 4.12: For any $x \in F$ there are integers $a_0, a_1, \ldots$ with $0 \leq a_i < b_i$ such that

$$a_0 + \frac{a_1}{b_1} + \frac{a_2}{b_1 b_2} + \ldots + \frac{a_{m-1}}{b_1 \cdots b_{m-1}} + \frac{a_m}{b_1 \cdots b_m} < x <$$

$$a_0 + \frac{a_1}{b_1} + \frac{a_2}{b_1 b_2} + \ldots + \frac{a_{m-1}}{b_1 \cdots b_{m-1}} + \frac{a_m + 2}{b_1 \cdots b_m}.$$

ii.   Define two notions of limit in $\mathcal{F}$ as follows:

$L = \lim_{n\to\infty}^{1} x_n$ iff for every positive $\epsilon \in F$ there is an $n_0 \in \mathbb{N}$ such that, for all $n > n_0, |x_n - L| < \epsilon$;

$L = \lim_{n\to\infty}^{2} x_n$ iff for every positive $q \in \mathbb{Q}$ there is an $n_0 \in \mathbb{N}^+$ such that, for all $n > n_0, |x_n - L| < q$.

Show: For any $L \in F$ and any sequence $x_0, x_1, \ldots$ of elements of F,

$$L = \lim_{n\to\infty}{}^{1} x_n \quad \text{iff} \quad L = \lim_{n\to\infty}{}^{2} x_n.$$

We may thus omit the superscripts and simply write lim.

iii.   Show: Limits in $\mathcal{F}$ are unique:

$$L_1 = \lim_{n\to\infty} x_n \ \& \ L_2 = \lim_{n\to\infty} x_n \ \Rightarrow \ L_1 = L_2.$$

iv.   Let $x_0, x_1, \ldots, \ y_0, y_1, \ldots$ be sequences of elements of F. Show: If $\lim x_n$ and $\lim y_n$ exist, then so do $\lim(x_n + y_n)$ and $\lim(x_n \cdot y_n)$ and

$$\lim(x_n + y_n) = \lim x_n + \lim y_n$$
$$\lim(x_n \cdot y_n) = \lim x_n \cdot \lim y_n.$$

v.   Prove Lemma 4.31.

vi.   Show: If $\mathcal{F}$ is complete, and $a_0, a_1, \ldots b_0, b_1, \ldots$ are as in part i, then

$$\sum_{i=0}^{\infty} \frac{a_i}{b_0 \cdots b_i}$$

converges to a limit in $\mathcal{F}$.

vii.   Prove Theorem 4.29.

A small remark: Letting $b_0 = b_1 = \ldots = 10$, parts i and vi show every element of a complete Archimedean ordered field $\mathcal{F}$ to have a decimal expansion and every decimal expansion to be an element of $\mathcal{F}$.

9.11 Exercise. (Dedekind's Example) We want a rational function $f(X)$ satisfying, for (say) positive x:
$\alpha.$ $x < \sqrt{2} \Rightarrow x < f(x) < \sqrt{2}$;
$\beta.$ $x > \sqrt{2} \Rightarrow \sqrt{2} < f(x) < x$;
$\gamma.$ $f(\sqrt{2}) = \sqrt{2}$.
It is natural to assume

$$f(X) = X \cdot \frac{P(X^2)}{Q(X^2)},$$

where $P(Y), Q(Y)$ have integral coefficients and $P(2)/Q(2) = 1$. Assume P, Q are as simple as possible: $P(Y) = aY + b, Q(Y) = cY + d$, with a, b, c, d positive.

i.a. Show: Condition γ holds iff $2a + b = 2c + d$.

b. Assuming γ, show that conditions α, β hold if f is strictly increasing and $g(x) = f(x)/x$ is strictly decreasing for positive x.

c. Show that $g'(x) < 0$ iff $ad < bc$. Show $f'(x) > 0$ if $bc < 3ad$.

d. Conclude that

$$f(X) = X \cdot \frac{aX^2 + b}{cX^2 + d} \tag{100}$$

satisfies conditions α, β, γ if $2a + b = 2c + d$ and $ad < bc < 3ad$. This does not hold for for $a = 1, b = 6, c = 3, d = 2$ as in Dedekind's Example 5.2, but we can use it to construct other functions that do work. For example, if $a = 1, b = 6$, we have $2a + b = 8$ and $\langle c, d \rangle$ must be on the line $2c + d = 8$ and lie between the lines $d = c$ and $d = c/3$. A glance at the graph suggests the choice $c = 3/2, d = 5$. Rationalising the denominator gives $a = 2, b = 12, c = 3, d = 10$

e. Dedekind's function f does not satisfy $bc < 3ad$. Show nonetheless that $f'(x) > 0$ for $x > 0$ by differentiating f.

ii. (No Calculus Allowed) Let f be as in (100) and suppose

δ. $2a + b = 2c + d$;

ε. $ad < bc$;

with a, b, c, d all positive.

a. Show: $a < c$.

b. Let $y = f(x)$ and show:

$$y - x = \frac{(c - a)x(2 - x^2)}{cx^2 + d}.$$

c. For $A = \{x \in \mathbb{Q}^+ | x^2 < 2\}, B = \{x \in \mathbb{Q}^+ | 2 < x^2\}$, show

$$y > x \text{ iff } x \in A$$
$$y < x \text{ iff } x \in B.$$

d. Show

$$y^2 - 2 = \frac{a^2 x^6 + (2ab - 2c^2)x^4 + (b^2 - 4cd)x^2 - 2d^2}{(cx^2 + d)^2}.$$

e. Divide the numerator of the fraction in part d by $x^2 - 2$ to get $Ax^4 + Bx^2 + C$ with remainder R, where

$$A: a^2$$
$$B: 2(a^2 + ab - c^2)$$
$$C: b^2 - 4cd + 4a^2 + 4ab - 4c^2$$
$$R: 2C - 2d^2.$$

f. Show by clever grouping and repeated replacement of $2a + b$ by $2c + d$ that $C = d^2$ and thus $R = 0$.

g. (Now we cheat) We have

$$y^2 - 2 = \frac{\left(a^2x^4 + 2(a^2 + ab - c^2)x^2 + d^2\right)\left(x^2 - 2\right)}{(cx^2 + d)^2}.$$

We can make the numerator equal to $(x^2 - 2)^3$ by choosing $a = 1, d = 2$ and then choosing b, c such that

$$a^2 + ab - c^2 = -2,$$

i.e. $1 + b - c^2 = -2$, i.e. $b = c^2 - 3$. Bearing in mind that $2a + b = 2c + d$, show that, if we want a, b, c, d positive, we must choose $c = 3, b = 6$.

h. The choices $1, 5, 2, 3$ and $1, 11, 5, 3$ for a, b, c, d also satisfy the crucial conditions δ, ε. Can you verify

$$y^2 - 2 < 0 \quad \text{iff} \quad x^2 - 2 < 0$$

for these choices without appealing to the Calculus or simple facts about the real numbers?

9.12 Exercise. (Applying Horner's Method) Given a function f and an approximation x to a zero of f, Newton's Method proposes

$$g(x) = x - \frac{f(x)}{f'(x)}$$

as the next approximation to a zero. William George Horner[214] proposes

$$h(x) = x - \frac{2f(x)f'(x)}{2(f'(x))^2 - f(x)f''(x)}.$$

Let $f(x) = x^2 - 2$. and find $h(x)$.

9.13 Exercise. (Dedekind's Example Revisited) Consider the graph (*Figure 7*) of Dedekind's rational function f defined by (47).

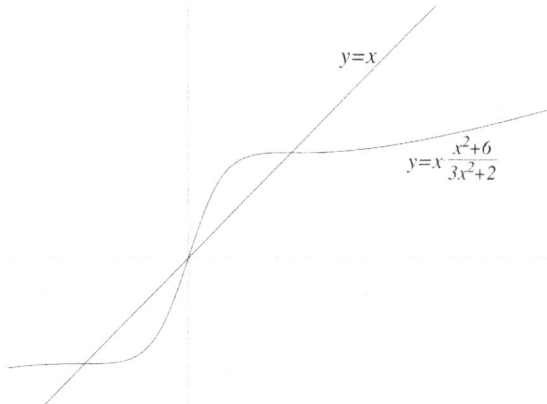

$y = x$

$$y = x\frac{x^2+6}{3x^2+2}$$

Figure 7

---

[214] I refer the reader to my *History of Mathematics; A Supplement* (Springer-Verlag, New York, 2007) for a discussion of Horner.

In addition to the fact that the graph of $y = f(x)$ intersects that of $y = x$ only in the points $0, \pm\sqrt{2}$, we ought to be impressed by two facts:

a. for $x \in (0, \infty)$, $f(x) \lessgtr \sqrt{2}$ iff $x \lessgtr \sqrt{2}$;

b. for $x \in (0, \infty)$, $x \lessgtr f(x)$ iff $x \lessgtr \sqrt{2}$.

These conditions, along with the rationality of $f(x)$ for x rational, are exactly what is needed to show his cut not to be determined by a rational number. If we replace $(0, \infty)$ by a smaller interval around $\sqrt{2}$, we can meet these conditions with a simple polynomial.

i.   Suppose f is differentiable at $\sqrt{2}$, $f(\sqrt{2}) = \sqrt{2}$, and $0 < f'(\sqrt{2}) < 1$. Show that in some interval $(\sqrt{2} - \delta, \sqrt{2} + \delta)$ one has

$$f(x) \lessgtr \sqrt{2} \ \text{ iff } \ x \lessgtr \sqrt{2}$$

$$x \lessgtr f(x) \ \text{ iff } \ x \lessgtr \sqrt{2}.$$

ii.   Assume $f(X) = aX^2 + bX + c$, $f(\sqrt{2}) = \sqrt{2}$, and $0 < f'(x) < 1$. Show: IF a, b, c are rational, $-\sqrt{2}/4 < a < 0$, $b = 1$ $c = -2a$.

iii.   Choose $a = \frac{-1}{4}, b = 1, c = \frac{1}{2}$, so that

$$f(x) = \frac{-x^2 + 4x + 2}{4}.$$

Without appeal to any special properties of the real numbers (i.e., dealing only with rational numbers) show that if x is rational and $0 < x < 2$,

$$x^2 \lessgtr 2 \ \text{ iff } \ x \lessgtr f(x).$$

iv.   For $f(x)$ as just defined, show

$$f(x)^2 = \frac{x^4 - 8x^3 + 12x^2 + 16x + 4}{16}.$$

Dealing only with rational numbers and simple properties of inequality, show that for rational x with $0 < x < 2$,

$$x^2 \lessgtr 2 \ \text{ iff } \ f(x)^2 \lessgtr 2.$$

v.   Klaus Mainzer[215] offers:

$$R(X) = \frac{2X + 2}{X + 2}$$

as another simple example. Compare its graph with those of the other functions of this exercise and give a proof that Mainzer's function does the trick.

9.14 Exercise. (Dedekind's Example Yet Again) Heinrich Weber took a different approach entirely to showing the cut of Example 5.2 not to correspond to any rational number. Choose $x \in \mathbb{Q}$ such that $x^2 < 2$ and write $x = p/q$ with p, q positive integers. He wants to construct y in the form

---

[215] In Ebbinghaus *et al.*, *op. cit.*, p. 31.

$$y = \frac{np + m}{nq} = \frac{p}{q} + \frac{m}{nq}.$$

Note that, for any positive m, n, we have $y > x$.

i.   Show that
$$y^2 - 2 = \frac{n^2(p^2 - 2q^2) + 2npm + m^2}{n^2 q^2}.$$

ii.   Show that if one chooses $n > m$ such that
$$n(2q^2 - p^2) > m(2p + 1),$$

then $y^2 - 2 < 0$, i.e. $y^2 < 2$.

iii.   Let m be any positive integer and
$$y = \frac{p}{q} + \frac{m}{nq}.$$

Find n_0 depending on p, q, m such that for $n > n_0$, we have $y^2 < 2$.

iv.   Handle the case in which $x^2 > 2$.

9.15 Exercise. (Order Awareness and Continuity, I) Let $(G, +, <, 0)$ be an ordered abelian group. Define absolute value on G by

$$|x| = \begin{cases} x, & 0 \le x \\ -x, & x < 0. \end{cases}$$

Note that $|-x| = |x|$ and, if f is a homomorphism of such groups, $|f(-x)| = |-f(x)| = |f(x)|$.

i.   Show: If f is a nonconstant order aware homomorphism of ordered abelian groups, then
$$|x| < |y| \Rightarrow |f(x)| < |f(y)|.$$

ii.   Show: If f is an order aware homomorphism of $(\mathbb{Q}, +, <, 0)$ or $(\mathbb{R}, +, <, 0)$ into $(\mathbb{R}, +, <, 0)$, then f is continuous at 0: for any $\epsilon > 0$ there is a $\delta > 0$ such that, for x in $\mathbb{Q}$ or $\mathbb{R}$, respectively,
$$|x| < \delta \Rightarrow |f(x)| < \epsilon.$$

iii.   Show: If f is an order aware homomorphism of $(\mathbb{Q}, +, <, 0)$ or $(\mathbb{R}, +, <, 0)$ into $(\mathbb{R}, +, <, 0)$, then f is continuous at all x_0 in $\mathbb{Q}$ or $\mathbb{R}$, respectively: for any $\epsilon > 0$ there is a $\delta > 0$ such that, for all appropriate x,
$$|x - x_0| < \delta \Rightarrow |f(x) - f(x_0)| < \epsilon.$$

[Hint: ii. $f(\mathbb{Q})$ is dense in $\mathbb{R}$.]

9.16 Exercise. (Order Awareness and Continuity, II) Let $f : \mathbb{R} \to \mathbb{R}$ be a homomorphism. Let us say that f maps positive to positive if, for all $x \in \mathbb{R}$,
$$x > 0 \Rightarrow f(x) > 0,$$

and f maps positive to negative if, for all $x \in \mathbb{R}$,

$$x > 0 \implies f(x) < 0.$$

i. Show: Let $f : \mathbb{R} \to \mathbb{R}$ be a homomorphism.
   a. If f maps positive to positive, then f is order preserving.
   b. If f maps positive to negative, then f is order reversing.
ii. Show: Let $f : \mathbb{R} \to \mathbb{R}$ be a continuous homomorphism. Let $x_0 > 0$.
   a. If $f(x_0) > 0$, then f maps positive to positive.
   b. If $f(x_0) < 0$, then f maps positive to negative.
iii. Show: If $f : \mathbb{R} \to \mathbb{R}$ is a continuous homomorphism, then f is order aware.
[Hint. ii.a. If $f(x_0) > 0$, then $f(q) > 0$ for some positive rational q close to x_0.
Let g_r be the restriction of f to $\mathbb{Q}$. Show g_r maps positive to positive. Use some
estimates to show f must do the same.]

9.17 Exercise. ($X^2 = X$) Consider the structures introduced by van Rootselaar and Laugwitz on page 253:

$$\mathcal{Q}_r : \quad (\mathbb{Q}^{\mathbb{N}}, +, \cdot, <, =)$$
$$\mathcal{Q}_l : \quad (\mathbb{Q}^{\mathbb{N}}, +, \cdot, <, \equiv).$$

i.   Show that $f \in \mathbb{Q}^{\mathbb{N}}$ satisfies the equation $X^2 = X$ in $\mathcal{Q}_r$ iff the range of f
consists solely of 0's and 1's.
ii.   Show that f satisfies $X^2 \equiv X$ in $\mathcal{Q}_l$ iff $f(n) \in \{0, 1\}$ for all but finitely
many $n \in \mathbb{N}$.
iii.   Show that $\mathcal{Q}_r$ contains "minimal" nonzero solutions f to $X^2 = X$, i.e.
solutions f which are not the sum of two nonzero solutions.
iv.   Show that every nonzero solution f to $X^2 \equiv X$ in $\mathcal{Q}_l$ can be written as the
sum of two nonzero solutions. Conclude that $\mathcal{Q}_r$ and $\mathcal{Q}_l$ are not isomorphic.
v.   Show that in neither structure is there a solution f to $X^2 = X$ or $X^2 \equiv X$
such that $0 < f < 1$.
vi.   Define $\equiv_E$ by $f \equiv_E g$ iff $f(n) = g(n)$ for all even $n \in \mathbb{N}$. Show that
$(\mathbb{Q}^{\mathbb{N}}, +, \cdot, <, =)$ is isomorphic to $(\mathbb{Q}^{\mathbb{N}}, +, \cdot, <, \equiv_E)$.

9.18 Exercise. (Rational Functions) Define a rational function over $\mathbb{Q}$ to be
a ratio of two polynomials P, Q with coefficients in $\mathbb{Q}$:

$$f(t) = \frac{P(t)}{Q(t)}.$$

A real number a is called a *pole* of f if $f(a)$ is undefined, i.e. if $Q(a) = 0$; a is a
zero of f if $f(a)$ is defined and $f(a) = 0$, i.e. if $P(a) = 0$ and $Q(a) \neq 0$. We let
$\Omega_0(t)$ denote the set of all rational functions over $\mathbb{Q}$, and define equality and
order on $\Omega_0(t)$ as follows:

$$f \equiv g \quad \text{iff} \quad \text{for all but finitely many } a \in \mathbb{R}, f(a) = g(a)$$
$$f < g \quad \text{iff} \quad \text{there is an } M \in \mathbb{R} \text{ such that for all } a > M, f(a) < g(a).$$

i.a. Show: Every $f \in \Omega_0(t)$ has only finitely many roots and only finitely many poles. Is $\sin(1/t) \in \Omega_0(t)$?

b. Show: $<$ is a linear order on $\Omega_0(t)$.

c. Show: $\Omega_0(t)$ is a non-Archimedean ordered field under the usual algebraic operations on rational functions. Characterise the infinitesimals.

ii. Define $f \in \Omega_0(t)$ to be *measurable* if for any positive integer q one can find an integer p such that

$$\frac{p}{q} < f < \frac{p+2}{q}.$$

Show: If f is measurable, then f differs from a rational number by an infinitesimal. Hence, if one identifies measurable rational functions differing only by an infinitesimal, the structure obtained is isomorphic to $\mathbb{Q}$.

iii. Let K be a field of functions extending $\Omega_0(t)$ and satisfying conclusions i.a - i.b. Let ω be any element of K and define K_ω to be the set of all functions of the form

$$f(t) = g(t) + h(t)\sqrt{1 + \omega(t)^2}, \quad g, h \in K.$$

a. Show that K_ω is a field (where equality is again taken to mean equality of all but finitely many values). [Hint: Consider the conjugate $\bar{f} = g - h\sqrt{1+\omega^2}$.]

b. Show that K_ω satisfies i.a - i.b.

c. Conclude that Hilbert's domain $\Omega(t)$ defined on page 324 is a non-Archimedean ordered field.

9.19 Exercise. (Non-Horn Formulæ and $\Omega\mathbb{R}$)

i. Show that Trichotomy fails in $\Omega\mathbb{R}$, i.e. there are $a, b \in \Omega\mathbb{R}$ such that none of

$$\bar{a} < \bar{b}, \quad \bar{a} = \bar{b}, \quad \bar{b} < \bar{a}$$

is true in $\Omega\mathbb{R}$.

ii. Let $R_\leq$ be the relation in $\mathbb{R}$ defined by:

$$R_\leq(x, y) : \ x < y \vee x = y.$$

The sentence

$$\forall x \forall y (\overline{R}_\leq(x, y) \to x < y \vee x = y)$$

is not a Horn formula. Show that it is not valid in $\Omega\mathbb{R}$: find $a, b \in \Omega\mathbb{R}$ such that

$$\Omega\mathbb{R} \models \overline{R}_\leq(\bar{a}, \bar{b}), \text{ but } \Omega\mathbb{R} \not\models \bar{a} < \bar{b}, \bar{a} = \bar{b}.$$

iii. Show that the following is not valid in $\Omega\mathbb{R}$:

$$\forall x (\overline{R}_\leq(\bar{0}, x) \wedge \neg x = \bar{0} \to \bar{0} < x)$$

(i.e., $0 \leq x$ and $x \neq 0$ do not imply $0 < x$).

9.20 Exercise. (Limits in $\mathbb{R}$ and $\Omega\mathbb{R}$) Show by constructing a counterexample that, if

$$\lim_{x \to a} f(x) = L \text{ in } \mathbb{R}$$

according to the non-Horn definition of limit in Examples 6.15.iv, it need not be the case that the non-Horn statement that L is the limit is true in $\Omega\mathbb{R}$. [Note: f must not be continuous at a.]

9.21 Exercise. (Induction in $^\Omega\mathbb{R}$) $\mathbb{N}$ is a unary relation in $\mathbb{R}$ and thus has a relation symbol $\overline{\mathbb{N}}$ in $L_\mathbb{R}$. The set of nonstandard natural numbers of $^\Omega\mathbb{R}$ is just

$$^\Omega\mathbb{N} = \{x \in {}^\Omega\mathbb{R} \,|\, {}^\Omega\mathbb{R} \vDash \overline{\mathbb{N}}(x)\}.$$

i. Show:

$$^\Omega\mathbb{N} = \{[x] \,|\, x_n \in \mathbb{N} \text{ for all but finitely many } n\}$$
$$= \{[x] \,|\, x_n \in \mathbb{N} \text{ for all } n \in \mathbb{N}\}.$$

ii. Since $\Omega \in {}^\Omega\mathbb{N}$ and $^\Omega\mathbb{N}$ is closed under addition of 1 [Quick proof: $\forall x(\overline{\mathbb{N}}(x) \to \overline{\mathbb{N}}(x+\overline{1}))$ is a Horn formula.], induction does not hold for $^\Omega\mathbb{N}$ for all properties. Show that it does not hold for all formulæ of the language $L_\mathbb{R}$. [Hint: Let $\varphi(x)$ assert x is not a 0-divisor.]

The following three exercises are only for those who know a little logic.

9.22 Exercise. (Horn Formulæ I) Let φ be a Horn formula of the form

$$Q_0 v_0 \ldots Q_{n-1} v_{n-1} (\psi_0 \wedge \ldots \wedge \psi_{m-1} \to \psi), \tag{101}$$

where each Q_i is a quantifier and ψ and each ψ_i is atomic. Show: $\neg\varphi$ is logically equivalent to a Horn formula. Conclude: If a sentence φ of $L_\mathbb{R}$ is of the form (101), then $\mathbb{R} \vDash \varphi$ iff $^\Omega\mathbb{R} \vDash \varphi$.

9.23 Exercise. (Horn Formulæ II) Let φ be a Horn formula,

$$Q_0 v_0 \ldots Q_{n-1} v_{n-1} \bigwedge (\psi_{ij_0} \wedge \ldots \wedge \psi_{ij_{m_0-1}} \to \psi_i),$$

and let $\overline{D}_0, \ldots, \overline{D}_{n-1}$ be unary relation symbols of the given language. Show:

$$Q_0 v_0 \in D_0 \ldots Q_{n-1} v_{n-1} \in D_{n-1} \bigwedge (\psi_{ij_0} \wedge \ldots \wedge \psi_{ij_{m_0-1}} \to \psi_i)$$

is logically equivalent to a Horn formula, where we use the abbreviations:

$$\forall v_i \in D_i \, \theta : \forall v_i (\overline{D}_i(v_i) \to \theta)$$
$$\exists v_i \in D_i \, \theta : \exists v_i (\overline{D}_i(v_i) \wedge \theta).$$

9.24 Exercise. (Cardinality in $^\Omega\mathbb{R}$) Recall from Chapter II that the cardinalities of $\mathbb{N}$ and $\mathbb{R}$ are denoted $\aleph_0$ and $\mathfrak{c}$, respectively.
i. Use the Cantor-Schröder-Bernstein Theorem to conclude that $^\Omega\mathbb{R}$ has cardinality $\mathfrak{c}$.
ii. Let $f : \mathbb{N} \to \mathbb{N} \times \mathbb{N}$ be one-to-one and onto. For $X \subseteq \mathbb{N}$, define $g_X : \mathbb{N} \to \mathbb{R}$ by

$$g_X(n) = \begin{cases} 1, & \text{for some } k, m, f(n) = \langle k, m\rangle \,\&\, k \in X \\ 0, & \text{otherwise.} \end{cases}$$

Show: The function $X \mapsto [g_X]$ maps the power set of $\mathbb{N}$ one-to-one into $\{N \in {}^\Omega\mathbb{N} \,|\, 0 \le N \le 1\}$. Conclude: There are as many integers "between" 0 and 1 in $^\Omega\mathbb{R}$ as there are real numbers.
iii. Show: There are $\mathfrak{c}$ infinite integers in $^\Omega\mathbb{R}$.
iv. Show: There are $\mathfrak{c}$ infinitesimals in $^\Omega\mathbb{R}$.

9.25 Exercise. (Trichotomy) Let $\mathcal{R}$ be a totally ordered ring extending $\mathbb{R}$ and satisfying Horn-preservation; i.e. assume

$$\mathcal{R} \vDash \forall x \forall y (x < y \lor x = y \lor y < x)$$

and, for any Horn sentence φ of $L_\mathbb{R}$,

$$\mathbb{R} \vDash \varphi \implies \mathcal{R} \vDash \varphi.$$

i.   Show: $\mathcal{R} \vDash \forall x \forall y \big((\neg x = y \to x \# y) \land (x \# y \to \neg x = y) \big)$.
ii. Show: $\mathcal{R} \vDash \forall x (\overline{0} < x \to |x| = x) \land \forall x (x < \overline{0} \to |x| = -x)$.
iii. Show: $\mathcal{R}$ is a field: $\mathcal{R} \vDash \forall x \exists y (\neg x = \overline{0} \to xy = \overline{1})$.
iv. Show: Every bounded element $x \in \mathcal{R}$ has a *standard part*, i.e. if $|x| < r$ for some $r \in \mathbb{R}$, then there is a real number $r' \in \mathbb{R}$ such that $x \approx r'$.
v.   Show: If the extension is proper (i.e. some $x \in \mathcal{R}$ for which $x \notin \mathbb{R}$ exists), then $\mathcal{R}$ has infinitesimal and infinitely large elements.
vi. Show: $\mathcal{R} \vDash \forall x \exists n (\overline{\mathbb{N}}(n) \land x < n)$. Conclude: If the extension is proper, then $\mathcal{R}$ has infinitely large natural numbers.

9.26 Exercise. (Cardinality in $\mathbb{H}$) Let $\mathbb{H}$ be a ring $\mathcal{R}$ as in the preceding exercise. Because $\mathcal{R}$ includes $\mathbb{R}$, its cardinality is at least $\mathfrak{c}$. It could be greater. The calculations of Exercise 9.24 do not apply here because

$$\{ x \in \mathcal{R} \mid \mathcal{R} \vDash \mathbb{N}(\overline{x}) \ \& \ 0 \le x \le 1 \} = \{0, 1\}.$$

However, $\mathcal{R}$ still has a lot of nonstandard elements.
i.   Let $q_0, q_1, \ldots$ be a sequence including all rational numbers. Show: Every real number is a limit point of this sequence.
ii. Choosing for each real $r \in \mathbb{R}$ an infinite integer $N_r \in \mathcal{R}$ such that $q_{N_r} \approx r$, conclude there to be at least $\mathfrak{c}$ infinite integers in $\mathcal{R}$.
iii. Show: There are at least $\mathfrak{c}$ infinitesimals in $\mathcal{R}$.

9.27 Exercise. (Undefinability in *$\mathbb{R}$) No infinite subset $X \subseteq \mathbb{R}$ is definable in *$\mathbb{R}$ by a formula $\varphi(v)$ of $L_\mathbb{R}(^*\mathbb{R})$. [Hint: Use the fact that induction,

$$\forall v_0 \ldots \forall v_{n-1} [\psi(\overline{0}) \land \forall v (\psi(v) \to \psi(v+\overline{1})) \to \forall v \psi(v)],$$

holds in *$\mathbb{R}$ to show that no formula ψ of $L_\mathbb{R}(^*\mathbb{R})$ defines $\mathbb{N}$ in *$\mathbb{R}$. Then consider any one-to-one function f mapping $\mathbb{N}$ into X.]

9.28 Exercise. (Infinitesimal Prolongation Lemma, aka Robinson's Sequential Lemma) Let $f : \mathbb{N} \times \mathbb{R} \to \mathbb{R}$. For each $x \in {}^*\mathbb{R}$, this defines a *sequence *$\mathbb{N} \to {}^*\mathbb{R}$: $n \mapsto {}^*f(n, x)$. Suppose for some fixed $x \in {}^*\mathbb{R}$, *$f(n, x)$ is infinitesimal for all finite n. Show: There is an infinite $M \in {}^*\mathbb{N}$ such that for all $N \in {}^*\mathbb{N}$, if $N < M$ then *$f(N, x)$ is infinitesimal. [A more intuitive way of expressing this is: If $a_0, a_1, \ldots$ is a sequence of hyperreals definable in *$\mathbb{R}$ such that a_n is infinitesimal for all finite n, then there is some infinite M such that a_N is infinitesimal for all infinite $N < M$. The result is quite useful in finding nonstandard replacements for some tricky limit arguments. As for the proof, note that one cannot reduce it to the preceding Exercise in the obvious way by showing that, if the result were false, one could define *$\mathbb{N}$ by

$$\varphi(n): \ \forall m \leq n \ \text{“}a_m \text{ is infinitesimal”}$$

for the simple reason that there is no formula in $L_\mathbb{R}(*\mathbb{R})$ defining the set of infinitesimals. Once must be a little more devious in choosing φ.]

9.29 Exercise. (Mean Value Theorem I) Let f be continuous on $[a,b]$ and differentiable on (a,b). Assume

$$h(x) = Af(x) + Bx + C,$$

for some $A, B, C \in \mathbb{R}$ with $A \neq 0$.
i. Show: $h(a) = h(b)$ iff

$$-\frac{B}{A} = \frac{f(b) - f(a)}{b - a}, \ \text{i.e., } B = -\frac{f(b) - f(a)}{b - a}A.$$

ii. Show: $h'(c) = 0 \ \Rightarrow \ f'(c) = -B/A$.
[Note that C can be chosen arbitrarily. Basically one chooses C so as to make the equation $h(b) = h(a)$ simple to recognise. What value of C works best for the choice $A = 1$? Repeat the Exercise using $h(x) = Af(x) + B(x - a) + C$.]

The algebraic exercise for finding the auxiliary function g reducing the proof of the Mean Value Theorem to that of Rolle's Theorem is fine for anyone who vaguely recalls the auxiliary function g to be a linear function of x and $f(x)$, but who doesn't recall the coefficients, or who guesses there may be some such function. A geometrically more intuitive approach can also be taken:

9.30 Exercise. (Mean Value Theorem II) Consider *Figure 8*, below, borrowed from the contribution of Yates to *Selected Papers in Calculus*[216].

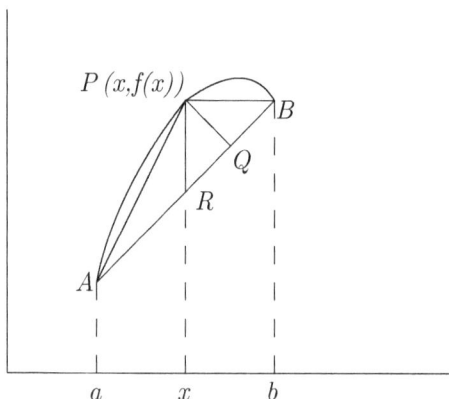

Figure 8

Intuitively, the tangent at P is parallel to AB when PQ is maximised, where PQ is perpendicular to AB.

---

[216] *Op. cit.*, p. 195.

i. Suppose the line AB has equation $Y = mX + \beta$.
   a. Show: The distance PQ is the absolute value of

$$h(x) = \frac{f(x) - mx - \beta}{\sqrt{m^2 + 1}}. \qquad (102)$$

   b. What is m?
   c. Show: $h(b) = h(a)$.
   d. Let $h'(c) = 0$ and calculate $f'(c)$.
ii. PQ is maximised when the area of the triangle ABP is maximised. The area of the triangle is

$$A(x) = \frac{1}{2} \begin{vmatrix} x & f(x) & 1 \\ a & f(a) & 1 \\ b & f(b) & 1 \end{vmatrix}.$$

Expand $A(x)$ and carry out the reduction of the Mean Value Theorem to Rolle's Theorem using it.
iii. Show: The angle $\angle PRQ$ does not depend on the choice of $P \neq A, B$ on the curve. Hence all the triangles PRQ are similar, and PQ is maximised when PR is maximised. Find an expression for the length of PR.

For the next three exercises, it is desired that the reader give nonstandard proofs based on the relation between $\mathbb{R}$ and $^*\mathbb{R}$.

9.31 Exercise. (Intermediate Value Theorem) Prove the Intermediate Value Theorem: If $f : [a, b] \to \mathbb{R}$ is continuous and $f(a) < 0 < f(b)$, then for some $c \in (a, b)$, $f(c) = 0$. [Hint. Consider the proof of the Extreme Value Theorem.]

9.32 Exercise. (Uniform Continuity) A function $f : D \to \mathbb{R}$ is continuous on a domain $D \subseteq \mathbb{R}$ if it is continuous at all points of D:

$$\mathbb{R} \vDash \forall x \in D \, \forall \epsilon > \overline{0} \, \exists \delta > \overline{0} \, \forall y \in D \big(|x - y| < \delta \to |\overline{f}(x) - \overline{f}(y)| < \epsilon \big).$$

f is *uniformly continuous* on D if the postulated δ depends only on ϵ and not on x:

$$\mathbb{R} \vDash \forall \epsilon > \overline{0} \, \exists \delta > \overline{0} \, \forall x \in D \, \forall y \in D \big(|x - y| < \delta \to |\overline{f}(x) - \overline{f}(y)| < \epsilon \big).$$

i. Show: f is uniformly continuous on D iff, for all $x, y \in {}^*D$,

$$x \approx y \;\Rightarrow\; {}^*f(x) \approx {}^*f(y).$$

ii. Show: If f is continuous on a closed, bounded interval $[a, b]$, then f is uniformly continuous on $[a, b]$.
iii. Show: The function $f(x) = 1/x$ is continuous, but not uniformly continuous on $(0, 1)$.

9.33 Exercise. (Uniform Convergence) Let I be a closed, bounded interval $[a, b]$. A sequence $f_0, f_1, f_2, \ldots$ of functions with domain I *converges pointwise* to a function f with domain I if, for every $x \in I$, the sequence $f_0(x), f_1(x), f_2(x), \ldots$ converges to $f(x)$:

$$\mathbb{R} \vDash \forall x \in I \, \forall \epsilon > \overline{0} \, \exists n_0 \in \mathbb{N} \, \forall n \in \mathbb{N}\big(n > n_0 \to |\overline{f}_n(x) - \overline{f}(x)| < \epsilon\big).$$

The convergence is *uniform* if n_0 does not depend on x:

$$\mathbb{R} \vDash \forall \epsilon > \overline{0} \, \exists n_0 \in \mathbb{N} \, \forall n \in \mathbb{N} \, \forall x \in I \, \big(n > n_0 \to |\overline{f}_n(x) - \overline{f}(x)| < \epsilon\big).$$

i. Show: $f_0, f_1, f_2, \ldots$ converges uniformly to f on I iff, for all $x \in {}^*I$ and all infinite $N \in {}^*\mathbb{N}$, ${}^*f_N(x) \approx f(x)$.

ii. Show: If $f_0, f_1, f_2, \ldots$ are continuous and converge uniformly to f on I, then f is continuous on I.

[After giving a nonstandard proof, the reader might like to consult the standard proof given in section 8 (Lemma 8.4).]

9.34 Exercise. (Cauchy's Derivative) Define, *á la* Cauchy, a function to be differentiable in an interval I if for each $x \in {}^*I$ there is some finite number y such that for all infinitesimals $\alpha \neq 0$ such that $x + \alpha \in {}^*I$

$$\frac{{}^*f(x + \alpha) - {}^*f(x)}{\alpha} \approx y. \tag{103}$$

i. Show that if f is differentiable in I in Cauchy's sense then the function

$$f'(x) = st\left(\frac{{}^*f(x + \alpha) - {}^*f(x)}{\alpha}\right), \text{ for } x \in I,$$

is continuous.

ii. Show this is not the case for ordinary differentiability where (103) is assumed only for standard $x \in I$.

[Hint: i. Let $x \in I$ and $x \neq 0$ be infinitesimal such that $x + \alpha \in {}^*I$. Compare the difference quotient (103) and

$$\frac{{}^*f(x + \alpha + \beta) - {}^*f(x + \alpha)}{\beta}$$

for $\beta = -\alpha$ to show ${}^*f(x + \alpha) - {}^*f(x) \approx 0$.]

9.35 Exercise. (Cauchy Inequality) Consider $\mathbb{R}^n$.

i. Prove the Cauchy inequality for $n = 1$: for all $a, b \in \mathbb{R}, |2ab| \leq a^2 + b^2$.

ii. Prove the Cauchy inequality for $n = 2$: for all $(a_1, b_1), (a_2, b_2) \in \mathbb{R}^2$,

$$(a_1 b_1 + a_2 b_2)^2 \leq (a_1^2 + a_2^2)(b_1^2 + b_2^2).$$

iii. Prove the Cauchy inequality on $\mathbb{R}^n$ by induction on n.

9.36 Exercise. (Product Metric) Let $(X, d_1), (Y, d_2)$ be metric spaces. Prove that one can define a distance function d_3 on $X \times Y$ by

$$d_3\big((x_1, y_1), (x_2, y_2)\big) = \sqrt{d_1(x_1, x_2)^2 + d_2(y_1, y_2)^2}.$$

9.37 Exercise. (Inner Product Spaces) Let V be a vector space over the reals. A function $\cdot : V \times V \to \mathbb{R}$ is called an *inner product* if it satisfies

a. for all $x \in V$, $x \cdot x \geq 0$ and $x \cdot x = 0$ iff $x = 0$;

b. for all $x, y \in V$, $x \cdot y = y \cdot x$;

c. for all $x, y, z \in V$, $x \cdot (y + z) = (x \cdot y) + (x \cdot z)$; and

d. for all $r \in \mathbb{R}$ and all $x, y \in V$, $(rx) \cdot y = r(x \cdot y)$.

Let $\cdot$ be an inner product on V.

i.   Show:

   a. for all $x, y, z \in V$, $(x + y) \cdot z = (x \cdot z) + (y \cdot z)$;

   b. for all $r \in \mathbb{R}$ and all $x, y \in V$, $x \cdot (ry) = r(x \cdot y)$.

Define the *norm* on V by

$$\|x\| = \sqrt{x \cdot x}.$$

ii. Show the following for $r \in \mathbb{R}$ and $x, y \in V$:

   a. $\|x\| \geq 0$ and $\|x\| = 0$ iff $x = 0$;

   b. $\|rx\| = |r| \cdot \|x\|$.

iii. Prove the Cauchy inequality: for all $x, y \in V$,

$$|x \cdot y| \leq \|x\| \cdot \|y\|.$$

iv. Prove the Triangle Inequality: for all $x, y \in V$,

$$\|x + y\| \leq \|x\| + \|y\|.$$

9.38 Exercise. (Bunyakovsky Inequality) Prove the Bunyakovsky Inequality: for all $f, g \in C[0, 1]$,

$$\left| \int_0^1 f(x)g(x)dx \right| \leq \left(\int_0^1 f(x)^2 dx \right)^{\frac{1}{2}} \left(\int_0^1 g(x)^2 dx \right)^{\frac{1}{2}}.$$

9.39 Exercise. (Normed Vector Spaces) Define a vector space V over $\mathbb{R}$ to be a *normed vector space* if it has a norm $\| \cdot \|$ satisfying the properties of Exercise 9.37.ii and 9.37.iv. Prove that the function d on $V \times V$ defined by

$$d(x, y) = \|x - y\|$$

is a distance function on V. If V is not complete, show how to define $+$ and $\| \cdot \|$ on its completion $\overline{V}$ to make $\overline{V}$ a normed vector space.

9.40 Exercise. (Uniform Limit Theorem) Show $C[0, 1]$ is a complete metric space under $\| \cdot \|_\infty$. Prove that the sequence $f_0, f_1, \ldots$ of (88) has a continuous limit.

9.41 Exercise. (A Bounded Norm) Define, for all $x \in X$ with X either $\mathbb{Q}$ or $\mathbb{R}$,

$$\|x\|_B = \frac{|x|}{1 + |x|}.$$

a. Show: $\|x\|_B \geq 0$ with equality only at $x = 0$;

b. Show: $|x| > |y| \Rightarrow \|x\|_B > \|y\|_B$ (i.e., show the function $f(x) = \frac{x}{1+x}$ is strictly increasing for nonnegative reals);

c. Show: $||x + y||_B \le ||x||_B + ||y||_B$ by considering the cases in which $||x + y||_B$ is or is not the maximum of $\{|x|, |y|, |x + y|\}$;

d. Show: $||x||_B < 1$ for all $x \in X$;

e. Show: $||xy||_B$ is not generally equal to $||x||_B \cdot ||y||_B$.

9.42 Exercise. (A Bounded Metric) Define for $X = \mathbb{Q}$ or $\mathbb{R}$,

$$d_B(x, y) = ||x - y||_B = \frac{|x - y|}{1 + |x - y|}.$$

i. Let $\iota : X \to X$ be the identity map $\iota(x) = x$. Show that

$$\iota : (X, d) \to (X, d_B)$$

is continuous, where d is the usual distance function $d(x, y) = |x - y|$. Show that

$$\iota : (X, d_B) \to (X, d)$$

is likewise continuous.

ii. What is the metric completion of $(\mathbb{Q}, d_B)$?

9.43 Exercise. (Another Bounded Metric) Repeat the previous exercise for

$$d_B(x, y) = \min\{1, |x - y|\}.$$

9.44 Exercise. (Norms in $\mathbb{Q}_g$) Let $g \ge 2$ be given.

i.   Show: for $q \in \mathbb{Q}$, $||q||_g = d_g(0, x)$.

ii. For $x \in \mathbb{Q}_g$, define $||x||_g = \overline{d}_g(0, x)$. Show that for all $x, y \in \mathbb{Q}_g$:
 a. $||x||_g \ge 0$, with equality only for $x = 0$;
 b. $||x + y||_g \le \max\{||x||_g, ||y||_g\} \le ||x||_g + ||y||_g$;
 c. Show: If g is a prime p, then $||x \cdot y||_p = ||x||_p \cdot ||y||_p$. iii. Let $x \in \mathbb{Q}_g$ and choose a Cauchy sequence $a : a_0, a_1, a_2, \dots$ of rational numbers converging to x. Show:

$$||x||_g = \lim_{n \to \infty} ||a_n||_g.$$

iv. For $x, y \in \mathbb{Q}_g$, show: $||xy||_g \le ||x||_g \cdot ||y||_g$.

v. Show: $\{||q||_g \,|\, q \in \mathbb{Q}\} = \{0\} \cup \{g^k \,|\, k \in \mathbb{Z}\}$.

vi. Let $a : a_0, a_1, a_2, \dots$ be a sequence of rational numbers that is Cauchy convergent under d_g and let $x = \lim_{n \to \infty} a_n$ in $\mathbb{Q}_g$. Show: If $x \ne 0$, then the sequence of g-norms $||a_0||_g, ||a_1||_g, ||a_2||_g, \dots$ is eventually constant: there is an n_0 such that, for all $m, n > n_0, ||a_m||_g = ||a_n||_g = ||x||_g$. Conclude $\{||x||_g \,|\, x \in \mathbb{Q}_g\} = \{0\} \cup \{g^k \,|\, k \in \mathbb{Z}\}$.

9.45 Exercise. (Ultrametric Inequality in $\mathbb{Q}_g$) Let g be given.

i. Let $a_0, a_1, a_2, \dots$ be a sequence of elements of $\mathbb{Q}_g$. Show:

$$\sum_{n=0}^{\infty} a_n \text{ converges in } \mathbb{Q}_g \text{ iff } \lim_{n \to \infty} a_n = 0.$$

ii. Let $x \in \mathbb{Q}_g$ with $\|x\|_g < 1$. Show: The geometric progression $\sum x^k$ converges in $\mathbb{Q}_g$.

iii. Let $x \in \mathbb{Q}_g$ with $\|x\|_g < 1$. Show: $1 - x$ has a multiplicative inverse in $\mathbb{Q}_g$.

9.46 Exercise. (1/2) i. Show: $\|1/2\|_6 \neq 1/\|2\|_6$.

ii. Show: $\|1/2\|_2 = 2$. Conclude that

$$\sum_{k=0}^{\infty} \left(\frac{1}{2} \right)^k$$

does not converge in $\mathbb{Q}_2$.

iii. Show: If $\|x\|_g < 1$ and x has a multiplicative inverse x^{-1} in $\mathbb{Q}_g$, then $\|x^{-1}\|_g > 1$ and

$$\sum_{k=0}^{\infty} x^{-k}$$

does not converge in $\mathbb{Q}_g$.

9.47 Exercise. (Grandi's Sequence) Let $G : 1, -1, 1, -1, \ldots$ be the Grandi sequence of Chapter I.

i. Show that

$$\sum_{n=0}^{\infty} G_n = \sum_{n=0}^{\infty} (-1)^n$$

does not converge in $\mathbb{Q}_g$ for any $g \geq 2$.

ii. Define $S_{H,1,g}(a)$ for a sequence $a : a_0, a_1, a_2, \ldots$ to be the limit in $\mathbb{Q}_g$ of the sequence of averages,

$$s_n = \frac{a_0 + \ldots + a_n}{n+1},$$

provided this limit exists. Show: $S_{H,1,g}(G)$ does not exist for any $g \geq 2$.

iii. The rings $\mathbb{Q}_g$ are not ordered[217], but the norms of the elements are. We can thus define

$$S_{L,g}(a) = \lim_{x \to 1^-} \sum_{n=0}^{\infty} a_n x^n,$$

provided it exists, where we define $\lim_{x \to 1^-} f(x) = L$ iff

$$\forall \epsilon \in \mathbb{R}^+ \exists \delta \in \mathbb{R}^+ \forall x \in \mathbb{Q}_g \left(\|x\|_g < 1 \, \& \, \|x - 1\|_g < \delta \Rightarrow \|f(x) - L\|_g < \epsilon \right).$$

Show: $S_{L,g}(G) = \frac{1}{2}$.

---

[217] They, in fact, cannot be ordered. This seems to be a nontrivial fact. At least, the simplest proof I can come up with goes as follows: If $p \geq 5$ is prime, there are $p - 1$ $(p-1)$-th roots of unity in $\mathbb{Q}_p$, i.e. there are $r_1 < r_2 < \ldots < r_{p-1} \in \mathbb{Q}_p$ such that $(r_i)^{p-1} = 1$. Half of these numbers must be positive. But for $0 < r_i < r_j$ one would have $(r_i)^{p-1} < (r_j)^{p-1}$, contradicting the equation $(r_i)^{p-1} = (r_j)^{p-1} = 1$. Thus $\mathbb{Q}_p$ cannot be ordered. For g containing two distinct prime factors, it can be shown $\mathbb{Q}_g$ has 0-divisors, whence 0 would be the product of two positive numbers, which should be positive. This still doesn't cover $\mathbb{Q}_2$ or $\mathbb{Q}_3$ (or for their isomorphic copies $\mathbb{Q}_g$ for $g = 2^n$ or 3^n, respectively).

9.48 Exercise. ($\mathbb{Q}_p$ I) Let p be a prime number. Show that one can define $\|q\|_p$ more simply by:

$$\left\|\frac{m}{n}\right\|_p = \frac{1}{p^k} \text{ iff } \frac{m}{n} = p^k \cdot \frac{r}{s},$$

where p does not divide r or s.

9.49 Exercise. ($\mathbb{Q}_p$ II) Show: $\mathbb{Q}_p$ is a field. [Hint: If $x \in \mathbb{Q}_p$ is not 0, then $x = \lim_{n\to\infty} a_n$ for some sequence $a_0, a_1, a_2, \ldots$ of rational numbers. Show that there is an n_0 such that for all $n > n_0$, $a_n \neq 0$.]

9.50 Exercise. (Mini-monster). Recall from footnote 65 (p. 132, above) that the Intermediate Value Theorem was often taken as characteristic of continuity. Certainly continuity implies the Intermediate Value Theorem (*Exercise* 9.31, above). The converse is not true: Define a function f by

$$f(x) = \begin{cases} \sin\frac{1}{x}, & x \neq 0, \\ 0, & x = 0. \end{cases}$$

Show that f satisfies the Intermediate Value Theorem: for any $a < b$, if $f(a) \neq f(b)$ and d lies between $f(a)$ and $f(b)$, then $f(c) = d$ for some c between a and b. Show too that f is discontinuous at $x = 0$. Can you construct a function satisfying the Intermediate Value Theorem for which the points of disconitnuity are not isolated? dense? Etc.

IV

The Axiomatic Method

1 What is the Axiomatic Method?

There are two views of the axiomatic method that are polar opposites of each other. In the traditional view, one has some theory, science, discipline or what-have-you to axiomatise. The axioms chosen are to be evident truths of the subject to be studied, and truth is preserved outward by the derivation of theorems. In the more modern view, the axioms are not about anything in particular and truth is borne from the theorems to the axioms; so long as there are no contradictions among the theorems, they are true about something and thus the axioms are true of these objects whatever they may be.

In addition to the traditional/modern opposition, there is also a dichotomy of description and practice. The description of the axiomatic method is usually a normative idealisation written by someone who may or may not actually apply the axiomatic method and, should he do so, may or may not meet his own norms. To further complicate matters, terminology is not constant or even consistent.

The classic example of the application of the axiomatic method is Euclid's axiomatisation of geometry in *The Elements*. This work is divided into 13 books, the first of which begins with 23 *definitions*, 5 *postulates*, and 5 *common notions*. Successive books usually begin with additional definitions. The common notions are basic principles of a general, not specifically geometric character. For example, the third one reads, "If equals be added to equals, the wholes are equal", and the fifth, "The whole is greater than the part". The postulates are of two kinds. The first three postulate our ability to perform various constructions: "to draw a line from any point to any point"; "to produce a finite straight line continuously in a straight line"; and "to describe a circle with any centre and distance". The other two postulates are more in the nature of our modern axioms: one asserts all right angles are equal to one another, and the other is his famous parallel postulate. The definitions are a mixed bag. They can define terms ("1. A *point* is that which has no part", or "2. A *line* is breadthless length"), or they can make additional axiomatic assumptions ("3. The extremities of a line are points"). They can even do both ("17. A *diameter* of the circle is any straight line drawn through the centre and terminated in both directions by the circumference of the circle, and such a straight line also bisects the circle"). Starting from these, Euclid proceeds to derive numerous consequences, the truths of which are secured by the derivations from the truths of the definitions, postulates, and common notions.[1]

Euclid practised the axiomatic method, but did not describe it. Aristotle had earlier done that and I should quote him. However, I have at hand the description by one of Euclid's commentators, Proclus, and shall quote him instead.

> First of all then, to repeat what I said, it was incumbent on him to set apart the principles from their consequences; and this is just what Euclid does in practically every book besides setting forth at the outset of his whole treatise the common principles of the science. Next he divides them into hypotheses, postulates, and axioms, for these are all different from each other. Axiom, postulate, and hypothesis are not the same thing, as the inspired Aristotle somewhere says[2]. When a proposition that is accepted into the rank of first principles is something both known to the reader and credible in itself, such a proposition is an axiom: for example, that things equal to the same thing are equal to each other. When the student does not have a self-evident notion of the assertion proposed but nevertheless posits it and concedes the point to his teacher, such an assertion is a hypothesis. That a circle is a figure of such-and-such a sort we do not know by a common notion in advance of being taught, but upon hearing it we accept it without a demonstration. Whenever, on the other hand, the statement is unknown and nevertheless is taken as true without the student's conceding it, then, he says, we call it a postulate: for example, that all right angles are equal... In this way axiom, postulate, and hypothesis are distinguished according to Aristotle's teaching.[3]

The terminology is not fixed and Proclus cites disagreement with other writers.[4] As to the basic principles as they appear in Euclid, he makes the identifica-

---

[1] Euclid did not do a perfect job of it. Already in Greek times commentators filled their commentaries with proofs of neglected cases, and in the 19th century, with the renewed attention to rigour, Euclid's unmentioned assumptions surfaced and new axiomatisations appeared. The culmination of all this activity was the publication in 1899 by David Hilbert of his *Grundlagen der Geometrie* [*Foundations of geometry*], a slim volume of his lectures on the subject.

[2] The reference is to a passage in Aristotle's *Posterior Analytics*, numbered 76a31 - 77a4 in the standard edition. 76a30 - 77a2 can be found in both Greek and English on pp. 418 - 423 of Ivor Thomas, *Greek Mathematical Works I: Thales to Euclid*, Harvard University Press, Cambridge (Mass.) and William Heinemann Ltd., London, 1939, with various reprints.

[3] Glenn R. Morrow (trans.), *Proclus: A Commentary on the First Book of Euclid's Elements*, Princeton University Press, Princeton, 1970, pp. 62 - 63.

[4] *Ibid.*, p. 63.

tions[5]:

Euclid	definition	postulate	common notion
Aristotle/Proclus	hypothesis	postulate	axiom.

The flow of truth from basic principles to theorems is not announced all at once by Proclus, but it is there:

> Principles must always be superior to their consequences in being simple, indemonstrable, and evident of themselves.[6]

> Science as a whole has two parts: in one it occupies itself with immediate premises, while in the other it treats systematically the things that can be demonstrated or constructed from these first principles, or in general are consequences of them.[7]

> The proof draws the proposed inference by reasoning scientifically from the propositions that have been admitted. The conclusion reverts to the enunciation, confirming what has been proved.[8]

This normative description of the axiomatic method has been refined over the ages, but had not essentially changed when David Hilbert published his nonconforming axiomatisation of geometry in 1899. Friedrich Ludwig Gottlob Frege immediately reacted and attempted to lay down the law in a letter to Hilbert of 27 December of that year:

> I should like to divide up the totality of mathematical propositions into definitions and all the remaining propositions (axioms, fundamental laws, theorems). Every definition contains a sign (an expression, a word) which had no meaning before and which is first given a meaning by the definition. Once this has been done, the definition can be turned into a self-evident proposition which can be used like an axiom. But we must not lose sight of the fact that a definition does not assert anything but lays down something... It is very essential for the strictness of mathematical investigations that the difference between definitions and all other propositions be observed in all strictness. The other propositions (axioms, fundamental laws, theorems) must not contain a word or sign whose sense and meaning, or whose contribution to an expression of a thought, was not already completely laid down, so that there is no doubt about the sense of the proposition and the thought it expresses. The only question can be whether this thought is true and what its truth rests on. Thus axioms and theorems can never try to lay down the meaning of a sign or word that occurs in them,

---

[5] *Ibid.*, p. 140 for the equation of definitions and hypotheses. Not being a linguist, I cannot rule out the possibility that Euclid and Proclus used the same words. I only note that this table is useful for one, like me, who relies on translations by different individuals.

[6] *Ibid.*, p. 141.

[7] *Ibid.*, p. 157.

[8] *Ibid.*, p. 159.

but it must already be laid down... I call axioms propositions that are true but are not proved because our knowledge of them flows from a source very different from the logical source, a source which might be called spatial intuition. From the truth of the axioms it follows that they do not contradict one another. There is therefore no need for a further proof.[9]

The second ellipsis concerned elucidatory propositions, which Frege considered as not properly belonging to mathematics itself. They are like definitions, but incomplete. I would think Euclid's definition of a point or a line to be good examples, unlike the first half of his definition of a diameter of a circle, which completely characterises the notion.

Frege's letter was a response to and criticism of Hilbert's axiomatisation of geometry in which he did not define his concepts in the Euclidean-Fregean fashion. Hilbert defined his terms implicitly through the axioms they satisfied and Frege disapproved. Nowadays, people say Frege was right if by "definition" one means *explicit* definition, and Hilbert was right if one accepts *implicit* definitions. The whole dispute is unimportant; the relevant point here is that Frege advocated the traditional view of axiomatics in a refined form as a normative idealisation, while Hilbert responded with a descriptive realisation:

> I was very much interested in your sentence: 'From the truth of the axioms it follows that they do not contradict one another', because for as long as I have been thinking, writing, lecturing about these things, I have been saying the exact reverse: If the arbitrarily given axioms do not contradict one another, then they are true, and the things defined by the axioms exist. This for me is the criterion of truth and existence.[10]

This backwards flow of truth is nicely illustrated in the history we have already discussed. When Cavalieri encountered a contradiction with his Principle (cf. page 106, above), he reformulated the Principle in such a way that it no longer yielded a contradiction, but gave the correct numerical results. History would repeat itself with Frege and Russell.

Recall the split by Proclus of science into two phases, a first phase occupied with the enumeration of immediate premises, and a second with the systematic working out of consequences of these premises. Where does Leibniz fit into this? Leibniz used infinitesimals, but unlike some of his contemporaries, he did not believe they existed. They were *ideal elements*, convenient fictions that led to quicker and easier proofs. If the elements do not exist their basic properties are not true and truth cannot flow from them to their consequences. Rather, one accepts them because of their consequences. The situation is like theoretical physics: one accepts electrons not because one can see them, but because of their explanatory powers, i.e., because of their consequences.

Proclus was a commentator— something of a mathematical analogue of a film critic— not a serious mathematician. Frege was a mathematician, though

---

[9] Gottlob, Frege, *Philosophical and Mathematical Correspondence*, University of Chicago Press, Chicago, 1980, pp. 36 - 37.

[10] *Ibid.*, p. 42.

not a particularly gifted one. He is better known to philosophers, who value him more highly than mathematicians do. This may tell us something. In any event, Proclus and Frege occupy themselves with the first phase of science, not discussing how one decides which propositions are immediate, but in establishing nomenclature for their respective classifications of immediate premises. Hilbert was a working mathematician, the greatest of his generation. His descriptions of the axiomatic method were also normative, but in a different way. He did not attempt to set up a form the neophyte axiomatiser would have to fill in and submitt for approval, but described what was to be accomplished. One point he emphasised was rigour: the starting principles, or "immediate premises", had to offer a complete description of the science being axiomatised. All theorems had to follow from them and only from them; additional properties (of the sort Euclid had silently drawn upon) were not to be appealed to. In a card to Frege of 22 September 1900, he wrote:

> In my opinion, a concept can be fixed logically only by its relations to other concepts. These relations, formulated in certain statements, I call axioms, thus arriving at the view that axioms (perhaps together with propositions assigning names to concepts) are the definitions of the concepts. I did not think up this view because I had nothing better to do, but I found myself forced into it by the requirements of strictness in logical inference and in the logical constructions of a theory. I have become convinced that the more subtle parts of mathematics and the natural sciences can be treated with certainty only in this way: otherwise one is only going around in a circle.[11]

Hilbert apologised for sending only a card and not a full letter, citing his preparation of a lecture on partial differential equations in physics as an excuse. I am inclined to read this card as his dismissal of Frege, finding in the cited paragraph a lack of patience with the hair-splitting criticisms Frege had been wasting his time with. Indeed, the only subsequent item in the published correspondence is a thank-you note from Hilbert acknowledging receipt of the second volume of Frege's *Grundgesetze der Arithmetik* [*Fundamental laws of arithmetic*], in which Frege had announced the discovery by Bertrand Russell of an inconsistency within his theory. Hilbert says:

---

[11] *Ibid.*, p. 51.

Your example at the end of the book (p. 253) was known to us here[12]; I found other even more convincing contradictions as long as four or five years ago; they led me to the conviction that traditional logic is inadequate and that the theory of concept formation needs to be sharpened and refined. As I see it, the most important gap in the traditional structure of logic is the assumption made by all logicians and mathematicians up to now that a concept is already there if one can state of any object whether or not it falls under it. This does not seem adequate to me. What is decisive is the recognition that the axioms that define the concept are free from contradiction.[13]

We have, in fact, already quoted Hilbert to the effect that consistency was for him the criterion of truth and existence. Indeed, in his *Grundlagen der Geometrie*, after introducing his axioms for geometry, he proved their consistency by modelling them in the arithmetic of the reals. And we read Frege's deriding the necessity of this by writing, "From the truth of the axioms it follows that they do not contradict one another. There is therefore no need for a further proof." *If* the axioms are true, then indeed he is correct. Unfortunately, he had neglected to verify the truth of his own axioms.

Frege's error, for the reader who may not have seen it before, is easily described. He assumed the full *Comprehension Axiom*: the collection of all sets having a given property exists, i.e., if $P(x)$ is a property of sets, then $\{x\,|\,P(x)\}$ is a set. Bertrand Russell wrote to him in June of 1902 to inform him that if one took,

$$R = \{x\,|\,x \notin x\},$$

then $R \in R$ iff $R \notin R$ and either possibility led to a contradiction.

The first volume of Frege's book was already published and the second was in press. All he could do was add a note acknowledging the inconsistency of his theory and admitting his helplessness before it. The ensuing attempts to replace Frege's system by a correct one were completely modern. There was no attempt to determine the true basic principles of set theory; the goal was a *consistent* theory, yet one still strong enough to carry out Frege's programme of developing mathematics within it.

Earlier paradoxes in set theory had not drawn much attention. Cantor, on discovering there could be no set of all cardinals, decided that some collections, like that of all cardinal numbers or the "set of all sets", were simply too

---

[12] Ernst Zermelo who habilitated in Göttingen in 1899 had already discovered Russell's paradox. That the result was certainly no surprise to Hilbert can also be gleaned from his lecture to the International Congress of Mathematicians in Paris in 1900, where he offhandedly remarks that there are inconsistent sets like the set of all Cantor's alephs, i.e., the collection of all infinite cardinal numbers. In 1897 already Burali-Forti had published his paradox of the largest ordinal number, and before 1900 Hilbert had discovered his own paradox, as he announces in the next phrase of his letter to Frege. I refer the reader to Volker Peckhaus and Reinhard Kahle, "Hilbert's paradox", *Historia Mathematica* 29 (2002), pp. 157 - 175 for further details.

[13] Frege, *op. cit.*, pp. 51 - 52.

large to exist as completed totalities. Hilbert, who considered his own paradox to be purely mathematical, concluded the subtlety of such matters and maintained every axiomatic theory had to be accompanied by a consistency proof. In *Grundlagen der Geometrie*, for example, he accompanied his axiomatisation with a consistency proof obtained by modelling the geometric axioms in the arithmetic of the real numbers. This, of course, raised the question of the consistency of an axiomatic theory of the arithmetic of the real numbers, which question he placed second on his famous list of problems posed at the 1900 International Congress of Mathematicians in Paris. The publication of Russell's Paradox, i.e., Frege's acknowledgment of the inconsistency of his theory, drew greater attention and a number of responses.

The responses of mathematicians were generally mild. Hilbert saw consistency as a necessary and sufficient condition an axiomatic theory had to satisfy, and regarded this consistency as just another problem to solve. Richard Dedekind, in whose slim monograph *Was sind und was sollen die Zahlen?*[14] [*What are numbers and what should they be?*] the natural numbers were treated set theoretically, delayed publication of the third edition for several years because of concerns about the possibility of deriving a contradiction within his system.

The French mathematician Henri Poincaré, who was equally at home in philosophy and mathematics, took up the baton from another French mathematician, Jules Richard, and ran with the latter's explanation that the paradoxes were caused by *vicious circles* lurking dangerously in *impredicative* definitions. In *Science and Method* he explains Richard's paradox:

> The Richard antinomy[15] is as follows: Consider all the decimal numbers definable by a finite number of words; these decimal numbers form an aggregate E, and it is easy to see that this aggregate is countable, that is to say we can *number* the different decimal numbers of this assemblage from 1 to infinity. Suppose the numbering effected, and define a number N as follows: If the nth decimal of the nth number of assemblage E is
>
> $$0, 1, 2, 3, 4, 5, 6, 7, 8, 9$$
>
> the nth decimal of N shall be:
>
> $$1, 2, 3, 4, 5, 6, 7, 8, 1, 1$$

---

[14] Published in 1888. It can be found in English translation by Wooster Woodruff Beman as "The nature and meaning of numbers" in: R. Dedekind, *Essays on the Theory of Numbers*, Open Court Publishing Company, Chicago, 1901, and reprinted by Dover Publications, New York, 1963.

[15] "Antinomy" is an old Kantian term resurrected by philosophers of the day to apply to what we now generally call paradoxes.

As we see, N is not equal to the nth number of E, and as n is arbitrary, N does not appertain to E and yet N should belong to this assemblage since we have defined it with a finite number of words.[16]

A couple of pages later, Poincaré offers the clarification of the paradoxes:

It seems to me that the solution is contained in a letter of M. Richard of which I have spoken above, to be found in the *Revue générale des sciences* of June 30, 1905. After having set forth the antinomy we have called Richard's antinomy, he gives its explanation. Recall what has already been said of this antinomy. E is the aggregate of *all* the numbers definable by a finite number of words, *without introducing the notion of the aggregate E itself.* Else the definition of E would contain a vicious circle; we must not define E by the aggregate E itself.

Now we have defined N with a finite number of words, it is true, but with the aid of the notion of the aggregate E. And this is why N is not part of E. In the example selected by M. Richard, the conclusion presents itself with complete evidence and the evidence will appear still stronger on consulting the text of the letter itself. But the same explanation holds good for the other antinomies, as is easily verified. Thus *the definitions which should be regarded as not predicative are those which contain a vicious circle.* And the preceding examples sufficiently show what I mean by that.[17]

With nothing else to fall back on, this was pretty compelling stuff in the first decade of the 20th century. Indeed, in the 1930's Alfred Tarski made a study of the notion of truth in formal languages and showed that, under some minimal adequacy conditions, truth for a language cannot be defined within that language. However, it can be defined in a new formal language— which cannot define its own truth, etc.[18] With respect to Richard's Paradox, when we replace English by a formal language, we discover that the class E of definable numbers is not definable. For, a real number is definable iff it is the unique number about which some predicate is true— but truth itself is undefinable in the language and without a definition of truth we cannot define "definability" and thus we cannot define E. We can go up to the next language and define E

---

[16] Henri Poincaré, *The Foundations of Science: Science and Hypothesis; The Value of Science; Science and Method*, The Science Press, New York, 1913; 1924 reprint, pp. 477 - 478.

[17] *Ibid.*, pp. 480 - 481. Richard's letter explaining his paradox and its solution, appears in English translation in Jean van Heijenoort, *From Frege to Gödel; A Source Book in Mathematical Logic, 1879 - 1931*, Harvard University Press, Cambridge (Mass.), 1967, pp. 142 - 144.

[18] Alfred Tarski, "Der Wahrheitsbegriff in den formalisierten Sprachen", *Studia Philosophica* 1 (1936), pp. 261 - 405. An English translation by J.H. Woodger appears in: Alfred Tarski, *Logic, Semantics, Metamathematics; Papers from 1923 to 1938*, Oxford University Press, Oxford, 1956. The "Wahrheitsbegriff" paper was a revised translation of an earlier paper written in Polish; it was, however, the version that attracted international attention.

and N. N is indeed both undefinable (in the original language) and definable (in the second language), but there is no contradiction here, hence no paradox.

Having said this, I note that the insistence on predicativity, that one not allow definitions of sets in terms of themselves or collections they might belong to, is too draconian. In Chapter III I have more than once defined sets inductively— in defining number expressions in discussing Bolzano's theory of real numbers (III.4.13, 4.14, 4.15) and in defining terms, formulæ, and truth in a formal language in discussing nonstandard analysis (III.6.5, 6.7, 6.8, 6.11). Such inductive definitions are necessary and commonplace in mathematics. Cutting out all impredicative definitions would cripple mathematics, as Russell would soon learn.

As seventh wrangler in the 1893 mathematical Tripos in Cambridge, Russell was no mathematical slouch. He was, however, not a mathematician at heart, but a philosopher. He, like Frege, wanted to show mathematics to be a part of logic, and this greatly complicated matters. He could not simply develop an axiomatic theory that encompassed all known mathematics, but had to include a theory of propositions and "propositional functions" as well. This necessitated avoiding not only the set theoretic paradoxes, but also linguistic ones, such as "I am lying" or "This sentence is false". Russell's final solution, the *ramified theory of types*, was a monstrosity nobody liked.

The *simple theory of types* is easy to describe. The universe of discourse is stratified into a hierarchy of numerically indexed levels, called *types*. The base, i.e., type 0, consists of a collection of *individuals* that are generally distinct from sets. The type 0 objects could, for example, be natural numbers, rational numbers, or real numbers. (Russell, of course, wanted to derive them logically, so he did not take the type 0 objects to be numbers, which he would construct set theoretically.) The type 1 objects are sets of type 0 objects, and generally type $n+1$ objects are sets of type n objects.

If one starts with the natural numbers as one's type 0 objects, one has sets of natural numbers as type 1 objects, and sets of sets of natural numbers as type 2 objects. This includes the pairs of natural numbers if we use the familiar set theoretic encoding[19] thereof:

$$\langle m, n \rangle = \{\{m\}, \{m, n\}\}.$$

---

[19] This definition was given by Kazimierz Kuratowski in 1921. Similar, but slightly less pleasing, definitions by Felix Hausdorff and Norbert Wiener were earlier given in 1914:

Wiener: $\langle x, y \rangle = \{\{\emptyset, \{x\}\}, \{\{y\}\}\}$

Hausdorff: $\langle x, y \rangle = \{\{x, 1\}, \{y, 2\}\}$,

where $1, 2$ are distinct objects differing from x, y. For natural numbers, one can also use a numerical encoding of ordered pairs such as the one given by Georg Cantor:

$$\langle x, y \rangle = \frac{(x+y)^2 + 3x + y}{2}.$$

Thus the integers, being viewed as in Chapter III as sets of ordered pairs (equivalence classes of the pairs) are of type 3. Again, sets of integers are type 4, pairs of type 5, and rational numbers (equivalence classes of the pairs) are objects of type 6. Sets of rationals are type 7 objects. Using Dedekind cuts, depending on one's choice, real numbers can come in at level 8 (identifying real numbers with their defining nonprincipal upper cuts, i.e., sets of rational numbers), 10 (identifying real numbers with pairs of upper and lower cuts, with no upper cuts being principal), or 11 (using equivalence classes of pairs of upper and lower cuts with no restriction on where the principal generator can lie).

The stratification into types resolved the set theoretic paradoxes. For, the form assumed by the Comprehension Axiom is simply that, for any type n and any property $P(x)$ of type n objects, the collection of all type n objects having the property P exists— as a type $n+1$ object. Superscripting variables by their types, this can be written symbolically:

$$\exists x^{n+1} \left(x^{n+1} = \{x^n \mid P(x^n)\} \right).$$

The $n+1$-st level Russell set,

$$R^{n+1} = \{x^n \mid x^n \notin x^n\},$$

is a set of type $n+1$; it does not contain itself, and there is no contradiction.

Starting with the natural numbers at type 0 and the basic arithmetical operations on them, the simple theory of types can be used to develop a large portion of mathematics. Avoiding the linguistic paradoxes, however, moved Russell to stratify the properties themselves into numerically indexed orders. One still starts with individuals of type 0. But now, in place of sets of types $0, 1, \ldots$, there are properties ("propositional functions" in Russell) of orders $1, 2, \ldots$ A first-order property is one in which all bound variables are individual variables. For example, if one's individuals are natural numbers,

$$P(n): \quad \exists m (n = 2m)$$

is the property of being even. Second-order properties allow quantification over first-order properties in their expression, e.g.

$$P^2(n): \quad \exists P^1 \forall m \left(P^1(m) \leftrightarrow \exists k (m = k^n) \right)$$

is a second-order property asserting the existence of a property singling out those numbers that are nth powers. Note that, identifying a property with the set of individuals possessing the property, we see that the collection of sets of type 1 splits into different orders. In fact, there is nothing to guarantee that type 1 exists as a completed set, i.e., property.

The Comprehension Axiom is replaced by the construction of formulæ expressing properties: If you write down a formula obeying the obvious order restrictions (e.g., you don't predicate a lower order property holding of a higher order one), the property described by the formula exists.

My description is over-simplified and one can readily produce elementary questions my description cannot supply the answers to. I recommend not worrying about this for the simple reason that nobody uses the ramified theory today. I offer the oversimplified description merely to give a taste of how Russell's theory goes, not to actually do anything with it. It is background, not subject matter for the present discussion. That said, I note that there are more detailed descriptions available in the literature.[20] The interest here in the ramified theory is that its Comprehension Axiom has been so severely restricted that one cannot derive within the theory some basic results of mathematics. Indeed, it is not even immediately clear how to state some of these results.

One sees the problem when one tries to prove the Least Upper Bound Principle. If X is a bounded set of real numbers, one defines a lower cut associated with X by

$$L = \{q \in \mathbb{Q} \mid \exists x \in X (q \leq x)\}. \tag{1}$$

If X is of order $i + 1$, the variable x ranging over X is of order i and we see L is of order $i + 1$. It is a real number of higher order than that of the reals of X. There is no obvious way of bringing the order down; and, indeed, there is no reason to suppose we can prove all real numbers to have a fixed order. There may well be real numbers (sets of rationals) of every order and thus no totality of all real numbers, no totality of all sets of real numbers, and no single statement asserting all bounded sets of real numbers possess least upper bounds. One can only assert schematically that, for each order i, every bounded set of order $i + 1$ of real numbers possesses an upper bound of order $i + 1$ that is less than or equal to every upper bound of order $i + 1$.

To get around this, Russell introduces the *Axiom of Reducibility*, an axiom schema asserting for any set X of a given level i, if a subset $Y \subseteq X$ can be defined at a level $j > i$, then there is a property of level $i + 1$ defining Y. Thus, if the set of rational numbers has order k, then every set of rational numbers— including L given by (1) above, has order $k + 1$, and the proof of the Least Upper Bound Principle goes through. (Also, the set of real numbers exists.)

But, is the Axiom of Reducibility true? I quote Bertrand Russell.

---

[20] The ramified theory is introduced in Bertrand Russell, "Mathematical logic as based on the theory of types", *American Journal of Mathematics* 30 (1908), pp. 222 - 262; reprinted in J. van Heijenoort, *op. cit.*, pp. 150 - 182. It forms the basis of *Principia Mathematica*, Russell's monumental 3 volume joint work with Alfred North Whitehead. The first edition was published in the years 1910 - 1913 by Cambridge University Press, Cambridge, and the second edition by the same publisher appeared in the years 1925 - 1927. In my student days both editions were prohibitively expensive, but a paperback, *Principia Mathematica to *56*, was published by Cambridge University Press in 1962. This volume covers the first 56 sections of the second edition, which is 50+ sections more than anyone would have the patience to slog through. Instead of reading Russell, I would recommend a commentary written by a very patient soul. In my personal library I have Irving M. Copi, *The Theory of Types*, Routledge & Kegan Paul, London, 1971. It is much more readable. The reader might also want to take a peek at the final chapter of David Hilbert and Wilhelm Ackermann, *Grundzüge der theoretischen Logik*, Springer Verlag, Berlin, 1928.

VII. *Reasons for Accepting the Axiom of Reducibility*

That the axiom of reducibility is self-evident is a proposition which can hardly be maintained. But in fact self-evidence is never more than a part of the reason for accepting an axiom, and is never indispensable. The reason for accepting an axiom, as for accepting any other proposition, is always largely inductive, namely that many propositions which are nearly indubitable can be deduced from it, and that no equally plausible way is known by which these propositions could be true if the axiom were false, and nothing which is probably false can be deduced from it. If the axiom is apparently self-evident, that only means, practically, that it is nearly indubitable; for things have been thought to be self-evident and have yet turned out to be false. And if the axiom itself is nearly indubitable, that merely adds to the inductive evidence derived from the fact that its consequences are nearly indubitable: it does not provide new evidence of a radically different kind. Infallibility is never attainable, and therefore some element of doubt should always attach to every axiom and to all its consequences. In formal logic, the element of doubt is less than in most sciences, but it is not absent, as appears from the fact that the paradoxes followed from premises which were not previously known to require limitations. . . [21]

Russell's reasoning seems a bit specious. It reminds me of the passage in Goethe's *Faust* where Mephistopheles explains to Faust that it is perfectly alright for the two of them to give a deposition on a matter of which they are (or, at least, Faust is) ignorant:

MEPHISTOPHELES

We've but to take a valid oath to tell
That her dead husband's limbs repose
At Padua in sacred ground.

FAUST

That's clever! First we'll go there, I suppose?

MEPHISTOPHELES

Sancta simplicitas! No need for that!
Swear, without knowing more than you do.

FAUST

Unless you devise a better plan— I'm through!

MEPHISTOPHELES

O saintly fellow! There you go!
Is this the first time that your task
Has been to bear false witness, may I ask?

---

[21] *Principia Mathematica to *56, op. cit.*, p. 59.

With brazen brow and dauntless breast,
Did you not expound with power and zest
Meanings of God and of the World— as well
As of all creatures that within it dwell?
Of Man, of all that stirs his heart and head,
While if you probe the matter to the core,
You must admit you knew but little more
Than of this Schwerdtlein's death, when all's been said!

FAUST

You are and remain a liar and a sophist too.[22]

The scene ends with Faust capitulating:

But come, I'm sick of all this useless chatter!
Since I must, I'll own you're right in this matter.[23]

And, of course, Faust is not capitulating because Mephistopheles is right in the matter, but because he must go along with Mephistopheles as a means to an end, just as Russell must assume some dubious axiom to derive basic mathematical theorems.

Rationalisation or not, Russell's bow to utilitarian necessity was a sound decision, and accords well with modern practice. Using the ramified type theory augmented by the Axiom of Reducibility, Russell and Whitehead were able to carry out the formalisation of a huge chunk of mathematics in a monumental case study in such formalisation within an apparently consistent system. With the introduction of the empirical consideration, however, Russell failed in his attempt to reduce mathematics to logic.

Russell's system is a monstrosity. The type structure already repels. The American philosopher Willard van Orman Quine wrote:

But the theory of types has unnatural and inconvenient consequences. Because the theory allows a class to have members only of uniform type, the universal class V gives way to an infinite series of quasi-universal classes, one for each type. The negation $-x$ ceases to comprise all nonmembers of x, and comes to comprise only those nonmembers of x which are next lower in type than x. Even the null class Λ gives way to an infinite series of null classes... Even arithmetic, when introduced by definitions on the basis of logic, proves to be subject to the same reduplication. Thus the numbers cease to be unique; a new 0 appears for each type, likewise a new 1, and so on, just as in the case of V and Λ. Not only are all these cleavages and reduplications intuitively repugnant, but they call continually for more or less elaborate technical maneuvers by way of restoring severed connections.[24]

---

[22] Johann Wolfgang von Goethe, *Faust*, Jonathan Cape and Harrison Smith, Inc., New York, 1930, translation by Alice Raphæl, pp. 163 - 164.

[23] *Ibid.*, p. 165.

[24] Willard van Orman Quine, "New foundations for mathematical logic", *American Mathematical Monthly* 44 (1937), pp. 70 - 80; reprinted with corrections and sup-

Quine restated this more simply in his autobiography:

> ... each of Russell's types had had its own empty class, its own universe class, its own number 1, its own 2 and so on... A further fault of the theory of types was that an infinitude of individuals had to be postulated to make the construction of arithmetic[25] come out right; and this assumption seemed to encroach on questions of fact beyond logic and mathematics.[26]

Quine's quoted criticism covers objections to the type theoretic framework and do not yet explicitly[27] mention the most criticised aspect of Russell's theory, namely the Axiom of Reducibility. For this, a couple of quotes from the logician Frank Plumpton Ramsey may suffice here. In a paper on the foundations of mathematics of 1925, Ramsey critiqued the *Principia* and described what he considered a more acceptable alternative. His criticism begins with the introduction of the notion of a *tautology*, a proposition that is always true regardless of the truth values of its component parts and distinguished tautologies and contradictory statements from *genuine propositions*, which actually assert something. He went on:

> So, as the process of deduction is such that from tautologies only tautologies follow, were it not for one blemish the whole structure would consist of tautologies. The blemish is of course the Axiom of Reducibility, which is... a genuine proposition, whose truth or falsity is a matter of brute fact, not of logic.[28]

He added:

> One thing is, however, clear: that mathematics does not consist of genuine propositions or assertions of fact which could be based on inductive

---

plement in: Quine, *From a Logical Point of View*, Harper Torchbooks, New York, 1963. The paragraph cited appears on pp. 78 - 79 of the original and on pp. 91 - 92 of the reprint.

[25] I.e., Russell had to assume the collection of type 0 objects formed an infinite set. Logicians of the period defined natural numbers to be the cardinal numbers of finite sets, a cardinal number being an equivalence class of sets of equal cardinality. If one does not assume infinitely many individuals, there are only finitely many sets of each type, hence no infinite totalities— in particular, no set of all natural numbers.

[26] Willard van Orman Quine, *The Time of My Life; An Autobiography*, MIT Press, Cambridge (Mass.), 1985, p. 126.

[27] The "more or less elaborate technical maneuvers" referred to at the end of the first quote obviously include the Axiom of Reducibility, so criticism of this axiom is implicit.

[28] Frank Plumpton Ramsey, "The foundations of mathematics", in: R.B. Braithwaite (ed.), *The Foundations of Mathematics and other Logical Essays by Frank Plumpton Ramsey*, Littlefield, Adams & Co., Totowa (New Jersey), 1965, pp. 11 - 12. The paper first appeared in the *Proceedings of the London Mathematical Society* (2), 25, part 5, pp. 338 - 384.

evidence as it was proposed to base the Axiom of Reducibility, but is in some sense necessary or tautologous.[29]

Ramsey, like Russell and Frege, wanted to reduce mathematics to logic by setting up one Master System in which to work. Russell replaced Frege's system because it had been inconsistent. To do so, he had to formally add an axiom of dubious truth. Ramsey replaced Russell's system because a Master System could not have dubious axioms. Unlike philosophers, mathematicians had more modest goals. David Hilbert, whose name would become more closely linked to the axiomatic method than anyone— with the possible exception of Euclid— had this to say in introducing the second problem of his famous list of problems at the International Congress of Mathematicians in Paris in 1900:

> When we are engaged in investigating the foundations of a science, we must set up a system of axioms which contains an exact and complete description of the relations subsisting between the elementary ideas of that science. The axioms so set up are at the same time the definitions of those elementary ideas: and no statement within the realm of the science whose foundation we are testing is held to be correct unless it can be derived from the axioms by means of a finite number of logical steps.[30]

Note the adjectives "a", "that", and "the" attached to the word "science". He exemplified the narrowness of his scope in referring to his own axiomatisations of Euclidean geometry and the arithmetic of the real numbers as examples. Seventeen years later he would give a talk in Zürich on the axiomatic method, a popular exposition in which he further illustrated the modesty of his goals in the opening remarks:

> ... The essence of these relations and the ground of their fertility will be made most distinct, I believe, if I sketch to you that general method of inquiry which appears to grow more and more significant in modern mathematics; the *axiomatic method*, I mean.
> If we collate the facts of a specific field of more or less comprehensive knowledge, then we shortly observe that these facts can be set in or- der. This order occurs invariably with the aid of certain *framework of concepts* such that there exists correspondence between the individual objects in the field of knowledge and a logical relation among concepts. The framework of concepts is nothing but the *theory* of the field of knowledge.
> The geometric facts thus order themselves into a geometry, the arith- metic facts into a number theory, the static, mechanic, electrodynamic facts into a theory of statics, mechanics, electrodynamics, or the facts

---

[29] *Ibid.*, p. 12.

[30] Hilbert's paper was first published in 1900 in the *Göttinger Nachrichten* and reprinted in the *Archive der Mathematik und Physik* in 1901. It is widely available. In my day the best source was Felix Browder (ed.), *Mathematical Developments Arising from Hilbert Problems*, American Mathematical Society, Providence, 1976.

out of the physics of gases into a gas theory...

If we examine a specific theory more closely, we then discern on all occasions that at the bottom of the construction of a framework of concepts are certain few prominent propositions of the field of knowledge, which alone are then sufficient for the building up the entire framework upon them in accordance with logical principles.

...

Viewed from a primary standpoint, these theorems may be looked upon as the *axioms of individual fields of knowledge*; the advancing development of individual fields of knowledge rests then on the more extensive logical enlargement of the completed framework of concepts.[31]

Hilbert peppered his talk with numerous examples and cited two things one had to prove about one's axiomatic system— the axioms should be *independent* and they should be *consistent*. Independence, or irredundancy, is an æsthetic requirement, the violation of which, however inelegant, is harmless— indeed, it might even have some pædogogical use. Consistency, however, is crucial. It is clearly a necessary condition an axiom system must satisfy. In his letter to Frege, he had also said it was sufficient.

Notice that in his lecture on mathematical problems Hilbert wanted an "exact and complete" axiomatic description. It is not clear how complete the axioms should be. Consider our axiomatic treatment of generalised sums back in Chapter I. The axioms did not completely capture any particular notion of infinite summation. They merely captured some common properties of the various sums considered. And in Chapter III, instead of carrying out some constructions for this structure or that structure, I presented them in some generality— for all structures satisfying certain "axioms". Whereas Hilbert would devote much energy to a variation of the traditional axiomatic method, describing a science axiomatically and developing the consequences of the axioms deductively, another— algebraic— axiomatic method was developing and, indeed, major work in the new algebraic practice of studying classes of structures all satisfying the same axioms would be carried out concurrently in Hilbert's university in Göttingen under the leadership of Emmy Noether, the greatest female mathematician to date. In algebraic axiomatics, one does not try to find an axiomatisation of a particular structure, which axiomatisation would completely determine the true properties of the given structure, be it the geometry of the plane, the arithmetic of the reals, or what have you; but, rather, one looks for an axiomatisation yielding the properties common to the structures under consideration— or, one starts with the axioms and studies the structural properties of the structures satisfying these axioms.

This, at any rate, is the ideal. The paradigm of deductive axiomatics is geometry and the derivation of geometric theorems from Euclid's or Hilbert's axioms, while that for algebraic axiomatics is group theory or field theory and

---

[31] David Hilbert, "Axiomatisches Denken", *Mathematische Annalen* 78 (1918), pp. 405 - 415. I quote from the unpolished English translation of Joong Fang, which appears as an appendix to his *Hilbert: Towards a Philosophy of Modern Mathematics*, Paideia, Hauppage (New York), 1970 (pp. 187 - 188 for the material quoted).

the study of structures defined by the group or field axioms, respectively. The Proclus-Frege description does not take into account the algebraic practice or the use of the axiomatic method in physics, where the axioms are tentative empirical laws, accepted as true so long as their consequences agree with experience. Hilbert's description is closer to practice, but his demands of completeness and the derivability of propositions from the axioms in finitely many steps are not unproblematic and need to be discussed, as we will do later. For now, let us leave the axiomatic method undefined— a mildly vague Type I notion if you will— and turn to a simpler question.

2 Where Do Axioms Come From?

Axiom systems have arisen in a number of different ways. Let us consider some of them.

2.1 Truth

The archetypical axiomatisation obtained by compiling a list of known truths about a science and pruning it of redundancies is, of course, Euclid's axiomatisation of geometry in *The Elements*. In the 1880's Moritz Pasch cited betweenness as a notion Euclid had omitted from his geometric definitions and postulates as well as from the common notions. Hilbert incorporated the notions of betweenness and order in his presentation of an axiomatisation of geometry in the *Grundlagen der Geometrie* of 1899. His axiomatisation was more complete, but used a true— but not immediately intuitive— axiom of completeness. Insofar as we have had an extensive discussion of the foundations of the real number system already in Chapter III, I prefer discussing Hilbert's related axiom system for the arithmetic of the reals. In his Parisian address on mathematical problems in 1900 Hilbert describes these axioms simply as follows:

> The axioms of arithmetic are essentially nothing else than the known rules of calculation, with the addition of the axiom of continuity. I recently collected them[32] and in so doing replaced the axiom of continuity by two simpler axioms, namely the well-known axiom of Archimedes, and a new axiom essentially as follows: that numbers form a system of things which is capable of no further extension, as long as all the other axioms hold (axiom of completeness).[33]

In the language of Chapter III, section 5, Hilbert axiomatised the reals as an archimedean ordered field with a maximality property. Specifically, he divided his axioms into four groups.

The first group of axioms are the *Axioms of Combination*:

---

[32] Hilbert here cites his paper, "Über den Zahlbegriff", *Jahresbericht der Deutschen Mathematiker-Vereinigung* 8 (1900), pp. 180 - 184.

[33] Browder, *op. cit.*, p. 9.

I1. (Addition) To each pair of numbers, a, b, a number $a + b$ is assigned.

I2. (Left- and Right-Subtraction) To each pair a, b, there are unique x, y such that

$$a + x = b \quad \text{and} \quad y + a = b.$$

I3. (Zero Element) There is a number 0 such that, for all a,

$$a + 0 = 0 + a = a.$$

I4. (Multiplication) To each pair of numbers, a, b, a number ab is assigned.

I5. (Left- and Right-Division) To each pair a, b with $a \neq 0$, there are unique x, y such that

$$ax = b \quad \text{and} \quad ya = b.$$

I6. (Unit Element) There is a number 1 such that, for all a,

$$a \cdot 1 = 1 \cdot a = a.$$

The second group of axioms are the *Axioms of Computation*:

II1. $a + (b + c) = (a + b) + c$,
II2. $a + b = b + a$,
II3. $a(bc) = (ab)c$,
II4. $a(b + c) = ab + ac$,
II5. $(a + b)c = ac + bc$,
II6. $ab = ba$.

Taken together, axiom groups *I* and *II* form a haphazardly arranged, redundant set of axioms for a field. They are followed by the *Axioms of Order*:

III1. For any a, b, if $a \neq b$ then $a > b$ or $b > a$; for no a is $a > a$.
III2. If $a > b$ and $b > c$, then $a > c$.
III3. If $a > b$, then for all c

$$a + c > b + c \quad \text{and} \quad c + a > c + b.$$

III4. If $a > b$ and $c > 0$, then

$$ac > bc \quad \text{and} \quad ca > cb.$$

Added to axiom groups *I* and *II*, group *III* yields a set of axioms for an ordered field. Hilbert characterises the real number system through a final group of axioms he calls the *Axioms of Continuity*. Exactly which axioms comprise the Axioms of Continuity varied, as noted in the quotation given above. For a while during the 1890's, in axiomatising geometry, Hilbert formulated the geometric counterpart of this axiom as a form of the Least Upper Bound Principle for increasing sequences of points. Arithmetically, this reads:

C. For any sequence $a_0 < a_1 < a_2 < \ldots$ of numbers, if there is a number b such that, for all n, $a_n < b$, then there is a least such b.

In the first edition of *Grundlagen der Geometrie*, Hilbert cited only the Archimedean Axiom:

IV1. (Archimedean Axiom) If a, b are two positive numbers, it is always possible to add a to itself a sufficient number of times so that

$$a + a + a + \ldots + a > b.$$

In "Über den Zahlbegriff", however, he added a maximality axiom he called the *Completeness Axiom*:

IV2. (Completeness Axiom) No proper extension of the real numbers satisfies axioms *I* - *III* and *IV1*.

All of Hilbert's axioms for the reals are true and those of axiom groups *I* - *III* are evident. The various formulations of group *IV* vary in their evidence.

I am inclined to accept *C* and *IV1* as evident enough on geometric grounds. The existence of "geometric evidences" for the former was testified to by Richard Dedekind in the quotation given on page 266, above. And *IV1* was accepted for linear magnitudes by the Greeks. But the non-extendability of the reals does not seem so evident. The history of the real number system has been one of expansion— from the positive integers to the positive rationals to some additional reals to... That there is an end to this expansion is not obvious. We accept it because it is a theorem (as proven back in Chapter III), not because it is evident.

As apparently first mentioned in print by Richard Baldus in 1928 [34], a quarter century after Hilbert published his axiomatisation, the Completeness Axiom has a different character from that of the other axioms:

> First, according to page 2 of the "Grundlagen" [35], the purpose of the axioms of geometry is to describe the relations like "between", "parallel", "congruent", etc. that hold between the things considered— the points, lines, and planes. This is to be understood so: that the axioms deal with the things considered *and only with these*, so that, for example, in investigating the workability of an interpretation one has to test only the things considered by the interpretation and not also other things for their formal logical conformity with the axioms. In opposition to this, the Completeness Axiom claims one can imagine no more things than the things considered; *it contains therefore a statement not only about the things considered, but also about all things imaginable at all.* In order to be able to maintain the Completeness Axiom as an axiom, one must thus *also permit among the axioms statements about things other than those considered in the relevant interpretation of the*

---

[34] Baldus states he is unaware of any prior publication concerning the matter. And when Paul Bernays, who had been Hilbert's assistant and chief collaborator on foundational work in Göttingen from about 1918 until his dismissal by the nazis in 1933, wrote on the Completeness Axiom in 1955, he cited only Baldus and a later paper on the subject by Rudolf Carnap and Friedrich Bachmann.

[35] The page reference is to Hilbert's *Grundlagen der Geometrie*. Baldus cites the 4th to 6th editions. The 7th and later editions had not yet appeared when the article was published.

axiom system, which would extend the concept of axioms in geometry in a questionable and superfluous manner.[36]

Baldus was referring to Hilbert's related axiomatisation of geometry, but his remark applies equally well to Hilbert's axiomatisation of arithmetic. Hilbert had said (cf. page 463, above) that, "when we are engaged in investigating the foundations of a science, we must set up a system of axioms which contains an exact and complete description of the relations subsisting between the elementary ideas of that science", yet now his axiomatisation of the science of arithmetic included in addition to a description of the relations subsisting between the "ideas" of the science, also a partial description of the relation subsisting between arithmetic and other possible arithmetics.

In a specifically geometric context, Baldus had another interesting criticism of the Completeness Axiom: If one drops the parallel axiom, one can still add the Completeness Axiom. There are maximal structures for both Euclidean and noneuclidean geometry. Thus Hilbert's Completeness Axiom does not always yield the logical completeness Hilbert desired. Indeed, using the Completeness Axiom to derive theorems may not be so easy. In the arithmetic case, for example, consider how one would prove that every positive real number r has a square root. This is fairly easy if one assumes axiom C or the Least Number Principle: One considers

$$Y = \{x \in \mathbb{R} \mid x^2 \leq r\}$$

and shows that $\mathrm{lub}(Y)$ is the root desired. Using Completeness one must start with $\mathbb{R}$ and show via some construction that an archimedean ordered extension $\mathbb{R}[\sqrt{r}]$ exists and then appeal to the Completeness Axiom to conclude $\sqrt{r}$ already to be in $\mathbb{R}$. The standard construction of such a structure $\mathbb{R}[\sqrt{r}]$ proceeds by assuming r to have no square root, so that the polynomial $X^2 - r$ has no factors. One then takes the set $\mathbb{R}[X]$ of all polynomials in one variable with real coefficients and defines two polynomials to be equivalent if $X^2 - r$ divides their difference. The set of equivalence classes under this equivalence relation forms an archimedean ordered field in which the equivalence class of the polynomial X is a square root of r. Except for the details dealing with the ordering relation, the construction is taught in the standard undergraduate course in Abstract Algebra and is no more difficult than the algebraic constructions of the first few sections of Chapter III, but it is still a detailed construction and offers a somewhat roundabout route to the desired result. [An alternative logical approach (i.e., an approach via Mathematical Logic) is to appeal to the Compactness Theorem cited in our discussion in Chapter III, section 6, of Nonstandard Analysis. One starts with $\mathbb{R}$ and a positive real number r and adds a new constant $\bar{c}$ and axioms

$$\bar{r} - \bar{\epsilon} < \bar{c}^2 < \bar{r} + \bar{\epsilon}$$

---

[36] Richard Baldus, "Zur Axiomatik der Geometrie. I. Über Hilberts Vollständigkeitsaxiom", *Mathematische Annalen* 100 (1928), pp. 321 - 333; here, p. 331.

for all $\epsilon > 0$. This has a model *$\mathbb{R}$ in which the square of the element c interpreting the new constant is infinitesimally close to r. Taking equivalence classes of the finite elements of *$\mathbb{R}$ under the infinitesimal closeness relation yields the extension, and then Completeness applies.]

I appear once again to have strayed off course. Criticism of the Completeness Axiom as an axiom recalls the unanswered question of the preceding section and anticipates the topic of the next chapter. Hilbert's axiomatisation of geometry is, however, one of the most famous axiomatisations in mathematics, second only to Euclid's, and I feel justified in my digression on that account: it is unthinkable not to have taken a brief look at its most questionable axiom. But let us get back on track and consider other ways in which axioms are given.

2.2 Proof-Generation

After Euclid and Hilbert, the most famous axiomatisation may well be that of set theory. This did not arise from an attempt to give a complete Hilbertian description of "a science", but appears instead to have been proof-generated.

Proof-generated concepts arise quite often in mathematics. One of the most famous examples is uniform convergence, which one discovers when one examines the proof of the false result that the limit of a convergent sequence of continuous functions is continuous. It can happen that a concept is not generated by a single proof, but by the repeated occurrence of some condition in a number of proofs. Without checking the history, I can only surmise, but my guess is that the notions of isomorphism and homomorphism arose in this way. After dealing with constructions like those of Chapter III in which one passed from structure to structure via embeddings and identifications of elements of equivalence classes, mathematicians found a need to give these maps names along with (Type II) definitions singling out their defining properties.

Axioms can also be generated by one or more proofs. The various properties (60) - (71), and (74) - (75) of the Dirac δ-function cited in Chapter III, section 5, are just the properties of the Dirac δ-function that P.A.M. Dirac used in computations— proofs— involving the function. These were certainly not evident properties of a known object as, indeed, Dirac knew there to be (at least at that time) no such object. Another example of proof-generated axioms is Ernst Zermelo's axiom system for set theory.

Georg Cantor had developed what is now called Naïve Set Theory. He did not explain what sets were or what basic properties they had (i.e., he did not lay down axioms for them), but worked with them in a naïve Type I fashion. Zermelo did the same in 1904 when he proved his Well-Ordering Theorem: Every nonempty set X can be ordered in such a way that every nonempty subset $Y \subseteq X$ has a least element. His proof used a newly formulated principle that would eventually be called the Axiom of Choice: For any nonempty set X there is a function f such that, for any $Y \subseteq X$, if Y is not empty then $f(Y) \in Y$. In words, the Axiom of Choice asserts that, given a set X, there is a function that chooses an element from each nonempty subset of X. The truth of the Axiom of Choice was quickly questioned and a shadow was cast on Zermelo's accomplishment. In 1908 he published two papers of importance here. The first was a new proof of the Well-Ordering Theorem and the second a detailed exposition of his new axiomatisation of set theory.[37]

The common idea behind the two proofs of the Well-Ordering Theorem is very simple and goes back to Cantor. If X is a nonempty set and f is the *choice function* posited by the Axiom, and if one writes

$$A \backslash B = \{x \in A \mid x \notin B\},$$

then one can define a sequence

$$x_0 = f(X), x_1 = f(X \backslash \{x_0\}), x_2 = f(X \backslash \{x_0, x_1\}), \dots$$

This is carried out into the transfinite until X has been exhausted and one has a well-ordering of X:

$$x_0 \prec x_1 \prec x_2 \prec \dots$$

In a sufficiently strong axiomatisation of set theory, after one has developed a bit of the theory of transfinite ordinals, this is very quickly turned into a correct formal proof. In 1904 Zermelo did not have any axiomatisation and in 1908 he did not have a sufficiently strong axiomatisation, whence his proofs are a bit more complicated than this. Both proofs, however, are based on this idea of generating a well-ordering by iterated application of f. In the 1904 proof, he built the well-ordering up from below, and, in the 1908 proof, he shrunk down from above.

The difference between the proofs is best illustrated by considering a simple inductive definition. Consider the set E of even natural numbers. It can be defined inductively from below as follows:

D1. A number $n \in \mathbb{N}$ is even if it can be generated after finitely many steps by application of the following rules:
i. 0 is even;
ii. if m is even, then $m + 2$ is even.

Alternatively, E can be defined from above:

---

[37] These papers are "Neuer Beweis für die Möglichkeit einer Wohlordnung" and "Untersuchungen über die Grundlagen der Mengenlehre, I" in *Mathematische Annalen* 65, pp. 107 - 128 and 261 - 281, respectively. They appear in English translation as "A new proof of the possibility of a well-ordering" and "Investigations in the foundations of set theory I" in: Jean van Heijenoort (ed.), *From Frege to Gödel; A Source Book in Mathematical Logic, 1879 - 1931*, Harvard University Press, Cambridge (Mass), 1967. The original 1904 paper (Ernst Zermelo, "Beweis daß jede Menge wohlgeordnet werden kann", *Mathematische Annalen* 59 (1904), pp. 514 - 516) can also be found translated in this source book. I also recommend Zermelo's biography: Heinz-Dieter Ebbinghaus (in cooperation with Volker Peckhaus), *Ernst Zermelo; An Approach to His Life and Work*, Springer-Verlag, Berlin-Heidelberg, 2007. Ebbinghaus includes a detailed exposition of the 1904 proof (pp. 58 - 59) along with background information and a hint for completing our discussion below of the 1908 proof.

D2. The set of even numbers is the smallest set $E \subseteq \mathbb{N}$ such that $0 \in E$ and whenever $m \in E$ it follows that $m + 2 \in E$.

Now, if E is the smallest such set, then

$$E = \bigcap \{F \subseteq \mathbb{N} \,|\, 0 \in F \ \& \ \forall m \in \mathbb{N}(m \in F \Rightarrow m + 2 \in F)\}.$$

Taking the intersection of all sets containing a given set, singleton or otherwise, and closed under some operations is a more abstract way of defining the closure of a set under the operations than defining it inductively in stages, but it yields the same set. This is the approach used by Dedekind in his set theoretic construction of the natural numbers in his famous monograph *Was sind und was sollen die Zahlen?*, and it is the approach Zermelo followed in his 1908 proof of the Well-Ordering Theorem.

Zermelo's 1908 proof starts with a nonempty set X and its choice function,

$$f : \mathcal{P}(X) \backslash \{\emptyset\} \to X,$$

where $\mathcal{P}(X)$ denotes the power set of X, $\mathcal{P}(X) = \{Y \,|\, Y \subseteq X\}$. He then considers the set $\mathcal{F}$ of final segments under the ordering to be defined. Now the set $\mathcal{F}$ will have the following properties:

i. $X \in \mathcal{F}$;
ii. if $\emptyset \neq Y \in \mathcal{F}$, then $Y \backslash \{f(Y)\} \in \mathcal{F}$;
iii. if $\mathcal{Y} \subseteq \mathcal{F}$, then $\bigcap \mathcal{Y} = \{x \in X \,|\, \forall Y \in \mathcal{Y}(x \in Y)\} \in \mathcal{F}$.

In words, i. the full set X is a final segment of itself, ii. if Y is a nonempty final segment then deleting its least element $f(Y)$ results in a final segment, and iii. the intersection of any collection of final segments is a (possibly empty) final segment. For lack of a better name, Zermelo calls any family $\mathcal{F} \subseteq \mathcal{P}(X)$ that satisfies i - iii a Θ-*chain*. A quick example of a Θ-chain is $\mathcal{P}(X)$ itself.

$\mathcal{P}(X)$ is obviously the largest possible Θ-chain. There is also a smallest possible one, namely the intersection $\mathcal{F}_0$ of all Θ-chains. To see that $\mathcal{F}_0$ is indeed a Θ-chain, note that

i. $X \in \mathcal{F}_0$ because it is in every Θ-chain, whence in their intersection;
ii. if $\emptyset \neq Y \in \mathcal{F}_0$, then Y is in every Θ-chain, whence so is $Y \backslash \{f(Y)\}$, whence $Y \backslash \{f(Y)\} \in \mathcal{F}_0$
iii. if $\mathcal{Y} \subseteq \mathcal{F}_0$, then $\mathcal{Y} \subseteq \mathcal{F}$ for every Θ-chain $\mathcal{F}$, whence $\bigcap \mathcal{Y} \in \mathcal{F}$ for all Θ-chains $\mathcal{F}$, whence $\bigcap \mathcal{Y} \in \mathcal{F}_0$.

Given $\mathcal{F}_0$, the well-ordering is easily defined: for any $x, y \in X$,

$$x \prec y \quad \text{iff} \quad \text{for some } Y \in \mathcal{F}_0, \ x \notin Y \ \& \ y \in Y.$$

The relation $\prec$ is indeed a well-ordering of X. The verification of this is, of course, the main part of the proof. As the page count of this book is already quite high, I shall forego the pleasure of presenting any of the details and simply refer the reader to Zermelo's original paper. The reader who reads that paper in its entirety will be treated to some delicious sarcasm as Zermelo follows his new proof with his response to critics of the Axiom of Choice and his earlier proof.

Zermelo accompanied his 1908 proof with the statement of the main set theoretic principles applied in the proof. In addition to a variant of the Axiom of Choice, they are as follows:

I. (Restricted Comprehension). For any set x and any property P "well-defined for every single element" the set $\{y \in x \mid P(y)\}$ exists.

II. (Power Set). For any set x, the power set $\mathcal{P}(x) = \{y \mid y \subseteq x\}$ exists.

III. (Intersection). For any nonempty set y, viewed as a family of sets, $\bigcap y = \{x \mid \forall z(z \in y \to x \in z)\}$ exists.

Immediate consequences of I are the existence of the difference sets, $Y \setminus Z = \{x \in Y \mid x \notin Z\}$ and, for $x \in X$, the singleton $\{x\} = \{y \in X \mid y = x\}$. In particular, one has the existence of $Y \setminus \{f(Y)\}$ mentioned in the second closure condition of the definition of a Θ-chain. The existence of the intersection is another consequence. For, let $\mathcal{Y}$ be a nonempty family of sets and let $Y_0 \in \mathcal{Y}$. Then

$$\bigcap \mathcal{Y} = \{x \mid \forall Y(Y \in \mathcal{Y} \to x \in Y)\} = \{x \in Y_0 \mid \forall Y(Y \in \mathcal{Y} \to x \in Y)\}.$$

The existence of the power set yields the existence of a Θ-chain. Moreover, every Θ-chain in X is a subset of $\mathcal{P}(X)$, whence an element of $\mathcal{P}(\mathcal{P}(X))$ and we have the existence of $\mathcal{F}_0$:

$$\mathcal{F}_0 = \bigcap \{\mathcal{F} \in \mathcal{P}(\mathcal{P}(X)) \mid \mathcal{F} \text{ is a } \Theta\text{-chain}\}.$$

Today[38], we think of a relation such as the well-ordering $\prec$ constructed by Zermelo as a set of ordered pairs, the ordered pairs themselves as sets. This would require a few more existential axioms such as[39]

V. (Unordered Pairs). For all sets x, y, the set $\{x, y\}$ exists.

Zermelo only uses the existence of such an unordered pair at one point in his proof and here both x and y are elements of the set X, whence his restricted Comprehension Axiom already suffices to yield the existence of $\{x, y\} = \{z \in X \mid z = x \lor z = y\}$.

Now conditions I - V go a long way toward axiomatising set theory and we have obtained them from the proof (so far as we have given it) of a single theorem. To complete the axiomatisation, one has but to examine a few more proofs in set theory and collect what is missing. Zermelo did just this in his second paper of 1908 and, to exemplify the use of his axioms, showed how the theory of cardinal equivalence could be developed in his axiomatic framework.

In modern notation, Zermelo's axioms are these:

Z1. (Axiom of Extensionality). Sets are determined by their elements, i.e., if two sets have the same elements, they are equal:

---

[38] Recall from footnote 19, that the coding of pairs of sets as sets did not appear in print until 1914— 6 years after Zermelo's paper.

[39] Zermelo listed the Axiom of Choice as number IV. Hence I skip to V for the axiom of unordered pairs.

$$\forall xy\big(\forall z(z \in x \leftrightarrow z \in y) \to x = y\big).$$

Note that the converse implication,

$$\forall xy\big(x = y \to \forall z(z \in x \leftrightarrow z \in y)\big),$$

is not an axiom of Set Theory, but one of logic: if two things are equal, anything true of the one is true of the other. Zermelo listed only the set theoretic axioms and not the logical ones or "common notions" as it were.

Z2. (Axiom of Elementary Sets). The empty set, singletons, and unordered pairs exist:

$$\exists x \forall y(y \notin x)$$
$$\forall a \exists x \forall y(y \in x \leftrightarrow y = a)$$
$$\forall ab \exists x \forall y(y \in x \leftrightarrow y = a \lor y = b).$$

Today we generally split this into two axioms, the existence of an empty set $\emptyset$ and the existence of unordered pairs. The existence of singletons follows from that of unordered pairs by noting that a, b need not be distinct to form $\{a, b\}$.

Z3. (Axiom of Separation). Whenever a "propositional function" $P(x)$ is "definite" for all elements of a set x, one can separate out of x those elements of x satisfying the propositional function:

$$\forall x \exists z \forall y\big(y \in z \leftrightarrow y \in x \land P(y)\big).$$

This is, of course, the restricted version of the Comprehension Axiom cited as *I*, above. The terms "propositional function" and "definite" are vague and various Type II explanations would later be proposed.

Z4. (Axiom of Power Set). The set of all subsets of a given set exists:

$$\forall x \exists y \forall z(z \in y \leftrightarrow z \subseteq x).$$

Z5. (Axiom of Unions). The union of any family of sets exists:

$$\forall x \exists y \forall z\big(z \in y \leftrightarrow \exists w(z \in w \land w \in x)\big).$$

Notice that the intersection of a family of sets is a subset of any element of the family and can thus be separated out of such an element by the Axiom of Separation. One therefore does not need a corresponding intersection axiom. The union of a family, however, can be larger than each of the elements of the family and no set from which it can be separated has been supplied. One thus needs a special axiom to produce the union. The existence of the union of a family pops up, for example, in the theory of cardinal numbers, where one defines, say, $\mathfrak{m}_0 + \mathfrak{m}_1 + \mathfrak{m}_2 + \ldots$ to be the cardinality of the union of a family of disjoint sets, $x_0, x_1, x_2, \ldots$ of cardinalities $\mathfrak{m}_0, \mathfrak{m}_1, \mathfrak{m}_2, \ldots$, respectively. Zermelo does just this in the second half of his paper when he discusses the theory of cardinal equivalence. Note too that the existence of the union of two sets follows from this axiom and *Z2*: $x \cup y = \bigcup \{x, y\}$.

Z6. (Axiom of Choice). If x is a collection of disjoint sets, there is a set y whose intersection with each element of x has exactly one element.

We will discuss this axiom in some detail after presenting Zermelo's final axiom.

Z7. (Axiom of Infinity). There is a set x such that $\emptyset \in x$ and, for all sets y, if $y \in x$ then $\{y\} \in x$:

$$\exists x \big(\emptyset \in x \wedge \forall y (y \in x \to \{y\} \in x) \big).$$

Today we present this differently.

The formulation of the Axiom of Infinity is based on Dedekind's 1888 monograph *Was sind und was sollen die Zahlen?* in which Dedekind characterised the infinite sets as those which could be mapped one-to-one onto proper subsets of themselves. One could then start with such a set X, a one-to-one function $x \mapsto x'$, and element 0 not in the range of the function, and define

$$N = \bigcap \{ Y \subseteq X \mid 0 \in Y \ \& \ \forall x \in X (x \in Y \to x' \in Y) \}.$$

N is unique up to isomorphism.[40] The fact that it is the smallest set containing 0 and closed under *successor* allows proof by induction and one can even define addition and multiplication inductively on N to obtain a structure isomorphic to $(\mathbb{N}, +, \cdot, 0, 1)$. However, one does need an infinite set and successor function $x \mapsto x'$ to begin with and Dedekind thought to prove the existence of such by letting X be the set of all his thoughts, 0 some thought not of the form "I am having the thought that...", and the successor function to be that which takes a thought θ to the thought "I am having the thought that θ". This sort of reasoning about thoughts is not universally acceptable and Zermelo wisely chose not to attempt such a dubious proof, but simply to assume the existence of such a set as an axiom.

For 0 Zermelo chose the empty set and as successor function he chose the map $y \mapsto \{y\}$. The most common choice today is again to choose the empty set for 0, but to take the more complicated $y \mapsto y \cup \{y\}$ as the successor function. If one takes the intersection of all sets yielded by Zermelo's axiom, and writes 0 for the empty set, his version yields

$$\{0, \{0\}, \{\{0\}\}, \{\{\{0\}\}\}, \dots\},$$

while the modern choice yields the more complicated

$$\omega = \{0, \{0\}, \{0, \{0\}\}, \{0, \{0\}, \{0, \{0\}\}\}, \dots\}.$$

The elements of ω have cardinalities $0, 1, 2, 3, \dots$, respectively, and, if we identify them with the natural numbers $0, 1, 2, 3, \dots$, we have

$$0 = \emptyset, \ 1 = \{0\}, \ 2 = \{0, 1\}, \ 3 = \{0, 1, 2\}, \ \dots, \ \omega = \{0, 1, 2, \dots\}.$$

---

[40] For more on Dedekind's work, cf. pp. 506 - 508, below.

Later, Zermelo would not always include the Axiom of Infinity in listing his axioms because he also wanted to consider "general" set theory without this assumption. And we have already[41] quoted Quine on his objection to the Axiom of Infinity as he considered it an "assumption [that] seemed to encroach on questions of fact beyond logic and mathematics", a curious remark that contrasts sharply with Hilbert's "mathematical analysis is but a single symphony of the infinite"[42].

We cannot ask if the Axiom of Infinity is true because we haven't explained what it is we want to know it might be true about. Zermelo was still two decades away from offering an explanation of what sets were supposed to be. He was merely offering a list of axioms— principles he and others had found useful in proofs. In any event, as questionable axioms go, the Axiom of Infinity was not heavily objected to. The real controversy was over the Axiom of Choice.

There are three things to be discussed about the Axiom of Choice: the equivalence of the formulation $Z6$ with the functional version cited earlier, the reason for the new function-free version, and the critique of the Axiom.

Zermelo was unaware of it, but Bertrand Russell had introduced $Z6$ two years earlier and proven its equivalence with the ordinary Axiom of Choice. It certainly follows easily from the functional form: If X is a nonempty set and $\mathcal{Y}$ is a family of disjoint nonempty subsets of X, let f be a choice function on X and define

$$Z = \{f(Y) \in X \mid Y \in \mathcal{Y}\} = \{x \in X \mid \exists Y \in \mathcal{Y}(x = f(Y))\}.$$

For any $Y \in \mathcal{Y}$, $Z \cap Y = \{f(Y)\}$, as the reader can easily verify.

The converse implication is a bit trickier. Let X be a nonempty set and consider, for each set $Y \in \mathcal{P}(X)\backslash\{\emptyset\}$, the set

$$A_Y = \{\langle Y, y\rangle \mid y \in Y\}.$$

and the collection

$$\mathcal{B} = \{A_Y \mid Y \in \mathcal{P}(X)\backslash\{\emptyset\}\}.$$

$\mathcal{B}$ is a collection of disjoint nonempty sets and by $Z6$ there is a set Z which intersects each A_Y in exactly one element. We use this fact to obtain the choice function f on X: Given $Y \in \mathcal{P}(X)\backslash\{\emptyset\}$, there is an element $y_0 \in Y$ such that

$$Z \cap A_Y = \{\langle Y, y_0\rangle\}.$$

Set $f(Y) = y_0$. Indeed, thinking of a function as a set of ordered pairs, f is in fact the set Z.

---

[41] Cf. page 462, above. Another quote from Quine: "... the basis of *Principia* is presumably adequate to the derivation of all codified mathematical theory, except for a fringe requiring the axiom of infinity and the axiom of choice as additional assumptions". (Willard van Orman Quine, "New foundations for mathematical logic", *American Mathematical Monthly* 44 (1937), pp. 70 - 80; here: p. 76.)

[42] Van Heijenoort, *op. cit.*, p. 373.

One might ask whether this proof can be carried out in Zermelo's set theory. The answer is "yes", and today we would verify this as follows. If we define, for $a \in A, b \in B$,

$$\langle a, b \rangle = \{\{a\}, \{a, b\}\},$$

we have

$$\langle a, b \rangle \in \mathcal{P}(\mathcal{P}(A \cup B)),$$

and

$$A \times B = \{\langle a, b \rangle \,|\, a \in A \ \& \ b \in B\} \in \mathcal{P}(\mathcal{P}(\mathcal{P}(A \cup P))).$$

Thus each

$$A_Y = \{\langle Y, y \rangle \,|\, y \in Y\} = \{\langle Y, y \rangle \in \text{ some power set} \,|\, Y \in \mathcal{P}(X) \ \& \ y \in Y\}$$

exists by appeal to the Axiom of Separation, and similarly the collection $\mathcal{B}$ of all these sets is a subset of another power set. Zermelo did not have Wiener's, Hausdorff's, or Kuratowski's definitions of ordered pairs[43] at his disposal, and used $Z6$ to define the Cartesian product of a family of disjoint sets. He then showed that, given any two sets A, B, one could find disjoint copies A', B' and take their product to serve as the product of A, B. A couple of pages of additional proofs and he derived the Axiom of Choice in its familiar formulation.

The choice of $Z6$ over the usual functional formulation of the Axiom of Choice can thus be simply explained: It can be expressed as an axiom before going through all the rigmarole of defining the product for disjoint sets, proving there to be a set disjoint from any two given sets, and using it to define the product of arbitrary sets— thus, finally allowing one to define functions by their graphs. Moreover, as a simple special case of the familiar Axiom, it makes an apparently (but *only* apparently) more modest axiomatic assumption. It is clearly the preferable Choice.

As for the critique of the Axiom, the crucial point is whether or not it is evident. The argument that it is not evident seems to depend on the changing meaning of the word "function". Functions were initially analytic expressions, which gave way to "discontinuous" functions defined piecewise using different analytic expressions on different intervals. Eventually, Dirichlet's discontinuous characteristic function of the rationals, and the "monsters" of Riemann, Weierstrass, and others led to a more general notion of a function as a *rule* assigning to each element of the domain a unique element of the range. With set theory and the identification of a function f with its graph (i.e., the set of ordered pairs $\langle x, y \rangle$ such that $y = f(x)$), the word "rule" became more nominal than descriptive. The existence of a rule telling one how to find y vanished, though the use of the word remained. Indeed, Zermelo's formulation of the Well-Ordering Theorem in his 1908 paper begins

> If with every nonempty subset of a set M an element of that subset is associated by some law as "distinguished element"...[44]

---

[43] Cf. footnote 19, above.

[44] Van Heijenoort, *op. cit.*, p. 184.

However, the Axiom of Choice as formulated in the 1904 paper,

> ...with every subset M' there is associated an arbitrary element m'_1 such that m'_1 occurs in M' itself,[45]

makes no mention of any law or rule for making the association. It simply posits the existence of such an association. The point is in fact the subject of a familiar English nursery rhyme:

> A preacher to logical flocks,
> Bert Russell was smart as a fox.
> He knew how to choose
> His left and right shoes,
> But wasn't decisive on socks.

The poem refers to a curious example of Russell's. To illustrate the dependence of cardinal arithmetic on the Axiom of Choice, which he referred to as the *Multiplicative Axiom*, Russell trotted out a collection of shoes and socks:

> This is illustrated by the millionaire who bought a pair of socks whenever he bought a pair of boots, and never at any other time, and who had such a passion for buying both that at last he had $\aleph_0$ pairs of boots and $\aleph_0$ pairs of socks. The problem is: How many boots had he, and how many socks? One would naturally suppose that he had twice as many boots and twice as many socks as he had pairs of each, and therefore he had $\aleph_0$ of each, since that number is not increased by doubling. But this is an instance of the difficulty already noted, of connecting the sum of ν classes each having μ terms with $\mu \times \nu$. Sometimes this can be done, sometimes it cannot. In our case it can be done with the boots, but not with the socks... The reason for the difference is this: Among the boots we can distinguish right and left, and therefore we can make a selection of one out of each pair, namely, we can choose all the right boots or all the left boots; but with socks no such principle of selection suggests itself, and we cannot be sure, unless we assume the multiplicative axiom, that there is any class consisting of one sock out of each pair. Hence the problem.[46] [47]

One usually finds the point more pithily expressed:

> Imagine an infinite collection of pairs of boots. One readily gives a rule telling which boot to choose so as to select exactly one boot from each pair: choose the right one. But what do you do if you have an infinite collection of pairs of socks?

[45] *Ibid.*, p. 140.

[46] Bertrand Russell, *Introduction to Mathematical Philosophy*, George Allen & Unwin, Ltd., London and The Macmillan Company, New York, 2nd printing, 1920, p. 126.

[47] *Essay Question*: In *My Philosophical Development* (Simon and Schuster, Inc., New York, 1959 and George Allen & Unwin, Ltd., London, 1959; p. 93), Russell says, "I once put this puzzle to a German mathematician to whom I happened to sit next at the High Table at Trinity, but his only comment was: 'Why a millionaire?'". Indeed, why a millionaire?

The argument that the Axiom of Choice is evident is historically compelling, if not logically convincing: everybody used the axiom without having realised it. Zermelo mentions this:

> That this axiom, even though it was never formulated in textbook style, has frequently been used, and successfully at that, in the most diverse fields of mathematics, especially in set theory, by Dedekind, Cantor, F. Bernstein, Schoenflies, J. König, and others is an indisputable fact, which is only corroborated by the opposition that, at one time or another, some logical purists have directed against it. Such an extensive use of a principle can be explained only by its self-evidence.[48]

Another argument for accepting the Axiom of Choice is its necessity. Zermelo cites a number of examples of theorems which were only known to be provable on assuming the Axiom of Choice. Russell's boots and socks example bears on the cardinal identity

$$\sum_{i=0}^{\infty} 2 = 2 \cdot \aleph_0 = \aleph_0.$$

Indeed almost every assertion about infinite cardinal arithmetic is equivalent to the Axiom of Choice. And, since the 1960's, a great many consequences of the Axiom of Choice have been proven to be independent of the remaining set theoretic axioms. The necessity of using an axiom to prove more theorems will not make it any truer or more evident, but it is a powerful argument for its acceptance. And, indeed, the Axiom of Choice is almost universally accepted today.

2.3 Necessity

Axioms can be found acceptable because they are needed in order to derive desired consequences. Both Russell and Zermelo appealed to the potency of their axioms to justify them. Necessity to derive certain results can also be a source for new axioms. Russell introduced the Axiom of Reducibility to provide sets at the appropriate levels in his hierarchy in order to carry out certain classical proofs within his axiomatic system. In set theory, Abraham Frænkel and Thoralf Skolem independently introduced an axiom overlooked by Zermelo.

It follows from the Axiom of Choice that every set can be well-ordered and one can order the infinite cardinal numbers by size in a transfinite sequence:

$$\aleph_0 < \aleph_1 < \ldots < \aleph_\omega < \ldots$$

Now it turns out that, even if one restricts the domain of this enumerating function to the natural numbers, a set whose existence is guaranteed by the Axiom of Infinity, Zermelo's axioms do not allow one to prove that the range of the function, so restricted, exists as a set. Thus, in Zermelo's theory, one cannot prove the existence of the sum,

---

[48] Van Heijenoort, *op. cit.*, p. 187.

$$\aleph_0 + \aleph_1 + \aleph_2 + \ldots = \sum_{n \in \mathbb{N}} \aleph_n.$$

To remedy the situation, Frænkel and Skolem proposed the Axiom of Replacement,

Z8. (Axiom of Replacement). If $\varphi(x, y)$ is a formula of the language of set theory defining a function on a set a, the range of the function is contained in a set:

$$\forall a \big[\forall x \in a \exists! y \varphi(x, y) \rightarrow \exists b \forall x \in a \forall y (\varphi(x, y) \rightarrow y \in b) \big],$$

where $\exists!$ abbreviates "there is a unique" and b does not occur in $\varphi(x, y)$:

$$\exists! y \psi(y) : \quad \exists y \psi(y) \wedge \forall z w \big(\psi(z) \wedge \psi(w) \rightarrow z = w \big).$$

Three quick remarks: First, that the range of the function is "contained in a set" implies via the Axiom of Separation that the range is indeed a set. The odd formulation is merely easier to put into symbols. Second, once the domain and range are sets, and one has ordered pairs, the Separation and Power Set Axioms can be appealed to to conclude that the restriction of the function to the domain is a set. Third, that the function whose range is posited to be a set is given by a formula of the language is a restriction applied by Skolem, who also used formulæ of a formal language to replace the "definite propositional functions" of Zermelo's Axiom of Separation.

Another axiom I would like to see adopted because it is needed in some proofs of useful results is the *Generalised Continuum Hypothesis*. This axiom asserts that, for every cardinal number $\mathfrak{m}$, there is no cardinal number strictly between $\mathfrak{m}$ and $2^{\mathfrak{m}}$. The argument-by-necessity for accepting the Generalised Continuum Hypothesis is two-fold. First, it is the only plausible axiom allowing calculations involving exponentiation with infinite cardinals. It says that such calculations are as trivial as possible and thus matches the Axiom of Choice, which renders such calculations involving addition and multiplication as trivial as possible.

The second argument-by-necessity concerns certain constructions of models in Mathematical Logic. It would take a bit too much space to describe them here, but there are constructions in that discipline in which one has to satisfy $2^{\mathfrak{m}}$ conditions in constructing an extension of a structure of cardinality $\mathfrak{m}$ and one can do this one condition at a time, still yielding a structure of cardinality $\mathfrak{m}$ at each step. If $2^{\mathfrak{m}}$ is the next cardinal after $\mathfrak{m}$, the cardinality of the successive extensions remains $\mathfrak{m}$ until the very end, when the final extension of the structure has cardinality $2^{\mathfrak{m}}$ and all $2^{\mathfrak{m}}$ conditions have been satisfied.[49]

Another reason for accepting the Generalised Continuum Hypothesis is that there are large classes of statements where its use in proofs of such a statement is not necessary but merely a convenience.

---

[49] Keisler's theorem cited in Chapter III, section 6, characterising the Horn formulæ as those preserved under reduced direct product (cf. footnote 102 on page 335) was first proven by such a construction.

2.4 Convenience

The use of the Generalised Continuum Hypothesis as a mere convenience is a technical matter depending on one of the great theorems of Mathematical Logic of the 20th century. In 1938 Kurt Gödel published a short announcement of a proof of the consistency of the Axiom of Choice and the Generalised Continuum Hypothesis relative to that of the other axioms of set theory.[50] His proof proceeded by defining a notion of *constructible* set and showing that the class of constructible sets satisfied all the axioms of set theory, plus the Axiom of Choice and the Generalised Continuum Hypothesis. The set ω of natural numbers turns out to be a constructible set. Hence, as first pointed out some years later by Georg Kreisel, if one can prove a statement about the natural numbers, not referring to arbitrary sets thereof, using the Axiom of Choice or the Generalised Continuum Hypothesis, then the truth of the statement in the constructible sets carries over to the full universe of all sets, i.e., the Axiom or Hypothesis was not needed after all. Hence, in number theory at least, one may as well assume the Axiom of Choice or the Generalised Continuum Hypothesis to prove a fact about the natural numbers. I note in this respect that the best results in number theory are proven using analysis and that many simple results of analysis rely on the Axiom of Choice, so this is not so far-fetched a scenario. Indeed, the language of arithmetic is very expressive and certain problems of interest to mathematical logicians can be expressed in the language and have been solved by constructions that can only be carried out if the Generalised Continuum Hypothesis is assumed. I shall have to leave it at this as these matters are too specialised and technical to be discussed in any depth here. What I can discuss in more detail are a couple of examples of the axiomatic use of *ideal elements* as proof-simplifying mathematical conveniences. A few examples that stand out are Leibniz's use of infinitesimals, Peacock's symbolical algebra, and Dirac's δ-function.

In papers published after his death, Leibniz is unambiguous in stating that infinitesimals do not exist, but they may be used nonetheless. In one paper responding to criticism by Bernard Nieuwentijt, he says:

> I take for granted the following postulate:
> *In any supposed transition, ending in any terminus, it is permissible to institute a general reasoning, in which the final terminus may also be included.*
>
> . . .
>
> Moreover, from this postulate arise certain expressions which are generally used for the sake of convenience, but seem to contain an absurdity, although it is one that causes no hindrance, when its proper meaning is substituted. For instance, we speak of an imaginary point of intersection as if it were a real point, in the same manner as in algebra imaginary roots are considered as accepted numbers. . .

---

[50] Kurt Gödel, "The consistency of the Axiom of Choice and the Generalised Continuum Hypothesis", *Proceedings of the National Academy of Sciences* 24 (1938), pp. 556 - 557.

Truly it is very likely that Archimedes, and one who seems to have surpassed him, Conon, found out their wonderfully elegant theorems by the help of such ideas; these theorems they completed with *reductio ad absurdum* proofs, by which they at the same time provided rigorous demonstrations and also concealed their methods. Descartes very appropriately remarked in one of his writings that Archimedes used as it were a kind of metaphysical reasoning...; in our time Cavalieri has revived the method of Archimedes, and afforded an opportunity for others to advance still further. Indeed Descartes himself did so, since at one time he imagined a circle to be a regular polygon with an infinite number of sides, and used the same idea in treating the cycloid; and Huygens too, in his work on the pendulum, since he was accustomed to confirm his theorems by rigorous demonstrations; yet at other times, in order to avoid too great prolixity, he made use of infinitesimals; as also quite lately did the renowned La Hire.

For the present, whether such a state of instantaneous transition from inequality to equality, from motion to rest, from convergence to parallelism, or anything of the sort, can be sustained in a rigorous or metaphysical sense, or whether infinite extensions successively greater and greater, or infinitely small ones successively less and less, are legitimate considerations, is a matter that I own to be possibly open to question; but for him who would discuss these matters, it is not necessary to fall back upon metaphysical controversies, such as the composition of the continuum, or to make geometrical matters depend thereon...

It will be sufficient if, when we speak of infinitely great (or more strictly unlimited), or infinitely small quantities (i.e., the very least of those within our knowledge), it is understood that we mean quantities that are indefinitely great or indefinitely small, i.e., as great as you please or as small as you please, so that the error that any one may assign may be less than a certain assigned quantity. Also, since in general it will appear that, when any small error is assigned, it can be shown that it should be less, it follows that the error is absolutely nothing; an almost exactly similar kind of argument is used in different places by Euclid, Theodosius and others; and this seemed to them to be a wonderful thing, although it could not be denied that it was perfectly true that, from the very thing that was assumed as an error, it could be inferred that the error was non-existent. Thus, by infinitely great and infinitely small, we understand something indefinitely great, or something indefinitely small, so that each conducts itself as a sort of class, and not merely as the last thing of a class. If any one wishes to understand these as the ultimate things, or as truly infinite, it can be done, and that too without falling back upon a controversy about the reality of extensions, or of infinite continuums in general, or of the infinitely small, ay, even though he think that such things are utterly impossible; it will be sufficient simply to make use of them as a tool that has advantages for the purpose of the calculation, just as the algebraists retain imaginary roots with great profit. For they contain a handy means of reckoning, as can manifestly be verified in every case in a rigorous manner by the method already stated.[51]

Leibniz also comments on the convenient use of infinitesimals in letters to Pinson, Pierre Varignon, and Guido Grandi. Henk Bos translates from the first two in his dissertation. From the letter to Pinson, he excerpts:

> For instead of the infinite or the infinitely small, one takes quantities as large, or as small, as necessary in order that the error be smaller than the given error, so that one differs from Archimedes's style only in the expressions, which are more direct in our method and more conform to the art of invention.[52]

And the excerpt from the letter to Varignon reads:

> And to this effect I have given once some lemmas on incomparables in the Leipzig Acts, which one may understand as one wishes, either as rigorous infinites, or as quantities only, of which the one does not count with respect to the other. But at the same time one has to consider that these ordinary incomparables are by no means fixed or determined; they can be taken as small as one wishes in our geometrical arguments. Thus they are effectively the same as rigorous infinitely small quantities, for if an opponent would deny our assertion, it follows from our calculus that the error will be less than any error which he will be able to assign, for it is in our power to take the incomparably small small enough for that, as one can always take a quantity as small as one wants.[53]

If one simply assumes infinitesimals are somewhere out there and starts using them, one is engaged in a Type I activity. When one justifies their use by presenting a model, one's use of infinitesimals is Type II. Leibniz's discussion of infinitesimals is Type III in that he offers the beginnings of an axiomatisation via his Continuity Law, the moderately unintelligible italicised statement of our first quotation. In his study of Leibniz, Bos remarks that Leibniz offered other formulations of the law and translates one of these:

> If any continuous transition is proposed terminating in a certain limit, then it is possible to form a general reasoning, which covers also the final limit.[54]

---

[51] J.M. Child, *The Early Mathematical Manuscripts of Leibniz; Translated from the Latin Texts Published by Carl Immanuel Gerhardt with Critical and Historical Notes*, Open Court Publishing Company, Chicago and London, 1920, pp. 147 - 150.

[52] Hendrik Jan Maarten Bos, *Differentials, Higher Order Differentials and the Derivative in the Leibnizian Calculus*, dissertation, Rijksuniversiteit te Utrecht, 1973, p. 73. A version of this dissertation was published in the *Archive for History of Exact Sciences* 14 (1974), pp. 1 - 89.

[53] *Ibid.*

[54] *Ibid.*, p. 74.

Bos remarks

> The law, not too clear in its formulation, was explained by some examples: in the case of intersecting lines, for instance, arguments involving the intersection could be extended (by introducing "imaginary" points of intersection and considering the angle between the lines "infinitely small") to the case of parallelism; also arguments about ellipses could be extended to parabolas by introducing a focus infinitely distant from the other, fixed, focus.[55]

Writing on Leibniz in the historical remarks at the end of his book on nonstandard analysis, Abraham Robinson quotes several passages from Leibniz in French and then states that

> ... although Leibniz assures us that the infinitely small and infinitely large numbers are ideal elements which are only introduced for convenience, and that arguments involving them are equivalent to the method of Archimedes (i.e., the so-called method of exhaustion), there is no attempt to justify this claim except for the appeal to the 'sovereign principle'. The same principle is supposed to serve as a justification for the assumption that the infinitely small and infinitely large numbers obey the same laws as ordinary (real) numbers. Elsewhere, Leibniz appeals to his principle of continuity to justify this assumption.[56]

Alas, Robinson doesn't explain what the "sovereign principle" is and, as near as I can make out with my limited linguistic ability, it isn't explained in the passage he quotes from Leibniz. But it doesn't matter here, as the point is that Leibniz does transcend Type I formalism in that he offers a justification for the use of infinitesimals by appeal to (perhaps not very precisely) stated principles.

In 1696, Guillaume François Antoine de L'Hôpital, Marquis de Sainte Mesme, Comte d'Entremont, usually referred to simply as the Marquis de L'Hospital, published the first textbook on the Differential Calculus. In it, following Leibniz, he laid down some definitions and axioms. According to Robinson,

> In laying down definitions and axioms as the point of departure of his theory, de l'Hospital of course, merely followed the classical examples of Euclid and Archimedes... We may also point out that in the approach of Euclid and Archimedes, which is also the approach of de l'Hospital, a definition frequently is an explication of a previously given and intuitively understood concept, and an axiom is a true statement from which later results are obtained deductively. Thus, the axiomatic approach implied a belief in the reality of the objects mentioned (e.g., differentials), which, as we have seen, Leibniz was unwilling to concede. It is interesting to contrast this with our present attitude according to

---

[55] *Ibid.*

[56] Abraham Robinson, *Non-Standard Analysis*, North-Holland Publishing Company, Amsterdam, 1966, p. 263.

which an axiom implies no ontological commitment (except possibly in set theory).[57]

Robinson went on to criticise the Leibniz-l'Hospital axiomatisation, stating that a

> weakness of Leibniz' theory was that neither he nor his successors were able to state *with sufficient precision* just what rules were supposed to govern their extended number system. Leibniz did say... that what succeeds for the finite numbers succeeds also for the infinite numbers and vice versa... But to what sort of laws was this principle supposed to apply?[58]

With Leibniz and his followers, there is not the Type I lack of a clear notion of infinitesimal, but a Type I lack of a clear notion of a law, i.e., a Type II definition of a *transfer principle*— a delineation of those properties which are assumed to succeed for the finite numbers iff they succeed for the infinite ones.

A much clearer, more sharply defined example of an axiomatisation of ideal elements is given by Peacock's *symbolical algebra*.

Peacock was very much a child of the times, a product of a particular time and place— early 19th century Cambridge. Helena Pycior informs us

> In the British universities of this period, mathematics was, after all, valued not so much for itself, but as an instrument for the training of logical minds. Possibly with this pedagogical purpose in mind, Woodhouse, in a letter of 1801 to Maseres, had agreed that "till the doctrines of negative and imaginary quantities are better taught than they are at present taught in the university of Cambridge... they had better not be taught".[59]

Francis Maseres, as the reader may recall, was one of the last great opponents of negative numbers. And we also met Robert Woodhouse briefly in our discussion thereof. Much of the debate on negative numbers was centred in Cambridge. Maseres and William Frend, the other great opponent of negatives, had graduated there; Woodhouse was a professor there; and Peacock earned his BA (1813), MA (1816), and DD (1839) there, becoming a lecturer in 1815 and professor in 1837, while serving as tutor at Trinity College from 1823 to 1839. In 1839 he was named dean of Ely College and ceased lecturing.

The first sentence in his entry in the *Dictionary of Scientific Biography* reads, "Peacock is known for his role in the reform of the teaching of mathemat-

---

[57] *Ibid.*, p. 265.

[58] *Ibid.*, p. 266.

[59] Helena Pycior, "George Peacock and the British origins of symbolical algebra", *Historia Mathematica* 8 (1981), pp. 23 - 45; here: pp. 32 - 33. My brief account of Peacock is primarily gleaned from this paper and Harvey A. Becher, "Woodhouse, Babbage, Peacock, and modern algebra", *Historia Mathematica* 7 (1980), pp. 389 - 400. I have also consulted the two editions of Peacock's algebra treatise, which are available online.

ics at Cambridge and his writings on algebra"[60]. Indeed, Peacock's mathematical fame rests mostly on his attempts at reform. As an undergraduate, he and fellow students, Charles Babbage, John Frederick William Herschel, Edward Ffrench Bromhead, etc., founded the Cambridge Analytical Society with the goal of jump-starting British mathematics, which had been stagnating under the reign of Newton's fluxions and its uninspired dot notation, by introducing the continental practice of analysis with its suggestive dy/dx notation— as some historian put it: from Newtonian dotage to Leibnizian d-ism. Likewise, in algebra Babbage, Herschel and Peacock showed a reform spirit. Herschel, however, "lost by 1821 the 'keen relish for abstract mathematical studies... he once felt'"[61] and Babbage, who had written a work on algebra by 1821, got interested in calculating machines and never published his treatment of the subject. It is not reported in my sources how far Bromhead progressed, but the end result of their common approach was the publication in 1830 of *A Treatise on Algebra* by Peacock.

One reason for the commonality of their approach is that they were all influenced by their older contemporary at Cambridge, Robert Woodhouse, whose *The Principles of Analytical Calculation* of 1803 anticipated Peacock's work. Indeed, Woodhouse had a notion of the *Extension of Demonstrated Forms*. What Woodhouse meant by this is not clarified in the source at hand. Woodhouse distinguished between arithmetical algebra— the algebra of the positive rationals— and more general algebra like the algebraic manipulation of power series *á la* Lagrange whose theory of analytic functions he expounded in his book. Woodhouse appears to have selectively applied the principle of the Extension of Demonstrated Forms as a transfer principle lifting results from arithmetical algebra such as the Binomial Theorem to his own general algebra[62]. On the other hand, in algebra he accepted infinitary calculation involving series expansions, e.g., the long divisions,

$$\frac{1}{1+x} = 1 - x + x^2 - x^3 + \ldots$$

$$\frac{1}{x+1} = \frac{1}{x} - \frac{1}{x^2} + \frac{1}{x^3} - \frac{1}{x^4} + \ldots$$

and thus did not extend the arithmetical identity[63]

$$\frac{1}{1+x} = \frac{1}{x+1}.$$

Peacock appears to have clarified matters better.

---

[60] Elaine Koppelman, "Peacock, George", *Dictionary of Scientific Biography*, volume 10, Charles Scribners Sons, p. 437.

[61] Becher, *op. cit.*, p. 396.

[62] Becher, *op. cit.*, pp. 390 - 391.

[63] *Ibid.*, p. 392.

Peacock's most important publications on the matter were his book, *A Treatise on Algebra*[64], and a report published a few years later[65]. Peacock reorganised his exposition, but broke no new ground, in a second, two-volume edition of his *Treatise* in the 1840's. The first volume, published in 1842, is subtitled *Arithmetical Algebra* and concerns the algebra of nonnegative numbers; and the second, published in 1845, bears the longer subtitle *On Symbolical Algebra and Its Applications to the Geometry of Position* and concerns a general algebra not based on any specific domain, but generally containing negatives.

Peacock's programme is succinctly described by Pycior:

> Influenced by Maseres and Frend, Peacock began his *Treatise on Algebra* of 1830 and "Report" of 1833 with the claim that there were within algebra no adequate foundations for the negative numbers and the unrestrained subtraction operation and, therefore, that algebra's standing as a deductive science was suspect. Together with the opponents of the negatives, he dismissed the popular idea that the principles of arithmetic were also the principles of algebra. He pointed out, for example, that it was erroneous to maintain that the subtraction operation of algebra could be justified by an appeal to that of arithmetic.[66]

> While he concurred with Maseres and Frend in diagnosing the problems of algebra, Peacock proposed a remedy very different from theirs. In order to establish algebra as a science, Maseres and Frend had scuttled all questionable algebraic entities and reduced algebra to universal arithmetic in the strictest sense. Peacock, on the other hand, shared the reluctance of Woodhouse and other defender[s] of the negatives to lose those parts of mathematics in which the negatives figured, and so sought to construct a deductive science of algebra in which the negatives, imaginaries, and an unrestricted subtraction operation were preserved.[67]

> Basic to Peacock's work on the foundations of algebra was his well-known distinction between arithmetical algebra and symbolical algebra, which he described as two independent sciences, distinguished from one another by the assumption in the latter of the existence of the negatives and an unrestricted subtraction operation. Arithmetical algebra was universal arithmetic in the strictest sense which... had been clearly and logically developed by Frend...; symbolical algebra... was above all traditional algebra.[68]

The laws of arithmetical algebra, or universal arithmetic, are true properties of the whole numbers. They are not true properties of all the integers or all the

---

[64] Cambridge University Press, Cambridge, 1830.

[65] "Report on the recent progress and present state of certain branches of analysis", in: *Report of the Third Meeting of the British Association for the Advancement of Science*, London, 1833.

[66] Pycior, *op. cit.*, p. 33

[67] *Ibid.*, pp. 32 - 33.

[68] *Ibid.*, p. 34.

rationals or all the reals or all the imaginaries because none of these totalities exists. The cause for the validity of the laws of symbolical algebra lies elsewhere. Pycior wishes to establish credit for Peacock for the "recognition of the freedom of mathematics" in choosing laws, a freedom usually ascribed to William Rowan Hamilton[69] and offers a couple of quotations from the "Report":

> In arithmetical algebra, the definitions of the operations determine the rules; in symbolical algebra, the rules determine the meaning of the operations, or more properly speaking, they furnish the means of interpreting them.[70]

And,

> ... we may *assume* any laws for the combination and incorporation of such symbols, so long as our assumptions are independent, and therefore not inconsistent with each other.[71]

Finally,

> ... in symbolical algebra, the rules determine the meaning of the operations... we might call them *arbitrary* assumptions, in as much as they are *arbitrarily* imposed upon a science of symbols and their combinations, which might be adapted to any other system of consistent rules.[72]

Peacock, of course, was not arbitrary in his assumptions. His goal was to justify then current mathematical practice— the algebra of negative and imaginary numbers— and the arbitrariness of his assumptions was limited to his arbitrary choice of the familiar rules. He backed them up with the Principle of the Permanence of Equivalent Forms. He offers a nice discussion of methodology in the latter portion of the third chapter of the 1830 edition of *A Treatise on Algebra*[73] I present here the beginning of this discussion:

> 131. Our great object in the very lengthened discussion which we have just concluded, has been to point out the distinction between the science of Algebra when considered with reference to its own principles, and when considered with reference to its applications, and to shew in what manner and to what extent the assumptions which regulate the combinations of general and arbitrary symbols in Algebra were suggested, and their interpretation limited by other and subordinate sciences: the principles which determine the connection between these sciences being once established, we shall be fully prepared to consider to what extent we can consider equivalent forms suggested or investigated upon the principles of a subordinate science, as equally true when expressed in

---

[69] *Ibid.*, p. 37.

[70] *Ibid.*, pp. 35 - 36.

[71] *Ibid.*, p. 36. Compare this remark with Hilbert's retort to Frege cited on page 452, above.

[72] *Ibid.*

[73] More specifically, I refer the reader to §§131 - 149 on pp. 103 - 113 of the *Treatise*.

general symbols.

Thus the principles of Arithmetical Algebra lead to the equation

$$a^n \times a^m = a^{n+m},$$

when n and m were whole numbers: it was the conversion of this conclusion in one science into an assumption in the other, which lead to the same equation,

$$a^n \times a^m = a^{n+m},$$

when n and m were general symbols.

If, however, we had commenced with the *assumption* that there existed some *equivalent* form for $a^n \times a^m$, when n and m were general symbols; and if we had discovered and *proved* that this form in Arithmetical Algebra was a^{n+m}, where n and m were such quantities as Arithmetical Algebra recognizes, then we might infer that such likewise must be the *equivalent* form in Symbolical Algebra: for this form can undergo no change, according to the assumptions which we have made, from any change in the nature of its symbols, and must therefore continue the same when the symbols are numbers: if, therefore, we discover this form in any one case, we discover it for all others.

132. Let us again recur to this principle or law of the *permanence of equivalent forms*, and consider it when stated in the form of a *direct* and *converse* proportion.

"Whatever form is Algebraically equivalent to another, when expressed in general symbols, must be true, whatever those symbols denote."

Conversely, if we discover an equivalent form in Arithmetical Algebra or any other subordinate science, when the symbols are general in form though specific in their nature, the same must be an equivalent form, when the symbols are general in their nature as well as in their form.

The direct proposition must be true, since the laws of the combinations of symbols by which such equivalent forms are deduced, have been assumed without any reference to their specific nature, and the forms themselves, therefore, are equally independent.

The converse proposition must likewise be true for the following reasons:

If there be an equivalent form when the symbols are general in form and in their nature also, it must coincide with the form discovered and proved in the subordinate science, where the symbols are general in form but specific in their nature: for in passing from the first to the second, no change in its form can take place by the first proposition.

Secondly, we may assume such an equivalent form in general symbols, since the laws of the combinations of symbols are assumed in such a manner as to coincide strictly with the corresponding laws in subordinate sciences such as Arithmetical Algebra: the conclusions, therefore, so far as their form is concerned, are necessarily the same in both; and the Algebraical *equivalence* which exists in one case must exist likewise in the other.[74]

I would hope that this proof is not entirely convincing. Now it must be admitted that my reading of the various volumes of Peacock's *Treatise* is very superficial: Using the table of contents I have selected those passages that seemed most likely to explain the Principle of the Permanence of Equivalent Forms, and not how to factor polynomials or solve the quadratic equation. Nonetheless, it seems to me that Peacock himself was unconvinced of the validity of his proof, for in the preface to the first volume of the second edition we find the passage:

> Numerical fractions, which have not a common denominator, are not homogeneous, and are incapable of addition and subtraction in arithmetic, and therefore in arithmetical algebra; and the multiplication and division of a number or fraction by a fraction is only admissible in arithmetic, and therefore in arithmetical algebra, in virtue of a convention which assumes the *permanence of forms*, which constitutes the great and fundamental principle of symbolical algebra.[75]

Thus, the Principle of the Permanence of Equivalent Forms is now a "convention", the validity of which is no longer demonstrated, but assumed.

The Principle of the Permanence of Equivalent Forms does double duty. On the one hand, it lifts the laws of arithmetical algebra from the whole numbers to the integers, to the rationals, and ultimately to the reals and imaginaries, thus giving us axioms for our ideal elements. Conversely, it tells us that any law derived in symbolical algebra when interpreted meaningfully in arithmetical algebra is valid there. Hence, the use of ideal numbers— negatives, irrationals, imaginaries— is merely a convenience.

Given our discussion of the Binomial Theorem in Chapter II, it might be of interest to see how Peacock applies the Principle to establish the result in his *Treatise*. To this end, let me first explain a few points. A form for Peacock is just an algebraic expression— polynomial, rational function, infinite series, etc. The word "value" applies in specific domains like the whole numbers, but not in general in symbolical algebra which deals only with the forms. And, in place of our current use of the word "exponent" he used "index"; the Law of Exponents,

$$a^m \times a^n = a^{m+n},$$

he called the "general principle of indices". Finally, I note that in the second edition, volume I consists of Chapters I - X concerning arithmetical algebra and volume II thus begins with Chapter XI. Chapter XXI begins as follows:

THE BINOMIAL THEOREM AND ITS APPLICATIONS

679. In Chapter VIII (Art. 486...), we have proved, when the index n is a whole number, that

---

[74] Peacock, *Treatise*, 1830 edition, pp. 103 - 105.
[75] Peacock, *Treatise*, 2nd edition, volume I, p. iv.

$$(1+x)^n = 1 + nx + n(n-1)\frac{x^2}{1.2} + n(n-1)(n-2)\frac{x^3}{1.2.3} + \ldots :$$

and it will be seen, from an examination of this series, and of the law of its formation, that the powers of x and their divisors are independent of n, and that the coefficients of

$$x, \quad \frac{x^2}{1.2}, \quad \frac{x^3}{1.2.3}, \quad \ldots\ldots, \quad \frac{x^r}{1.2\ldots r},$$

are $n, n(n-1), n(n-1)(n-2), \ldots, n(n-1)\ldots(n-r+1),$

being, for the $(1+r)^{\text{th}}$ term, the continued product of the descending series of natural numbers from n to $n-r+1$.

680. The series for $(1+x)^n$ is perfectly general in its form, though n is specific in its value, and it will continue therefore, by "the principle of the permanence of equivalent forms" (Art. 631) to be equivalent to $(1+x)^n$, when n is general in value as well as in form: and it will consequently admit, in virtue of this equivalence, of being immediately translated into the whole series of propositions respecting indices and their interpretation, which are given in Chapter XVI.

681. Thus "the general principle of indices" (Art. 635) shows that

$$(1+x)^n (1+x)^{n'} = (1+x)^{n+n'}$$

for all values of n and n', and consequently the product of the series for $(1+x)^n$ or

$$(1+x)^n = 1 + nx + n(n-1)\frac{x^2}{1.2} + n(n-1)(n-2)\frac{x^3}{1.2.3} + \&c.,$$

and of the series for $(1+x)^{n'}$, or

$$1 + n'x + n'(n'-1)\frac{x^2}{1.2} + n'(n'-1)(n'-2)\frac{x^3}{1.2.3} + \&c.$$

will be equivalent to the series for $(1+x)^{n+n'}$, or

$$1 + (n+n')x + (n+n')(n+n'-1)\frac{x^2}{1.2}$$

$$+(n+n')(n+n'-1)(n+n'-2)\frac{x^3}{1.2.3} + \&c,$$

under the same circumstances.[76]

At this point, the reader might review Euler's proof of the Binomial Theorem given in Chapter II, Section 4, Subsection 1. In that proof the result of Peacock's Article 681 was a lemma to the proof of the Theorem and not a

---

[76] Peacock, *Treatise*, 2nd edition, volume II, pp. 110 - 111.

corollary to the Theorem. If we accept Peacock's proof, we now indeed obtain an easy proof of the Vandermonde Identity (formula (37) of that Chapter on page 136), demonstrating once again the usefulness of ideal elements.

I don't know if anyone has attempted to justify Peacock's algebra in any way analogous to the justification of the use of infinitesimals in nonstandard analysis. Pycior's paper is concerned with Peacock's formalistic approach and the acceptance of his formalism rather than the validity of his inferences. On the subject of the acceptance of this formalism, she cites two letters of Hamilton's on the subject. These letters were originally published in the second volume of Robert Perceval Graves's biography of Hamilton[77]. The first letter, dated 11 July 1835, is to John T. Graves and mentions Hamilton's forthcoming paper on algebra, which paper we discussed in Chapter III, and the reasons for his distaste for symbolical algebra:

> My Paper on *Conjugate Functions* is in the press; at least a *Preliminary and Elementary Essay* which I am trying to incorporate with, or to prefix to it, *On Algebra as the Science of Pure Time*, and of which I received the seventh proof-sheet this morning. Perhaps I have not ever talked to you about this crochet of mine, for I know that with all our personal and intellectual ties we belong to opposite poles in Algebra; since you, like Peacock, seem to consider Algebra as a "System of Signs and of their combinations," somewhat analogous to syllogisms expressed in letters; while I am never satisfied unless I think that I can look beyond or through the signs to the things signified. I habitually desire to find or make in Algebra a system of demonstrations resting at last on intuitions, analogous in some way or other to Geometry as presented by Euclid— for I own that Geometry itself might be presented in a merely logical or symbolical form, though I for one would not thank him who should so present it.[78]

Hamilton's view changed and one finds in a draft of a letter of 13 October 1846 to Peacock,

> I indulge myself once more by enclosing to you, that you may read or burn, a batch of manuscript which the Cambridge printers lately returned to me, with proof-sheets to match, of my Paper on *Symbolical Geometry* for the *Cambridge and Dublin Mathematical Journal...*
> My views respecting the nature, extent, and importance of symbolical science may have approximated gradually to yours; and that approximation may be due chiefly to the influence of your writings and conversation; while you may still refuse, perhaps, and with justice, to recognise me as belonging to your school. At least I am sure that the school which produced, with such admitted ability, many articles on

---

[77] Robert Perceval Graves, *The Life of Sir William Rowan Hamilton*, Dublin University Press, Dublin, 1885. The biography has 3 volumes published in 1882, 1885, 1889, respectively.

[78] *Ibid.*, p. 143.

Symbolical Algebra, or on subjects connected therewith, in the *Cambridge Mathematical Journal*, would not concede to me the honour of their fellowship. For I still look more and more habitually *beyond* the symbols than they would choose to do...

[When I first read that work, now many years ago, and indeed for a long time afterwards, it seemed to me, I own— so hard is it for even a candid reader to enter at once into the whole spirit of an original work— that the author designed to reduce algebra to a mere system of symbols, and *nothing more;* an affair of pothooks and hangers, of black strokes upon white paper, to be made according to a fixed but arbitrary set of rules: and I refused, in my own mind, to give the high name of *Science* to the results of such a system; as I should, even now, think it a stretch of courtesy, however it may be allowed by custom, to speak of chess as a "science", though it may well be called a "scientific game".][79]

Hamilton's remarks anticipate the two major criticisms of the formalist enterprise, first that it banishes intuition from mathematics, and second that it trivialises mathematics by reducing it to a game. I will have more to say on this in the final chapter.

Peacock's symbolical algebra was itself anticipatory and may be viewed as a dress rehearsal for David Hilbert's yet more sharply defined mathematical programme undertaken in the 1920's. It is a fascinating story, but, as I know from having written about it elsewhere[80], it takes quite a number of pages to discuss with any degree of completeness and I shall not repeat it here.

2.5 System Refinement

Another source of new axioms is the tweaking of an existing theory. The individual new axioms may or may not be of the same sort as the old ones. We have seen that Zermelo's axioms for set theory were *proof-generated*. Later Fraenkel and Skolem added the Axiom of Replacement as *necessary* for certain constructions. Still later, in 1930, Zermelo added yet another *convenient* axiom, known alternately as the Axiom of Foundation or the Axiom of Regularity. The disparate names both point to the fact that regular sets, the ones that pop up in practice, are *well-founded*. A set may contain sets, which may contain sets, and so on— but only up to a point. We do not get an infinite regress,

$$\ldots x_{n+1} \in x_n \in \ldots \in x_1 \in x_0. \tag{2}$$

Formally, we may state the axiom as follows:

---

[79] *Ibid.*, pp. 527 - 528. The square brackets are Hamilton's and do not indicate an editorial insertion by Graves or the present author.

[80] Craig Smoryński, "Hilbert's programme", *CWI Quarterly* 1, no. 4 (1988), pp. 3 - 59. This article is most accessible in its reprinting, accompanied by a new introduction, in Eckart Menzler-Trott, *Logic's Lost Genius; The Life of Gerhard Gentzen*, American Mathematical Society, Providence, 2007.

Z9. (Axiom of Foundation). Every nonempty set contains an $\in$-minimal element:

$$\forall a\big(a \neq \emptyset \rightarrow \exists x(x \in a \wedge x \cap a = \emptyset)\big).$$

To see that this statement rules out chains like (2), suppose $x_0, x_1, \ldots$ are given and let $a = \{x_0, x_1, \ldots\}$. Choosing x_n in accordance with *Z9*, $x_n \cap a = \emptyset$; in particular $x_{n+1} \notin x_n$ and (2) cannot hold. Similarly, *Z9* rules out the existence of strange singletons $x = \{x\}$ (let $a = \{x\}$) and cycles $x \in y \in x$ (let $a = \{x, y\}$) or $x \in y \in z \in x$, etc. In the presence of the Axiom of Choice, the Axiom of Foundation can, in fact, be rewritten as the impossibility of (2):

Z9′. (Axiom of Foundation). There is no infinite descending $\in$-chain:

$$\forall f \neg \forall n\big(n \in \omega \rightarrow f(n') \in f(n)\big),$$

where $'$ denotes the successor function on ω.

The Axiom of Foundation has a different character from the other axioms of set theory. Aside from the Axiom of Extensionality, which tells us when two sets are equal, the remaining axioms are all existence axioms, telling us that certain collections exist as sets. The Axiom of Foundation is essentially a nonexistence axiom asserting strange sets do not exist.

One can compare the adoption of the Axiom of Foundation with the following situation. Suppose we want to axiomatise the Calculus and we start setting up axioms about functions. We might start with an extensionality axiom,

$$\forall f g \big(\forall x (f(x) = g(x)) \rightarrow f = g\big),$$

then add some function existence axioms,

$$\forall x \, \exists f \, \forall y \big(f(y) = x\big)$$
$$\forall f g \, \exists h \, \forall x \, \big(h(x) = f(x) + g(x)\big)$$
$$\text{etc.}$$

At some point we decide we don't like discontinuous functions and so we add an axiom of continuity:

$$\forall f (f \text{ is continuous}).$$

Now, the axiom of continuity is false if we interpret our function variables to range over all functions, but it is true if we let our function variables range only over continuous functions. It is convenient in that we do not have constantly to write "f is continuous" as a premise in such theorems as "all functions have a definite integral on any closed bounded interval". The Axiom of Foundation is similarly an axiom of convenience, restricting our universe of discourse from all sets to all "well-behaved" sets. Note that it violates the Euclidean-Fregean norm of truth of the axioms. We do not know if it is true of all sets or not, only that it is assumed true of the sets we are going to be interested in. Following Hilbert, the Axiom of Foundation becomes thus part of the definition of the notion "set".

Indeed, that the axioms define the objects they axiomatise is more literally true of the axioms of set theory than Hilbert could have imagined. A year before Zermelo officially proclaimed Foundation as a new axiom, Johann von Neumann showed that the universe of sets included a subuniverse of well-founded sets in which all the axioms of set theory, including the Axiom of Foundation, held. Moreover, assuming all the other axioms, the Axiom of Foundation implies that von Neumann's subuniverse is the universe of sets. And, it is this von Neumann universe that finally gives us a clear notion of set about which set theory deals.

It is easy to give an intuitive description of von Neumann's subuniverse. It is based on the idea that a collection cannot be formed until all of its elements have been formed, and thus that the universe V of all sets is built up in stages. At stage 0 one has nothing

$$V_0 = 0 = \emptyset.$$

But then at the next stage, we can collect all the subsets of V_0:

$$V_1 = \{\emptyset\},$$

and then all subsets of this:

$$V_2 = \{\emptyset, \{\emptyset\}\}.$$

Then the subsets of this:

$$V_3 = \{\emptyset, \{\emptyset\}, \{\emptyset, \{\emptyset\}\}\}.$$

In general,

$$V_{n+1} = \mathcal{P}(V_n).$$

After one has gone through all the finite stages, one collects the results,

$$V_\omega = \bigcup\{V_n \mid n \in \omega\},$$

and starts over:

$$V_{\omega+1} = \mathcal{P}(V_\omega), \quad V_{\omega+2} = \mathcal{P}(V_{\omega+1}), \ldots$$

When one has finished creating these, one again collects all the results:

$$V_{\omega+\omega} = \bigcup\{V_{\omega+n} \mid n \in \omega\}.$$

One continues in this way to generate all the well-founded sets. Of course, to make this description definite, we need a Type II definition or a Type III description of what is meant by "continues". We are, in essence, counting the stages and to do this we need *numbers*. The necessary numbers in this case are called *ordinal numbers* and I have carefully avoided being very specific about them until now, occasionally referring vaguely to one's somehow iterating a procedure "into the transfinite".

Back in Chapter II we discussed cardinal numbers in connexion with one aspect of counting. Our definition was basically Type III. We did not say what a cardinal number was, but that it was something abstracted from sets of equal cardinality. Just as Dedekind imagined us using our divine powers of creation to create irrational numbers where his nonprincipal cuts told him where to put them, Cantor created cardinal numbers for sets. He also created *order types* for any ordered set.[81] An *ordinal number* is the order type of a well-ordered set.

In set theory one proves basic properties of well-ordered sets. First among these is that, given any two well-ordered sets, one can be embedded as an initial segment of the other. For, if $(a, <)$ and $(b, <)$ are two well-ordered sets, one can try to define a function $f : a \to b$ by setting

$$f(x) = \text{ least element of } b \text{ not in } \{f(y) \mid y \in a \wedge y < x\}.$$

If this does not define a function on a, there is a least element $x_0 \in a$ for which $f(x_0)$ fails to be defined. By the minimality of x_0, $f(y)$ is defined for $y < x_0$, whence $\{f(y) \mid y \in a \wedge y < x_0\}$ exists and its complement in b is either empty or has a least element— whence $f(x_0)$ is defined. Hence, either f is defined for all $x \in a$ and $(a, <)$ is embedded into $(b, <)$ as an initial segment, or there is an x_0 such that f maps the initial segment $\{y \in a \mid y < x_0\}$ onto b, i.e., the inverse map embeds $(b, <)$ into $(a, <)$ as an initial segment. (*Exercise.* Fill in all the details.)

The embedding just described is easily seen to be unique, whence it follows more-or-less quickly that the ordinal numbers can be ordered. I should like to say that the ordering is a well-ordering, that is, that any set of ordinal numbers has a least-element. However, this is a bit difficult without first stating what an ordinal number is. In discussing Quine's objections to Russell's theory of types, I mentioned fleetingly (in footnote 25) that Russell and his contemporaries defined a cardinal number to be the equivalence class of all sets of a given cardinality. Such a class, unfortunately, is not a set: For any set a, the product $\omega \times \{a\}$ is countable. Hence $\omega \times \{a\}$ would be in the equivalence class comprising $\aleph_0$. If this were a set, so too would be its union, $\cup \aleph_0$, and its union. Now $\langle 0, a \rangle = \{\{0\}, \{0, a\}\} \in \cup \aleph_0$, whence $\{0, a\} \in \cup \cup \aleph_0$, and $a \in \cup \cup \cup \aleph_0$, i.e., $\cup \cup \cup \aleph_0$ contains all sets and the Russell paradox quickly yields a contradiction.

By the same token, an ordinal number defined as an equivalence class of well-ordered sets of a common order type is not a set and cannot belong to a set. There are, under this definition of an ordinal number, no sets of ordinal numbers and talk of them being well-ordered is vacuous.

However, in addition to the free creation of ordinal numbers by abstraction and the formation of equivalence classes of well-ordered sets sharing common order types, there is a third possibility: one can choose representatives of the given classes to be the ordinal numbers. And the key to this choice is simple: Every element in an ordered set is determined by the set of all elements less than it. An ordinal number would thus be a well-ordered set in which every element equals the set of all its predecessors. Every well-ordered set is order-isomorphic

---

[81] In fact, given an ordered set A, he wrote $\overline{A}$ for the order type abstracted from it, and $\overline{\overline{A}}$ for the cardinal number further abstracted therefrom.

to such a set. Indeed, if $(a, <)$ is well-ordered, one defines[82]

$$f(x) = \{f(y) \mid y \in a \wedge y < x\}.$$

Suppose a is given by

$$a_0 < a_1 < a_2 < \ldots < a_\omega < \ldots$$

Then one sees

$$f(a_0) = \{f(y) \mid y < a_0\} = \emptyset = 0$$
$$f(a_1) = \{f(a_0)\} = \{0\} = 1$$
$$f(a_2) = \{f(a_0), f(a_1)\} = \{0, 1\} = 2$$

$$\vdots$$

$$f(a_\omega) = \{f(a_0), f(a_1), \ldots\} = \{0, 1, 2, \ldots\} = \omega$$

$$\vdots,$$

using the definition of $0, 1, 3, \ldots, \omega$ given earlier in discussing the Axiom of Infinity.

We are only one intermediate definition away from a Type II formal definition of ordinal number.

2.1 Definition. *A set x is* transitive *if it contains all the elements of all of its elements:*

$$x \text{ is transitive: } \forall yz(y \in x \wedge z \in y \rightarrow z \in x).$$

2.2 Definition. *A set x is an* ordinal number *if it satisfies*
i. x is transitive;
ii. $(x, \in)$ is a linear ordering; and
iii. every nonempty subset y of x has an $\in$-minimal element:

$$\forall y (y \subseteq x \rightarrow \exists z(z \in y \wedge z \cap y = \emptyset)).$$

The third condition is that $\in$ not only linearly orders x but that the ordering is a well-ordering. When one assumes the Axiom of Foundation, it is automatically true of all sets and one can drop it from the definition. The transitivity clause is needed because one wants unique representatives of every well-ordered order type. The sets

$$\{0, 1, 2\} \quad \text{and} \quad \{0, 1, 3\},$$

for example, are well-ordered by $\in$ and have the same order type. The second set, however, has a gap: $2 \in 3$, but 2 is not in the set. Transitivity rules out such duplication of order types among the ordinal numbers.

---

[82] The existence of f can be justified rigorously in set theory by the fact that $(a, <)$ is a well-ordering. If f were not defined, there would be a least element x_0 of a for which $f(x_0)$ was not defined. By Replacement, $\{f(y) \mid y \in a \wedge y < x_0\}$ exists, so $f(x_0)$ exists after all.

My discussion is rather sketchy and there are many details to be verified and gaps to be filled in. I refer the reader to any good textbook on elementary set theory for these. What is important here is to have an intuitive grasp of the von Neumann hierarchy, officially called the *cumulative hierarchy*. There is an extension of the natural number sequence into the transfinite, and we can define functions on them recursively just as we can define functions on the natural numbers. In particular, we can iterate the application of the power set:

$$V_0 = \emptyset$$
$$V_{\alpha+1} = \mathcal{P}(V_\alpha), \text{ where } \alpha + 1 = \alpha \cup \{\alpha\}$$
$$V_\alpha = \bigcup_{\beta \in \alpha} V_\beta, \text{ if } \alpha \text{ has no greatest element.}$$

The union of all the sets V_α as α ranges over all the ordinals is not a set, but a subuniverse of the universe of all sets. It satisfies all the axioms *Z1 - Z8* of set theory, plus the axiom *Z9* of Foundation. Conversely, from the Axiom of Foundation it can be shown that all sets belong somewhere in this hierarchy.

Our modern notion of set is of a member of this hierarchy, i.e., it is a collection of sets that arise in this inductive generation. Somewhat after the fact, we can see the axioms of set theory as a collection of truths about these sets. But are they *evident* truths á la Proclus and Frege? Do they capture the science completely á la Hilbert?

The structure $(V_\omega, \in)$,

$$V_\omega = \emptyset \cup \mathcal{P}(\emptyset) \cup \mathcal{P}(\mathcal{P}(\emptyset)) \cup \ldots,$$

is a model of all the axioms of set theory other than the Axiom of Infinity: ω is a subset of V_ω, but all elements of V_ω are finite.

The structure $(V_{\omega+1}, \in)$ satisfies the Axiom of Infinity, but not the Axiom of Power Set: $V_\omega \in V_{\omega+1}$, but $\mathcal{P}(V_\omega) = V_{\omega+1} \notin V_{\omega+1}$. Indeed, no structure $(V_{\alpha+1}, \in)$ satisfies the power set axiom.

The next non-successor ordinal after ω is

$$\omega + \omega = \{0, 1, 2, \ldots; \omega, \omega + 1, \omega + 2, \ldots\}.$$

$(V_{\omega+\omega}, \in)$ satisfies all the axioms of set theory, except the Replacement Axiom: For, one can define F by $F(n) = V_{\omega+n}$ and $V_{\omega+\omega} = \bigcup\{F(n) \mid n \in \omega\}$, whence $\{F(n) \mid n \in \omega\}$ does not exist in $V_{\omega+\omega}$.

For no countable ordinal α can $(V_\alpha, \in)$ be a model of all the axioms of set theory. For, by the Well-Ordering Theorem, every set in V_α could be well-ordered. Moreover, there would then be an ordinal number with that order type. This holds for the uncountable set $\mathcal{P}(\omega)$. But the ordinal numbers in V_α are all initial segments of α and are thus countable.

The Axiom of Replacement is, in fact, a very strong closure condition on the class of ordinals. It cannot be proven from axioms *Z1 - Z9* that there exists a set $(V_\kappa, \in)$ that satisfies all of *Z1 - Z9*.[83] In fact, if $(V_\kappa, \in)$ is a model of *Z1 -*

---

[83] If such a κ exists, then there is a least such κ. But if the existence is a theorem, there is an $\alpha \in V_\kappa$ such that $(V_\alpha, \in)$ satisfies *Z1 - Z9*, contrary to the minimality of κ.

Z9, then κ is what is called an *inaccessible cardinal*, the smallest of a series of what are called large cardinal numbers.

I do not deem it appropriate here to get into a discussion of large cardinals, but I should explain the following. By the Well-Ordering Theorem, every set can be well-ordered, and, as we have discussed, every well-ordered set shares its order type with an ordinal number. Hence every set can be put into one-to-one correspondence with at least one ordinal number. The least such ordinal number is called the *cardinal number* of the set. That is, instead of having to define cardinal numbers as equivalence classes of equicardinal sets, we can choose as representatives of these classes the least ordinal numbers in them, i.e., we can use the least ordinal numbers of given cardinalities as a Type II definition of the notion of a cardinal number. Now, if $(V_\kappa, \in)$ is a model of *Z1* - *Z9*, then κ is a cardinal number and, in fact, κ is the cardinal of V_κ. If $\mathfrak{m} < \kappa$ is a cardinal number, then $\mathfrak{m}$ is an ordinal number in V_κ, whence the cardinality $2^{\mathfrak{m}}$ of $\mathcal{P}(\mathfrak{m})$ is in V_κ, whence $2^{\mathfrak{m}} < \kappa$. Moreover, if $\mathfrak{m}_0, \mathfrak{m}_1, \ldots$ is a transfinite sequence $\mathfrak{m}_\beta$ of length $\alpha < \kappa$, then

$$\sum_{\beta < \alpha} \mathfrak{m}_\beta \in V_\kappa, \text{ whence } \sum_{\beta < \alpha} \mathfrak{m}_\beta < \kappa.$$

Thus κ is inaccessible in that it cannot be reached from below by cardinal exponentiation or the summation of fewer than κ cardinal numbers smaller than κ. κ is very large indeed.

In the 1930's Zermelo started introducing so-called large cardinal axioms[84] to complete set theory. Such a programme became quite popular in the latter half of the 20th century. However, the rationale behind the study of such axioms is not much explained and one gets the impression that it is just a contest to see who can come up with the largest cardinal number whose existence is not refuted by the axioms of set theory. For the purposes of the present discussion, however, they serve as a nice example of axioms introduced not because they are evident truths, assumptions necessary[85] to derive desired results, or even convenient tools, but are axioms introduced to complete a theory by answering questions left unanswered by the remaining axioms. In the present case, the question would be: how far does the cumulative hierarchy extend?

2.6 Pure Formalism

A final category could be called "Miscellaneous" or "Other". I have decided to call it "Pure Formalism" because the examples that immediately come to mind are examples of axioms chosen purely formally. They are not attempts to capture a notion by listing evident truths about it, or to encapsulate existing

---

[84] The cardinals themselves predate Zermelo. Paul Mahlo had studied classes of inaccessible cardinals as early as 1911.

[85] Actually, inaccessible cardinals have been used. Keisler's construction mentioned in footnote 102 on page 335 can be carried out if, instead of assuming the Generalised Continuum hypothesis, one assumes the existence of an inaccessible cardinal.

practice by collecting frequently used principles or established results. Indeed, initially they had no semantic interpretations. There are four such theories that come to mind— the various noneuclidean geometries of the 18th and 19th centuries, the set theories of Alonzo Church and Willard van Orman Quine, and Jean Yves Girard's system of linear logic. I propose, in keeping with the general emphasis on set theory in this chapter, to discuss the middle two of these. The story of noneuclidean geometry does not need yet another exposition (at least not such as I could supply), and, besides, now that we have models of the noneuclidean geometries, they are no longer purely formal. Girard's linear logic is too complex to discuss here.

Church's paper was published in 1932[86] and Quine's in 1937[87], and, ignoring culture lag, both papers were behind the times. Throughout the 1920's a system of logic known as the "restricted functional calculus" and now termed "first-order logic" was carefully developed. By 1930 it was so well understood that Kurt Gödel was able in his dissertation to prove the completeness of this system— the equivalence between the syntactic notion of derivability within the system and the semantic notion of truth in all structures. In pure logic, it ought to have been the system by which all other systems of logic were measured, and which ought to have served as a paradigm for any new system. Church largely ignored it. Similarly, Zermelo's axiomatisation of set theory and his explication of set as a member of the cumulative hierarchy were in place by the time Quine was postulating his variant of the *old* foundations of mathematics. His main concerns were Russell's paradox and the outdated system of *Principia Mathematica* and the inconsistent set theory of Frege.

Church had two goals in setting up his system. One was to create a system using no free variables:

> One reason for avoiding the use of the free variables is that we require that every combination of symbols belonging to our system, if it represent a proposition at all, shall represent a particular proposition, unambiguously, and without the addition of verbal explanations. That the use of free variables involves violation of this requirement, we believe is readily seen.[88]

The obvious violation would be that, while propositions like

$$2 + 2 = 4 \quad \text{or} \quad 2 + 3 = 4$$

are particular and have definite truth values, statements like

$$2 + x = 4$$

do not. That this is not his main objection to free variables becomes clear if one reads on to the example he gives:

---

[86] Alonzo Church, "A set of postulates for the foundations of logic", *Annals of Mathematics* (2) 33 (1932), pp. 346 - 366.

[87] "New foundations. . .", *op. cit.*

[88] Church, *op. cit.*, p. 346.

For example, the identity

$$a(b + c) = ab + ac \qquad (3)$$

in which a, b, and c are used as free variables, does not state a definite proposition unless it is known what values may be taken on by these variables, and this information, if not implied in the context, must be given by a verbal addition.[89]

The point is that one has to state that a, b, c range over, say, real numbers and one should write something like

$$\forall abc\big(R(a) \wedge R(b) \wedge R(c) \rightarrow a(b + c) = ab + ac\,\big). \qquad (4)$$

This example tells us that the free variables range over all conceivable objects and that Church, like Frege and Russell before him, wants to build an all-encompassing master system, the logic of everything.

Church's second goal was the avoidance of the paradoxes. To do this he dropped the basic tenet of Aristotelian logic— the Law of Excluded Middle— whereby every definite proposition is either true or false:

> Rather than adopt the methods of Russell for avoiding the familiar paradoxes of mathematical logic, or that of Zermelo, both of which appear somewhat artificial, we introduce for this purpose, as we have said, a certain restriction on the law of excluded middle.[90]

Both Russell and Zermelo, via the former's type hierarchy and the latter's Axiom of Separation, restricted Comprehension to the existence of a set of all sets having the given property and "already existing" in some sense, and I don't think too many would find this as artificial as restricting the Law of Excluded Middle. In a constructivist approach, where one is allowed to assert a disjunction $A \vee B$ only when one can assert one of the disjuncts A or B, the Law of Excluded Middle is not universally valid and restricting it is quite natural. But in Church's context? It smacks of a formal device, not an evident truth.

Church's system is annoyingly novel in too many ways to allow ready comparison with more familiar systems, which comparison might allow one to pinpoint with ease where the restriction of the Law of Excluded Middle is and how extensive it is. In not allowing free variables and insisting that expressions like (3) be replaced by expressions like (4), he guarantees a more complicated system right from the start. Second, instead of dealing with sets or objects, he considers all to be functions and develops a theory thereof— a theory with no natural semantics as his functions have no definite domains and are not the familiar functions of everyday mathematics. And, to top it off, he introduces a new symbolism, some of which looks to be duplication, but isn't. For example, he assumes a propositional function Π such that $\Pi(F, G)$ denotes "$G(x)$ is a

---

[89] *Ibid.*

[90] *Ibid.*, p. 347.

true proposition for all values of x for which $F(x)$ is a true proposition" and distinguishes it from $\forall x\big(F(x) \to G(x)\big)$ ("for every x, $F(x)$ implies $G(x)$"). There is no clarification offered. The resulting system has five rules of inference and no fewer than 37 axioms, which he proceeds to use in detailed formal derivations of I forget what.

The only thing of lasting importance to come out of his paper is his method of term construction via λ-abstraction:

2.3 Definition. *The class of* λ-terms *is generated inductively as follows:*
i. *variables* $x, y, z, \ldots$ *are* λ-terms;
ii. *if* x *is a variable and* M *is a* λ-term, then $\lambda x[\mathsf{M}]$ *is a* λ-term; and
iii. *if* M, N *are* λ-terms, then $\{\mathsf{M}\}(\mathsf{N})$ *is a* λ-term.

Variables range over propositional functions. The term $\lambda x[\mathsf{M}]$ is the functional abstraction operator analogous to the set theoretic abstraction operation $\{x \mid P(x)\}$: $\lambda x[\mathsf{M}]$ is a function whose argument is x and whose value, given x, is M. And $\{\mathsf{M}\}(\mathsf{N})$ is application or evaluation.

The analogy extends to include the Russell set, which is rendered as the propositional function $\lambda x[\{\sim\}(\{x\}(x))]$, more simply written $\lambda x[\sim x(x)]$, where $\sim$ is an old-fashioned notation for negation, $\neg$. Asking if the Russell set is a member of itself translates to evaluating this function when applied to itself. Let $R = \lambda x[\sim x(x)]$ and observe for $P = \{R\}(R)$:

$$P = \{R\}(R) = \{\lambda x[\sim x(x)]\}(R) = \sim R(R) = \sim P.$$

So P and $\sim P$ are equivalent as in Frege's theory. However, without assuming one of P and $\sim P$ to be true, we do not get a contradiction. The paradox is avoided by Church's not accepting the Law of Excluded Middle.

Unfortunately, Church's precaution only ruled out the most direct proof of a contradiction and the following year saw Church publishing a patch to his system. The patched system again proved to be inconsistent and Church soon abandoned the programme, keeping only the λ-notation and developing the λ-calculus, which later acquired great importance in the theory of computability.

Quine was more fortunate than Church. Although he was visited by an inconsistency in a later refinement of his set theory, his original basic system is not known to be inconsistent. Like Church, Quine offered no attempted explication of the notion of set under which his axioms are evident truths, but simply offered another formal trick to avoid the paradoxes. His view was that one did not have to stratify sets into a hierarchy of types, but that, to avoid impredicative definitions, one had only to restrict Comprehension to those properties which could be represented by *stratifiable* formulæ in a formal language, i.e., $\{x|P(x)\}$ is asserted to exist if P is expressible in a formal language for set theory by a formula to which types may consistently be assigned to the variables. This means one can assign natural numbers to the variables of the formula expressing P in such a way that, whenever $x \in y$ occurs as a subformula, the number assigned to x is one less than that assigned to y. A few examples may explain more than a formal definition.

2.4 Examples. The following are stratifiable:

i. $x = a$: $\forall z(z \in x \leftrightarrow z \in a)$
ii. $x = a \lor x = b$: $\forall z(z \in x \leftrightarrow z \in a) \lor \forall z(z \in x \leftrightarrow z \in b)$;
iii. $x = x$: $\forall z(z \in x \leftrightarrow z \in x)$
iv. $x \neq x$: $\exists z(z \in x \land \neg z \in x)$
v. $x \subseteq a$: $\forall z(z \in x \to z \in a)$.

The stratifications are easy: Assign 0 to z and 1 to x, a, b. Note that Quine's Comprehension Axiom, when applied to these formulæ respectively, yields the existence of

i.   singletons $\{a\}$;
ii. unordered pairs $\{a, b\}$;
iii. the set V of all sets;
iv. the empty set $\emptyset$, denoted Λ by Quine; and
v.   the power set $\mathcal{P}(a)$.

2.5 Examples. The following are not stratifiable:

i. $x \in x$
ii. $x \notin x$: $\neg x \in x$.

By the second of these, Quine's restricted Comprehension Axiom does not apply and we cannot immediately conclude the existence of the Russell set. In fact, the derivation of the Russell paradox shows that $R = \{x | x \notin x\}$ does not exist as a set (unless, of course, Quine's system is inconsistent), and, since the set V of all sets exists, the nonexistence of $R = \{x | x \notin x\} = \{x \in V | x \notin x\}$ shows the failure of Zermelo's Axiom of Separation in Quine's system.

Quine was quick to point out the advantages of his system. It did not require separate axioms postulating the existence of the empty set, singletons, unordered pairs, or power sets. He also pointed to V as an example of an infinite set[91]. The disadvantages of his system slowly emerged and some are discussed in a supplement to his paper added to the anthologised version[92]. Chief among these is the problem his system has with mathematical induction.

In set theory, one establishes mathematical induction as follows: By the Axiom of Infinity there is a set containing 0 $(= \emptyset)$ and closed under the successor function $(x' = \{x\})$. One defines the set ω of natural numbers to be the intersection of all such sets:

$$\omega = \{x \mid \forall y(0 \in y \land \forall z(z \in y \to z' \in y) \to x \in y)\}. \tag{5}$$

Induction for sets follows immediately:

$$\forall y(0 \in y \land \forall z(z \in y \to z' \in y) \to \omega \subseteq y). \tag{6}$$

Induction on properties P,

---

[91] Given his remark cited in footnote 41, it is surprising to see the derivability of the Axiom of Infinity touted in the same paper when he remarks that "the deductive power of the system outruns that of *Principia*. A more striking inference, however, is the axiom of infinity". ("New foundations. . .", *op. cit.*, p. 80.)

[92] In: Quine, *From a Logical Point of View, op. cit.*

$$P(0) \wedge \forall z \big(P(z) \to P(z') \big) \to \forall x \big(x \in \omega \to P(x) \big), \tag{7}$$

follows from (6) by some form of Comprehension by considering the set $y = \{x | P(x)\}$ or $\{x \in \omega | P(x)\}$. In Quine's system, this argument breaks down in two places, namely the applications of Comprehension to conclude (5) and to reduce (7) to (6). The latter failure is clear: we cannot conclude induction on P from (6) for unstratifiable P because we cannot produce the intermediate set y.

The existence or nonexistence of ω must be considered more closely. First, to go back a step, Quine unambiguously states the Axiom of Infinity to be derivable in his system and cites V as an example of a set containing 0 and closed under successor. A couple of years later, J. Barkley Rosser states equally forcefully that it is not known whether the Axiom of Infinity is provable or not in Quine's system. Presumably Rosser refers to a different formulation of the Axiom of Infinity than Quine does— he doesn't say and I didn't feel like looking up all the papers cited by Rosser in his paper's bibliography. The relevant point can be made by noting the failure of the obvious attempt to define ω by (5): The formula defining ω is not stratifiable. The culprit is the closure clause $\forall x (x \in y \to \{x\} \in y)$:

$$\forall x w (x \in y \wedge w = \{x\} \to w \in y)$$
$$\leftrightarrow \forall x w \big(x \in y \wedge \forall v (v \in w \leftrightarrow \forall t (t \in v \leftrightarrow t \in x)) \to w \in y \big). \tag{8}$$

If one tries to assign numbers stratifying the variables of the last clause, one will start by assigning a natural number n_0 to t. v, x must, by virtue of the subformulæ $t \in v, t \in x$, be assigned the number $n_0 + 1$, w the number $n_0 + 2$ (because of the occurrence of $v \in w$), and y the values $n_0 + 2$ (because of $x \in y$) and $n_0 + 3$ (because of $w \in y$). Thus, a unique number cannot be consistently assigned to y in order to stratify the formula.

One might ask at this point why the ranks assigned to terms in subformulæ $x \in y$ must be consecutive. Numerical types and numerical stratification in Quine's Comprehension Axiom were intended to avoid impredicativity. Since the sets themselves are not stratified and one can introduce sets like V containing elements like $a, \{a\}, \{a, \{a\}\}$, where one cannot consistently stratify the assertions,

$$a \in \{a\}, \quad a \in \{a, \{a\}\}, \quad \{a\} \in \{a, \{a\}\},$$

using consecutive numbers, why then must the numbers be consecutive in stratifying formulæ in the Comprehension Axiom? Shouldn't it be sufficient that the number assigned to y when $x \in y$ occurs in the comprehension formula be strictly larger than the number assigned to x? If one does this, the formula (8) can be stratified by assigning t the number 0, v and x the number 1, w the number 2, and y the number 3. There would seem to be no less justification for the Comprehension Axiom so generalised than for Quine's original restricted version. But we cannot relax the restriction on choosing types: the system, so modified, is inconsistent. What does this do for our confidence in Quine's original system?

2.6 Remark. The inconsistency mentioned is an easy matter. Let φ be the formula,

$$\varphi(x): \quad \forall a(x \in a \to \exists y \in a(y \cap a = \emptyset)).$$

Written in full, $\varphi(x)$ reads

$$\forall a(x \in a \to \exists y \in a \forall z \neg(z \in y \wedge z \in a))$$

and can be stratified under the looser stratification requirement by assigning the types 0 to x and z, 1 to y, and 2 to a. The extended Comprehension Axiom thus yields the existence of

$$R = \{x|\varphi(x)\} = \{x|\forall a(x \in a \to \exists y \in a(y \cap a = \emptyset))\}.$$

The set R leads fairly easily to a contradiction.

If $R \in R$, we have

$$\forall a(R \in a \to \exists y \in a(y \cap a = \emptyset)).$$

This holds in particular for $a = \{R\}$:

$$R \in \{R\} \to \exists y \in \{R\}(y \cap \{R\} = \emptyset),$$

i.e., $R \cap \{R\} = \emptyset$. But $R \in R$ and $R \in \{R\}$, whence $R \in R \cap \{R\} \neq \emptyset$, a contradiction.

Thus $R \notin R$, whence

$$\exists a(R \in a \wedge \forall y \in a(y \cap a \neq \emptyset)).$$

Let A be such a set a:

$$R \in A \wedge \forall y \in A(y \cap A \neq \emptyset). \tag{9}$$

Thus $R \cap A \neq \emptyset$. Choose $B \in R \cap A$. Because $B \in R$,

$$\forall a(B \in a \to \exists y \in a(y \cap a = \emptyset)).$$

In particular,

$$B \in A \to \exists y \in A(y \cap A = \emptyset),$$

i.e., $\exists y \in A(y \cap A = \emptyset)$, contrary to (9). [For those more familiar with set theory, one can offer a partial motivation for the choice of φ as follows. Note that the conclusion of the implication of φ,

$$\psi(a): \quad \exists y \in a(y \cap a = \emptyset),$$

asserts a to have an $\in$-minimal element. The Axiom of Foundation asserts $\forall a \psi(a)$ and tells us that all sets are well-founded. And, already in 1917, Dmitry Mirimanov (1861 - 1945) proved there to be no set of all well-founded sets. This is fairly simple: let

$$W = \{x|x \text{ is well-founded}\}.$$

If $W \in W$, then W is well-founded by the definition of W, but it is not well-founded by the definition of well-founded. To apply this argument in the present context, one needs to define well-foundedness via a stratified formula. Now, although $\forall a \psi(a)$ implies all sets to be well-founded, $\psi(a)$ doesn't do the trick. For a to be well-founded, ψ must hold for the *transitive closure* of a, i.e., for

$$a \cup \bigcup a \cup \bigcup \bigcup a \cup \ldots$$

Now $\varphi(x)$ implies this for, e.g., $\{x\}$, and thus for x. Sets satisying φ are called *grounded* and the proof given is called the *Paradox of Grounded Sets*.[93]]

In 1940 Quine published an extension of his system in a book titled *Mathematical Logic*[94]. Quine's original system has become known as NF, an abbreviated acronym for the title of his paper "*New foundations*...", and the new system is correspondingly called ML after the book. ML extends NF by the addition of *proper classes*, i.e., collections that could not be elements of sets or classes. In ML mathematical induction is derivable. Unfortunately, in the initial formulation of the system, everything else also turned out to be derivable, as shown by Rosser, who seems to have devoted quite a bit of time early in his career to proving the inconsistency of ill-conceived formal systems. Fortunately for Quine, Hao Wang semi-saved the day by reformulating the theory ML and proving its consistency relative to that of NF.

As Quine notes in his autobiography,

> Though superior to "New foundations", the system of *Mathematical Logic* is not ideal. Rosser has shown that the class of natural numbers cannot be proved in that system to be a set, unless the system is inconsistent. One must postulate that it is a set in order to found the theory of real numbers, and the postulate is an unwelcome artificiality.[95]

I hate to kick a man when he is down, but I must report one more problem with Quine's theories: In 1953 Ernst Specker proved that the Axiom of Choice is refuted in NF.

I must confess to having a strong, negative, knee-jerk reaction to Quine's theories. Aside from their purely formal nature unsupported by any model to give them meaning, the pretension of the names "new foundations for mathematical logic" and "mathematical logic" for systems which cannot prove the existence of the set of natural numbers (or even induction in the case of NF) and which refute the Axiom of Choice (a "law of thought" if ever there were one) strikes me as being particularly deserving of scorn. Perhaps I would have felt the same about some of Russell's hyperbolic pronouncements on *Principia Mathematica* had I been as knowledgeable about mathematics when I learned of that system as I was when I came across Quine's set theories, but the book

---

[93] Richard Montague, "On the paradox of grounded classes", *Journal of Symbolic Logic* 20 (1955), p. 140.

[94] Norton Publishing Company, New York, 1940; later revised editions published by Harvard University Press.

[95] Quine, *The Time of My Life, op. cit.*, p. 146.

of Russell and Whitehead did accomplish something. It was a monumental case study in the formalisation of mathematics and, if one drops the ramification and the Axiom of Reducibility, its simple type theory can be construed as an honest attempt at the foundations of mathematics. Quine eschews foundational work in favour of a formal system based not on ideas but on a formal device, a system which has to be augmented before any serious mathematical development can begin. Better men than I have taken Quine's set theories seriously and studied them, so I should not dismiss these theories out of hand. But these theories have been around now for three quarters of a century and have yet to justify their existence through some notable achievement, either ease of use (like nonstandard analysis), some amazing application, or a clearer perspective on mathematics in general.

*   *   *

The axiomatic enterprise consists, Proclus tells us, of two parts: first, the isolation of the axioms of a science, and, second, the derivations of the consequences of the axioms once chosen. We have just seen *ad nauseum* that in practice axioms are not chosen for their evident truth, but for a variety of reasons. Indeed, the variety is great enough that we might well ask if there is anything in common to these choices other than that the choice is made, i.e., it would seem that Peacock's and Hilbert's uses of the word "arbitrary" to describe the choice of axioms offers a most accurate description of this part of the undertaking. What is left for us, then, is the consideration of the following question.

3 What are the Consequences of the Axioms?

That the title question of this section is genuinely problematic can nicely be illustrated by comparing Dedekind's and Peano's respective axiomatisations of arithmetic. The two gave essentially the same axiomatisation for the arithmetic of the positive integers, but they offer fundamentally different notions of consequence, Dedekind implicitly and Peano explicitly.

As briefly noted a few pages back (cf. p. 474), Dedekind eventually accompanied his set theoretic construction of the real numbers with a set theoretic construction of the positive integers. The key to this is his characterisation of infinite sets as those which can be put into one-to-one correspondence with proper subsets of themselves. That is, he starts with an infinite set X, a one-to-one function $f : X \to X$, and an element $x_0 \in X$ not in the range of f, and defines N to be the intersection of all subsets $Y \subseteq X$ containing x_0 and closed under f:

$$N = \bigcap \{Y \subseteq X \mid x_0 \in Y \land \forall x (x \in Y \to f(x) \in Y)\}$$
$$= \bigcap \{Y \in \mathcal{P}(X) \mid x_0 \in Y \land \forall x (x \in Y \to f(x) \in Y)\}$$
$$= \{n \in X \mid \forall Y (x_0 \in Y \land \forall x (x \in Y \to f(x) \in Y) \to n \in Y)\}.$$

Note that the Principle of Mathematical Induction is an immediate consequence of this identity:

$$\forall n \in N\big(x_0 \in Y \wedge \forall x \in N(x \in Y \rightarrow f(x) \in Y) \rightarrow n \in Y\big).$$

This will take on a more familiar appearance if we follow Dedekind by writing 1 for x_0, and x' for $f(x)$:

$$\forall n \in N\big(1 \in Y \wedge \forall x \in N(x \in Y \rightarrow x' \in Y) \rightarrow n \in Y\big).$$

Dedekind next observes that one can define the closure of any element of N under successor:

$$N_n = \big\{x \in N \,\big|\, \forall Y\big(n \in Y \wedge \forall y(y \in Y \rightarrow y' \in Y) \rightarrow x \in Y\big)\big\}$$
$$= \bigcap\big\{Y \subseteq N \,\big|\, n \in Y \wedge \forall y(y \in Y \rightarrow y' \in Y)\big\},$$

and that an ordering can be defined from this[96]:

$$m < n:\ N_n \subsetneq N_m.$$

Then, for fixed m, he shows by induction

$$\exists f\big(\mathrm{domain}(f) = \{x \in N \,|\, x \leq n\} \wedge f(1) = m' \wedge \forall y < n(f(y') = f(y)')\big).$$

This proof is very simple: For $n = 1$, we can take $f_1 = \{\langle 1, m'\rangle\} = \{\{1\}, \{1, m'\}\}$. Assuming the existence of f_n for n, define

$$f_{n'} = f_n \cup \{\langle n', f_n(n)'\rangle\}.$$

Having done this, he defines

$$m + n = f_n(m).$$

Thus he has a function satisfying the recursion equations,

$$m + 1 = m'$$
$$m + n' = (m + n)' \quad \text{(or: } m + (n + 1) = (m + n) + 1).$$

By induction, there is only one function satisfying this recurrence, whence he has indeed defined the familiar addition function. In a similar way he defines multiplication. Indeed, Dedekind proved a general theorem asserting the existence of recursively defined functions on the positive integers.

One of Dedekind's recursively defined functions is used to show that he has characterised $(N, ', 1)$ up to isomorphism: If (X, f, x_0) is such that

i. $f : X \rightarrow X$ is one-to-one,

ii. $x_0 \neq f(x)$ for any $x \in X$, and

---

[96] Compare this with the definition of the well-ordering in Zermelo's second proof of his Well-Ordering Theorem cited on page 471, above.

iii. $X = \bigcap \{Y \subseteq X \mid x_0 \in Y \wedge \forall x(x \in Y \rightarrow f(x) \in Y)\}$,

then there is a unique function $g : N \rightarrow X$ which is one-to-one and onto and satisfies:

$$g(1) = x_0, \qquad g(n') = f(g(n)).$$

The sketch of Dedekind's construction just given merely outlines the basic structure of the proof. There is a lot of detail to be filled in and the reader can find it fairly readably presented in Dedekind's monograph. His terminology and notation are a bit archaic, and the reader with no taste for decipherment can readily find presentations more conformable to modern usage.[97]

Let us now turn to Giuseppe Peano.

Peano is famous in mathematics for three things: his space-filling curve (1890) mentioned in passing in the previous chapter, his general existence theorem for first-order differential equations (1886), and his axiomatisation of the arithmetic of the positive integers (1889). Peano, who named the budding discipline Mathematical Logic, was keenly interested in symbolism and the precision it offers.

> Mathematical logic has interested me for many years. In the preliminary chapter to my *Calcolo geometrico* [*Geometric calculus*] (1888) I explained in summary fashion the studies of Mr Schröder, in *Der Operationskreis des Logikkalküls* [*The operation-circle of the logical calculus*] (1887), of Boole, and other authors. I pointed out there the identity of the calculus of classes, given by these authors, with the calculus of propositions, as found in the writings of Peirce, MacColl, etc.
>
> Continuing this research, in my *Arithmetices principia, nova methodo*[98] (1889) I was fortunate enough to arrive at a complete analysis of the ideas of logic, reducing them to a quite limited number, which are expressed by the symbols $\varepsilon, \supset, =, \cap, \cup, \sim, \Lambda$.
>
> A result of this analysis was the construction of a graphic symbolism, or ideography, capable of representing all the ideas of logic, so that by introducing symbols to represent the ideas of the other sciences, we may

---

[97] For example: Paul R. Halmos, *Naive Set Theory*, D. van Nostrand Company, Inc., Princeton, 1960, chapters 11 - 13; Klaus Mainzer, "Natürliche, ganze und rationale Zahlen", in: H.-D. Ebbinghaus, H. Hermes, F. Hirzebruch, M. Koecher, K. Mainzer, A. Prestel, R. Remmert, *Zahlen*, Springer-Verlag, Berlin, 1983. The modern preference is to base the construction on the Axiom of Infinity, choosing $0 = \emptyset$ for x_0, the map $x \mapsto x \cup \{x\}$ for f, and ω for N. The modern preference offers the psychological advantage of providing a definite set, function, and starting element as N, f, and x_0, respectively.

[98] Bocca, Torino, 1889. This booklet of 36 pages appears in full in English translation as "The principles of arithmetic, presented by a new method" in: Hubert C. Kennedy, *Selected Works of Giuseppe Peano*, George Allen & Unwin, Ltd., London, 1973, and University of Toronto Press, Toronto, 1973.

express every theory symbolically. For the first time, in that booklet, a complete theory was expressed in symbols.[99]

The *Arithmetices principia* divides into two parts. The first is a 16 page preface, written in Latin, explaining the logical notation and presenting a list of 66 logical propositions which, for the most part, Peano had gleaned from the works of George Boole, Ernst Schröder, and others. One may take them all to be logical axioms, though presumably there is a great deal of redundancy among them. The second part of the booklet, titled "The principles of arithmetic", is a 20 page axiomatic development of the arithmetics of the positive integers, positive rationals taken as pairs of integers, and positive reals taken as sets of rationals. The treatment is purely symbolic with words used only in a few elucidatory comments (to borrow the term from Frege— or his translator) such as, "The sign N means *number* (positive integer); 1 means *unity*; $a + 1$ means the *successor of a*, or *a plus* 1; and = means *is equal to* (this must be considered as a new sign, although it has the appearance of a sign of logic)", and some labels ("Axioms", "Definitions", etc.). He presents axioms for arithmetic which, in modern notation, would read:

P1. $1 \in N$
P2. $a \in N \rightarrow a = a$
P3. $a, b \in N \rightarrow (a = b \leftrightarrow b = a)$
P4. $a, b, c \in N \rightarrow (a = b \wedge b = c \rightarrow a = c)$
P5. $a = b \wedge b \in N \rightarrow a \in N$
P6. $a \in N \rightarrow a + 1 \in N$
P7. $a, b \in N \rightarrow (a = b \leftrightarrow a + 1 = b + 1)$
P8. $a \in N \rightarrow \neg a + 1 = 1$
P9. $\left(k \in K \wedge 1 \in k \wedge \forall x (x \in N \wedge x \in k \rightarrow x + 1 \in k) \right) \rightarrow N \subseteq k.$

Here, N denotes the set of numbers and K the collection of sets of numbers.

Nowadays, *P2* - *P4* would be called equality axioms and would be considered logical axioms rather than arithmetical ones, but this is a matter of classification: they would still be considered axioms. Peano follows these axioms with definitions of specific numerals:

P10. $2 = 1 + 1; 3 = 2 + 1; 4 = 3 + 1$: etc.

Then comes his first theorem:

P11. $2 \in N$.

Proof.

---

[99] Kennedy, *op. cit.*, p. 190. This quotation is from Kennedy's translation of the introduction to: Giuseppe Peano, "Studii di logica matematica", *Atti Accad. sci. Torino* 32 (1896 - 97), pp. 565 - 583.

$$P1 \to \qquad\qquad\qquad 1 \in N \qquad\qquad\qquad (1)$$

$$P6(1) \to \qquad\qquad (1 \in N \to 1+1 \in N) \qquad (2)$$

$$(1) \wedge (2) \to \qquad\qquad 1+1 \in N \qquad\qquad (3)$$

$$P10 \to \qquad\qquad\qquad (2 = 1+1) \qquad\qquad (4)$$

$$(4) \wedge (3) \wedge P5(2, 1+1) \to \quad 2 \in N. \qquad (\text{Theorem})$$

He then says that "for the sake of brevity" we can rewrite the proof as

$$P1 \wedge P6 \to (1+1 \in N \wedge P10 \wedge P5 \to \text{Th}),$$

"Th" abbreviating "Theorem", i.e., the conclusion. I'm not sure what to make of this. I think it is simply that Peano's symbol for implication, $\supset$, which he reads as "one deduces that", does double duty as a logical connective (our arrow $\to$) in the language, and a deducibility assertion outside the formal language (usually denoted $\vdash$). Today we woud make the distinction and add a rule of inference to the system:

$$\text{from} \vdash A \text{ and } \vdash A \to B, \text{ conclude } \vdash B, \qquad (modus\ ponens)$$

i.e., if you can prove A and $A \to B$, then you have proven B. If one does this, all of Peano's proofs become correct with minor modification. The proof of *P11* now reads:

Proof.

$$
\begin{aligned}
P1: &\quad \vdash 1 \in N \\
P6: &\quad \vdash 1 \in N \to 1+1 \in N \\
&\quad \vdash 1+1 \in N, \text{ by (1), (2), } modus\ ponens & (10) \\
P10: &\quad \vdash 2 = 1+1 & (11) \\
P5: &\quad \vdash 2 = 1+1 \wedge 1+1 \in N \to 2 \in N.
\end{aligned}
$$

A little extra logical argumentation combines (10) and (11),

$$\vdash 2 = 1+1 \wedge 1+1 \in N,$$

and *modus ponens* yields $\vdash 2 \in N$. [Many early logical systems were extremely awkward tools for the derivation of logical theorems.]

To help us make our promised comparison of Dedekind's and Peano's axiomatisations, recall from our quote on page 463, above, the two desiderata Hilbert proposed for what an axiomatisation of a science had to accomplish:

i. it had to "contain an exact and complete description of the relations subsisting between the elementary ideas of that science"; and

ii. all consequences of the axioms had to be capable of being "derived from the axioms by means of a finite number of logical steps".

These two demands are distinct.

I wish to examine the two axiomatisations in the light of these desiderata. This is not as easy as it sounds; for, practically all the terms used in expressing them are vague: "exact", "complete", "description of the relations subsisting between", "elementary ideas", "consequence".

The first step in attempting to set up an "exact and complete description", one which comes even before enumerating axioms, is the enumeration of the concepts involved. Only after this has been done can one begin to set up axioms concerning the "relations subsisting between" the concepts. Traditional axiomatisations are not complete here, isolating only the notions of central interest and listing axioms for them, but allowing the use of other concepts— common notions as it were— and their properties in proofs. Thus, Euclid, who seems not to have been axiomatising a theory of space so much as a theory of figures in space, included points, lines, planes, circles and solids among his primitive notions and listed axioms for them. It did not occur to him to cite non-figurative spatial concepts like betweenness, order, or continuity. He could have included them among his common notions[100], but demanding such an inclusion on his part is an anachronism. Indeed, that he included such common notions as "equals added to equals are equal" redounds much to his credit.

As late as 1930 in the 7th edition of Hilbert's *Grundlagen der Geometrie*, the last edition Hilbert himself had contributed to, there are unaxiomatised common notions and other blatant violations of his first desideratum. On page 29 we read

> Explanation. If M is an arbitrary point in a plane α, then the totality of all such points A in α, for which the segments MA are congruent, is called a *circle*; M is called the *centre of the circle*.

Up to this point, sets have not been introduced. They were not cited in any of Hilbert's axioms nor even listed among his primitive notions. Yet now some form of Comprehension Axiom is about to be called upon.[101] And, if one turns the page, one is confronted with the Archimedean Axiom which seems to refer to number:

> VI (Axiom of Measure or Archimedean Axiom). If AB and CD are any segments, then on the line AB there is a number of points $A_1, A_2, A_3, \ldots, A_n$ such that the segments $AA_1, A_1A_2, A_2A_3, \ldots,$ $A_{n-1}A_n$ are congruent to the segment CD and B lies between A and A_n.

---

[100] I am allowing the words "common notion" here and below to serve double duty, both denoting the concept and, in Euclidean fashion, a postulated property of the concept. I think the intended meaning in any instance will be clear from the context.

[101] The Explanation itself does not assume existence, but Hilbert follows it with the remark that, on the basis of the Explanation, with the help of his congruence axioms and the parallel postulate, "the familiar theorems on the circle, in particular the possibility of the construction of a circle through any three points not lying in a straight line", are easily proven.

Is the notion of number a common notion and does this axiom assert the existence of a number n and a function f mapping numbers to points in such a way that, for each $1 \leq k \leq n - 1$, if $A_k = f(k)$ and $A_{k+1} = f(k + 1)$, then $A_k A_{k+1}$ is congruent to CD? Or is this an infinite disjunction

$$\exists A_1 (CD \text{ is congruent to } AA_1 \text{ and } B \text{ lies between } A \text{ and } A_1)$$
$$\vee \exists A_1 A_2 (CD \text{ is congruent to } AA_1, AA_2 \text{ and } B \text{ lies between } A \text{ and } A_2)$$
$$\vee \ldots \qquad\qquad ?$$

This I cannot answer because logic itself is an untreated common notion in Hilbert's early axiomatisations.

Hilbert was not alone in not mentioning logic explicitly as part of his axiomatisation. It may be that logic was overlooked, like Euclidean betweenness, or that it was something applied to the system rather than a part of the system. Frege, Russell, Ramsey, and Quine included logical axioms, but then logic was the focus of their interest and one of their goals was to reduce mathematics to logic. For the most part, however, everyone else ignored the issue.

Dedekind cannot, of course, be faulted for not citing logical axioms as he was not proceeding axiomatically, but constructively: he was constructing the set of positive integers. However, we can read an axiomatisation off his characterisation. The natural numbers form the smallest structure containing the number 1 and closed under a specific one-to-one function, called the successor function ′, and for which 1 is not in the range. Symbolically,

$$\forall xy (x' = y' \to x = y)$$
$$\forall x (1 \neq x')$$
$$\forall n \forall Y \big(1 \in Y \wedge \forall x (x \in Y \to x' \in Y) \to n \in Y\big).$$

These are Dedekind's axioms. They agree with Peano's *P7 - P9*. Axioms corresponding to Peano's *P1 - P6* would be considered logical axioms, thus among the unmentioned common notions and not axioms of arithmetic *per se*.

Do Dedekind's axioms satisfy Hilbert's desiderata? The answer will, of course, depend on what precisely Hilbert meant— assuming he had something precise in mind. As we are familiar from our study of Type I Formalism, mathematicians often deal with imprecise concepts and can speak in vague generalities. Informally, one would have to say Dedekind, in characterising the positive integers up to isomorphism by his axioms, gave as exact and complete an axiomatisation as possible; and we ought therefore to credit him with satisfying the first desideratum. As for the second desideratum, Dedekind does not explain at all what "logical steps" are to be used, or even what a consequence of his axioms is.

Peano, on the other hand, offers roughly the same axiomatisation, plus an axiomatisation of logic, and, if we add *modus ponens* and the *generalisation rule*,

$$\text{from } A(x), \text{ deduce } \forall x A(x),$$

where x is a *free variable* of A, then we have a means of deriving consequences by means of a finite number of logical steps. And, indeed, Peano gave a number of examples. *If* by a proposition being a consequence of the axioms we mean it can be derived from the axioms by a finite number of logical steps, then Peano's axiomatisation satisfies Hilbert's second desideratum. Can we conclude thus, since Dedekind and Peano offer essentially the same arithmetic axioms, that Hilbert's two desiderata have been satisfied? The answer is "No". For, there is no guarantee that the logical derivations are sufficient to generate all "consequences" of the axioms.

The matter may sound rather subtle and the reader may be thinking of throwing his hands up in defeat, but it really is no more subtle than any other case of the use of vaguely conceived notions we have encountered in the past. The situation merely calls for Type II definitions of the undefined terms used by Hilbert in stating his desiderata:

Elementary ideas. I am inclined to think this would mean the notions taken to be primitive. In arithmetic, this would generally mean the operations of addition and multiplication, and the distinguished element 1 (as well as 0 if it is included). Subtraction and division are inverse to these and are thus definable therefrom and need not be assumed outright. Dedekind and Peano go back a step and consider 1 and the successor function as basic, defining addition and multiplication inductively. Equality is usually placed among the common notions these days, but can be considered an elementary idea of the science. Order may be taken as an elementary or a derivative idea, as it is definable: $x < y$ holds just in case there is a positive number z such that $x + z = y$.

Description of the relations subsisting between the elementary relations. I take this to mean the proper axioms of the science, i.e., those axioms other than the logical ones.

Exact. There are degrees of exactness. In formulating his Axiom of Separation, Zermelo referred vaguely to a "definite" property. Frænkel replaced this by a vague reference to some hypothetical enumeration of properties. Eventually, Skolem suggested "definite property" meant a property definable in a formal language. For reasons to be discussed in the next section, Zermelo did the same for a more powerful language than that suggested by Skolem. I take exactness today to mean "formal", or, as Peano said in describing his *Arithmetices principia*, "symbolical".

Complete. There is an ambiguity here. Is an axiomatisation of arithmetic supposed to completely describe the practice of arithmetic, thus allowing the formal derivation of every theorem one would ordinarily prove? Or, is it supposed to completely capture the structure of the set of all positive integers, thus allowing the derivation of every true statement? These two tasks are by no means equivalent and Hilbert does not make clear in his Congress address which of the two meanings is intended. However, in further publications he makes it clear that, at least for arithmetic, he means the latter— on the basis of Dedekind's isomorphism result: If φ is a true arithmetical proposition, since every model of the arithmetical axioms is isomorphic to the set of positive integers, $\neg \varphi$ has no model. Hence $\neg \varphi$ is contradictory, i.e., $\neg \neg \varphi$ is provable and thus so is φ.

Consequence. There are two notions of logical consequence— *semantic* and *syn-*

tactic— and they do not necessarily agree. The semantic notion is this: Given a set Γ of propositions serving as axioms, a proposition φ is a consequence of Γ just in case φ is true in all models of Γ. The syntactic notion means that φ can be derived from Γ by means of some precisely determined rules of inference. Generally speaking, the semantic notion comes first and the problem is to find a syntactic notion, i.e., a set of axioms and rules of inference, that completely captures the semantic notion. Usually the rules take one from finitely many premises to a single conclusion, thus guaranteeing that a derivation will consist of a finite number of logical inferences from the axioms. It is not *a priori* evident that such a syntactic equivalent to a semantic notion will exist. Moreover, some leeway in determining which notions not specific to the theory are "common notions" results in a multiplicity of semantics.

4 The Logical Solution(s)

4.1 First-Order Logic

The questions we have concerned ourselves with can all be answered logically by introducing Type II definitions for the various vague and ambiguous terms. Just as was the case with infinite sums in Chapter I, there are multiple possibilities to choose from. And, just as in Chapter III, once definite choices have been made, surprising and counterintuitive results can arise.

The most successful logical explication of the axiomatic method is given via *first-order logic*. We are not totally unfamiliar with this, having encountered it back in Chapter III in our discussion of Nonstandard Analysis. We shall now take a closer look at it.

First-order logic is not intended *á la* Frege, Russell, *et al.* to be a universal system. Rather, it is intended to describe a single structure, say $\mathfrak{M}$, which, in its simplest manifestation, is given by a single set M and some designated elements, functions, and relations thereon. In discussing Nonstandard Analysis, the structure considered was the set $\mathbb{R}$ of all real numbers, and all numbers, functions, and relations on the reals were designated. In Abstract Algebra, one would not be so all-inclusive but would consider only the ordered field of real numbers, i.e., the structure $(\mathbb{R}; +, \cdot, 0, 1, <)$. In discussing Dedekind and Peano, one would initially consider $(\mathbb{N}^+; {}', 1)$, or, as logicians prefer, $(\mathbb{N}; {}', 0)$. In discussing the arithmetic of the natural numbers, the favoured structure today is $(\mathbb{N}; +, \cdot, 0, 1)$, while that for the integers would be $(\mathbb{Z}; +, \cdot, 0, 1, <)$, because the order relation turns out not to be as simply definable in $(\mathbb{Z}; +, \cdot, 0, 1)$ in a first-order language as it is in $(\mathbb{N}; +, \cdot, 0, 1)$.

Formally, we define a structure simply as follows:

4.1 Definition. *A structure*[102] $\mathfrak{M}$ *is a 4-tuple* $(M; I, F, R)$, *where*

i. M *is a nonempty set;*

---

[102] In the present chapter I am using a semicolon instead of a comma to separate the domain from the structural elements. This will come in handy in Example 4.28 on page 550, below.

ii. *I is a subset of M;*
iii. *F is a set of functions of one or more variables from M to itself; and*
iv. *R is a set of relations on M.*

Equality may or may not be listed in R. Obviously, it is a relation on every structure, but, because of this universality, it is not generally listed among the relations in R, but is considered a part of logic, or a "common notion". Its inclusion in the internal description of a structure is thus optional.

If I ("I" for "individual"), F, R are finite, we can represent M in familiar form by simply enumerating the elements of the sets. There is no hard and fast rule as to the ordering of the enumerated items. However, if considering two structures with matching I's, F's, and R's, one should list these in corresponding order. Thus, one could write $(\mathbb{N}^+; ', 1, <)$ or $(\mathbb{N}^+; ', <, 1)$, but in comparing the structure with, say, $(\mathbb{N}; ', 0, <)$, the first structure should be written $(\mathbb{N}^+; ', 1, <)$.

Each structure $\mathfrak{M}$ has a language $L_{\mathfrak{M}}$ associated with it which includes as *primitives*, or *primitive symbols*, names for the designated individuals, functions, and relations. These names are called *constants*, *function symbols*, and *relation symbols*, respectively. When one wants to distinguish an object from its name, one overlines the object to indicate the name, or superscripts the name with "$\mathfrak{M}$" to indicate the object, as we did in Chapter III. Thus $\bar{0}$ is the constant naming 0, $\overline{+}$ would the function symbol for $+$ (though, because of confusion with the minus-plus symbol $\mp$, one would not actually write this), and $\overline{<}$ the relation symbol for $<$. Officially, function and relation symbols are all assumed to be prefix. When there is no danger of confusion, one reverts to familiar usage— one drops the overlining, writes some function symbols as infix operations (e.g., $+, \cdot$) and some as postfix (e.g., $'$), and does the same with such infix relation symbols as $<$ and $=$.

Two points should be mentioned.

The language $L_{\mathfrak{M}}$ should not be confused with the language $L(\mathfrak{M})$ defined in Definition III.6.9 on page 333, above. $L(\mathfrak{M})$ was obtained by starting with some language L fitting the structure $\mathfrak{M}$ and adding constants naming all individuals in M. $L_{\mathfrak{M}}$ also fits the structure $\mathfrak{M}$ but only has constants naming the designated individuals in I.

In most cases it is necessary to include a symbol for equality among the primitive symbols of the language. There are three exceptional cases. First, one may simply not be interested in talking about equality in a particular structure. This usually happens when one is discussing some technical question concerning a fragment of the language, for example in introducing the subject in stages— first pure logic, then logic with equality, then a system with nonlogical axioms. Second, one may have under consideration a structure with an equivalence relation respecting all the designated functions and relations. In this case, one doesn't want to distinguish equivalent, but distinct individuals and thus does not include a symbol for equality. (Or, one includes the symbol, but interprets it as the equivalence relation and not as identity. Alternatively, one can replace the structure by a new one consisting of the equivalence classes and the functions and relations imposed upon them by the original functions and relations.)

The third exception occurs when a primitive symbol for equality is unnecessary because equality is expressible in terms of the other primitives. E.g., in linear orderings one has

$$x = y \leftrightarrow \neg(x < y \vee y < x),$$

and in set theory one has

$$x = y \leftrightarrow \forall z(z \in x \leftrightarrow z \in y).$$

In these cases, one can dispense with the use of a primitive symbol for equality and replace each equation by the more complicated defining formula. When deriving theorems, one must also add replacements for the missing equality axioms to the stock of nonlogical axioms or verify that they are already derivable. In the sequel, I shall ignore all these complications and assume all languages to have a primitive symbol for equality.

As illustrated by the alternative definitions of equality cited for linear orderings and sets, the language $L_{\mathfrak{M}}$ contains more than just the primitive symbols naming objects of $\mathfrak{M}$. It must also contain primitive logical symbols and allow the construction of complex terms and formulæ. In first order logic, other than equality, the primitive logical symbols are the familiar logical connectives $\neg, \wedge, \vee, \rightarrow$, quantifiers $\forall, \exists$, and variables $v_0, v_1, v_2, \ldots$ intended to range over elements of M.

The rest of the description of first-order logic and its semantics is now the almost verbatim repetition of Definitions III.6.5 - 6.11 given on pages 331 - 334, above. The only modification needed is to replace the occurrences of "$\mathbb{R}$" and "$L_{\mathbb{R}}$" in Definitions III.6.5 - 6.7 by "$\mathfrak{M}$" and "$L_{\mathfrak{M}}$", respectively.

As Proclus told us, the axiomatic enterprise consists of two parts— choosing the axioms and working out their consequences. One cannot say much in general about the first task. Ostensibly, one has a fixed structure $\mathfrak{M}$ in mind and the axioms being about $\mathfrak{M}$ must be true in $\mathfrak{M}$. There are no general guidelines beyond that, and, as we saw in §2, even this does not cover all choices of axioms. From a logical point of view, we might as well consider any set of sentences to be a set of axioms. This leaves us with only one task: given a set of axioms, to work out the consequences.

As announced at the end of the preceding section, there are two notions of consequence— semantic and syntactic. The semantic notion is easy to define:

4.2 Definition. *Let L be a given language and Γ a set of sentences of L. We say that a sentence φ of L is a* semantic consequence *of Γ, or that φ is* semantically entailed *by Γ, written $\Gamma \vDash \varphi$, just in case φ is true in every structure in which all sentences of Γ are true.*

This definition meets the goal of defining in a precise manner what it means for a sentence φ to be a consequence of a set Γ of axioms. It is not much use, however, in determining if φ is such a consequence. For that one usually needs a proof of φ from the axioms given by Γ. And for this we need a formal definition of proof— or, *derivation*, as formal proofs are called. Formal systems of logical derivation come in three flavours: Hilbert-style systems, systems of natural deduction, and sequent calculi. Each has its own advantages and disadvantages.

Hilbert-style systems, named after Hilbert, who perfected them, are the simplest to describe, if not the easiest to use right out of the box. Natural deduction systems, introduced independently by Gerhard Gentzen and Stanisław Jaśkowski, are probably the easiest to work with for neophyte logicians deriving logical theorems within the system. And sequent calculi, introduced in the 1920's by Paul Hertz, a student of Hilbert's, and perfected by Gentzen in a two-part paper[103] of 1934 - 1935, are most useful for proving theorems about the logical system itself. This book not being a textbook in Mathematical Logic, I shall opt for simplicity of presentation rather than ease of use and discuss Hilbert-style systems.[104]

Now, Hilbert-style systems are heavy on the axioms and light on rules of inference. To cut down on the number of axioms needed, one often takes only a few of the logical symbols as primitive and treats the rest as abbreviations. In his influential first book on logic[105], Hilbert chose to take $\neg, \vee, \forall, \exists$ as primitive, treating $\wedge, \rightarrow, \leftrightarrow$ as abbreviations. Thus a formula $\varphi \wedge \psi$ is taken as shorthand for $\neg(\neg\varphi \vee \neg\psi)$ and $\varphi \rightarrow \psi, \varphi \leftrightarrow \psi$ are taken as such for $\neg\varphi \vee \psi, (\varphi \rightarrow \psi) \wedge (\psi \rightarrow \varphi)$, respectively[106]. If one takes this route, one would normally choose only one of the quantifiers $\forall, \exists$ as primitive, defining the other via one

[103] Gerhard Gentzen, "Untersuchungen über das logische Schliessen", *Mathematische Zeitschrift* 39 (1934 - 1935), pp. 136 - 210 and 405 - 431. An English translation can be found in: Manfred E. Szabo (ed.), *Collected Papers of Gerhard Gentzen*, North-Holland Publishing Company, Amsterdam, 1969.

[104] Any of the textbooks on Mathematical Logic of the 1950's and 1960's will include the development of a Hilbert-style system. I cite in particular the textbook I first learned logic from: Elliott Mendelson, *Introduction to Mathematical Logic*, D. van Nostrand Company, Inc., Princeton, 1964. In addition to one official choice to base his discussion on, several alternative systems are introduced and treated in the exercises. The book is currently in its 4th edition (1997). A good source for natural deduction is Dirk van Dalen, *Logic and Structure*, Springer-Verlag, Heidelberg, 1980; 4th edition, 2004. My favoured approach is via the sequent calculi of Gentzen and I used such as the basis of the formal systems discussed in: Craig Smoryński, *Logical Number Theory I*, Springer-Verlag, Heidelberg, 1991. The system I used was overly redundant and I would recommend Gentzen's original paper or another specialised account of sequent calculi. Jean H. Gallier, *Logic for Computer Science; Foundations of Automatic Theorem Proving*, Harper & Row, New York, 1986 surveys a variety of such calculi and their applications. More recent expositions are: Anne Troelstra and Helmut Schwichtenberg, *Basic Proof Theory*, Cambridge University Press, Cambridge, 1996 (2nd edition, 2000) and Sara Negri and Jan von Plato, *Structural Proof Theory*, Cambridge University Press, Cambridge, 2001.

[105] David Hilbert and Wilhelm Ackermann, *Grundzüge der theoretischen Logik*, Springer-Verlag, Berlin, 1928.

[106] This last is, of course, itself an abbreviation for the somewhat unreadable

$$\neg\big(\neg(\neg\varphi \vee \psi) \vee \neg(\neg\psi \vee \varphi)\big).$$

of the equivalences,

$$\exists x \varphi(x) \leftrightarrow \neg \forall x \neg \varphi(x)$$
$$\forall x \varphi(x) \leftrightarrow \neg \exists x \neg \varphi(x).$$

For axioms, the practice is often to choose a large set of logically valid sentences and whittle them down to as small a set as possible by eliminating those derivable from the others. Hilbert's axioms for the fragment without quantifiers are:

H1. $\varphi \vee \varphi \rightarrow \varphi$

H2. $\varphi \rightarrow \varphi \vee \psi$

H3. $\varphi \vee \psi \rightarrow \psi \vee \varphi$

H4. $(\varphi \rightarrow \psi) \rightarrow (\chi \vee \varphi \rightarrow \chi \vee \psi)$,

where φ, ψ, χ are any formulæ. And his rule of inference[107] for this fragment is *modus ponens*:

R1. from φ and $\varphi \rightarrow \psi$ conclude ψ,

for any formulæ φ, ψ. The axioms are certainly simple enough, and they are true. Their adequacy can also be established. However,

> The rule of inference is, from $A \supset B$ and A to deduce B. In the axioms and the rule, implications $A \supset B$ are just abbreviations for $\sim A \vee B$, and negation and disjunction are the only primitive connectives of the language. What guarantees that the axioms, especially the last one, are "simple logical truths"? What is worse, should not a formula such as $A \supset A$ be such a truth? In Hilbert's logic, it has a rather complicated proof. The overall conclusion is that Hilbert's axiomatisation captures propositional logic, but the price for having just one rule of inference is that the system is next to useless for the actual proving of theorems of propositional logic.[108]

The proof of $\varphi \rightarrow \varphi$ referred to goes as follows:

---

[107] Actually, he wrote *H1 - H4* using variables for formulæ and derived instances via a substitution rule. Today we dispense with the variable formulæ and simply declare all instances to be axioms.

[108] Jan von Plato, "From Hilbert's Programme to Gentzen's Programme", appendix in: Eckart Menzler-Trott, *Logic's Lost Genius; The Life of Gerhard Gentzen*, American Mathematical Society, Providence, 2008; here, p. 376. I note that $\supset$ is a descendent of Peano's $\supset$ and is our familiar $\rightarrow$, while $\sim$ is $\neg$, and "propositional logic" is the quantifier-free fragment of first-order logic.

1. $\varphi \vee \varphi \to \varphi$ *H1*

2. $(\varphi \vee \varphi \to \varphi) \to \left(\neg \varphi \vee (\varphi \vee \varphi) \to \neg \varphi \vee \varphi\right)$ *H4*

3. $\neg \varphi \vee (\varphi \vee \varphi) \to \neg \varphi \vee \varphi$ 1, 2, *modus ponens*

4. $(\varphi \to \varphi \vee \varphi) \to (\varphi \to \varphi)$ 3, definition of $\to$

5. $\varphi \to \varphi \vee \varphi$ *H2*

6. $\varphi \to \varphi$ 5, 4, *modus ponens.*

We can go a step farther and note that since $\varphi \to \varphi$ abbreviates $\neg \varphi \vee \varphi$ we have derived a form of the Law of Excluded Middle. Using *H3* and *modus ponens*, we quickly obtain the Law in its more familiar form $\varphi \vee \neg \varphi$.

To call the proof of $\varphi \to \varphi$ "rather complicated" borders on exaggeration, but the formal derivation is more complicated than one feels a proof of so basic a result should be.[109] This seems to be an inherent difficulty with Hilbert-style formalisms. However, we need not concern ourselves with this as our goal is simply to offer some formal device for deriving theorems from axioms. It need not be convenient; it only needs to *be*.

In fact, if we assume already given the logical axioms and rules of inference, the definition we seek is readily at hand. First, a couple of auxiliary definitions:

4.3 Definitions. *Let L be a given language and Γ a set of formulæ of L. A derivation from Γ is a finite sequence $\varphi_0, \varphi_1, \ldots, \varphi_{n-1}$ of formulæ such that, for each $k < n$,*
i. φ_k is an element of Γ; or
ii. φ_k is a logical axiom; or
iii. there are $i_0, \ldots, i_{j-1} < k$ such that φ_k follows directly from $\varphi_{i_0}, \ldots, \varphi_{i_{j-1}}$ by a rule of inference.
The last formula of a derivation $\varphi_0, \varphi_1, \ldots, \varphi_{n-1}$ is the conclusion *of the derivation.*

This immediately allows us to define a syntactic notion of consequence:

4.4 Definition. *Let L be a given language and Γ a set of formulæ of L. We say that a formula φ of L is a* syntactic consequence *of Γ, or that φ is* syntactically entailed *by Γ, written $\Gamma \vdash \varphi$, if φ is the conclusion of a derivation from Γ.*

For the sake of definiteness, let me quickly list the remaining axioms of Hilbert's own system. They are quite simple:

H5. $\forall x \varphi(x) \to \varphi(t)$
H6. $\varphi(t) \to \exists x \varphi(x)$,

where t is any term of L. In addition to these, one has more two rules of

---

[109] And the proof is a bit devious. Replacing φ by $\varphi \vee \varphi$, ψ by φ, and χ by $\neg \varphi$ in *H4* in step 2 is perhaps comparable to integrating the secant by multiplying it by $(\sec x + \tan x)/(\sec x + \tan x)$.

inference:

R2. from $\varphi \to \psi(x)$ conclude $\varphi \to \forall x \psi(x)$

R3. from $\psi(x) \to \varphi$ conclude $\exists x \psi(x) \to \varphi$,

where x does not occur free in φ. Hilbert does not include equality axioms among the logical ones as is now generally done. These would be

E1. $x = x$

E2. $x = y \to y = x$

E3. $x = y \wedge y = z \to x = z$

E4. $x = y \wedge \varphi(x) \to \varphi(y)$

where φ is atomic and $\varphi(y)$ results from $\varphi(x)$ by replacing some, but not necessarily all, occurrences of x in $\varphi(x)$ by y. [φ need not be atomic, but if it is not, one has to state carefully that the substitution not involve any conflicts involving the scopes of quantifiers.]

The exact details are not important here. What does matter is that we have some list of axioms and rules of inference. Were we to actually want to carry out formal proofs within the system, the fact that the axioms are recognisable instances of finitely many schemata and that one can quickly determine if a formula is an immediate consequence of others via a rule of inference would be important in that it means we can determine whether or not a given sequence $\varphi_0, \varphi_1, \ldots, \varphi_{n-1}$ is a derivation.

Of course, none of this is any good if we have stupidly chosen false axioms or invalid rules which allow us to draw false conclusions from true premises, or if our rules are too weak to allow us to formalise a lot of accepted mathematical proofs. Now, the logical axioms are easily seen to be valid and the rules of inference also evidently preserve truth. So that is no problem. The adequacy of the system is another matter. It is by no means obvious that the given axioms are powerful enough to derive all logically valid formulæ (i.e., all formulæ valid in all structures for a given language). Barring that, it is not even clear these are adequate for actual practice, i.e., that they capture at least the valid reasoning that actually occurs in practice. Recall the roundabout derivation of $\varphi \to \varphi$ necessitated by the choice of axioms. Or, if you are really feeling adventurous, try deriving associativity from the axioms:

$$\varphi \vee (\psi \vee \chi) \to (\varphi \vee \psi) \vee \chi.$$

Such an enterprise will demonstrate that the adequacy of the axiomatisation cannot be taken for granted.

One of the major results of 20th century Mathematical Logic asserts that the axiomatisation of logic given is adequate:

4.5 Theorem (Completeness Theorem). *Let L be a first order language. For any set Γ of sentences of L and any sentence φ of L,*

$$\Gamma \vdash \varphi \quad \text{iff} \quad \Gamma \vDash \varphi.$$

The proof of this theorem, which, in the case of a countable language (i.e., a language with only countably many constants, function symbols, and relation symbols) was first given by Kurt Gödel in his doctoral dissertation in 1930, must lie outside the scope of this book. It is maybe one level of abstraction higher than the construction of the real numbers from the rationals, and thus is not really all that difficult; but it involves a lot of messy little details and would occupy a disproportionate amount of space. I can, however, say a thing or two about how the proof goes— or, better: how the most popular proof goes.

There are two implications in the statement of the Theorem and they are proven separately. That $\Gamma \vdash \varphi$ implies $\Gamma \vDash \varphi$ is a fairly straightforward induction on the number of steps in a derivation. Because derivations can involve formulæ with free variables, one has to take formulæ as well as sentences into account in this induction and prove that every formula occurring in a derivation from Γ is *valid* in every structure $\mathfrak{M}$ making all sentences of Γ true, i.e., if $\varphi(v_0, \ldots, v_{m-1})$ is a formula occurring in the derivation and $\mathfrak{M} \vDash \Gamma$, then $\mathfrak{M} \vDash \varphi(\bar{a}_0, \ldots, \bar{a}_{m-1})$ for all $a_0, \ldots, a_{m-1} \in M$ and their constants $\bar{a}_0, \ldots, \bar{a}_{m-1}$ in $L(\mathfrak{M})$.

The interesting part of the proof is the converse and this is the part I wish to discuss, albeit most briefly. The argument is by contraposition. One assumes $\Gamma \nvdash \varphi$ and constructs a *model* of $\Gamma \cup \{\neg\varphi\}$, i.e., a structure $\mathfrak{M} \vDash \Gamma \cup \{\neg\varphi\}$. There are three parts to the construction.

First, one adds to the language an infinite set C of new constants, as many as there are formulæ in the language L. We call the new language $L(C)$. If L has cardinality κ, i.e., if there are κ many formulæ of L, then $L(C)$ will also have cardinality κ. Choose enumerations $\{c_\alpha | \alpha < \kappa\}, \{\varphi_\alpha(v) | \alpha < \kappa\}$, and $\{\psi_\alpha | \alpha < \kappa\}$ of the set C, the set of all formulæ $\varphi(v)$ of $L(C)$ with only one free variable, and the set of all sentences ψ of $L(C)$, respectively.

The next step is to expand $\Gamma \cup \{\neg\varphi\}$ inductively. One starts with $\Gamma_0 = \Gamma \cup \{\neg\varphi\}$. At stage $\alpha + 1$, we add two sentences of $L(C)$ to Γ_α. By adding only two sentences at a time, we can guarantee that Γ_α has fewer than κ constants of C occurring in its formulæ. Let c_β be the first in the enumeration that does not occur in any formula of Γ_α and define

$$\Gamma'_\alpha = \Gamma_\alpha \cup \{\exists v \varphi_\alpha(v) \to \varphi_\alpha(c_\beta)\}.$$

The consistency of Γ'_α can be shown, on the basis of $R3$, to follow from that of Γ_α.

One finishes the construction of $\Gamma_{\alpha+1}$ by adding ψ_α to Γ'_α if $\Gamma'_\alpha \cup \{\psi_\alpha\}$ is consistent and adding $\neg\psi_\alpha$ otherwise. So defined, $\Gamma_{\alpha+1}$ is consistent. (Otherwise, Γ'_α is seen to be inconsistent by the derivability of

$$(\psi \to \chi) \to ((\neg\psi \to \chi) \to \chi)$$

from $H1$ - $H4$. (*Exercise.*))

In the countable case, this completes the description of the sets Γ_α. In the uncountable case, there are also ordinals α which are not successors. For such an ordinal α, one first defines $\Gamma^*_\alpha = \bigcup_{\beta<\alpha} \Gamma_\beta$ and then treats $\varphi_\alpha, \psi_\alpha$ as before, adding the two new sentences to Γ^*_α.

Having all of the sets Γ_α defined, one goes one step farther and defines $\overline{\Gamma} = \bigcup_{\alpha < \kappa} \Gamma_\alpha$. The final part of the construction is to use $\overline{\Gamma}$ to define a structure $\mathfrak{M}$ in which $\Gamma \cup \{\neg\varphi\}$ is true. The structure is simple to describe. Its domain consists of equivalence classes of constants in C:

$$[c] = \{c' \in C \mid c = c' \text{ is in } \overline{\Gamma}\}.$$

For any relation symbol R, one defines

$$R^{\mathfrak{M}}([c_{i_0}], \dots, [c_{i_{m-1}}]) \quad \text{iff} \quad R(c_{i_0}, \dots, c_{i_{m-1}}) \text{ is in } \overline{\Gamma}.$$

The equality axioms guarantee that the relation is well-defined. For any constants $c_{i_0}, \dots, c_{i_{m-1}}$ and any m-ary function symbol f, choose c such that

$$\exists v \big(f(c_{i_0}, \dots, c_{i_{m-1}}) = v \big) \to f(c_{i_0}, \dots, c_{i_{m-1}}) = c$$

is in $\overline{\Gamma}$, and define

$$f^{\mathfrak{M}}([c_{i_0}], \dots, [c_{i_{m-1}}]) = [c].$$

Again, $f^{\mathfrak{M}}$ is well-defined.

We have thus defined a structure $\mathfrak{M}$. It now remains only to verify that $\Gamma \cup \{\neg\varphi\}$ is true in $\mathfrak{M}$. This is done by showing, by induction on the number of connectives and quantifiers in any sentence ψ, that

$$\mathfrak{M} \vDash \psi \quad \text{iff} \quad \psi \in \overline{\Gamma}.$$

The reader can find all the gory details in any textbook on Mathematical Logic (albeit, probably with a different choice of axioms) and I leave it to him to look them up. I note only that the treatment of the quantifiers is made possible by the addition of all the axioms

$$\exists v \varphi_\alpha(v) \to \varphi(c_\beta)$$

in the construction of $\Gamma_{\alpha+1}$.

4.2 Consequences of the Completeness Theorem

The Completeness Theorem is simultaneously the doing and the undoing of first-order logic; it accounts for both the usefulness and limitations of this logic. On the positive side, it shows that for the first-order language and pure logic both of Hilbert's desiderata can be met. The axiomatisation captures fully the first-order logical laws valid in all structures and these are exactly those which follow from the axioms in finitely many logical steps.

Another positive result is the Compactness Theorem (Theorem 6.31, p. 353) of Chapter III:

4.6 Theorem (Compactness Theorem). *A set Γ of sentences has a model iff every finite subset $\Gamma_0 \subseteq \Gamma$ has a model.*

Proof. Let $\perp$ be any convenient refutable sentence, e.g., $\varphi \wedge \neg\varphi$. Observe

$$\begin{array}{lll} \Gamma \text{ has a model} & \text{iff} & \Gamma \not\vdash \perp \\ & \text{iff} & \Gamma \not\vdash \perp, \text{ by the Completeness Theorem} \\ & \text{iff} & \forall \text{ finite } \Gamma_0 \subseteq \Gamma \, (\Gamma_0 \not\vdash \perp) \end{array}$$

by the finiteness of any possible derivation of $\perp$,

$$\begin{array}{lll} & \text{iff} & \forall \text{ finite } \Gamma_0 \subseteq \Gamma(\Gamma_0 \not\vdash \perp), \text{ by Completeness} \\ & \text{iff} & \forall \text{ finite } \Gamma_0 \subseteq \Gamma, \Gamma_0 \text{ has a model.} \end{array} \qquad \square$$

The great utility of the Compactness Theorem was demonstrated in Chapter III when we referred to it as the source of a model $^*\mathbb{R}$ of nonstandard analysis. Many simpler applications may be found in any introductory textbook on Mathematical Logic. Standard examples are:

4.7 Applications. *i. Every nonempty set can be linearly ordered.*
ii. If every finite subset of an infinite graph can be k-coloured (i.e., the nodes can be coloured by k colours with no two adjacent nodes receiving the same colour), then the graph itself can be k-coloured.
iii. Every abelian group can be embedded into a divisible abelian group.

The first two applications are fairly trivial. For the first, let X be a nonempty set and consider the language L with constants $\overline{x}$ naming all individuals $x \in X$, the equality symbol $=$, and a relation symbol for order $<$. Let Γ be the union of the sets of formulæ,

$$\{\neg \overline{x} = \overline{y} \mid x, y \in X \ \& \ x \neq y\}, \quad \{\overline{x} = \overline{x} \mid x \in X\},$$

and the axioms of order:

$$\forall x(\neg x < x)$$
$$\forall xyz(x < y \wedge y < z \rightarrow x < z)$$
$$\forall xy(x < y \vee x = y \vee y < x).$$

Any finite subset $\Gamma_0 \subseteq \Gamma$ names only finitely many elements $x_0, x_1, \ldots, x_{n-1}$ of X and has a model iff $\{x_0, x_1, \ldots, x_{n-1}\}$ can be ordered. Now, that any finite set can be linearly ordered is easily proven by induction. Hence Γ_0 has a model. This holding for all such Γ_0, Compactness applies and Γ has a model $\mathfrak{Y} = \{Y; \{\overline{x}^{\mathfrak{Y}} \mid x \in X\}, <^{\mathfrak{Y}}, =\}$. Obviously X embeds into Y via the map $x \mapsto \overline{x}^{\mathfrak{Y}}$ and thus inherits an ordering from $\mathfrak{Y}$:

$$x < y \quad \text{iff} \quad \overline{x}^{\mathfrak{Y}} <^{\mathfrak{Y}} \overline{y}^{\mathfrak{Y}}.$$

The second application is simpler in that the existence of the models of the relevant Γ_0's is assumed and does not have to be proven. The third application requires greater knowledge of abelian groups than has been presented in this book and I shall merely state that it is a simple matter in Abstract Algebra to explicitly embed any finitely generated abelian group into a divisible abelian group, thus establishing the existence of models for the appropriate sets Γ_0.

While these results may not be all that impressive (with respect to 4.7.i, for example, we already know in fact that every set can be well-ordered and not just linearly ordered), the simplicity of the proof ought to be. The Compactness Theorem is a powerful tool of first-order logic. Ultimately, however, its power derives from a fundamental weakness in the expressive power of the first-order language. The notion of finiteness and many notions dependent thereon are not expressible in a first-order language.

4.8 Examples. There is no set of sentences Δ in a first-order language such that the truth of Δ in $\mathfrak{M}$ characterises the following:

i. $\mathfrak{M}$ is finite;
ii. $\mathfrak{M}$ is archimedean ordered;
iii. $\mathfrak{M}$ is well-ordered.

I leave the verification of these as exercises to the reader. I note that one can characterise infinite structures as those in which an infinite set Δ is true; namely one can take Δ to be the set of sentences

$$\exists v_0 \ldots v_{n-1} \Big(\bigwedge_{i<j} \neg\, v_i = v_j \Big),$$

for $n \geq 1$. Δ cannot be replaced by any finite set of sentences. (Why?) It is also interesting to note that in Chapter III we obtained a nonarchimedean field $^*\mathbb{R}$ by adding a constant c and axioms $c > \bar{0}, c > \bar{1}, \ldots$ to the set of sentences true in $\mathbb{R}$. We cannot guarantee the nonarchimedean property without so expanding the language. For, if Δ defined nonarchimedean-ness, we would have

$$^*\mathbb{R} \vDash \Delta \text{ since } {}^*\mathbb{R} \text{ is nonarchimedean}$$

whence

$$\mathbb{R} \vDash \Delta \text{ since } \mathbb{R} \text{ and } {}^*\mathbb{R} \text{ satisfy the same sentences,}$$

but $\mathbb{R} \nvDash \Delta$ because $\mathbb{R}$ is archimedean ordered.

Mentioning the nonstandard reals and the fact that they are not archimedean points to a major weakness of first-order logic as a tool of the axiomatic method. Hilbert's first desideratum cannot be satisfied by any first-order axiomatisation of the real numbers in the sense that no structure $\mathfrak{R} = (\mathbb{R}; \ldots)$ for any choice of individuals, functions and relations can be characterised uniquely up to isomorphism the way the field of real numbers was characterised as *the* complete, nonarchimedean ordered field back in Chapter III. The same holds for Dedekind's characterisation of the natural numbers, a result that can be more concretely presented. For those interested I shall discuss this in the next subsection. I wish first, however, to discuss another consequence of the Completeness Theorem and a surprising application thereof. The consequence is called the Löwenheim-Skolem Theorem and the application is known as Skolem's Paradox.

The Löwenheim-Skolem Theorem was initially proven by Leopold Löwenheim in 1915 and was subsequently given a simplified proof as well as a generalisation by Thoralf Skolem in 1920[110], and has since been generalised further. The basic result is the following:

4.9 Theorem (Löwenheim-Skolem Theorem). *Let Γ be a set of sentences in a countable first-order language L. If Γ has any infinite model at all, then Γ has a countable model.*

In this form, the Löwenheim-Skolem Theorem— and, indeed, its generalisation to uncountable languages— is an immediate application of the proof of the Completeness Theorem: if L has cardinality κ, then the model constructed in that proof had cardinality at most κ— for, its domain consisted of equivalence classes of new constants, of which there were exactly κ introduced.

A sharper version due to Skolem asserts that if $\mathfrak{M}$ is an infinite model of Γ, then some countable substructure of $\mathfrak{M}$ is a model of Γ.

Like the Compactness Theorem, the Löwenheim-Skolem Theorem is a mainstay of Mathematical Logic. While it is not as easy to point to any impressive application comparable to the use of the Compactness Theorem to found nonstandard analysis, it does have its use. For example, a generalisation of the sharper version to all infinite cardinals provides a key step in Gödel's proof that the Generalised Continuum Hypothesis holds in his subuniverse of constructible sets mentioned in subsection 2.4 above.

The Löwenheim-Skolem Theorem also yields the *Skolem Paradox*: There is a countable model of set theory in which there are uncountable sets.

To prove this, Skolem had to choose a first-order formulation of set theory. Now the original version of the Axiom of Separation referred vaguely to a "definite" property. This axiom becomes first-order as soon as we declare "definite" to mean "given by a first-order formula of the language":

Z3. If $\varphi(x)$ is a formula of the first-order language of set theory and x is a set, one can separate out of x those elements of x satisfying φ:

$$\forall x \exists z \forall y \big(y \in z \leftrightarrow y \in x \wedge \varphi(y)\big),$$

where the variable z does not occur in $\varphi(y)$. The same should, of course, be done with the Axiom of Replacement *Z8*, but, as the reader can look back and verify, I already did this in writing down *Z8*. Thus we have set theory presented to us as a first-order theory to which the Löwenheim-Skolem Theorem applies: There is a countable model $\mathfrak{M}$ of set theory which contains an infinite set $\omega^{\mathfrak{M}}$ the power set of which, $\mathcal{P}^{\mathfrak{M}}(\omega^{\mathfrak{M}})$, is of necessity uncountable in $\mathfrak{M}$ and yet must be countable because its members are all elements of the countable set comprising the domain of $\mathfrak{M}$.

I cannot improve on Gentzen's remarks on this paradox:

---

[110] Both papers can be found in English translation in van Heijenoort, *op. cit.* The terminology of Löwenheim's paper is sufficiently archaic as to have required four pages of explanation prior to the paper's presentation and his paper is not recommended reading.

One could probably say that this result is not especially agreeable for axiomatic set theory. It asserts that all uncountable powers one speaks about in set theory are, in a certain sense, only appearances, insofar as one can substitute certain *countable* sets for them without altering the validity of any theorems.

One's first impression is that an *inconsistency* must arise from this. One proves in axiomatic set theory, for example, that the set of all real numbers is not countable. To be precise, one proves the theorem: there is no one-to-one correspondence between the natural numbers and the real numbers. Consider what this means in Skolem's countable model of set theory. This model contains objects which represent the natural numbers of the axiom system, other objects which represent the real numbers, and still others representing the *correspondences* which are possible on the basis of the axiom system; and each sort consists of at most countably many objects. Nevertheless the theorem mentioned remains valid in this model, for, among the *correspondences* available in the model, there are none mapping the countably many representatives of the "natural numbers" one-to-one onto the countably many representatives of the "real numbers". If we assume given a one-to-one correspondence, it will not be represented among the correspondences available in the model.

Perhaps these not easily understood facts become somewhat clearer if one gives them the following expression, whereby I restrict myself to the continuum of real numbers as the prototype of an "uncountable set": one puts oneself in the standpoint of the in-itself conception, that the continuum in- and of-itself is given beforehand, say as the set of all arbitrary infinite decimal fractions. Then following *Cantor* one can prove the uncountability of this system. Now, however, the following may be said: every *axiom system* for analysis that one may propose is in a certain sense *inadequate* for the complete capture of the continuum so conceived. For, the theorem of Skolem tells us that, taking a definite axiom system, this continuum can be *replaced* by a countable model, which similarly satisfies all the properties of the continuum laid down *in the axiom system*. By this conception, Skolem's result would so to speak be demonstrated not to be a defect of the continuum or of higher cardinalities, but rather a defect of human thought with respect to characterising these powers.[111]

Zermelo was not at all pleased with the Skolem Paradox. In early 1930 Zermelo was writing an important paper[112].

---

[111] Menzler-Trott, *op. cit.*, pp. 358 - 359. These remarks are from a lecture given by Gentzen in 1937 on "The current situation in research in the foundations of mathematics".

[112] Ernst Zermelo, "Über Grenzzahlen und Mengenbereiche. Neue Untersuchungen über die Grundlagen der Mengenlehre", *Fundamenta mathematicæ* 16 (1930), pp. 29 - 47.

During the process of publication of the paper Zermelo was confronted with Skolem's first-order approach which admitted countable models of set theory, a fact totally alien to his idealistic point of view. As a reaction he cancelled the condition of definiteness that was still present in the axiom of separation as formulated in the submitted version, now allowing "*arbitrary* propositional functions." Putting aside his former scepticism, he also allowed arbitrary functions in the axiom of replacement, altogether arriving at an axiom system of second order. Although he was informed then that Skolem had formulated the axiom of replacement nearly as early and for similar reasons as Fraenkel, he stuck to the names ZF and ZF′. He thus followed von Neumann who used the name "Zermelo-Fraenkel set theory" for the first time.[113]

Zermelo's snub of Skolem would be followed in other papers by anti-skolemical polemics. He would also consider variations of ZF in languages allowing infinite conjunctions and disjunctions.

4.3 Dedekind and Peano Revisited

Recall the Dedekind-Peano axioms for the structure[114] $\mathfrak{N}_0 = (\mathbb{N}; \prime, 0)$. Other than the logical axioms, these are

D1. $\forall x(\neg 0 = x')$

D2. $\forall xy(x' = y' \to x = y)$

D3. $\forall x \forall X \big(0 \in X \wedge \forall y(y \in X \to y' \in X) \to x \in X\big).$

The third axiom is not first-order. It contains a variable X ranging, not over individuals, but over *sets* of individuals. Following our discussion of Skolem's modification of the Axiom of Separation, the obvious accommodation is to replace the axiom *D3* by a first-order schema,

D3¹. $\forall x\big(\varphi(0) \wedge \forall y(\varphi(y) \to \varphi(y')) \to \varphi(x)\big),$

for all first-order formulæ $\varphi(x)$ in the language with 0, $\prime$, $=$ as primitives. Does this subsitution work, i.e., do *D1, D2, D3¹* have all the first-order consequences that *D1 - D3* do? Or, do we have to expand our logic to handle structures like $(\mathbb{N}; \mathcal{P}(\mathbb{N}); \in, \prime, =, 0)$? And, if so, precisely how do we do this?

   Let us first consider the structure $\mathfrak{N}_0 = (\mathbb{N}; \prime, 0)$ and its first-order theory. It turns out that very little induction is needed to determine this theory. To see this, consider an arbitrary structure $\mathfrak{X} = (X; f)$ where $f : X \to X$ is one-to-one. What can $\mathfrak{X}$ look like? Well, $\mathfrak{X}$ does not have a whole lot of structure to it.

---

[113] Ebbinghaus, *op. cit.*, p. 189. I have deleted several parenthetical references to the literature from this passage. Ignoring fine points about the choice of the language, ZF is the system of set theory based on axioms *Z1 - Z9* other than the Axiom of Choice; ZF′ is that system minus the Axiom of Infinity.

[114] I know: Dedekind and Peano considered $(\mathbb{N}^+; \prime, 1)$. As a logician my preference is for starting with 0. Peano himself switched over from $\mathbb{N}^+$ to $\mathbb{N}$ in "Sul concetta di numero", *Revista di Matematica* 1 (1891), pp. 87 - 102.

Given an element $x \in X$, one can follow its progress as one iterates application of the map:

$$x, f(x), f(f(x)), \dots$$

And, since f is one-to-one, a partial inverse f^{-1} with the range of f as its domain exists and one can follow the backwards progress of x as one iterates the application of the inverse as long as this is possible:

$$\dots, f^{-1}(f^{-1}(x)), f^{-1}(x), x.$$

The full course of this progress we call the *trace* of the element x,

$$Tr(x) = \{f^n(x) \mid n \in \mathbb{Z}\},$$

where $f^0(x) = x$, $f^n(x)$ is the n-fold application of f to x for $n \in \mathbb{N}^+$, and f^n is the $|n|$-fold application of f^{-1} to x for negative integers n.

There are three kinds of traces:

i.   There is an element $a \in Tr(x)$ which is not in the range of f. In this case,

$$Tr(x) = Tr(a) = \{a, f(a), f^2(a), f^3(a), \dots\}$$

is isomorphic to $(\mathbb{N}; ', 0)$.

ii. $Tr(x)$ is finite with m elements for some $m \in \mathbb{N}^+$. Then, for any $a \in Tr(x)$, $f^m(a) = a$, i.e.

$$Tr(x) = \{a, f(a), \dots, f^{m-1}(a)\}.$$

iii. Every element of $Tr(x)$ is in the range of f and there is no looping. Then, for any $a \in Tr(x)$,

$$Tr(x) = Tr(a) = \{\dots, f^{-2}(a), f^{-1}(a), a, f(a), f^2(a), \dots\}$$

and $(Tr(x); f, a)$ is isomorphic to $(\mathbb{Z}; ', 0)$.

If Y is another set with a one-to-one function g defined on it, then $\mathfrak{Y} = (Y; g)$ is isomorphic to $\mathfrak{X}$ just in case $\mathfrak{X}$ and $\mathfrak{Y}$ have the same number of traces of kind i, the same number of traces of cardinality m for each $m \in \mathbb{N}^+$, and the same number of traces of kind iii.

Writing DP for the theory given by axioms *D1, D2, D3*[1], we can prove in DP that there is exactly one trace of the first kind and there are none of the second.

4.10 Lemma. *i.* $\mathsf{DP}\vdash \forall x\big(\neg x = 0 \to \exists z(x = z')\big)$.
ii. $\mathsf{DP}\vdash \forall x(\neg x = x'^{\dots'})$, for $n \geq 1$ *applications of the successor function.*

Proof. Both proofs are by induction within DP.
i. Let $\varphi(x)$ be the formula

$$\neg x = 0 \to \exists z(x = z').$$

Observe that $\mathsf{DP}\vdash \varphi(0)$ since $\neg 0 = 0$ logically implies any sentence. To prove $\forall y\big(\varphi(y) \to \varphi(y')\big)$, it suffices to prove $\varphi(y') : \neg y' = 0 \to \exists z(y' = z')$. But

$$\mathsf{DP} \vdash y' = y' \Rightarrow \mathsf{DP} \vdash \exists z(y' = z')$$
$$\Rightarrow \mathsf{DP} \vdash \neg\, y' = 0 \rightarrow \exists z(y' = z').$$

ii. Let $\varphi(x)$ be $\neg\, x = x'^{\cdots\prime}$. $\varphi(0)$ follows directly from *D1*. And the induction step follows from *D2*:

$$\mathsf{DP} \vdash y' = (y'^{\cdots\prime})' \rightarrow y = y'^{\cdots\prime},$$

i.e., $\mathsf{DP} \vdash \neg\,\varphi(y') \rightarrow \neg\,\varphi(y)$, whence contraposition yields $\mathsf{DP} \vdash \varphi(y) \rightarrow \varphi(y')$.

$\square$

And what about the third kind of trace? DP is noncommittal. It is not that some additional axioms have to be added, but that the language cannot distinguish between models of DP which have traces of the third kind and those which don't. To state this formally, we need a definition.

4.11 Definition. *Let L be a first-order language and $\mathfrak{M}_0, \mathfrak{M}_1$ two structures for the language. We say $\mathfrak{M}_0, \mathfrak{M}_1$ are* elementarily equivalent, *written $\mathfrak{M}_0 \equiv \mathfrak{M}_1$, if we have*

$$\mathfrak{M}_0 \vDash \varphi \quad iff \quad \mathfrak{M}_1 \vDash \varphi$$

for all sentences φ of the language L.

Let us further write DP_0 for the subtheory of DP given by choosing as nonlogical axioms the axioms *D1, D2*, and the consequences of Lemma 4.10:

D4. $\forall x\big(\neg\, x = 0 \rightarrow \exists z(x = z')\big)$.
D5$_n$. $\forall x(\neg\, x = x'^{\cdots\prime})$, for $n \geq 1$ applications of the successor.

4.12 Theorem. *Let $\mathfrak{M}$ be a model of DP_0. There is a countable model $\mathfrak{M}^* \vDash \mathsf{DP}_0$ which has infinitely many distinct traces of the third kind and which is elementarily equivalent to $\mathfrak{M}$.*

We will prove this by a compactness argument. First, however, let us derive the following corollary from it.

4.13 Corollary. *i. Any two models of DP_0 are elementarily equivalent.*
ii. DP_0 is complete: *for any sentence φ of the language,*

$$\mathsf{DP}_0 \vdash \varphi \quad or \quad \mathsf{DP}_0 \vdash \neg\varphi.$$

iii. For any sentence φ of the language,

$$\mathsf{DP}_0 \vdash \varphi \quad iff \quad \mathfrak{N}_0 \vDash \varphi.$$

iv. DP_0 and DP have the same theorems.

Proof. i. Let $\mathfrak{M}_0, \mathfrak{M}_1$ be models of DP_0. By the Theorem, there are countable models $\mathfrak{M}_0^*, \mathfrak{M}_1^*$ of DP_0 elementarily equivalent to $\mathfrak{M}_0, \mathfrak{M}_1$, respectively, which have infinitely many traces of the third kind. As they each have exactly one trace of the first kind, none of the second, and countably many of the third,

$\mathfrak{M}_0^*, \mathfrak{M}_1^*$ are isomorphic and thus satisfy the same sentences of L, i.e., they are elementarily equivalent. But

$$\mathfrak{M}_0 \equiv \mathfrak{M}_0^* \equiv \mathfrak{M}_1^* \equiv \mathfrak{M}_1,$$

and the obvious transitivity of $\equiv$ yields the result.

ii - iii. Note

$$\mathsf{DP}_0 \vdash \varphi \quad \text{iff} \quad \text{for all } \mathfrak{M} \vDash \mathsf{DP}_0, \ \mathfrak{M} \vDash \varphi$$
$$\text{iff} \quad \mathfrak{N}_0 \vDash \varphi.$$

This establishes iii. The same equivalence holds for $\neg\varphi$, whence from

$$\mathfrak{N}_0 \vDash \varphi \quad \text{or} \quad \mathfrak{N}_0 \vDash \neg\varphi,$$

the conclusion ii follows.

iv. Since the axioms of DP_0 are provable in DP, it follows that

$$\mathsf{DP}_0 \vdash \varphi \Rightarrow \mathsf{DP} \vdash \varphi.$$

The converse implication follows from completeness:

$$\mathsf{DP}_0 \nvdash \varphi \Rightarrow \mathsf{DP}_0 \vdash \neg\varphi$$
$$\Rightarrow \mathsf{DP} \vdash \neg\varphi$$
$$\Rightarrow \mathsf{DP} \nvdash \varphi. \qquad \qquad \square$$

Proof of Theorem 4.12. Let $\mathfrak{M} \vDash \mathsf{DP}_0$ and let

$$\Delta = \{\varphi \text{ in the language } L \mid \mathfrak{M} \vDash \varphi\}.$$

We wish to add a sequence $c_0, c_1, \ldots$ of constants to the language and axioms to Δ asserting that the elements interpreting the constants in any model of the new theory belong to distinct traces of the third kind.

To guarantee that each c_i denotes an element with the third kind of trace, we must add axioms asserting we can always find predecessors, i.e., we must add all axioms of the form

$$\exists y(y'^{\cdots'} = c_i).$$

For later discussion, let us denote by $t + n$ the term $t'^{\cdots'}$ with n applications of the successor function. Thus, we will add to Δ the set of axioms

$$\Pi_i = \{\exists y(y + n = c_i) \mid n \in \mathbb{N}^+\}.$$

To guarantee that, for $i \neq j$, the constants denote elements of disjoint traces, we must add axioms asserting no predecessor or successor of c_i equals any such of c_j:

$$\neg c_i \pm m = c_j \pm n,$$

where $c - m$ denotes the m-fold predecessor of c. A moment's reflexion should convince the reader that these axioms contain a lot of redundancy and can be replaced by the simpler set of axioms of the form,

$$\neg c_i + n = c_j.$$

For, if c_i and c_j are in the same trace, either they are equal ($c_i + 0 = c_j$) or one is obtained from the other by iterated application of the successor function ($c_i + n = c_j$ or $c_j + n = c_i$). Thus we take as axioms of incomparability of traces for $i \neq j$ the sets I_{ij}, I_{ji}, where

$$I_{ij} = \{\neg c_i + n = c_j \mid n \in \mathbb{N}\}.$$

Let $\Gamma = \Delta \cup \bigcup\{\Pi_i \mid i \in \mathbb{N}\} \cup \bigcup\{I_{ij} \mid i, j \in \mathbb{N}\ \&\ i \neq j\}$. By the Compactness Theorem, Γ will have a model iff every finite subset $\Gamma_0 \subseteq \Gamma$ has a model. It turns out that, for each such Γ_0, we can interpret the constants c_i occurring in Γ_0 within $\mathfrak{M}$, so that $\mathfrak{M}$ itself models Γ_0. To see this, let Γ_0 be given and choose n_0 so large that only constants c_i for $i < n_0$ occur in Γ_0 and axioms $\exists y(y + n = c_i), \neg c_i + n = c_j$ occur in Γ_0 only for $n < n_0$. To make the interpretation, note that $\mathfrak{M}$ has a trace of the first kind, the elements of which we can identify with the natural numbers. We interpret the constants $c_0, c_1, \ldots, c_{n_0 - 1}$ in the natural numbers as follows:

$$c_i^{\mathfrak{M}} \text{ is } (i + 1)n_0,$$

i.e., $c_0^{\mathfrak{M}}$ is n_0, $c_1^{\mathfrak{M}}$ is $2n_0$, etc.

The axioms of $\Pi_i \cap \Gamma_0$ are true because they are of the form $\exists y(y + n = c_i)$ for $n < n_0$ and $c_i^{\mathfrak{M}}$ has all the necessary predecessors because $c_i^{\mathfrak{M}} = (i + 1)n_0 > n_0 > n$.

Similarly, the axioms of $I_{ij} \cap \Gamma_0$ are true because they assert $c_i^{\mathfrak{M}} + n \neq c_j^{\mathfrak{M}}$ for $i \neq j$ and $n < n_0$, while it takes n_0 applications of successor to get from one $c_k^{\mathfrak{M}}$ to the next.

Thus Γ_0 has a model and by Compactness, the full set Γ has such a model. The Löwenheim-Skolem Theorem tells us this model $\mathfrak{M}^*$ can be taken to be countable. Finally, I note that $\mathfrak{M}^* \equiv \mathfrak{M}$ because $\Delta \subseteq \Gamma$ and Δ was the set of all sentences of the original language true in $\mathfrak{M}$.   □

I believe I promised some pages back that our discussion of nonstandard models of DP would be more concrete than that of nonstandard models of analysis, and yet here I've again applied Compactness and, additionally, the Löwenhem-Skolem Theorem. Be that the case, we can see however that the models themselve are more concretely defined.

4.14 Example. Let $\mathfrak{M}$ result by affixing to $\mathfrak{N}_0 = (\mathbb{N}; ', 0)$ a copy of $(\mathbb{Z}; ')$, i.e., define

$$M = \{\langle 0, n \rangle \mid n \in \mathbb{N}\} \cup \{\langle 1, n \rangle \mid n \in \mathbb{Z}\},$$

choose $\langle 0, 0 \rangle$ for $0^{\mathfrak{M}}$, and define successor by

$$\langle i, n \rangle' = \langle i, n + 1 \rangle.$$

Then: $\mathfrak{M} = (M; ', 0) \equiv (\mathbb{N}; ', 0) = \mathfrak{N}_0$. For, obviously, $\mathfrak{M} \models \mathsf{DP}_0$.

Actually, the whole treatment can be replaced by a more concrete syntactic one. The theory DP_0 is so simple and the language so inexpressive that the theory admits an *elimination of quantifiers*: For every formula φ one can find in the same language a formula ψ with no quantifiers such that $\mathsf{DP}_0 \vdash \varphi \leftrightarrow \psi$. The procedure is quite effective and has other applications besides yielding anew Theorem 4.12 and its corollary. Most notably, by analysing the quantifier-free formulæ of the language with primitives $', 0, =$, one can easily show that the only sets definable in the structure $\mathfrak{N}_0$ are finite or *co-finite* (i.e., possessing a finite complement). Thus, for example, the set of even numbers is not definable in this structure— and therefore neither of the operations of addition and multiplication is so definable. For, evenness is definable in terms of each of these operations:

$$x \text{ is even iff } \exists y(y + y = x) \text{ iff } \exists y(2y = x).$$

The definable binary relations also have a relatively simple description from which one can quickly see that the ordering relation and the graphs of the doubling and squaring functions are not definable. From these latter two non-definabilities one again sees the nondefinability of addition and multiplication.

Nothing is to be gained here by presenting the quantifier elimination and its application to the analysis of the definable sets and binary relations. The most important nondefinability results are readily established by simple, albeit *ad hoc*, arguments.

4.15 Lemma. *The set of even numbers is not definable in* $\mathfrak{N}_0$.

Proof. Suppose to the contrary that a formula $\varphi(x)$ defined the set of even numbers in $\mathfrak{N}_0$. Now

$$\mathfrak{N}_0 \vDash \forall x\big(\varphi(x) \leftrightarrow \neg\,\varphi(x')\big)$$

whence completeness yields

$$\mathsf{DP}_0 \vdash \forall x\big(\varphi(x) \leftrightarrow \neg\,\varphi(x')\big),$$

whence

$$\mathfrak{M} \vDash \forall x\big(\varphi(x) \leftrightarrow \neg\,\varphi(x')\big), \qquad (12)$$

for the structure $\mathfrak{M}$ of Example 4.14.

Define $f : M \to M$ by

$$f(i, j) = \begin{cases} \langle i, j \rangle, & i = 0 \\ \langle i, j + 1 \rangle, & i = 1. \end{cases}$$

f is an isomorphism of $\mathfrak{M}$ onto itself. Because isomorphisms obviously preserve truth, we have

$$\mathfrak{M} \vDash \varphi(\overline{m}) \text{ iff } \mathfrak{M} \vDash \varphi(\overline{f(m)}),$$

for any $m \in M$. In particular, for $m = \langle 1, 0 \rangle$, say,

$$\mathfrak{M} \vDash \varphi(\overline{m}) \text{ iff } \mathfrak{M} \vDash \varphi(\overline{m'}),$$

contrary to (12).   □

4.16 Corollary. *Addition and multiplication are not definable in* $\mathfrak{N}_0$.

One can similarly use an isomorphism argument applied to a structure with two traces of the third kind to verify that the ordering relation on $\mathbb{N}$ is not definable in $\mathfrak{N}_0$.

So we see quite dramatically that the theory DP_0, whence DP, does not capture Dedekind's and Peano's axiomatisation of arithmetic. The question now is how one goes about recapturing what has been lost in replacing the second-order axiom *D3* by the first-order schema *D3*[1].

Two approaches come to mind. The first is to stick with first-order logic, but expand the language to include primitives for some functions or relations not already definable and add new axioms codifying their basic properties. The second is to expand the logic to allow the existence of the missing functions and relations to be consequences of axioms *D1* - *D3*. The Compactness Theorem and the Löwenheim-Skolem Theorem, which we have exploited as tools, each precludes the possibility of any first-order set of axioms characterising the natural numbers up to isomorphism. On the other hand, Hilbert's second desideratum, that all consequences be derivable by means of a finite number of logical steps, would seem to make Compactness inevitable. This would again mean nonstandard models. So there is no compelling reason to rule out first-order logic.

There are several natural additions to the language we have been considering. I wish to report quickly and shallowly on the theories of the natural numbers formulated in these languages, not getting into any details. Including the language we have been using, the languages are the following:

L_0: $0, ', =$

L_1. $0, ', <, =$

L_2: $0, 1, +, <, =$

L_3: $0, 1, +, \cdot, =$.

The theory DP_0 of the structure $(\mathbb{N}; ', 0)$ for the language L_0 was treated syntactically by Jacques Herbrand in his thesis, *Recherches sur la théorie de la démonstration*, submitted to the Sorbonne in 1929.[115] Herbrand presents the quantifier elimination in the 4th chapter of this thesis[116] and remarks without proof that he has checked that the treatment extends to the theory of the structure $(\mathbb{Z}; ', \mathrm{pd}, <, =)$, where pd is the predecessor function, $\mathrm{pd}(x) = x - 1$.[117]

A quantifier elimination for this latter theory, as well as for the theory of the structure $\mathfrak{N}_1 = (\mathbb{N}; 0, ', <, =)$ for L_1 had already been given by the American mathematician Cooper Harold Langford[118]. The theory of the structure $\mathfrak{N}_1$ is

---

[115] An English translation, "Investigations in proof theory", can be found in: Warren D. Goldfarb (ed.), *Jacques Herbrand: Logical Writings*, Harvard University Press, Cambridge (Mass.), 1971.

[116] *Ibid.*, pp. 112 - 132.

[117] *Ibid.*, p. 131.

[118] C.H. Langford, "Theorems on deducibility (second paper)", *Annals of Mathematics*, 2nd series, 28 (1926/27), pp. 459 - 471. A first paper bearing the name "The-

just the theory DiLO of a discrete linear ordering with first, but no last element, the nonlogical axioms of which are:

A1. $\forall x(\neg\, x < x)$

A2. $\forall xyz(x < y \land y < z \to x < z)$

A3. $\forall xy(x < y \lor x = y \lor y < x)$

A4. $\forall x(x = 0 \lor 0 < x)$

A5. $\forall xy(x < y \to x' < y')$

A6. $\forall xy(x < y' \leftrightarrow x = y \lor x < y)$

A7. $\forall x\big(0 < x \to \exists y(x = y')\big)$

Axioms *A1* - *A3* assert that the relation $<$ is a linear ordering, *A4* that 0 is the first element of the ordering, *A5* that successor preserves order, *A6* that the ordering is discrete, and *A7* that every nonzero element is a successor. Of the axioms for DP_0, we have retained only *D4* as *A7*, the rest being derivable from *A1* - *A7*.

Our semantic treatment of DP_0 can be extended to DiLO. There is a slight new twist in that the models of DiLO are no longer characterised up to isomorphism by the number of traces of the third kind they have. However, the ordering of a model $\mathfrak{M} \models \mathsf{DiLO}$ induces an ordering on these traces and any two countable models of the theory for which the ordering of the traces of the third kind is dense with no first nor last elements are isomorphic. The analogue to Theorem 4.12 will thus assert the structure $\mathfrak{M}^*$ to be of this sort. With that the analogue to the corollary to 4.12 also carries over and we see that DiLO is indeed the theory of the structure $\mathfrak{N}_1$.[119]

Adding order to the model $\mathfrak{M}$ of Example 4.14, one again sees that the set of even numbers is not definable in $\mathfrak{N}_1$, whence addition and multiplication are not definable in $\mathfrak{N}_1$.

The most cited quantifier elimination in the literature is that for the theory PSA (for *Presburger-Skolem Arithmetic*) of the structure $\mathfrak{N}_2 = (\mathbb{N}; +, 0, 1, <)$. This theory does not admit an elimination of quantifiers in the strict sense that every formula φ of L_2 is equivalent to a quantifier-free formula of L_2, but one does get such an elimination if one adds new primitive relation symbols for some definable relations. In 1929, Mojżesz Presburger proved such an elimination for the structure $(\mathbb{Z}; +, 0, 1)$ after expanding the language by adding new primitives for the relations of congruence modulo n for $n = 2, 3, 4, \ldots$ He also announced

---

orems on deducibility" appeared on pp. 16 - 40 of the same volume and presented a quantifier elimination for the theory of dense linear orderings.

[119] The interested reader will find some hints to the proof in Smoryński, *Logical Number Theory I*, op. cit., Exercise III.3.4, pp. 306 - 307. Some additional exercises:

i.   If $(M; ', 0)$ is a model of DP_0, there is a relation $<$ on M for which $(M; ', 0, <)$ is a model of DiLO.

ii. Every set of natural numbers definable by a formula of the language L_1 in $\mathfrak{N}_1$ is finite or co-finite.

iii. Every set of natural numbers definable by a formula of the language L_0 in $\mathfrak{N}_0$ is finite or co-finite.

without proof that the result carried over to the structure $(\mathbb{Z}; +, 0, 1, <)$.[120] In 1930, Skolem independently provided a quantifier elimination using a radically different choice of new primitives.[121] The elimination procedures for the integers apply also to the natural numbers and are quite effective, if not efficient. They yield a complete axiomatisation for the theory PSA of $\mathfrak{N}_2$, a decision procedure (i.e., an effective procedure for determining the truth or falsity of any sentence of L_2 in $\mathfrak{N}_2$), and a characterisation of the definable sets in the language as the *ultimately periodic* sets: $X \subseteq \mathbb{N}$ is definable in $\mathfrak{N}_2$ iff there are positive integers p, n_0 such that, for all $n > n_0$, $n \in X$ iff $n + p \in X$. Noting that the set of squares is not ultimately periodic, we can conclude that the set of squares, whence also the multiplication operation, is not definable in the language.

The simple, induction-free axiomatisation DP_0 of the theory of the structure $\mathfrak{N}_0$ gave us a simple description of all models of the theory and allowed us to construct simple nonstandard models thereof in Example 4.14. The induction-free axiomatisation DiLO does the same for the theory of the structure $\mathfrak{N}_1$: models of DiLO are obtained from models of DP_0 by ordering the elements of the individual traces so that $x < x + n$ for all positive n, and then ordering the traces, placing the trace containing 0 at the beginning. In like manner, the theory PSA of the structure $\mathfrak{N}_2$ has a complete induction-free axiomatisation:

A1. $\forall xyz\big(x + (y + z) = (x + y) + z\big)$

A2. $\forall xy(x + y = y + x)$

A3. $\forall x(x + 0 = x)$

A4. $\forall xy\big(x < y \rightarrow \exists z(x + z = y)\big)$

O1. $\forall x(\neg\, x < x)$

O2. $\forall xyz(x < y \wedge y < z \rightarrow x < z)$

O3. $\forall xy(x < y \vee x = y \vee y < x)$

O4. $\forall x(x = 0 \vee 0 < x)$

O5. $\forall xyz(x < y + 1 \leftrightarrow x = y \vee x < y)$

OA. $\forall xyz(x < y \rightarrow x + z < y + z)$

D_n. $\forall x \exists y \left(\bigvee_{k=0}^{n-1} x = ny + \overline{k} \right)$,

where n is an integer > 1, nt denotes the n-fold sum $t + \ldots + t$ for any term t, and $\overline{k}$ the term $k1$ or 0 according to whether $k > 0$ or $k = 0$.

The induction-free axiomatisation again allows us to exhibit relatively simple nonstandard models.

---

[120] M. Presburger, "Über die Vollständigkeit eines gewissen Systems der Arithmetik ganzer Zahlen, in welchen die Addition als einzige Operation hervortritt", in: *Compte Rendus I Congrès des Mathématiciens des Pays Slaves*, Warsaw, 1930.

[121] Thoralf Skolem, "Über einige Satzfunktionen in der Arithmetik", *Skrifter Vitenskapsakademiet i Oslo* I, No. 7 (1931), pp. 1 - 28; reprinted in: Thoralf Skolem, *Selected Works in Logic*, Universitetsforlaget, Oslo, 1970. A presentation of Skolem's method in English is given in Smoryński, *Logical Number Theory I, op. cit.*, pp. 307 - 329.

4.17 Examples. Let $n \in \mathbb{N}$. Define P_n to be the set of polynomials of degree $m \leq n$,

$$a_0 + a_1 X + \ldots + a_{m-1} X^{m-1} + a_m X^m,$$

for which

i. $a_0 \in \mathbb{N}$
ii. $a_1, \ldots, a_m \in \mathbb{Q}$
iii. $a_m > 0$.

$0, 1, +$ are defined on P_n in the usual way, and the polynomials of P_n are ordered by eventual dominance:

$$P < Q : \quad \exists x_0 \forall x > x_0 \big(P(x) < Q(x)\big).$$

The structure $\mathfrak{P}_n = (P_n; 0, 1, +, <)$ is a model of PSA. (*Exercise.*[122])

Finally, we consider the language L_3 and the corresponding structure $\mathfrak{N}_3 = (\mathbb{N}; +, \cdot, 0, 1)$.

4.4 Arithmetic with $+$ and $\cdot$

Once one has addition and multiplication together, the expressive power of the language changes dramatically. To begin with, one has Cantor's pairing function referred to earlier in footnote 19:

$$\langle x, y \rangle = \frac{(x + y)^2 + 3x + y}{2},$$

i.e.

$$x = \langle x, y \rangle \text{ iff } 2z = (x + y)^2 + 3x + y.$$

Kurt Gödel observed that using the Chinese Remainder Theorem from elementary Number Theory, one can in fact code arbitrary finite sequences.

The Chinese Remainder Theorem, for those not familiar with it, makes its appearance in the *Shùshū jiǔzhāng* of Qín Jiǔsháo of 1247 and asserts that if $d_0, d_1, \ldots, d_{n-1}$ are pairwise relatively prime positive integers and $x_0, x_1, \ldots, x_{n-1}$ are any nonnegative integers, then there is an integer x such that, for each i, x is congruent to x_i modulo d_i, i.e., d_i divides $x - x_i$. Thus the sequence $x_0, x_1, \ldots, x_{n-1}$ is coded by the sequence $x, d_0, d_1, \ldots, d_{n-1}$. This is useful because there are lots of sequences $d_0, d_1, \ldots, d_{n-1}$ which can be coded by simple pairs of numbers.

4.18 Lemma (Gödel's β-Function). *Define*

$$\beta(c, d, i) = \text{the remainder of } d \text{ after division by } 1 + (i + 1)c.$$

For any sequence $x_0, x_1, \ldots, x_{n-1}$ of nonnegative integers there are c, d such that, for $i = 0, 1, \ldots, n-1$, $\beta(c, d, i) = x_i$.

---

[122] Another exercise: Show that none of the structures $\mathfrak{P}_n$ for $n > 0$ admits a multiplication operation on it which satisfies the usual algebraic laws and preservation of order: $\forall xyz(x < y \wedge 0 < z \rightarrow xz < yz)$. Conclude that multiplication is not definable on $\mathfrak{N}_2$ by any formula of L_2.

Proof. Let $m > n, x_0, x_1, \ldots, x_{n-1}$ and choose $c = m! = m(m-1)\cdots 1$ The numbers

$$1 + 2c, 1 + 3c, \ldots, 1 + nc \tag{13}$$

are pairwise relatively prime. For, if $1 + (i+1)c$ and $1 + (j+1)c$ have a common factor where $i < j$, they would have a common prime factor, say, p. If p divides two numbers, it divides their difference, in this case:

$$1 + (j+1)c - (1 + (i+1)c) = (j-i)c.$$

Now, p cannot divide c since a similar subtraction,

$$1 + (i+1)c - (i+1)c = 1,$$

shows p would then divide 1. Because p is prime, does not divide c, and divides $(j-i)c$, it follows that p divides $j - i$. Recalling that $c = m!$ for $m > n > j > 0$, it follows that $m > p$ and thus p divides $m!$, i.e., p divides c, a contradiction. Thus the numbers in the sequence (13) are pairwise relatively prime.

To finish the proof, apply the Chinese Remainder Theorem to find a number d_0 such that, for $i = 0, 1, \ldots, n-1$,

$$d_0 \equiv x_i \mod 1 + (i+1)c,$$

i.e., $1 + (i+1)c$ divides $d_0 - x_i$. Now choose

$$d = d_0 + \prod_{i=0}^{n-1}(1 + (i+1)c).$$

For $i = 0, 1, \ldots, n-1$, we clearly have

$$d \equiv d_0 \mod 1 + (i+1)c$$
$$\equiv x_i \mod 1 + (i+1)c.$$

Moreover, $d > 1 + (i+1)c > c > m > x_i$, whence x_i is the remainder of d on division by $1 + (i+1)c$, i.e., $\beta(c, d, i) = x_i$ ☐

The β-function is easily shown to be expressible in the language L_3. First, note that congruence is expressible:

$$x \equiv y \mod z: \quad \exists w(x + wz = y \vee y + wz = x).$$

Then observe

$$\beta(c, d, i) = x \text{ iff } (d \equiv x \mod 1 + (i+1)c) \wedge (x < 1 + (i+1)c).$$

As already remarked, the ability to refer to finite sequences yields a dramatic increase in expressive power in the step from L_2 to L_3. For example, one can define the graph of the exponential function:

$$z = x^y: \quad \exists cd\Big(\beta(c, d, 0) = 1 \wedge$$

$$\forall i\big(i < y \to \beta(c,d,i+1) = x \cdot \beta(c,d,i)\big) \wedge \beta(c,d,y) = z\big).$$

That one should find this impressive is made manifest by comparison with L_2. The simple exponential function $y = 2^x$ grows more rapidly than any polynomial, i.e., than any function representable by a term of the language L_3. In L_2, on the other hand, if $f(x)$ is a function expressible in the language, its range is ultimately periodic, whence the function itself will be bounded by a linear function $y = mx + b$, i.e., it will be bounded by a term $(x + \ldots + x) + (1 + \ldots + 1)$ of L_2 itself.

The language is expressive enough to transcend exponentiation:

$$z = x^{x^{\cdot^{\cdot^{\cdot^{x}}}}}\Big\}^{y} \;:\; \exists cd\Big(\beta(c,d,0,) = 1 \wedge$$

$$\forall i\big(i < y \to \beta(c,d,i+1) = x^{\beta(c,d,i)}\big) \wedge \beta(c,d,y) = z\big).$$

And this function can be iterated, and the next, and the next, ...

The ability to define functions recursively within the language L_3 is what separates arithmetic in L_3 from that in the weaker languages L_0, L_1, and L_2. The theory TA (for *True Arithmetic*) of the structure $\mathfrak{N}_3$ is more elusive. TA cannot be *effectively* axiomatised, and *effective* procedures cannot construct its nonstandard models.

There is a subfield of Mathematical Logic known variously as Recursion Theory, the Theory of Recursive Functions, and the Theory of (Effective) Computability. It emerged in the mid-1930's when Alonzo Church and Alan Turing independently proposed Type II definitions of the notion of a computable function. Other definitions of the notion by various mathematicians[123] soon appeared and the various definitions were all proven equivalent. Indeed, the definitions all involve the notion of a computation being a finite sequence of states, each of which can be coded by a finite list of symbols, and the equivalence proofs routinely proceed by showing that the functions computable by one definition can computably recognise the computations of functions that are computable according to a second definition. In this way, the functions computable according to the first definition can simulate the computations generated by the second definition, and the functions computable according to this second definition are seen to be computable according to the first definition. In

---

[123] An earlier definition had been proposed by Jacques Herbrand and cited, with minor modification, by Gödel in lectures at Princeton in 1934. Gödel, however, did nothing with the definition and was not ready to commit himself on the question of the complete generality of his definition, and Herbrand had already died in a mountaineering accident. Thus, credit generally goes to Church and Turing, whose work will be discussed in slightly more detail below. The other early major definitions of computability are due to Emil Post and A.A. Markov, Jr. A number of the very readable seminal papers of the field are anthologised in Martin Davis, *The Undecidable; Basic Papers on Undecidable Propositions, Unsolvable Problems and Computable Functions*, Raven Press Books, Ltd., Hewlett (New York), 1965. The collection has been republished by Dover.

like manner, with its ability to discuss finite sequences, L_3 has formulæ defining the graphs of all computable functions $f : \mathbb{N}^n \to \mathbb{N}$. This is not difficult to verify. The details can be found in any introductory textbook on Mathematical Logic.[124]

Now the computable functions are definable by, say, programs in the reader's preferred programming language. Like the computations, the programs themselves can be coded as sequences of symbols, hence sequences of natural numbers, and there are, thus, only countably many computable functions. One can diagonalise on a list $f_0, f_1, \ldots$ of them and obtain a noncomputable function,

$$f(n) = f_n(n) + 1.$$

This diagonalisation is expressible in L_3, i.e., there is a formula $\varphi(x, y)$ of L_3 such that, for all $m, n \in \mathbb{N}$,

$$f(m) = n \ \text{ iff } \ \mathfrak{N}_3 \vDash \varphi(m, n).$$

Thus, there are noncomputable functions definable in $\mathfrak{N}_3$. Indeed, the computable functions and relations form only the bottom level of a hierarchy[125] of functions and relations definable in $\mathfrak{N}_3$.

What does all of this expressive power have to tell us about TA, its possible axiomatisability, and the construction of its models? Let me first address the semantic issue.

Every discussion in mathematics begins with some new terminology. A structure $\mathfrak{M}$ that models TA or some axiomatic theory approximating TA is called a *model of arithmetic*. $\mathfrak{M}$ is *nonstandard* if it is not isomorphic to $\mathfrak{N}_3$. $\mathfrak{M}$ is a *strong* model of arithmetic if $\mathfrak{M} \vDash$ TA. The existence of nonstandard models of arithmetic was first proven by an infinitistic construction by Skolem in 1933 for any finitely axiomatised subtheory of TA and in 1934 for TA itself.[126]

---

[124] For the sake of definiteness, I cite the logic texts of Mendelson, van Dalen, and Smoryński mentioned in footnote 104, above, but any textbook on Mathematical Logic will be more than adequate. An alternative, of course, is the source book by Davis mentioned in the preceeding footnote.

[125] This hierarchy was first independently studied in the 1940's by Stephen Cole Kleene and Andrzej Mostowski. It is not quite as ubiquitous in the logic textbooks as the computable functions are and I refer the reader to my text cited in the preceding footnote or to a dedicated textbook on Recursion Theory.

[126] The first proof appeared in a paper, "Über die Unmöglichkeit einer Charakterisierung der Zahlenreihe mittels eines endlichen Axiomensystems" [On the impossibility of a characterisation of the number series by means of a finite axiom system], *Norsk Matematisk Forening, Skrifter*, series 2 (1933), pp. 73 - 82. The second proof appeared in the paper, "Über die Nichtcharakterisierbarkeit der Zahlenreihe mittels endlich oder abzählbar unendlich vieler Aussagen mit ausschliesslich Zahlenvariablen" [On the noncharacterisability of the number series by means of finitely or infinitely many statements with only number variables], *Fundamenta Mathematicæ* 23 (1934), pp. 150 - 161. The construction was repeated, in English, at a conference in Amsterdam in 1954: Th. Skolem, "Peano's axioms and models of arithmetic", in: Th. Skolem, G. Hasenjaeger, G. Kreisel, A. Robinson, Hao Wang, L. Henkin, and

Skolem's construction is not hard to describe and might give us a feeling of *déjà vu*. There is, in fact, a certain inevitability to it. Suppose we are already given a strong nonstandard model of arithmetic $\mathfrak{M} = (M; +, \cdot, 0, 1)$ and $a \in M$ is a nonstandard natural number. Let $\mathcal{F}$ be the family of all arithmetically definable unary functions, i.e., all functions $f : \mathbb{N} \to \mathbb{N}$ definable by formulæ of L_3. It can be shown that the structure $\mathfrak{M}_0 = (M_0; +, \cdot, 0, 1)$ given by

$$M_0 = \{ f(a) \,|\, f \in \mathcal{F} \}$$

is also a strong nonstandard model of arithmetic. Now the elements of M_0 can almost be identified with the functions of $\mathcal{F}$. They would more properly be identified with equivalence classes of these functions under the relation

$$f \equiv g : \quad \mathfrak{M} \vDash f(a) = g(a).$$

As we saw in our discussion of Nonstandard Analysis in Chapter III, where we presented Laugwitz's construction of $^{\Omega}\mathbb{R}$, it is not enough in constructing a nonstandard model elementarily equivalent to a given standard structure to just take the functions themselves. The equivalence relation is a vital part of the construction and it is fully as infinitistic as the class $\mathcal{F}$ of functions itself. Skolem showed, however, that, for any sequence $f_0, f_1, f_2, \ldots$ of functions from $\mathbb{N}$ to $\mathbb{N}$, one could find a strictly increasing function g such that for any pair f_i, f_j of functions in the list, a number n_{ij} can be found so that one of the following holds:

$$\forall n > n_{ij} \big(f_i(g(n)) < f_j(g(n)) \big),$$
$$\forall n > n_{ij} \big(f_i(g(n)) = f_j(g(n)) \big),$$
$$\forall n > n_{ij} \big(f_i(g(n)) > f_j(g(n)) \big).$$

In words, f_i is eventually dominated by f_j on the range of g, f_i is eventually equal to f_j on the range of g, or f_i eventually dominates f_j on the range of g. Skolem's nonstandard model of arithmetic is then given by taking as $\mathfrak{M}$ the set of equivalence classes of arithmetically definable functions $f : \mathbb{N} \to \mathbb{N}$ given by the equivalence relation,

$$f_i \equiv f_j : \quad \exists n_{ij} \forall n > n_{ij} \big(f_i(g(n)) = f_j(g(n)) \big).$$

That Skolem's construction works is verified in two stages. First, because g is strictly increasing, the identity function is not equivalent to any constant

---

J. Łoś, *Mathematical Interpretations of Formal Systems*, North-Holland Publishing Company, Amsterdam, 1955. All three papers can be found in Skolem's selected logical works, *op. cit.*, as can a French treatment, a couple of Skolem's papers anticipating the result, and some informative remarks by Skolem's editor, Jens Erik Fenstad (pp. 40 - 42). Incidentally, the paper by Łoś in the conference proceedings is a generalisation of Skolem's construction to arbitrary first-order theories and is the modern *ultraproduct* construction on which expositions of nonstandard analysis are often based.

function and it will follow that the model $\mathfrak{M}$ is not isomorphic to $\mathfrak{N}_3$. Second, one proves by induction on the construction of any formula $\varphi(x_0, \ldots, x_{m-1})$ of L_3 that, for any equivalence classes $[f_0], \ldots, [f_{m-1}]$ of arithmetical functions,

$$\mathfrak{M} \vDash \varphi\big(\overline{[f_0]}, \ldots, \overline{[f_{m-1}]}\big) \quad \text{iff}$$

$$\text{for sufficiently large } n, \mathfrak{N}_3 \vDash \varphi\big(\overline{f_0(g(n))}, \ldots, \overline{f_{m-1}(g(n))}\big).$$

I refer the reader to Skolem's papers for the proof. It is no more difficult than the proofs given in the opening sections of Chapter III and makes a nice exercise for graduate students or advanced undergraduate students with some background in writing proofs.[127]

In the 1950's the question of whether less complex models of TA could be constructed was looked into and various negative results were established. It was shown, for example, that one cannot replace $\mathcal{F}$ in Skolem's construction by the family of all computable functions. More generally it was shown that, if $\mathcal{F}_n$ is the family of all functions belonging to the first n levels of the arithmetical hierarchy fleetingly referred to above, then the structure $\mathfrak{M}_n$ obtained by replacing $\mathcal{F}$ by $\mathcal{F}_n$ in Skolem's construction was not a model of TA: There is a function belonging to the $n+1$-st level of the hierarchy with graph defined by a formula $\varphi(x, y)$ and such that

$$\mathfrak{N}_3 \vDash \forall x \exists y\, \varphi(x, y), \quad \mathfrak{M}_n \vDash \exists x \forall y\, \neg\, \varphi(x, y).$$

Moreover, the totality of the function exhibited is provable in the axiomatic theory PA of *Peano Arithmetic* shortly to be introduced, whence $\mathfrak{M}_n$ is not only not a *strong* nonstandard model of arithmetic, but it is also not a model of arithmetic— the instance of induction used to prove the totality of the function must fail in $\mathfrak{M}_n$.

Divorcing oneself from Skolem's construction, one can also apply the Compactness Theorem to prove the existence of nonstandard models of TA. By the Löwenheim-Skolem Theorem, there are models of TA that are countable. Identifying the domain of such a model with the set of natural numbers, one sees there are nonstandard models of TA of the form $\mathfrak{M} = (\mathbb{N}; \oplus, \odot, 0, 1)$, where $\oplus, \odot$ are binary operations on $\mathbb{N}$. The question becomes: how complicated must $\oplus, \odot$ be? This was answered in 1959 by Stanley Tennenbaum:

4.19 Theorem (Tennenbaum's Theorem, Version 1). *Let* $\mathfrak{M} = (M; \oplus, \odot, 0, 1)$ *be a nonstandard model of* TA. *Neither function* $\oplus$ *nor* $\odot$ *is arithmetically definable. In particular, neither* $\oplus$ *nor* $\odot$ *is a computable function.*

Much of this complexity is due to the demand that the nonstandard model be strong, i.e., that it satisfy all true sentences of arithmetic. When we replace TA by its most common axiomatic counterpart PA one can construct models in which $\oplus$ and $\odot$ are arithmetically definable functions. However, we still have

---

[127] Skolem's proof is not that widely presented in textbooks because it has been supplanted by the ultraproduct construction of Jerzy Łoś cited in the preceding footnote. This construction is expounded upon in most textbooks on Model Theory.

4.20 Theorem (Tennenbaum's Theorem, Version 2). *Let* $\mathfrak{M} = (M; \oplus, \odot, 0, 1)$ *be a nonstandard model of* PA. *Neither function* $\oplus$ *nor* $\odot$ *is a computable function.*

With this we have finished our discussion of semantics. With respect to syntax, our first order of business is to present the axioms of Peano Arithmetic, PA. These are, of course, not the axioms Peano used in defining his system earlier described. But they are embedded in his *Arithmetices principia* and, except for the definition of exponentiation, they are the principles used by him in his symbolical development of elementary number theory and the system was thus named in his honour. Basically the axioms of PA consist of the axioms of successor postulated by him, the recursion equations for addition and multiplication assumed by Peano as definitions, and the first-order induction schema replacing his single axiom *P9*:

PA1. $\forall x(\neg 0 = x + 1)$

PA2. $\forall xy(x + 1 = y + 1 \rightarrow x = y)$

PA3. $\forall x(x + 0 = x)$

PA4. $\forall xy\big(x + (y + 1) = (x + y) + 1\big)$

PA5. $\forall x(x \cdot 0 = 0)$

PA6. $\forall xy\big(x \cdot (y + 1) = (x \cdot y) + x\big)$

PA7. $\forall x\big(\varphi(0) \wedge \forall y(\varphi(y) \rightarrow \varphi(y + 1)) \rightarrow \varphi(x)\big)$, for all formulæ $\varphi(x)$ of L_3.

One often formulates PA in a slight variant of L_3 in which the constant 1 is replaced by a function symbol for the successor function. The intended meanings of the axioms are perhaps a little clearer in such a formulation:

PA1. $\forall x(\neg 0 = x')$

PA2. $\forall xy(x' = y' \rightarrow x = y)$

PA3. $\forall x(x + 0 = x)$

PA4. $\forall xy\big(x + y' = (x + y)'\big)$

PA5. $\forall x(x \cdot 0 = 0)$

PA6. $\forall xy\big(x \cdot y' = (x \cdot y) + x\big)$

PA7. $\forall x\big(\varphi(0) \wedge \forall y(\varphi(y) \rightarrow \varphi(y')) \rightarrow \varphi(x)\big)$, for all formulæ $\varphi(x)$ of the appropriate language.

Peano Arithmetic is a very powerful theory, a fact that takes some effort to establish. Peano borrowed proofs from Hermann Grassmann's *Lehrbuch der Arithmetik* of 1861, expressing them symbolically. He first established the usual algebraic laws (commutativity, associativity, distributivity) by various inductions. Then he defined exponentiation recursively and presented the Laws of Exponentiation. This is, of course, a sticking point in PA, which is first-order and does not immediately have the mechanism to handle such recursive definitions. Peano, the reader will recall, worked in a theory including sets as well as numbers, in which theory the existence of functions defined by recursion is easily expressed and proven by induction. In PA one must first exert some effort to show that one can express such things. For this Gödel introduced the β-function.

It is a curious fact, but for many years this extremely important part of the development of Peano Arithmetic was ignored by the textbook writers. Authors would present the proof of Lemma 4.18, develop material informally, and then, when they needed the formal derivability of some of the consequences, would blithely say "formalising the treatment just given". However, the "treatment just given" does not formalise: Lemma 4.18 cannot be expressed in L_3 without already having the means to express finite sequences— which, in turn, requires some form of the Lemma. Lemma 4.18 must, in fact, be replaced by another lemma asserting that, given c, d, n, a one can find c^*, d^*, such that

i. $\beta(c^*, d^*, n) = a$

ii. for $i = 0, \ldots, n - 1$, $\beta(c^*, d^*, i) = \beta(c, d, i)$.

Such a lemma was first provided by Paul Bernays in 1934 in the first volume of the *Grundlagen der Mathematik*, his joint exposition with David Hilbert[128] of the work in mathematical logic arising from the latter's programme in proof theory, and subsequently ignored by the textbook writers for over three decades until Joseph Shoenfield included the proof in his textbook[129].

Once one has proven the extendible version of Lemma 4.18, the definability of functions by simple recursions becomes a routine matter and we can continue to follow Peano's progress.[130] He proceeds at this point to list, without proofs, a number of arithmetic propositions, culminating in Fermat's theorem asserting that an integer $p > 1$ is prime iff, for $0 < n < p$,

$$n^{p-1} \equiv 1 \mod p.$$

He then points out that, basically, positive rational numbers are ordered pairs of integers and develops within his formalism the algebra of the rational numbers. Finally, considering irrational numbers as sets of rational numbers he develops the rudiments of the theory of real numbers. This last part cannot be carried out within the first-order theory PA, but various approximations can be made. The whole family $\mathcal{F}$ of arithmetically definable functions cannot be de-

---

[128] This work was published in two volumes in 1934 and 1939, respectively, by Springer-Verlag, Berlin. The designation of Hilbert and Bernays as authors, with Hilbert's name listed first, followed the German tradition of the Professor providing the grand sweep of a book and having his assistant do the actual writing. By the early 1930's, however, Hilbert obviously had little grasp of the details of the then current foundational work and it is generally acknowledged to be Bernays's book. A second expanded edition was published by Springer-Verlag, Heidelberg, in 1968 (volume I) and 1970 (volume II).

[129] Joseph R. Shoenfield, *Mathematical Logic*, Addison-Wesley Publishing Company, New York, 1967, pp. 115 - 116.

[130] An alternative approach is to add exponentiation to the language and the recursion equations for exponentiation to the axioms. With exponentiation, the formal development of the theory of finite sequences is greatly simplified. Cf., e.g., van Dalen, *op. cit.*, pp. 239 - 241, for details.

fined within the language L_3, but the family of computable functions can be[131] and one can develop a theory of computable real numbers within PA. One can, in fact, do the same at each level of the arithmetical hierarchy to get a sequence of successive approximations to analysis. In 1972, Gaisi Takeuti gave a course of lectures on the analytic strength of Peano Arithmetic[132] in which he showed, by developing analysis in an extension of PA through the addition of variables for sets and a weak comprehension axiom, that Peano Arithmetic was sufficiently strong to derive all arithmetic theorems provable in Analytic Number Theory. Thus, for example, the Prime Number Theorem on the distribution of primes, easily translatable into L_3 by replacing the analytic equation in the formula by an equivalent statement concerning explicitly definable sequences of rational approximations, is provable in PA.

I stress the strength of PA because, strong as it is, it is yet incomplete: PA is not TA. And the reason for this is *not* that anything obvious has been overlooked, as was the case with Zermelo's having missed the Axiom of Replacement. This incompleteness is, in fact, a symptom of an inherent weakness in first-order logic: No consistent effectively enumerable set of axioms in L_3 can yield all of TA. This is known as Gödel's Incompleteness Theorem, although it is not really a theorem so much as a cluster of theorems stated in varying degrees of generality over the years. The first version was proven by Gödel in 1930 for his habilitation thesis[133]. A second major variant was proven indepen-

---

[131] What does it mean for $\mathcal{F}$ or some subfamily thereof to be definable in $\mathfrak{N}_3$? The language L_3 has no variables for functions, so a definition

$$f \text{ is definable iff } \mathfrak{N}_3 \vDash \varphi(f)$$

is ruled out. Another possibility would be an arithmetical enumeration of unary arithmetical functions, i.e., a binary arithmetical function $f(m, n)$ for which the sequence $f_0, f_1, \ldots$ given by $f_m(n) = f(m, n)$ enumerates all unary arithmetically definable functions. If f is arithmetically definable, however, then so too is $f(n) = f(n, n) + 1$, which differs from each function in the enumeration. The family of computable functions, however, does have such an arithmetically definable enumeration, albeit not a computable one.

[132] Gaisi, Takeuti, "A conservative extension of Peano Arithmetic", in: Gaisi Takeuti, *Two Applications of Logic to Mathematics*, Iwanami-Shoten, Publishers, and Princeton University Press, Princeton, 1978.

[133] Kurt Gödel, "Über formal unentscheidbare Sätze der Principia Mathematica und verwandter Systeme I" [On formally undecidable sentences of Principia Mathematica and related systems I], *Monatshefte für Mathematik und Physik* 38, pp. 173 - 198. The paper remains today one of the best expositions of the proof of the basic result and I recommend it highly to the reader. It can be found in the original German as well as in English translation in: Kurt Gödel, *Collected Works, I*, Oxford University Press, Oxford, 1986. Standalone English translations by various hands can also be found in Davis, *op. cit.*; van Heijenoort, *op. cit.*; and Stephen Hawking, *God Created the Integers: The Mathematical Breakthroughs that Changed History*, Running Press, Philadelphia, 2005. I haven't read all the translations and cannot say if one stands out from the crowd on merit; however, I find the version in van Heijenoort's source book to be particularly readable on account of its typography

dently in 1936 by Church and Turing[134], but is usually simply called Church's Theorem.

Gödel started it all by showing that a simplified version PM of Russell and Whitehead's *Principia Mathematica* could discuss its own syntax via an encoding using *primitive recursive* functions— basically, those functions built up from the constant functions, $+$, and $\cdot$ by composition and the simplest form of recursion. PM correctly computed primitive recursive functions and most syntactic operations were simulated by these functions. Playing with this and mimicking the paradoxes, he constructed a sentence φ_G asserting its own unprovability from the axioms of PM. Using the β-function, φ_G could in fact be expressed in L_3. More generally he defined a theory T to be primitive recursive if the set of numerical codes, now commonly called *gödel numbers*, of its axioms has a primitive recursive characteristic function, and proved:

4.21 Theorem (Gödel's First Incompleteness Theorem). *Let* T *be a primitive recursive theory extending* PM. *Let* φ_G^T *be the sentence asserting its own unprovability from* T,

$$\text{PM} \vdash \varphi_G^T \leftrightarrow \text{``}\varphi_G^T \text{ is not provable in } T\text{''}.$$

Then:

i. *if* T *is consistent,* $T \nvdash \varphi_G^T$;

ii. *if* T *is ω-consistent,* $T \nvdash \neg \varphi_G^T$.

Thus, if T *is ω-consistent,* T *is incomplete:* T *does not decide the sentence* φ_G^T.

The notion of ω-consistency referred to is a proof-generated concept, a soundness assertion stronger than consistency and weaker that the assertion that the axioms of T are true. Without reference to hierarchical refinement, it is just the condition needed to establish 4.21.ii. One could replace it by the simpler condition that T be a true theory:

i. if T is consistent, $T \nvdash \varphi_G^T$;

ii. if the axioms of T are true, $T \nvdash \neg \varphi_G^T$.

In very short order, Gödel had replaced PM by PA.

The second part of Gödel's paper was to have included the proof of a second theorem:

4.22 Theorem (Gödel's Second Incompleteness Theorem). *Let* T *be a primitive recursive theory extending* PA. *If* T *is consistent, then* T *cannot prove its own consistency.*

---

and layout. On the other hand, the collected works and Davis's source book contain Gödel's lectures on the matter, and Davis also includes subsequent basic papers by Rosser and Kleene on the subject.

[134] Their fundamental papers, "An unsolvable problem of elementary number theory", *American Journal of Mathematics* 58 (1936), pp. 345 - 363, and "A note on the Entscheidungsproblem", *Journal of Symbolic Logic* 1 (1936), pp. 40 - 41 (correction: pp. 101 - 102), by Church, and "On computable numbers, with an application to the Entscheidungsproblem", *Proceedings of the London Mathematical Society*, series 2, vol. 42 (1936 -1937), pp. 230 - 265 (correction: vol. 43 (1937), pp. 544 - 546), by Turing, are perhaps most accessible in Davis, *op. cit.*

This consequence had also been noticed independently by Johann von Neumann and its truth was so readily accepted that Gödel never bothered to write up the second part of his paper. The proof proceeds by formalising a key step in the proof of Theorem 4.21.i to show

$$\mathsf{T} \vdash \text{``T is consistent''} \to \text{``T} \nvdash \varphi_G^{\mathsf{T}}\text{''}$$

$$\vdash \text{``T is consistent''} \to \varphi_G^{\mathsf{T}}, \ \text{by choice of } \varphi_G^{\mathsf{T}},$$

whence Theorem 4.21.i yields

$$\mathsf{T} \nvdash \text{``T is consistent''}.$$

The first detailed exposition of the proof of Gödel's Second Incompleteness Theorem was given by Paul Bernays in 1939 in the second volume of *Grundlagen der Mathematik*.

In the spring of 1934, Gödel lectured on his incompleteness theorems at Princeton and the lecture notes were written up by two of Church's students, Stephen Cole Kleene and J. Barkley Rosser. It was at the end of these notes that Gödel tentatively put forth a definition of computable function. Church, Kleene and Rosser had continued to develop the λ-calculus which Church now recognised as a fully general notion of computability. In 1936 he published a paper[135] in which he used this calculus to obtain a Type II formal definition of computability, stated the thesis (known as *Church's Thesis*) that this definition did indeed capture the intuitive notion of computable function, presented a proof, due mainly to Kleene but also in part to Church and Rosser, of the equivalence of his notion of computability with that put forth by Gödel, gave examples of problems which are not algorithmically solvable— more briefly called *unsolvable problems*— and applied this to the *Entscheidungsproblem* to conclude the theory of *Principia Mathematica* to be undecidable.

The Entscheidungsproblem, or *decision problem*, was one of the logical problems worked on intensely in Germany in the 1920's. Put quite simply, it was the problem of finding a method of deciding the validity of formulæ of first-order logic, or of deciding the derivability of formulæ in various axiomatic theories. For pure logic, the decision problem had been solved in special cases, and for theories there was the method of quantifier elimination introduced by Skolem in 1919 and applied by various researchers in the 1920's, most notably by Alfred Tarski and his school in Warsaw. Presburger's quantifier elimination for PSA was the best result at the time. In his paper under discussion, Church noted that the unsolvable problems concerning λ-terms he had come up with could be represented within PM and thus PM had an unsolvable Entscheidungsproblem, i.e., PM is undecidable.[136]

---

[135] "An unsolvable problem of elementary number theory", cited in footnote 134, above.

[136] The word "undecidable" does double duty. A single sentence φ is undecidable in a theory T if T proves neither φ nor $\neg\varphi$; a theory T is undecidable if there is no effective procedure for determining which sentences φ are provable in T.

In his second paper, "A note on the Entscheidungsproblem"[137], Church observed that one did not need the full strength of PM to represent his given unsolvable problems, but only a little arithmetic and finitely many defining equations as axioms for the few functions explicitly used in setting up the representation. Thus, a finitely axiomatised theory was undecidable, whence first-order logic itself was undecidable.[138]

4.23 Theorem (Church's Theorem). *First-order logic is undecidable: there is no effective procedure that decides, for every sentence φ of a first-order language, whether or not φ is logically valid.*

Alan Turing had also come up with the idea of characterising computability and giving a negative solution to the Entscheidungsproblem. His approach was somewhat different. He described an abstract machine operating on strings in a finite alphabet, declared the functions computable by such machines to be the computable functions (*Turing's Hypothesis*), produced unsolvable problems, showed his definition equivalent to Church's, and demonstrated the unsolvability of the decision problem by directly describing a language and axioms for each of his machines.

The question now was mainly one of generality. The First Incompleteness Theorem yielded the incompleteness of any ω-consistent extension of PA with a primitive recursive set of axioms. Church's argument showed the undecidability of any ω-consistent extension of PA with a primitive recursive set of axioms. And his proof of the undecidability of first-order logic depended on the ω-consistency (better: a special case of the ω-consistency) of his finitely axiomatised theory. Rosser[139] cleverly exploited the ordering of the natural numbers to eliminate the appeal to ω-consistency, and he also weakened the effectiveness requirement on the set of axioms. Recall that a set of axioms is called primitive recursive if the characteristic function of the set of (codes of, gödel numbers of) axioms is primitive recursive. Following Rosser, we now call a set of axioms *recursively enumerable* if the set of (codes of, gödel numbers of) axioms is the range of a computable function.[140]

4.24 Theorem (Rosser's Theorem). *Let* T *be a consistent recursively enumerable theory extending* PM.
i. T *is incomplete: there is a sentence φ such that*

---

[137] *Op. cit.* Cf. footnote 134.

[138] Let χ be the conjunction of the axioms of the undecidable theory. If first-order logic were decidable, one could decide the validity of any sentence $\chi \to \varphi$, i.e., one could decide the derivability of φ from the axioms making up χ, contrary to the assumed undecidability of the finitely axiomatised theory.

[139] J.B. Rosser, "Extensions of some theorems of Gödel and Church", *Journal of Symbolic Logic* 1 (1936), pp. 87 - 91; reprinted in Davis, *op. cit.*

[140] Computable functions are also called *general recursive functions*, or simply *recursive* functions, in honour of the Herbrand-Gödel definition of computability which generalised the use of recursion equations in defining the primitive recursive functions. Hence the name "recursively enumerable" instead of "computably enumerable" or "effectively enumerable".

$$T \nvdash \varphi \ \ and \ \ \mathsf{T} \nvdash \neg \varphi.$$

ii. T *is undecidable.*

In terms of generality, there remained the question of replacing PM by as weak a theory as possible. Rosser could already easily have replaced PM by PA with no change to the rest of his paper. There is probably no definitively weakest theory for which Rosser's Theorem remains true, but there is a weakest theory for which the standard proof works without any additional argument. This is the system R introduced by Raphæl M. Robinson[141]. In addition to a few order properties, the axioms of R are all the instances of true sentences of the forms

$$m + n = p, \ \ m \cdot n = p, \ \ \neg m = n.$$

Each such instance is derivable without the use of induction from the recursion equations for $+, \cdot$ and the axioms of successor listed in $\mathsf{DP_0}$. Thus, for the sake of finite axiomatisability, he also introduced a stronger theory Q which is essentially PA minus induction, plus $D4$, which is no longer redundant without induction.

The question of how general Gödel's Second Incompleteness Theorem is has not been ignored. Solomon Feferman relaxed the requirement of the primitive recursiveness of the axioms to that of their recursive enumerability in what is probably the most important paper on the subject since Gödel's in 1960.[142] As for replacing PA by a weaker theory, it will have to suffice here to note that the full power of induction is scarcely needed.

4.5 Second-Order Logic

The inability of first-order logic to characterise any infinite structure up to isomorphism through some set of axioms did not sit well with everyone in the 1930's. Skolem jettisoned quantifiers and developed a modest, safe theory of recursive arithmetic. In the opposite direction, Zermelo expanded logic in a couple of different ways. One was to allow infinite conjunctions and disjunctions, and one was to allow quantification over sets. In 1931 Hilbert, not liking the limitations placed on formal theories by Gödel's Theorems, made no change in the language, but expanded the mode of reasoning by adding to PA the ω-rule:

$$\text{from } \varphi(0), \varphi(1), \varphi(2), \dots \text{ conclude } \forall x \varphi(x).$$

Not surprisingly, the theory (formulated in L_3) is complete and Gödel's First Incompleteness Theorem fails. Gödel's Second Incompleteness Theorem simply

---

[141] In: Alfred Tarski, Andzej Mostowski, and Raphæl M. Robinson, *Undecidable Theories*, North-Holland Publishing Company, Amsterdam, 1953. The exposition in this book forms the basis of most modern treatments of incompleteness and undecidability.

[142] Solomon Feferman, "Arithmetization of metamathematics in a general setting", *Fundamenta Mathematicae* 49 (1960), pp. 35 - 92.

does not apply because the consistency of the new theory cannot be stated in the language: provability of arithmetic statements under the ω-rule equals truth, and truth, as Tarski would famously demonstrate in 1936[143], is not expressible in the language.

Today, a number of extensions of first-order logic have been introduced and studied. The simplest are obtained by adding new quantifiers, for example

$$\exists_\alpha x \varphi(x) : \text{ there are at least } \aleph_\alpha \text{ } x\text{'s such that } \varphi(x).$$

$\exists_0$ ("there are infinitely many") is the most obvious of these, but $\exists_1$ ("there are uncountably many") has the smoothest theory. Another example is the quantifier/connective

$$\exists_\equiv x(\varphi(x), \psi(x)) : \text{ the sets } \{x | \varphi(x)\}, \{x | \psi(x)\} \text{ have the same cardinality.}$$

The most natural extension, however, is second-order logic wherein one adds variables ranging over sets of individuals and allows quantification over these variables. We shall finish this chapter with a few words on second-order logic.

Syntactically there is not much to distinguish second-order logic from first-order logic. Indeed, it is not a bad idea to consider the second-order language to be the first-order language associated with a particular kind of two-sorted structure. To this end, I digress briefly to introduce two-sorted first-order logic.

Ordinary first-order logic is concerned with describing a structure with a single domain and some operations and relations defined on it. In many-sorted logic, a structure might consist of several domains, relations on and among them, and functions of one or more variables from one or more of the domains to one of them. In the case to be considered, namely a two-sorted structure, there are just two of these domains.

4.25 Example. A vector space V over the real numbers may be considered to be a two-sorted structure with domains $\mathbb{R}$ of scalars and V of vectors. There are three designated elements, namely $0, 1 \in \mathbb{R}$ and the zero vector $\mathbf{0} \in V$. There are four designated functions— real addition and multiplication mapping $\mathbb{R} \times \mathbb{R}$ into $\mathbb{R}$, vector addition mapping $V \times V$ into V, and scalar multiplication mapping $\mathbb{R} \times V$ into V. The only designated relations are the relations of equality on $\mathbb{R}$ and V.

4.26 Example. The Dedekind-Peano structure $\mathfrak{N}_0^2 = (\mathbb{N}; \mathcal{P}(\mathbb{N}); \in, ', 0)$ is a structure with two sorts of objects— numbers and sets of numbers. The only designated individual is the number 0, the only designated function the successor, and the only designated relation besides the two equality relations is the membership relation on $\mathbb{N} \times \mathcal{P}(\mathbb{N})$.

4.27 Example. Let $(G_i; \oplus_i, 0_i)$ be abelian groups for $i = 0, 1$, and suppose $f : G_0 \to G_1$ is a group homomorphism. The structure $(G_0; G_1; \oplus_0, 0_0, \oplus_1, 0_1, f)$ is a two-sorted structure.

---

[143] "Der Wahrheitsbegriff...", *op. cit.*

4.28 Example. $(\mathbb{Q}; \mathbb{Z}; +, \cdot, 0, 1)$ is a two-sorted structure with the rational numbers and the integers as the two sorts. Since $\mathbb{Z} \subseteq \mathbb{Q}$, we can also view it as a single-sorted structure $(\mathbb{Q}; \mathbb{Z}, +, \cdot, 0, 1)$ in which $\mathbb{Z}$ is now considered not one of two independent domains, but as a unary relation on the domain $\mathbb{Q}$. Though very similar and functionally interchangeable, the structures are distinct with distinct languages: the former has variables ranging over integers, but no relation symbol for the set of integers; while it is just the opposite with the latter.

To give a reasonably intelligible formal definition of what we've been discussing, it will help to make some simplifying assumptions. First, we assume the domains to be disjoint. This loses the last Example, but as we are interested in structures like $\mathfrak{N}_0^2$, this is no essential loss for us.[144] As there are two sorts we may as well give them the clever names α and β and denote by A the set of individuals of sort α and by B the set of individuals of sort β. We assume each designated function f to be of one of the forms,

$$f : A^m \times B^k \to A \quad \text{or} \quad f : A^m \times B^k \to B,$$

where $m, k \in \mathbb{N}$ and $m + k > 0$. The first kind of function has *type* (m, k, α) and the second has type (m, k, β). Similarly, each designated relation is assumed to be of the form

$$R \subseteq A^m \times B^k,$$

for some $m, k \in \mathbb{N}$ with $m + k > 0$, and is said to be of type (m, k).

The rules of term formation differ from those of first-order logic only in that terms are now assigned sorts, and sorts and types must match before a term can be formed:

i. a constant naming an individual of sort α or β is a term of sort α or β, respectively;

ii. variables $x_0, x_1, \ldots$ are terms of sort α and variables $y_0, y_1, \ldots$ are terms of sort β; and

iii. letting γ denote α or β, if f is a function symbol of type (m, k, γ), $t_0, \ldots, t_{m-1}$ are terms of sort α, and $u_0, \ldots, u_{k-1}$ are terms of sort β, then $f(t_0, \ldots, t_{m-1}, u_0, \ldots, u_{k-1})$ is a term of sort γ.

Atomic formulæ are likewise defined to be expressions of the form $R(t_0, \ldots, t_{m-1}, u_0, \ldots, u_{k-1})$, where R has type (m, k), the t_i's are terms of sort α, and the u_i's are terms of sort β. General formulæ are then constructed by applying the logical connectives and quantifiers applied to both sorts of variables.

The definition of a structure $\mathfrak{M} = (A; B; I, F, R)$ for a two-sorted language is as before; and the truth of a sentence in such a structure is basically the same as for a single-sorted structure, only now the quantifier clauses split:

---

[144] In ordinary mathematics numbers are numbers and sets are sets, with no overlap between them. In ZF, however, one identifies $\mathbb{N}$ with the set ω of finite ordinals and it happens that $\omega \subseteq \mathcal{P}(\omega)$. Let us ignore this little difficulty.

$$\mathfrak{M} \vDash \exists x \varphi(x) \quad \text{iff} \quad \text{for some } a \in A, \mathfrak{M} \vDash \varphi(a)$$
$$\mathfrak{M} \vDash \exists y \varphi(y) \quad \text{iff} \quad \text{for some } b \in B, \mathfrak{M} \vDash \varphi(b),$$

where x, y are variables of sorts α, β, respectively, and

$$\mathfrak{M} \vDash \forall x \varphi(x) \quad \text{iff} \quad \text{for all } a \in A, \mathfrak{M} \vDash \varphi(a)$$
$$\mathfrak{M} \vDash \forall y \varphi(y) \quad \text{iff} \quad \text{for all } b \in B, \mathfrak{M} \vDash \varphi(b),$$

where again x, y are variables of sorts α, β, respectively.

Once one has defined truth, the definition of semantic consequence (Definition 4.2) carries over word for word. The definition of syntactic consequence differs from that of the single-sorted case only in the statements of axioms *H5* - *H6*, where the match up of sorts must now be stipulated:

H5′. $\forall x \varphi(x) \to \varphi(t)$, provided x and t are of the same sort,

H6′. $\varphi(t) \to \exists x \varphi(x)$, provided x and t are of the same sort.

The Completeness Theorem and its consequences carry over from the single-sorted to the two-sorted case with no difficulty.

Syntactically, second-order logic is just such a two-sorted first-order logic, where the individuals of the second sort are taken to be the sets of individuals of the first sort. There are variants— the individuals of the second sort can be functions or relations of the first sort instead of sets. Thus, one often refers to the variant under discussion as the *monadic* second-order logic, because one only quantifies over monadic relations.

The idea behind second-order logic is to start with a first-order structure $\mathfrak{M}$ with a specific first-order language L, such as our familiar $\mathfrak{N}_0, \mathfrak{N}_1, \mathfrak{N}_2, \mathfrak{N}_3$ with languages L_0, L_1, L_2, L_3, respectively, and consider the two-sorted structures obtained from $\mathfrak{M}$ by adjoining the power set of its domain as the second sort, and adding the membership relation $\in$ to the list of designated relations to obtain a new language L^2. This yields a second-order structure $\mathfrak{M}^2$. Thus:

$$\mathfrak{N}_0^2 = (\mathbb{N}; \mathcal{P}(\mathbb{N}); \in, \prime, 0)$$
$$\mathfrak{N}_1^2 = (\mathbb{N}; \mathcal{P}(\mathbb{N}); \in, \prime, 0, <)$$
$$\mathfrak{N}_2^2 = (\mathbb{N}; \mathcal{P}(\mathbb{N}); \in, +, 0, 1, <)$$
$$\mathfrak{N}_3^2 = (\mathbb{N}; \mathcal{P}(\mathbb{N}); \in, +, \cdot, 0, 1).$$

The syntactic notion of consequence requires new axioms. There are really only two obvious ones about sets in structures $(A; \mathcal{P}(A); \in, \ldots)$. These are Extensionality and Comprehension:

Ext. $\quad \forall XY \big(\forall x(x \in X \leftrightarrow x \in Y) \to X = Y \big)$

Comp. $\exists X \forall x \big(x \in X \leftrightarrow \varphi(x) \big)$, where X does not occur free in φ.

[Note that I now follow the convention of using lower case letters for individual variables and upper case letters for set variables.] In general these are insufficient and do not yield all sentences valid in all second-order structures $(A; \mathcal{P}(A); \in, \ldots)$. This is a fairly easy consequence of Gödel's, Church's or

Rosser's Theorems discussed in the preceding section. Derivability in second-order logic is actually derivability from extra axioms in two-sorted first-order logic, but semantic consequence, defined again by copying Definition 4.2, does not agree with the two-sorted first-order concept because the quantification in the defining implication,

$$\text{for all structures } \mathfrak{M}, \quad \mathfrak{M} \vDash \Gamma \ \Rightarrow \ \mathfrak{M} \vDash \varphi, \tag{14}$$

is taken over a much, much narrower class of structures.

At this point, it might be instructive to consider the structures $\mathfrak{N}_i^2$ and their axioms.

We start with $\mathfrak{N}_0^2$ and the Dedekind-Peano theory DP^2 given by $D1$ - $D3$, Ext and $Comp$.[145] In first-order logic, the structures $\mathfrak{M}$ of (14) would be of the form $(A; B; E, \ldots)$ for some sets A, B, and a relation $E \subseteq A \times B$ for which the axioms are true. Among these are countable structures. In second-order logic, one considers only structures $(A; \mathcal{P}(A); \in, \ldots)$, none of which is countable. Thus the quantification in (14) is indeed over a smaller class of structures. Indeed, as Dedekind showed, all second-order structures modelling DP^2 are isomorphic to $\mathfrak{N}_0^2$ and the quantification is essentially over a singleton for $\Gamma = \mathsf{DP}^2$: $\mathsf{DP}^2 \vDash \varphi$ iff $\mathfrak{N}_0^2 \vDash \varphi$.

DP^2 does not illustrate, however, the difference between the syntactic and semantic notions of consequence for second-order logic. For, DP^2 happens to be syntactically complete. As was the case with DP_0, the language L_0^2 of $\mathfrak{N}_0^2$ is inexpressive enough to allow a complete analysis of the theory. In 1960, J. Richard Büchi applied the methods of finite automata theory to the the theory of $\mathfrak{N}_0^2$ and showed that every formula of the language of the structure could be simulated by a finite automaton operating on infinite strings of inputs. This yielded a decision procedure and a characterisation of definable relations, the definable sets being exactly the ultimately periodic sets, i.e., those definable in PSA.[146] A decade later Dirk Siefkes completed the analysis, producing DP^2 as an axiomatisation of Büchi's theory.[147]

It follows from the existence of a decision procedure for the theory of the structure $\mathfrak{N}_0^2$ that addition and multiplication are not both definable in the language L_0^2. The characterisation of the definable sets as the ultimately periodic ones shows once again that the set of squares— hence multiplication— is not definable. We rule out the definability of addition by observing that multiplication is definable in the structure $\mathfrak{N}_2^2$. To see this, first observe that the set

---

[145] Note that DP^2 uses the original second-order induction axiom and not merely the first-order schema $D3\,^1$ of DP or the non-inductive replacements $D4$ - $D5$ of DP_0.

[146] J. Richard Büchi, "On a decision method in restricted second-order arithmetic", in: Ernst Nagel, Patrick Suppes, and Alfred Tarski (eds.), *Logic, Methodology and Philosophy of Science*, Stanford University Press, Stanford, 1962.

[147] Dirk Siefkes, *Büchi's Monadic Second Order Successor Arithmetic*, Springer-Verlag, Heidelberg, 1970. As Siefkes pointed out, Büchi's paper, like most papers published in a conference proceedings, is rather abbreviated and "leaves a lot of work to the reader". His exposition is much easier on the reader and is the preferred reference for an introduction to the subject.

$\{x, 2x, 3x, \ldots\}$ of nonzero multiples of x is definable:

$$y \in M(x): \quad \forall X\big(x \in X \land \forall z(z \in X \leftrightarrow z + x \in X) \to y \in X\big).$$

Then successively define first $y = x^2 + x$:

$$x = 0 \land y = 0 \lor 0 < x \land y \in M(x) \land y \in M(x+1) \land$$
$$\forall z\big(z \in M(x) \land z \in M(x+1) \to y \le z\big)$$

and then

$$\begin{array}{ll}
y = (x+1)^2: & \exists z(z = x^2 + x \land y = z + x + 1) \\
y = x^2: & x = 0 \land y = 0 \lor \exists z\big(x = z + 1 \land y = (z+1)^2\big) \\
z = xy: & \exists uvw\big(w = (x+y)^2 \land u = x^2 \land v = y^2 \land w = u + z + z + v\big).
\end{array}$$

It follows that Dedekind's proof of the existence of addition and multiplication operations on $\mathfrak{N}_0^2$ transcends the monadic second-order theory of successor as there are not the means of expressing the graphs of these functions within the language.

4.29 Remark. Dedekind's result can be established, however, in a more general *dyadic* second-order theory allowing quantification over binary relations, or even over unary functions. One defines addition and multiplication in $(\mathbb{N}; \mathbb{N}^{\mathbb{N}}; \mathrm{App}, ', 0)$ (where $\mathrm{App}: \mathbb{N}^{\mathbb{N}} \times \mathbb{N} \to \mathbb{N}$ is defined by $\mathrm{App}(f, n) = f(n)$):

$$\begin{array}{ll}
z = x + y: & \forall f\big(f(0) = y \land \forall v(f(v') = f(v)') \to f(x) = z\big) \\
z = x \cdot y: & \forall f\big(f(0) = 0 \land \forall v(f(v') = f(v) + y) \to f(x) = z\big);
\end{array}$$

and in $(\mathbb{N}; \mathcal{P}(\mathbb{N} \times \mathbb{N}); \in, ', 0)$ one defines these operations respectively by:

$$\forall R\big(\forall x(R(0, w) \leftrightarrow w = y) \land \forall vw(R(v, w) \leftrightarrow R(v', w')) \to R(x, z)\big)$$
$$\forall R\big(\forall x(R(0, w) \leftrightarrow w = 0) \land \forall vw(R(v, w) \leftrightarrow R(v', w + y)) \to R(x, z)\big).$$

The language L_0^2, however, is expressive enough to allow the ordering relation to be defined:

$$x < y: \quad \neg x = y \land \forall X\big(x \in X \land \forall z(z \in X \to z' \in X) \to y \in X\big). \tag{15}$$

It follows that L_1^2 is no more expressive than L_0^2 and the structure $\mathfrak{N}_1^2$ is not essentially different from $\mathfrak{N}_0^2$. Thus, other than perhaps finding a more elegant axiomatisation for the theory than that obtained by adding the equivalence (15) to DP^2, there is no particular reason to look more closely at $\mathfrak{N}_1^2$.

Similarly, by the definability of multiplication on $\mathfrak{N}_2^2$, we need only consider one of the structures $\mathfrak{N}_2^2$ and $\mathfrak{N}_3^2$. I prefer to consider $\mathfrak{N}_3^2$ and its language L_3^2: $\in, +, \cdot, 0, 1, =$. The axioms of second-order Peano Arithmetic, PA^2, are the familiar non-inductive axioms *PA1* - *PA6* of Peano Arithmetic, the induction axiom *D3*,

$$\forall x \forall X \big(0 \in X \wedge \forall y (y \in X \to y + 1 \in X) \to x \in X \big),$$

in place of the schema $PA\,7$, and the set theoretic axioms Ext and $Comp$. The syntactic consequences of PA^2 do not coincide with the semantic ones. For, PA^2 has up to isomorphism only the single second-order model $\mathfrak{N}_3^2$, whence for any sentence φ of the language L_3^2,

$$\mathsf{PA}^2 \vDash \varphi \quad \text{iff} \quad \mathfrak{N}_3^2 \vDash \varphi.$$

But, syntactically PA^2 is just a primitive recursively axiomatised first-order theory extending PA, and, say, Rosser's Theorem applies: there is a sentence φ such that $\mathsf{PA}^2 \nvdash \varphi$ and $\mathsf{PA}^2 \nvdash \neg\varphi$. Now one of these, say φ, is true in $\mathfrak{N}_3^2$ and we have

$$\mathsf{PA}^2 \vDash \varphi, \quad \text{but} \quad \mathsf{PA}^2 \nvdash \varphi.$$

We thus see that the semantic and syntactic notions of consequence do not agree in second-order logic. We have satisfied Hilbert's first desideratum for the structure $\mathfrak{N}_3^2$: PA^2 completely determines the structure and thus every true statement is a semantic consequence of its axioms. But not every true sentence is a syntactic consequence of the axioms of PA^2 and the axiomatisation does not satisfy Hilbert's second desideratum that every statement we hold to be true in the structure be derivable by means of finitely many purely logical inferences. Is this a fault of the particular choice of axioms and rules of inference, or is it indeed the case that Hilbert's second desideratum cannot be met?

This last question is, of course, a vague Type I affair that has no definite answer. We can ask, for any precise Type II formal definition of an axiomatisation, such as the one given, whether it satisfies the desideratum or not. Or, we could take a Type III, axiomatic approach and postulate certain properties of an axiomatic system and ask if these properties rule out this satisfaction.

Now two reasonable assumptions to make of an axiomatisation are that we be able to recognise of a given statement whether or not it is an axiom and, given formulæ $\varphi_0, \ldots, \varphi_{n-1}, \varphi$, we should be able to recognise if φ is an immediate consequence of $\varphi_0, \ldots, \varphi_{n-1}$ via one of the axiomatisation's given rules of inference. Now, in Rosser's paper it is proven that, assuming a recursive, or even just recursively enumerable, set of axioms, if the problem of recognising the applicability of a rule of inference is recursively recognisable, then the set of derivable formulæ is recursively enumerable.

We have already remarked that there is a noncomputable function definable by a formula of L_3. Its graph, considered as a set of ordered pairs and thus, via the pairing function of Cantor (footnote 19 on page 457, above), as a set of natural numbers, is not recursively enumerable. Let $\varphi(x)$ be a formula defining such a non-recursively enumerable set and observe that the recursively enumerable set $\{n \,|\, \vdash \varphi(n)\}$ cannot equal the non-recursively enumerable set $\{n \,|\, \vDash \varphi(n)\}$ and completeness fails. No effective set of axioms and rules for second-order logic, or any other logic powerful enough to allow a semantically complete theory of arithmetic in a language including L_3 has a syntactically complete axiomatisation allowing the derivation of all semantic consequences in finite sequences of purely logical steps. For a sufficiently expressive theory we can either hope for a powerful axiomatisation like PA from which we can derive theorems in finitely many logical steps, but which is incomplete, or for a complete axiomatisation like PA^2 for which there is no adequate accompanying notion of derivation. We cannot have both.

5 Essay Questions

5.1 Essay Question. Consider Zermelo's argument (page 478, above) for the self-evidence of the Axiom of Choice. If everyone uses a principle is it self-evident? If so, does that make it true? Discuss some examples from the history of mathematics cited in this book.

5.2 Essay Question. Seriously consider Russell's millionaire (page 477, above). Did the German mathematician of footnote 47 miss the point? Did Russell miss the German mathematician's point? Read Chapter I, section 10, "Imprecise limits" of John Crossley, *The Emergence of Number*[148], before answering this.

5.3 Essay Question. Computer programmers, when faced with a program that isn't working correctly, are supposed to sit back and analyse where the program went wrong. Impatience often leads them instead to the practice of *hacking*: sitting at he computer and "hacking" away at the code in a seemingly random fashion in the hopes that one of their changes will fix the program. Discuss, citing examples from the text, the analogous axiomatic practice. Who are the hackers of the axiomatic method?

5.4 Essay Question. Drawing inspiration from the poem about Russell, I have made up my own authentic old English nursery rhyme. The following, touching on the preceding chapter, is called "Bolzano's Monster":

Bolzano, with some regularity,
Repeated a sole singularity.
And now he is feted,
For thus he created
A function of great angularity.

Give a line-by-line explanation of this poem. Is the last line true? False? Meaningful? Make up your own authentic old English nursery rhyme and write a short commentary explaining it and, if needs be, confessing to where you should have applied for a poetic licence.

5.5 Essay Question. The "NF" of Quine's theory is more accurately considered an acronym for "*no* foundations" than for "new foundations". Discuss. (By the way: Just what are foundations?)

---

[148] John N. Crossley, *The Emergence of Number*, World Scientific, Singapore, 1987, pp. 25 - 27.

5.6 Essay Question. Poincaré had said, "Logic sometimes makes monsters". Are Gödel's Theorems analogous to the monsters of Chapter III? What are the similarities? The differences?

5.7 Essay Question. What is the axiomatic method?

V

The Crisis of Intuition

Critics are wont to exclaim
That formalists bear all the blame
For following paths
That make of pure maths
No more than a meaningless game.
—W.R. du Bois-Weyl

I am at a loss on how to proceed with this final chapter. The responses to formalism are many and varied. And they depend on the type of formalism being responded to. Type I formalism is simply not rigorous. One can respond to it by merely noting a Type I argument establishes nothing and moving on, by taking the supposed result as a strong possibility and looking for a proof (possibly along different lines), or, best of all, by trying to justify the use of the Type I method perhaps through some Type II or Type III device. Types II and III formalism are correct and rigorous, but they may still be objected to. If one does not believe in the truth of Type III axioms, what is one to make of the results derived from them? Hasn't one reduced mathematics to a meaningless game of symbolic manipulation? And Type II mathematics, by allowing the construction of monsters, banishes geometric, spatial intuition from mathematics. That one can no longer rely on intuition and must formally deal with formally defined concepts, and one must pretty much follow formal rules as in some sort of game, is a bone of contention for many mathematicians. The main theme of this chapter is thus the "great gulf fixed" between intuition and formalism— largely, Type II formalism.

The trauma of the "crisis of intuition", as Hahn, named it, was not limited to mathematics. Its occurrence in physics is perhaps more widely known to the general public through Einstein's celebrated aphorism about God's refusal to play dice with the universe, or Russell McCormmach's depressing narrative[1] of an aging classical physicist not quite able to accept the modern physics of the beginning 20th century. McCormmach described the period up to the end of

---

[1] Russell McCormmach, *Night Thoughts of a Classical Physicist*, Harvard University Press, Cambridge (Mass.), 1982.

the First World War. By the beginning of the Second World War, nostalgia[2] had given way to anger and in Nazi Germany the Jews were blamed for all that was bad in both fields. That episode has passed, but the crisis continues to erupt from time to time.

1 Criticism of Formalism

So far in this book we have used the word "formalism" to describe various mathematical practices. In the 19th century it was used to describe a certain perspective on mathematics—a mathematical *Weltanschauung* or philosophy if you will. In the 20th century it also described a specific mathematical research programme, and today it has become a cliché, a name for an imagined philosophy that no one really believes in. It is the formalist perspective that drew the first fire.

We saw in Chapter IV that the axiomatic method has two stages, first a setting up stage in which the subject matter is designated and axioms are chosen, and then a developmental stage in which consequences of the axioms are worked out. When one reaches this latter stage it is no longer necessary to consider the meanings of the terms involved, the intuition behind one's conjectures, or the grounds for the truth of one's theorems; all that matters is that the rules for derivation are strictly followed. This is the concern for rigour. Formalism is a perspective on, or philosophy of, not all of mathematics but of this second phase of deriving consequences from given axioms.

When one is deriving such consequences it shouldn't matter where the axioms come from. Indeed, to maintain rigour, one should banish all connexions to the external world from one's mind in checking derivations. The objects one is studying might just as well be the symbols themselves. Modern philosophers now call this view that the objects of mathematical study are marks on paper *term formalism*: the representations are the objects themselves, e.g., numerals rather than numbers. The phrase *game formalism* is applied to the attitude that mathematics is merely a game following certain rules. The game pieces may or may not represent real objects, either physical or abstract entities.

The phraseology of "term" and "game formalism" is modern. In the 19th century one spoke only of "formalism", though likening the practice to playing a game certainly occurred.

It should be remembered that before the beginning of the age of specialisation in the 20th century, mathematicians wore many hats: many were also physicists or astronomers or both. And a few were philosophers. Mathematics was considered a science and had to be serious. Recall William Frend's remark concerning the interpretation of negative numbers as debits:

---

[2] Recall the beer drinking songs written for the International Congress of Mathematicians in Heidelberg in 1904, cited on pp. 428*f*., above.

Now, when a person cannot explain the principles of a science without reference to metaphor, the probability is, that he has never thought accurately upon the subject.[3]

That extreme formalism robs mathematics of its scientific standing was remarked on in a letter from William Rowan Hamilton to John T. Graves concerning Peacock's algebra:

When I first read that work, now many years ago, and indeed for a long time afterwards, it seemed to me, I own— so hard is it for even a candid reader to enter at once into the whole spirit of an original work— that the author designed to reduce algebra to a mere system of symbols, and *nothing more;* an affair of pothooks and hangers, of black strokes upon white paper, to be made according to a fixed but arbitrary set of rules: and I refused, in my own mind, to give the high name of *Science* to the results of such a system; as I should, even now, think it a stretch of courtesy, however it may be allowed by custom, to speak of chess as a "science", though it may well be called a "scientific game".[4]

The name most commonly associated with criticism of formalism as a game is that of Paul David Gustav du Bois-Reymond (1831 - 1889) whom we have already briefly encountered in connexion with Weierstrass's nowhere differentiable function. It was he who first published the result of his teacher in 1875. He is also known for introducing the diagonal argument, most famously applied by Georg Cantor in showing the reals to be uncountable. The younger brother of the famous physiologist Emil Heinrich du Bois-Reymond, Paul almost followed his brother into medicine, but got diverted into mathematical physics. The two shared an interest in philosophical matters and Paul discusses such matters as relate to mathematics in some depth in *Die allgemeine Functionentheorie*. In chapter I of the first volume he describes the various forms of number and remarks on formalism, drawing the analogy between the moves in chess and the combinatorial nature of arithmetic calculation. In Article 14 we read

Out of this observation, which should mediate between the real and the mathematical quantities, one does not want to conclude, in my view, that the boundaries between the higher games and mathematics are also to be erased, and that the combinations of the game of chess are to be accorded the same scientific status as those, e.g., of algebra... The generally felt difference of status between the combinations of chess and those of mathematics appear to me, apart from the so-called practical uses, to rest on the following.

It cannot be denied that the knight problems, particularly in the so-called end games of games of chess, with their pre-arranged beginning

---

[3] Fuller quote on page 212, above.

[4] *Cf.* page 492 and the preceding for a fuller citation and more of its context.

positions of a few pieces and necessarily ensuing conclusions, exhibit the character of genuine mathematical exercises...[5]

Du Bois-Reymond has here drawn the analogy between following rules in chess and doing the same in mathematics. He hasn't yet said that it turns mathematics into no more than a game, but he hasn't finished with the subject. A few pages later, in Article 18, we read

A purely formalistic-literal framework of Analysis which amounts to the separation of the number together with the analytic signs of the quantities, would ultimately devalue this science, which is in truth a natural science,... to a mere symbol game, where the written marks will have arbitrary meanings attached, like chess pieces and playing cards.[6]

He continues, warning that the symbol game can be so delightful that "this literal mathematics, if it becomes completely separated from the ground on which it grew, soon enough amounts to no more than fruitless inclinations" [7]

Another critic of formalism, whom I mention for the sake of any philosophers reading this, was Gottlob Frege. A mathematician, Frege has always had a higher reputation among philosophers than among his mathematical colleagues. In 1884 he published a slim volume[8] on the nature of number which is considered a classic in the philosophy of mathematics. The translator/editor of the English edition was a philosopher; mathematicians, as Frege noted in the first volume of his later book on the subject[9], ignored it: the 1884 volume was not cited in any of several works on the concept of number written in the interim by mathematicians. In the 1884 volume, *Die Grundlagen der Arithmetik*, Frege does indeed have critical things to say about formalism, but no really excitingly quotable remarks as far as I could tell before becoming bored with the book. A first snippet,

but it is precisely in this respect that mathematics aspires to surpass all other sciences, even philosophy[10],

tells us that he agrees mathematics is a science[11]. A second quote,

---

[5] Paul du Bois-Reymond, *Die allgemeine Functionentheorie. Erster Theil. Metaphysik und Theorie der mathematischen Grundbegriffe: Grösse, Grenze, Argument und Function*, Verlag der H. Laupp'schen Buchhandlung, Tübingen, 1882, p. 40.

[6] *Ibid.*, pp. 53 - 54.

[7] *Ibid.*, p. 54.

[8] G. Frege, *Die Grundlagen der Arithmetik. Eine logisch mathematische Untersuchung über den Begriff der Zahl*, Verlag von Wilhelm Koebner, Breslau, 1884. Dual German-English edition: J.L. Austin (trans.), *The Foundations of Arithmetic; A logico-mathematical enquiry into the concept of number*, Basil Blackwell & Mott, Ltd., Oxford, 1950; 2nd revised edition, 1953.

[9] G. Frege, *Grundgesetze der Arithmetik. Begriffsschriftlich abgeleitet. I*, Verlag von Hermann Pohle, Jena, 1893, p. XI.

[10] Frege, *Grundlagen...*, 2nd ed., *op.cit.*, p. IVe.

[11] The German word for science is more inclusive than the English, including philosophy, economics, and other serious disciplines, the key thing here is the serious and disciplined nature of the enterprise.

As to the third point[12], it is a mere illusion to suppose that a concept can be made an object without altering it. From this it follows that a widely-held formalist theory of fractional, negative, etc., numbers is untenable.[13]

telegraphs his disagreement with the formalist enterprise. And a few pages later he brings up the game comparison:

> MILL, of course, would explain 0 as something that has no sense, a mere manner of speaking; calculations with 0 would be a mere game, played with empty symbols, and the only wonder would be that anything rational could come of it.[14]

From this we can read dissatisfaction with the formal enterprise on two counts—first, the reduction of mathematics in status from a science to a game, and, second, the epistemological headache it gives rise to: how can a meaningless game be so useful?

Frege repeats this objection more quotably in his *Grundgesetze der Arithmetik* of 1893:

> From time to time it seems the number marks are like chess pieces and the so-called definitions like the rules of play. The mark then denotes nothing, but rather is the thing itself. A trifle one overlooks this way: that namely with ,$3^2 + 4^2 = 5^2$' we express a thought, while a position of chess pieces says nothing. Where one is satisfied with such superficialities [there] is certainly no ground for a deeper understanding.[15]

Frege has been credited by philosophers with demolishing term formalism. This, of course, had no effect on mathematical practice. Types II and III formalism grew. By 1900 David Hilbert had wholeheartedly embraced the axiomatic method. But paradoxes had also arisen in set theory and Hermann Weyl (1885 - 1955), a student of Hilbert's, feared that contradiction in Analysis was imminent. His response was extreme retrenchment, or, as Hilbert phrased it, "to throw overboard everything that has an uncomfortable appearance"[16]. Weyl's initial criticism was not of formalism itself, but of mathematics and its reasoning about the infinite. Along with a demand for predicativity he saw the use of classical reasoning as a sort of misguided Type I extension to the infinite case of principles accepted in the finite case. Discussing the problem with classical logic, he described the use of quantifiers as follows:

> *An existential assertion*—e.g., "there is an even number"—*is not at all a judgement in the proper sense, which asserts a fact.*

---

[12] "never lose sight of the distinction between concept and object".

[13] *Ibid.*, p. X^e.

[14] *Ibid.*, p. 11^e.

[15] Frege, *Grundgesetze...*, *op.cit.*, p. XIII.

[16] David Hilbert, "Neubegründung der Mathematik, erste Mitteilung", *Abhandlungen aus dem mathematischen Seminar der Hamburger Universität* 1, (1922) pp. 157 - 177.

Rather, it was a "judgement abstract" to be compared with a slip of paper offering a treasure but not telling where the treasure was to be found.[17] Continuing,

> The general "Every number has the property $\mathfrak{E}$"—e.g., "for each number $m, m + 1 = 1 + m$"—is equally less a genuine judgement; rather [it is] a general instruction on judgement.[18]

Thus, Weyl more-or-less considers

$$\exists : \qquad \text{an I.O.U.}$$

$$\forall : \qquad \text{a payment slip.}$$

Combining these observations he says

> In this light mathematics appears as a dreadful paper-economy.[19]

Weyl also announced various beloved results of mathematics that were lost under his programme and declared his new allegiance to the more promising programme of L.E.J. Brouwer.

Hilbert's reaction was immediate. As we saw in Chapter IV in discussing Hilbert and Frege, Hilbert had been thinking about foundational matters and the axiomatic method for some time. Weyl's defection was the stimulus that caused his thoughts to crystalise and over the next few years Hilbert developed an axiomatic programme he called *Beweistheorie* [proof theory], but which more popularly co-opted the name *formalism*.

Hilbert's approach was to acknowledge that large parts of mathematics were without meaning, but that some was meaningful. He would encode the meaningless bits in a formal logical system (like the first-order logic discussed in Chapter IV) and use the meaningful portion to prove the consistency of the meaningless theory. This would allow mathematicians to work with the meaningless bits (e.g., negative and imaginary numbers) *as if* they actually existed. Hilbert would eventually produce a convincing reason as to why this would justify the use of ideal elements to prove results about real objects, but he tended to speak most forcefully about the requirements for rigour, the meaninglessness of theoretical mathematics, the fact that contentual mathematics reasoned about the combinatorial manipulation of symbols, and how the all-important goal of his programme was to prove consistency. In short, he sounded like the archetypal game formalist. Everyone confused his programme with a philosophy of game formalism, Weyl even referring to it as a "formula-game", a description that angered Hilbert, who really should have anticipated the epithet.

Although Hilbert's name is nearly synonymous with the modern use of the word "formalism", I don't think there is a need here to give a more detailed

---

[17] Hermann Weyl, "Über die neue Grundlagenkrise der Mathematik. (Vorträge, gehalten im mathematischen Kolloquium Zürich.)", *Mathematische Zeitschrift* 10 (1921), pp. 39 - 79; here: p. 54.

[18] *Ibid.*, p. 55.

[19] *Ibid.*

account[20]. The immediate outcome of Hilbert's programme was that in 1930 Kurt Gödel proved his Incompleteness Theorems, which proved the untenability of Hilbert's programme.

Parallel to the criticism that formalism turns mathematics into a game was, of course, the dissatisfaction reported on in Chapter III with the invention of Type II monsters. Poincaré's use of words like "monster", "bizarre", "tera-tologic", as well as his complaint that "heretofore when a new function was invented, it was for some practical end; today they are invented expressly to put at fault the reasonings of our fathers" [21], reflects the similar attitude that mathematics was veering away from being a serious scientific enterprise. With the rise of the nazis, criticism of the new situation in mathematics brought on by the increase in fromalism, would rise[22] to new levels, as we will see in the next section.

2 Intuition and Logic

In a letter to Sonia Kovalevskaya of 27 August 1883, Carl Weierstrass wrote

> Among the older mathematicians there are different sorts of men, a trivial statement, which however still explains a lot. My dear friend Kummer, for example, in the time when he set his entire energy on discovering the proofs of the higher reciprocity laws, did not concern himself then or later, after he had exhausted himself thereon, with what had happened in mathematics; he was, if not opposed, indifferent to it. If you said to him the Euclidean geometry rests on an unproven basic principle, he would grant you that; from this insight, however, now put the question so: What form does geometry take without this basic principle? That is against his nature; the endeavours built thereon and the ensuing general investigations of the empirical givens or assumptions are to him idle speculations or, indeed, a horror. Kronecker is different; he quickly makes himself familiar with everything that is new; his ready ability to grasp enables him to do so, but not in a penetrating manner. He does not possess the talent to engage himself in a good, but unfamiliar work with the same scientific interest that he pursues his own studies.
>
> Beyond this he shares the shortcoming that one finds in many intelligent people, especially those of Semitic stock: he does not possess sufficient fantasy (intuition I would prefer to say). And it is true, a mathematician who is not something of a poet will never be a complete mathematician. Comparisons are instructive: an all-embracing vision

---

[20] I refer the curious reader to my earlier exposition of this subject: Craig Smoryński, "Hilbert's programme", *CWI Quarterly* 1, no. 4 (1988), pp. 3 - 59; reprinted in Eckart Menzler-Trott, *Logic's Lost Genius; The Life of Gerhard Gentzen*, American Mathematical Society, Providence, 2007.

[21] *Cf.* page 392, above.

[22] Or, should I say "sink"?

focused on the loftiest of ideals distinguishes Abel from Jacobi, Riemann from his contemporaries (Eisenstein, Rosenhain), and Helmholtz from Kirchhoff (although the latter is without a drop of Semitic blood) in an altogether splendid manner.[23]

A decade after Weierstrass wrote these words, Felix Klein was delivering a series of lectures in Evanston, Illinois, as part of an international congress of mathematicians (a forerunner to the series of international congresses bearing the acronym ICM) held in conjunction with the Chicago World's Fair. Among his comments in Lecture I one reads

Among mathematicians in general, three main categories may be distinguished; and perhaps the names *logicians, formalists,* and *intuitionists* may serve to characterize them. (1) The word *logician* is here used, of course, without reference to the mathematical logic of Boole, Peirce, etc.; it is only intended to indicate that the main strength of the men belonging to this class lies in their logical and critical power, in their ability to give strict definitions, and to derive rigid deductions therefrom. The great and wholesome influence exerted in Germany by *Weierstrass* in this direction is well known. (2) The *formalists* among the mathematicians excel mainly in the skilful formal treatment of a given question, in devising for it an "algorithm." *Gordan,* or let us say *Cayley* and *Sylvester,* must be ranged in this group. (3) To the *intuitionists,* finally, belong those who lay particular stress on geometrical intuition (*Anschauung*), not in pure geometry only, but in all branches of mathematics. What Benjamin Peirce has called "geometrizing a mathematical question" seems to express the same idea.[24]

In Lecture VI, we read

Finally, it must be said that the degree of exactness of the intuition of space may be different in different individuals, perhaps even in different races. It would seem as if a strong naïve space-intuition were an attribute pre-eminently of the Teutonic race, while the critical, purely logical sense is more fully developed in the Latin and Hebrew races. A full investigation of this subject on the lines suggested by Francis Galton in his researches on heredity, might be interesting.[25]

---

[23] Quoted in: W. Ahrens, *Scherz und Ernst in der Mathematik; Geflügelte und Ungeflügelte Worte,* Verlag von B.G. Teubner, Leipzig, 1904, pp. 324 - 325. That part of the quotation about Kronecker and Semitic blood is repeated on page 442 of: David E. Rowe, "'Jewish mathematics' at Göttingen in the era of Felix Klein", *Isis* 77 (1986), pp. 422 - 449. I have cheated and used Rowe's translation for this portion of the quotation. The book by Ahrens, which I found online, is an entertaining collection of mathematical quotations culled from the literature and I recommend it highly.

[24] Felix Klein, *The Evanston Colloquium; Lectures on Mathematics,* Macmillan and Co., New York, 1894; here, p. 2.

[25] *Ibid.,* p. 46. Galton was an early pioneer in the application of statistics in his study of heredity. This paragraph is also quoted in Ahrens, *op. cit.,* p. 148.

Klein's trichotomy has long since collapsed into a dichotomy, the formalists and logicians being lumped together. In his posthumously published history of mathematics in the 19th century, Klein distinguishes between periods of rigour and periods lacking rigour, which correspond to the respective dominances of logic and intuition and he compares this to a dichotomy of types of researchers:

The ideal of "rigour" has historically not always had the same significance for the development of our science, rather each different according to circumstances. In times of great, violent productivity it often recedes into the background in favour of the richest and most rapid possible growth, and in one of the succeeding critical periods, which examines the treasures won, will again be stressed all the more so. One thinks of the development of the Differential and Integral Calculus in the 18th century, in which next to many insufficiently founded results also some directly false results were furthered through the liveliest imagination and the desire for discovery of the day, or in the creation of the theory of algebraic curves in the 19th century. Contrary to this I'd like to remember the time of the scholastics, which united the most extreme sharpness of critical and dialectical reasoning with narrower productivity. Scholasticism is frequently very unjustly judged scornfully as an intellectual direction losing itself in sterile hairsplitting. Our time especially should not agree with this superficial view as pure theological hairsplitting, for it reveals itself frequently as the most correct formulations of that which we designate today as "set theory". If something like the question, whether God could have created the infinite universe in one hour, is raised, the considerations related thereto would in fact be handled no differently from the things which our present mathematicians would be led to through the problem of the infinite set of points of the unit interval. In fact, this is because Georg Cantor, the creator of set theory, has gone to school with the scholastics. Looking back we must say that seldom has the spirit of criticism, the striving for meticulous analysis of every conceivable step, the "ideal of rigour" been so alive as in scholastic times.

The same antithesis which we have here between the different scientific periods can also be recognised in the individual types of researchers. There are the bold conquerors who with the strongest intuition, but with disorderly concepts, who through instinct and empathy find and unearth new treasures; and there are the careful organisers and custodians of the winnings, who appreciate each thing correctly and are able to put it in its place with the clear, sure critique of a sharp mind. The union of these two mutually contradictory gifts is found in only very rare cases... [26]

---

[26] Felix Klein, *Vorlesungen über die Entwicklung der Mathematik im 19. Jahrhundert, I*, Springer-Verlag, Berlin, 1926, pp. 52 - 53. A second volume appeared the following year. In 1979 Springer-Verlag published a single-volume reprint.

Somewhat later in the book, Klein exemplifies the two types by Weierstrass and Riemann:

> Riemann is the man of brilliant intuition. Through his comprehensive genius he towers above all his contemporaries. Wherever his interest is aroused, he begins anew, without letting himself be swayed by tradition and without accepting any systematic constraints.
>
> Weierstraß is first of all a logician; he goes forth slowly, systematically, step-by-step. Where he works, he strives for the final form.[27]

Klein was not alone in his assessment. The distinction between intuition and logic was discussed in detail by Henri Poincaré in Chapter I of *The Value of Science*:

> It is impossible to study the works of the great mathematicians, or even those of the lesser, without noticing and distinguishing two opposite tendencies, or rather two entirely different kinds of minds. The one sort are above all preoccupied with logic: to read their works, one is tempted to believe they have advanced only step by step, after the manner of a Vauban who pushes on his trenches against the place besieged, leaving nothing to chance. The other sort are guided by intuition and at the first stroke make quick but sometimes precarious conquests, like bold cavalrymen of the advance guard.
>
> The method is not imposed by the matter treated. Though one often says of the first that they are *analysts* and calls the others *geometers*, that does not prevent the one sort from remaining analysts even when they work at geometry, while the others are still geometers even when they occupy themselves with pure analysis. It is the very nature of their mind which makes them logicians or intuitionalists, and they can not lay it aside when they approach a new subject.
>
> Nor is it education which has developed in them one of the two tendencies and stifled the other. The mathematician is born, not made, and it seems he is a born geometer or an analyst. I should like to cite examples and there are surely plenty; but to accentuate the contrast I shall begin with an extreme example, taking the liberty of seeking it in two living mathematicians.
>
> M. Méray wants to prove that a binomial equation always has a root, or, in ordinary words, that an angle may always be subdivided. If there is any truth that we think we know by direct intuition, it is this. Who could doubt that an angle may always be divided into any number of equal parts? M. Méray does not look at it that way; in his eyes this proposition is not at all evident and to prove it he needs several pages. On the other hand, look at Professor Klein: he is studying one of the most abstract questions of the theory of functions: to determine whether on a given Riemann surface there always exists a function admitting of given singularities. What does the celebrated German geometer do?

---

[27] *Ibid.*, p. 246.

He replaces his Riemann surface by a metallic surface whose electric conductivity varies according to certain laws. He connects two of its points with the two poles of a battery. The current, says he, must pass, and the distribution of this current on the surface will define a function whose singularities will be precisely those called for by the enunciation. Doubtless Professor Klein well knows he has given here only a sketch; nevertheless he has not hesitated to publish it; and he would probably believe he finds in it, if not a rigorous demonstration, at least a kind of moral certainty. A logician would have rejected with horror such a conception, or rather he would not have had to reject it, because in his mind it would never have originated.

. . .

Among the German geometers of this century, two names above all are illustrious, those of the two scientists who founded the general theory of functions, Weierstrass and Riemann. Weierstass leads everything back to the consideration of series and their analytic transformations; to express it better, he reduces analysis to a sort of prolongation of arithmetic; you may turn through all his books without finding a figure. Riemann, on the contrary, at once calls geometry to his aid; each of his conceptions is an image that no one can forget, once he has caught the meaning.

More recently, Lie was an intuitionalist; this might have been doubted in reading his books, no one would doubt it after talking with him; you saw at once that he thought in pictures. Madame Kovalevski was a logician.

Among our students we notice the same differences; some prefer to treat their problems 'by analysis,' others 'by geometry.' The first are incapable of 'seeing in space,' the others are quickly tired of long calculations and become perplexed.

The two sorts of minds are equally necessary for the progress of science; both the logicians and the intuitionalists have achieved great things that others would not have done. Who would venture to say whether he preferred that Weierstrass had never written or that there had never been a Riemann? Analysis and synthesis have then both their legitimate rôles.[28]

In the next section of his chapter, Poincaré discusses the necessity of logic, that intuition can mislead. E.g.,

We know there exist continuous functions lacking derivatives. Nothing is more shocking to the intuition than this proposition which is imposed upon us by logic. Our fathers would not have failed to say: "It is evident

---

[28] Henri Poincaré, *The Foundations of Science*, The Science Press, New York, 1913, pp. 210 - 212. I quote from the 1929 reprinting of George Bruce Halsted's omnibus translation of three popular works by Poincaré: *Science and Hypothesis*, *The Value of Science*, and *Science and Method*. The particular passage is from *The Value of Science*, which was originally published in French in 1905.

that every continuous function has a derivative, since every curve has a tangent."[29]

And again:

> Intuition, therefore, does not give us certainty. This is why the evolution had to happen; let us see how it happened.
>
> It was not slow being noticed that rigor could not be introduced into reasoning unless first made to enter into the definitions. For the most part the objects treated of by mathematicians were long ill defined; they were supposed to be known because represented by means of the senses or the imagination; but one had only a crude image of them and not a precise idea on which reasoning could take hold. It was there first that the logicians had to direct their efforts.
>
> So, in the case of incommensurable numbers. The vague idea of continuity, which we owe to intuition, resolved itself into a complicated system of inequalities referring to whole numbers.[30]

Two sections later, Poincaré explains why mathematics needs intuition:

> The philosophers make still another objection: "What you gain in rigor," they say, "you lose in objectivity. You can rise toward your logical ideal only by cutting the bonds which attach you to reality. Your science is infallible, but it can only remain so by imprisoning itself in an ivory tower and renouncing all relation with the external world. From this seclusion it must go out when it would attempt the slightest application."
>
> For example, I seek to show that some property pertains to some object whose concept seems to me at first indefinable, because it is intuitive. At first I fail or must content myself with approximate proofs; finally I decide to give to my object a precise definition, and this enables me to establish this property in an irreproachable manner.
>
> "And then," say the philosophers, "it still remains to show that the object which corresponds to this definition is indeed the same made known to you by intuition; or else that some real and concrete object whose conformity with your intuitive idea you believe you immediately recognize corresponds to your new definition. Only then can you affirm that it has the property in question. You have only displaced the difficulty."
>
> This is not exactly so; the difficulty has not been displaced, it has been divided. The proposition to be established was in reality composed of two different truths, at first not distinguished. The first was a mathematical truth, and it is now rigorously established. The second was an experimental verity. Experience alone can teach us that some real and concrete object corresponds or does not correspond to some abstract definition. This second verity is not mathematically demonstrated, but

---

[29] *Ibid.*, p. 213.
[30] *Ibid.*, p. 214.

neither can it be, no more than can the empirical laws of the physical and natural sciences. It would be unreasonable to ask more.

Well, is it not a great advance to have distinguished what long was wrongly confused? Does this mean that nothing is left of this objection of the philosophers? That I do not intend to say; in becoming rigorous, mathematical science takes a character so artificial as to strike every one; it forgets its historical origins; we see how the questions can be answered, we no longer see how and why they are put.

This shows us that logic is not enough; that the science of demonstration is not all science and that intuition must retain its rôle as complement, I was about to say as counterpoise or as antidote of logic. I have already had occasion to insist on the place intuition should hold in the teaching of the mathematical sciences. Without it young minds could not make a beginning in the understanding of mathematics; they could not learn to love it and would see in it only a vain logomachy; above all, without intuition they would never become capable of applying mathematics. But now I wish before all to speak of the rôle of intuition in science itself. If it is useful to the student it is still more so to the creative scientist.[31]

And, in section V, he offers the following:

Pure analysis puts at our disposal a multitude of procedures whose infallibility it guarantees; it opens to us a thousand different ways on which we can embark in all confidence; we are assured of meeting there no obstacles; but of all these ways, which will lead us most promptly to our goal? Who shall tell us which to choose? We need a faculty which makes us see the end from afar, and intuition is this faculty. It is necessary to the explorer for choosing his route; it is not less so to the one following his trail who wants to know why he chose it.

If you are present at a game of chess, it will not suffice, for the understanding of the game, to know the rules for moving the pieces. That will only enable you to recognize that each move has been made conformably to these rules, and this knowledge will truly have very little value. Yet this is what the reader of a book on mathematics would do if he were a logician only. To understand the game is wholly another matter; it is to know why the player moves this piece rather than that other which he could have moved without breaking the rules of the game. It is to perceive the inward reason which makes of this series of successive moves a sort of organized whole. This faculty is still more necessary for the player himself, that is, for the inventor.[32]

Poincaré finishes the section with the words,

---

[31] *Ibid.*, pp. 216 - 217.
[32] *Ibid.*, p. 218.

Thus logic and intuition have each their necessary rôle. Each is indispensable. Logic, which alone can give certainty, is the instrument of demonstration; intuition is the instrument of invention.[33]

Poincaré does not ascribe to any race a particular leaning toward intuition or logic, but he is otherwise in agreement on the dichotomy noted by Weierstrass and Klein. His remarks are more expansive and he brings up the rôle of intuition in creativity and its necessity in pædagogy, two points that would be emphasised in years to come. And like Klein, he appreciates the importance and necessity of logic, an appreciation not always enthusiastically expressed.

The pædagogical point is also raised by Hans Hahn in a review published in 1919 of Alfred Pringsheim's *Vorlesungen über Zahlen- und Funktionenlehre* [*Lectures on number theory and the theory of functions*]:

> Indeed we seem to be dealing here with a *psychological* or *pedagogical* question: it is feared that the use of geometrical concepts might bring with it an irresistible temptation to use (perhaps unconsciously) intuitive and alogical means of proof. That there is such a danger must definitely be admitted. But whether this advantage is outweighed by the disadvantages, which to my mind are quite significant, will always remain a matter of opinion. I regard these disadvantages as so significant because I am convinced that all progress in the more subtle parts of analysis, as for instance in the theory of real functions, is first achieved subjectively, in an intuitive way, and hence that the use of intuition is an indispensable means of investigation. Of course, I do not mean the kind of crude and uncritical use by which analysis was led astray at one time, but a use of intuition purified and refined by the knowledge gained over a century. It seems to me an important task of mathematical instruction, in the classroom as well as in textbooks, to go on strengthening and refining this means of investigation— without damage to the logical rigour which is an inexorable requirement of all proof under all circumstances— instead of letting it atrophy by casting it aside.[34]

I should also mention Pierre Léon Boutroux, mathematician, philosopher, historian, and cousin to Henri Poincaré. In 1920 Boutroux published *L'idéal scientifique des mathématiciens dans l'antiquité et dans les temps modernes* [*The scientific ideal of the mathematicians in antiquity and in modern times*]. Ronald Calinger states in his article on Boutroux in the *Dictionary of Scientific Biography*,

> The main purpose of Boutroux in writing this book, however, was not to investigate the constituent elements of progress. He had two didactic goals in mind. After showing that the different sciences do not progress

---

[33] *Ibid.*, p. 219.
[34] Hans Hahn, "Review of Alfred Pringsheim, *Vorlesungen über Zahlen- und Funktionenlehre*", in: Hans Hahn, *Empiricism, Logic, and Mathematics; Philosophical Papers*, D. Reidel Publishing Co., Dordrecht, 1980; p. 68.

independently, he first asserted that the history of science should be a study of the continuous interactions between the various sciences. He opposed the view of the history of science as consisting only of narrow, technical studies. Second, he told teachers and researchers that no one type of solution exists for all problems. He felt that the nature of the problem best dictated the methods needed for its solution.[35]

This doesn't bear directly on the intuition/formalism divide, so I quote Herbert Mehrtens:

> The other translation initiated by Bieberbach was of Pierre Boutroux's *L'idéal scientifique des mathématiciens*[36]... According to his own statements..., Bieberbach was deeply influenced by Boutroux's book, adopting many of its ideas and even phrasings. Boutroux had maintained the existence of two conflicting "orientations of thought", described alternately as synthetic vs. analytic, intuitive vs. discursive, or as order of invention vs. order of proof. He postulated an "intrinsic" objectivity of mathematics, a realm of mathematical facts independent of constructions. The motives of mathematicians, he wrote, "have almost nothing in common with the algebraico-logical presentation"... which is, instead, an
>> edifice of symbols which we pile up to infinity— like a clever juggler would do, who finds joy in multiplying the difficulties of his exercises.[37]

The mention of Bieberbach brings us to Nazi Germany and Nazi attitudes toward mathematics. The three names that must be mentioned are those of Ludwig Bieberbach, Erich Rudolf Jaensch, and Hugo Dingler. Jaensch was a psychologist who eventually claimed to be influenced by Klein's lectures on the psychology of mathematics given in the winter semester of 1909 - 1910, a claim much to be doubted[38]. Jaensch developed a typology, a theory of psychological types.

> Jaensch had postulated two basic psychological types, the "S-type" with unstable psychic functions, internally generated synaesthetic (hence the "S") perceptions, and a tendency towards disintegration; in contrast was the "J-type" with stable psychic functions, in whom perceptual im-

---

[35] *Dictionary of Scientific Biography*, volume 2, Charles Scribner's Sons, New York, p. 358.

[36] Pierre Boutroux, *Das Wissenschaftsideal der Mathematiker*, Teubner, Leipzig, 1927. The original French text is available online.

[37] Herbert Mehrtens, "Ludwig Bieberbach and 'Deutsche Mathematik'", in: Esther R. Phillips, ed., *Studies in the History of Mathematics*, Mathematical Association of America, 1987; here, p. 206.

[38] Rowe, *op. cit.*, p. 441. In discussing Jaensch and Bieberbach I rely heavily on Rowe, Mehrtens, *op. cit.*, and Eckart Menzler-Trott, *Logic's Lost Genius: The Life of Gerhard Gentzen*, American Mathematical Society, Providence, 2007.

agery and conceptual thinking were strongly integrated (the 'J", earlier
"I", is related to the term "integration type.").[39]

Bieberbach was a mathematician, but, like most mathematicians of those
days, a man of broad, if occasionally dilletantish, interests. He claimed Poincaré
and Boutroux as influences, but developed Jaensch's typology, refining it into
subtypes. I quote Mehrtens:

> In general the J-type was, in Bieberbach's words
>> open to reality, with all sense and psychic functions, so that for
>> him *Anschauung*[40] and thinking merge into a harmonic unity.
>
> The differentiation of types appeared later in Bieberbach's Heidelberg
> lecture:
>> The J_1-type does not turn the world into a problem, rather the
>> problem comes to him out of the world... He is attracted by
>> the colorful richness of reality; he is interested in coherence, in
>> the grand scale of events; while thinking he has to see or feel
>> the relation to reality.
>
> For Bieberbach, Felix Klein was the paradigmatic J_1-type, while Gauss
> represented the J_2-type, which
>> does not so much long for a wealth of knowledge but rather for
>> its meaning and range. He approaches reality with fixed values
>> and ideals. He tries to form cognitions into a world-view. His
>> aim of work is a perfect harmonious construction. He loves truth
>> for its beauty.
>
> The third type, J_3, is illustrated by Weierstrass:
>> It is the type for whom knowledge must have command over
>> the world... The scientist and his subject matter are standing
>> face to face like fighters struggling for power. Cognition is a
>> struggle with reality. In pure mathematics these are the critics,
>> the systematists, who carve out clear rules for the control of
>> the subject, who clarify the basic concepts and deprive them
>> of their mysteries, who form the accumulated results into a
>> system.
>
> In contrast, there is the S-type:
>> None of these types runs the risk of accepting no criterion other
>> than the inner coherence of the edifice of his thought or losing

---

[39] Mehrtens, *op.cit.*, pp. 228 - 229. My copy of *The Concise Oxford Dictionary* defines
synæsthesia as
> *n.* Sensation produced in part of body by stimulus elsewhere; production of
> mental sense-impression by stimulation of another sense.

Some people, for example, see colours when music is played. As for "I" and "J",
the identification of the two letters lasted much longer in German than in English.

[40] *Anschauung* is an untranslatable word usually rendered as "intuition". According
to Mehrtens, "This notion touches on visual perception, geometrical intuition or
visualization, as well as 'intuition' in the sense of Brouwer or Kant. It remained
sufficiently vague to have different meanings in different contexts, and it set a sharp
contrast to formal operations and reasoning." (p. 205)

sight of the natural place of things in science. This is in fact the case with the fourth type, which Jaensch called the S-type, the *Strahltypus* [ray type], who beams his autistic thought into reality. At best he tries to recover his ideas within reality, but not as a confirmation of his thinking, rather as an *epitheton ornans* of reality. Among the great German mathematicians— I emphasize the word "great"— no case of this intellectualist type can be found. Among Germans, however, juveniles and also mathematicians frequently remind one of this type. Namely, among the strangers who took up certain studies of Hilbert, there are some who belong to this intellectualist type.

("Strangers" in this quote refers not only to foreigners but also— and in a stronger sense— to the racial "strangers.")[41]

Hilbert, the greatest mathematician of the first quarter of the 20th century and the pride of German mathematics, posed a bit of a problem with his emphasis on consistency, i.e., the "inner coherence of the edifice of his thought". Bieberbach solved this difficulty admirably:

The struggle over the foundations of mathematics is dependent on races or, expressed differently, one's position in this struggle corresponds to certain types of intellectual creativity. As such the J-type will tend towards intuitionism or towards the ways of Klein, while formalism appears to belong to the S-type. This seems to be contradicted by the East-Prussian heritage of the founder of formalism[42]. Indeed Hilbert cannot possibly be taken as an S-type regarding his other accomplishments. In the psychology of types, however, a form of J-type is known which tends to be open to influences of the S-type.[43]

And later

The difference is quite compatible with the fact that both Hilbert and Brouwer should be classified under the psychological type J_3/J_2. The fact that two men approach their science with an ideal norm does not necessarily imply that it has to be the same norm in both cases.[44]

It is hard to imagine anyone taking all of this seriously, but Bieberbach apparently actually believed what he wrote. Mathematicians abroad were horrified. It did not help his reputation any when Bieberbach publicly defended nazi students who had boycotted the classes of Edmund Landau, a Jewish mathematician whose excessively formal style they detested. Certainly the *Landau*

---

[41] Mehrtens, *op. cit.*, pp. 229 - 230. The quotation is almost verbatim. I have deleted references like "[Bieberbach, 1934d, 5]".

[42] "Formalism" is used here in the narrow sense of Hilbert's programme of encapsulating mathematics within formal deductive systems.

[43] *Ibid.*, p. 228.

[44] *Ibid.*, p. 228.

style, as the now familiar totally rigorous Definition-Theorem-Proof enumeration was called, leaves a lot to be desired pædagogically, as is still remarked on
today:

> The writing in the Bourbaki books is crisp, clean and precise. Bourbaki
> has a very strict notion of mathematical rigor. For example, *no Bourbaki
> books contain any pictures!* That is correct. Bourbaki felt that pictures
> are an intuitive device, and have no place in a proper mathematics text.
> If the mathematics is written correctly then the ideas should be clear—
> at least after sufficient cogitation. The Bourbaki books are written in a
> strictly logical fashion, beginning with definitions and axioms and then
> proceeding with lemmas and propositions and theorems and corollaries.
> Everything is proved rigorously and precisely. There are few examples
> and little explanation. Mostly just theorems and proofs. There are no
> "proofs omitted", no "sketches of proofs", and no "exercises left for the
> reader".
>
> The Bourbaki books have had a considerable influence in modern math-
> ematics. For many years, other textbook writers sought to mimic the
> Bourbaki style. Walter Rudin was one of these, and he wrote a number
> of influential texts without pictures and adhering to a strict logical for-
> malism. Certainly, in the 1950s and 1960s and 1970s, Bourbaki ruled
> the roost. This group of dedicated French mathematicians with the fic-
> titious name had set a standard to which everyone aspired. It can safely
> be said that an entire generation of mathematics texts danced to the
> tune that was set by Bourbaki.
>
> But fashions change. It is now a commonly held belief in France that
> Bourbaki caused considerable damage to the French mathematics en-
> terprise.[45][46]

Refusal to take courses under Landau is any student's right and would
have been reasonable and defensible, as would have been even protests and
petitions were the courses taught by Landau necessary for further progress
in their studies. The Göttingen students far exceeded this: they blocked his
entrance to the classroom and forced his resignation. Bieberbach defended these
actions in a series of lectures given in April 1934. On the 8th of the month, a
summary of the lectures appeared in the journal *Deutsche Zukunft* and on 1
May the Danish mathematician Harald Bohr, brother to the now more famous
physicist, criticised Bieberbach's views in a Danish newspaper. Bieberbach used
his position in the German Mathematicians Union to publish a response in the
official organ of the society:

> In the framework of my lecture the statement referred to had the clear
> intention to show the inner reason that the valiant appearance of the

---

[45] Steven G. Krantz, "The history and concept of mathematical proof", draft of a
paper dated 5 February 2007, posted online; pp. 28 - 29.

[46] The book before you has pictures, omits some proofs, sketches others, and leaves
some to the reader. It also has a chapter of non-rigorously proven results. It is thus
a paragon of pædagogically sound exposition!

Göttinger students put a limit on the further effectiveness of Mr. Landau as a teacher of German youth... A people, which wishes to develop its own gifts, needs teachers of its own kind.[47]

Bieberbach's point was that students should be taught by those with the same psychological style of thinking they had. It is nicely expressed in a summary of Bieberbach's lectures published in May of that year:

> In the series of mathematical lectures Professor Dr. Bieberbach spoke about personality structures and mathematical creativity. To this end he points out that the Göttingen student league had rejected the activity of the number-theoretician E. Landau because his presentations were strange, and justified this with the fact that the mathematics itself is international in its results but not in the manner of its thinking and representation.[48]

In 1936 Bieberbach founded his own journal *Deutsche Mathematik* [*German mathematics*] so that there would be at least one Jew-free mathematics journal. It was not a resounding success, but contained enough solid mathematical work that it was reprinted after the war— in censored form: blank pages were substituted for some of the more offensive papers.

Bieberbach would later claim that he was not anti-semitic. Already in his open letter to Bohr, he cited Klein, Weierstrass and Poincaré[49] in noting racial differences in psychological styles and he added that he simply felt that German mathematicians should develop their own style. Writing to Bohr, he maintained he was not opposed to other styles and indeed he co-edited Brouwer's journal *Compositio Mathematica* which was internationalist in outlook. Even if one believes this defence, however, one is reminded of the old adage: The road to Hell is paved with good intentions.

Two nazis whose anti-semitic activities have never raised the suspicion of good intentions are Max Steck and Hugo Dingler.

> Steck would label Hilbert a formalist whose success in "foundations" was based upon a Jewish perspective of philosophy which had been laid down by Cohen, Vaihinger, Schlick et al. Formalism changes mathematics into a "system of science from pure tautologies." In this Steck stuck closely to Tornier [who had written]:
>> It is namely the typical Jewish-liberalistic thesis that the criterion of right to existence of a mathematical theory is its "æsthetic beauty", whereby a logically closed... construction from definitions will be meant.

---

[47] Ludwig Bieberbach, "Die Kunst des Zitierens. Ein offener Brief an Herrn Harald Bohr in København" [The art of citation. An open letter to Mr. Harald Bohr in Copenhagen], *Jahresbericht der Deutschen Mathematiker-Vereinigung* 44, 2nd division (1934), pp. 1 - 3; here, p. 2.

[48] Quoted in Menzler-Trott, *op. cit.*, p. 155.

[49] In *Science and Hypothesis* (pp. 175 - 176 in the omnibus edition cited in footnote 28, above), Poincaré had compared Maxwell's approach with that of the French— which latter approach he explicitly attributed to French education, not race.

Formalism begets, ever more complex and varied, a subtlety which is accessible only to a circle of esoterics who ban the German mathematical idea and in the end work only with arbitrarily set names. Steck described formalism as mathematical futurism and cubism:

> The pinnacle of absurd development is to be found in the books and works of E. Landau, A. Rosenthal, H. Minkowski, M. Dehn, I. Schur, S. Bochner and many other Jews in mathematics.

The correct mathematics, however, proceeds from Euclid, where intuition plays a decisive rôle:

> ... the intuition, as the highest intellectual power of man, to seek in pure ideas of the mathematical, to examine and symbolise the ideal adequately.

The consistency of an axiom system is to be given through geometry, because if the basic assertions of the axioms of a system were to be logically or intuitively compelling and illuminating, they would also be considered to be true.[50]

The nazi I wish most to quote, however, is Hugo Dingler. This philosopher/physicist who fancied himself an expert on the foundations of mathematics is responsible for what has to be one of the most disgusting documents in the history of mathematics[51]. I would love to reproduce the entire thing here, but it is rather long and not all of it is relevant. The most relevant parts, however, express most dramatically the dissatisfaction of those oriented by geometric intuition with formal mathematics. Dingler begins with an attack on Klein, who he claimed had Jewish ancestry[52]. Klein, according to Dingler, was solely interested in acquiring power over others and mathematics was his tool. After explaining how Klein set out to conquer Germany, Dingler continues:

> ... In the end no one in Germany could hold more than the smallest position without Klein's consent. Soon nearly all the mathematical chairs in Germany were occupied by his people.
>
> The persons whom Klein chose for these were exclusively such who proved themselves worthy as collaborators in his "Betrieb" [business, or concern] (a favourite word of his). But his ambition reached farther. He wished also to subjugate mathematics abroad. So he drew foreigners as students to Göttingen, where each arbitrary foreign student finds the highest interest, while the young German, in case he is not Jewish,

---

[50] Menzler-Trott, *op. cit.*, p. 205.

[51] This was a memorandum composed in two parts in April and September 1933 and sent to Philipp Lenard, who combined the two parts and forwarded it to various ministries in December 1933 and January 1934. It bore a title which translates to "Absurd developments in the area of pædagogy of mathematics and the exact sciences in the last half century".

[52] Klein, with his support of and emphasis on applied mathematics, embodied a mathematical ideal of the nazis. It wasn't long before proof that Klein was not Jewish was produced. Indeed, in the opening lines of Rowe, *op. cit.* (p. 422), it is noted that his Aryan standing was proudly proclaimed in the right-wing *Göttinger Tageblatt*.

must lead a modest existence. The atmosphere is not only somewhat international and pacifistic, but is already since the 90s outspokenly anti-german. A hint of nationalist convictions by a young German leads automatically to a professional transfer to the provinces [Kaltstellung]. The Jewish students, on account of their skillfulness, quickness, and rapid absorption, are everywhere in the foreground (This agility in the intellectual handling of the— for the moment— immediately vital and important facts exists only because of a lack of depth; the non-Jewish mentality therefore only appears slow, because many more secondary associations will be made— associations which, to be sure, can appear less important for the purely superficial material consideration in the moment, but in fact carries with it an, in many ways, larger breadth of perspective and of possibilities and which are preconditions of the extraction of new inventions.); the young Germans are so impressed thereby that they, not to be too different, emulate them in speech, bearing, and gesture with all eagerness, so that a specific Göttingen mathematical behavioural style has formed and become widespread German mathematical fashion. Only those German nonjews who have adapted themselves in nature and attitude as completely as possible have a modest prospect to arrive in the general struggle for a permanent job.

Through Cauchy's reform an entirely analogous situation has developed in mathematics as in physics. With the turn to pure approximatives, the incentive is lost to search for large lines and categories in functionals. This situation is exceedingly favourable for purely casuistically oriented mentalities like the Jews have been ever since their classical literature, and so there developed then, especially in the so-called function theory and its neighbouring areas (analytic number theory, set theory), a vast, extensive casuistic research, which lies mainly in Jewish hands. Countless small and smallest little results were inflationarily complicated[53], and thereby published in high scientific appearing presentation, and were accepted by the like-striving with the greatest expenditure of enthusiasm and a praise that previously was reserved only for genuinely fundamental advances of science. So there soon developed a uniform protected public opinion in mathematics, which soon became the more uniform and could be the more uniformly controlled, ever more the Klein dictatorship resulted in the creation of a uniform-minded math-

---

[53] A good example is Lemma 3.4 of Chapter I (page 19, above), by which we proved that if L_1 is the limit of a sequence and L_2 is the limit of the same sequence, then $L_1 = L_2$. Bolzano's calculation of the limit of a geometric progression, discussed in Chapter 1, section 4, and again in Chapter III, section 4, is another example of over-elaboration wrought by Type II formalism and the requirement of rigour. The most famous example, which I have not discussed here, is the multipage proof given in *Principia Mathematica* (1910) by Bertrand Russell and Alfred North Whitehead that $1 + 1 = 2$. Some decades earlier, Richard Dedekind claimed, on the basis of his construction of the reals, to be the first to actually prove $\sqrt{6} = \sqrt{2} \cdot \sqrt{3}$.

ematicianship, which finally became a heavily visibly closed organisation. (Already in the first decade of the 20th century there was at
every German high school a, usually Jewish, "Göttingen envoy", who
reported occurrences involving the mathematical and physical person-
nel to Göttingen, and through the Göttingen influence almost ruled.)
Through this thus created public opinion this then appeared: Sheaves
of such little theorems soon acquired larger than life significance and
would work for the given candidate for an open position, whereby the
Göttingen organisation acquired further increase in power.

A very far reaching side effect of this development was the following: the
outstanding mathematically gifted German nonjews are, in the direction of their talent, intent mostly very intuitively on form-ideas, as these
shaped the nature of all pure science since the Greeks and make up the
proper nature of this science. Through the casuistic type of treatment,
however, such talents no longer find a proper field of work. Opposed
to that this circle focuses more narrowly on the dialectic mastery of
hundreds of casuistic special cases, which must not seem meaningful
enough to them (and correctly so), to devote one's life to these. The
Jewish powers were, through the casuistic gift, an advantage here. At
the same time it seems for them the lack of an inner sense is less of a
hindrance there, where a biological-utilitarian sense of an occupation
is given. Through the ever more widening casuistic treatment of mathematics in books and lectures the appearance arose, that the Jewish
ability at mathematics was a greater "gift" than the nonjewish. This
in turn automatically had an effect on the selection of the supposedly
gifted, where the lectures and exams would be held by casuistically
oriented teachers. Thus the Jewish-casuistic element penetrated ever
further into mathematics, so that finally only Jewish teachers appeared
suitable to occupy the higher mathematical positions. Which then occurred.

Another characteristic of the Jewish talent comes in, namely the denial
of all that is intuitive, which so to speak is a corollary to the predilection
for the casuistic-dialectic treatment of pure word-laws. ("You should
make yourself no image nor any likeness at all.") So it came to pass
that in German mathematics all that is *intuitive* (on which in the shape
of form-ideas all pure science rests), thus in particular pure geometry,
would be pushed aside ever more. This circumstance has worked out so
that today it is very difficult to find by the above methods among the
younger generation selected representatives who are suitable to occupy
a chair in pure geometry. Almost only at the technical high schools,
where an absolute material necessity forces intuition, there lingers yet
a remnant of real geometry. But through the above treatment of public opinion this remnant leads only a patient existence. Its works and
problems will, through the general cheap propaganda, be considered as
of little value and they have difficulties with publication in professional
journals, which are entirely in the hands of the non-intuitive direction.

The invisible organisation created by Klein has full power in its hands, uses it mercilessly, and admits only such non-jews who have subjugated themselves body and soul. *Everything that exists in the talents in German non-jewish circles has from the beginning with all means been put off* and attempted to be kept aside, whereby the mechanism described proves itself restlessly useful and effective. Now let us turn to physics...[54]

Let us not turn to physics, but consider what is already at hand. A number of points have been raised, mixed up and confused. And there is more! So it may be a good idea to pause and try to sort out the issues. They are:

•P s y c h o l o g i c a l. There are two types of mathematical minds, *intuitive* and *logical*. The intuitive mathematician, as exemplified by Riemann, thinks visually, looks for the overall picture, proceeds by leaps and bounds, and is creative and inventive. The logical mathematician is good at formal manipulation and proceeds step-by-step. He provides rigour.

•P æ d a g o g i c a l. The logical development favoured by the second type of mathematician is a detriment to intuition-oriented students. This has nothing to do with race. Augustin Louis Cauchy's excessively formal detail drew protest from his students and criticism from his fellow French colleagues.[55] Likewise, the Italian Giuseppe Peano had been an excellent teacher to Italian students until he allowed himself to get carried away by his love of symbolism to introduce ever more formalism into his courses. The Military Academy at which he taught for 15 years had to fire him.[56] And I have already mentioned the French judgment of Bourbaki-inspired mathematical education. The turn away from intuitive mathematics to an excessive formalism in the mid-20th century mathematical instruction was decried by many, most vehemently, if not most notably, by Morris Kline in his book *Why Johnny Can't Add*[57].

•P h y s i o l o g i c a l. The traditional teutonic view that different races have different "personality types" when it comes to mathematics has not gone away. The complaint, not restricted to mathematics, is that this or that race would profit more from intuition-oriented education and suffers under the logico-

---

[54] Menzler-Trott, *op. cit.*, pp. 104 - 105.

[55] James R. Hofmann, *André-Marie Ampère: Enlightenment and Electrodynamics*, Cambridge University Press, 1996, p. 172; Hans Niels Jahnke, *A History of Analysis*, AMS, Providence, 2003, p. 160.

[56] Hubert C. Kennedy, *Peano; Life and Works of Giuseppe Peano*, D. Reidel Publishing Company, Dordrecht, 1980, pp. 100 - 102.

[57] Morris Kline, *Why Johnny Can't Add; The Failure of the New Math*, St. Martin's Press, New York, 1973. I acquired a copy with the intention of citing some passages. The book is written with a heavy hand and each page drips of sarcasm, but I found Kline's repetitive rhetoric boring after a while. Beyond that he makes many good points, but wordily. Nonetheless, I recommend this book to the reader. Aside from its prolonged discussion of the pædagogical shortcomings of too much formalism, it is a superb document on the distinction *cum* antagonism between intuition and rigour.

formalistic education aimed at white males. Gender differences are also declared. The differences are explained by the known physiological asymmetry in the functioning of the left and right hemispheres of the brain. How much of this is science and how much pseudoscience is not clear.

•M a t h e m a t i c a l. The prevailing view is that the use of intuition— geometrical, spatial intuition, or the right hemisphere of the brain— obtains new results, while logic— the left hemisphere— provides rigour in the demonstrations. The pædagogical point is that instruction aimed at the left brain is insufficient. Aside from turning students away from mathematics, it leads to confusing a multitude of trivialities with good mathematics. Of those we quoted here, only Dingler was explicit on this point, but he said it forcibly enough for everyone. Formal mathematics is open to criticism on two counts. First, its excessive attention to detail focusses attention on small points and elevates trivialities in status to that of genuine results. Second, by banishing intuition, formalism turns mathematics into a game, thus trivialising it further.

The psychological, pædagogical, and physiological issues have become intertwined. The left brain/right brain divide explains the psychological dichotomy, that it is a dichotomy *á la* Weierstrass and Poincaré and not a trichotomy *á la* Klein. The traditional educational choice between the historical approach to the subject, teaching it as it evolved, and the logical approach, building it up step-by-step on a firm foundation, is no longer a matter of personal preference, but a physiological dictate. I have little to say on the subject, but it is important enough today that some coverage should be given. We will encounter this in the immediately following section.

3 Physiological Digression: Left Brain/Right Brain

I first learned of the differentiation between the two hemispheres of the brain in my freshman psychology course taken in the mid-1960's. Thereafter, aside from snickering at the idea of a friend of mine teaching a "Math for Girls" course based on the supposed fact that men and women favour different hemispheres of the brain, I had no other dealing with the subject and had forgotten all about it until thinking about Weierstrass, Klein, Poincaré *et alia*. I thought the left/right distinction would provide a nice underpinning for the discussion. So I looked up a few references and found a great deal of contradictory material: solid results, highly speculative overinterpretation, harsh criticism, *non sequiturs*, and so on. I would throw up my hands in despair and drop the subject were it not so pervasive in the discussion of mathematical pædagogy. The left/right distinction, however much surrounded by pseudoscience, has a certain heuristic value and, so long as one does not mistake its suggestions for firmly established results, thinking in terms of it is harmless enough.

Much of the literature on the left brain/right brain divide comes not from the fields of physiology or psychology, but from the business world and the education establishment, where implications of the basic research are imaginatively and enthusiastically explored.

I am unfamiliar with the literature and cannot say how representative the ensuing examples are. From the world of education, there is a 49 page booklet by Ronald L. Rubenzer put out by the Council for Exceptional Children in 1982. The work starts out speculatively enough, and then, like a bad term paper, lists every fact, relevant or irrelevant, that the author could find about the brain— citing sources for every find. His history of left brain/right brain research I reproduce in full (minus the bibliographic references):

History of Brain Research

Although the relationship between damage to the left side of the head and resultant language impairment was noted several thousand years ago by the Egyptians, the first clearly documented account of the idea of hemispheric specialization of function was attributed to Goethe in 1796 in his description of the association between lesions in the left hemisphere and the inability to speak. Marc Dax is credited with presenting the first formal theory of localization of speech in the left hemisphere in 1811, followed by Broca in 1865, who also forwarded a theory of specialization of the left hemisphere for speech. In 1866, Jackson performed the first medical investigation to determine the relationship between brain lesions and aphasia. Roger Sperry's most significant "split brain" experiments in the 1950's marked an upsurge in hemisphericity research that has continued to grow over the last three decades. Sperry's studies revealed that when the neural bridge (corpus collasum [*sic*, the reference should be to the *corpus callosum*]), physically connecting and relaying neural signals between the two hemispheres, was cut (to reduce seizure activity in epileptic patients), the left and right brains functioned nearly independently in receiving and controlling various types of stimuli. Stimuli received from the body's right sensory/motor fields were processed and controlled by the left side of the brain, while the right side of the brain processed the information and controlled the responses of the left sensory/motor fields. Further research indicated that hemispheric processing preferences varied according to the types of information received (verbal, spatial, etc.) and the cognitive style demands of the task (convergent, divergent, etc.). Evidence concerning specific involvement in cognitive/affective functioning has been collected for over a century. [58]

This is followed by a passage of roughly the same length describing briefly the methods used to study the brain. The point is that the techniques are very scientific and the findings to be cited are not to be doubted. The attitude is conveyed by the tone of the closing sentence:

---

[58] Ronald L. Rubenzer, *Educating the Other Half: Implications of Left/Right Brain Research*, ERIC Clearinghouse on Handicapped and Gifted Children and The Council for Gifted Children, Reston (Va.), 1982, p. 3.

The technique is called positron emission transaxial tomography or PETT III and possibly represents as significant a technological advance in brain research as the development of the telescope in astronomy.[59]

Then comes the important statement of the basic left brain/right brain hypothesis:

Dominance

The left hemisphere has been considered historically to be the dominant or major hemisphere because it was held that this hemisphere was primarily responsible for the processing of language and planning— the two functions that clearly distinguish man from the rest of the animal kingdom. Interestingly, it has been found that the left hemisphere is, in fact, anatomically larger than the right hemisphere. Furthermore, it has been found that the left hemisphere is predominantly more active than the right in most adults. The dominance relationship between the left and right hemispheres in daily functioning can be compared to that of the coexistence of the sun and stars. Although right brain functions are always occurring, they are dominated by left brain functions, just as the sun obscures the presence of the other stars in the sky during the day.

Divisions of Labor

Although anatomical differences in brain structure between individuals vary as much as facial features, generalizations concerning the divisions of labor of the bifunctional brain can be made based on extensive research findings. In order to avert the field of brain research from becoming merely a "modern phrenology," it is imperative that conclusions be drawn from scientifically reputable studies. The generalizations regarding the functional organization of the brain presented herein are based on a review of over 150 studies conducted within the last 30 years. The majority of these studies dealt primarily with right-handed, English-speaking adults; therefore, all statements made will be limited to that population unless stated otherwise. Variations in brain processes for left-handed and non-English speaking individuals, children, and learning/behaviorally handicapped students will be addressed separately.[60]

Rubenzer then proceeds in several pages to mine his earlier (1978) review of over 150 studies to cite the conclusions to be drawn therefrom. These findings are summarised in a copyrighted table from another earlier (1981) report of his reproduced as *Table 1*.

This sort of table is common. A few days before writing these words I found a variant of *Table 2* on *Wikipedia* online. The table as presented on the next page is today's version[61]. A few days ago, "analytical" was matched up with "holistic", and "verbal" with "imagistic". Moreover, there were extra rows

---

[59] *Ibid.*, p. 4.
[60] *Ibid.*, pp. 4 - 5.
[61] I double checked the format and found the difference.

TABLE 1
BRAIN FUNCTION MATRIX

Predominant Brain Processing Mode			
FUNCTIONS	LEFT	INTEGRATED OR ALTERNATING	RIGHT
General	Facts	Intuition	Feelings
Thinking Style	Verbal		Spatial
	Orderly		Random
	Temporal	Problem Finding/Solving	Simultaneous
	Convergent		Creative
Emotions	Positive	Well-adjusted	Negative
Education	Reading	Singing	Music
	Writing	Poetry	Art (Logos)
	Arithmetic	Geometry	Tangrams
	Computer Programming	Computer Graphics	Computer Games

Table 2

Left hemisphere functions	Right hemisphere functions
analytical	holistic
verbal	prosodic
logical	intuitive
mathematics (exact calculation, numerical comparison, estimation) left hemisphere only: direct fact retrieval	mathematics (exact calculation, numerical comparison, estimation)
present and past	present and future
language: grammar/vocabulary, literal	language: intonation/accentuation, prosody, pragmatic, contextual

with "sequential" matched with "simultaneous" and with "linear" and "holistic algorithmic processing" opposing each other on the left and right, respectively.

But let us return to the 1980's. Rubenzer was an administrator, the State Coordinator of Gifted Education in the Special Education Unit of the New Mexico State Department of Education. Terence Hines, by 1985, was an Assistant Professor of Psychology at the University of Oregon when he published his own review. His well-written article comes straight to the point:

> Research on the differences in function between the left and right sides of the human brain once was discussed only in professional medical and psychology journals. Now, neuroscience research has captured the popular imagination. References to it appear regularly in the popular press. Further, claims are made frequently that the differences between the two sides of the brain have important practical implications. This seems to be especially true in the area of training and personnel development. Several articles have appeared in this *Journal* over the past few years claiming that understanding left-brain/right-brain differences is important for people in the field of training and development. The basic thesis of these articles is that the two hemispheres of the human brain differ greatly in their mode of cognitive processing or thinking, that Western society emphasizes "left brain" modes of thought, that the left brain represses the right brain and its natural mode of thought, and that, if only the right brain could be better trained to express its mode of thought, we'd all be better managers, salespersons, artists or whatever... The left brain is said to be "Conscious, inductive, logical, linear thinking and questions why and how." The right brain is said to be "Subconscious or even unconscious, deductive, intuitive and non-linear thinking."
>
> All this may sound very scientific, especially when presented with the patina of modern neuroscience. However, it is an astonishingly uninformed and simplistic view of the brain. None of this left-brain/right-brain "mythology" is supported by the actual research on the differences between the left and right human cerebral hemispheres. In fact, the research literature flatly contradicts most of the mythology.[62]

The article then splits into three sections, tellingly titled "Left-brain/right-brain mythology", "Little distinction", and "Reliable or valid?" Citing several studies in which the proclaimed hemispheric differences are statistically significant but minor (e.g., differences in reaction times measured in tens of milliseconds), he concludes

> The real research findings on hemispheric differences have shown, in sharp contrast to the claims of left-brain/right-brain mythology, that differences in function between the hemispheres, while very real and extremely interesting are, with the exception of vocal control, rather small

---

[62] Terence Hines, "Left brain, right brain: Who's on first?", *Training and Development Journal*, November 1985, pp. 32 - 34; here, p. 32.

and matters of degree. There is no evidence to support the claims that, for example, the left hemisphere is "logical" and the right "intuitive" or that the left hemisphere is "conscious" while the right is "unconscious." Harnad and Steklis have pointed out that such simplistic dichotomies bear "about as much relation to the known facts about hemisphere functioning as astrology does to astronomy." In view of this, attempts to improve performance and training by relying on nonexistent left-brain/right-brain differences are unlikely to be productive.[63]

These references are about a quarter century old, but the overall picture hasn't changed. A bit more online searching will uncover more exaggerated claims and more debunking of exaggerated claims. With respect to mathematics, the claims appear to be that this or that minority is right brain dominated and its children require a different educational approach than the left brain dominated instruction aimed at white children.[64]

In the popular imagination, women are better at logical/linguistic tasks than men and are thus left brain dominated, while men are more geometrically inclined and thus are right brain dominated. Thus we have explained for us the age old conflict between men who prefer to read maps and women who would rather ask for directions. I am toying with the idea of proposing a theory that men leave the toilet seat up because they are comfortable with the 3-dimensional configuration, while women prefer it down because the 2-dimensional configuration is closer to the orderly, 1-dimensional, sequential world view of the female mind. And I wonder where the vaunted woman's intuition comes from if the *Wikipedia* is correct in placing intuition in the right hemisphere— where proponents of right hemispheric dominance in minorities often place it.

Come to think of it, if women are left-brained and mathematical instruction is aimed at the left brain, why did we need those "Math for Girls" courses? It would seem we already had them.

4 Public Dismay

I confess I haven't tied all the dissatisfaction reported on in this final chapter to the rise of Types II and III formalism as tightly as I would have liked. Perhaps I have not given the subject enough thought. Or perhaps it is a reflexion of the fact that I am better at writing about simple technical matters and am lacking as a writer when it comes to less clear-cut issues. But I think it is

---

[63] *Ibid.*, p. 34.

[64] Marjorie Lee, "The match: Learning styles of Black children and microprocessor programming", *Journal of Negro Education* 55 (1986), pp. 78 - 93, does not specialise on mathematics, but is a good example of the call for race-based education. In an appendix to the second edition of Steven Krantz's book *How to Teach Mathematics* (American Mathematical Society, Providence, 1999), David Klein cites the *Journal of American Indian Education* for both the claim and criticism of the claim that American Indians are right brain dominated.

important to raise these issues, however imperfectly. For, the dissatisfaction is real and unbalanced proposals for reform—of education and practice—are occasionally put forwad rather too forcefully for comfort. Twice in the pages of *The Mathematical Intelligencer*, having seen a "clear and present danger", I have launched my own counterattack in what must surely have been perceived as a young unknown trying to make a name for himself by going on the offensive against more famous targets.[65] I tend to pride myself in my self-knowledge and am fairly confident that this was not my motive. In the early years of *The Intelligencer*, the editor Roberto Minio had to scramble for material and, as he was an all-round great guy and a valued acquaintance, I tried to oblige. And, I confess, I like the sound of my own voice—metaphorically: my physical voice is a source of embarrassment, but I do love to read and reread everything I write[66]. But, granted all of this, I believe in what I wrote and, given the importance I place on the issues involved, should repeat myself here.

I hasten to preface these remarks with the disclaimer that I do not regard these men with the same moral repugnance I feel toward Steck and Dingler. But they share the same dissatisfaction with the new that Dingler did and their solution was condemnation, *à la* Dingler. Where Dingler sought redress through official Nazi channels, Morris Kline and Saunders Mac Lane used popular literature to exert their respective wills through influencing public opinion. There is no mistaking the political intent of these two men and though they are both dead and can no longer act on their own, others can act on their behalf and they should be opposed at every opportunity.

Morris Kline (1908 - 1992) was a competent applied mathematician, educator, and populariser of mathematics. His two great contributions to the popularisation of mathematics were his *Mathematics in Western Culture*[67] and *Mathematical Thought from Ancient to Modern Times*[68]. The former is a celebration of mathematics in western culture, celebrating both its westernness and its contribution to culture. In today's politically correct climate, the book might appear reactionary, but it is a fact, however uncomfortable to some, that although every culture has its mathematics, the discipline as we know it today was primarily shaped by (and helped shape) western culture. The second book cited is a massive summary of the history of mathematics through the ages and is probably as good as it gets for a single-volume history of mathematics by an amateur historian of the field.

Kline also made solid contributions to mathematical literature stressing applications and intuitive understanding. However, he is most famous for his polemical, somewhat political, works. His *Why Johnny Can't Add*[69] is a scathing indictment of the failure of the New Math. The solid analysis and

---

[65] Indeed, one of them said as much.

[66] A fact which explains why the reader has found no typographical errors in the present work!

[67] Oxford University Press, New York, 1953.

[68] Oxford University Press, New York, 1972

[69] Morris Kline, *Why Johnny Can't Add: The Failure of the New Mathematics*, St. Martin's Press, New York, 1973.

criticism he offered was, however, masked by a heavy sarcasm that quickly ceases to amuse. It is reported that his follow-up, *Why the Professor Can't Teach*[70], bore a title he neither chose nor wanted but which was forced on him by a publisher with a mind to the publicity the earlier book had received. I have not studied either book carefully. With respect to the former, I simply tired of his prose; and the latter book is not in my collection. However, I have read one of his attacks on a field he believed had spiralled out of control, and it happens to concern an area I had studied fairly deeply. I began my review thus:

> I have just completed a rapid reading of Morris Kline's new book *Mathematics; The Loss of Certainty*[71] and I am angry. I am not angry because the book is badly written—it is nicely done. I am not angry because I disagree with him—I think I agree with much of what he says. I am angry because he wasted all that masterly research and his many years of scholarship: Instead of compiling his impressive history and drawing conclusions from it, he simply related this history to prove a point. How else is the lack of perspective explicable? A fair study of history should lend perspective, not block it.[72]

Kline's contention, in a nutshell, is that mathematics, both in its teaching and its choice of research topics loses sight of intuition. In explaining this, one can do no better than to steal from J.J. O'Connor and E.F. Robertson's mini-biography of Morris Kline at the Math Tutor web site. First, they quote Kline's "A proposal for the high school mathematics curriculum" (1966):

> Instead of presenting mathematics as rigorously as possible, present it as intuitively as possible. Accept and use without mention any facts that are so obvious that students do not recognize that they are using them. Students will not lose sleep worrying about whether a line divides the plane into two parts. Prove only what the students think requires proof. The ability to appreciate rigor is a function of the age of the student and not of the age of the mathematics.

This seems reasonable enough, although the word "instead" makes me nervous. Here, however, he is restricting his attention to the high school curriculum. As to higher education, at some stage the student will either have developed a good mathematical intuition or he will not. The former student is ready for rigour and the latter might best be served by being directed to another field. I've not read Kline's textbook on the Calculus—there comes a point when one has seen too many and, besides, one already has a favourite. O'Connor and Robertson report:

---

[70] Morris Kline, *Why the Professor Can't Teach: Mathematics and the Dilemma of University Education*, St. Martin's Press, New York, 1977.

[71] Oxford University Press, New York, 1980.

[72] C. Smoryński, "Morris Kline; the DeKline from grace", *The Mathematical Intelligencer* 4, No. 3 (1982), pp. 125 - 129; here: p. 125.

In 1967 he published a two-volume calculus text *Calculus, An Intuitive and Physical Approach* which teaches calculus through physical problems but also tries to develop the student's intuition by approaching problem solving as a beginner might, making false starts and changing tack.

Kline has certainly put his money where his mouth is, so to speak, and cannot be faulted for hypocrisy. Perhaps one should take a careful look at this book to see how rigorous it is. In my own experience I find it easier to teach the basic ϵ-δ technique to first-year Calculus students in a course where it is probably an excess of rigour than to teach it to upper division students in a Real Analysis course. The American practice of postponing difficulties does have its deleterious effect, in this case, an inability to accept new directions in one's study.

Of course, Kline is not to be blamed for this. His emphasis on intuition and physical applications may have been just the right antidote for the high level of rigour in American textbooks of the early 1960,s that came on the heels of American embarrassment over Sputnik and the resulting desire to improve science education. The new Calculus textbook adopted where I was a student was far more rigorous than the textbook being used at the time in Advanced Calculus. By the time Kline wrote *The Loss of Certainty*, however, he had gone overboard, decrying modern mathematical research as a waste of time, and citing the evolution of mathematical concepts and the continued sharpening of the notion of proof as a demonstration of his own uncertainty principle:

> I emphasized applied mathematics in the *Loss of Certainty* because, in view of current disagreements on what is correct mathematics, application to the physical world may be our only check.[73]

Kline shares with Dingler the disdain for what the latter called casuistic mathematics and the love of the intuitive. This is fair enough. Dingler blamed everything on the Jews and Kline attacks pure mathematical research. Dingler was something of an outsider who thought he was more of an expert in the subject than he actually was, and with regard to foundational matters, the same can be said of Kline. In fact, I regard Kline as much the same as Russell McCormmach's classical physicist—an aging mathematician who has been left behind by developments and resentful and embittered by that fact. I am not the first to say this. O'Connor and Robertson cite Ian Stewart's review of the same book as probably being the view of many mathematicians:

> I think three quarters of it is superb, and the other quarter is outrageous nonsense; and the reason is that Morris Kline really doesn't understand what today's mathematics is about, although he has an enviable grasp of yesterday's.

If this is the view of Morris Kline among his fellow mathematicians, why is it so important to be wary of him? The answer is that, like Dingler, he is not

---

[73] Quoted from Kline's comment on my attack, *The Mathematical Intelligencer* 4, No. 3 (1982), p. 129.

trying to bring about reform in mathematics education or research from the inside by convincing his fellow mathematicians via reasoned argumentation. Rather he is trying to impose his will, *à la* Dingler, from the outside by going over the heads of the mathematicians to the general public and, via them, school boards, legislators, and funding agencies. And this is a very real danger. In reviewing *Why the Professor Can't Teach*, Peter Hilton finishes with the lines

> To sum up we must conclude that Kline's book is an intemperate emotional outburst, and that the author has not only missed an opportunity to devote his great talents to the improvement of the quality of mathematical education but has also placed in the hands of the enemies of education in general and mathematics education in particular a potent if unreliable weapon.[74]

Lest this sound like overstated alarmism, let me point out that the enemies of science in the United States have continued passing legislation attempting to blunt the teaching of evolution by requiring the concomitant teaching of creationism in science courses even in the present century. And United States senators with no medical or scientific background have jumped on the anti-vaccination bandwagon because of claims by those with fewer relevant credentials than Morris Kline, a full professor and author of books published by reputable publishers. As I write, there is universal concern in the United States with the failure of the public school system, with calls for more teaching of mathematics and science, creating a ripe climate for hastily legislated "reforms" by such easily misled and ill-informed legislators.

The case of Saunders Mac Lane (1909 - 2005) is different. True, he was an old, old man when he published his article on the health of mathematics in the *Intelligencer*[75], but he was still in touch with modern mathematics, as much as could be expected of anyone in the latter half of the 20th century. His complaint was that mathematicians had overspecialised to the point that much of their research was pointless drivel. The complaint is reminiscent of the dissatisfaction with the monsters that threatened to overrun mathematics after Weierstrass's nowhere differentiable function arrived on the scene. Where the construction of such a counterexample is a necessary antidote to too heavy a reliance on intuition, the elaboration of such constructions certainly seemed to be carrying things too far. This sort of behaviour has repeated itself in all areas of mathematics and few mathematicians would disagree with Mac Lane in principle. They would differ in specifics, in their level of alarm at the situation, and on what—if anything—needs to be done about it. What is galling about Mac Lane's article is that he chose to illustrate bad examples of what he termed *persistent specialisation* in a subfield of mathematics he clearly did

---

[74] Once again I borrow from O'Connor and Robertson. These references are to a single web page at the Math Tutor web site and I cannot pinpoint their locations more sharply than this.

[75] Saunders Mac Lane, "The health of mathematics", *The Mathematical Intelligencer* 5, No. 4 (1983), pp. 53 - 55

not have as firm a grasp of as he thought. This happened to be my own field of specialisation, Mathematical Logic.

Mac Lane was not absolutely and totally ignorant of the field. He had studied under David Hilbert's logical partner Paul Bernays in Göttingen and even served as advisor to a couple of doctoral theses in logic. His remarks on logic show some knowledge of the overall situation:

> This effect of the isolation of a specialty is especially strong in mathematical logic. This field started in a study of the foundations of mathematics, but its practitioners were soon ostracized by other mathematicians. This led perforce to the isolation of mathematical logicians. Subsequently, despite splendid progress and many specific results connected with classical mathematics, the isolation has tended to continue—and at the same time mathematical logic has almost completely lost track of its original concern with foundations. Some of its practitioners are less concerned with concepts than with the demonstration that they too can solve hard problems. This they do, for example, by new axions [*sic*] setup in the never-never land of large cardinals. Or given that one can prove the continuum hypothesis to be independent of the axioms of set theory, let us prove the independence of all sorts of combinatoric notions. Or, given that recursive functions arise in Gödel's incompleteness theorem, and that recursive functions suggest a hierarchy of degrees, let us explore all the technical difficulties in the elaborate fine structure of this hierarchy. Or, given that the axioms of set theory *and* the continuum hypothesis can all be satisfied in Gödel's constructible sets, let us explore the fine structure of these sets, no matter how deep the morass which they form.[76]

Mac Lane's description here is accurate. I only take exception with his judgment that there is something wrong with all of this. To some extent he allows for this by continuing

> Now in each of these illustrative examples of mathematics gone wild, the specialist may well see a purpose which has eluded me. But I still contend that there are far too many such cases of elaboration without illumination, and that this is the price of persistent specialization.[77]

I have never found much interest in those particular topics in Mathematical Logic he chose to condemn, especially the recursion theoretic research alluded to in what are called *degrees of unsolvability*. And the fact that a leader in the field once acknowledged that the interest was in the techniques and not in the results proves to my satisfaction that Mac Lane was spot on in his judgment in this case.

But... as to the other examples, I think the intrinsic interest in other independence results for set theory requires no explanation, and I am not sufficiently familiar with the work on the fine structure of Gödel's constructible sets

---

[76] *Ibid.*, pp. 54 - 55.
[77] *Ibid.*, p. 55.

to say why or why not they should not be classed as an example of persistent overspecialisation. I have a vague remembrance that some portions of the fine structure were called morasses and thus Mac Lane's use of the word was jocular and not judgmental *à la* Poincaré's use of "teratologic" and "monster". I know the ostensible justification for the study of large cardinals[78], and published a letter in the *Intelligencer*[79] explaining this.

Heavy criticism of Mac Lane came later in the same issue as my letter in an article by Morris W. Hirsch, whose opening remarks underscored the danger that rash opinions like Mac Lane's posed:

> The thoughtful and provocative articles by Borel, "Mathematics: Art and Science" and Mac Lane, "The Health of Mathematics" in Volume 5, No. 4 present an interesting contrast. Both discuss, among other things, which branches of mathematics should or should not be pursued. A few years ago this question was no more than an interesting topic for coffee-hour conversation; today, however, we rely increasingly on governmental financial support for education and research, and our answers will therefore have very practical consequences.[80]

With respect to Mathematical Logic, Hirsch cites some of Mac Lane's remarks and offers some kind words in defence:

> I must admit that to me, not a practitioner of mathematical logic, these seem like perfectly natural and reasonable enterprises. While "concepts" are nice to develop, solving "hard problems" is also a good occupation, particularly when the problems arise naturally from fine pieces of research. Mac Lane, however, sees no purpose in them and considers them cases of "elaboration without illumination".[81]

So there are now two issues, the economic consequences for researchers of the public condemnation of their work by an influential mathematician like Mac Lane and the nasty things he had said about various subjects in Mathematical Logic.

In the next volume of the *Intelligencer* William Browder responded and, concerning Mac Lane's objection to narrow, persistent specialisation in fields he did not approve of, wrote

> How to judge the current value of directions of mathematical research is a question of great importance, from the personal view of the researcher,

---

[78] The *real* reason for such a study was probably the fact that the combinatorial definitions of these cardinals gave one a handle on dealing with them. Derek da Solla Price once said something to the effect that it more often happens that science chooses its problems by the tools at hand than it develops the tools to deal with the problems at hand.

[79] C. Smoryński, letter to the editor, *The Mathematical Intelligencer* 6, No. 3 (1984), p. 5.

[80] Morris W. Hirsch, "The health of mathematics—a second opinion", *The Mathematical Intelligencer* 6, No. 3 (1984), pp. 61 - 62.

[81] *Ibid.*, p. 61.

and from the point of view of an administrator allocating resources. The pull of fashion is very strong, and fashions, after all, are created at least partly by exciting developments. But if all effort and resources are concentrated in a few "hot" subjects, how will the groundwork be created for the next change of direction?[82]

It is in Mac Lane's response to Browder that the first hint of deeper motives is seen when he remarks that

> I can't help noting that the conference at princeton in honor of John Moore was not extended to honor his major accomplishments (with Eilenberg) in category theory (triples!).[83]

There is not enough in this quote to raise any suspicion of Mac Lane's motives yet. That would come with his response to the next article in the thread.

The set theorist Frank Drake of Leeds was compelled to justify himself and his persistent specialisation in the next volume of the *Intelligencer*[84]. It was a nice little exposition and that it arose in response to Mac Lane's mild assault on the field is in part to Mac Lane's credit; that Drake felt it necessary to justify his existence, however, is much to Mac Lane's discredit. Mac Lane's response[85] could only further diminish him. Unwilling to acknowledge he may have been wrong in this instance, he redoubled his attack on logic, once again bringing up Category Theory:

> There are by now proposals of alternative foundations; almost all logicians cheerfully ignore these proposals. Any real study of foundations, and this is what I would urge, would involve active comparison of such different foundations.[86]

The alternate proposals for foundations are Category Theory, in the 1980's Topos Theory. Category Theory provides a useful language for describing mathematics globally. It offers a good algebraic organisation for some aspects of mathematics, but it did not at the time offer foundations in any sense compatible with the meaning of the word in describing the foundational work in analysis in the 19th century as reported on in Chapter III, above, or even that in the later developments out of which Mathematical Logic emerged, as reported on in Chapter IV. Yet Mac Lane would dictate that logic devote resources to the algebraic field in the name of foundational research. That is not how science works. Should Category Theory come up with something interesting, logicians,

---

[82] William Browder, "Mathematical judgment", *The Mathematical Intelligencer* 7, No. 1 (1985), pp. 51 - 52.

[83] Saunders Mac Lane, "Whose mathematical judgment?", *The Mathematical Intelligencer* 7, No. 1 (1985), pp. 52, 76; here: p. 76.

[84] F.R. Drake, "How recent work in mathematical logic relates to the foundations of mathematics", *The Mathematical Intelligencer* 7, No. 4 (1986), pp. 27 - 35.

[85] Saunders Mac Lane, "Are we all just specialists?", *The Mathematical Intelligencer* 8, No. 4 (1986), pp. 74 - 75.

[86] *Ibid.*, p. 75.

among others, will take an active interest in the field. And, indeed, some logicians have involved themselves in the study of Category Theory. I am not familiar with this, but I would hazard a guess that it is in spite of and not because of Mac Lane's demands. Back in the 1980's it was easier to "cheerfully ignore" category theoretic foundations because much of it simply translated into well-travelled paths and the category theorists were saying nothing new. Moreover, the global approach of Category Theory precluded its applicability to the fine structure so popular among logicians and criticised by Mac Lane who, nonetheless, criticised logicians for ignoring it in the one case that he wanted it[87]. This, of course, elicited a response from me[88], the details of which need not concern us here.

5 Closing Remarks

Mac Lane's change in direction from expressing his dissatisfaction with parts of mathematics, the purposes of which had eluded him and which he thus denigrated as "persistent specialization", to an attempt to co-opt the field of Mathematical Logic and press it into the service of Category Theory can quickly sidetrack our underlying discussion. This discussion, I remind the reader who may feel I have wandered away from, is the crisis of intuition. For a long time mathematics dealt with things that could be readily visualised. An intuition developed. Every so often, however, intuition leads one astray—irrational numbers are discovered, theorems have "exceptions", and monsters are unleashed. The rise of Type II formalism may mitigate the existence of irrational numbers by giving us the tools to deal with them, and it can help explain the exceptions to theorems, but it also gives the tools for the construction of monsters, complicates proofs by requiring even the obvious to be stated and verified, leading to the much hated casuistic practice of Felix Klein and his imaginary Zionist conspiracy. One spends one's life developing an understanding of mathematics until it becomes a matter of intuition and suddenly the rules change. At the end of the 19th century the increasing emphasis on general functions unlike anything that had come up in practical applications provoked a reaction—one described them with colourful negative adjectives, shrank in horror from such functions, and ridiculed them in beer drinking songs. The increased rigour and its lack of immediacy to the goals at hand brought protests from students— witness Cauchy, Peano, and Landau, the last of whom was not only disliked by the Nazi students because he was Jewish. His Definition-Theorem-Proof style became the standard for textbooks, but is still much maligned today.

Dissatisfaction with the nature and direction of mathematics is now almost two centuries old, and the one thing common to those who proposed too force-

---

[87] "This sort of weakening of the Zermelo axioms is an example of a real question about foundations neglected by logicians." (*Ibid.*, p. 75.) Actually, the question he raised was already an active subject of research at the time.

[88] C. Smoryński, "Against Mac Lane", *The Mathematical Intelligencer* 10, No. 3 (1988), pp. 12 - 16.

fully to take over the direction of the field is that they have damaged their own reputations. Who remembers William Frend and Francis Maseres today for anything other than their opposition to negative numbers? Is not the first thing one thinks of at the mention of Leopold Kronecker his opposition to Georg Cantor and his opinion that Cantor didn't know when to stop, an opinion that led him to direct action? Paul Gordan, the "king of the invariants" is remembered by the general mathematical public for his initial opposition to a proof of Hilbert's not for his own positive accomplishments or his subsequent change of heart. Had Kline limited himself to his works on the application and popularisation of mathematics his reputation among mathematicians might not have been high, but it would not have been as low as it is. It may be too soon to say whether Mac Lane's complaints were broadly enough known to have damaged his reputation, but I doubt any conference in his honour at the University of Chicago will be extended to include his major achievements in fighting against persistent specialisation.

There will always be differences in style and perspective in mathematics and any researcher can be guaranteed to find some approach not to his liking. When he does, like mathematicians facing the problems of Type I formalism, he can respond in various ways. If he chooses rejection, simply choosing to ignore the offending mathematics is the safest course; it will not harm others or his reputation the way actively campaigning against unwelcome mathematical subdisciplines will.

Index